Optimal Decision Making in Operations Research and Statistics

Methodologies and Applications

Editors

Irfan Ali
Department of Statistics & Operations Research
Aligarh Muslim University, Aligarh, India

Leopoldo Eduardo Cárdenas-Barrón
Department of Industrial and Systems Engineering
School of Engineering and Sciences
Technológico de Monterrey, México

Aquil Ahmed
Department of Statistics & Operations Research
Aligarh Muslim University, Aligarh, India

Ali Akbar Shaikh
Department of Mathematics
The University of Burdwan, Burdwan, India

CRC Press
Taylor & Francis Group
Boca Raton London New York

CRC Press is an imprint of the
Taylor & Francis Group, an **informa** business
A SCIENCE PUBLISHERS BOOK

First edition published 2021
by CRC Press
6000 Broken Sound Parkway NW, Suite 300, Boca Raton, FL 33487-2742

and by CRC Press
4 Park Square, Milton Park, Abingdon, Oxon OX14 4RN

ISBN: 978-0-367-61875-9 (hbk)
ISBN: 978-0-367-61881-0 (pbk)
ISBN: 978-1-003-10695-1 (ebk)

Typeset in Times New Roman
by Radiant Productions

Preface

Operations Research (OR) has become a powerful technique for optimal decision-making. New techniques and sophisticated analysis tools are required to resolve the challenges arising from modern problems. It leads to the emergence of OR for efficiently determining optimal solutions to problems of real world. Although there are many types of conceivable problems, OR practitioners and researchers have found several problems in different circumstances. Thus, a challenge problem may be in the manufacturing industry area while another may be in the service sector. However, their essential features are the same. Thus, it is possible to describe these problems by naming the general categories into which they fall irrespective of their physical descriptions. A common analytical technique can be used to find the optimal solution to problems belonging to the same general category. In this direction,OR helps make better decision and solve problems in the real world. It uses mathematical relations, statistical computations, engineering techniques, economics and management methodologies to know the consequences of deciding for any possible alternative actions.

The decision-making techniques can be used in industries and services for making business decisions under risk and uncertainty. Furthermore, the decision-making techniques are also applied successfully to almost every possible sphere of human activity. Moreover, decision-making techniques are widely applied in different fields, ranging from almost every branch of science, engineering, industrial management, management planning, medical sciences, social sciences and economics, among others.

The book "Optimal Decision Making in Operations Research & Statistics: Methodologies and Applications" has been written by unified authors with a diverse background expertise from the faculties of Operations Research, Management, Applied Statistics and Mathematics. The contributed chapters are based on the vast research experiences of the authors in real-world decision-making problems.

The book is on the recent developments and contributions in optimal decision-making using optimization and statistical techniques. Mathematical modelling of cost-effective management policies are also part of the book.

The book presents challenging and practical real-world applications based on decision-making problems in various fields. The modelling and solution procedures of such real-world problems are provided concisely. This book provides readers a valuable compendium of several decision-making problems as a reference for this field's researchers and industrial practitioners. After reading this book, the readers will understand the formulations of decision-making problems and their solution procedures using appropriate optimization and statistical techniques.

This book broadly covers applications of applied statistics and optimization techniques in decision making in the various areas such as—estimation, control charts, econometric, regression, sampling, stochastic modelling, inventory control and management, transportation problem and optimization.

Finally, this book benefits the teachers, students, researchers, and industrialists working in material science, especially Operations Research and Applied Statistics, as a valuable reference handbook for teaching, learning, and research.

Contents

CHAPTER 1

A New Version of the Generalized Rayleigh Distribution with Copula, Properties, Applications and Different Methods of Estimation

M Masoom Ali[1], *Haitham M Yousof*[2,*] and *Mohamed Ibrahim*[3]

1. Introduction

A random variable (RV) Z is said to have the generalized Rayleigh (GR) distribution if its probability density function (PDF) and hence the CDF are given by

$$\pi_\beta(z) = 2\beta z e^{-z^2}(1-e^{-z^2})^{\beta-1}, \tag{1}$$

and

$$H_\beta(z) = (1-e^{-z^2})^\beta, \tag{2}$$

respectively, for $z > 0$, $\beta > 0$. Alizadeh et al. (2016) generalized the Odd G (O-G) family and the proportional reversed hazard rate family (PRHR) by proposing a new broader family called the Odd Burr (OB) family. The CDF of the OB family is given by

$$F_{\delta,\theta,\underline{\xi}}(z) = 1 - \frac{\overline{H}_{\underline{\xi}}(z)^{\delta\theta}}{\left[\overline{H}_{\underline{\xi}}(z)^\delta + H_{\underline{\xi}}(z)^\delta\right]^\theta}, \tag{3}$$

where $\overline{H}_{\underline{\xi}}(z) = 1 - H_{\underline{\xi}}(z)$. The PDF corresponding to (3) is given by

$$f_{\delta,\theta,\underline{\xi}}(z) = \frac{\delta\theta\pi_{\underline{\xi}}(z) H_{\underline{\xi}}(z)^{\delta-1}\overline{H}_{\underline{\xi}}(z)^{\delta\theta-1}}{\left[\overline{H}_{\underline{\xi}}(z)^\delta + H_{\underline{\xi}}(z)^\delta\right]^{1+\theta}}. \tag{4}$$

For $\theta = 1$, the OB-G family reduces to O-G family (see Gleaton and Lynch (2006)). For $\delta = 1$, the OB-G family reduces to the PRHR (see Gupta and Gupta (2007)).

In this paper, we propose and study a new version of the GR called the Odd Burr GR (OBGR) model. Some of its properties are derived and numerically analyzed. The usefulness and flexibility of the OBGR distribution is illustrated by means of a real data set related to failure times. Many bivariate and multivariate type distributions are derived based on Farlie Gumbel Morgenstern (FGM) Copula, modified FGM Copula, Clayton Copula and Renyi's entropy Copula. We briefly describe and consider different estimation methods namely, the maximum likelihood estimation (MLE), Cramér-von-Mises estimation (CVM), ordinary least square estimation (OLS), weighted least square estimation (WLSE), Anderson Darling estimation (ADE), right tail Anderson Darling estimation (RTADE), and the left tail Anderson Darling estimation (LTADE) method. These methods are used in the estimation process of the

[1] Department of Mathematical Sciences, Ball State University, Muncie, IN, USA.
Email: mali@bsu.edu
[2] Department of Statistics, Mathematics and Insurance, Benha University, Egypt.
[3] Department of Applied, Mathematical and Actuarial Statistics, Faculty of Commerce, Damietta University, Damietta, Egypt.
Email: mohamed_ibrahim@du.edu.eg
* Corresponding author: haitham.yousof@fcom.bu.edu.eg

unknow parameters. Monte Carlo simulation experiments are performed for comparing the performances of the proposed methods of estimation for both small and large samples.

The OBGR survival function (SF) is given by

$$S_{\underline{\Theta}}(z)=\frac{\left[1-(1-e^{-z^2})^{\beta}\right]^{\delta\theta}}{\left\{(1-e^{-z^2})^{\beta\delta}+\left[1-(1-e^{-z^2})^{\beta}\right]^{\delta}\right\}^{\theta}}, \tag{5}$$

where $S_{\underline{\Theta}}(z)=1-F_{\underline{\Theta}}(z)|_{(\underline{\Theta}=\delta,\theta,\beta)}$. For $\theta=1$, the OBGR reduces to the O-F. For $\delta=1$, the OBGR reduces to the PRHR-R. The PDF corresponding to (5) is given by

$$f_{\underline{\Theta}}(z)=2\delta\theta\beta ze^{-z^2}\frac{(1-e^{-z^2})^{\beta\delta-1}\left[1-(1-e^{-z^2})^{\beta}\right]^{\delta\theta-1}}{\left\{(1-e^{-z^2})^{\beta\delta}+\left[1-(1-e^{-z^2})^{\beta}\right]^{\delta}\right\}^{1+\theta}}. \tag{6}$$

The hazard rate function (HRF) for the new model can be obtained from $f_{\underline{\Theta}}(z)/S_{\underline{\Theta}}(z)$. The asymptotics of the CDF, PDF and hazard rate function (HRF) as $z\longrightarrow 0$ are given by

$$F_{\underline{\Theta}}(z)|_{(z\to 0)}\sim\theta(1-e^{-z^2})^{\beta\delta},\ f_{\underline{\Theta}}(z)|_{(z\to 0)}\sim 2\delta\theta\beta ze^{-z^2}(1-e^{-z^2})^{\beta\delta-1},$$

and

$$h_{\underline{\Theta}}(z)|_{(z\to 0)}\sim 2\delta\theta\beta ze^{-z^2}(1-e^{-z^2})^{\beta\delta-1}.$$

The asymptotics of CDF, PDF and HRF as $z\longrightarrow\infty$ are given by

$$1-F_{\underline{\Theta}}(z)|_{(z\to\infty)}\sim\delta^{\theta}\left[1-(1-e^{-z^2})^{\beta}\right]^{\theta},$$

$$f_{\underline{\Theta}}(z)|_{(z\to\infty)}\sim\frac{2\delta^{\theta}\theta\beta ze^{-z^2}(1-e^{-z^2})^{\beta-1}}{\left[1-(1-e^{-z^2})^{\beta}\right]^{1-\theta}},$$

and

$$h_{\underline{\Theta}}(z)|_{(z\to\infty)}\sim\frac{2\theta\beta ze^{-z^2}(1-e^{-z^2})^{\beta-1}}{1-(1-e^{-z^2})^{\beta}}.$$

Figure 1 gives some plots of the OBGR PDF for selected parameter values. Figure 2 gives some plots of the OBGR HRF for selected parameter values.

Based on Figure 1, the new PDF can be "right skewed" with "bimodal" and "unimodal" shapes. Based on Figure 2, the new HRF can be "increasing", " **U**-shape or (bathtub)", "**J**-shape", "upside-down-increasing", "decreasing", "upside-down" or "increasing-constant-increasing".

For simulation of this new model, we obtain the quantile function (QF) of Z (by inverting the CDF), say $z_u=F^{-1}(u)$, as

$$z_u=\left(-ln\left\{1-\left[\frac{u(\delta,\theta)}{(1-u)^{\frac{1}{\delta\theta}}+u(\delta,\theta)}\right]^{\frac{1}{\beta}}\right\}\right)^{\frac{1}{2}}, \tag{7}$$

where $u(\delta,\theta)=\left[1-(1-u)^{\frac{1}{\theta}}\right]^{\frac{1}{\delta}}$. Equation (7) is used for simulating the new model.

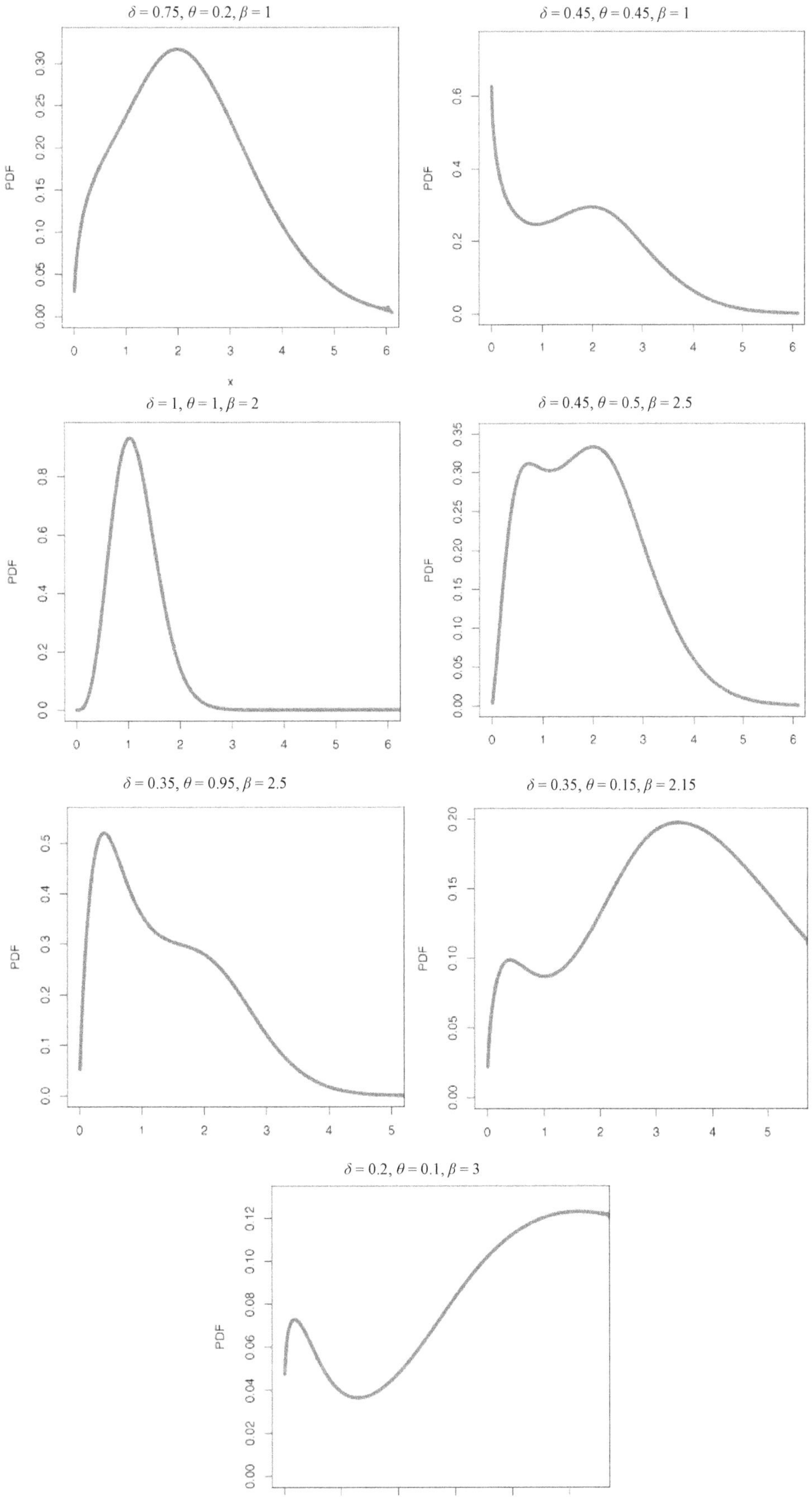

Figure 1: Plots of the OBGR PDF for selected parameter values.

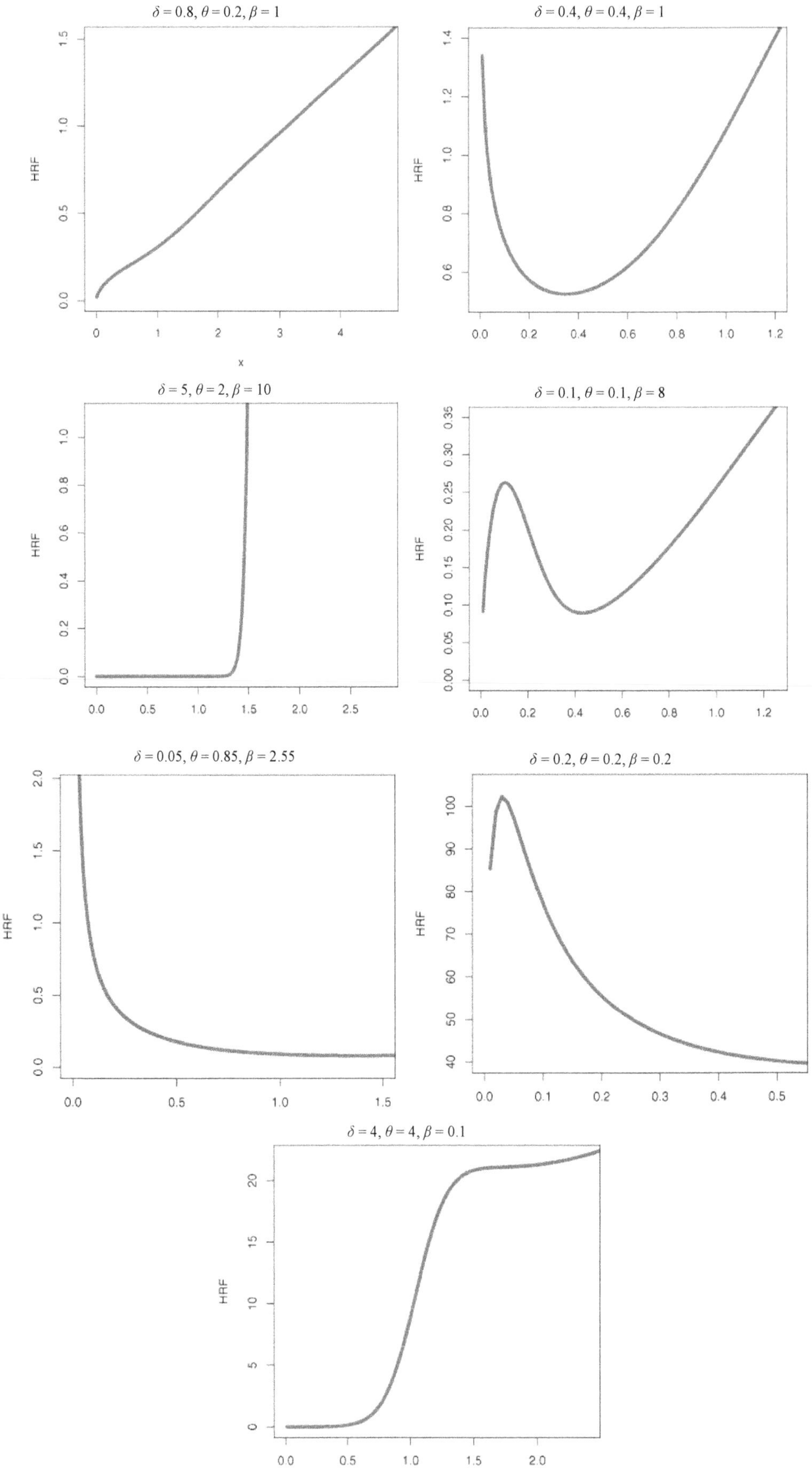

Figure 2: Plots of the OBGR HRF for selected parameter values.

2. Simple Type Copula

In this section, we derive some new bivariate OBGR (BvOBGR) and Multivariate OBGR (MvOBGR) types. However, future works may be allocated to study these new models. For more details about Copula and distributions see Mansour et al. (2020a,b,c).

2.2 BvOBGR Type via FGM Copula

Consider the joint CDF of the FGM family (see Morgenstern (1956), Gumbel (1960), Farlie (1960), Gumbel (1961), Johnson and Kotz (1975) and Johnson and Kotz (1977))

$$C_{\mathbf{\Delta}}(w,\varpi) = w\varpi(1+\mathbf{\Delta}\overline{w\varpi}),$$

where the continuous marginal function $w = 1 - \overline{w} \in [0,1]$, $\varpi = 1 - \overline{\varpi} \in [0,1]$, $\mathbf{\Delta} \in [-1,1]$ is a dependence parameter and for every $C_{\mathbf{\Delta}}(w,0) = C_{\mathbf{\Delta}}(0,\varpi) = 0 \mid_{(w,\,\varpi \,\in\, [0,1])}$ which is "grounded minimum" and $C_{\mathbf{\Delta}}(w,1) = w$ and $C_{\mathbf{\Delta}}(1,\varpi) = \varpi$ which is "grounded maximum". Then, setting $\overline{w} = \overline{w}_{\underline{\Theta}_1}|_{\underline{\Theta}_1=\delta_1,\theta_1,\beta>0}$, and $\overline{\varpi} = \overline{\varpi}_{\underline{\Theta}_2}|_{\underline{\Theta}_2=\delta_2,\theta_2,a>0}$, we have

$$F(z_1,z_2) = \left\{1 - \frac{\left(1-\varrho_{z_1}^{\beta}\right)^{\delta_1\theta_1}}{\left[\varrho_{z_1}^{\beta\delta_1} + \left(1-\varrho_{z_1}^{\beta}\right)^{\delta_1}\right]^{\theta_1}}\right\}\left\{1 - \frac{\left(1-\varrho_{z_2}^{a}\right)^{\delta_2\theta_2}}{\left[\varrho_{z_2}^{a\delta} + \left(1-\varrho_{z_2}^{a}\right)^{\delta_2}\right]^{\theta_2}}\right\}$$

$$\times\left(1+\mathbf{\Delta}\left\{\frac{\left(1-\varrho_{z_1}^{\beta}\right)^{\delta_1\theta_1}\left(1-\varrho_{z_2}^{a}\right)^{\delta_2\theta_2}}{\left[\varrho_{z_1}^{\beta\delta_1} + \left(1-\varrho_{z_1}^{\beta}\right)^{\delta_1}\right]^{\theta_1}\left[\varrho_{z_2}^{a\delta_2} + \left(1-\varrho_{z_2}^{a}\right)^{\delta_2}\right]^{\theta_2}}\right\}\right),$$

where $F(z_1, z_2) = C(F_{\underline{\Theta}_1}(z_1), F_{\underline{\Theta}_2}(z_2))$ and $\varrho_y^{\beta} = (1 - e^{-y^2})^{\beta}$. The joint PDF can be derived from

$$C_{\mathbf{\Delta}}(w,\varpi) = 1 + \mathbf{\Delta}w^{*}\varpi^{*} \mid_{(w^{*} = 1-2w \text{ and } \varpi^{*} = 1-2\varpi)},$$

or from

$$f(z_1,z_2) = f_{\underline{\Theta}_1}(z_1) f_{\underline{\Theta}_2}(z_2) C(F_{\underline{\Theta}_1}(z_1), F_{\underline{\Theta}_2}(z_2)).$$

2.3 BvOBGR Type via Modified FGM Copula

Consider the following modified FGM copula defined as (see Rodriguez-Lallena and Ubeda-Flores (2004)) $C_{\mathbf{\Delta}}(u,v) = uv + \mathbf{\Delta}\widetilde{\mathbf{\Phi}(u)}\,\widetilde{\psi(v)}$, $\widetilde{\mathbf{\Phi}(u)} = u\overline{\mathbf{\Phi}(u)}$ and $\widetilde{\psi(v)} = v\overline{\psi(v)}$, where $\mathbf{\Phi}(u)$ and $\psi(v)$ are two absolutely continuous functions on $(0,1)$ with the following conditions:

1-The boundary condition:

$$\mathbf{\Phi}(0) = \mathbf{\Phi}(1) = \psi(0) = \psi(1) = 0.$$

2-Let

$$\alpha = inf\left\{\frac{\partial}{\partial u}\widetilde{\mathbf{\Phi}(u)} : A_1\right\} < 0, \beta = sup\left\{\frac{\partial}{\partial u}\widetilde{\mathbf{\Phi}(u)} : A_1\right\} < 0,$$

$$\xi = inf\left\{\frac{\partial}{\partial v}\widetilde{\psi(v)} : A_2\right\} > 0, \eta = sup\left\{\frac{\partial}{\partial v}\widetilde{\psi(v)} : A_2\right\} > 0.$$

Then, $min(\alpha\beta, \xi\eta) \geq 1$, where

$$\frac{\partial}{\partial u}\widetilde{\mathbf{\Phi}(u)} = \mathbf{\Phi}(u) + u\frac{\partial}{\partial u}\mathbf{\Phi}(u)$$

$$A_1 = \left\{u \in (0,1) : \frac{\partial}{\partial u}\widetilde{\mathbf{\Phi}(u)} \text{ exists}\right\},$$

and

$$A_2 = \left\{ v \in (0,1) : \frac{\partial}{\partial v} \widetilde{\psi(v)} \ \text{ exists} \right\}.$$

2.3.1 BvOBGR-FGM (Type I)

Here, we consider the following functional form for both $\Phi(u)$ and $\psi(v)$ where $\widetilde{\Phi(u)} = u \frac{\left(1-\varrho_u^{\beta}\right)^{\delta_1\theta_1}}{\left[\varrho_u^{\beta\delta_1} + \left(1-\varrho_u^{\beta}\right)^{\delta_1}\right]^{\theta_1}} |_{\Theta_1 > 0}$, and $\widetilde{\psi(v)} = v \frac{\left(1-\varrho_v^{a}\right)^{\delta_2\theta_2}}{\left[\varrho_v^{a\delta_2} + \left(1-\varrho_v^{a}\right)^{\delta_2}\right]^{\theta_2}} |_{\Theta_2 > 0}$. Then using $C_\Delta(u,v) = uv + \Delta\widetilde{\Phi(u)}\,\widetilde{\psi(v)}$, the BvOBGR-FGM (Type I) can be obtained.

2.3.2 OBGR-FGM (Type II)

Consider the following functional form for both $\Phi(u)$ and $\psi(v)$ which satisfy all the conditions stated earlier where $\Phi(u)|_{(\Delta_1>0)} = u^{\Delta_1}(1-u)^{1-\Delta_1}$ and $\psi(v)|_{(\Delta_2>0)} = v^{\Delta_2}(1-v)^{1-\Delta_2}$.

The corresponding bivariate copula (henceforth, BvOBGR-FGM **(**Type **II)** copula) can be derived from $C_{\Delta,\Delta_1,\Delta_2}(u,v) = uv[1 + \Delta u^{\Delta_1} v^{\Delta_2}(1-u)^{1-\Delta_1}(1-v)^{1-\Delta_2}]$.

2.3.3 OBGR-FGM (Type III)

Consider the following functional form for both $\Phi(u)$ and $\psi(v)$ which satisfy all the conditions stated earlier where $\Phi(u) = u\,[log(1+\bar{u})]$ and $\psi(v) = v\,[log(1+\bar{v})]$. In this case, one can also derive a closed form expression for the associated CDF of the BvOBGR-FGM (Type **III**).

2.3.4 OBGR-FGM (Type IV)

Using Ghosh and Ray (2016) the CDF of the BvOBGR-FGM (Type **IV**) model can be derived from $C(u, v) = uF^{-1}(v) + vF^{-1}(u) - F^{-1}(u)\,F^{-1}(v)$ where $F^{-1}(u)$ and $F^{-1}(v)$ are derived before.

2.3 BvOBGR Type via Clayton Copula

The Clayton Copula can be considered as $C(v_1, v_2) = (v_1^{-\Delta} + v_2^{-\Delta} - 1)^{-\frac{1}{\Delta}}|_{\Delta\in[0,\infty]}$. Let us assume that $T \sim$ OBGR $(\delta_1, \theta_1, \beta)$ and $X \sim$ OBGR (δ_2, θ_2, a). Then, setting $v_1 = v(t)|_{\Theta_1 > 0}$ and $v_2 = v(x)|_{\Theta_2 > 0}$, the BvOBGR type distribution can be derived from $F(t, x) = C(F_{\Theta_1}(t), F_{\Theta_2}(x))$.

2.4 BvOBGR Type via Renyi's Entropy

Consider the theorm of Pougaza and Djafari (2011) where $R(w, \varpi) = z_2 w + z_1\varpi - z_1 z_2$. Then, the associated BvOBGR will be

$$R(z_1, z_2)|_{(\alpha=\alpha_1=\alpha_2)} = R(F_{\Theta_1}(z_1), F_{\Theta_2}(z_1)) = -z_1 z_2$$
$$+ z_2 \left\{ 1 - \frac{\left(1-\varrho_{z_1}^{\beta}\right)^{\delta_1\theta_1}}{\left[\varrho_{z_1}^{\beta\delta_1} + \left(1-\varrho_{z_1}^{\beta}\right)^{\delta_1}\right]^{\theta_1}} \right\} + z_1 \left\{ 1 - \frac{\left(1-\varrho_{z_2}^{a}\right)^{\delta_2\theta_2}}{\left[\varrho_{z_2}^{a\delta} + \left(1-\varrho_{z_2}^{a}\right)^{\delta_2}\right]^{\theta_2}} \right\}.$$

A straightforward Multivariate OBGR $\hbar$-dimensional extension can be derived from

$$H(\varpi_i) = \left[\sum_{i=1}^{\hbar} \varpi_i^{-\Delta} + 1 - \hbar \right]^{-\frac{1}{\Delta}}.$$

3. Mathematical Properties

3.1 Useful Representations

Due to Alizadeh et al. (2016), the PDF in (6) can be expressed as

$$f(z)=\sum_{k=0}^{\infty}\Omega_k\,\pi_{\beta^*}(z)\,|_{(\beta=\beta(1+k))}, \tag{8}$$

where

$$\Omega_k=\frac{\delta\theta}{1+k}\sum_{j,i=0}^{\infty}\sum_{l=k}^{\infty}(-1)^{i+k+l}\binom{-(1+\theta)}{j}\binom{-[\delta(1+j)+1]}{i}\binom{\delta(1+j)+i+1}{l}\binom{l}{k},$$

and $\pi_\beta(z)$ is the PDF of the EW model with power parameter β. By integrating Equation (8), the CDF of Z becomes

$$F(z)=\sum_{k=0}^{\infty}\Omega_k H_{\beta^*}(z), \tag{9}$$

where $H_{\beta^*}(z)$ is the CDF of the EW distribution with power parameter β^*.

3.2 Moments and Incomplete Moments

The rth ordinary moment of Z is given by $\mu_r'=E(Z^r)=\int_{-\infty}^{\infty}z^r f(z)dz$. Then we obtain

$$\mu_r'|_{(r>-2)}=\Gamma\left(1+\frac{r}{2}\right)\sum_{k,h=0}^{\infty}\Omega_{k,h}^{(r,\beta^*)}, \tag{10}$$

where $\Omega_{k,h}^{(r,\beta^*)}=\Omega_k\dfrac{\beta^*(-1)^h}{(h+1)^{(r+2)/2}}\dbinom{\beta^*-1}{h}$ and $\Gamma(1+\psi_1)|_{(\psi_1\in R^+)}=\prod_{r=0}^{\psi_1-1}(\psi_1-r)$, where $E(Z)=\mu_1'$ is the mean of Z. The variance (V(Z)), skewness (S(Z)) and kurtosis (K(Z)) can be derived easily using the well-known relationships. The rth incomplete moment, say $I_r(\tau)$, of Z can be expressed, from (9), as

$$I_r(\tau)=\int_{-\infty}^{\tau}z^r f(z)dz=\sum_{k=0}^{\infty}\Omega_k\int_{-\infty}^{\tau}z^r\pi_{\beta^*}(z)dz.$$

Then

$$I_r(\tau)|_{(r>-2)}=\gamma\left(1+\frac{r}{2},t^2\right)\sum_{k,h=0}^{\infty}\Omega_{k,h}^{(r,\beta^*)},$$

where $\gamma(\psi_1,\psi_2)$ is the incomplete gamma function.

$$\gamma(\psi_1,\psi_2)=\int_0^{\psi_2}z^{\psi_1-1}e^{-z}dz=\frac{\psi_2^{\psi_1}}{\psi_1}\{1F_1[\psi_1;\psi_1+1;-\psi_2]\}=\sum_{k=0}^{\infty}\frac{(-1)^k}{k!(\psi_1+k)}\psi_2^{\psi_1+k},$$

and $1F_1[\cdot,\cdot,\cdot]$ is a confluent hypergeometric function. The first incomplete moment given by

$$I_1(\tau)=\gamma\left(\frac{3}{2},t^2\right)\sum_{k,h=0}^{\infty}\Omega_{k,h}^{(1,\beta^*)}.$$

The dispersion index (DisIx) or the variance to mean ratio is a measure used to quantify whether a set of observed occurrences are clustered or dispersed compared to a standard statistical model. A numerical analysis for the DisIx (Z) for the new OBGR is presented in Table 2 with useful comments.

3.3 Moment Generating Function (MGF)

The MGF of Z can be derived from Equation (8) as

$$M_z(\tau)=\sum_{k=0}^{\infty} v_k M_{\beta^*}(\tau),$$

where $M_\beta^*(\tau)$ is the MGF of the GR model, then

$$M_z(\tau)|_{(r>-2)}=\sum_{r=0}^{\infty}\sum_{k,h=0}^{\infty}\frac{\tau^r}{r!}\Gamma\left(1+\frac{r}{2}\right)\Omega_{k,h}^{(r,\beta^*)}.$$

3.4 Residual Life and Reversed Residual Life Functions

The rth moment of the residual life $A_r(\tau) = E[(Z-\tau)^r|_{z>\tau,\ r=1,2,...}]$. The rth moment of the residual life of Z is given by

$$A_r(\tau)=\frac{1}{1-F(\tau)}\int_{\tau}^{\infty}(Z-\tau)^r\,dF(z).$$

Therefore,

$$A_r(\tau)=\frac{1}{1-F(\tau)}\sum_{k,h=0}^{\infty}a_{k,h}^{(r,\beta^*)}\Gamma\left(1+\frac{r}{2},t^2\right)|_{(r>-2)},$$

where $a_{k,h}^{(r,\beta^*)} = \Omega_k \sum_{m=0}^{r}\binom{r}{m}(-\tau)^{r-m}, \Gamma(\psi_1,r)|_{r>0} = \int_r^{\infty} z^{\psi_1-1}e^{-z}dz$ and $\Gamma(\psi_1, r) = \Gamma(\psi_1) - \gamma(\psi_1, r)$.

The rth moment of the reversed residual life, say $Z_r(\tau)= E[(\tau-Z)^r|_{z\le\tau,\tau>0\,\text{and}\,r=1,2,...}]$ uniquely determines $F(z)$. Then, we obtain $Z_r(\tau)=\frac{1}{F(\tau)}\int_0^\tau(\tau-Z)^r\,dF(z)$. Then, the rth moment of the reversed residual life of Z becomes

$$Z_r(\tau)=\frac{1}{F(\tau)}\sum_{k,h=0}^{\infty}b_{k,h}^{(r,\beta^*)}\gamma\left(1+\frac{r}{2},t^2\right)|_{(r>-2)},$$

where

$$b_{k,h}^{(r,\beta^*)}=\Omega_k\sum_{m=0}^{r}\binom{r}{m}\tau^{m-r}(-1)^m.$$

4. Numerical Analysis

Table 1 gives some numerical calculations for the mean, V(Z), S(Z), K(Z) and DisIx(Z). Based on Table 1, we note that the skewness of the OBGR model can be positive and negative as well. The spread for the OBGR kurtosis is much larger ranging from –16.774 to 75224271. The DisIx(Z) can be "between 0 and 1" and "more than 1".

5. Estimation Methods

In this Section, we briefly describe and consider different classical estimation methods namely, the MLE method, CVM method, OLS method, WLSE method, ADE method, RTADE method, left tail LTADE. All these methods are discussed in the statistical literature with more details. In this work, we may ignore some of its derivation details for avoiding the repetition.

5.1 The MLE Method

Let $Z_1, Z_2,..., Z_m$ be any RS from the new OBGR. The log likelihood function $\left(\ell_{[\Theta]}^{(m)}\right)$ for $\underline{\Theta}$ may be expressed as

$$\begin{aligned}\ell_{[\Theta]}^{(m)} &= m\,log(2\delta\theta\beta z)-\sum_{k=1}^{m}z_{[m,k]}^2+(\beta\delta-1)\sum_{k=1}^{m}log\left(1-e^{-z_{[m,k]}^2}\right)\\ &+(\delta\theta-1)\sum_{k=1}^{m}log\left[1-\left(1-e^{-z_{[m,k]}^2}\right)^{\beta}\right]\\ &-(1+\theta)\sum_{k=1}^{m}log\left\{\left(1-e^{-z_{[m,k]}^2}\right)^{\beta\delta}+\left[1-\left(1-e^{-z_{[m,k]}^2}\right)^{\beta}\right]^{\delta}\right\}.\end{aligned}$$

Table 1: Mean, variance, skewness, kurtosis and dispersion index.

δ	θ	β	E(Z)	V(Z)	S(Z)	K(Z)	DisIx(Z)
0.1	10	2	0.017204	0.00566576	14.73350	327.3479	0.3293370
0.5			0.301815	0.03094014	1.029830	4.555105	0.1025136
1			0.551176	0.02997869	0.159010	2.853232	0.0543905
5			0.961227	0.00418211	–0.809210	4.195831	0.0043508
10			1.032716	0.00118227	–0.940050	4.706613	0.0011448
20			1.069991	0.00031101	–0.993379	4.940258	0.0002907
30			1.082614	0.00013998	–0.774423	–16.77394	0.0001293
40			1.088960	7.9560×10^{-5}	–1.093857	14.74018	7.31×10^{-5}
50			1.092778	5.1102×10^{-5}	–1.026987	6.243116	4.68×10^{-5}
60			1.095328	3.5584×10^{-5}	–1.022265	5.143180	3.25×10^{-5}
100			2.490×10^{-7}	2.6199×10^{-7}	2055.689	42258570	1.052221
2.5	0.5	5	1.567618	0.05841878	0.8909679	4.746629	0.037265960
	1		1.436109	0.02756367	0.3076453	3.891405	0.019193290
	5		1.250278	0.01171343	–0.4948790	3.503164	0.009368659
	10		1.189912	0.00969201	–0.5873196	3.568369	0.008145148
	25		1.118369	0.00790280	–0.6376401	3.620711	0.007066365
	50		1.068920	0.00687203	–0.6543228	3.646666	0.006428947
	100		1.022832	0.00600896	–0.6648330	3.670132	0.005874830
	200		0.979754	0.00527184	–0.6737164	3.694863	0.005380779
	500		0.926985	0.00445288	–0.6858751	3.731091	0.004803618
5	5	0.1	0.012519	7.00243×10^{-5}	0.7876379	6.968286	0.005593269
		0.5	0.418374	0.005015261	–0.4656010	3.404089	0.011987510
		1	0.713840	0.005464616	–0.6484830	3.864594	0.007655137
		5	1.335607	0.003546402	–0.7279617	4.128052	0.002655274
		10	1.559558	0.002855629	–0.7220810	4.117927	0.001831051
		50	2.001435	0.001860090	–0.6989719	4.054250	0.000929378
		100	2.001435	0.001860090	–0.6989719	4.054250	0.000929378
		500	2.510104	0.001197058	–0.6745922	3.987213	0.000476896
		1000	$3.073 \times e^{-8}$	7.10490×10^{-8}	8673.1930	75224271	2.311848000

Following the normal routine of parameter estimation for the MLE of δ, θ and β we differentiate $\ell_{[\Theta]}^{(m)}$ with respect to δ, θ and β to obtain the score vector $\left(\frac{\partial \ell_{[\Theta]}^{(m)}}{\partial \delta}, \frac{\partial \ell_{[\Theta]}^{(m)}}{\partial \theta}, \frac{\partial \ell_{[\Theta]}^{(m)}}{\partial \beta}\right)^T$ as follows

$$U_{(\delta)} = \frac{\partial}{\partial \delta} \ell_{[\Theta]}^{(m)},\ U_{(\theta)} = \frac{\partial}{\partial \theta} \ell_{[\Theta]}^{(m)},\ U_{(c)} = \frac{\partial}{\partial \beta} \ell_{[\Theta]}^{(m)}.$$

Setting the nonlinear system of equations $U_{(\delta)} = U_{(\theta)} = U_{(\beta)} = 0$ and solving them simultaneously yields the MLE of $\Theta = (\delta, \theta, \beta)^T$. These equations cannot be solved analytically. So, statistical software can be used to solve them numerically using iterative methods such as the Newton-Raphson type algorithms.

5.2 The CVME Method

The CVME of the parameters δ, θ and β are obtained via minimizing the following expression with respect to (w.r.t) the parameters δ, θ and β respectively, where

$$CVME_{(\Theta)} = \frac{1}{12} m^{-1} + \sum_{k=1}^{m} \left[F_{\delta,\theta,\beta}(z_{[m,k]}) - c_{(k,m)} \right]^2,$$

and $c_{(k,m)} = [(2k-1)/2m]$ and

$$CVME_{(\underline{\Theta})} = \sum_{k=1}^{m}\left(1-\frac{\left[1-\left(1-e^{-z_{[m,k]}^{2}}\right)^{\beta}\right]^{\delta\theta}}{\left\{\left(1-e^{-z_{[m,k]}^{2}}\right)^{\beta\delta}+\left[1-\left(1-e^{-z_{[m,k]}^{2}}\right)^{\beta}\right]^{\delta}\right\}^{\theta}}-c_{(k,m)}\right)^{2}.$$

The CVME of the parameters δ, θ and β are obtained by solving the following non-linear equations

$$\sum_{k=1}^{m}\left(1-\frac{\left[1-\left(1-e^{-z_{[m,k]}^{2}}\right)^{\beta}\right]^{\delta\theta}}{\left\{\left(1-e^{-z_{[m,k]}^{2}}\right)^{\beta\delta}+\left[1-\left(1-e^{-z_{[m,k]}^{2}}\right)^{\beta}\right]^{\delta}\right\}^{\theta}}-c_{(k,m)}\right)\nabla_{(\delta)}\left(z_{[m,k]};\underline{\Theta}\right)=0,$$

$$\sum_{k=1}^{m}\left(1-\frac{\left[1-\left(1-e^{-z_{[m,k]}^{2}}\right)^{\beta}\right]^{\delta\theta}}{\left\{\left(1-e^{-z_{[m,k]}^{2}}\right)^{\beta\delta}+\left[1-\left(1-e^{-z_{[m,k]}^{2}}\right)^{\beta}\right]^{\delta}\right\}^{\theta}}-c_{(k,m)}\right)\nabla_{(\theta)}\left(z_{[m,k]};\underline{\Theta}\right)=0,$$

and

$$\sum_{k=1}^{m}\left(1-\frac{\left[1-\left(1-e^{-z_{[m,k]}^{2}}\right)^{\beta}\right]^{\delta\theta}}{\left\{\left(1-e^{-z_{[m,k]}^{2}}\right)^{\beta\delta}+\left[1-\left(1-e^{-z_{[m,k]}^{2}}\right)^{\beta}\right]^{\delta}\right\}^{\theta}}-c_{(k,m)}\right)\nabla_{(\beta)}\left(z_{[m,k]};\underline{\Theta}\right)=0,$$

where

$$\nabla_{(\delta)}(z_{[m,k]};\delta,\theta,\beta)=\partial F_{\delta,\theta,\beta}\left(z_{[m,k]}\right)/\partial\delta,$$

$$\nabla_{(\theta)}(z_{[m,k]};\delta,\theta,\beta)=\partial F_{\delta,\theta,\beta}\left(z_{[m,k]}\right)/\partial\theta,$$

and

$$\nabla_{(\beta)}(z_{[m,k]};\delta,\theta,\beta)=\partial F_{\delta,\theta,\beta}\left(z_{[m,k]}\right)/\partial\beta.$$

5.3 The OLSE Method

Let $F_{\underline{\Theta}}(z_{[m,k]})$ denotes the CDF of OBGR model and let $z_1 < z_2 < \cdots < z_m$ be the m ordered RS. The OLSEs are obtained upon minimizing

$$OLSE(\delta,\theta,\beta)=\sum_{k=1}^{m}\left[F_{\underline{\Theta}}(z_{[m,k]})-b_{(k,m)}\right]^{2}.$$

Then, we have

$$OLSE(\delta,\theta,\beta)=\sum_{k=1}^{m}\left(1-\frac{\left[1-\left(1-e^{-z_{[m,k]}^{2}}\right)^{\beta}\right]^{\delta\theta}}{\left\{\left(1-e^{-z_{[m,k]}^{2}}\right)^{\beta\delta}+\left[1-\left(1-e^{-z_{[m,k]}^{2}}\right)^{\beta}\right]^{\delta}\right\}^{\theta}}-b_{(k,m)}\right)^{2},$$

where $b_{(k,m)}=\frac{k}{m+1}$. The LSEs are obtained via solving the following non-linear equations

$$0=\sum_{k=1}^{m}\left(1-\frac{\left[1-\left(1-e^{-z_{[m,k]}^{2}}\right)^{\beta}\right]^{\delta\theta}}{\left\{\left(1-e^{-z_{[m,k]}^{2}}\right)^{\beta\delta}+\left[1-\left(1-e^{-z_{[m,k]}^{2}}\right)^{\beta}\right]^{\delta}\right\}^{\theta}}-b_{(k,m)}\right)\nabla_{(\delta)}\left(z_{[m,k]};\underline{\Theta}\right),$$

$$0=\sum_{k=1}^{m}\left(1-\frac{\left[1-\left(1-e^{-z_{[m,k]}^{2}}\right)^{\beta}\right]^{\delta\theta}}{\left\{\left(1-e^{-z_{[m,k]}^{2}}\right)^{\beta\delta}+\left[1-\left(1-e^{-z_{[m,k]}^{2}}\right)^{\beta}\right]^{\delta}\right\}^{\theta}}-b_{(k,m)}\right)\nabla_{(\theta)}\left(z_{[m,k]};\underline{\Theta}\right),$$

and

$$0=\sum_{k=1}^{m}\left(1-\frac{\left[1-\left(1-e^{-z_{[m,k]}^{2}}\right)^{\beta}\right]^{\delta\theta}}{\left\{\left(1-e^{-z_{[m,k]}^{2}}\right)^{\beta\delta}+\left[1-\left(1-e^{-z_{[m,k]}^{2}}\right)^{\beta}\right]^{\delta}\right\}^{\theta}}-b_{(k,m)}\right)\nabla_{(\beta)}\left(z_{[m,k]};\underline{\Theta}\right),$$

where $\nabla_{(\delta)}(z_{[m,k]};\underline{\Theta})$, $\nabla_{(\theta)}(z_{[m,k]};\underline{\Theta})$ and $\nabla_{(\beta)}(z_{[m,k]};\underline{\Theta})$ are defined before.

5.4 The WLSE Method

The WLSE are obtained by minimizing the function **WLSE** (δ, θ, β) w.r.t δ, θ and β where

$$WLSE(\delta,\theta,\beta)=\sum_{k=1}^{m}\omega_{(k,m)}\left[F_{\delta,\theta,\beta}(z_{[m,k]})-b_{(k,m)}\right]^{2},$$

and

$$\omega_{(k,m)}=[(1+m)^{2}(2+m)]/[k(1+m-k)].$$

The WLSEs are obtained by solving

$$0=\sum_{k=1}^{m}\left(1-\frac{\left[1-\left(1-e^{-z_{[m,k]}^{2}}\right)^{\beta}\right]^{\delta\theta}}{\left\{\begin{array}{c}\left(1-e^{-z_{[m,k]}^{2}}\right)^{\beta\delta}\\+\left[1-\left(1-e^{-z_{[m,k]}^{2}}\right)^{\beta}\right]^{\delta}\end{array}\right\}^{\theta}}-b_{(k,m)}\right)\omega_{(k,m)}\nabla_{(\delta)}\left(z_{[m,k]};\underline{\Theta}\right),$$

$$0=\sum \omega_{(k,m)}\left(1-\frac{\left[1-\left(1-\quad {}_{[m,k]}\right)\quad\right]^{\delta\theta}}{\left\{\begin{array}{l}\left(\quad {}_{[m,k]}\right)^{\beta\delta}\\ +\left[1-\left(1-\quad {}_{[m,k]}\right)\quad\right]\end{array}\right\}}-b_{(k,m)}\right)\nabla_{(\;)}\left(z_{[m,k]};\underline{\Theta}\right),$$

and

$$0=\sum_{k=1}^{m} \omega_{(k,m)}\left(1-\frac{\left[1-\left(1-e^{-z^{2}_{[m,k]}}\right)^{\beta}\right]^{\delta\theta}}{\left\{\begin{array}{l}\left(1-e^{-z^{2}_{[m,k]}}\right)^{\beta\delta}\\ +\left[1-\left(1-e^{-z^{2}_{[m,k]}}\right)^{\beta}\right]^{\delta}\end{array}\right\}^{\theta}}-b_{(k,m)}\right)\nabla_{(\beta)}\left(z_{[m,k]};\underline{\Theta}\right).$$

5.5 The ADE Method

The ADE of δ, θ and β are obtained by minimizing the function

$$ADE_{(z_{[m,k]},z_{[-k+1+m\,:\,m]})}(\delta,\theta,\beta)=-m-m^{-1}\sum_{k=1}^{m}(2k-1)\left\{\begin{array}{c}log\,F_{\underline{\Theta}}(z_{[m,k]})\\ +\,log\,[1-F_{\underline{\Theta}}(z_{[-k+1+m\,:\,m]})]\end{array}\right\}.$$

The parameter estimates of δ, θ and β follow by solving the nonlinear equations

$$0=\partial\left[ADE_{(z_{[m,k]},z_{[-k+1+m\,:\,m]})}(\underline{\Theta})\right]/\partial\delta,$$

$$0=\partial\left[ADE_{(z_{[m,k]},z_{[-k+1+m\,:\,m]})}(\underline{\Theta})\right]/\partial\theta,$$

and

$$0=\partial\left[ADE_{(z_{[m,k]},z_{[-k+1+m\,:\,m]})}(\underline{\Theta})\right]/\partial\beta.$$

5.6 The RTADE Method

The RTADE of δ, θ and β are obtained by minimizing

$$\mathrm{RTADE}_{(z_{[m,k]},z_{[-k+1+m\,:\,m]})}(\delta,\theta,\beta)=\frac{1}{2}m-2\sum_{k=1}^{m}F_{\underline{\Theta}}(z_{[m,k]})$$
$$-\frac{1}{m}\sum_{k=1}^{m}(2k-1)\left\{log\left[1-F_{(\delta,\theta,\beta)}(z_{[-k+1+m\,:\,m]})\right]\right\}.$$

The estimates of δ, θ and β are obtained by solving the nonlinear equations

$$0=\partial\left[\mathrm{RTADE}_{(z_{[m,k]},z_{[-k+1+m\,:\,m]})}(\underline{\Theta})\right]/\partial\delta,$$

$$0 = \partial\left[\text{RTADE}_{(z_{[m,k]}, z_{[-k+1+m\,:\,m]})}(\underline{\Theta})\right] / \partial\theta,$$

and

$$0 = \partial\left[\text{RTADE}_{(z_{[m,k]}, z_{[-k+1+m\,:\,m]})}(\underline{\Theta})\right] / \partial\beta.$$

5.7 The LTADE Method

The LTADE of δ, θ and β are obtained by minimizing

$$\text{LTADE}_{(z_{[m,k]})}(\delta,\theta,\beta) = -\frac{3}{2}m + 2\sum_{k=1}^{m} F_{\underline{\Theta}}(z_{[m,k]}) - \frac{1}{m}\sum_{k=1}^{m}(2k-1)\log F_{\underline{\Theta}}(z_{[m,k]}).$$

The parameter estimates of δ, θ and β are obtained by solving the nonlinear equations

$$0 = \partial\left[\text{LTADE}_{(z_{[m,k]})}(\underline{\Theta})\right] / \partial\delta,$$

$$0 = \partial\left[\text{LTADE}_{(z_{[m,k]})}(\underline{\Theta})\right] / \partial\theta,$$

Table 2: Simulation results for parameters $\delta = 1.5$, $\theta = 1.2$, $\beta = 0.6$.

Methods	m	**Bias**			**RMSE**			**D**	
		δ	θ	β	δ	θ	β	**abs**	**max**
MLE	50	0.01733	0.01462	0.00692	0.18690	0.17700	0.06148	0.00326	0.00604
OLS		0.00587	0.02778	0.02002	0.23111	0.19671	0.06918	0.01881	0.02701
WLS		0.15120	0.03727	0.03172	0.28615	0.17838	0.07225	0.03267	0.05784
CVM		0.01435	0.01284	0.00674	0.23040	0.20309	0.06500	0.00132	0.00228
ADE		0.03671	0.00639	0.00571	0.19604	0.18912	0.06281	0.00318	0.00585
RTADE		0.02503	0.00090	0.00992	0.22661	0.18012	0.07007	0.00553	0.00901
LTADE		0.01352	0.02381	0.00355	0.27197	0.22863	0.06221	0.00301	0.00601
MLE	100	0.01611	0.01197	0.00354	0.12770	0.12313	0.04202	0.00232	0.00364
OLS		0.01121	0.01143	0.00949	0.15661	0.13793	0.04626	0.00887	0.01312
WLS		0.10673	0.01317	0.01794	0.18434	0.12787	0.04852	0.01863	0.03469
CVM		0.00069	0.00884	0.00297	0.15601	0.14051	0.04484	0.00065	0.00102
ADE		0.01159	0.00590	0.00247	0.13343	0.13102	0.04337	0.00074	0.00117
RTADE		0.00235	0.00477	0.00360	0.16239	0.12544	0.04788	0.00130	0.00198
LTADE		0.01434	0.00940	0.00264	0.18054	0.15361	0.04305	0.00198	0.00305
MLE	150	0.00824	0.00894	0.00116	0.10411	0.09707	0.03410	0.00139	0.00276
OLS		0.00556	0.00446	0.00510	0.12838	0.11094	0.03658	0.00442	0.00659
WLS		0.07542	0.00821	0.01268	0.14125	0.10039	0.03842	0.01308	0.02443
CVM		0.00158	0.00910	0.00079	0.12819	0.11273	0.03594	0.00142	0.00206
ADE		0.00918	0.00626	0.00064	0.11079	0.10508	0.03483	0.00116	0.00215
RTADE		0.00311	0.00468	0.00138	0.12951	0.09941	0.03803	0.00063	0.00115
LTADE		0.00409	0.01250	0.00005	0.14916	0.12634	0.03514	0.00244	0.00404
MLE	300	0.00133	0.00034	0.00138	0.07307	0.06743	0.02462	0.00087	0.00138
OLS		0.00022	0.00560	0.00349	0.08828	0.07930	0.02598	0.00344	0.00490
WLS		0.04689	0.00801	0.00946	0.09546	0.07007	0.02757	0.00945	0.01719
CVM		0.00377	0.00115	0.00132	0.08830	0.07963	0.02568	0.00053	0.00094
ADE		0.00740	0.00012	0.00112	0.07778	0.07477	0.02500	0.00076	0.00149
RTADE		0.00188	0.00178	0.00132	0.09320	0.07182	0.02750	0.00019	0.00027
LTADE		0.00213	0.00330	0.00065	0.10257	0.08854	0.02476	0.00041	0.00079

and

$$0 = \partial\left[\text{LTADE}_{(z_{[m,k]})}(\underline{\Theta})\right] / \partial\beta.$$

6. Simulations for Comparing Methods

A numerical simulation is performed to compare the classical estimation methods. The simulation study is based on $N = 1000$ generated data sets from the OBGR version where $m = 50$, 100, 150 and 300 and

	δ	θ	β
I	1.5	1.2	0.6
II	2.5	2.5	0.9
II	2	0.5	0.5

The estimates are compared in terms of their biases, root mean square errors (RMSE). The mean of the absolute difference between the theoretical and the estimates (D-abs) and the maximum absolute difference between the true parameters and estimates (D-max) are also reported.

Tables 2, 3 and 4 give the simulation results. From Tables 2, 3 and 4 we note that the RMSE ($\underline{\Theta}$) tend to zero when m increases which means the incidence of consistency property.

Table 3: Simulation results for parameters $\delta = 2.9$, $\theta = 2.5$, $\beta = 0.9$.

Methods	m	Bias			RMSE			D	
		δ	θ	β	δ	θ	β	abs	max
MLE		0.03361	0.04490	0.00240	0.27553	0.38545	0.04710	0.00266	0.00429
OLS		0.10336	0.03186	0.00954	0.40371	0.41314	0.04707	0.02093	0.03238
WLS		0.19892	0.05335	0.01547	0.38032	0.37650	0.04860	0.03660	0.05730
CVM	50	0.03153	0.05360	0.00002	0.37792	0.43211	0.04566	0.00317	0.00649
ADE		0.00082	0.03925	0.00010	0.32346	0.39797	0.04444	0.00390	0.00578
RTADE		0.00215	0.01373	0.00377	0.34235	0.37911	0.04743	0.00222	0.00335
LTADE		0.02252	0.05947	0.00045	0.32158	0.48701	0.04807	0.00359	0.00693
MLE		0.02016	0.03802	0.00073	0.19294	0.25899	0.03148	0.00276	0.00536
OLS		0.07096	0.03498	0.00710	0.27542	0.28953	0.03382	0.01622	0.02459
WLS		0.15163	0.03916	0.01146	0.26252	0.26729	0.03589	0.02759	0.04334
CVM	100	0.03547	0.00702	0.00234	0.26385	0.29312	0.03298	0.00471	0.00804
ADE		0.01809	0.00258	0.00234	0.22553	0.27752	0.03254	0.00348	0.00556
RTADE		0.00854	0.00411	0.00241	0.24893	0.27002	0.03408	0.00252	0.00387
LTADE		0.01410	0.02165	0.00083	0.22100	0.32009	0.03342	0.00113	0.00230
MLE		0.01851	0.00960	0.00179	0.15703	0.21778	0.02713	0.00242	0.00421
OLS		0.03715	0.01218	0.00343	0.21363	0.23408	0.02680	0.00765	0.01181
WLS		0.10575	0.01764	0.00693	0.20685	0.21351	0.02842	0.01745	0.02797
CVM	150	0.01386	0.01598	0.00026	0.20845	0.23736	0.02649	0.00101	0.00175
ADE		0.00421	0.01297	0.00016	0.18401	0.22277	0.02598	0.00079	0.00149
RTADE		0.00619	0.01098	0.00028	0.19702	0.21171	0.02645	0.00140	0.00203
LTADE		0.00950	0.01976	0.00053	0.18210	0.26235	0.02731	0.00111	0.00196
MLE		0.00158	0.01471	0.00082	0.10669	0.14459	0.01813	0.00206	0.00305
OLS		0.02204	0.01078	0.00223	0.14725	0.16406	0.01890	0.00506	0.00769
WLS		0.06838	0.01156	0.00459	0.13949	0.15153	0.02017	0.01142	0.01831
CVM	300	0.01046	0.00317	0.00064	0.14514	0.16481	0.01875	0.00126	0.00220
ADE		0.00445	0.00184	0.00061	0.12726	0.15611	0.01845	0.00077	0.00127
RTADE		0.00181	0.00384	0.00039	0.14730	0.15067	0.01910	0.00020	0.00036
LTADE		0.00587	0.00455	0.00015	0.12671	0.18272	0.01935	0.00063	0.00115

Table 4: Simulation results for parameters $\delta = 2$, $\theta = 0.5$, $\beta = 0.5$.

Methods	m	Bias			RMSE	RMSE	RMSE	D	
		δ	θ	β	δ	θ	β	abs	max
MLE		0.02848	0.00725	0.00909	0.26069	0.07386	0.05986	0.00385	0.00639
OLS		0.03190	0.00785	0.01864	0.36466	0.08230	0.06854	0.01767	0.02565
WLS		0.12240	0.01222	0.03121	0.31694	0.07352	0.07041	0.01907	0.03656
CVM	50	0.03009	0.00918	0.00559	0.37646	0.08581	0.06414	0.00428	0.00808
ADE		0.01793	0.00667	0.00349	0.29050	0.07912	0.06067	0.00028	0.00049
RTADE		0.01939	0.00460	0.00765	0.33671	0.07583	0.07181	0.00300	0.00569
LTADE		0.03993	0.01368	0.00282	0.28938	0.09751	0.05895	0.00363	0.00484
MLE		0.01160	0.00295	0.00340	0.17582	0.05041	0.03932	0.00152	0.00236
OLS		0.00057	0.00103	0.00659	0.24948	0.05577	0.04361	0.00474	0.00718
WLS		0.10444	0.00436	0.01784	0.22326	0.05027	0.04571	0.01023	0.02105
CVM	100	0.03179	0.00750	0.00025	0.25542	0.05734	0.04243	0.00570	0.01018
ADE		0.00799	0.00613	0.00015	0.20615	0.05376	0.04056	0.00364	0.00579
RTADE		0.01857	0.00370	0.00400	0.24179	0.05449	0.05137	0.00216	0.00350
LTADE		0.01240	0.00865	0.00018	0.19663	0.06350	0.03884	0.00245	0.00415
MLE		0.01175	0.00470	0.00160	0.14364	0.04281	0.03270	0.00226	0.00426
OLS		0.01589	0.00507	0.00871	0.21893	0.04950	0.03863	0.00912	0.01318
WLS		0.07835	0.00589	0.01670	0.18488	0.04465	0.03980	0.00974	0.01974
CVM	150	0.00479	0.00053	0.00445	0.22075	0.04989	0.03750	0.00232	0.00408
ADE		0.00978	0.00028	0.00376	0.17852	0.04669	0.03573	0.00318	0.00469
RTADE		0.00635	0.00142	0.00256	0.18387	0.04169	0.03987	0.00212	0.00312
LTADE		0.01609	0.00180	0.00265	0.16793	0.05396	0.03312	0.00184	0.00304
MLE		0.00130	0.00069	0.00139	0.09969	0.02819	0.02173	0.00132	0.00191
OLS		0.00348	0.00086	0.00272	0.14325	0.03718	0.02559	0.0039	0.00349
WLS		0.05488	0.00133	0.00886	0.12405	0.03001	0.02588	0.00508	0.01030
CVM	300	0.00686	0.00196	0.00061	0.14412	0.03350	0.02533	0.00113	0.00213
ADE		0.00131	0.00150	0.00035	0.11671	0.03133	0.02421	0.00044	0.00066
RTADE		0.00543	0.00089	0.00207	0.12841	0.02906	0.02736	0.00101	0.00195
LTADE		0.00790	0.00226	0.00039	0.11140	0.03323	0.02318	0.00240	0.00081

Table 5: Comparing estimation methods via an application.

Methods	$\hat{\delta}$	$\hat{\theta}$	$\hat{\beta}$	CVM*	AD*
MLE	0.72446	0.18093	1.2458	**0.05531**	**0.55916**
OLS	0.78606	0.17720	3.38609	0.17462	1.47304
WLS	0.81674	0.20699	4.95758	0.26054	2.02331
CVM	0.77836	0.18221	3.39813	0.17096	1.44932
ADE	0.74950	0.17803	2.38404	0.10569	0.99061
RTADE	0.72604	0.19766	3.66245	0.16890	1.44117
LTADE	0.78181	0.15782	1.89213	0.08869	0.85082

7. Modeling Failure Times Data and Comparing Methods

An application to real data set is considered for comparing the estimation methods. The data consist of 84 aircraft windshield observations (see Murthy et al. (2004)). The required computations are carried out using the MATHCAD software. In order to compare the estimation methods, we consider the Cramér-von Mises (CVM) and the Anderson-Darling (AD) statistics. These two statistics are widely used to determine how closely a specific CDF fits the empirical distribution of a given data set. These statistics are given by

$$\mathrm{CVM}^* = \left[\frac{1}{12m} + \sum_{s=1}^{m}\left(z_{\hbar} - \frac{2s-1}{2m}\right)^2\right]\left(1 + \frac{1}{2m}\right),$$

and

$$AD^{*} = \left(1+\frac{9}{4m^{2}}+\frac{3}{4m}\right)\left\{m+\frac{1}{m}\sum_{s=1}^{m}(2s-1)log[z_{\hbar}(1-z_{m-s+1})]\right\},$$

respectively, where $z_{\hbar} = F(z_s)$ and the z_s's values are the ordered observations. The smaller these statistics are, the better the fit. From Table 6 we conclude that the MLE method is the best method with CVM = 0.05531 and AD = 0.55916. However, all other methods performed well.

Table 6: MLEs and standard errors (SEs) for failure times data set.

Distribution	Estimates (SEs)				
GR (β,α)	1.181876	0.377525			
	(0.17060)	(0.02532)			
OBGR (δ,θ,β)	**0.72167**	**0.1817**	**1.25096**		
	(0.19306)	**(0.0568)**	**(0.5024)**		
OLEW (θ,β,α)	0.15935	0.7322	0.7650		
	(0.3712)	(1.778)	(0.041)		
OLGR (θ,β,α)	1.45406	0.7543	0.2379		
	(0.9018)	(0.2530)	(0.0317)		
GREW (θ,β,α)	0.63684	4.2622	0.5364		
	(0.356)	(1.757)	(0.0997)		
PTLW (θ,β,α)	−5.78175	4.22865	0.65801		
	(1.395)	(1.167)	(0.039)		
MOEW (θ,β,α)	488.899	0.2832	1261.97		
	(189.358)	(0.013)	(351.07)		
GamW (θ,β,α)	2.37697	0.84809	3.5344		
	(0.378)	(0.00053)	(0.665)		
KumW (δ,θ,β,α)	14.4331	0.2041	34.6599	81.8459	
	(27.095)	(0.042)	(17.527)	(52.014)	
WFr (δ,θ,c,α)	630.938	0.3024	416.097	1.1664	
	(697.94)	(0.032)	(232.359)	(0.357)	
Beta-W (δ,θ,β,α)	1.360	0.2981	34.1802	11.496	
	(1.002)	(0.06)	(14.838)	(6.73)	
TrMW (δ,θ,β,α)	0.2722	1	4.6×10^{-6}	0.4685	
	(0.014)	(5.2×10^{-5})	(1.9×10^{-4})	(0.165)	
KumTrW (δ,θ,λ,β,α)	27.7912	0.178	0.4449	29.5253	168.06
	(33.401)	(0.017)	(0.609)	(9.792)	(129.17)
MBW (δ,θ,λ,β,α)	10.1502	0.1632	57.4167	19.3859	2.0043
	(18.697)	(0.019)	(14.063)	(10.019)	(0.662)
MacW (δ,θ,λ,β,α)	1.9401	0.306	17.686	33.6388	16.7211
	(1.011)	(0.045)	(6.222)	(19.994)	(9.722)
TrEGW (δ,θ,λ,β,α)	4.2567	0.1532	0.0978	5.2313	1173.33
	(33.401)	(0.017)	(0.609)	(9.792)	(6.999)

8. Modeling Failure Times Data and Comparing Models

In this section based on the data of Murthy et al. (2004), we provide an application of the OBGR distribution to show empirically its potentiality. The required computations are carried out using the R software.

Here, we shall compare the fits of the OBGR distribution with those of other competitive models, namely, GR, Odd Lindley EW (OLEW) (Silva et al. (2017)), Burr X EW (GREW) (Khalil et al. (2019)), Poisson Topp Leone-W (PTLW), MO extended-W (MOEW) (Ghitany et al. (2005)), Gamma-W (GamW) (Provost et al. (2011)), Kumaraswamy-W (KumW) (Cordeiro et al. (2010)), Beta-W (Lee et al. (2007)), Transmuted modified-W (TrMW) (Khan and King (2013)), W-Fréchet (WFr) (Afify et al. (2016b)),

Kumaraswamy transmuted-W (KumTrW) (Afify et al. (2016a)), Modified beta-W (MBW) (Khan (2015)) Mcdonald-W (MacW) (Cordeiro et al. (2014)), transmuted exponentiated generalized W (TrEGW) (Yousof et al. (2015)) distributions. The MLEs and the corresponding standard errors (in parentheses) of the model parameters are given in Table 6. The numerical values of the statistics CVM* and AD* are listed in Table 7. Figure 3 gives the TTT, Q-Q, box, and Estimated HRF (EHRF) plots for failure times data. Figure 4 gives the EPDF, ECDF, Probability-Probability (P-P) and Kaplan-Meier survival plots for the failure times data.

Table 7: CVM and AD statistics.

Distribution	CVM*	AD*
OBGR	**0.0553**	**0.5589**
OLEW	0.0723	0.6086
OLGR	0.0792	0.5910
GR	0.0690	0.6916
GREW	0.0744	0.6420
PTLW	0.1397	1.1939
MOEW	0.3995	4.4477
GamW	0.2553	1.9489
KumW	0.1852	1.5059
WFr	0.2537	1.9574
Beta-W	0.4652	3.2197
TrMW	0.8065	11.205
KumTrW	0.1640	1.3632
MBW	0.4717	3.2656
MacW	0.1986	1.5906
TrEGW	1.0079	6.2332

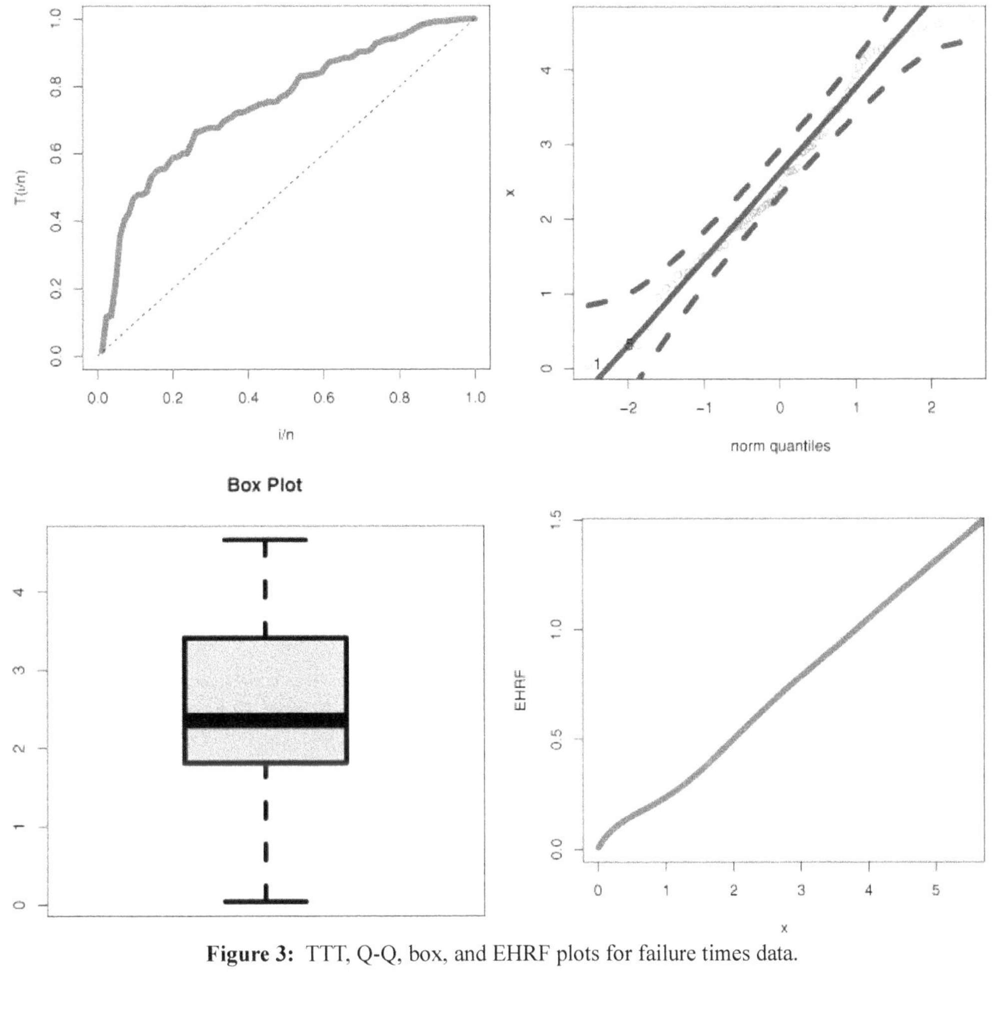

Figure 3: TTT, Q-Q, box, and EHRF plots for failure times data.

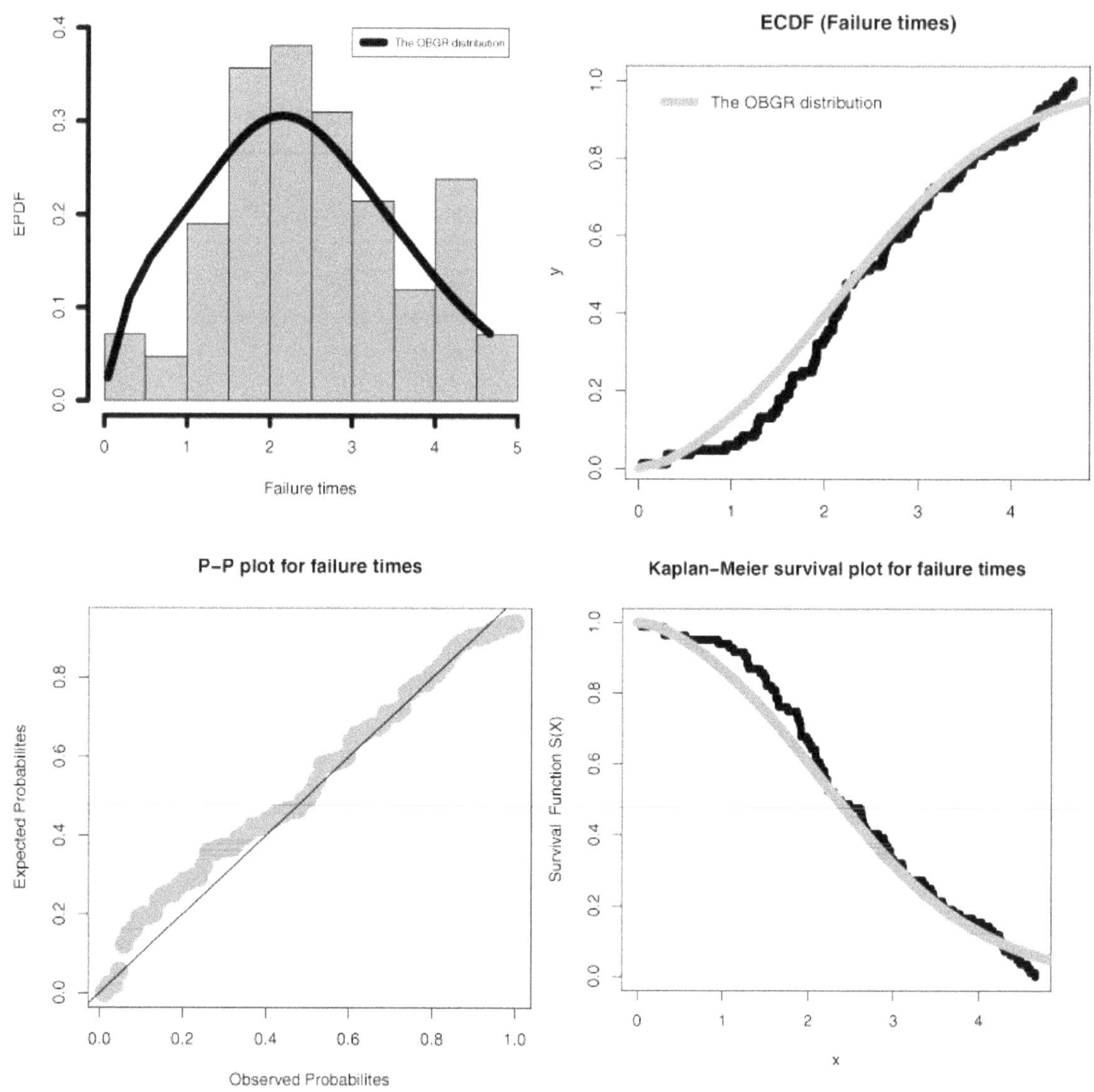

Figure 4: EPDF, ECDF, P-P and Kaplan-Meier plot plots for the failure times data.

Some other extensions of the Rayleigh distribution can also be used in this comparison, but are not limited to Alizadeh et al. (2016), Yousof et al. (2016a,b), Cordeiro et al. (2017a,b), Yousof et al. (2017), Brito et al. (2017), Aryal et al. (2017a,b), Korkmaz et al. (2017), Yousof et al. (2018), Ibrahim and Yousof (2020), Mansour et al. (2020d) and Hamedani et al. (2018 and 2019). Based on the figures in Table 3 we conclude that the new lifetime model provides adequate fits as compared to other W models with small values for CVM and AD. The proposed OBGR lifetime model is much better than the GREW, PTLW, MOEW, GamW, KumW, WFr, Beta-W, TrMW, KumTrW, MBW, MacW, TrEGW models, and a good alternative to these models.

9. Concluding Remarks

We introduced and studied a new version of the Generalized Rayleigh distribution. Some of its properties are derived and numerically analyzed. The usefulness and flexibility of the new distribution are illustrated by means of a real data set. The new PDF can be "right skewed" with "bimodal" and "unimodal" shapes. The new HRF can be "increasing", " **U**-shaped or(bathtub)", "**J**-shaped", "upside-down-increasing", "decreasing", "upside-down" or "increasing-constant-increasing". Many bivariate and multivariate type distributions have also been derived based on FGM Copula, Clayton Copula, modified FGM Copula and Renyi's entropy. We briefly describe different estimation methods namely, the maximum likelihood method, Cramér-von-Mises method, ordinary least square method, weighted least square method, Anderson Darling method, right tail Anderson Darling method, left tail Anderson Darling method which are used in the estimation process. Monte Carlo simulation experiments are performed for comparing the performances of the proposed methods of estimation for both small and large samples.

References

Afify, A. Z., Cordeiro, G. M., Yousof, H. M., Alzaatreh, A. and Nofal, Z. M. 2016a. The Kumaraswamy transmuted-G family of distributions: properties and applications, 14: 245–270.

Afify, A. Z., Yousof, H. M., Cordeiro, G. M., Ortega, E. M. M. and Nofal, Z. M. 2016b. The Weibull Fréchet distribution and its applications. Journal of Applied Statistics, 43(14): 2608–2626.

Ahmad, K. E., Fakhry, M. E. and Jaheen, Z. F. 1997. Empirical Bayes estimation of Pr (Y < X) and characterization of Burr-type X model. Journal of Statistical Planning and Inference, 64: 297–308.

Al-Babtain, A. A., Elbatal, I. and Yousof, H. M. 2020a. A new exible three-parameter model: properties, Clayton Copula, and modeling real data, 12(13): 1–26. doi: 10.3390/sym1203044017.

Al-Babtain, A. A., Elbatal, I. and Yousof, H. M. 2020b. A new three parameter Fréchet model with mathematical properties and applications. Journal of Taibah University for Science, 14(1): 265–278.

Alizadeh, M., Ghosh, I., Yousof, H. M., Rasekhi, M. and Hamedani, G. G. 2017. The generalized odd generalized exponential family of distributions: properties, characterizations and applications. J. Data Sci., 15: 443–466.

Alizadeh, M., Rasekhi, M., Yousof, H. M. and Hamedani, G. G. 2018. The transmuted Weibull G family of distributions. Hacettepe Journal of Mathematics and Statistics, 47(6): 1–20.

Almamy, J. A., Ibrahim, M., Eliwa, M. S., Al-mualim, S. and Yousof, H. M. 2018. The two-parameter odd Lindley Weibull lifetime model with properties and applications. International Journal of Statistics and Probability, 7(4): 1927–7040.

Alzaatreh, A., Lee, C. and Famoye, F. 2013. A new method for generating families of continuous distributions. Metron, 71: 63–79.

Aryal, G. R., Ortega, E. M., Hamedani, G. G. and Yousof, H. M. 2017a. The ToppLeone Generated Weibull distribution: regression model, characterizations and applications. International Journal of Statistics and Probability, 6: 126–141.

Aryal, G. R. and Yousof, H. M. 2017b. The exponentiated generalized-G Poisson family of distributions. Economic Quality Control, 32(1): 1–17.

Bjerkedal, T. 1960. Acquisition of resistance in guinea pigs infected with different doses of virulent tubercle bacilli. American Journal of Hygiene, 72: 130–148.

Bourguignon, M., Silva, R. B. and Cordeiro, G. M. 2014. The Weibull-G family of probability distributions. Journal of Data Science, 12: 53–68.

Brito, E., Cordeiro, G. M., Yousof, H. M., Alizadeh, M. and Silva, G. O. 2017. Topp-Leone odd log-logistic family of distributions. Journal of Statistical Computation and Simulation, 87(15): 3040–3058.

Burr, I. W. 1942. Cumulative frequency distribution. Annals of Mathematical Statistics, 13: 215–232.

Cordeiro, G. M., Ortega, E. M. and Nadarajah, S. 2010. The Kumaraswamy Weibull distribution with application to failure data. Journal of the Franklin Institute, 347: 1399–1429.

Cordeiro, G. M., Hashimoto, E. M., Edwin, E. M. M. Ortega. 2014. The McDonald Weibull model. Statistics: A Journal of Theoretical and Applied Statistics, 48: 256–278.

Cordeiro, G. M., Yousof, H. M., Ramires, T. G. and Ortega, E. M. M. 2017b. The Burr XII system of densities: properties, regression model and applications. Journal of Statistical Computation and Simulation, 88(3): 432–456.

Elbatal, I. and Aryal, G. 2013. On the transmuted additive Weibull distribution. Austrian Journal of Statistics, 42(2): 117–132.

Farlie, D. J. G. 1960. The performance of some correlation coefficients for a general bivariate distribution. Biometrika, 47: 307–323.

Gleaton, J. U. and Lynch, J. D. 2006. Properties of generalized log-logistic families of lifetime distributions. Journal of Probability and Statistical Science, 4: 51–64.

Ghosh, I. and Ray, S. 2016. Some alternative bivariate Kumaraswamy type distributions via copula with application in risk management. Journal of Statistical Theory and Practice, 10: 693–706.

Gumbel, E. J. 1960. Bivariate exponential distributions. Journ. Amer. Statist. Assoc., 55: 698–707.

Gumbel, E. J. 1961. Bivariate logistic distributions. Journal of the American Statistical Association, 56(294): 335–349.

Gupta, R. C. and Gupta, R. D. 2007. Proportional reversed hazard rate model and its applications. J. Statist Plan Inference. 137: 3525–3536.

Hamedani, G. G., Yousof, H. M., Rasekhi, M., Alizadeh, M. and Najibi, S. M. 2018. Type I general exponential class of distributions. Pak. J. Stat. Oper. Res., 14(1): 39–55.

Hamedani, G. G., Rasekhi, M., Najibi, S. M., Yousof, H. M. and Alizadeh, M. 2019. Type II general exponential class of distributions. Pak. J. Stat. Oper. Res., 15(2): 503–523.

Ibrahim, M. and Yousof, H. M. 2020. Transmuted Topp-Leone Weibull lifetime distribution: Statistical properties and different method of estimation. Pakistan Journal of Statistics and Operation Research, 16(3): 501–515.

Jaheen, Z. F. 1995. Bayesian approach to prediction with outliers from the Burr type X model. Microelectronic Reliability, 35: 45–47.

Jaheen, Z. F. 1996. Empirical Bayes estimation of the reliability and failure rate functions of the Burr type X failure model. Journal of Applied Statistical Sciences, 3: 281–288.

Johnson, N. L. and Kotz, S. 1975. On some generalized Farlie-Gumbel-Morgenstern distributions. Commun. Stat. Theory, 4: 415–427.

Johnson, N. L. and Kotz, S. 1977. On some generalized Farlie-Gumbel-Morgenstern distributions-II: Regression, correlation and further generalizations. Commun. Stat. Theory, 6: 485–496.

Khalil, M. G., Hamedani, G. G. and Yousof, H. M. 2019. The Burr X exponentiated Weibull model: characterizations, mathematical properties and applications to failure and survival times data. Pak. J. Stat. Oper. Res., 15(1): 141–160.

Khan, M. N. 2015. The modified beta Weibull distribution. Hacettepe Journal of Mathematics and Statistics, 44: 1553–1568.

Khan, M. S. and King, R. 2013. Transmuted modified Weibull distribution: a generalization of the modified Weibull probability distribution. European Journal of Pure and Applied Mathematics, 6: 66–88.

Kilbas, A. A., Srivastava, H. M. and Trujillo, J. J. 2006. Theory and Applications of Fractional Difierential Equations. Elsevier, Amsterdam.

Korkmaz, M. C., Yousof, H. M. and Hamedani, G. G. 2018. The exponential Lindley odd log-logistic G family: properties, characterizations and applications. Journal of Statistical Theory and Applications, 17(3): 554–571.

Lee, C., Famoye, F. and Olumolade, O. 2007. Beta-Weibull distribution: some properties and applications to censored data. Journal of Modern Applied Statistical Methods, 6: 17.

Mansour, M. M., Ibrahim, M., Aidi, K., Shafique Butt, N., Ali, M. M. et al. 2020a. A new log-logistic lifetime model with mathematical properties, Copula, modified goodness-of-fit test for validation and real data modeling. Mathematics, 8(9): 1508.

Mansour, M., Korkmaz, M. Ç., Ali, M. M., Yousof, H. M., Ansari, S. I. and Ibrahim, M. 2020b. A generalization of the exponentiated Weibull model with properties, Copula and application. Eurasian Bulletin of Mathematics, 3(2): 84–102.

Mansour, M., Rasekhi, M., Ibrahim, M., Aidi, K., Yousof, H. M. et al. 2020c. A new parametric life distribution with modified Bagdonavičius–Nikulin goodness-of-fit test for censored validation, properties, applications, and different estimation methods. Entropy, 22(5): 592.
Mansour, M. M., Butt, N. S., Yousof, H. M., Ansari, S. I. and Ibrahim, M. 2020d. A generalization of reciprocal exponential model: Clayton Copula, statistical properties and modeling skewed and symmetric real data sets. Pakistan Journal of Statistics and Operation Research, 16(2): 373–386.
Morgenstern, D. 1956. Einfache beispiele zweidimensionaler verteilungen. Mitteilingsblatt fur Mathematische Statistik, 8: 234–235.
Murthy, D. N. P., Xie, M. and Jiang, R. 2004. Weibull Models. John Wiley and Sons, Hoboken, New Jersey.
Nadarajah, S., Cordeiro, G. M. and Ortega, E. M. M. 2013. The exponentiated Weibull distribution: A survey. Statistical Papers, 54: 839–877.
Pougaza, D. B. and Djafari, M. A. 2011. Maximum entropies copulas. Proceedings of the 30th international workshop on Bayesian inference and maximum Entropy methods in Science and Engineering, 329–336.
Provost, S. B., Saboor, A. and Ahmad, M. 2011. The gamma-Weibull distribution. Pak. Journal Stat., 27: 111–131.
Raqab, M. Z. 1998. Order statistics from the Burr type X model. Computers Mathematics and Applications, 36: 111–120.
Rezaei, S., Nadarajah, S. and Tahghighnia, N. A. 2013. New three-parameter lifetime distribution. Statistics, 47: 835–860.
Rinne, H. 2009. The Weibull Distribution: A Handbook. CRC Press, Boca Raton, Florida.
Ristic, M. M. and Balakrishnan, N. 2012. The gamma-exponentiated exponential distribution. Journal of Statistical Computation and Simulation, 82: 1191–1206.
Rodriguez-Lallena, J. A. and Ubeda-Flores, M. 2004. A new class of bivariate copulas. Statistics and Probability Letters, 66: 315–25.
Sartawi, H. A. and Abu-Salih, M. S. 1991. Bayes prediction bounds for the Burr type X model. Communications in Statistics—Theory and Methods, 20: 2307–2330.
Silva, F. S., Percontini, A., de Brito, E., Ramos, M. W., Venancio, R. and Cordeiro, G. M. 2017. The odd Lindley-G family of distributions. Austrian Journal of Statistics, 46(1): 65–87.
Surles, J. G. and Padgett, W. J. 2001. Inference for reliability and stress-strength for a scaled Burr Type X distribution. Lifetime Data Analysis, 7: 187–200.
Tian, Y., Tian, M. and Zhu, Q. 2014. Transmuted linear exponential distribution: a new generalization of the linear exponential distribution. Communications in Statistics—Simulation and Computation, 43(10): 2661–2677.
Yadav, A. S., Goual, H., Alotaibi, R. M., Rezk, H., Ali, M. M. et al. 2020. Validation of the Topp-Leone-Lomax model via a modified Nikulin-Rao-Robson goodness-of-fit test with different methods of estimation. Symmetry, 12: 1–26. doi: 10.3390/sym12010057.
Yousof, H. M., Alizadeh, M., Jahanshahi, S. M. A., Ramires, T. G., Ghosh, I. et al. 2017. The transmuted Topp-Leone G family of distributions: theory, characterizations and applications. Journal of Data Science, 15: 723–740.
Yousof, H. M., Majumder, M., Jahanshahi, S. M. A., Ali, M. M. and Hamedani G. G. 2018. A new Weibull class of distributions: theory, characterizations and applications. Journal of Statistical Research of Iran, 15: 45–83.

Appendix

0.040, 1.866, 2.385, 3.443, 0.301, 1.876, 2.481, 3.467, 0.309, 1.899, 2.610, 3.478, 0.557, 1.911, 2.625, 4.570, 1.652, 2.300, 3.344, 4.602, 1.757, 3.578, 0.943, 1.912, 2.632, 3.595, 1.070, 1.914, 2.646, 3.699, 1.124, 1.981, 2.661, 3.779, 1.248, 2.010, 2.224, 3.117, 4.485, 1.652, 2.229, 3.166, 2.688, 3.924, 1.281, 2.038, 2.823, 4.035, 1.281, 2.085, 2.890, 4.121, 1.303, 2.089, 2.902, 4.167, 1.432, 4.376, 1.615, 2.223, 3.114, 4.449, 1.619, 2.097, 2.934, 4.240, 1.480, 2.135, 2.962, 4.255, 1.505, 2.154, 2.964, 4.278, 1.506, 2.190, 3.000, 4.305, 1.568, 2.194, 3.103, 2.324, 3.376, 4.663.

CHAPTER 2

Expanding the Burr X Model:

Properties, Copula, Real Data Modeling and Different Methods of Estimation

M Masoom Ali[1], *Mohamed Ibrahim*[2,*] and *Haitham M Yousof*[3]

1. Introduction and Genesis

A random variable Y is said to have the Burr type X (BX) distribution if its cumulative distribution function (CDF) and probability density function (PDF) are given by $\Pi_a(y) = (1-e^{-y^2})^a$ and $h_a(y) = 2aye^{-y^2}(1-e^{-y^2})^{a-1}|_{(y>0,a>0)}$ respectively. We write $Y \sim \text{BX}(a)$. The particular case for $a = 1$ is the Rayleigh (R) distribution. Based on Lehmann (1953), consider a baseline CDF of the Lehmann Burr type X (LBX) distribution

$$G_{\theta,a}(y) = 1-\left[1-(1-e^{-y^2})^a\right]^\theta|_{(y\geq 0 \,\text{and}\, \theta,a>0)}, \tag{1}$$

the corresponding PDF of (1) can be derived as

$$g_{\theta,a}(y) = 2\theta ay\frac{e^{-y^2}(1-e^{-y^2})^{a-1}}{\left\{1-(1-e^{-y^2})^a\right\}^{1-\theta}}|_{(y\geq 0 \,\text{and}\, \theta,a>0)}, \tag{2}$$

For $\theta = 1$, the LBX model reduces to the BX model. For $\theta = a = 1$, the LBX model reduces to the Rayleigh model. Marshall and Olkin (1996) pioneered a simple method of adding a positive shape parameter into a family of distributions and many authors used this method to extend many well-known models in the last few years. The reliability function (RF) of the Marshall-Olkin-G (MO-G) family of distributions is defined by

$$\overline{F}_{\delta,\underline{\psi}}(y) = \delta\overline{\Pi}_{\underline{\psi}}(y)\left[1-\overline{\delta}\,\overline{\Pi}_{\underline{\psi}}(y)\right]^{-1}|_{(y\in\Re,\delta>0)}, \tag{3}$$

where $\overline{F}_{\delta,\underline{\psi}}(y) = 1-F_{\delta,\underline{\psi}}(y)$ and $\overline{\Pi}_{\underline{\psi}}(y) = 1-\Pi_{\underline{\psi}}(y)$ is the RF of the baseline model, δ is a positive shape parameter, $\overline{\delta} = 1-\delta$ and $\underline{\psi}$ refers to the parameters vector of the baseline model. For $\delta \in (0,1)$, MOL-G family reduces to the complementary geometric-G ($\overline{\text{C}}$G-G) family. For $\delta = 1$, MO-G family reduces to the standard G family. The corresponding PDF of (3) is given by

$$f_{\delta,\underline{\psi}}(y) = \delta h_{\underline{\psi}}(y)\left[1-\overline{\delta}\,\overline{\Pi}_{\underline{\psi}}(y)\right]^{-2}|_{(y\in\Re,\delta>0)}, \tag{4}$$

where $h_{\underline{\psi}}(y) = \frac{d}{dx}\Pi_{\underline{\psi}}(y)$ is the PDF of the baseline model. By inserting (1) in (3), we obtain the CDF of the Marshall-Olkin Lehmann BX (MOLBX) class

[1] Department of Mathematical Sciences, Ball State University, Muncie, IN, USA.
Email: mali@bsu.edu
[2] Department of Applied, Mathematical and Actuarial Statistics, Faculty of Commerce, Damietta University, Damietta, Egypt.
[3] Department of Statistics, Mathematics and Insurance, Benha University, Egypt.
Email: haitham.yousof@fcom.bu.edu.eg
* Corresponding author: mohamed_ibrahim@du.edu.eg

$$F_{\underline{V}}(y) = \frac{1-[1-\eta_{(y,a)}]^{\theta}}{1-\bar{\delta}[1-\eta_{(y,a)}]^{\theta}}\Big|_{(y\geq 0 \text{ and } \delta,\theta,a>0)}, \quad (5)$$

where $\eta_{(y,\nabla)} = (1-e^{-y^2})^{\nabla}$ and $\underline{V} = (\delta, \theta, a)$. The corresponding PDF of (5) is given by

$$f_{\underline{V}}(y) = 2\delta\theta a y \frac{e^{-y^2}\eta_{(y,a-1)}}{\{1-[1-\eta_{(y,a)}]^{\theta}\}^{-1}}\{1-\bar{\delta}[1-\eta_{(y,a)}]^{\theta}\}^{-2}\Big|_{(y\geq 0 \text{ and } \delta,\theta,a>0)}. \quad (6)$$

Table 1 below gives some sub-models from the MOLBX model. Equations (5) and (6) can be also derived based on Yousof et al. (2018a).

Table 1: Some sub-models from the MOLBX model.

δ	θ	a	Reduced model	CDF
$\delta^*\vert_{[\delta^*\in(0,1)]}$			CGLBX	$\frac{1-[1-\eta_{(y,a)}]\theta}{1-\bar{\delta}^*[1-\eta_{(y,a)}]^{\theta}}\vert_{[\delta^*=1-\delta]}$
$\delta^*\vert_{[\delta^*\in(0,1)]}$	1		CGBX	$\frac{\eta_{(y,a)}}{1-\bar{\delta}^*[1-\eta_{(y,a)}]}\vert_{[\delta^*=1-\delta]}$
$\delta^*\vert_{[\delta^*\in(0,1)]}$	1	1	CGR	$\frac{\eta_{(y,1)}}{1-\bar{\delta}^*[1-\eta_{(y,1)}]}\vert_{[\delta^*=1-\delta]}$
	1		MOBX	$\frac{\eta_{(y,a)}}{1-\bar{\delta}[1-\eta_{(y,a)}]}$
1			LBX	$1-[1-\eta_{(y,a)}]^{\theta}$
1	1		BX	$\eta_{(y,a)}$
1	1	1	R	$\eta_{(y,1)}$

Figure 1 gives some plots of the MOLBX PDF. Figure 2 gives some plots of the MOLBX hazard rate function (HRF). From Figure 1 we conclude that the PDF MOLBX distribution can exhibit many important shapes with different skewness and kurtosis which can

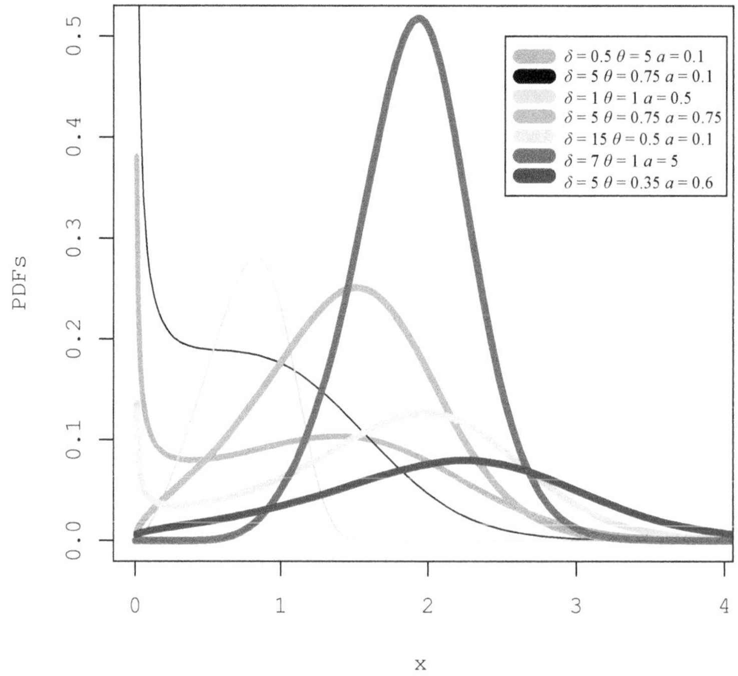

Figure 1: Plots of the MOLBX PDF for selected parameter values.

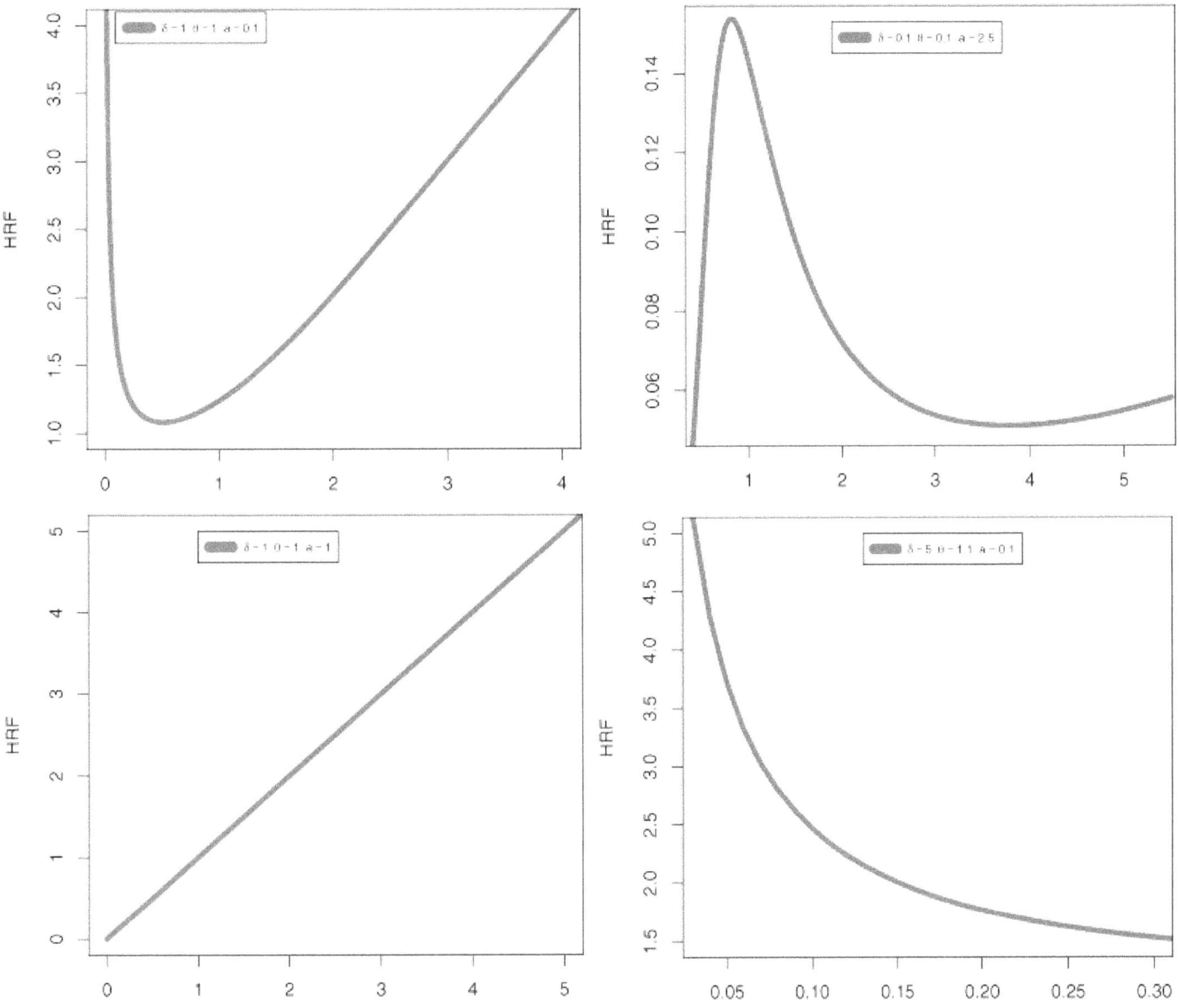

Figure 2: Plots of the MOLBX HRF for selected parameter values.

be unimodal and bimodal. From Figure 2 we conclude that the HRF distribution can be decreasing hazard rate ($\delta = 1, \theta = 0.5, a = 0.5$), J-hazard rate ($\delta = 5, \theta = 2, a = 5$), bathtub shape ($\delta = 1, \theta = 0.5, a = 0.5$), constant hazard rate ($\delta = 1, \theta = 1, a = 1$), increasing hazard rate ($\delta = 20, \theta = 2.85, a = 1$), upside down (reversed bathtub shape) ($\delta = 1, \theta = 1, a = 2.5$), and increasing-constant hazard rate ($\delta = 1, \theta = 1, a = 3$).

2. Properties

2.1 Moments

Let $A_a(y; \theta) = [1 - \eta_{(y,a)}]^{\theta}$. Then, expanding $A_a(y; \theta)$ we have

$$A_a(y;\theta) = 1 + \sum_{i_1=0}^{\infty} (-1)^{1+i_1} \binom{\theta}{i_1} A_a(y;\theta i_1) \sum_{i_1=0}^{\infty} \xi_{i_1} A_a(y;\theta i_1), \tag{7}$$

where $\xi_0 = 2$ and $\xi_{i_1} = (-1)^{1+i_1} \binom{\theta}{i_1}|_{(i_1 \geq 1)}$. Similarly, the quantity $1 - \bar{\delta}(1 - A_a(y; \theta))$ can be expanded as

$$1 - \bar{\delta}(1 - A_a(y;\theta)) = 1 - \bar{\delta} - \sum_{i_1=0}^{\infty} (-1)^{i_1} \binom{\theta}{i_1} A_a(y;\theta i_1) \sum_{i_1=0}^{\infty} \varphi_{i_1} A_a(y;\theta i_1), \tag{8}$$

where $\varphi_0 = \delta$ and $\varphi_{i_1} = (1-\delta)(-1)^{1+i_1}\binom{\theta}{i_1}$. Using (7) and (8) the CDF in (5) can be expressed as

$$F_{\underline{V}}(y) = \sum_{i_1=0}^{\infty} \frac{\xi_{i_1} A_a(y;\theta i_1)}{\varphi_{i_1} A_a(y;\theta i_1)} = \sum_{i_1=0}^{\infty} C_{i_1} A_a(y;\theta i_1),$$

where

$$C_0 = \frac{\xi_0}{\varphi_0},$$

and

$$C_{i_1} = \frac{1}{\varphi_0}\left(\xi_{i_1} - \frac{1}{\varphi_0}\sum_{r=1}^{i_1}\varphi_r C_{i_1-r}\right)|_{(i_1\geq 1)}.$$

The PDF of the MOLBX model can also be expressed as a mixture of exponentiated BX (ExpBX) PDF. By differentiating $F_{\underline{V}}(y)$, we obtain the same mixture representation

$$f_{\underline{V}}(y) = \sum_{i_1=0}^{\infty} C_{i_1} g_{1+\theta i_1,a}(y), \tag{9}$$

where $g_{1+\theta i_1,a}(y)$ is the ExpBX PDF with power parameter $(1+\theta i_1)$. Equation (9) reveals that the MOLBX density function is a linear combination of ExpBX densities. Thus, some structural properties of the new family such as the ordinary and incomplete moments and generating function can be immediately obtained from well-established properties of the EW distribution. The rth ordinary moment of Y is given by

$$\mu_r' = E(Y^r) = \int_{-\infty}^{\infty} y^r f(y)dy,$$

and then we obtain

$$\mu_r' = a^{-r}\Gamma\left(\frac{r}{2}+1\right)\sum_{i_1,i_2=0}^{\infty} \nabla_{i_1,i_2}^{(1+\theta i_1,r)}|_{(r>-2)}, \tag{10}$$

where

$$\nabla_{i_1,i_2}^{(1+\theta i_1,r)} = C_{i_1} \nabla_{i_2}^{(1+\theta i_1,r)},$$

and

$$\nabla_m^{(v,r)} = v\frac{(-1)^m}{(m+1)^{-\left(\frac{r}{2}+1\right)}}\binom{v-1}{m},$$

setting $r = 1,2,3,4$ in (10) we get

$$E(Y) = \mu_1' = a^{-1}\Gamma\left(\frac{3}{2}\right)\sum_{i_1,i_2=0}^{\infty} \nabla_{i_1,i_2}^{(1+\theta i_1,1)},$$

$$E(Y^2) = \mu_2' = a^{-2}\Gamma(2)\sum_{i_1,i_2=0}^{\infty} \nabla_{i_1,i_2}^{(1+\theta i_1,2)},$$

$$E(Y^3) = \mu_3' = a^{-3}\Gamma\left(\frac{5}{2}\right)\sum_{i_1,i_2=0}^{\infty} \nabla_{i_1,i_2}^{(1+\theta i_1,3)},$$

and

$$E(Y^4) = \mu_4' = a^{-4}\Gamma(3)\sum_{i_1,i_2=0}^{\infty} \nabla_{i_1,i_2}^{(1+\theta i_1,4)}.$$

The skewness and kurtosis measures can be calculated from the ordinary moments using well-known relationships. The moment generating function (MGF) is $M_Y(t) = E(e^{tY})$ of Y. Clearly, the first one can be derived from Equation (9) as

$$M_y(t) = \Gamma\left(\frac{r}{2}+1\right) \sum_{i_1,i_2,r=0}^{\infty} a^{-r} \nabla_{i_1,i_2,r}^{(1+\theta i_1,r)} |_{(r>-2)},$$

where

$$\nabla_{i_1,i_2,r}^{(1+\theta i_1,r)} = \frac{t^r}{r!} \nabla_{i_1,i_2}^{(1+\theta i_1,r)}.$$

The sth incomplete moment, say $I_s(t)$, of y can be expressed from (9) as

$$I_s(t) = \int_{-\infty}^{t} y^s f(y) dy = a^{-s} \gamma\left(\frac{r}{2}+1, t^2\right) \sum_{i_1,i_2=0}^{\infty} \nabla_{i_1,i_2}^{(1+\theta i_1,r)} |_{(s>-2)}, \tag{11}$$

setting s = 1,2,3,4 in (11) we get

$$I_1(t) = \int_{-\infty}^{t} y f(y) dy = a^{-1} \gamma\left(\frac{3}{2}, t^2\right) \sum_{i_1,i_2=0}^{\infty} \nabla_{i_1,i_2}^{(1+\theta i_1,1)},$$

$$I_2(t) = \int_{-\infty}^{t} y^2 f(y) dy = a^{-2} \gamma\left(2, t^2\right) \sum_{i_1,i_2=0}^{\infty} \nabla_{i_1,i_2}^{(1+\theta i_1,2)},$$

$$I_3(t) = \int_{-\infty}^{t} y^3 f(y) dy = a^{-3} \gamma\left(\frac{5}{2}, t^2\right) \sum_{i_1,i_2=0}^{\infty} \nabla_{i_1,i_2}^{(1+\theta i_1,3)},$$

and

$$I_4(t) = \int_{-\infty}^{t} y^4 f(y) dy = a^{-4} \gamma\left(3, t^2\right) \sum_{i_1,i_2=0}^{\infty} \nabla_{i_1,i_2}^{(1+\theta i_1,4)}.$$

2.2 Order Statistics

Suppose $Y_1,\ldots,Y_m$ is any random sample from any MOLBX distribution. Let $Y_{\hbar:m}$ denote the ith order statistic. The PDF of $Y_{\hbar;m}$ can be expressed as

$$f_{\hbar:m}(y) = \frac{f(y)}{B(\hbar, m-\hbar+1)} \sum_{s=0}^{m-\hbar} (-1)^s \binom{m-\hbar}{s} F(y)^{s+\hbar-1}.$$

Then,

$$f_{\hbar:m}(y) = \sum_{r,i_1=0}^{\infty} \Upsilon_{r,i_1} g_{1+r+\theta i_1}(y),$$

where $\Upsilon_{r,i_1} = \frac{m!(r+1)(\hbar-1)!C_{(r+1)}}{(1+r+\theta i_1)} \sum_{s=0}^{m-\hbar} \frac{(-1)^s \xi_{s+h-1,i_1}}{(m-h-s)!s!}$ C_{i_1} is given before and the quantities $\xi_{s+\hbar-1,i_1}$ can be determined with $\xi_{s+\hbar-1,0} = w_0^{s+\hbar-1}$ and recursively for $i_1 \geq 1$

$$\xi_{s+h-1,i_1} = (i_1 t_0)^{-1} \sum_{m=1}^{i_1} t_m [m(s+\hbar) - i_1] \xi_{s+h-1,i_1-m}.$$

Based on (12) we have

$$E\left(y_{\hbar:m}^{\xi}\right) = a^{-\xi} \Gamma\left(\frac{\xi}{2}+1\right) \sum_{r,i_1,i_2=0}^{\infty} \nabla_{r,i_1,i_2}^{(1+r+\theta i_1,\xi)} |_{(\xi>-2)},$$

where

$$\nabla_{r,i_1,i_2}^{(1+r+\theta i_1,\xi)} = \Upsilon_{r,i_1} \nabla_{i_2}^{(1+r+\theta i_1,\xi)}.$$

2.3 Residual Life and Reversed Residual Life Functions

The mth moment of the residual life

$$q_{m,t} = E\left[(Y-t)^m \mid_{(y>t \text{ and } m=1,2,\ldots)}\right].$$

The mth moment of the residual life of Y

$$q_{m,t} = \frac{1}{1-F(t)}\int_t^{\infty}(Y-t)^m \, dF(y).$$

Therefore,

$$q_{m,t} = a^{-m}\gamma\left(\frac{m}{2}+1, t^2\right)\frac{1}{\overline{F}(t)}\sum_{i_1,i_2=0}^{\infty}\sum_{r=0}^{m} v_{i_1,i_2,r}^{(1+\theta i_1,m)} \mid_{(m>-2)},$$

where $\overline{F}(t) = [1 - F(t)]$ and

$$v_{i_1,i_2,r}^{(1+\theta i_1,m)} = (1-t)^m \nabla_{i_1,i_2}^{(1+\theta i_1,m)}.$$

The mean residual life (MRL) at age t can be defined as

$$q_{1,t} = E[(y-t)|_{(y>t \text{ and } m=1)}],$$

which represents the expected additional life length for a unit which is alive at age t. The mth moment of the reversed residual life is given by

$$\Lambda_{m,t} = E[(t-Y)^m|_{(y \le t, t>0 \text{ and } m=1,2,\ldots)}],$$

or

$$\Lambda_{m,t} = \frac{1}{F(t)}\int_0^t (t-Y)^m \, dF(y).$$

Then, the mth moment of the reversed residual life of y becomes

$$\Lambda_{m,t} = a^{-m}\gamma\left(\frac{m}{2}+1, t^2\right)\frac{1}{F(t)}\sum_{i_1,i_2=0}^{\infty}\sum_{r=0}^{m} v_{i_1,i_2,r}^{(1+\theta i_1,m)} \mid_{(m>-2)},$$

where

$$v_{i_1,i_2,r}^{(1+\theta i_1,m)} = (-1)^r \binom{m}{r} t^{m-r} \nabla_{i_1,i_2}^{(1+\theta i_1,m)}.$$

The mean inactivity time (MIT) or mean waiting time (MWT) also called the mean reversed residual life function is given by

$$\Lambda_{1,t} = E[(t-y)|_{(y \le t\, m=1)}],$$

and it represents the waiting time elapsed since the failure of an item on the condition that this failure had occurred in $(0,t)$.

3. Copula

3.1 Bivariate MOLBX (BivMOLBX) Type viaRenyi's Entropy

The joint CDF of the Renyi's entropy Copula can be expressed as (Pougaza and Djafari (2011)) $C(u,v) = y_2u + y_1v - y_1y_2$. Then, the associated bivariate MOLBX will be

$$C(t_1, t_2) = C(F_{\underline{V_1}}(y_1), F_{\underline{V_2}}(y_2)).$$

3.2 BivMOLBX Type via Farlie Gumbel Morgenstern (FGM) Copula

Consider the joint CDF of the FGM family, then

$$C_{\Delta}(u,\ v) = uv\ (1 + \Delta\bar{u}\bar{v}),$$

where the marginal function $u = F_1(y_1) \in [-1,1]$, $v = F_2(y_2) \in [-1,1]$, $\Delta \in [-1,1]$ is a dependence parameter and for every $u, v \in [-1,1]$, $C_{\Delta}(u,0) = C_{\Delta}(0,v) = 0$ which is a "grounded minimum" and $C_{\Delta}(u,1) = u$ and $C_{\Delta}(1,v) = v$ which is a "grounded maximum". Then, setting $\bar{u} = \bar{u}_{\underline{V_1}} = F_{\underline{V_1}}(y_1)$, and $\bar{v} = \bar{v}_{\underline{V_2}} = F_{\underline{V_2}}(y_2)$, we have

$$F(y_1, y_2) = C(F_{\underline{V_1}}(y_1),\ F_{\underline{V_2}}(y_2)).$$

The joint PDF can bederived from

$$C_{\Delta}(u,v) = 1 + \Delta u^{*}v^{*}|_{(u^{*} = 1-2u \text{ and } v^{*} = 1-2v)}$$

or from

$$f(y_1,y_2) = f_{\underline{V_1}}(y_1) f_{\underline{V_2}}(y_2) c(F_{\underline{V_1}}(y_1),\ F_{\underline{V_2}}(y_2)).$$

For more details see Morgenstern (1956), Farlie (1960), Gumbel (1960), Gumbel (1961), Johnson and Kotz (1975) and Johnson and Kotz (1977).

3.3 BivMOLBXType via Modified FGM Copula

The modified joint CDF of the bivariate FGM copula can be expressed as

$$C_{\Delta}(u, w) = uv + \Delta\widetilde{\Theta(u)}\widetilde{\varphi(w)},$$

where $\widetilde{\Theta(u)} = u\overline{\Theta(u)}$, and $\widetilde{\varphi(w)} = w\overline{\varphi(w)}$ and $\Theta(u) = 1 - \overline{\Theta(u)}$ and $\varphi(w) = 1 - \overline{\varphi(w)}$ are two absolutely continuous functions on with the following conditions:

(I)-The boundary condition:

$$\Theta(0) = \Theta(1) = \varphi(0) = \varphi(1) = 0.$$

(II)-Let

$$a = inf\left\{\frac{\partial}{\partial u}\widetilde{\Theta(u)} : A_1\right\} < 0, c = sup\left\{\frac{\partial}{\partial u}\widetilde{\Theta(u)} : A_1\right\} < 0,$$

$$b = inf\left\{\frac{\partial}{\partial w}\widetilde{\varphi(w)} : A_2\right\} > 0, d = sup\left\{\frac{\partial}{\partial w}\widetilde{\varphi(w)} : A_2\right\} > 0.$$

Then,

$$min(ac,\ bd) \geq 1,$$

where

$$\frac{\partial}{\partial u}\widetilde{\Theta(u)} = \Theta(u) + u\frac{\partial}{\partial u}\Theta(u),$$

$$A_1 = \left\{u \in (0,1)\ :\ \frac{\partial}{\partial u}\widetilde{\Theta(u)}, \text{exists}\right\},$$

and

$$A_2 = \left\{w \in (0,1)\ :\ \frac{\partial}{\partial w}\widetilde{\varphi(w)}, \text{exists}\right\}.$$

3.3.1 BivMOLBX-FGM (Type I) Model

Here, we consider the following functional form for both $\mathbf{\Theta}(u)$ and $\boldsymbol{\varphi}(w)$ as

$$C_{\Delta}(u,w)=\left\{1-\frac{1-[1-\eta_{(u,a_1)}]^{\theta_1}}{1-\bar{\delta}_1[1-\eta_{(u,a_1)}]^{\theta_1}}\right\}\left\{1-\frac{1-[1-\eta_{(v,a_2)}]^{\theta_2}}{1-\bar{\delta}_2[1-\eta_{(v,a_2)}]^{\theta_2}}\right\}+\Delta[\widetilde{\mathbf{\Theta}(u)}\widetilde{\boldsymbol{\varphi}(w)}],$$

where,

$$\widetilde{\mathbf{\Theta}(u)}=u\left\{1-\frac{1-[1-\eta_{(u,a_1)}]^{\theta_1}}{1-\bar{\delta}_1[1-\eta_{(u,a_1)}]^{\theta_1}}\right\}|_{\underline{V_1}>0},$$

and,

$$\widetilde{\boldsymbol{\varphi}(w)}=v\left\{1-\frac{1-[1-\eta_{(v,a_2)}]^{\theta_2}}{1-\bar{\delta}_2[1-\eta_{(v,a_2)}]^{\theta_2}}\right\}|_{\underline{V_2}>0}.$$

3.3.2 BivMOLBX-FGM (Type II) Model

Consider the following functional form for both $\mathbf{\Theta}(u)$ and $\boldsymbol{\varphi}(w)$ which satisfy all the conditions stated earlier where,

$$\mathbf{\Theta}(u)|_{(\Delta_1>0)}=u^{\Delta_1}(1-u)^{1-\Delta_1} \text{ and } \boldsymbol{\varphi}(w)|_{(\Delta_2>0)}=v^{\Delta_2}(1-w)^{1-\Delta_2}.$$

The corresponding BivMOLBX-FGM (Type **II**) copula can be derived from,

$$C_{\Delta,\Delta_1,\Delta_2}(u,w)=uw[1+\Delta\upsilon^{\Delta_1}w^{\Delta_2}(1-u)^{1-\Delta_1}(1-w)^{1-\Delta_2}].$$

3.3.3 BivMOLBX-FGM (Type III) Model

Consider the following functional form for both $\mathbf{\Theta}(u)$ and $\boldsymbol{\varphi}(w)$ which satisfy all the conditions stated earlier where,

$$\mathbf{\Theta}(u)=u[log(1+\bar{u})] \text{ and } \boldsymbol{\varphi}(w)=w[log(1+\bar{w})].$$

In this case, one can also derive a closed form expression for the associated CDF of the BivMOLBX-FGM (Type **III**).

3.3.4 BivMOLBX-FGM (Type IV) Model

The CDF of the BivMOLBX-FGM (Type **IV**) model can be derived from,

$$C(u,w)=uF^{-1}(w)+wF^{-1}(u)-F^{-1}(u)F^{-1}(w).$$

3.4 Bivariate MOLBX Type via Clayton Copula

The Clayton Copula can be considered as,

$$C(v_1,v_2)=(v_1^{-\Delta}+v_2^{-\Delta}-1)^{-\frac{1}{\Delta}}|_{\Delta\in[0,\infty]}.$$

Let us assume that T ~ MOLBX (δ_1, θ_1, a_1) and W ~ MOLBX (δ_2, θ_2, a_2). Then, setting $v_1=v(t)=F_{\underline{V_1}}(t)|_{\underline{V_1}>0}$ and $v_2=v(w)=F_{\underline{V_2}}(w)|_{\underline{V_2}>0}$ the bivariate MOLBX type distribution can be derived as $F(t,w)=C(F_{\underline{V_1}}(t)|F_{\underline{V_2}}(w))$. A straightforward m-dimensional extension from the above will be, $H(v_i)=\left[\sum_{i=1}^{m}v_i^{-\Delta}+1-m\right]^{-\frac{1}{\Delta}}$.

4. Estimation

4.1 The MLE Method

Let $Y_1,\dots Y_m$ be a random sample from the MOLBX distribution with parameters δ, θ and a. Let $\underline{V}$ be the 3 × 1 parameter vector. For determining the MLE of $\underline{V}$, we have the log-likelihood function,

$$\ell=\ell(\underline{V})=m\log\delta+m\log\theta+2m\log a$$

$$+\sum_{k=1}^{m} log(y_{[m,k]})+(a-1)\sum_{k=1}^{m} log\,(1-e^{-y_{[m,k]}^{2}})$$
$$+\sum_{k=1}^{m} log\left\{1-\left[1-\eta_{(y_{[m,k]},a)}\right]^{\theta}\right\}-2\sum_{k=1}^{m} log\left\{1-\bar{\delta}\left[1-\eta_{(y_{[m,k]},a)}\right]^{\theta}\right\}.$$

The components of the score vector can be easily derived.

4.2 CVME method

The CVME of the parameters δ, θ and a are obtained via minimizing the following expression with respect to the parameters δ, θ and a respectively, where,

$$CVME_{(\underline{V})}=\frac{1}{12}m^{-1}+\sum_{k=1}^{m}[F_{\underline{V}}(y_{[m,k]})-\Upsilon_{(k,m)}]^{2},$$

where $\Upsilon_{(k,m)} = [(2k-1)/2m]$ and

$$CVME_{(\underline{V})}=\sum_{k=1}^{m}\left(\frac{1-[1-\eta_{(y_{[m,k]},a)}]^{\theta}}{1-\bar{\delta}[1-\eta_{(y_{[m,k]},a)}]^{\theta}}-\Upsilon_{(k,m)}\right)^{2}.$$

Then, CVME of the parameters δ, θ and a are obtained by solving the following three non-linear equations,

$$\sum_{k=1}^{m}\left(\frac{1-[1-\eta_{(y_{[m,k]},a)}]^{\theta}}{1-\bar{\delta}[1-\eta_{(y_{[m,k]},a)}]^{\theta}}-\Upsilon_{(k,m)}\right)C_{(\delta)}(y_{[m,k]};\underline{V})=0,$$
$$\sum_{k=1}^{m}\left(\frac{1-[1-\eta_{(y_{[m,k]},a)}]^{\theta}}{1-\bar{\delta}[1-\eta_{(y_{[m,k]},a)}]^{\theta}}-\Upsilon_{(k,m)}\right)C_{(\theta)}(y_{[m,k]};\underline{V})=0,$$

and

$$\sum_{k=1}^{m}\left(\frac{1-[1-\eta_{(y_{[m,k]},a)}]^{\theta}}{1-\bar{\delta}[1-\eta_{(y_{[m,k]},a)}]^{\theta}}-\Upsilon_{(k,m)}\right)C_{(a)}(y_{[m,k]};\underline{V})=0,$$

where

$$C_{(\delta)}(y_{[m,k]};\underline{V})=\partial F_{\underline{V}}(y_{[m,k]})/\partial\delta,$$
$$C_{(\theta)}(y_{[m,k]};\underline{V})=\partial F_{\underline{V}}(y_{[m,k]})/\partial\theta,$$

and

$$C_{(a)}(y_{[m,k]};\underline{V})=\partial F_{\underline{V}}(y_{[m,k]})/\partial a.$$

4.3 OLSE Method

Let $F_{\underline{V}}(y_{[m,k]})$ denote the CDF of MOLBX model and let $Y_1 < Y_2 < \cdots < Y_m$ be the m ordered RS. The OLSEs are obtained upon minimizing

$$OLSE(\underline{V})=\sum_{k=1}^{m}[F_{\underline{V}}(y_{[m,k]})-\psi_{(k,m)}]^{2},$$

where $\psi_{(k,m)}=\dfrac{k}{m+1}$. Then, we have

$$OLSE(\underline{V})=\sum_{k=1}^{m}\left(\frac{1-[1-\eta_{(y_{[m,k]},a)}]^{\theta}}{1-\bar{\delta}[1-\eta_{(y_{[m,k]},a)}]^{\theta}}-\psi_{(k,m)}\right)^{2},$$

The LSEs are obtained via solving the following non-linear equations

$$0=\sum_{k=1}^{m}\left(\frac{1-[1-\eta_{(y_{[m,k]},a)}]^{\theta}}{1-\bar{\delta}[1-\eta_{(y_{[m,k]},a)}]^{\theta}}-\psi_{(k,m)}\right)C_{(\delta)}(y_{[m,k]};\underline{V}),$$

$$0=\sum_{k=1}^{m}\left(\frac{1-[1-\eta_{(y_{[m,k]},a)}]^{\theta}}{1-\bar{\delta}[1-\eta_{(y_{[m,k]},a)}]^{\theta}}-\psi_{(k,m)}\right)C_{(\theta)}(y_{[m,k]};\underline{V}),$$

and

$$0=\sum_{k=1}^{m}\left(\frac{1-[1-\eta_{(y_{[m,k]},a)}]^{\theta}}{1-\bar{\delta}[1-\eta_{(y_{[m,k]},a)}]^{\theta}}-\psi_{(k,m)}\right)C_{(a)}(y_{[m,k]};\underline{V}),$$

where $C_{(\delta)}(y_{[m,k]};\underline{V})$, $C_{(\theta)}(y_{[m,k]};\underline{V})$ and $C_{(a)}(y_{[m,k]};\underline{V})$ are defined above.

4.4 WLSE Method

The WLSE are obtained by minimizing the function **WLSE** with respect to δ, θ and a, where

$$WLSE(\underline{V})=\sum_{k=1}^{m} w_{(k,m)}[F_{\underline{V}}(y_{[m,k]})-\psi_{(k,m)}]^{2},$$

and

$$w_{(k,m)}=[(1+m)^{2}(2+m)]/[k(1+m-k)].$$

The WLSEs are obtained by solving

$$0=\sum_{k=1}^{m}w_{(k,m)}\left(\frac{1-[1-\eta_{(y_{[m,k]},a)}]^{\theta}}{1-\bar{\delta}[1-\eta_{(y_{[m,k]},a)}]^{\theta}}-\psi_{(k,m)}\right)C_{(\delta)}(y_{[m,k]};\underline{V}),$$

$$0=\sum_{k=1}^{m}w_{(k,m)}\left(\frac{1-[1-\eta_{(y_{[m,k]},a)}]^{\theta}}{1-\bar{\delta}[1-\eta_{(y_{[m,k]},a)}]^{\theta}}-\psi_{(k,m)}\right)C_{(\theta)}(y_{[m,k]};\underline{V}),$$

and

$$0=\sum_{k=1}^{m}w_{(k,m)}\left(\frac{1-[1-\eta_{(y_{[m,k]},a)}]^{\theta}}{1-\bar{\delta}[1-\eta_{(y_{[m,k]},a)}]^{\theta}}-\psi_{(k,m)}\right)C_{(a)}(y_{[m,k]};\underline{V}).$$

4.5 The ADE Method

The ADE of δ, θ and a are obtained by minimizing the function.

$$ADE_{(y_{[m,k]},y_{[1-k+m:m]})}(\underline{V})=-m$$

$$-m^{-1}\sum_{k=1}^{m}(2k-1)\left\{\begin{matrix}\log F_{(\underline{V})}(y_{[m,k]})\\+\log[1-F_{(\underline{V})}(y_{[1-k+m:m]})]\end{matrix}\right\}.$$

The parameter estimates of δ, θ and a follow by solving the nonlinear equations

$$0=\partial[ADE_{(y_{[m,k]},y_{[1-k+m:m]})}(\underline{V})]/\partial\delta,$$

$$0=\partial[ADE_{(y_{[m,k]},y_{[1-k+m:m]})}(\underline{V})]/\partial\theta,$$

and

$$0 = \partial[ADE_{(y_{[m,k]}, y_{[1-k+m:m]})}(\underline{V})] / \partial a.$$

4.6 The RTADE Method

The RTADE of δ, θ and a are obtained by minimizing

$$\text{RTADE}_{(y_{[m,k]}, y_{[1-k+m:m]})}(\underline{V}) = \frac{1}{2}m - 2\sum_{k=1}^{m} F_{(\underline{V})}(y_{[m,k]})$$

$$-\frac{1}{m}\sum_{k=1}^{m}(2k-1)\{log\,[1 - F_{(\underline{V})}(y_{[1-k+m:m]})]\}.$$

The estimates of δ, θ and a follow by solving the nonlinear equations,

$$0 = \partial[\text{RTADE}_{(y_{[m,k]}, y_{[1-k+m:m]})}(\underline{V})] / \partial\delta,$$

$$0 = \partial[\text{RTADE}_{(y_{[m,k]}, y_{[1-k+m:m]})}(\underline{V})] / \partial\theta,$$

and

$$0 = \partial[\text{RTADE}_{(y_{[m,k]}, y_{[1-k+m:m]})}(\underline{V})] / \partial a.$$

4.7 The LTADE Method

The LTADE of δ, θ and a are obtained by minimizing

$$\text{LTADE}_{(y_{[m,k]})}(\underline{V}) = -\frac{3}{2}m + 2\sum_{k=1}^{m} F_{(\underline{V})}(y_{[m,k]})$$

$$-\frac{1}{m}\sum_{k=1}^{m}(2k-1)\,log\,F_{(\underline{V})}(y_{[m,k]}).$$

The parameter estimates of δ, θ and a follow by solving the nonlinear equations,

$$0 = \partial[\text{LTADE}_{(y_{[m,k]})}(\underline{V})] / \partial\delta,$$

$$0 = \partial[\text{LTADE}_{(y_{[m,k]})}(\underline{V})] / \partial\theta,$$

and

$$0 = \partial[\text{LTADE}_{(y_{[m,k]})}(\underline{V})] / \partial a.$$

5. Simulations for Comparing Estimation Methods

A Markov chain Monte Carlo simulation is performed to compare the classical estimation methods. The simulation study is based on $M = 1000$ generated data sets from the MOLBX model where m = 50,150,300 and 500 and

	δ	θ	a
I	2.5	2	1.2
II	0.9	3	0.9
II	1.6	5	3

The estimates are compared in terms of their "biases" and "root mean square errors (RMSE)". The "mean of the absolute difference between the theoretical and the estimates (D-ABS)" and the "maximum absolute difference between the true parameters and estimates (D-MAX)" are also reported in Tables 2, 3 and 4. From Tables 2,3 and 4 we note that the biases and the RMSE tend to zero when m increases which means the incidence of consistency property and the average estimated values (AEVs) tend to initial values when increases.

Table 2: Simulation results for parameters $\delta = 2.5$, $\theta = 2$, $a = 1.2$.

Methods	m	Bias			RMSE			D	
		δ	θ	a	δ	θ	a	ABS	MAX
MLE		0.05548	0.03603	0.00934	0.60219	0.23070	0.13926	0.00241	0.00360
OLS		0.18763	0.02504	0.03920	0.65758	0.24468	0.15428	0.02654	0.03855
WLS		0.27258	0.03036	0.06432	0.74908	0.23168	0.16409	0.09070	0.05675
CVM	50	0.05679	0.02801	0.00706	0.60242	0.25312	0.14536	0.00170	0.00308
ADE		0.05548	0.02102	0.00499	0.60219	0.23774	0.14144	0.01860	0.00437
RTADE		0.07442	0.01732	0.01060	0.63550	0.22858	0.15441	0.00507	0.00784
LTADE		0.04845	0.03827	0.00293	0.62327	0.28661	0.14309	0.00206	0.00424
MLE		0.02296	0.00948	0.00534	0.36312	0.13073	0.08629	0.00162	0.00268
OLS		0.06421	0.00878	0.01363	0.35767	0.14313	0.08691	0.00930	0.01353
WLS		0.11865	0.01257	0.03243	0.40599	0.13311	0.09252	0.01840	0.02676
CVM	150	0.02184	0.00881	0.00456	0.34671	0.14473	0.08509	0.00113	0.00173
ADE		0.02191	0.00595	0.00423	0.34380	0.13562	0.08220	0.00300	0.00330
RTADE		0.02745	0.00445	0.00422	0.35966	0.12882	0.08944	0.00227	0.00343
LTADE		0.01985	0.01164	0.00180	0.35379	0.16227	0.08239	0.00046	0.00080
MLE		0.01825	0.00362	0.00279	0.24459	0.08990	0.05689	0.00061	0.00111
OLS		0.03581	0.00585	0.00754	0.25893	0.10459	0.06346	0.00561	0.00814
WLS		0.07301	0.00741	0.02198	0.28290	0.09356	0.06547	0.01187	0.01727
CVM	300	0.02139	0.00294	0.00304	0.25474	0.10505	0.06276	0.00093	0.00162
ADE		0.02103	0.00287	0.00248	0.25215	0.09833	0.06067	0.00132	0.00197
RTADE		0.01358	0.00332	0.00201	0.26004	0.09366	0.0651	0.00093	0.00145
LTADE		0.01052	0.00576	0.00091	0.26217	0.11866	0.06143	0.00024	0.00037
MLE		0.00905	0.00317	0.00152	0.19919	0.07183	0.04659	0.00052	0.00086
OLS		0.03401	0.00798	0.00773	0.20038	0.08049	0.04905	0.00533	0.00776
WLS		0.05514	0.00429	0.01726	0.21844	0.07315	0.05076	0.00898	0.01305
CVM	500	0.01474	0.00274	0.00225	0.19763	0.08039	0.04851	0.00030	0.00045
ADE		0.01475	0.00124	0.00197	0.19621	0.07606	0.04698	0.00120	0.00193
RTADE		0.00960	0.00114	0.00157	0.19810	0.07127	0.04983	0.00089	0.00133
LTADE		0.00654	0.00277	0.00050	0.19640	0.08907	0.04620	0.00017	0.00032

6. Modeling Real Data and Comparing Methods

Two real data applications are introduced for comparing the estimation methods. The first data consists of 84 aircraft windshield observations (see Murthy et al. (2004)) and called "failure times data". This second data consists of 63 observations of the strengths of 1.5 cm glass fibers, originally obtained by workers at the UK National Physical Laboratory. This data is given in Smith and Naylor (1987) and called "glass fibers data". The required computations are carried out using the MATHCAD software. In order to compare the estimation methods, we consider the Cramér-von Mises (CVM*) and the Anderson-Darling (AD*). These two statistics are widely used to determine how closely a specific CDF fits the empirical distribution of a given data set. These statistics are given by

$$CVM^* = \left[\frac{1}{12}m^{-1} + \sum_{q=1}^{n}\left(z_q - \frac{1}{2}m^{-1}(2q-1)\right)^2\right]\left(\frac{1}{2}m^{-1}+1\right),$$

and

$$\text{AD}^* = \left(1+\frac{9}{4}m^{-2}+\frac{3}{4}m^{-1}\right)\left\{m+\frac{1}{m}\sum_{q=1}^{m}(2q-1)\log\left[z_q(1-z_{m-q+1})\right]\right\},$$

Table 3: Simulation results for parameters $\delta = 0.9$, $\theta = 3$, $a = 0.9$.

Methods	m	Bias			RMSE			D	
		δ	θ	a	δ	θ	a	ABS	MAX
MLE		0.02469	0.05512	0.00592	0.23117	0.45547	0.08046	0.00304	0.00517
OLS		0.07754	0.03993	0.02361	0.26854	0.52771	0.09486	0.02845	0.04120
WLS		0.11581	0.07487	0.04054	0.29640	0.47275	0.09750	0.04562	0.06590
CVM	50	0.02993	0.06507	0.00580	0.24630	0.55207	0.09002	0.00320	0.00548
ADE		0.03104	0.04606	0.00545	0.24408	0.51540	0.08652	0.00461	0.00734
RTADE		0.03790	0.02763	0.00833	0.24897	0.46759	0.09290	0.00867	0.01304
LTADE		0.01668	0.10911	0.00075	0.24757	0.63191	0.08523	0.00602	0.00912
MLE		0.00807	0.02925	0.00137	0.13145	0.26379	0.04660	0.00274	0.00399
OLS		0.02182	0.01163	0.00644	0.13342	0.28774	0.04980	0.00806	0.01174
WLS		0.03692	0.01598	0.01588	0.14143	0.25042	0.04950	0.01568	0.02274
CVM	150	0.00654	0.02318	0.00111	0.12963	0.29235	0.04902	0.00085	0.00128
ADE		0.00513	0.01954	0.00045	0.12804	0.27399	0.04703	0.00086	0.00134
RTADE		0.01336	0.00519	0.00331	0.13420	0.26212	0.05194	0.00365	0.00541
LTADE		0.01406	0.00975	0.00284	0.13594	0.32139	0.04758	0.00317	0.00478
MLE		0.00754	0.00562	0.00209	0.08632	0.17351	0.03084	0.00051	0.00108
OLS		0.01266	0.00783	0.00404	0.09430	0.20804	0.03550	0.00494	0.00716
WLS		0.02654	0.01544	0.01178	0.10287	0.18572	0.03616	0.01182	0.01710
CVM	300	0.00509	0.00949	0.00081	0.09281	0.20951	0.03515	0.00072	0.00118
ADE		0.00483	0.00667	0.00076	0.09142	0.19688	0.03382	0.00072	0.00115
RTADE		0.00775	0.00219	0.00197	0.09328	0.18123	0.03630	0.00256	0.00375
LTADE		0.00546	0.00803	0.00087	0.09277	0.22468	0.03291	0.00078	0.00125
MLE		0.00164	0.00415	0.00011	0.06917	0.13811	0.02482	0.00021	0.00034
OLS		0.00812	0.00779	0.00257	0.06929	0.15463	0.02632	0.00339	0.00490
WLS		0.01702	0.01297	0.00817	0.07548	0.13878	0.02699	0.00814	0.01177
CVM	500	0.00359	0.00260	0.00055	0.06856	0.15508	0.02615	0.00046	0.00102
ADE		0.00198	0.00257	0.00002	0.06819	0.14581	0.02529	0.00015	0.00026
RTADE		0.00494	0.00050	0.00131	0.07360	0.14599	0.02882	0.00159	0.00233
LTADE		0.00141	0.01014	0.00004	0.07057	0.17404	0.02521	0.00054	0.00083

respectively, where $z_q = F(z_q)$ and the z_q's values are the ordered observations.

6.1 Modeling Failure Times Data and Comparing Methods

From Table 5 we conclude that all methods performed well except the OLS method.

6.2 Modeling Glass Fibers Data and Comparing Methods

From Table 6 we conclude that LTADE method is better that all other methods with CVM* = 0.05794 and AD* = 0.39993. However, the CVM is the worst method CVM* = 0.47182 and AD* = 3.09173.

7. Modeling Real Data and Comparing Models

In this section, we provide two real data applications of the MOLBX distribution to show its potentiality. In order to compare the fits of the MOLBX model with other competing distributions, we consider the Cramér-von Mises (CVM) and the Anderson-Darling (AD). The MLEs and the corresponding standard errors (SEs) are given in Table 7 (for failure times data) and Table 9 (glass fibers data). The numerical values of the statistics CVM and AD are listed in Table 8 (failure times data) and Table 10 (glass fibers data).

Table 4: Simulation results for parameters $\delta = 1.6$, $\theta = 5$, $a = 3$.

Methods	*m*	**Bias**			**RMSE**			**D**	
		δ	θ	a	δ	θ	a	**ABS**	**MAX**
MLE		0.06640	0.04761	0.02050	0.42108	0.64998	0.23116	0.00767	0.01165
OLS		0.14065	0.09375	0.06444	0.44934	0.70422	0.24740	0.03048	0.04435
WLS		0.16594	0.08032	0.08168	0.48271	0.66370	0.25241	0.03489	0.05103
CVM	50	0.04371	0.10083	0.00414	0.44074	0.79420	0.25285	0.00456	0.00664
ADE		0.04694	0.05576	0.00746	0.41108	0.70377	0.23351	0.00304	0.00499
RTADE		0.07302	0.01771	0.02156	0.45087	0.64854	0.24867	0.01017	0.01514
LTADE		0.06559	0.05487	0.01693	0.42018	0.83587	0.23599	0.00658	0.01012
MLE		0.01668	0.03438	0.00385	0.22561	0.37484	0.12916	0.00204	0.00300
OLS		0.04878	0.02792	0.02249	0.24893	0.43268	0.14538	0.01051	0.01529
WLS		0.08113	0.03694	0.04567	0.27108	0.38747	0.14748	0.01822	0.02656
CVM	150	0.01634	0.05225	0.00779	0.21964	0.41009	0.13141	0.00298	0.00436
ADE		0.02087	0.01143	0.00564	0.23466	0.39831	0.13642	0.00251	0.00378
RTADE		0.01680	0.01374	0.00357	0.23283	0.36974	0.13817	0.00130	0.00247
LTADE		0.01257	0.03939	0.00018	0.24350	0.48892	0.14154	0.00092	0.00170
MLE		0.01174	0.00774	0.00553	0.15961	0.24741	0.09310	0.00066	0.00117
OLS		0.02080	0.01926	0.01018	0.16928	0.29990	0.10026	0.00491	0.00713
WLS		0.04590	0.02095	0.02853	0.18017	0.26187	0.10080	0.01085	0.01579
CVM	300	0.00499	0.00644	0.00664	0.16401	0.29372	0.09766	0.00124	0.00253
ADE		0.00766	0.00652	0.00113	0.16217	0.28133	0.09562	0.00063	0.00099
RTADE		0.00724	0.01445	0.00066	0.16931	0.27472	0.10172	0.00020	0.00031
LTADE		0.00924	0.01258	0.00206	0.16446	0.32461	0.09577	0.00062	0.00103
MLE		0.00690	0.00261	0.00200	0.12304	0.19520	0.07033	0.00063	0.00100
OLS		0.01980	0.01001	0.00918	0.12572	0.22251	0.07469	0.00432	0.00630
WLS		0.02439	0.00071	0.01656	0.13996	0.21302	0.07969	0.00545	0.00796
CVM	500	0.00293	0.00219	0.00113	0.12450	0.22626	0.07486	0.00036	0.00059
ADE		0.00429	0.00574	0.00053	0.12514	0.21717	0.07373	0.00022	0.00038
RTADE		0.00444	0.00724	0.00045	0.13238	0.21164	0.07955	0.00016	0.00029
LTADE		0.00329	0.01486	0.00038	0.13459	0.26943	0.07901	0.00055	0.00088

Table 5: Comparing estimation methods via an application.

Methods	$\hat{\delta}$	$\hat{\theta}$	$\hat{a}$	**CVM***	**AD***
MLE	2.67333	0.19220	0.60788	0.07696	0.57216
OLS	0.88898	0.12779	3.03425	0.24723	1.91613
WLS	2.73706	0.21360	1.27347	0.08684	0.55084
CVM	1.68589	0.16529	1.25096	0.06220	0.90125
ADE	2.03866	0.17590	1.00613	0.06006	0.55485
RTADE	1.37699	0.15675	1.86577	0.06118	0.91163
LTADE	2.84665	0.20578	0.81160	0.07245	0.56561

Figures 3 and 5 give the TTT, Q-Q, box, and Estimated HRF (EHRF) plots for failure times data and glass fibers data respectively. Figures 4 and 6 give the EPDF, ECDF, Probability-Probability (P-P) and Kaplan-Meier survival plots for the failure times data and glass fibers data, respectively.

Table 6: Comparing estimation methods via an application.

Methods	$\hat{\delta}$	$\hat{\theta}$	$\hat{a}$	CVM*	AD*
MLE	0.01114	0.05282	15.29034	0.07174	0.44446
OLS	0.00728	0.03011	14.69405	0.06898	0.43106
WLS	0.00486	0.02458	16.02191	0.07672	0.46692
CVM	55.0615	1.84835	1.25096	0.47182	3.09173
ADE	0.03095	0.13764	14.82671	0.06314	0.41266
RTADE	0.00791	0.04170	16.46871	0.07650	0.46559
LTADE	**0.04545**	**0.17950**	**14.10260**	**0.05794**	**0.39993**

8. Modeling Failure Times Data

The data consists of 84 observations. Here, we shall compare the fits of the MOLBX distribution with those of other competitive models, namely: Poisson Topp Leone-W (PTLW), MO extended-W (MOEW) (Ghitany et al., (2005)), Odd Lindley BX (OLBX) (Silva et al. (2017)), Gamma-W (GamW) (Provost et al., (2011)), Kumaraswamy-W (KumW) (Cordeiro et al., (2010)), Beta-W (Lee et al., (2007)), Transmuted modified-W (TrMW) (Khan and King, (2013)), W-Fréchet (WFr) (Afify et al., (2016b)), Kumaraswamy transmuted-W (KumTrW) (Afify et al., (2016a)), Modified beta-W (MBW) (Khan, (2015)) Mcdonald-W (MacW) (Cordeiro et al., (2014)), transmuted exponentiated generalized W (TrEGW) (Yousof et al., (2015)) distributions. Some other extensions of the W distribution can also be used in this comparison, but are not limited to Yousof et al. (2015), Alizadeh et al. (2016), Yousof et al. (2016a,b), Brito et al. (2017), Cordeiro et al. (2017a,b), Korkmaz et al. (2017), Yousof et al. (2017a-d), Aryal et al. (2017a,b), Yousof

Table 7: MLEs and standard errors (SEs) for failure times data set.

Distribution	Estimates (SEs)				
MOLBX(δ,θ,a)	3.31573	0.20480	0.5303		
	(1.42272)	(0.0322)	(0.2576)		
PTLW(θ,β,α)	−5.78175	4.22865	0.65801		
	(1.395)	(1.167)	(0.039)		
MOEW(θ,β,α)	488.899	0.2832	1261.97		
	(189.358)	(0.013)	(351.07)		
GamW(θ,β,α)	2.37697	0.84809	3.5344		
	(0.378)	(0.00053)	(0.665)		
OLBX(θ,β,α)	1.45406	0.7543	0.2379		
	(0.9018)	(0.2530)	(0.0317)		
KumW(δ,θ,β,α)	14.4331	0.2041	34.6599	81.8459	
	(27.095)	(0.042)	(17.527)	(52.014)	
WFr(δ,θ,c,α)	630.94	0.3024	416.097	1.1664	
	(697.94)	(0.032)	(232.359)	(0.357)	
Beta-W(δ,θ,β,α)	1.36	0.2981	34.1802	11.4956	
	(1.002)	(0.06)	(14.838)	(6.73)	
TrMW(δ,θ,β,α)	0.2722	1	4.6×10^{-6}	0.4685	
	(0.014)	(5×10^{-5})	(2×10^{-4})	(0.165)	
KumTrW(δ,θ,λ,β,α)	27.7912	0.178	0.4449	29.5253	168.06
	(33.401)	(0.017)	(0.609)	(9.792)	(129.17)
MBW(δ,θ,λ,β,α)	10.1502	0.1632	57.4167	19.3859	2.0043
	(18.697)	(0.019)	(14.063)	(10.019)	(0.662)
MacW(δ,θ,λ,β,α)	1.9401	0.306	17.69	33.639	16.721
	(1.011)	(0.045)	(6.222)	(19.99)	(9.72)
TrEGW(δ,θ,λ,β,α)	4.2567	0.1532	0.0978	5.2313	1173.33
	(33.401)	(0.017)	(0.609)	(9.792)	(6.999)

et al. (2018) and Hamedani et al. (2018 and 2019). Based on the results in Table 8 we conclude that the new lifetime model provides adequate fits as compared to other competitive models with small values for CVM* and AD*. The proposed MOLBX lifetime model is much better than all mentioned models and a good alternative to these models with CVM* = 0.0769 and AD* = 0.5722.

Table 8: CVM and AD statistics for failure times data set.

Distribution	CVM*	AD*
MOLBX	**0.0769**	**0.5722**
OLBX	0.0792	0.5910
PTLW	0.1397	1.1939
MOEW	0.3995	4.4477
GamW	0.2553	1.9489
KumW	0.1852	1.5059
WFr	0.2537	1.9574
Beta-W	0.4652	3.2197
TrMW	0.8065	11.205
KumTrW	0.1640	1.3632
MBW	0.4717	3.2656
MacW	0.1986	1.5906
TrEGW	1.0079	6.2332

9. Modeling Glass Fibers Data

For this data set, we shall compare the fits of the new distribution with some competitive models like OLEW and E-W. Based on Table 10 we conclude that the proposed MOLBX model is much better than OLEW and E-W models with CVM* = 0.0717 and AD* = 0.4445.

Table 9: MLEs (SEs) for glass fibers data.

Distribution	Estimates (SEs)		
MOLBX(δ,θ,a)	973.88	1.9405	0.3099
	(9.001)	0.2179	(0.001)
OLEW(δ,θ,a)	0.5088	2.534	1.7122
	(0.397)	(1.830)	(0.096)
E-W(δ,θ,a)	0.6710	7.2850	1.7180
	(0.249)	(1.707)	(0.086)

Table 10: CVM and AD statistics for glass fibers data.

Distribution	CVM*	AD*
MOLBX	**0.0717**	**0.4445**
OLEW	0.2711	1.4965
E-W	0.6361	3.4842

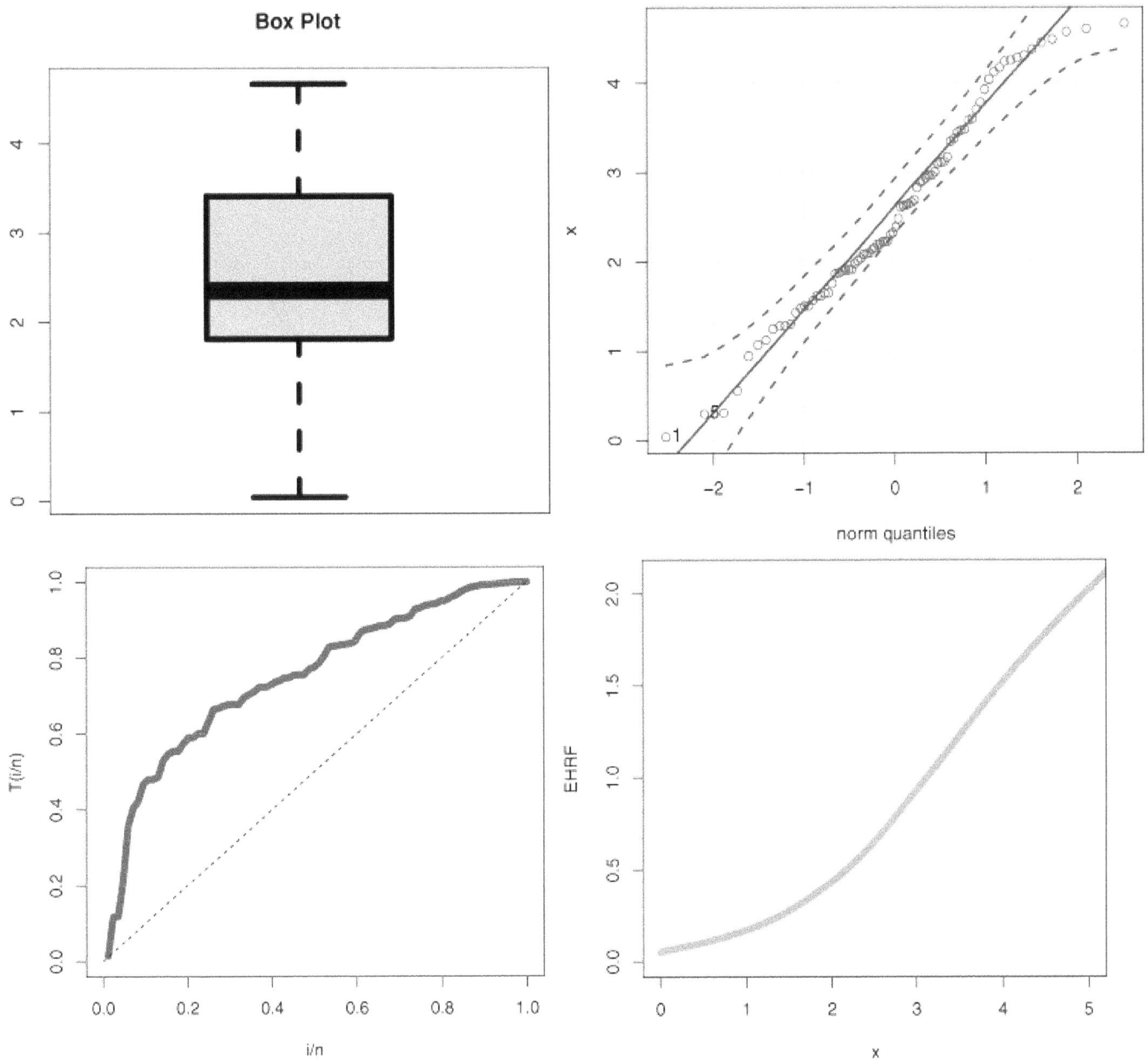

Figure 3: TTT, Q-Q, box, and EHRF plots for failure times data.

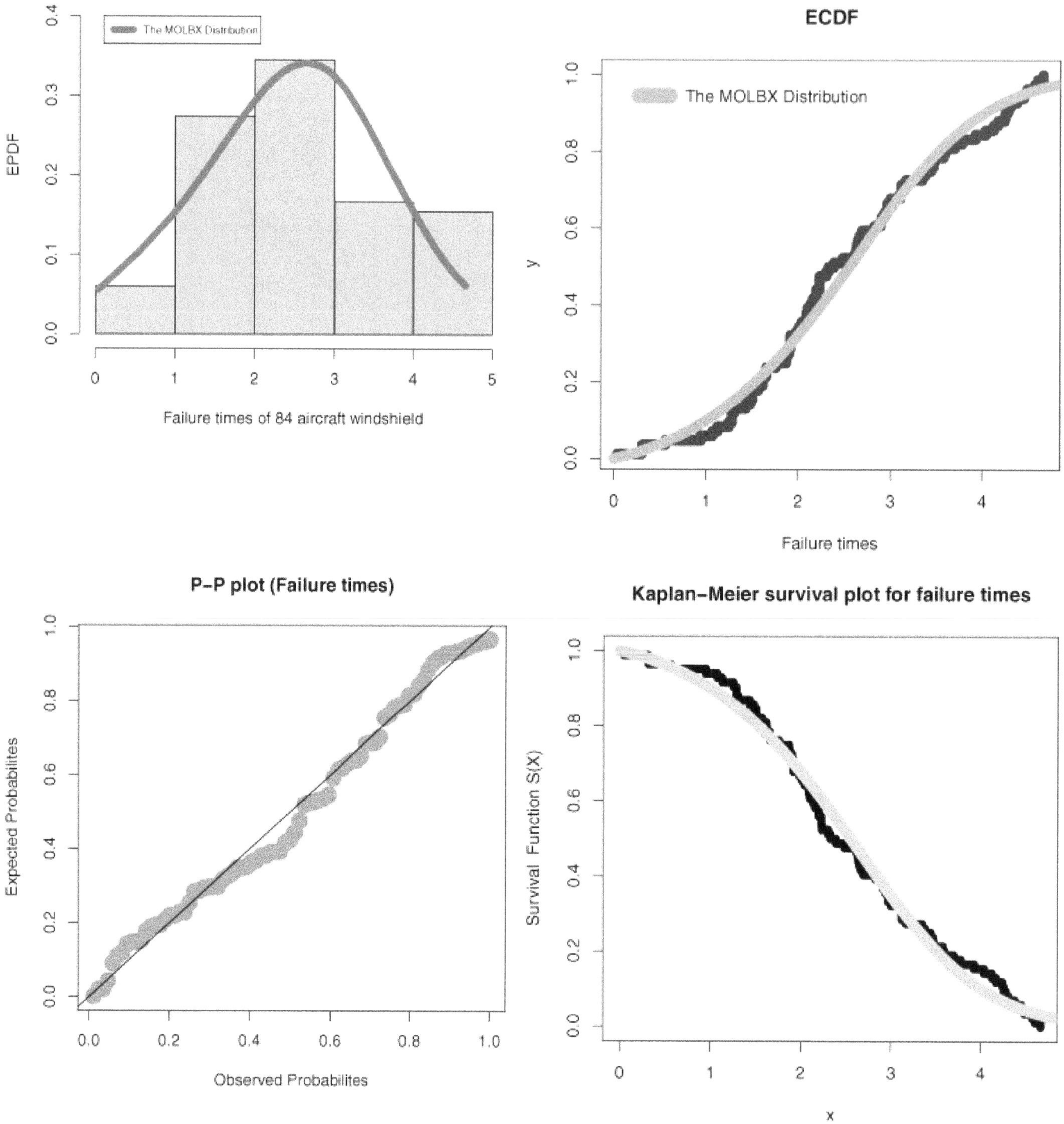

Figure 4: EPDF, ECDF, P-P and Kaplan-Meier plots for the failure times data.

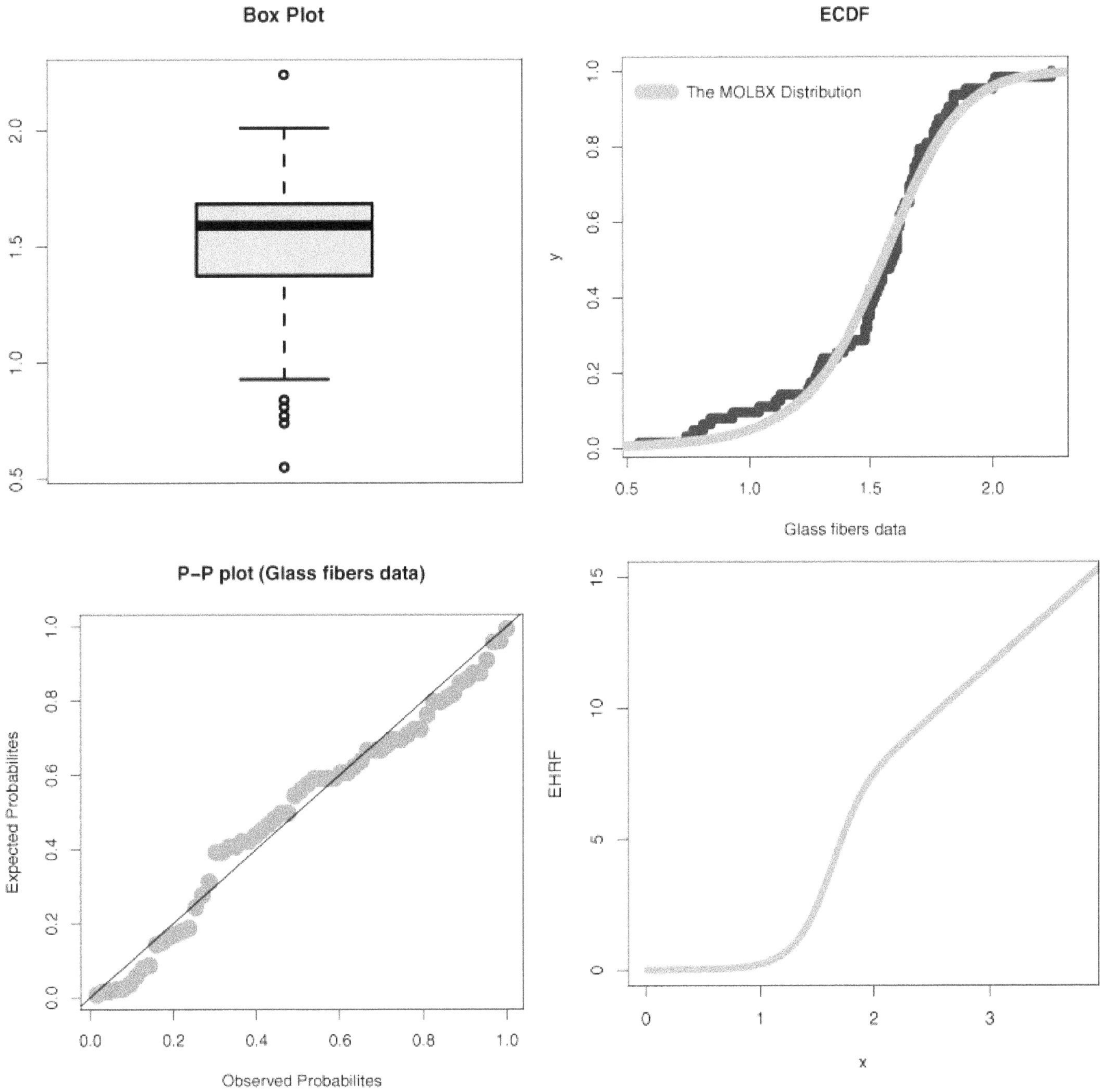

Figure 5: TTT, Q-Q, box, and EHRF plots for glass fibers data.

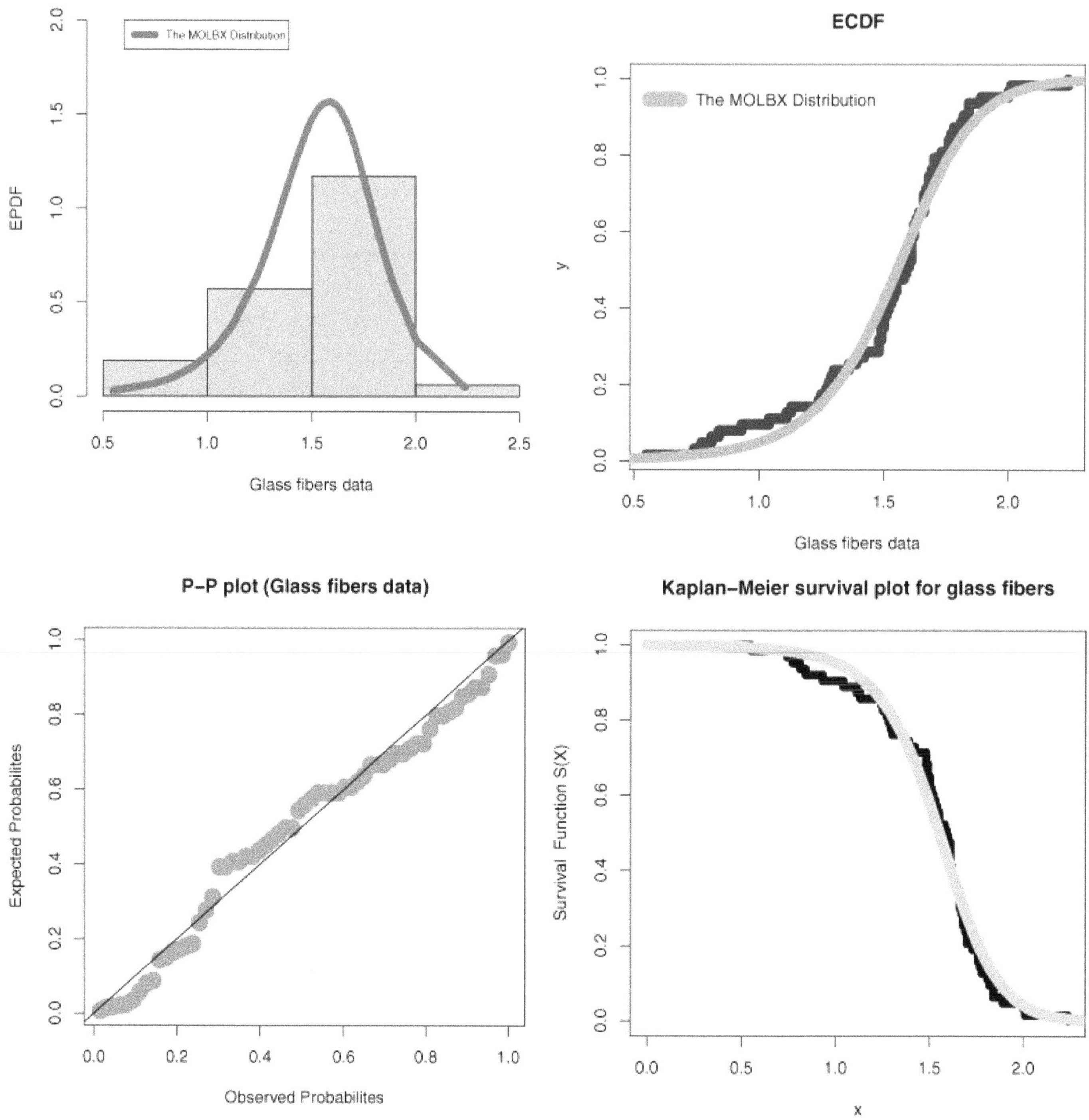

Figure 6: EPDF, ECDF, P-P and Kaplan-Meier plot plots for the glass fibers data.

10. Concluding Remarks

A new generalization of Burr type X distribution is proposed and studied. Some bivariate and multivariate type models are derived based on "Clayton Copula", "Farlie Gumbel Morgenstern Copula", "modified Farlie Gumbel Morgenstern Copula" and "Renyi's entropy Copula". Some of its mathematical properties are derived and analyzed. Different estimation methods such as "Cramér-von-Mises" method, "maximum likelihood method", "weighted least square" method, "Anderson Darling" method, "ordinary least square" method, "right and left tail Anderson Darling" methods are briefly described and compared using two real-life data sets. Comprehensive Monte Carlo simulation studies are performed for comparing the performances of the estimation methods. The usefulness and flexibility of the new Burr X distribution is illustrated by means of two real data sets.

References

Afify, A. Z., Cordeiro, G. M., Yousof, H. M. Alzaatreh, A. and Nofal, Z. M. 2016a. The Kumaraswamy transmuted-G family of distributions: properties and applications, 14: 245–270.

Afify, A. Z., Yousof, H. M., Cordeiro, G. M., Ortega, E. M. M et al. 2016b. The Weibull Fréchet distribution and its applications. Journal of Applied Statistics, 43(14): 2608–2626.

Ahmad, K. E., Fakhry, M. E. and Jaheen, Z. F. 1997. Empirical Bayes estimation of P(Y < X) and characterization of Burr-type X model. Journal of Statistical Planning and Inference, 64: 297–308.

Al-Babtain, A. A., Elbatal, I. and Yousof, H. M. 2020a. A new exible three-parameter model: properties, Clayton Copula, and modeling real data, 12(13): 1–26. doi: 10.3390/sym1203044017.

Al-Babtain, A. A., Elbatal, I. and Yousof, H. M. 2020b. A new three parameter Fréchet model with mathematical properties and applications. Journal of Taibah University for Science, 14(1): 265–278.

Alizadeh, M., Ghosh, I., Yousof, H. M., Rasekhi, M. and Hamedani, G. G. 2017. The generalized odd generalized exponential family of distributions: properties, characterizations and applications. J. Data Sci., 15: 443–466.

Alizadeh, M., Rasekhi, M., Yousof, H. M. and Hamedani, G. G. 2018. The transmuted Weibull G family of distributions. Hacettepe Journal of Mathematics and Statistics, 47(6): 1–20.

Alzaatreh, A., Lee, C. and Famoye, F. 2013. A new method for generating families of continuous distributions. Metron, 71: 63–79.

Aryal, G. R., Ortega, E. M., Hamedani, G. G. and Yousof, H. M. 2017a. The ToppLeone Generated Weibull distribution: regression model, characterizations and applications. International Journal of Statistics and Probability, 6: 126–141.

Aryal, G. R. and Yousof, H. M. 2017b. The exponentiated generalized-G Poisson family of distributions. Economic Quality Control, 32(1): 1–17.

Bjerkedal, T. 1960. Acquisition of resistance in guinea pigs infected with different doses of virulent tubercle bacilli. American Journal of Hygiene, 72: 130–148.

Bourguignon, M., Silva, R. B. and Cordeiro, G. M. 2014. The Weibull--G family of probability distributions. Journal of Data Science, 12: 53–68.

Brito, E., Cordeiro, G. M., Yousof, H. M., Alizadeh, M. and Silva, G. O. 2017. Topp-Leone odd log-logistic family of distributions. Journal of Statistical Computation and Simulation, 87(15): 3040–3058.

Burr, I. W. 1942. Cumulative frequency distribution, Annals of Mathematical Statistics, 13: 215–232.

Cordeiro, G. M., Hashimoto, E. M., Edwin and E. M. M. Ortega. 2014. The McDonald Weibull model. Statistics: A Journal of Theoretical and Applied Statistics, 48: 256–278.

Cordeiro, G. M., Ortega, E. M. and Nadarajah, S. 2010. The Kumaraswamy Weibull distribution with application to failure data. Journal of the Franklin Institute, 347: 1399–1429.

Cordeiro, G. M., Yousof, H. M., Ramires, T. G. and Ortega, E. M. M. 2017b. The Burr XII system of densities: properties, regression model and applications. Journal of Statistical Computation and Simulation, 88(3): 432–456.

Elbatal, I. and Aryal, G. 2013. On the transmuted additive Weibull distribution. Austrian Journal of Statistics, 42(2): 117–132.

Farlie, D. J. G. 1960. The performance of some correlation coefficients for a general bivariate distribution. Biometrika, 47: 307–323.

Gleaton, J. U. and Lynch, J. D. 2006. Properties of generalized loglogistic families of lifetime distributions. Journal of Probability and Statistical Science, 4: 51–64.

Gumbel, E. J. 1961. Bivariate logistic distributions. Journal of the American Statistical Association, 56(294): 335–349.

Gumbel, E. J. 1960. Bivariate exponential distributions. Journ. Amer. Statist. Assoc., 55: 698–707.

Gupta, R. C. and Gupta, R. D. 2007. Proportional reversed hazard rate model and its applications. J. Statist Plan Inference, 137: 3525–3536.

Hamedani, G. G., Yousof, H. M., Rasekhi, M., Alizadeh, M. and Najibi, S. M. 2018. Type I general exponential class of distributions. Pak. J. Stat. Oper. Res., XIV(1): 39–55.

Hamedani, G. G., Rasekhi, M., Najibi, S. M., Yousof, H. M. and Alizadeh, M. 2019. Type II general exponential class of distributions. Pak. J. Stat. Oper. Res., XV(2): 503–523.

Jaheen, Z. F. 1995. Bayesian approach to prediction with outliers from the Burr type X model. Microelectronic Reliability, 35: 45–47.

Jaheen, Z. F. 1996. Empirical Bayes estimation of the reliability and failure rate functions of the Burr type X failure model. Journal of Applied Statistical Sciences, 3: 281–288.

Johnson, N. L. and Kotz, S. 1975. On some generalized Farlie- Gumbel- Morgenstern distributions. Commun. Stat. Theory, 4: 415–427.

Johnson, N. L. and Kotz, S. 1977. On some generalized Farlie- Gumbel- Morgenstern distributions- II: Regression, correlation and further generalizations. Commun. Stat. Theory, 6: 485–496.

Khalil, M. G., Hamedani, G. G. and Yousof, H. M. 2019. The Burr X exponentiated Weibull model: characterizations, mathematical properties and applications to failure and survival times data. Pak. J. Stat. Oper. Res., 15(1): 141–160.

Khan, M. N. 2015. The modified beta Weibull distribution. Hacettepe Journal of Mathematics and Statistics, 44: 1553–1568.

Khan, M. S. and King, R. 2013. Transmuted modified Weibull distribution: a generalization of the modified Weibull probability distribution. European Journal of Pure and Applied Mathematics, 6: 66–88.

Kilbas, A. A., Srivastava, H. M. and Trujillo, J. J. 2006. Theory and Applications of Fractional Difierential Equations. Elsevier, Amsterdam.

Korkmaz, M. C., Yousof, H. M., Rasekhi, M. and Hamedani, G. G. 2017. The exponential Lindley odd log-logistic G family: properties, characterizations and applications. Journal of Statistical Theory and Applications, forthcoming.

Lee, C., Famoye, F. and Olumolade, O. 2007. Beta-Weibull distribution: some properties and applications to censored data. Journal of Modern Applied Statistical Methods, 6: 17.

Lehmann, E. L. 1953. The power of rank tests. Annals of Mathematical Statistics, 24: 23–43.

Marshall, A. W. and Olkin, I. 1996. A new methods for adding a parameter to a family of distributions with application to the exponential and weibull families. Biometrika, 84: 641–652.

Morgenstern, D. 1956. Einfache beispiele zweidimensionaler verteilungen. Mitteilingsblatt fur Mathematische Statistik, 8: 234–235.

Murthy, D. N. P., Xie, M. and Jiang, R. 2004. Weibull Models. John Wiley and Sons, Hoboken, New Jersey.

Nadarajah, S., Cordeiro, G. M. and Ortega, E. M. M. 2013. The exponentiated Weibull distribution: A survey, Statistical Papers, 54: 839–877.

Pougaza, D. B. and Djafari, M. A. 2011. Maximum entropies copulas. Proceedings of the 30th international workshop on Bayesian inference and maximum Entropy methods in Science and Engineering, 329–336.

Provost, S. B., Saboor, A. and Ahmad, M. 2011. The gamma--Weibull distribution, Pak. Journal Stat., 27: 111–131.

Raqab, M. Z. 1998. Order statistics from the Burr type X model, Computers Mathematics and Applications, 36: 111–120.

Rezaei, S., Nadarajah, S. and Tahghighnia, N. A. 2013. New three-parameter lifetime distribution, Statistics, 47: 835–860.

Rinne, H. 2009. The Weibull Distribution: A Handbook. CRC Press, Boca Raton, Florida.

Ristic, M. M. and Balakrishnan, N. 2012. The gamma-exponentiated exponential distribution. Journal of Statistical Computation and Simulation, 82: 1191–1206.

Sartawi, H. A. and Abu-Salih, M. S. 1991. Bayes prediction bounds for the Burr type X model, Communications in Statistics - Theory and Methods, 20: 2307–2330.

Surles, J. G. and Padgett, W. J. 2001. Inference for reliability and stress-strength for a scaled Burr Type X distribution. Lifetime Data Analysis, 7: 187–200.

Tian, Y., Tian, M. and Zhu, Q. 2014. Transmuted linear exponential distribution: a new generalization of the linear exponential distribution. Communications in Statistics - Simulation and Computation, 43(10): 2661–2677.

Yadav, A. S., Goual, H., Alotaibi, R. M. Rezk, H., Ali, M. M. et al. 2020. Validation of the Topp-Leone-Lomax model via a modified Nikulin-Rao-Robson goodness-of-fit test with di□erent methods of estimation. Symmetry, 12: 1–26. doi: 10.3390/sym12010057.

Yousof, H. M., Alizadeh, M., Jahanshahi, S. M. A., Ramires, T. G., Ghosh, I. et al. 2017. The transmuted Topp-Leone G family of distributions: theory, characterizations and applications. Journal of Data Science, 15: 723–740.

Yousof, H. M., Majumder, M., Jahanshahi, S. M. A., Masoom Ali, M. and Hamedani, G. G. 2018. A new Weibull class of distributions: theory, characterizations and applications. Journal of Statistical Research of Iran, 15: 45–83.

CHAPTER 3

Transmuted Burr Type X Model with Applications to Life Time Data

Tabassum Naz Sindhu,[1] *Zawar Huassian*[2] and *Muhammad Aslam*[3,*]

1. Introduction

In 1942 Irving W. Burr created a system of distributions. In consideration of immense varieties of shape, this system of distributions is applicable for approaching histograms, especially when a simple mathematical model for the fitted cumulative distribution function is needed. Various standard theoretical distributions are restricting types of Burr distributions. Burr Type X (B-X) distribution holds a dominant role in modeling the life time of random phenomena. The B-X model can be utilized quite adequately to model strength as well as general survival data. B-X model with one parameter is a limiting form of EW (Exponentiated Weibull) model suggested by Mudholkar and Srivastava (1993). Various features of one parameter ($\phi = 1$) B-X model were considered by Surles and Padgett (1998) and Raqab (1998). The relation of B-X with Weibull, Generalized Exponential (GE) and EW models are considered by Raqab and Kundu (2006).

Several scholars have proposed and considered a variety of transmuted families of distributions in the literature. In 2009, Aryal and Tsokos examined the transmuted Gumbel model (TGM) for climate data. In 2011, Aryal and Tsokos established the TWM (transmuted Weibull model) and its features to explore real life applications. More recently Khan et al. (2017) developed the transmuted GE model. The transmuted Weibull Lomax model with a discussion on some features of this family is designed by Afify et al. (2015). Merovic (2013) proposed the transmuted Rayleigh model. Ahmed et al. (2014) investigated transmuted complementary Weibull geometric model. Oguntunde et al. (2014) considered the transmuted inverse exponential model. In 2017, Nofal et al. developed the transmuted exponentiated generalized-G family of models. Khan et al. (2016) studied the transmuted Kumaraswamy (TK) model and discussed a range of mathematical properties. Characterization of transmuted inverse Weibull model was investigated by Khan et al. (2014).

Transmuted models are advantageous for a clear insight of foundation models and can be expanded to other models by introducing a pliability parameter. This article commences a novel three parametric model which is a generalization of B-X model, called TB-X model using the QRTM (see Shaw et al. (2009)). Currently a generalization of distributions by transforming suitable distribution, into a supplementary general distribution, with the aid of a shape variable has been thoroughly investigated for various unlike families of survival models. A random variable ($r.v.$) X is considered to be a transmuted model if its function is of the form

$$F(x) = G(x)(1+\lambda) - \lambda\{G(x)\}^2, \tag{1}$$

where $G(x)$ is a cumulative distribution function of the foundation model. It is significant to observe that at $\lambda = 0$ the foundation model $r.v.$ is obtained.

The article is structured according to the following. The analytical shapes of subject density, reliability, distribution and hazard functions are shown in Section 2. Section 3 includes the range of mathematical characteristics specifically quantile functions, median, estimation of moment and graphical representation of the model under study. MGF and entropy are demonstrated in Sections 4 and 5 respectively. Maximum Likelihood estimates of unknown parameters and asymptotic confidence intervals of TB-X model are

[1] Department of Statistics, Quaid-i-Azam University 45320 Islamabad Pakistan; Department of Science & Humanities, National University of Computer and Emerging Sciences Islamabad 44000, Pakistan.
Email: sindhuqau@gmail.com
[2] Department of Social & Allied Sciences, Cholistan University of Veterinary & Animal Sciences, Bahawalpur 63100, Pakistan.
Email: zhlangah@yahoo.com
[3] Department of Statistics, Faculty of Science, King Abdulaziz University, Jeddah 21551, Saudi Arabia.
* Corresponding author: aslam_ravian@hotmail.com

examined in Section 6. The evaluation of performance of MLEs using simulation is considered in Section 7. Bayesian Inference of unknown parameters of the model under study is presented in Section 8. An application of the TB-X distribution to the breaking stress of glass fibers is demonstrated in Section 9. The conclusions are presented in the last Section.

2. Transmuted Burr Type X (TB-X) Model

Let $\Phi = (x;\alpha,\phi,\lambda)^T$ a *r.v.* X is said to have a TB-X probability model if its cdf and pdf are

$$\begin{aligned} F_{TB-X}(x;\ \alpha,\phi,\lambda) &= (1+\lambda)\left(1-e^{-(\phi x)^2}\right)^{\alpha} - \lambda\left\{\left(1-e^{-(\phi x)^2}\right)^{\alpha}\right\}^2, \\ &= \left\{1+\lambda-\lambda\left(1-e^{-(\phi x)^2}\right)^{\alpha}\right\}\left(1-e^{-(\phi x)^2}\right)^{\alpha}, \end{aligned} \tag{2}$$

and

$$f_{TB-X}(x;\ \alpha,\phi,\lambda) = 2\alpha x\phi^2 e^{-(\phi x)^2}\left(1-e^{-(\phi x)^2}\right)^{\alpha-1}\left\{1+\lambda-2\lambda\left(1-e^{-(\phi x)^2}\right)^{\alpha}\right\}, \tag{3}$$

respectively, where x, α, $\phi > 0$ and $|\lambda| \le 1$.

Here α and ϕ represent shape and scale parameters, respectively and λ represents the transmuting variable showing different structures of the TB-X distribution. Then $X \sim TB-X(\alpha, \phi, \lambda)$ denotes the *r.v.* along the pdf given in (3). The $R(x;\alpha,\phi,\lambda)$ reliability and $h(x;\alpha,\phi,\lambda)$ hazard function (hf) of X are specified in (4 – 5) respectively.

$$R_{TB-X}(x;\ \alpha,\ \phi,\ \lambda) = 1-\left(1-e^{-(\phi x)^2}\right)^{\alpha}\left\{1+\lambda-\lambda\left(1-e^{-(\phi x)^2}\right)^{\alpha}\right\}. \tag{4}$$

and

$$h_{TB-X}(x;\ \alpha,\ \phi,\ \lambda) = \frac{2xe^{-(\phi x)^2}\left\{1+\lambda-2\lambda\left(1-e^{-(\phi x)^2}\right)^{\alpha}\right\}\alpha\phi^2}{\left(1-e^{-(\phi x)^2}\right)^{1-\alpha}-\left(1-e^{-(\phi x)^2}\right)\left\{1+\lambda-\lambda\left(1-e^{-(\phi x)^2}\right)^{\alpha}\right\}}. \tag{5}$$

The reversed hf and chf of X are:

$$h^*_{TB-X}(x;\ \alpha,\ \phi,\ \lambda) = \frac{2\alpha x\phi^2 e^{-(\phi x)^2}\left\{1+\lambda-\left(1-e^{-(\phi x)^2}\right)^{\alpha}2\lambda\right\}}{\left(1-e^{-(\phi x)^2}\right)\left\{1+\lambda-\lambda\left(1-e^{-(\phi x)^2}\right)^{\alpha}\right\}}. \tag{6}$$

and

$$H_{TB-X}(x;\ \alpha,\ \phi,\ \lambda) = -\ln\left[1-\left(1-e^{-(\phi x)^2}\right)^{\alpha}\left\{1+\lambda\left(1-\left(1-e^{-(\phi x)^2}\right)^{\alpha}\right)\right\}\right]. \tag{7}$$

Figures (1–4) illustrate the various patterns of the density curve, cumulative function, reliability and hfs for some nominated parameter values. According to the chosen values, the density, cumulative function, reliability and hf can take many behaviors.

A variation of feasible forms of the pdf and cdf of the TB-X distribution for α, ϕ and λ are disposed in Figures 1 to 2. The various patterns of the rf and hf of the TB-X distribution are shown in Figures 3 to 4. The hazard rates of the TB-X distribution have a constant, monotonically increasing and gradually decreasing bathtub shaped function. For varying parameters failure rates are defined. It is particularly noteworthy from Figure 4, that the model unveils a constant behavior of hazard rates for $\lambda = 0.03$, $\alpha = 0.5$, and $\phi = 1$. Further the distribution displays a strictly increasing and steadily decreasing structures of hazard rates.

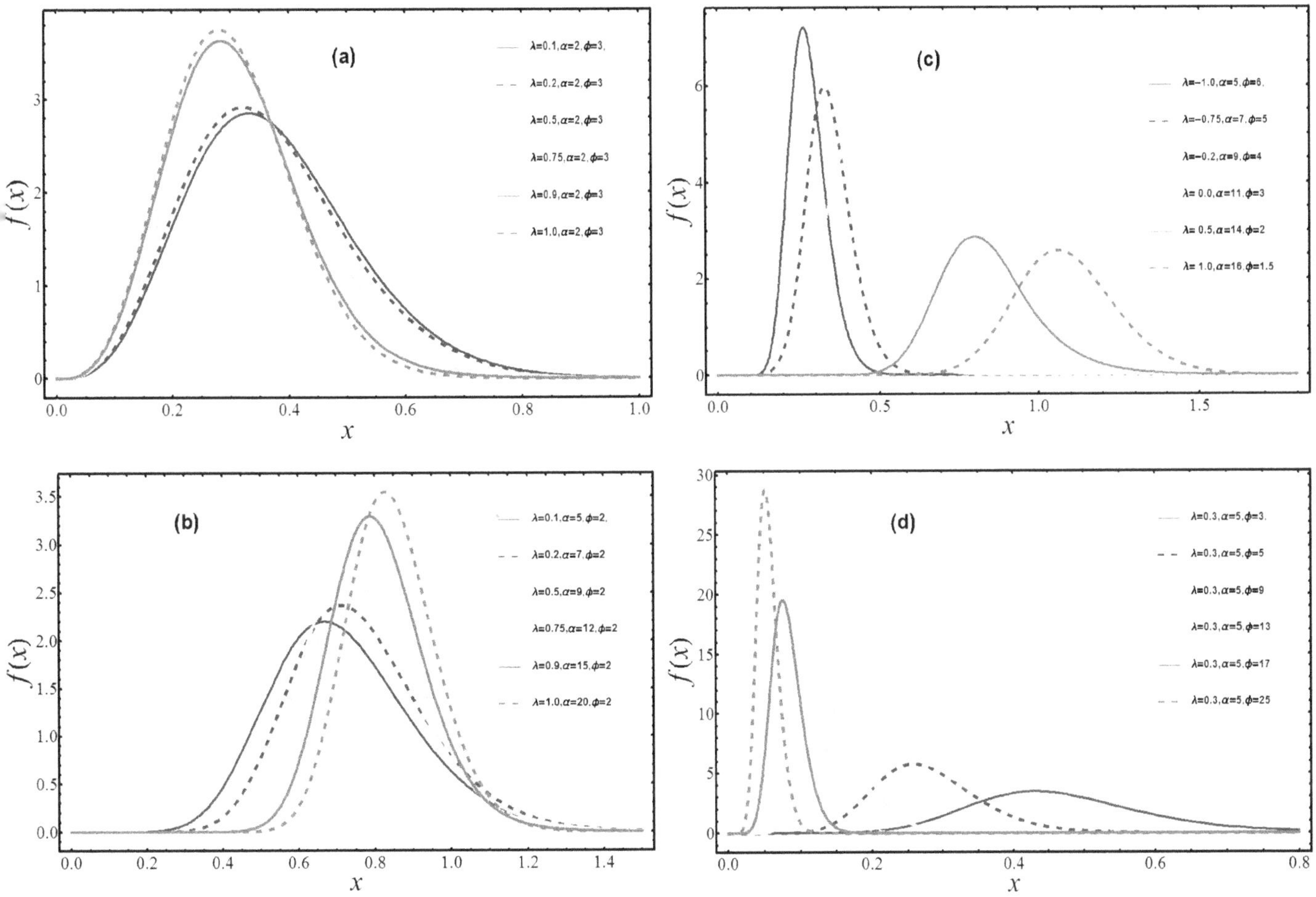

Figure 1: Graphs of TB-X model for some parameter values.

Limiting behavior of pdf of TB-X distribution

Now we examine the performance of the TB-X model as $x \to 0$ and $x \to \infty$.

$$\text{The } \lim_{x\to 0} f_{TB-X}(x;\ \alpha,\phi,\lambda) = \lim_{x\to 0} 2\alpha\phi^2 x e^{-(\phi x)^2} \frac{\left(1-\exp\left\{-(\phi x)^2\right\}\right)^{\alpha}}{\left(1-\exp\left\{-(\phi x)^2\right\}\right)} \tag{8}$$

$$\times\left\{1+\lambda-2\lambda\left(1-\exp\left\{-(\phi x)^2\right\}\right)^{\alpha}\right\}=0\Big),$$

and the

$$\lim_{x\to\infty} f_{TB-X}(x;\ \alpha,\phi,\lambda) = \lim_{x\to\infty} 2\alpha\phi^2 x e^{-(\phi x)^2} \frac{\left(1-\exp\left\{-(\phi x)^2\right\}\right)^{\alpha}}{\left(1-\exp\left\{-(\phi x)^2\right\}\right)} \left\{1+\lambda-2\lambda\left(1-\exp\left\{-(\phi x)^2\right\}\right)^{\alpha}\right\}=0,\Big) \tag{9}$$

$$\text{as} \quad \lim_{x\to\infty}\left(1-\exp\left\{-(\phi x)^2\right\}\right)=0.$$

The significance of the TB-X distribution is that it accommodates particular sub models as B-X models with two parameters, when $\lambda = 0$ and when $\lambda = 0$ and $\alpha = 1$, a subject model turns out to a B-X model with one parameter.

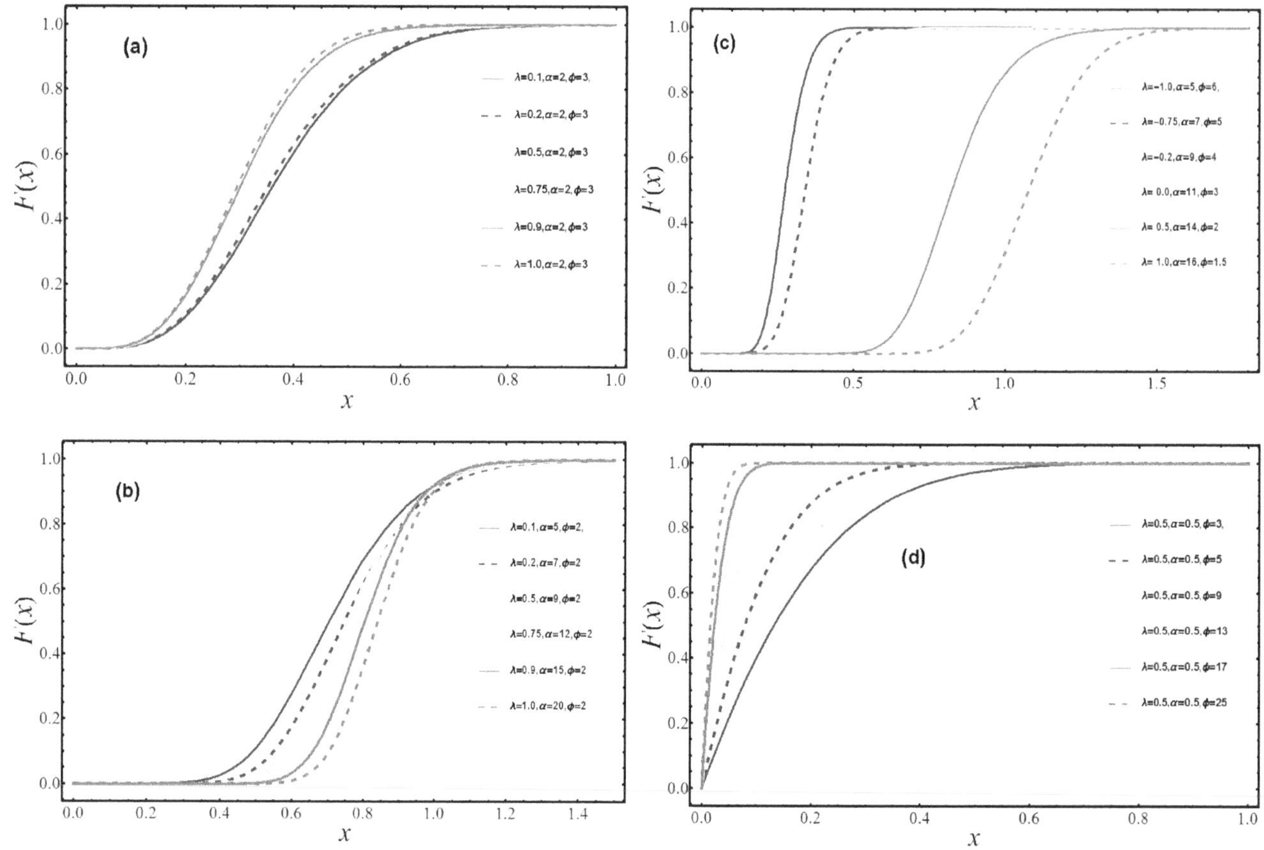

Figure 2: Graphs of the cumulative function of TB-X model.

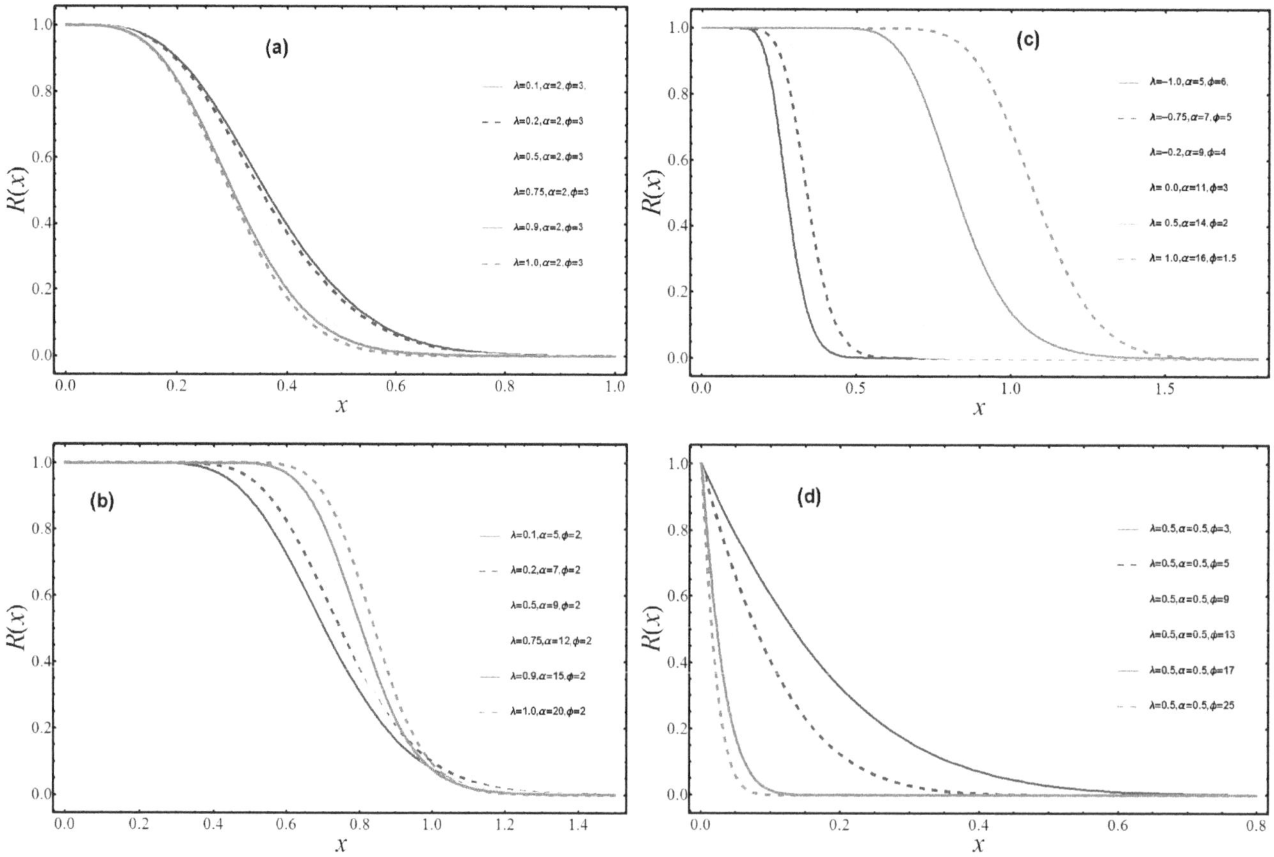

Figure 3: Graphs of the reliability function of TB-X model.

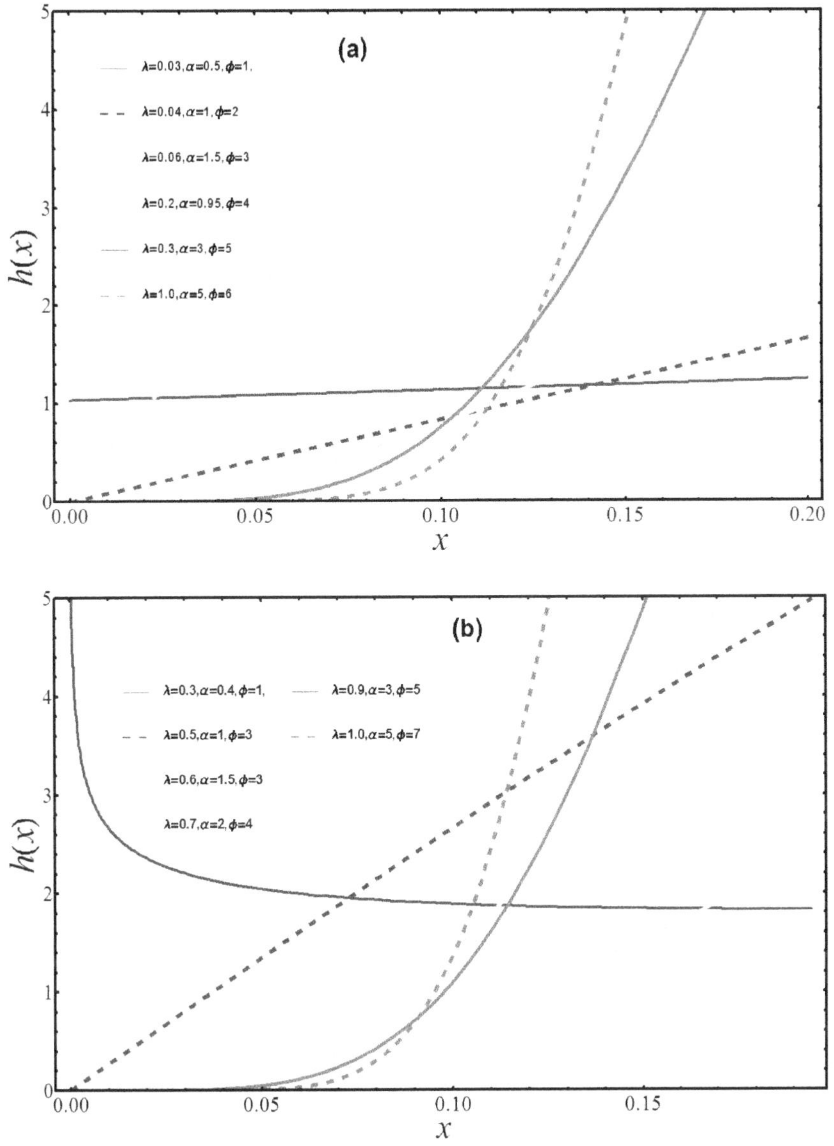

Figure 4: Graphs of hf of the TB-X model showing increasing decreasing and constant behaviour.

3. Mathematical Features

Some fundamental properties of the TB-X model are studied here. These properties are quantile function (qf.), median, the rth moment, mgf and entropy.

4. Quantile Function (qf)

In practice qf of a model is functional in several applications. The measures based on quantiles are considered to show skewness and kurtosis. The cdf of th TB-X model is inversed to get the quantile function that is stated as follows:

$$x = F^{-1}(u) = \sqrt{\frac{-1}{(\phi)^2}\ln\left\{1-\left(\frac{(1+\lambda)-\sqrt{(1+\lambda)^2-4\lambda u}}{2\lambda}\right)^{1/\alpha}\right\}}. \tag{10}$$

Simulating the r. v. for the TB-X model it is implemented by generating uniform numbers $U(0,1)$ and the concerned qf given in (10) when the parameters α, ϕ and λ are known. We study the Bowley Skewness because classical kurtosis is known to have limitations. Bowley Skewness (for details see Kenney and Keeping 1954) one of the primary measures of skewness is defined as:

$$S_k = \frac{Q_{3/4} + Q_{1/4} - 2Q_{1/2}}{Q_{3/4} - Q_{1/4}} \tag{11}$$

The Moors kurtosis (1988) established on octile of the TB-X model which can be obtained by applying the following formula:

$$M_k = \frac{-Q_{1/8} + Q_{3/8} + Q_{7/8} - Q_{5/8}}{Q_{3/4} - Q_{1/4}} \tag{12}$$

5. Median

Letting $u = 0.5$ in (10), we derive the median of the TB-X model. The median with chosen values of parameters depending upon λ is displayed in Figure 5. It is worth noting that as λ increases the median decreases asymptotically.

$$x = \sqrt{\frac{-1}{(\phi)^2} \ln\left\{1 - \left(\frac{(1+\lambda) - \sqrt{(1+\lambda^2)}}{2\lambda}\right)^{1/\alpha}\right\}}. \tag{13}$$

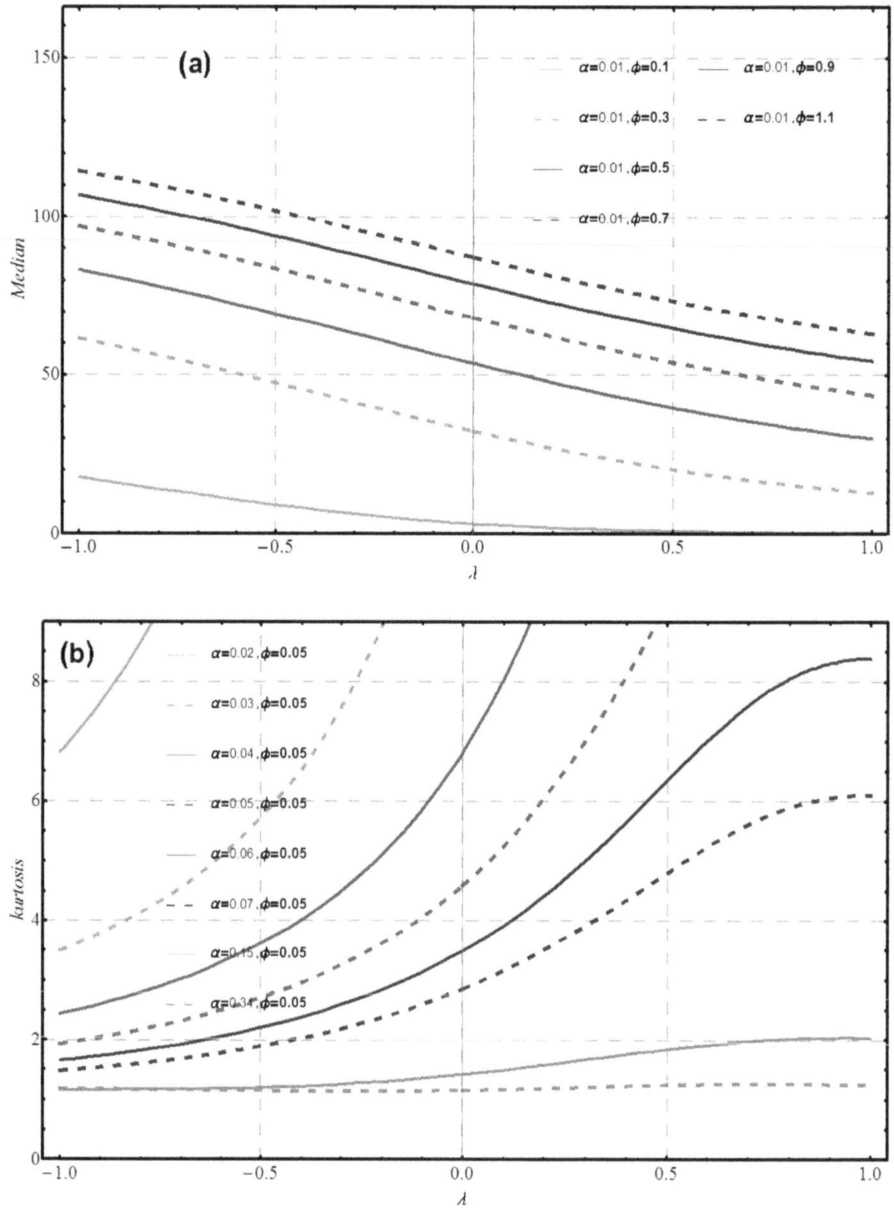

Figure 5: Graphs of median and kurtosis of the TB-X distribution.

6. Moments

Here, we study the moment and mgf for the TB-X model. Most of the characteristics of the model can be examined from moments like measures of dispersion and central tendency, and the shape and peak of the distribution. To exemplify out the impact of λ on skewness, median and kurtosis, we study the estimate depending on quantiles. Skewness and kurtosis are evaluated utilizing the relation of Bowley and Moors. Graphical representation of median and kurtosis as a function of λ under the various values of α and ϕ is sketched in Figure 5, whereas Skewness as a function of λ under the various values of α and ϕ are illustrated in Figure 6.

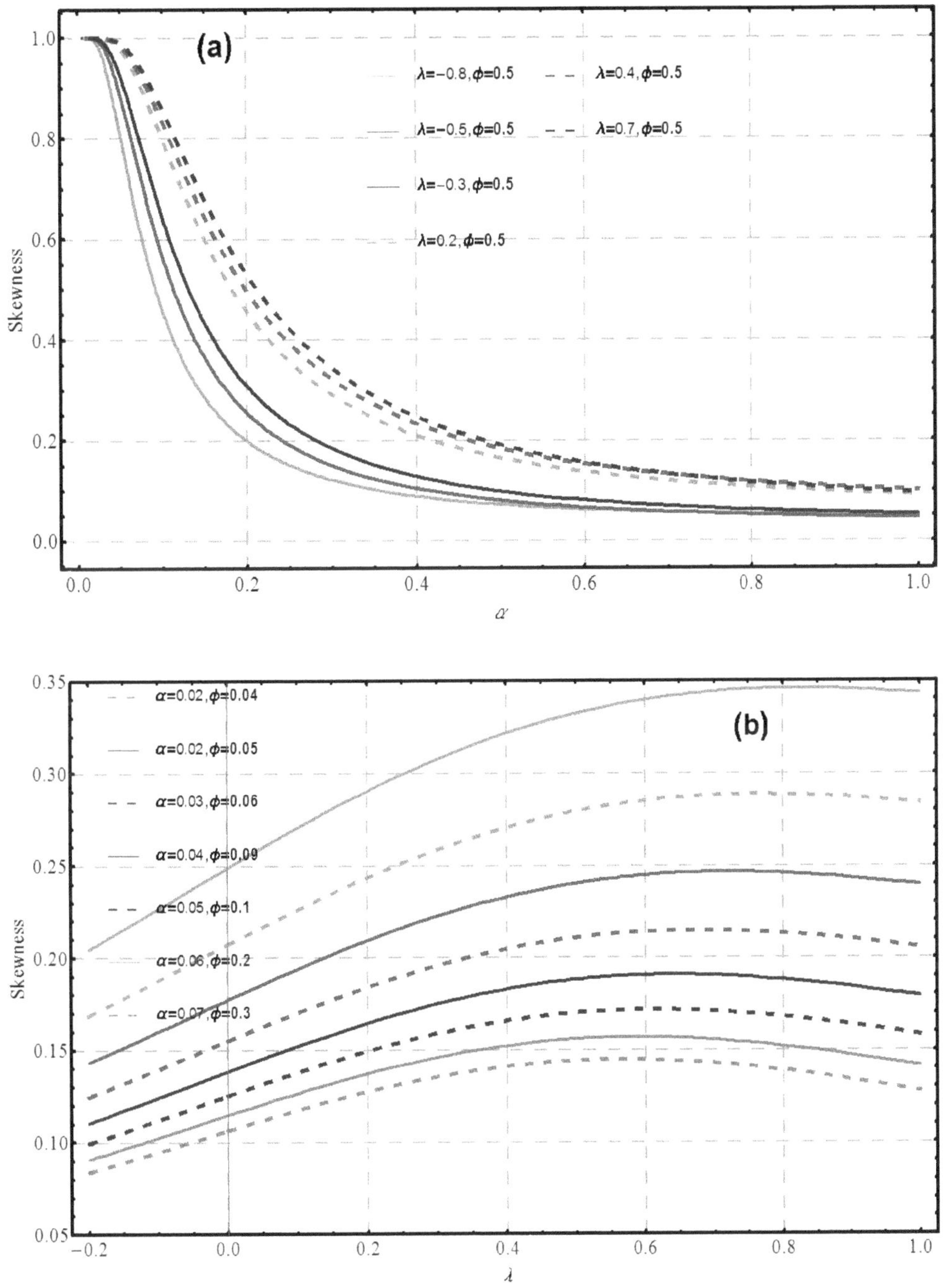

Figure 6: Graphs of skewness of the TB-X distribution.

Theorem 1:

If X has TB-X $(x;\alpha,\phi,\lambda)$ with $|\lambda|\le 1$, then ρ_r moment of X is as follows

$$\rho_r=\frac{\Gamma\left(\frac{r}{2}+1\right)}{\phi^r}\sum_{j=0}^{\infty}(-1)^j\left\{\frac{\alpha(1+\lambda)}{(j+1)^{\frac{r}{2}+1}}-\sum_{i=0}^{\infty}(-1)^i\,2\lambda\alpha\frac{\Gamma(\alpha+1)}{\Gamma(\alpha+1-1)\,i!\,(i+j+1)^{\frac{r}{2}+1}}\right\}\frac{\Gamma(\alpha)}{\Gamma(\alpha-j)\,j!}.\tag{14}$$

Proof:
First $0<\left(1-\exp\left\{^{-(\varphi x)^2}\right\}\right)^{\alpha}<1$ for $x>0$, we have the power series representation

$$(1-\theta)^{\alpha-1}=\sum_{j=0}^{\infty}(-1)^j\frac{\Gamma(\alpha)}{j!\Gamma(\alpha-j)}\theta^j \tag{15}$$

We want to find

$$\rho_r=\int_0^{\infty}x^r\,2\alpha\phi^2xe^{-(\phi x)^2}\left(1-e^{-(\phi x)^2}\right)^{\alpha-1}\left\{1+\lambda-2\lambda\left(1-e^{-(\phi x)^2}\right)^{\alpha}\right\}dx$$

$$=\int_0^{\infty}x^{r+1}2\alpha\phi^2e^{-(\phi x)^2}\left\{1+\lambda-2\lambda\left(1-e^{-(\phi x)^2}\right)^{\alpha}\right\}\frac{\left(1-e^{-(\phi x)^2}\right)^{\alpha}}{\left(1-e^{-(\phi x)^2}\right)}dx$$

$$=2\alpha\phi^2(1+\lambda)\sum_{j=0}^{\infty}(-1)^j\frac{\Gamma(\alpha)}{j!\Gamma(\alpha-j)}\int_0^{\infty}x^{r+1}\left\{e^{-(\phi x)^2}\right\}^{j+1}dx$$

$$-4\alpha\phi^2\lambda\sum_{j=0}^{\infty}(-1)^j\frac{\Gamma(\alpha)}{\Gamma(\alpha-j)j!}\int_0^{\infty}x^{r+1}\left\{e^{-(\phi x)^2}\right\}^{j+1}\left(1-e^{-(\phi x)^2}\right)^{\alpha}dx$$

$$=2\alpha\phi^2(1+\lambda)\sum_{j=0}^{\infty}\frac{(-1)^j\Gamma(\alpha)}{\Gamma(\alpha-j)j!}\int_0^{\infty}x^{r+1}\left\{e^{-(\phi x)^2}\right\}^{j+1}dx-$$

$$-4\alpha\phi^2\lambda\sum_{i=j=0}^{\infty}(-1)^{i+j}\frac{(-1)^{i+j}\Gamma(\alpha+1)\Gamma(\alpha)}{\Gamma(\alpha+1-i)\Gamma(\alpha-j)i!j!}\int_0^{\infty}x^{r+1}\left\{e^{-(\phi x)^2}\right\}^{i+j+1}dx$$

After integration and some simplification the *rth* moment can be noted as:

$$\rho_r=\frac{\Gamma(\frac{r}{2}+1)}{\varphi^r}\sum_{j=0}^{\infty}(-1)^j\frac{\Gamma(\alpha)}{\Gamma(\alpha-j)j!}\left\{\alpha(1+\lambda)(j+1)^{-\frac{r}{2}-1}-2\lambda\alpha\sum_{i=0}^{\infty}(-1)^i\frac{\Gamma(\alpha+1)}{\Gamma(\alpha+1-i)(i+j+1)^{\frac{r}{2}+1}}\right\}.$$

If $\lambda=0$ then

$$\rho_r=\frac{\Gamma(\frac{r}{2}+1)\alpha}{\phi^r}\sum_{j=0}^{\infty}\binom{\alpha-1}{j}(-1)^j(j+1)^{\frac{-r}{2}-1}. \tag{16}$$

The aforementioned is the rth moment for B-X model (see Surles and Padgett (2001)) with parameters α and ϕ.

7. Moment Generating Function (mgf)

Theorem 2:
The mgf for TB-X (x; α, ϕ, λ) model with $|\lambda|\le 1$, is:

$$M_x(t)=\sum_{j=0}^{\infty}\left\{\frac{(-1)^j}{(j+1)^{\frac{m}{2}+1}}\right\}\left[(1+\lambda)\binom{\alpha-1}{j}-2\lambda\binom{2\alpha-1}{j}\right]\alpha_1\Psi_0\left\{\binom{1,\frac{1}{2}}{-};\,t|\phi\right\}, \tag{17}$$

where

$${}_p\Psi_q\left[\begin{matrix}(\alpha_1,A_1),\ldots,(\alpha_p,A_p);\,z\\(\beta_1,B_1),\ldots,(\beta_p,B_p);\end{matrix}\right]=\sum_{k=0}^{\infty}\frac{\prod_{i=1}^{p}\Gamma(\alpha_i+A_ik)z^k}{\prod_{j=1}^{q}\Gamma(\beta_j+B_jk)k_!}. \tag{18}$$

Proof: By definition $M_x(t)$ for the r.v. X with TB-X $(x; \alpha, \phi, \lambda)$ as:

$$M_x(t)=\int_0^\infty e^{tx}\,2\alpha\phi^2 x e^{-(\phi x)^2}\frac{\left(1-e^{-(\phi x)^2}\right)^\alpha}{\left(1-e^{-(\phi x)^2}\right)}\left\{1+\lambda-2\lambda\left(1-e^{-(\phi x)^2}\right)^\alpha\right\}dx,$$

so that

$$M_x(t)=2\alpha\phi^2(1+\lambda)\int_0^\infty xe^{tx}e^{-(\phi x)^2}\left(1-e^{-(\phi x)^2}\right)^{\alpha-1}dx-4\lambda\alpha\phi^2\int_0^\infty e^{tx}xe^{-(\phi x)^2}\left(1-e^{-(\phi x)^2}\right)^{2\alpha-1}dx.$$

Using (15) and after some simplification we obtain

$$M_x(t)=2\alpha\phi^2(1+\lambda)\sum_{j=0}^\infty(-1)^j\binom{\alpha-1}{j}\int_0^\infty xe^{tx}e^{-(\phi x)^2(1+j)}dx-$$
$$4\lambda\alpha\phi^2\sum_{j=0}^\infty(-1)^j\binom{2\alpha-1}{j}\int_0^\infty e^{tx}xe^{-(\phi x)^2(1+j)}dx.$$

The application of the Taylor series extends the above integral shortening to

$$M_x(t)=\alpha(1+\lambda)\sum_{j=0}^\infty\sum_{m=0}^\infty(-1)^j\binom{\alpha-1}{j}\frac{(t/\phi)^m\,\Gamma\left(\frac{m}{2}+1\right)}{m!(1+j)^{\frac{m}{2}+1}}-$$
$$2\lambda\alpha\sum_{j=0}^\infty\sum_{m=0}^\infty(-1)^j\binom{2\alpha-1}{j}\frac{(t/\phi)^m\,\Gamma\left(\frac{m}{2}+1\right)}{m!(1+j)^{\frac{m}{2}+1}}$$

Using the Wright-generalized hyper geometric formula (Srivastava and Manocha (1984)), the above expression provide the following mgf

$$M_x(t)=\left[\sum_{j=0}^\infty(1+\lambda)\frac{(-1)^j}{(j+1)^{\frac{m}{2}+1}}\binom{\alpha-1}{j}-2\lambda\right.$$
$$\left.\sum_{j=0}^\infty\frac{(-1)^j}{(j+1)^{\frac{m}{2}+1}}\binom{2\alpha-1}{j}\right]\alpha\,{}_1\Psi_0\left\{\begin{matrix}(1,\frac{1}{2})\\ -\end{matrix};\ t|\phi\right\}.$$

This completes the proof.

8. Entropy

The entropy is a degree of variation of information or uncertainty of r.v. X with probability model TB-X $(x; \alpha, \phi, \lambda)$. In many field of science like engineering, quantum information, theory of communication, etc. Rényi entropy is extensively used. The more uncertainty the data contains, the greater the entropy value. The Rényi entropy (1961) for a r.v. X with probability distribution TB-X is:

$$I_R(\tau)=(1-\tau)^{-1}\ln\left\{\int_0^\infty f^{\tau}_{f_{TB-X}}(x)\,dx\right\},\qquad \tau\neq 1,\qquad \tau>0. \tag{19}$$

Let X has TB-X $(x; \alpha, \phi, \lambda)$ then by replacing (3) in (19), we obtain

$$I_R(\tau)=\frac{2^\tau\alpha^\tau\phi^{2\tau}}{1-\tau}\log\int_0^\infty x^\tau\left\{1+\lambda-2\lambda\left(1-e^{-(\phi x)^2}\right)^\alpha\right\}^\tau e^{-\tau(\phi x)^2}\left(1-e^{-(\phi x)^2}\right)^{\tau(\alpha-1)}dx.$$

The TB-X Rényi entropy shortens to

$$I_R(\tau)=\frac{1}{1-\tau}\log\left\{\left(\alpha\phi^2\right)^\tau H_{i,j,k}\left\{2^{\tau+i}(1+\lambda)^{\tau-i}\right\}\int_0^\infty e^{-(\phi x)^2(j+k+\tau)}x^\tau dx\right\},$$

where

$$H_{i,j,k} = \sum_{i,j,k=0}^{\infty} (-1)^{i+j+k} \lambda^{i} \frac{\Gamma(\tau+1)\Gamma(\alpha i+1)\Gamma(\tau(\alpha-1)+1)}{\Gamma(\tau+1-i)\,\Gamma(\tau(\alpha-1)+1-k)\Gamma(\alpha i+1-j)(i!j!k!)}.$$

The above integral can be computed as

$$I_R(\tau) = \frac{\tau}{1\text{-}\tau}\log(\alpha f^2) + \frac{\tau\text{-}i}{1\text{-}\tau}\log(1+\lambda) + \frac{\tau}{1\text{-}\tau}\log 2 + \frac{i}{1\text{-}\tau}\log 2\lambda + \frac{1}{1\text{-}\tau}\log\left\{\left(2f^{\tau+1}(j+k+\tau)^{\frac{\tau+1}{2}}\right)^{-1}\Gamma\left(\tfrac{\tau+1}{2}\right)H_{i,j,k}\right\}.$$

The ξ-entropy is developed as (see Havrda and Charvat (1967))

$$I_H(\xi) = \left\{1 - \int_0^{\infty} f_{f_{TB-X}}^{\xi}(x)\,dx\right\}\frac{1}{1-\xi}, \qquad \xi \neq 1,\ \xi > 0. \tag{20}$$

Let X has TB-X (x; α, ϕ, λ) then using (3) in (20), we get

$$I_H(\xi) = \frac{1}{1-\xi}\left\{1 - 2^{\xi}\alpha^{\xi}\phi^{2\xi}\int_0^{\infty} x^{\xi} e^{-\xi(\phi x)^2}\left\{\frac{1+\lambda-2\lambda\left(1-e^{-(\phi x)^2}\right)^{\alpha}}{\left(1-e^{-(\phi x)^2}\right)}\right\}^{\xi}\left(1-e^{-(\phi x)^2}\right)^{\xi\alpha} dx\right\},$$

after some simplification we have,

$$I_H(\xi) = Q_{i,j,k}\left\{2^{\xi+i}(1+\lambda)\right\}^{\xi-i}\lambda^{i}\left(\alpha\phi^2\right)^{\xi}\int_0^{\infty}\frac{x^{\xi}}{e^{(\phi x)^2(j+k+\xi)}}dx,$$

where

$$Q_{i,j,k} = \sum_{i,j,k=0}^{\infty}(-1)^{i+j+k}\frac{\Gamma(\xi+1)\Gamma(\alpha i+1)\Gamma(\xi(\alpha-1)+1)}{\Gamma(\xi+1-i)\Gamma(\alpha i+1-j)\Gamma(\xi(\alpha-1)+1-k)(i!j!k!)}.$$

The above integral can be calculated as

$$I_H(\xi) = \frac{1}{1-\xi}\left[1 - Q_{i,j,k}\frac{\Gamma\left(\frac{\tau+1}{2}\right)\left\{2^{\xi+i}(1+\lambda)\right\}^{\xi-i}\lambda^{i}\left(\alpha\phi^2\right)^{\xi}}{2\phi^{\tau+1}(j+k+\tau)^{\frac{\tau+1}{2}}}\right]. \tag{21}$$

This completes the proof.

9. Estimation

Here, MLEs of TB-X model are determined. Let $x_1, x_2, \ldots, x_3$ be a random sample of size n from TB-X distribution where $\Phi = (\alpha, \phi, \lambda)^T$. The log likelihood function for the vector of parameters Φ can be framed as

$$l(\Phi) = \sum_{i=1}^{n}\ln\left\{1+\lambda\left(1-2\left(1-\exp\left[-(\phi x_i)^2\right]\right)^{\alpha}\right)\right\} + (\alpha-1)\sum_{i=1}^{n}\ln\left\{1-\exp\left[-(\phi x_i)^2\right]\right\} + \sum_{i=1}^{n}\ln x_i - \phi^2\sum_{i=1}^{n}x_i^2 + n\ln 2\alpha + 2n\ln\phi. \tag{22}$$

The relative score function as:

$$U(\Phi) = \left(\frac{\partial l}{\partial\alpha}, \frac{\partial l}{\partial\phi}, \frac{\partial l}{\partial\lambda}\right)^T \tag{23}$$

The log likelihood function can be maximized by solving nonlinear likelihood equations derived from (22), namely

$$\frac{\partial l}{\partial \alpha}=\frac{n}{\alpha}+\sum_{i=1}^{n}\ln w(x_i,\phi)-\sum_{i=1}^{n}\frac{2\lambda\left(w(x_i,\phi)\right)^{\alpha}\ln w(x_i,\phi)}{1+\lambda-2\lambda\left(w(x_i,\phi)\right)^{\alpha}}=0, \tag{24}$$

$$\frac{\partial l}{\partial \phi}=\frac{2n}{\phi}-2\phi\sum_{i=1}^{n}x_i^2-4\sum_{i=1}^{n}\frac{\lambda\alpha\left(w(x_i,\phi)\right)^{\alpha-1}\phi x_i^2\exp\left[-(\phi x_i)^2\right]}{1+\lambda-2\lambda\left(w(x_i,\phi)\right)^{\alpha}} \tag{25}$$

$$\left.+2(\alpha-1)\sum_{i=1}^{n}\frac{\phi x_i^2\exp\left[-(\phi x_i)^2\right]}{w(x_i,\phi)}=0,\right\}$$

$$\frac{\partial l}{\partial \lambda}=\sum_{i=1}^{n}\frac{\left(1-2w(x_i,\phi)^{\alpha}\right)}{1+\lambda-2\lambda\left(w(x_i,\phi)\right)^{\alpha}}=0, \tag{26}$$

where $w(x_i,\phi)=-\left\{e^{-(\phi x_i)^2}-1\right\}$. The above equations cannot be analytically solved but using a variety of statistical packages, one can maximize the function of likelihood. In R-language, for example, package (Adequacy Model), in Mathematica (NMaximize procedure), or some iterative method like NR (Newton-Raphson). For statistical inference of $\left(\frac{\partial l}{\partial\alpha},\ \frac{\partial l}{\partial\phi},\ \frac{\partial l}{\partial\lambda}\right)^T$ and testing of parameters one can get observed information matrix ($J(\Phi)$) since its expectation bears numerical integration. The 3×3 $J(\Phi)$ is

$$J(\Phi)=-E\begin{pmatrix} L_{\alpha\alpha} & L_{\alpha\phi} & L_{\alpha\lambda}\\ . & L_{\phi\phi} & L_{\phi\lambda}\\ . & . & L_{\lambda\lambda}\end{pmatrix} \tag{27}$$

The elements of $J(\Phi)$ for the parameters $\left(\frac{\partial l}{\partial\alpha},\ \frac{\partial l}{\partial\varphi},\ \frac{\partial l}{\partial\lambda}\right)^T$ are

$$L_{\alpha\alpha}=\frac{\partial^2 l}{\partial\alpha^2}=-\frac{n}{\alpha^2}-\sum_{i=1}^{n}\frac{2\lambda(1+\lambda)\{w(x_i,\phi)\}^{\alpha}\ln\{w(x_i,\phi)\}^2}{\left\{1+\lambda-2\lambda\{w(x_i,\phi)\}^{\alpha}\right\}^2},$$

$$L_{\phi\phi}=\frac{\partial^2 l}{\partial\phi^2}=\frac{2n}{\phi^2}-2\sum_{i=1}^{n}x_i^2+4\sum_{i=1}^{n}x_i^2e^{-(\phi x_i)^2}\frac{\lambda\alpha\{w(x_i,\phi)\}^{\alpha-2}}{\left\{1+\lambda-2\lambda\{w(x_i,\phi)\}^{\alpha}\right\}^2}[w(x_i,\phi)\left(-1+\left(-1+2\{w(x_i,\phi)\}^{\alpha}\right)\lambda\right)$$

$$+2\left\{1+\lambda-2\lambda\{w(x_i,\phi)\}^{\alpha}\right\}-\alpha(1+\lambda)+e^{(\phi x_i)^2}w(x_i,\phi)\left\{1+\lambda-2\lambda\{w(x_i,\phi)\}^{\alpha}\right\}(\phi x_i)^2]$$

$$+(\alpha-1)\sum_{i=1}^{n}\frac{e^{-2(\phi x_i)^2}}{\{w(x_i,\phi)\}^2}\left[-4\phi^2x_i^4-2e^{(\phi x_i)^2}x_i^2w(x_i,\phi)\left\{-1+2(\phi x_i)^2\right\}\right],$$

$$L_{\lambda\lambda}=\frac{\partial^2 l}{\partial\lambda^2}=\sum_{i=1}^{n}\frac{\left(1-2\{w(x_i,\phi)\}^{\alpha}\right)^2}{\left\{1+\lambda-2\lambda\{w(x_i,\phi)\}^{\alpha}\right\}^2},$$

$$L_{\alpha\phi}=\frac{\partial^2 l}{\partial\alpha\partial\phi}=2\sum_{i=1}^{n}\left(\frac{\phi x_i^2e^{-(\phi x_i^2)}}{w(x_i,\phi)}\right)+4\lambda\phi\sum_{i=1}^{n}\left\{1+\lambda-2\lambda\{w(x_i,\phi)\}^{\alpha}\right\}^{-2}x_i^2e^{-(\phi x_i^2)_2}$$

$$\times\left(2\lambda\{w(x_i,\phi)\}^{\alpha}-(1+\lambda)(1+\alpha\ln w(x_i,\phi))\right),$$

$$L_{\alpha\lambda}=\frac{\partial^2 l}{\partial\alpha\partial\lambda}=-2\sum_{i=1}^{n}\left\{1+\lambda-2\lambda\{w(x_i,\phi)\}^{\alpha}\right\}^{-2}\{w(x_i,\phi)\}^{\alpha}\ln w(x_i,\phi),$$

$$L_{\phi\lambda}=\frac{\partial^2 l}{\partial\phi\partial\lambda}=-4\phi\alpha\sum_{i=1}^{n}\frac{x_i^2\{w(x_i,\phi)\}^{\alpha-1}e^{-(\phi x_i)^2}}{\left\{1+\lambda-2\lambda\{w(x_i,\phi)\}^{\alpha}\right\}^2}. \tag{28}$$

Translating the inverse matrix for the observed information matrix (27), provides the asymptotic variance and co-variance of the ML estimators $\hat{\lambda}$, $\hat{\phi}$ and $\hat{\alpha}$ By applying (27) approximate 100 $(1-\gamma)$% asymptotic confidence intervals for λ, ϕ and α are

$$\hat{\lambda} \pm z_{\frac{\gamma}{2}}\sigma_{\hat{\lambda}},\ \hat{\phi} \pm z_{\frac{\gamma}{2}}\sigma_{\hat{\phi}},\ \hat{\alpha} \pm z_{\frac{\gamma}{2}}\sigma_{\hat{\alpha}},$$

where $z_{\frac{\gamma}{2}}$ is upper γ^{th} percentile of SND.

10. Simulation Study

Simulation is widely used to evaluate the performance of estimators of parameters of the model under study. Here, we performed a simulation study depend on 5000 replications to assess efficiency of MLEs of parameters of TB-X model. The inverse transform method is used to generate the random sample from TB-X model of the sizes of 10, 30, 50, 100 and 500. We evaluated the mean of 5000 replications of the estimates, mean squared errors (MSEs) and biases for the parameters sets $\lambda = 0.2, 0.4$, $\alpha = 1.5$ and $\phi = 0.5$, $\lambda = 0.7, 0.5$, $\alpha = 1.5, 2.5$, $\phi = 0.5, 3.0$ and $\lambda = 0.7, 0.9$, $\alpha = 2.5$, $\phi = 3.0$. Simulation findings are entered in Tables 1–3. We notice that as sample size increases, the bias and MSEs of the estimate parameters reduce to approximately zero. Furthermore estimates tend to be close to the theoretical value of the parameters as sample size increases.

Table 1: MLEs, MSEs and Bias for $\lambda = 0.2, 0.4$, $\alpha = 1.5$ and $\phi = 0.5$.

n	$\lambda = 0.2$	$\alpha = 1.5$	$\phi = 0.5$	$\lambda = 0.4$	$\alpha = 1.5$	$\phi = 0.5$
10	0.29523 (1.73202)	1.68044 (0.51402)	0.40121 (0.01967)	0.44626 (0.34754)	1.45228 (0.37919)	0.61693 (0.01757)
Bias	0.09523	0.18044	0.09879	0.04626	0.04772	0.11693
30	0.25936 (0.65421)	1.64110 (0.08924)	0.45679 (0.00953)	0.41501 (0.17502)	1.45540 (0.14557)	0.61120 (0.00641)
Bias	0.05936	0.14110	0.04321	0.01501	0.04460	0.11120
50	0.21561 (0.40694)	1.63361 (0.07835)	0.46883 (0.00535)	0.41475 (0.16093)	1.56678 (0.10267)	0.54432 (0.00596)
Bias	0.01561	0.13361	0.03117	0.01475	0.06678	0.04432
100	0.20595 (0.25860)	1.57419 (0.05749)	0.50661 (0.00339)	0.41210 (0.12425)	1.56343 (0.03007)	0.47949 (0.00294)
Bias	0.00595	0.07419	0.00661	0.01210	0.06343	0.02051
500	0.19546 (0.07552)	1.54127 (0.01218)	0.50431 (0.00092)	0.40955 (0.04656)	1.52453 (0.00845)	0.48559 (0.00120)
Bias	0.00454	0.04127	0.00431	0.00955	0.02453	0.01441

Table 2: MLEs, MSEs and Bias for $\lambda = 0.7, 0.5$, $\alpha = 1.5, 2.5$ and $\phi = 0.5, 3.0$.

n	$\lambda = 0.7$	$\alpha = 1.5$	$\phi = 0.5$	$\lambda = 0.5$	$\alpha = 2.5$	$\phi = 3.0$
10	0.59181 (0.30520)	1.32515 (0.29025)	0.61725 (0.02699)	0.68179 (0.18256)	2.06982 (0.19987)	2.11671 (1.08481)
Bias	00.10819	0.17485	0.11725	0.18179	0.43018	0.88329
30	0.66560 (0.15101)	1.39639 (0.26111)	0.60318 (0.01020)	0.67098 (0.18069)	2.37330 (0.14485)	2.54201 (0.051249)
Bias	0.03444	0.10361	0.10318	0.17098	0.12670	0.45799
50	0.66632 (0.06624)	1.40382 (0.06303)	0.56483 (0.00349)	0.56915 (0.09575)	2.47027 (0.07167)	2.85203 (0.49404)
Bias	0.03368	0.09618	0.06483	0.06915	0.02973	0.14797
100	0.67732 (0.03834)	1.44258 (0.03444)	0.52983 (0.00272)	0.54322 (0.08675)	2.48769 (0.05358)	2.93042 (0.30352)
Bias	0.02268	0.05742	0.02983	0.04322	0.01231	0.06958
500	0.71548 (0.02222)	1.51275 (0.02574)	0.51310 (0.00185)	0.50274 (0.02828)	2.51175 (0.017695)	3.04887 (0.06432)
Bias	0.01547	0.01275	0.01310	0.00274	0.01175	0.04887

Table 3: MLEs, MSEs and Bias for $\lambda = 0.7, 0.9$, $\alpha = 2.5$ and $\phi = 3.0$.

n	$\lambda = 0.7$	$\alpha = 2.5$	$\phi = 3.0$	$\lambda = 0.9$	$\alpha = 2.5$	$\phi = 3.0$
10	0.62995 (0.146125)	2.15104 (0.21641)	2.24587 (1.21669)	0.53643 (0.58788)	3.28771 (0.74995)	4.25094 (2.90154)
Bias	0.07005	0.34896	0.75413	0.36357	0.78771	1.25094
30	0.67392 (0.12146)	2.28265 (0.14270)	2.54902 (0.51253)	0.54203 (0.12336)	3.20553 (0.15657)	4.21946 (2.33425)
Bias	0.02608	0.21735	0.45098	0.35797	0.70553	1.21946
50	0.68712 (0.04972)	2.44532 (0.05438)	2.80672 (0.44214)	0.60607 (0.27162)	3.14405 (0.15379)	3.32061 (0.71089)
Bias	0.01288	0.05468	0.19328	0.29393	0.64405	0.32061
100	0.69223 (0.03903)	2.46293 (0.04572)	2.88876 (0.31054)	0.87468 (0.02492)	2.41176 (0.03505)	2.71118 (0.17533)
Bias	0.00777	0.03707	0.11124	0.02532	0.08824	0.28882
500	0.70556 (0.02534)	2.48592 (0.03540)	2.98721 (0.07957)	0.91944 (0.00665)	2.47482 (0.00932)	3.06893 (0.05596)
Bias	0.00556	0.01408	0.01279	0.01944	0.02518	0.06893

10.1 Bayesian Inference

Let a sample $x_1, x_2, \ldots, x_3$ of size n is taken from TB-X distribution where unknown parameters $\Phi = (\alpha, \phi, \lambda)$ which are supposed to be independent possessing non informative priors to the extent that $g(\alpha) \propto k_1$, $g(\phi) \propto k_2$, and $g(\lambda) \propto k_3$. Then the joint prior distribution for Φ can be noted by $g(\Phi) \propto k$, where $\alpha, \phi > 0$ and $|\lambda| \leq 1$. By multiplying the joint prior distribution with equation (#), the joint posterior density for the vector Φ given the data develop into

$$g(\Phi \mid \mathbf{x}) = \frac{\exp\left\{\sum_{i=1}^{n} \ln\left\{1+\lambda-2\lambda\left(w(x_i,\phi)\right)^{\alpha}\right\}\right\}\alpha^2\phi^{2n} \exp\Big) \left\{\sum_{i=1}^{n} \ln x_i - \phi^2 \sum_{i=1}^{n} x_i^2 + (\alpha-1)\sum_{i=1}^{n} \ln\left\{w(x_i,\phi)\right\}\right\}}{\iiint_{\alpha\,\phi\,\lambda} \left[\exp\left\{\sum_{i=1}^{n} \ln\left\{1+\lambda-2\lambda\left(w(x_i,\phi)\right)^{\alpha}\right\}\right\}\alpha^2\phi^{2n} \exp\left\{\sum_{i=1}^{n} \ln x_i - \phi^2 \sum_{i=1}^{n} x_i^2 + (\alpha-1)\sum_{i=1}^{n} \ln\left\{w(x_i,\phi)\right\}\right\}\right] d\lambda\, d\phi\, d\alpha}. \tag{34}$$

The full posterior joint distribution is quantified and use over a set of random variables under the execution of Bayesian inference. But this often involves calculating complex integrals. In such examples, one may give up on resolving the analytical equations, and continuing with sampling techniques rested upon Markov Chain Monte Carlo (MCMC) methods. Under implementing MCMC methods, we estimate the posterior distribution and the complex integrals utilizing simulated samples from the posterior distribution.

The most accepted example of a MCMC method is the Metropolis-Hastings algorithm (MHA). The primary issue that it unfolds is to yield a technique for sampling from some non specific distribution. The object is that in many occurrences, the probability density function is known but we don't know how to generate a random number from this probability density. This is the case where MCMC is functional. Actually, for the execution MHA we don't even require to know how to evaluate the probability distribution completely. Bayes Estimates are obtained using Metro Hasting function in R and reported in Table 4.

Table 4: Posterior Mean and 95% C.I for TB-X.

Parameter Estimate	Posterior Mean	95% Confidence Interval	
		Lower	Upper
$\hat{\lambda}$	0.25355	0.161708	0.381779
$\hat{\alpha}$	0.596351	0.411966	0.819905
$\hat{\phi}$	0.571629	0.389896	0.762178

11. Application

This section provides the data analysis in order to assess the goodness of fit of the TB-X distribution. The data set is from Smith and Naylor (1987); with respect to the strengths of 1.5 cm glass fibers with 63 observations

0.55	0.74	0.77	0.8	0.84	0.93	1.04	1.11	1.13	1.24	1.25	1.27	1.28
1.36	1.39	1.42	1.48	1.48	1.49	1.49	1.50	1.50	1.51	1.52	1.53	1.54
1.58	1.59	1.60	1.61	1.61	1.61	1.61	1.62	1.62	1.63	1.64	1.66	1.66
1.68	1.69	1.70	1.70	1.73	1.76	1.76	1.77	1.78	1.81	1.82	1.84	1.84
2.00	2.01	2.24	1.30	1.55	1.68	1.89	1.29	1.55	1.67	1.84		

The summary statistics for TB-X distribution are given in Table 5. The summary statistics for TB-X, BX and BX-1 distributions for strength fibers data are given in Table 5.

Table 5: Summary Statistics for TB-X, BX and BX-1 distributions for strengths of glass fibers data.

Distribution	Median	Quartile Deviation	Bowley Skwness	Moors Kurtosis
TB-X	1.49760	0.23611	0.04409	5.99922
BX	1.24680	0.59794	0.05683	3.86337
BX-1	1.35479	0.26275	0.05689	4.84979

The MLEs of the unknown parameters with the corresponding standard errors and 95% confidence intervals for TB-X, BX and BX-1 for strength fibers data are given in Table 6. The standard error estimates gained applying the observed information matrix seem to be smaller than the parameters estimates.

Table 6: Estimated parameters of TB-X, BX and BX-1 distributions for strengths of glass fibers data.

Model	Parameters Estimate	ML Estimate	Standard Error	95% Confidence Interval	
				Lower	Upper
TB-X	$\hat{\lambda}$	–0.653432	0.19636067	–1.0382991	–0.26856529
	$\hat{\alpha}$	4.719131	1.26349582	2.24267939	7.195583007
	$\hat{\phi}$	1.040204	0.05482823	0.93274027	1.147666931
BX	$\hat{\alpha}$	5.416223	1.1674457	3.12802973	7.704416872
	$\hat{\phi}$	0.982542	0.0538427	0.87701041	1.088073792
BX-1	$\hat{\phi}$	5.731362	0.7220837	4.31607795	7.146646052

In order to choose the best model we use criteria like Log Likelihood (l), Akaike information criterion (AIC), Corrected Akaike Information Criterion ($CAIC$), Bayesian Information Criterion (BIC) and Hannan-Quinn Information Criterion ($HQIC$) statistics which are described as fellows,

$$AIC = -2l + 2\psi$$

$$CAIC = AIC + \frac{2\psi(\psi+1)}{n-\psi-1}$$

$$BIC = -2l + \psi\log(n),$$

$$HQIC = -2l + 2\psi\log\{\log(n)\},$$

where ψ is the number of parameters in the statistical model, $l(.)$ is the maximized value of the log-likelihood function and n is the sample size. In additional we consider Kolmogorov-Smirnov (K-S) test, modified Cramér-von Mises (W*) and modified Anderson-Darling (A*) goodness of fit statistics and statistics W* and A* are given by

$$W^* = \left(\frac{1}{2n}+1\right)\left\{\sum_{i=1}^{n}\left(z_i - \frac{2i-1}{2n}\right)^2 + \frac{1}{12n}\right\},$$

and

$$A^* = \left(\frac{9}{4n^2} + \frac{3}{4n} + 1\right)\left\{-n - \frac{1}{n}\sum_{i=1}^{n}(2i-1)\log\left(z_i\left(1 - z_{n-i+1}\right)\right)\right\},$$

respectively, $z_i = F(y_{(i)})$, and the $y_{(i)}$'s are the ordered observations. The lesser these statistics are, the more appropriate the fit. The findings of model fitting are reported in Table 7. The Kolmogorov-Smirnov (K-S) distance between the empirical and fitted TB-X distribution is the smallest among other of two distributions. Table 7 also states that the modified Cramér-von Mises (W*) and modified Anderson-Darling (A*) goodness of fit statistics have the smallest values for the strength fibers data with view to the TB-X distribution. According to the above cited benchmarks AIC,CAIC,BIC, HQIC, W* and A* we have supporting evidence that the TB-X model provides the best fit among other two sub models. On the basis of these goodness of fit measures we infer that the TB-X distribution yields a better fit than the sub models.

Table 7: Goodness-of-fit statistics for strengths of glass fibers data.

Distributions	l	AIC	CAIC	BIC	HQIC	K-S	W*	A*
TB-X (α, ϕ, λ)	21.84830	49.69659	50.10337	56.126	52.22531	0.19117	0.450326	2.501158
BX (α, ϕ)	24.36439	52.72878	52.92878	57.01505	54.41459	0.21257	0.532857	2.95641
BX-1 (α)	24.41755	50.8351	50.90067	52.97823	51.67801	0.22092	0.538946	2.989869

The estimated PDFs, CDFs and the histogram of the fibers strengths data set for the fitted model are appended in Figures 5 and 7. The plots of estimated densities show that the relative histogram of strengths fibers data recommends that the fit of proposed model preferable to baseline distributions.

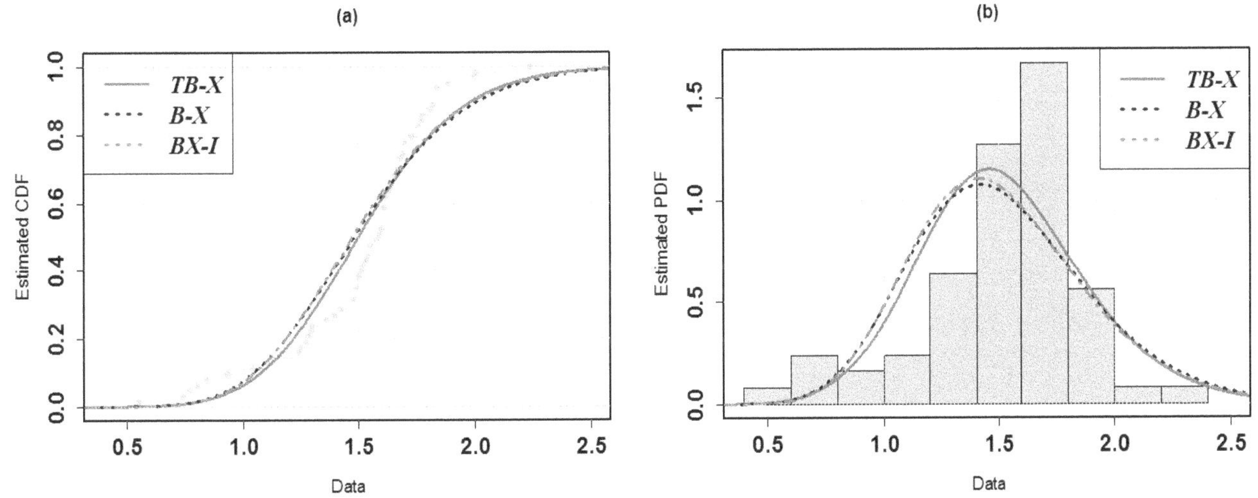

Figure 7: The estimated CDF, PDF and histogram for fibers strength data.

12. Conclusions

In this article, generalization of Burr type X (TB-X) distribution is developed. This generalization is attained by transmuting the two parameters Burr distribution via the quadratic rank transmuted map execution. We explore its different characteristics including explicit form of reliability function, hazard function, median, moments, moment generating function, entropies and maximum likelihood estimator. We discuss the maximum likelihood estimation inside the structure of asymptotic log-likelihood estimation including confidence intervals. Moreover Bayesian estimation of unknown parameters are also discussed via Metropolis- Hasting algorithm. The parameters of TB-X yield monotonically increasing, decreasing and constant hazard rates. The TB-X is practiced to fit real data set. The importance of TB-X is demonstrated by real data set. Concerning the statistical weight of model adequacy, the TB-X shows to a better fit than its competitor's models.

Acknowledgements

The authors are deeply thankful to the editors and reviewers for their valuable suggestions to improve the quality of the chapter.

References

Afify, A. Z., Nofal, Z. M., Yousof, H. M., El Gebaly, Y. M. and Butt, N. S. 2015. The transmuted Weibull Lomax distribution: properties and application. Pakistan Journal of Statistics and Operation Research, 11(1): 1–10.

Ahmed, Z. A., Nofal, Z. M. and Butt, N. S. 2014. Transmuted complementary Weibull geometric distribution. Pakistan Journal of Statistics and Operation Research, 10(4): 1–15.

Aryal, G. R. and Tsokos, C. P. 2009. On the transmuted extreme value distribution with application. Nonlinear Analysis: Theory, Methods & Applications, 71(12): e1401–e1407.

Aryal, G. R. and Tsokos, C. P. 2011. Transmuted Weibull distribution: A generalization of the Weibull probability distribution. European Journal of Pure and Applied Mathematics, 4(2): 89–102.

Burr, I. W. 1942. Cumulative Frequency Functions. The Annals of Mathematical Statistics, 13: 215–232.

Havrda, J. and Charvát, F. 1967. Quantification Method of Classification Processes. Concept of Structural α-Entropy. Kybernetika, 3: 30–35.

Kenney, F. and Keeping, E. S. 1951. Mathematics of statistics-part one. 3rd Edn., D. Van Nostrand Company Inc; Toronto; Princeton; New Jersey.

Khan, M. S., King, R. and Hudson, I. 2014. Characterizations of the transmuted inverse Weibull distribution. Anziam Journal, 55: 197–217.

Khan, M. S., King, R. and Hudson, I. L. 2016. Transmuted Kumaraswamy distribution. Statistics in Transition New Series, 17(2): 183–210.

Khan, M. S., King, R. and Hudson, I. L. 2017. Transmuted generalized exponential distribution: A generalization of the exponential distribution with applications to survival data. Communications in Statistics-Simulation and Computation, 46(6): 4377–4398.

Merovci, F. 2013. Transmuted Rayleigh distribution. Austrian Journal of Statistics, 42(1): 21–31.

Moors, J. J. A. 1988. A quantile alternative for kurtosis. The statistician, 25–32.

Mudholkar, G. S. and Srivastava, D. K. 1993. Exponentiated Weibull family for analyzing bathtub failure-rate data. IEEE Transactions on Reliability, 42(2): 299–302.

Nofal, Z. M., Afify, A. Z., Yousof, H. M. and Cordeiro, G. M. 2017. The generalized transmuted-G family of distributions. Communications in Statistics-Theory and Methods, 46(8): 4119–4136.

Oguntunde, P. and Adejumo, O. 2014. The transmuted inverse exponential distribution. International Journal of Advanced Statistics and Probability, 3(1): 1–7.

Raqab, M. Z. 1998. Order statistics from the Burr type X model, Computers Mathematics and Applications, 36: 111–120.

Raqab, M. Z. and Kundu, D. 2006. Burr type X distribution: revisited. Journal of Probability and Statistical Sciences, 4(2): 179–193.

Rényi, A. 1961. On Measures of Entropy and Information. In Proceedings of the Fourth Berkeley Symposium on Mathematical Statistics and Probability, 1: Contributions to the Theory of Statistics. The Regents of the University of California.

Shaw, W. T. and Buckley, I. R. 2009. The alchemy of probability distributions: beyond Gram-Charlier expansions and a skew-kurtotic-normal distribution from a rank transmutation map. arXiv preprint arXiv: 0901.0434.

Smith, R. L. and Naylor, J. C. 1987. A comparison of maximum likelihood and Bayesian estimators for the three-parameter Weibull distribution. Applied Statistics, 358–369.

Srivastava, H. M. and Manocha, H. L. 1984. A Treatise on Generating Functions, John Wiley and Sons/Ellis Horwood, New York/Chichester.

Surles, J. G. and Padgett, W. J. 1998. Inference for P(Y < X) in the Burr Type X model, Journal of Applied Statistical Science, 7: 225–238.

Surles, J. G. and Padgett, W. J. 2001. Inference for reliability and stress-strength for a scaled Burr Type X distribution. Lifetime Data Analysis, 7(2): 187v200.

CHAPTER 4

Monitoring Patients Blood Level through Enhanced Control Chart

Muhammad Aslam,[1,*] *Khushnoor Khan*[1] and *Nasrullah Khan*[2]

1. Introduction

Some variation in healthcare is desirable, even essential since each patient is different and should be cared for uniquely. Variation if remains within some prescribed limits are not treated as harmful for instance, a person of 50 years of age having blood pressure 130/85 can be thought of having normal blood pressure whereas the normal rate is 120/80. Variation in the clinical variables like blood pressure, blood sugar, and temperature are measured by specific gadgets and are the first step towards medical diagnostics. Effective monitoring clinical variables can serve both as a bulwark and an early warning alarm for all the stakeholders in health care. Continuous monitoring of clinical variables provides a safe base for future complications that may lead to unnecessary complications or other adverse endings. However, early detection of unusual changes in clinical variables may contain the acuteness of a disease. Two clinical variables (temperature and blood pressure) are incessantly measured as a patient approaches a medical doctor. Blood pressure usually refers to the pressure of the circulating blood on the walls of the blood vessels. Both high and low blood pressures lead to multiple complications, if not monitored continuously. Patients can become partners in health care by self-monitoring the clinical variables. (McCray, 2005) stated that the illiteracy of the patients is directly related to the understanding and monitoring of blood pressure. Health illiteracy is a major obstacle to the monitoring of blood pressure (Bosworth and Oddone, 2002; Williams et al., 1998). Blood pressure is measured in millimeters of mercury (mmHg) and expressed in systolic blood pressure (maximum, the pressure when your heart pushes blood out), and diastolic blood pressure (minimum, the pressure when your heart rests between beats). It is used to measure or to diagnose oxygen saturation, heart rate, respiratory rate, or/and the body temperature in a person. Normal values in a healthy adult must be 120/80 mmHg. Conventionally it is measured using a mercury manometer, which is considered as the gold standard among the medical community. Individuals will be considered to be hypertensive with a mean blood pressure of more than 135/85 mmHg Vilaplana, J, J (2007).

The use of mercury manometer in current hospital settings has been replaced by automatic electronic devices which gained much-augmented popularity (Ogedegbe and Pickering, 2010). Medical science has developed many successful treatments and solved much-complicated ailments of human beings but the monitoring and diagnosing of the clinical variables remains a challenge for medical practitioners and academia. Control charts are an important technique for monitoring unusual changes in the variable of interest, particularly in the manufacturing industry. Control charts are low cost, fast, and easy, and have high statistical power. The statistical power of control charts is based on multiple measures over time.

During the first part of the last century, the application of statistical control charts in monitoring clinical variables has attracted the attention of medical researchers. (Steiner and Jones, 2010) used the exponentially weighted moving average (EWMA) control chart technique to monitor the risk-adjusted survival times. The cumulative sum (CUSUM) control chart procedure was used by (Steiner et al., 2000) for monitoring the patient's pre-operative risk of surgical failure rate. The CUSUM control chart and the Shewhart p-chart were compared based on the average run length (ARL) for the cardiac surgery patients by (Grigg and Farewell, 2004). A thorough discussion about the use of CUSUM charts, EWMA charts, Risk-adjusted charts, and prospective detection of a cluster of diseases using control chart schemes in the health care and public surveillance problems was presented by (Woodall et al., 2006). Statistical process control charts for monitoring the infections, nosocomial infections, needle stick injuries, intensive care units,

[1] Department of Statistics, Faculty of Science, King Abdulaziz University, Jeddah, 21551, Saudi Arabia.
Email: khushnoorkhan64@gmail.com
[2] Department of Statistics, College of Veterinary and Animal Sciences, Jhang, University of Veterinary and animal sciences Lahore (Pakistan).
Email: nasrullah.khan@uvas.edu.pk
* Corresponding author: aslam_ravian@hotmail.com

and bacteria have been used by (Sellick, 1993) and (Morton et al., 2001). Chaabane et al., 2019 presented a multivariate chart for monitoring the health system. Aslam et al., 2019 applied the EWMA chart for monitoring blood glucose levels. Ali et al., 2020 used a control chart for monitoring cardiac patents. Carmo-Silva et al., 2020 applied a control chart using hypertension data. For more application of control chart techniques in the medical field see (Buckeridge et al., 2005; Harrison et al., 2004; Kirkham and Bouamra, 2008; Marshall et al., 2004; Maruthappu et al., 2014; Tekkis et al., 2003).

Sherman (1965) originally developed the paradigm of a repetitive group sampling plan (RGS). This new technique could considerably reduce the average numbers of samples for disposition of the lot under inspection. The operational procedure of the repetitive sampling is the same as in the sequential sampling. In the latter sampling, the process is continued until the decision is made. The repetitive sampling is simple in implanting than the sequential sampling. Although, sequential sampling is more efficient than repetitive sampling but required administrative work, time, and inspection cost. The work by (Balamurali and Jun, 2006) have also compared single and double sampling plans with the repetitive acceptance sampling plans and have found that using the latter technique less number of samples are required for inspection of the lot. Aslam et al., 2014a applied the repetitive sampling technique on control charts. The application of RGS in the construction of control charts (both for attributes and variables) has been discussed in (Aslam et al., 2014a). Control charts can be divided into attributes and variables control charts, depending on whether the quality characteristic is attributed or measurable. In the present study, the means of the blood pressure for five samples are used as statistics. Control charts for variables using RGS will be considered. The proposed control chart consists of a pair of inner control limits (LCL_1 & LCL_2) and a pair of outer control limits (UCL_2 & UCL_1) a repetitive sample is selected if the plotted statistics falls between the pairs of inner or outer control limits. Nevertheless, if the statistic is located between UCL_2 & UCL_1 or between LCL_1 & LCL_2 then a new sample is selected and inspected. Visual depiction for the proposed chart showing the acceptance, rejection (out-of-control), and indecisive regions are exhibited in Figure 1.

The criteria to evaluate the statistical performance of both the control charts (traditional and proposed) average run lengths for in-control and out-of-control processes will be compared. In traditional Shewhart control for blood pressure, the decision about the state

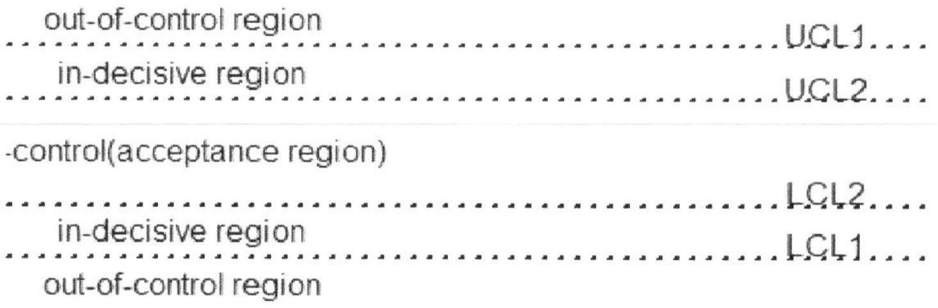

Figure 1: Operational procedure for RGS source Aslam et al. (2014).

of the disease is taken on the information from the single sample only. In some situations, medical practitioners are in-decision using single sample information. In this case, the process is repeated by taking a new sample to make a decision. Besides, the control charts using repetitive sampling are more efficient than charts using the single sampling in average run length (ARL). The main purpose of proposing repetitive sampling charts for blood pressure is to develop an early warning or early detection mechanism. Early detection of out-of-control clinical variables will result in the avoidance of health hazards before they get out of hand. After going through relevant literature, the authors have not come across any study applying control charts based on RGS to monitor clinical variables.

The rest of the paper is organized as follows in Section 2 and 3 the designing of the Shewhart and proposed RGS charts are discussed. Section 4 gives a brief description of the data, and the diagnostics applied for checking the assumptions of normality of data. In Section 5 performance of the proposed RGS control charts is compared with traditional control charts, and the statistical performance of the proposed and traditional techniques are also compared in terms of ARL. The application of the proposed RGS charts on a real data set is explained in Section 6. An elaborate discussion of the results is done in Section 7. Section 8 concludes the study with some future research options.

2. Design of the Traditional Shewhart chart for Monitoring Systolic Blood Pressure

Suppose that the variable blood pressure, say X follows the normal distribution with mean m and standard deviation σ and let $\bar{X}$ be the sample mean of the variable X. Here $\bar{X} \sim N(\mu, \sigma/\sqrt{n})$ In a traditional control chart (Shewhart chart) only two lines are drawn, the upper control limit and the lower control limit, the limits are drawn using the following equations:

$$UCL = \mu + k\sigma/\sqrt{n} \tag{1}$$

$$LCL = \mu - k\sigma/\sqrt{n} \tag{2}$$

where the value of 'k' is a control limit coefficient and will be determined by simulation which is traditionally taken as '3'.

3. Design of Proposed RGS Chart for Monitoring Systolic Blood Pressure

The control chart using RGS is constructed under the assumption that the quality of characteristic or statistic understudy is normally distributed with mean μ and standard deviation σ. It is further assumed that the target mean is m when the process is in control. The operational procedure exhibited in Figure 1 and narrated below is taken from (Aslam et al., 2014b):

Step 1: Select a sample of size 'n'. Then measure the required quality characteristic X and calculate $\overline{X}$.

Step 2: Declare out-of-control if $\overline{X} >$ UCL1 or $\overline{X} <$ LCL1. Declare in-control if LCL1 $< \overline{X} <$ UCL1 Otherwise, go to Step 1 and repeat the process.

Since the proposed control charts consist of a pair of inner or outer control limits and they were constructed using the following equations:

The upper and lower outer control limits of the proposed chart will be constructed using (see (Aslam et al., 2014b):

$$UCL_1 = \mu + k_1\sigma/\sqrt{n} \tag{3}$$

$$LCL_1 = \mu - k_1\sigma/\sqrt{n} \tag{4}$$

The upper and lower inner control limits of the proposed chart will be constructed using:

$$UCL_2 = \mu + k_2\sigma/\sqrt{n} \tag{5}$$

$$LCL_2 = \mu - k_2\sigma/\sqrt{n} \tag{6}$$

where k_1 and k_2 are positive constants whose values are to be determined by the simulation process.

4. Data Description and Checking the Assumption of Normality for Systolic Blood Pressure readings

For the current study, two data sets were used to illustrate the working of both the traditional and proposed RGS control charts for monitoring systolic blood pressure. The first data set as shown in Table 1 of Appendix "A" are 50 simulated readings using a normal distribution with 5 samples for each individual/patient of which the first 20 readings are taken from an in-control process with mean 100 and sigma 6.66 next 30 readings are simulations from a process with a change in the shift, i.e., c = 0.50. The simulations are performed using R software. The second data set shown in Table 2 of Appendix "A" contains original systolic blood pressure readings from each of the 50 patients at 5 different timings. The assumption of normality for the original data set was diagnosed through Kolmogorov Smirnov test.

Probability plots for the data are shown in Figure 2. Since the p-value in both cases is greater than 0.01 so we do not reject the null hypothesis of data being normal and conclude that there is sufficient evidence that the data follows a normal distribution.

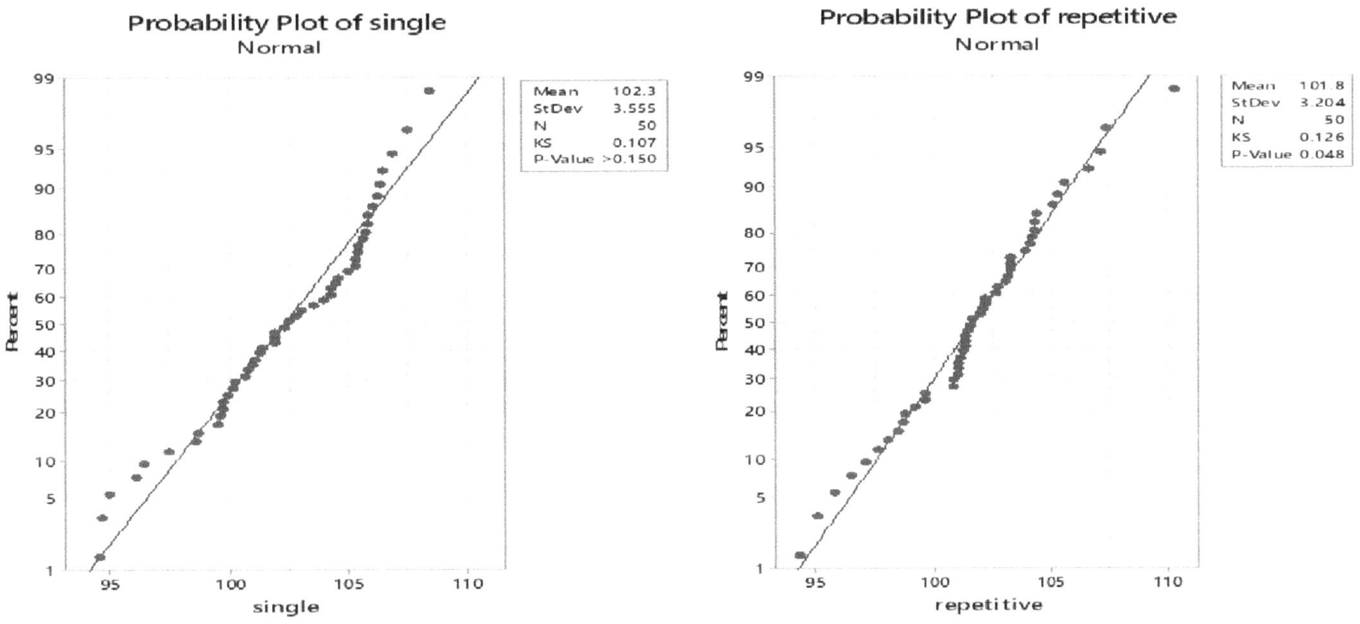

Figure 2: Normal probability plots for single and repetitive samples.

5. Comparing Traditional and Proposed RGS Control Charts of Systolic Blood Pressure

In this section, we will discuss the application of the chart using the simulated data and real data.

5.1 Using Simulated Systolic Blood Pressure Data

The traditional control chart is shown in panel (a) of Figure 3, clearly shows that all average systolic blood pressure readings for 50 patients lie within the ±3S.D limits; thus the readings are in-control. However, applying the control limits produced by RGS as shown in panel (b) of Figure 3 the systolic blood pressure reading for 48th patient is out-of-control, and readings for the 32nd, 43rd, and 44th patient lie in indecisive regions thus, warranting the attention of health practitioners. Hence we can say that Shewhart/traditional control charts were unable to detect any out-of-control points, but when RGS control charts are applied, it detected the out-of-control points. In other words, Shewhart control charts were unable to detect the problem when in fact the problem was present this in statistical parlance is known as committing a type-II error and the probability of committing a type-II error is called 'β'. However, with the application of the proposed chart 'β' is contained, which may warrant effective remedies on the part of health practitioners.

5.2 Using a Real Systolic Blood Pressure Dataset

Data are displayed in Table 2, the control charts using traditional and proposed RGS schemes are shown in panels (c) and (d) of Figure 3. The same pattern as discussed in 5.1 emerged with traditional control charts exhibiting all readings as in-control but proposed RGS chart detected 5 patients in the indecisive region (two above UCL_2 and three below LCL_2) but no readings were found to be out-of-control. Though, all the readings are in-control but still, the readings for the three patients in the indecisive region warrant the attention of health practitioners. These points indicate that the health practitioners cannot take the decision either the patients have blood pressure or not. For these patients, new blood pressure information should be taken to make the decision.

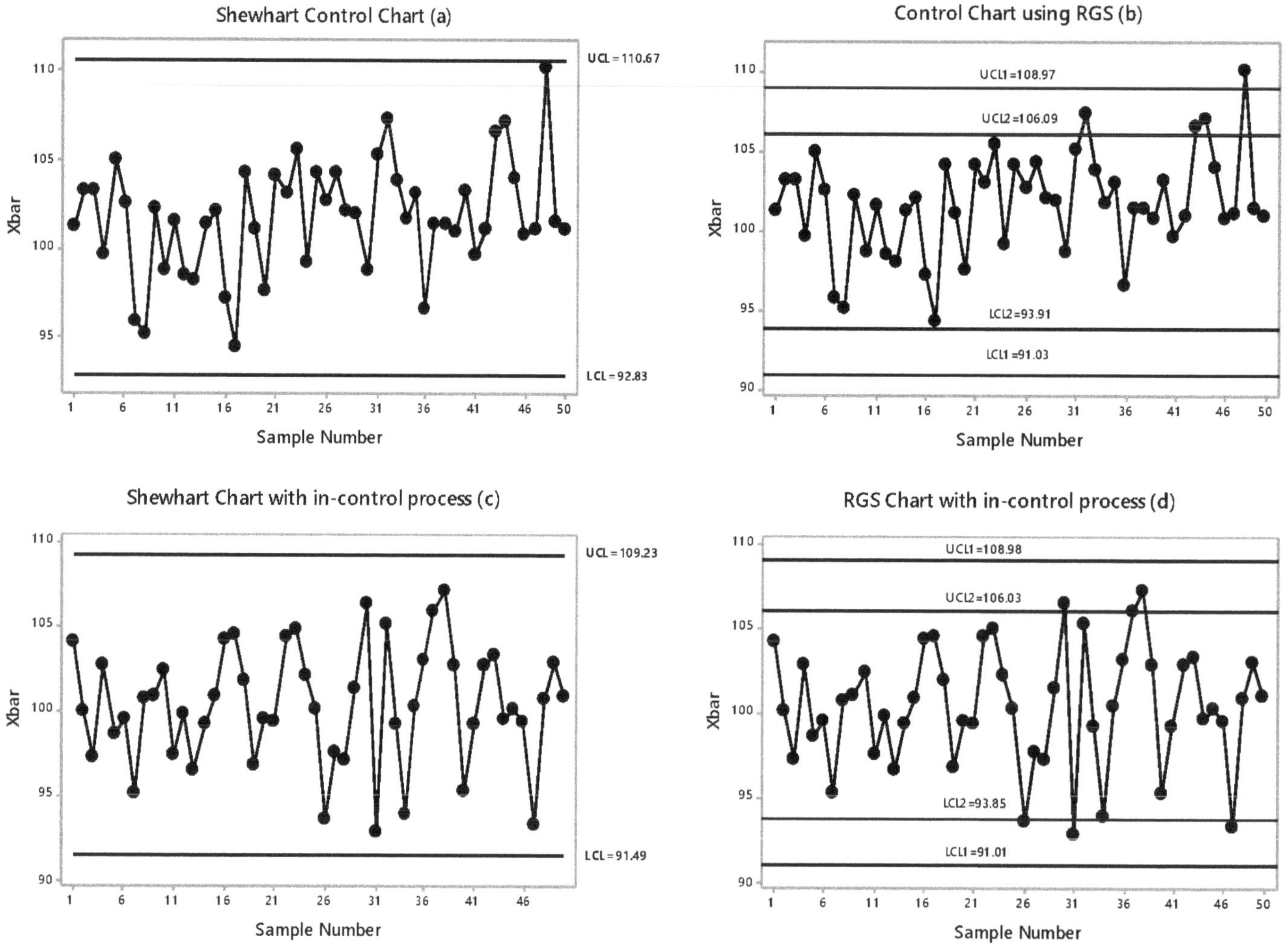

Figure 3: Control Charts using Shewhart and RGS on Simulated and Real dataset.

6. Evaluation Performance of Traditional and Proposed RGS Blood Pressure schemes using ARLs

The performance of the proposed control charts can also be assessed in terms of the average run length (ARL). For comparison purposes, two types of ARLs have to be calculated-ARLo and ARL1 commonly referred to as in-control and out-of-control ARLs. ARLo is the expected number of samples until a control chart signals, under the condition that the actual process is genuinely in-control which means that if the quality of characteristic follows a normal distribution, then there is 99.73% (1-α) chance that the values will lie within the Upper and Lower control limits and only 0.27% (α) chance that any value will lie outside the control limits. To put it in other words if the process is in-control, the process will give a false alarm after every (1/α=1/0.0027=370) sample.

ARL1 is the expected number of samples until a control chart signals, under the condition that the actual process is out-of-control. We would expect ARLo to be as large as possible and ARL1 as small as possible. For instance, if ARL1 = 10 this means that it will take 10 samples on average, to detect the occurrence of a change in shift. As compared to 370 samples, 10 is significantly a small sample to detect a problem. Hence we need a big ARLo and a small ARL1.
The equations for ARLs are taken from (Aslam et al., 2014b)

$$ARLo = \frac{1}{1-P_{in}} \tag{7}$$

$$ARL1 = \frac{1}{1-P^{*}_{in}} \tag{8}$$

Where P_{in} is the probability that the process is declared as in-control under the proposed X bar chart is given as (see (Aslam et al., 2014b)):

$$P_{in} = \frac{2\Phi(k_2)-1}{1-2[\Phi(k_1)-\Phi(k_2)]} \tag{9}$$

and, P^{*}_{in} is the probability that the process is declared as in-control when the process means has shifted from μ to $\mu + c\sigma$ is obtained by (see (Aslam et al., 2014b):

$$P^{*}_{in} = \frac{\Phi(k_2-c\sqrt{n})+\Phi(k_2+c\sqrt{n})-1}{\Phi(k_2+c\sqrt{n})-\Phi(k_1+c\sqrt{n})-\Phi(k_1-c\sqrt{n})+\Phi(k_2-c\sqrt{n})+1} \tag{10}$$

Table 3 compares the ARLs for single sampling (using Shewhart rules) and RGS, with n=5 for different values of the constants (k and k1, k2) for different values of c = shift in the process, on a real data set. If there is no process shift, i.e., c = 0 the values of ARLs

Table 3: Comparison of ARLs for both Shewhart and RGS Systolic Blood Pressure charts for n = 5.

	Single Sampling Plan			Repetitive Group Sampling			
				k_1	2.808532	3.049411	3.011835
k	2.8071	2.9352	2.9996	k_2	2.591891	1.006548	2.045961
Shift	ARLs			Shift	ARLs		
	$r_0 = 200$	$r_0 = 300$	$r_0 = 370$		$r_0 = 200$	$r_0 = 300$	$r_0 = 370$
0	200.0655	300.0149	370.0042	0	200.0144	300.1173	370.4018
0.01	199.6305	299.3069	369.0953	0.01	199.5777	299.3072	369.4683
0.03	196.2112	293.7492	361.9663	0.03	196.145	292.9583	362.1471
0.05	189.6817	283.1742	348.4260	0.05	189.5902	280.9237	348.2483
0.07	180.5943	268.5382	329.7402	0.07	180.4683	264.3687	329.0815
0.1	163.6805	241.5501	295.4489	0.1	163.4926	234.1551	293.9515
0.13	144.8876	211.9475	258.0819	0.13	144.6346	201.4997	255.7405
0.15	132.2667	192.2932	233.4160	0.15	131.9722	180.1148	230.5617
0.17	120.0469	173.4375	209.8613	0.17	119.7145	159.8369	206.5545
0.2	103.0126	147.4404	177.5641	0.2	102.6309	132.2974	173.7048
0.25	78.95581	111.3120	133.0388	0.25	78.51382	94.97286	128.5828
0.3	60.31604	83.81958	99.45888	0.3	59.83801	67.52932	94.73639
0.4	35.71480	48.30887	56.54790	0.4	35.21541	34.03003	51.91085
0.5	21.90919	28.88507	33.37620	0.5	21.42731	17.41396	29.22203
0.7	9.333199	11.71620	13.20523	0.7	8.930657	5.110121	10.24875
0.8	6.481997	7.947599	8.850906	0.8	6.125388	3.067051	6.434107
0.9	4.686012	5.615827	6.181654	0.9	4.374934	2.040571	4.232458

Table 4: Showing Percentage Decrease in ARLs using RGS.

Shift	Percentage Decrease ARLs (n = 300)
0.13	5%
0.5	39%
0.9	63%

for both Single sampling and RGS should be greater than r. From Table 3 it can be seen that at c = 0 ARLo ≥ ro. It is also witnessed that the proposed RGS chart can detect early shift/change in the readings as compared to traditional single sampling control charts. For example, at c = 0.13 and ro = 300, the value ARL from the proposed RGS chart is 201 while it is 212 from the traditional control charts also at c = 0.5 and ro = 300 the proposed RGS control chart would require 17 samples before a shift in the process are detected, and the traditional control would use 29 samples before a shift in the process is detected. Hence comparing the ARLs one concludes that the proposed RGS needs significantly fewer samples to detect a shift in the process and so is more economical. Table 4 shows the percentage decrease in the ARLs for three different shifts, and it is quite clear that the ARLs using RGS significantly decrease as the shift increases from 0.15 to 0.9.

7. Discussion

High blood pressure is the main cause leading to hypertension which in turn finally leads to myocardial infarction. Continuously monitoring blood pressure is the first step in diagnosing hypertension. However, if unwarranted variation in the blood pressure can be detected at an earlier stage, it can help in preventing a health upheaval. The present study focused on the application of control charts using the RGS technique to monitor blood pressure. Results were compared with the traditional Shewhart control chart; it was witnessed that using traditional Shewhart control charts shifted measurements lied within the ±3SD limits, i.e., in-control for visual presentation see panel (a) of Figure 3. However, on the other hand, RGS control charts effectively detected the out-of-control measurements, as shown in panel (b) of Figure 3. Moreover, the efficacy of the proposed control chart is gauged through ARLs. ARLs for the shifted process in the case of RGS decreased more rapidly as compared to traditional Shewhart control charts, as shown in Table 4. Further elaborating ARLo and ARL1 we can say that by using traditional Shewhart control chart a false alarm will occur after 370 samples which means the detection of an unusual measurement will occur after 370/5=74 days of readings whereas using RGS procedure if the process is shifted to 0.7S.D it will take only 10.24875/5 = 2.04 days of reading to detect an out-of-control reading. Keeping in view the results of the present study it can be posited that early detection of variation in the blood pressure measurements can serve both as a bulwark against a possible health upheaval and also result in the reduction of the medical cost. The findings of the study are in line with Hebert, and Neuhauser (2004). Control charts that are simple to comprehend and easy to put to practice have so far not been able to gain currency in the medical field. Procedure and the results of the present study will provide the control charts much-needed impetus in the medical field, in general, and monitoring clinical variables, in particular. All stakeholders (clinical researchers, healthcare managers, and individual patients) will benefit from the proposed application in terms of early detection of out-of-control measurements, reduction in effort, and recurring medical cost. The detecting of a small shift becomes more important, especially when the patient is in a critical situation and the non-detection of out-of-control clinical variables can be life-threatening. Proposed repetitive sampling charts are an attempt to develop an early warning/detection mechanism of clinical variables of which blood pressure is the most vital one. Just by merely looking at the RGS charts and specially focusing on the measurements lying in the indecisive region health practitioners and even patients will know beforehand the health hazard that is going to ensue. This will guide both the practitioner and patient to look for preemptive measures and will also provide patients with substantial relief from physical and financial burden. Another outcome of the present study was to demonstrate an easy-to-follow procedure for application of the RGS charts for monitoring clinical variables which will not only help the medical practitioners to adopt the procedure but also assist patients to become partners in monitoring clinical variables. For establishing the external generalizability of the proposed RGS charts, further work can be developed using a much larger sample with different clinical variables. The proposed chart can be developed using rank set sampling as future research.

8. Conclusion and Recommendation

Traditional control charts can be used to detect a massive shift in the process. But detecting a small shift like in the patient's heartbeat, respiratory rate, or blood pressure is not within the ambit of the traditional control chart. Therefore, the present study focused on the application of a new control chart based on the RGS plan, which helps in detecting small shifts in the processes. Results and the analysis of the proposed chart revealed the efficient detecting ability of the out-of-control readings of systolic blood pressures at different shift levels as compared with the traditional Shewhart control charts. The ease of use and early detection ability will make the proposed control chart a value addition in monitoring clinical variables in the future. RGS may be applied to study behaviour patterns of more clinical variables like blood sugar, temperature, and asthma.

Acknowledgements

The authors are deeply thankful to the editor and reviewers for their valuable suggestions to improve the quality of the chapter.

References

Ali, S., Altaf, N., Shah, I., Wang, L. and Raza, S. M. M. 2020. On the Effect of Estimation Error for the Risk-Adjusted Charts. Complexity, 2020.

Aslam, M., Azam, M. and Jun, C.-H. 2014a. New attributes and variables control charts under repetitive sampling. Industrial Engineering & Management Systems, 13(1): 101–106.

Aslam, M., Azam, M. and Jun, C.-H. 2014b. New attributes and variables control charts under repetitive sampling. Industrial Engineering & Management Systems, 13(1): 101–106.

Aslam, M., Rao, G. S., Khan, N. and Al-Abbasi, F. A. 2019. EWMA control chart using repetitive sampling for monitoring blood glucose levels in type-II diabetes patients. Symmetry, 11(1): 57.

Balamurali, S. and Jun, C.-h. 2006. Repetitive group sampling procedure for variables inspection. Journal of Applied Statistics, 33(3): 327–338.

Bosworth, H. B. and Oddone, E. Z. 2002. A model of psychosocial and cultural antecedents of blood pressure control. Journal of the National Medical Association, 94(4): 236.

Buckeridge, D. L., Burkom, H., Campbell, M., Hogan, W. R. and Moore, A. W. 2005. Algorithms for rapid outbreak detection: a research synthesis. Journal of Biomedical Informatics, 38(2): 99–113.

Carmo-Silva, S., Fernandes, N., Thürer, M. and Ferreira, L. P. 2020. Extending the POLCA production control system with centralized job release. Production, 30.

Chaabane, M., Mansouri, M., Ben Hamida, A., Nounou, H. and Nounou, M. 2019. Multivariate statistical process control-based hypothesis testing for damage detection in structural health monitoring systems. Structural Control and Health Monitoring, 26(1): e2287.

Grigg, O. and Farewell, V. 2004. An overview of risk-adjusted charts. Journal of the Royal Statistical Society: Series A (Statistics in Society), 167(3): 523–539.

Harrison, W. N., Mohammed, M. A., Wall, M. K. and Marshall, T. P. 2004. Analysis of inadequate cervical smears using Shewhart control charts. BMC Public Health, 4(1): 25.

Kirkham, J. J. and Bouamra, O. 2008. The use of statistical process control for monitoring institutional performance in trauma care. Journal of Trauma and Acute Care Surgery, 65(6): 1494–1501.

Marshall, T., Mohammed, M. A. and Rouse, A. 2004. A randomized controlled trial of league tables and control charts as aids to health service decision-making. International Journal for Quality in Health Care, 16(4): 309–315.

Maruthappu, M., Carty, M. J., Lipsitz, S. R., Wright, J., Orgill, D. and Duclos, A. 2014. Patient-and surgeon-adjusted control charts for monitoring performance. BMJ Open, 4(1): e004046.

McCray, A. T. 2005. Promoting health literacy. Journal of the American Medical Informatics Association, 12(2): 152–163.

Morton, A. P., Whitby, M., McLaws, M. L., Dobson, A., McElwain, S., Looke, D., . . . Sartor, A. 2001. The application of statistical process control charts to the detection and monitoring of hospital-acquired infections. Journal of Quality in Clinical Practice, 21(4): 112–117.

Ogedegbe, G. and Pickering, T. 2010. Principles and techniques of blood pressure measurement. Cardiology Clinics, 28(4): 571–586.

Sellick, J. A. 1993. The use of statistical process control charts in hospital epidemiology. Infection Control & Hospital Epidemiology, 14(11): 649–656.

Steiner, S. H., Cook, R. J., Farewell, V. T. and Treasure, T. 2000. Monitoring surgical performance using risk-adjusted cumulative sum charts. Biostatistics, 1(4): 441–452.

Steiner, S. H. and Jones, M. 2010. Risk-adjusted survival time monitoring with an updating exponentially weighted moving average (EWMA) control chart. Statistics in Medicine, 29(4): 444–454.

Tekkis, P. P., McCulloch, P., Steger, A. C., Benjamin, I. S. and Poloniecki, J. D. 2003. Mortality control charts for comparing performance of surgical units: validation study using hospital mortality data. Bmj, 326(7393): 786.

Williams, M. V., Baker, D. W., Parker, R. M. and Nurss, J. R. 1998. Relationship of functional health literacy to patients' knowledge of their chronic disease: a study of patients with hypertension and diabetes. Archives of Internal Medicine, 158(2): 166–172.

Woodall, W. H., Mohammed, M. A., Lucas, J. M. and Watkins, R. 2006. The Use of Control Charts in Health-Care and Public-Health Surveillance/Discussion/Discussion/Discussion/Discussion/Discussion/Discussion/Rejoinder. Journal of Quality Technology, 38(2): 89.

Appendix A1

Table 1: Simulated data of Systolic Blood Pressures with process shift at c = 0.50 after 20 readings.

Sr#	Sample					X-bar
	1	2	3	4	5	
1	100.7385	102.2734	94.6612	103.0965	105.7119	101.63
2	97.83624	102.5221	107.8113	96.97438	111.0983	103.24
3	110.27	99.33251	106.516	102.4628	97.60766	103.23
4	91.78635	101.6098	102.8323	101.8947	100.2099	99.666
5	106.4972	102.2223	107.6683	100.2894	108.8075	105.09
6	110.6655	95.46205	107.6529	96.96528	102.3688	102.62
7	94.63982	102.0321	92.7982	93.67574	95.88415	95.806
8	106.5309	102.3121	92.87655	88.66825	85.28101	95.133
9	101.7937	101.1751	108.1471	87.84557	112.0228	102.19
10	93.02251	104.9352	108.711	94.05091	93.16621	98.777
11	106.0698	106.8852	107.576	93.79566	93.69032	101.60
12	110.2967	91.88698	103.0292	94.86668	92.24868	98.465
13	97.87486	94.54178	103.599	89.87386	104.6859	98.115
14	100.3956	100.4486	107.5911	95.6549	102.7758	101.37
15	95.59126	104.4267	102.0556	115.2831	93.42409	102.15
16	102.6204	91.5505	94.86818	99.24844	97.56791	97.171
17	86.53896	99.64029	103.6689	91.15329	90.67284	94.334
18	108.6953	103.5092	106.6964	102.4805	99.8022	104.23
19	96.39254	116.8282	90.4565	102.6486	99.36415	101.13
20	91.19734	95.6254	102.1309	89.40949	109.8598	97.644
21	100.8754	107.9001	94.72875	118.6112	98.78346	104.17
22	100.1533	105.4268	104.6778	100.075	105.0521	103.07
23	105.6686	110.9464	110.6396	100.7768	99.92239	105.59
24	97.7555	96.5203	106.0686	96.52734	99.1936	99.213
25	99.09783	103.9376	113.2101	109.1923	95.91227	104.27
26	102.1886	102.704	100.4174	103.5873	104.6329	102.70
27	102.8076	109.4424	103.9446	96.29374	109.272	104.35
28	97.6741	110.1638	94.35566	100.2316	108.4616	102.17
29	96.7007	112.3402	90.80676	111.7776	98.419	102.00
30	99.3479	104.2852	105.5087	94.41017	89.95204	98.700
31	95.6858	115.3021	102.2462	99.72916	113.3205	105.25
32	111.273	114.3441	108.1555	105.285	98.01963	107.41
33	97.2827	105.5666	99.29938	105.1219	112.0786	103.86
34	101.0683	102.7298	103.5851	107.3169	93.99339	101.73
35	109.8759	101.6261	106.9515	99.02943	97.98045	103.09
36	94.1171	92.84766	100.8692	88.67016	106.3928	96.579
37	102.0881	91.09228	103.1282	98.85657	111.7608	101.38
38	99.65058	98.45035	102.201	95.95615	110.6881	101.38
39	100.2571	107.9399	110.1239	95.55587	90.51469	100.87
40	105.347	100.5891	99.32045	105.9669	104.9497	103.23
41	97.77464	105.8433	94.9357	101.7957	98.00207	99.670
42	93.04817	114.6297	108.4192	91.50512	97.56873	101.03
43	100.6554	112.8825	102.2596	109.8157	107.4406	106.61
44	98.42453	107.0448	103.5635	108.4263	118.4221	107.17
45	104.3796	99.36517	99.17847	113.5168	103.9135	104.07
46	106.7304	104.1398	98.84753	97.41218	96.9872	100.82
47	94.76094	114.735	97.93556	108.3842	89.57198	101.07
48	102.5532	111.1191	115.4306	108.3154	113.7208	110.22
49	91.5696	108.6762	94.49935	100.0848	112.5474	101.47
50	104.1254	109.9207	96.16842	95.73897	99.22109	101.03

Table 2: Blood Pressure readings of 50 Patients with n = 5.

Sr#	Sample					
	1	2	3	4	5	X-bar
1	108.1408	106.0261	102.2834	108.3822	95.85502	104.1375
2	106.0158	95.87905	94.59415	112.5115	91.16965	100.034
3	89.30123	103.5636	111.1814	89.36414	92.96581	97.27523
4	102.0911	116.171	94.9434	103.4601	97.3643	102.806
5	107.9255	98.91923	101.1612	87.76988	97.30268	98.61569
6	93.92313	101.166	101.4831	109.0759	91.91271	99.51217
7	86.49222	97.17847	109.1286	93.20071	89.93768	95.18753
8	105.1543	100.9217	94.66072	99.46109	103.2403	100.6876
9	105.5901	100.7221	98.21327	106.0969	94.26356	100.9772
10	100.9665	101.9212	108.5324	96.41088	104.0827	102.3827
11	93.79724	104.5559	96.4482	90.31736	102.062	97.43615
12	98.60252	102.4993	95.46994	104.4101	98.12953	99.82226
13	89.95764	103.7956	98.73211	90.62833	99.73884	96.5705
14	96.22114	101.3612	97.95194	101.2116	99.53282	99.25574
15	108.9432	92.25044	102.6333	102.48	98.0893	100.8792
16	95.58205	104.998	115.7695	112.1029	93.09923	104.3103
17	93.44128	99.4983	111.1326	108.5134	109.9364	104.5044
18	106.0243	96.49521	98.5419	102.0768	105.9937	101.8264
19	101.0227	99.60589	100.4084	97.51797	85.48709	96.80842
20	86.26896	102.8447	108.4962	101.7577	98.24691	99.52289
21	94.13993	91.80717	106.7484	95.95409	108.1909	99.36809
22	102.0717	113.8884	105.2354	100.4067	100.8023	104.4809
23	109.8234	104.3039	106.7581	102.1589	101.6791	104.9447
24	95.05288	99.58406	105.7905	103.1173	107.1836	102.1457
25	103.4545	95.45521	96.01003	102.5217	103.6843	100.2252
26	96.08035	96.58612	94.82887	90.49069	89.74615	93.54643
27	97.54538	98.27692	102.2939	92.69679	97.25074	97.61275
28	104.252	93.53836	84.36582	109.1857	94.48108	97.16459
29	111.9776	100.7552	102.8408	102.4413	88.80523	101.364
30	113.9289	108.6201	112.6002	95.26037	101.4651	106.3749
31	89.67786	91.00594	100.5388	90.18289	92.68399	92.8179
32	106.5712	110.011	107.2001	100.5873	101.3711	105.1481
33	95.88769	98.63501	99.60716	103.852	98.15262	99.2269
34	74.02221	100.4998	109.0632	91.8349	94.2858	93.94119
35	92.04026	96.8461	112.6373	94.13244	105.895	100.3102
36	103.4026	103.7816	97.63245	110.374	100.3076	103.0997
37	103.7772	112.7894	105.4058	103.8172	104.2264	106.0032
38	117.503	116.9446	104.8352	95.55063	101.0931	107.1853
39	95.55775	106.2544	109.5741	107.3046	95.24202	102.7866
40	101.1591	94.73998	93.85267	90.91292	95.52575	95.23808
41	97.95058	96.84893	109.5362	93.15954	98.46497	99.19204
42	102.4448	98.57065	107.2964	103.5564	102.194	102.8125
43	98.00142	96.53036	101.1893	112.7749	108.1268	103.3246
44	99.19671	101.165	96.01435	98.58104	102.9681	99.58503
45	100.2563	90.86267	100.9015	100.1078	108.9025	100.2061
46	101.6538	102.2814	90.88722	95.95553	106.4674	99.44907
47	100.5916	97.07469	91.34126	87.96762	89.67231	93.32950
48	108.7162	106.7887	95.17716	86.40084	107.0093	100.8184
49	116.1274	92.6772	107.4484	96.91483	101.2982	102.8932
50	103.7453	102.1566	99.8606	97.56155	101.2317	100.9112

CHAPTER 5

Goodness of Fit in Parametric and Non-parametric Econometric Models

Shalabh, [1,*] ***Subhra Sankar Dhar*** [1] **and** ***N Balakrishna*** [2]

1. Introduction

Statistical tools play a key role in deriving a model from the observed data obtained from any study. Various types of parametric and nonparametric modelling procedures have been proposed in the literature. An important feature of any statistical model is that it is constructed based only on a set of sample data but it is expected to remain valid for the entire population. For example, the efficacy of a drug is tested over a sample of patients but the outcome is expected to remain valid for the entire population. Statisticians experiment with the data using different functional forms of relationship between input and output variables including various transformations on input and output variables. The objective of the statistician is to find out a suitable model which closely matches to the real data obtained through the experiment. The success of the modelling depends upon how much the data obtained from the fitted model is well matched to the data obtained from the experiment. Various statistical measures are available in the literature to quantitatively measure the degree of such closeness between the data obtained from the fitted statistical model and the observed data. Such measures have been constructed based on various philosophies concerning how to measure the "goodness" of a fitted model. The model obtained by using statistical procedures over the given set of data is usually termed as "fitted model". A user is always interested in quantitatively measuring the "goodness of fit" of the model obtained from the given data. Usually the statistics measuring the goodness of fit are constructed such that they have a capability of capturing the features of data generating process and are compatible with the fitted model. Whenever any statistical measure of goodness of fit is developed, it is based on certain assumptions which are required for the validity and correct interpretation. Such assumptions are required due to the constraints posed by the rules of statistics with a view to obtain a form that is easy to use and interpret from a practitioner's point of view. Moreover, a lot of improved and more efficient estimators have been developed in the literature but a practitioner is usually more interested in knowing how good is the resultant model that is being obtained by using those estimators. This will help the practitioner to have an idea about the performance of this model when it is used in further applications like forecasting based on the available observations.

Various types of statistical and econometric models are used in different real data situations under a variety of statistical assumptions. A goodness of fit measure developed for a given model may not necessarily have capability to judge the goodness of fit in other models and even may not remain a valid measure in other models. For example, the coefficient of determination (CoD), popularly known as R^2, is a measure of goodness of fit in multiple linear regression model which is applicable only when certain conditions are satisfied like as all the observations are recorded error-free, an intercept term is present in the model, explanatory variables are independent of random errors, etc. In practice, suppose that the observations are not observable, and they are recorded with measurement error, and accordingly, the measurement error models are used. The usual coefficient of determination cannot be used in such a situation. Similarly, there can be many such situations where different type of measures of goodness of fit are required, and the practitioners are not aware of them. A modest effort is being made in this article to present a discussion on various goodness of fit measures available in parametric and non-parametric modelling.

The goodness of fit of a model depends upon the variables selected in the modelling. In fact, the two aspects or goodness of fit and variable selection are interrelated. The choice and form of the variables affect the goodness of fit. Various techniques have been suggested for variable selection. Whenever the modelling of a phenomenon starts, the statisticians usually start with a set of chosen variables. The number of variables are usually large in the initial set as

[1] Department of Mathematics and Statistics, Indian Institute of Technology Kanpur, Kanpur 208016 (India).
Email: subhra@iitk.ac.in
[2] Department of Statistics, Cochin University of Science and Technology, Cochin 682022 (India).
Email: nb@cusat.ac.in
* Corresponding author: shalab@iitk.ac.in

the experimenter might be over enthusiastic to make the model as realistic as possible etc. A large number of variables pose their own challenges in any statistical analysis from the point of view of computation and interpretation of results. Moreover, it is possible that out of large number of variables, many of them may have little effect or practically no effect on the output variables. So another important objective of any statistical analysis is to identify the important and relevant input variables. The importance of variables is measured in terms of their explanatory powers in terms of their contribution in explaining the variability in the outcomes. The goodness of fit measures help them in choosing relevant and important variables. We make an attempt in this article to present how to choose such explanatory variables using the proposed measures of goodness of fit in various kinds of statistical models.

The research work in obtaining the different suitable forms of the coefficient of determination (CoD) for various situations has been addressed in the literature by several researchers. Eshima and Tabata (2010, 2011) suggested the CoD in entropy form for generalized linear models while a robust coefficient of determination in regression was proposed by Renaud and Victoria-Feser (2010). Tjur (2009) proposed a coefficient of determination for the logistic regression model, see also Hong, Ham and Kim (2005), Liao and McGee (2003); the CoD in the local polynomial model was suggested by Huang and Chen (2008), the CoD in the mixed regression model was developed by Hössjer (2008), and the CoD in the case of association in a regression framework is discussed in van der Linde and Tutz (2008). A family of the CoD in the linear regression model was suggested by Srivastava and Shobhit (2002). The point estimation of the CoD is considered in Marchand (2001), see also Marchand (1997). The partial correlation coefficient and the CoD for the multivariate normal repeated measures data is discussed in Lipstiz et al. (2001). A general CoD for the covariance structure models under arbitrary generalized least squares estimation was suggested by Tanaka and Hahn (1998) whereas Nagelkerke (1991) presented a generalization of the CoD. The CoD for the least absolute deviation analysis is discussed in McKean and Sievers (1987). The role of CoD in the simultaneous equation models is explained in Knight (1980) and Hilliard and Llyod (1980). Ohtani (1994) derived the density of CoD and its adjusted version. He also analyzed their risk performance under an asymmetric loss function in the misspecified linear regression model. A systematic study of the properties of CoD and its adjusted version under the normality of disturbances was conducted by Cramer (1987).

Although the various methodologies related to goodness of fit have been extensively studied for multiple linear regression and its variants, these topics are not much studied in other models, e.g., measurement error models, restricted regression, non-parametric regression, time series models etc. which have enormous utility in the real data analysis and an attempt is made in this direction in this chapter. The statistical measures of goodness of fit in such models under nonstandard statistical assumptions and the guidelines to use them in real data applications are presented. We have described how to use the proposed goodness of fit measures under respective models for variable screening and selection in various models.

The plan of the paper is as follows. The Section 2 presents the goodness of fit in classical multiple linear regression model. The goodness of fit in measurement error model, restricted measurement error model and shrinkage estimation are discussed in the Sections 3, 4 and 5 respectively followed by a brief details about AIC, BIC and Mallows' C_p in Section 6. The goodness of fit and variable selection in non-parametric regression model are discussed in Sections 7 and 8 respectively. The goodness of fit in time series models is discussed in 9 followed by a brief description on model selection using goodness of fit statistics in any of the earlier proposed models and conclusions in Section 10. We have tried our best to avoid the repeating of symbols but the symbols used in any section may be considered to be restricted to that section only to avoid any confusion,

2. Goodness of Fit in Multiple Linear Regression Model

Consider the following multiple linear regression model

$$y = \alpha \mathbf{e}_n + X\beta + u, \tag{2.1}$$

where y is the $(n \times 1)$ vector of n observations on the study variable, X is the $(n \times p)$ matrix of n observations on each of the p explanatory variables $X_1, X_2, \ldots, X_p$, α is the intercept term, β is the $(p \times 1)$ vector of regression coefficients $\beta_1, \beta_2, \ldots, \beta_p$ associated with the explanatory variables or covariates $X_1, X_2, \ldots, X_p$, respectively, u is the $(n \times 1)$ vector of random disturbances, and $\mathbf{e}_n$ is the $(n \times 1)$ vector with all the elements unity (**1s**).

2.1 Coefficient of Determination (CoD) based on OLSE

Assume that random disturbances u in (2.1) follow a distribution with mean vector 0 and covariance matrix $\sigma^2 I$, i.e., $u_1, u_2, \ldots, u_n$ are identically and independently distributed. The ordinary least squares estimator (OLSE) of α and β are obtained by minimizing $(y - \alpha\mathbf{e}_n - X\beta)'(y - \alpha\mathbf{e}_n - X\beta)$. The OLSE of β and α are obtained as

$$b = (X'PX)^{-1}X'Py = (b_1, b_2, \cdots, b_p)' \tag{2.2}$$

and

$$\hat{\alpha} = \bar{y} - b_1\bar{x}_1 - b_2\bar{x}_2 - \cdots - b_p\bar{x}_p, \tag{2.3}$$

respectively, where $P = I_n - n^{-1}\mathbf{e}_n\mathbf{e}_n'$ and $\bar{x}_j = \frac{1}{n}\sum_{i=1}^n x_{ij}$.

The next issue is that once the unknown parameter is estimated by OLSE, how to know if using this estimator, the resultant model will be good and also how to define "good". The goodness of fit in a multiple linear regression model is judged by the coefficient of determination (CoD). The concept, idea and philosophy behind the definition of CoD are based on the multiple correlation coefficient between the study variable y and all the explanatory variables $X_1, X_2, \ldots, X_p$. The square of population multiple correlation coefficient between the y and $X_1, X_2, \ldots, X_p$ is defined as

$$\theta = \frac{\beta'\Sigma\beta}{\beta'\Sigma\beta + \sigma^2}, \quad 0 \le \theta \le 1, \tag{2.4}$$

where $\Sigma = \text{plim}_{n\to\infty} n^{-1}X'PX$. Here the notation plim denotes the "probability in limit" defined as follows:

Let $\{Q_n\}$ be a sequence of random variables defined on some probability space. Then $\{Q_n\}$ is said to converge to the random variable Q if for every $\vartheta > 0$ if $P\{|Q_n - Q| > \vartheta\} \to 0$ as $n \to \infty$. We say that $\{Q_n\}$ converges to Q in probability as n goes to infinity and is denoted as $\text{plim}_{n\to\infty} Q_n = Q$. The probability in limit also indicates the consistency property of an estimator.

Under this set up of classical multiple linear regression model (2.1), the coefficient of determination (CoD), denoted as R^2, is defined as

$$R^2 = \frac{b'X'PXb}{y'Py} = \frac{y'PX(X'PX)^{-1}X'Py}{y'Py}, \quad 0 \le R^2 \le 1. \tag{2.5}$$

The CoD R^2 is the sample based multiple correlation coefficient between y and $X_1, X_2, \ldots, X_p$ and can be viewed as an estimator of θ. The R^2 measures the explanatory power of the model, which in turn reflects the goodness of fit or capability of the model to explain the variation in y of the model in terms of the explanatory power of the explanatory variables, see Rao et al. (2008).

The R^2 can be considered as an estimator of θ, i.e., the square of population multiple correlation coefficient between the study variable y and all the explanatory variables $X_1, X_2, \ldots, X_p$. Assuming $E(u) = 0$, covariance matrix $V(u) = \sigma^2 I$ with the assumptions $\text{plim}_{n\to\infty}\frac{X'PX}{n} = \Psi > 0$ and $\text{plim}_{n\to\infty}\frac{X'Pu}{n} = 0$ of the classical linear regression analysis, it can be shown that

$$\text{plim}_{n\to\infty} R^2 = \theta, \tag{2.6}$$

i.e., R^2 is a consistent estimator of θ.

The CoD can also be defined using the concept of analysis of variance in multiple linear regression model. Using the analysis of variance in the model (2.1), the total sum of squares (SS_{total}) can be partitioned into two orthogonal components, sum of squares due to regression ($SS_{regression}$) and sum of squares due to error (SS_{error}) as

$$SS_{total} = SS_{regression} + SS_{error}, \tag{2.7}$$

where $SS_{total} = y'Py$, $SS_{regression} = y'PX(X'PX)^{-1}X'Py$, and $SS_{error} = y'[I - X(X'X)^{-1}X']y$, the R^2 can be defined as

$$R^2 = \frac{SS_{regression}}{SS_{total}} = 1 - \frac{SS_{error}}{SS_{total}}, \quad 0 \le R^2 \le 1, \tag{2.8}$$

which is indicating the proportion of the total variation being explained by the fitted model.

Another representation of R^2 is based on the residuals. The ith residual is defined as the difference between the observed and fitted values of y, i.e., $y_i \sim \hat{y}_i$ where $\hat{y}_i = \hat{\alpha} - b_1x_{i1} - b_2x_{i2} - \ldots - b_px_{ip}$, $i = 1, 2, \ldots, n$. The R^2 is defined as

$$R^2 = 1 - \frac{\sum_{i=1}^n (y_i - \hat{y}_i)^2}{\sum_{i=1}^n (y_i - \bar{y})^2}, \quad 0 \le R^2 \le 1, \tag{2.9}$$

where $\bar{x}_j = \frac{1}{n}\sum_{j=1}^n x_{ij}$ and $\bar{y} = \frac{1}{n}\sum_{i=1}^n y_i$. This is based on quantifying the error in the fitted model.

2.1.1 Interpretation of R^2

First we observe the interpretation about the goodness of fit from the expression of θ. The model is said to be best fitted if the variation in y is ideally zero. In such a case, $\sigma^2 = 0$ which implies $\theta = 1$ and is interpreted as the model is best fitted in the sense that the model obtained from the sample of data matches exactly with the true model in the population. The model is considered as worst fitted when no explanatory variable contributes in explaining the variation in y in the modelling in the sense that $\beta_1 = \beta_2 = \ldots = \beta_p = 0$. In such a situation, $\beta = 0$ which implies $\theta = 0$. Similarly any other value of $0 < \theta < 1$ is considered as indicating the degree of goodness of fitted model explained by θ. So, θ measures the contribution of explanatory variables in explaining the variability among the values of study variable obtained by the model and can be considered as a measure of goodness of fit.

Since R^2 is an estimator of θ, it can also be considered to have similar interpretations for the goodness of fit. The lower and upper limits of R^2 are 0 and 1, respectively, and the interpretation of R^2 is as follows. The value of $R^2 = 0$ indicates the worst fit of the model and $R^2 = 1$ indicates the best fit of the model. Any other value of R^2 between 0 and 1 similarly indicates the degree of adequacy of the fitted obtained by the concerned estimator. For example, $R^2 = 0.7$ indicates that 70% of the variation in y is explained by the involved explanatory variables. In general, one can say that the model is approximately 70% good. Similarly, $R^2 = 0.1$ indicates that 10% of the variation in y is explained by the involved explanatory variables, and hence, it may be considered as not a good model.

2.1.2 Adjusted R^2

The CoD R^2 has a problem that its value increases as more explanatory variables are added to the model. As long as the relevant and meaningful explanatory variables are added, this indicates that the model is improving and getting better. On the other hand, in case the added variables are irrelevant, even then R^2 will increase and thus giving an overly optimistic picture. To correct such an inadequacy so that the R^2 corrects and does not give an overly optimistic picture and incorrect inferences, a variant of R^2 is suggested which is called as adjusted R^2 and denoted as $\bar{R}^2$. The adjusted R^2 is also used as a measure of goodness of fit, just like R^2 and is defined as

$$\bar{R}^2 = 1 - \frac{SS_{error}/(n-p)}{SS_{total}/(n-1)} = 1 - \left(\frac{n-1}{n-p}\right)(1 - R^2)\ , \ 0 \leq \bar{R}^2 \leq 1. \tag{2.10}$$

The lower and upper limits of $\bar{R}^2$ are 0 and 1, respectively. The interpretation of values of $\bar{R}^2$ is just like the same as of R^2. The value of $\bar{R}^2$ declines with the addition of an explanatory variable that produces too small reduction in $(1 - R^2)$ to compensate for the increase in $\left(\frac{n-1}{n-p}\right)$.

2.1.3 Limitations of R^2 and $\bar{R}^2$

The CoD and its adjusted version are the most popular measures of goodness of fit available in the literature but they have several limitations also, see Hahn (1973), but in spite of them, they remain a popular choice among practitioners. We discuss a few such features about which the practitioners must be aware of.

1. The R^2 and $\bar{R}^2$ work correctly only when the relationship between study and explanatory variables is linear. If the relationship between study and explanatory variables is nonlinear, even then R^2 and $\bar{R}^2$ will provide the numerical value but it will be inappropriate and incorrect to make any interpretation. The R^2 and $\bar{R}^2$ cannot be used to measure the degree of nonlinear relationship.

2. If the linear model is without an intercept term, then R^2 cannot be defined as in (2.5). If (2.5) is used to judge the goodness of fit in such a case, then R^2 can be negative which is meaningless. How to measure the goodness of fit in such a situation is not known. Some ad-hoc measures based on the structure of R^2 have been proposed in the literature but their dependability can always be challenged, see Rao et al. (2008).

3. The R^2 lacks robustness. The R^2 is sensitive to extreme values, i.e., if an extreme value irrespective of being an outlier or not is present in the data, the value of R^2 is disturbed.

4. As discussed earlier, the value of R^2 always increases when more explanatory variables are included in the model. The added explanatory variables may be relevant or irrelevant. In case the explanatory variables added in the model are irrelevant even then the R^2 increases which presents a faulty picture of the model as this will indicate that the model is getting better, which may not really be correct.

5. When the experimenter provides the data on study and explanatory variables, the statistician assumes a form of the model and/or variables and uses various transformation to linearizes the model so that the tools of multiple linear regression models including CoD can be used. The aim is only to judge that which form of the model gives a better fit. Comparison of goodness of fit among such model poses different challenges. For example, consider a situation where we have the following two models:

$$y_i = \beta_1 + \beta_2 X_{i2} + \ldots + \beta_p X_{ip} + u_i, \quad i = 1, 2, \ldots, n \tag{2.11}$$

$$\log y_i = \gamma_1 + \gamma_2 X_{i2} + \ldots + \gamma_p X_{ip} + u_i^*. \tag{2.12}$$

A question arises how to judge which of the model is better fitted? For the first model (2.11), the R^2 is defined as

$$R_1^2 = 1 - \frac{\sum_{i=1}^n (y_i - \hat{y}_i)^2}{\sum_{i=1}^n (y_i - \bar{y})^2}, \tag{2.13}$$

where $\hat{y}_i = \hat{\alpha} - b_1 x_{i1} - b_2 x_{i2} - \ldots - b_p x_{ip}$.

For the second model (2.12), the R^2 can be defined as

$$R_2^2 = 1 - \frac{\sum_{i=1}^n (\log y_i - \log \hat{y}_i)^2}{\sum_{i=1}^n (\log y_i - \log \bar{y})^2}. \tag{2.14}$$

The CoDs R_1^2 and R_2^2 in (2.13) and (2.14), respectively are not really comparable. If the two models are to be compared, a possible ad hoc proposition is to define the CoDs as follows:

$$R_3^2 = 1 - \frac{\sum_{i=1}^n (y_i - \text{antilog } \hat{y}_i^*)^2}{\sum_{i=1}^n (y_i - \bar{y})^2}, \tag{2.15}$$

where $\hat{y}_i^* = \widehat{\log y_i}$. Now possibly R_1^2 and R_3^2 can be compared but such a comparison can always be questioned.

6. A limitation of $\bar{R}^2$ is that it can be negative also. For example, if $p = 3$, $n = 10$, $R^2 = 0.15$ then $\bar{R}^2 = -0.09$, and it has no interpretation. Although mathematically correct but this is not a meaningful argument and statistical inference because $R^2 = 0.15$ itself indicates that the model fitting is not so good. The fitting with the currently used explanatory variables is itself unsatisfactory, so its not advisable to add more explanatory variables or use this fitted model.

3. Goodness of Fit in Multiple Measurement Error Model

A fundamental assumption defining the CoD or any goodness of fit statistic in multiple linear model based on the observations on study and explanatory variables is that all the observations are correctly observed. It may happen in many real life data that the observations on study and explanatory variables are not correctly observed. There can be several reasons for this. For example, sometimes the variable is clearly defined but it is hard to take correct observations on them, the variable is qualitative in nature, the variable is well understood but the observations are available only on their proxy variables etc. A few examples of such variables are, e.g., intelligence which is well understood but observations are obtained on IQ scores, the education level is measured by the number of years of schooling, taste is measured by an indicator variable etc. In such practical situations, such an assumption of observing the data correctly is difficult to be satisfied and the observations on the variables are not correctly observable. Rather measurement errors enter into the data and the true observations on the variable are not available. This disturbs the optimal properties of the OLSEs and other statistical tools. It may be noted that the OLSEs are the best linear unbiased estimators of the regression coefficients but this assumption does not continue to remain valid when the data has measurement errors in it. The OLSE becomes biased and inconsistent estimators of regression coefficients in the presence of measurement errors in the data. Consequently, the procedures like estimation of parameters, testing of hypothesis etc. cannot be used as they will provide invalid inferences. Since the CoD R^2 is a function of OLSE, so if the same R^2 is used to judge the goodness of fit of the model, ignoring the measurement errors, the conclusions for goodness of fit will become misleading and provide incorrect statistical inferences. Under such circumstances, the question is how to judge the goodness of fit of the measurement error models?

Another reason that why CoD R^2 cannot be used in measurement error models is based on the analysis of variance in multiple linear regression model. The CoD is based on the analysis of variance where the total sum of squares is partitioned into two orthogonal components- sum of squares due to regression and sum of squares due to errors, see (2.7) and (2.8). Such partitioning of total sum of squares in orthogonal components is not possible in measurements error models.

A fundamental challenge in the measurement error models is to first obtain the estimators of the regression coefficients which are at least consistent and then the goodness of fit can be measured. The consistent estimators can be obtained by using some additional information from outside the sample which removes the problem of identification concerning the estimation of parameters, see Rao et al. (2008) for more details. There are only two possible forms of such information which can be utilized in the set up of multiple measurement error models. These additional information can be either the knowledge of the covariance matrix of measurement errors associated with explanatory variables or the knowledge of the reliability matrix of explanatory variables. Using both types of additional knowledge, Cheng et al. (2014) suggested the measures of goodness of fit in measurement error models and have analyzed their asymptotic properties under an ultrastructural form of the measurement error models. It may be noted that there are two possible forms of measurement error models- functional form and structural form. Another form, *viz.*, the ultrastructural form is a synthesis of the functional and structural forms along with the classical regression model, see Dolby (1976).

3.1 Measurement Error Model

We first consider the following true exact relationship of measurement error free observations between the study and explanatory variables. Let η be the $(n \times 1)$ vector of true but unobservable values of study variable and T be the $(n \times p)$ matrix of n true but unobservable values on each of the p explanatory variables $T_1, T_2, \ldots, T_p$:

$$\eta = \alpha \mathbf{e}_n + T\beta, \tag{3.1}$$

where α is the intercept term, $\mathbf{e}_n$ is the $(n \times 1)$ vector of elements unity ($\mathbf{1}'$s), and β is the $(p \times 1)$ vector of regression coefficients. The presence of intercept term in the model is the basic requirement for the CoD statistics in multiple linear regression model without measurement errors. It is difficult to verify the need for such a condition for the validity of goodness of fit statistic proposed later under the measurement error models but to be on safe side, without loss of generality, we assume the presence of an intercept term in the model (3.1).

Assuming the presence of measurement errors in the observations, the observations on the study and explanatory variables η and T, respectively cannot be obtained but we assume that they are observed with additive measurement errors as

$$y = \eta + \epsilon \tag{3.2}$$

and

$$X = T + \Delta, \tag{3.3}$$

respectively. Here, $\epsilon = (\epsilon_1, \epsilon_2, \ldots, \epsilon_n)'$ is the $(n \times 1)$ vector of measurement errors associated with η and $\Delta = (\delta_1, \delta_2, \ldots, \delta_n)'$ is the $(n \times p)$ matrix of measurement errors associated with the explanatory variables in T, respectively. To consider a general form of the measurement error model, *viz.*, an ultrastructural model, we assume that T is expressible as the sum of matrix $M = E(T) = (\mu_1, \mu_2, \ldots, \mu_n)'$ which is the $(n \times p)$ of unknown means (constants) of T and the matrix of random errors $\Phi = (\phi_1, \phi_2, \ldots, \phi_n)'$ associated with T as

$$T = M + \Phi. \tag{3.4}$$

The $(p \times 1)$ random vectors $\phi_1, \phi_2, \ldots, \phi_n$ of matrix Φ are assumed to be independently and identically distributed with mean vector 0 and covariance matrix Σ_ϕ. Similarly, $(p \times 1)$ random vectors $\delta_1, \delta_2, \ldots, \delta_n$ of matrix Δ are assumed to be independently and identically distributed with mean vector 0 and covariance matrix Σ_δ. Further, the measurement errors $\epsilon_1, \epsilon_2, \ldots, \epsilon_n$ are assumed to be independently and identically distributed with mean zero and variance σ_ϵ^2.

There are two forms of measurement error models - functional and structural, which depends upon the nature of the explanatory variables in the sense whether they are fixed or stochastic, respectively. The setup of an ultrastructural model (see Dolby (1976)) is described by (3.1)-(3.4). The ultrastructural model comprises of functional form, structural form as well as the classical linear regression model without measurement errors as its particular cases. If the results are derived under the ultrastructural model, then the results for the functional and structural forms as well as the classical regression model can be derived as the special cases from them as follows. It may be observed that when all the rows of M are identically distributed, i.e., the rows of X are random and independent following some multivariate distribution, then the structural form of the measurement error model is specified. When $\Phi = 0$ (null matrix) which is indicated by $\Sigma_\phi = 0$ (null matrix), this means that X is fixed but measured with error, then the functional form of the measurement error model is specified. When $\Delta = \Phi = 0$ (null matrix) which is indicated by $\Sigma_\phi = \Sigma_\delta = 0$ (null matrix), i.e., X is fixed and without measurement errors, then the classical multiple regression model is specified. The advantage of considering the ultrastructural model is that the properties of estimators and goodness of fit statistics under these three models can be derived directly from the results developed and obtained under the ultrastructural model.

We further assume that $\lim_{n\to\infty} n^{-1}M'PM = \Sigma_\mu$ which is a symmetric and positive definite matrix where $P = I_n - n^{-1}\mathbf{e}_n\mathbf{e}_n'$. This assumption is needed for the validity of asymptotic results and avoids the possibility of any trend in the data, see Schneeweiss (1991). The goodeness of fit statistics under measurement error model have been obtained by Cheng et al. (2014) and an analysis of their asymptotic properties under ultrastructural model is presented.

3.2 Goodness of Fit Statistics in Measurement Error Model

The OLSE of the regression coefficient becomes inconsistent in the presence of measurement errors in the data. The consistent estimators of regression coefficients in the multiple measurement error models can be obtained using the additional information in following two forms of information from outside the sample, *viz.*, the known covariance matrix of measurement errors (Σ_δ) associated with the explanatory variables and the known reliability matrix (K_x) associated with the explanatory variables. Such information can be available from the past experience of the researcher, from some similar kind of studies done in the past, pilot survey etc. In case, such information is unavailable, an alternative is to estimate them using the repeated observations under the setup of repeated multiple measurement error model and replace Σ_δ or K_x by their respective estimated values.

We observe that the probability in limit of CoD $R^2 = \frac{y'PX(X'PX)^{-1}X'Py}{y'Py}$ under the ultrastructural model (3.1)-(3.4) is

$$\text{plim}_{n\to\infty} R^2 = \frac{\beta'(\Sigma - \Sigma_\delta)\Sigma^{-1}(\Sigma - \Sigma_\delta)\beta}{\beta'(\Sigma - \Sigma_\delta)\beta + \sigma_\epsilon^2} \neq \theta,$$

in general, and hence, R^2 is an inconsistent estimator of θ under the measurement error models. Consequently, use of R^2 to judge the goodness of the fit with measurement error ridden data will provide incorrect, invalid and misleading conclusions. Therefore, it is not recommended to use CoD meant for multiple regression model to judge the goodness of fit in the measurement error models.

The statistic for measuring the goodness of fit in measurement error models is expected to be developed which is at least a consistent estimator of θ in the presence of measurement errors in the data which will be based on the utilization of some additional information from outside the sample. Such additional information is required to consistently estimate β. We consider the goodness of fit statistic for estimating θ under the measurement error model.

3.2.1 Goodness of Fit Statistics when Σ_δ is known

When the covariance matrix Σ_δ is known, the consistent estimator of β in measurement error model is given by

$$b_\delta = (S - \Sigma_\delta)^{-1} Sb, \tag{3.5}$$

where $S = n^{-1}X'PX$ and $b = (X'PX)^{-1}X'Py$ is the OLSE of β. The goodness of fit statistic using Σ_δ is given as

$$R_\delta^2 = \frac{b_\delta' S b_\delta}{b_\delta' S b_\delta + \{n^{-1}y'Py - b_\delta'(S - \Sigma_\delta)b_\delta\}} = \frac{b_\delta' S b_\delta}{n^{-1}y'Py + b_\delta'\Sigma_\delta b_\delta}, \quad 0 \leq R_\delta^2 \leq 1, \tag{3.6}$$

provided $b_\delta' S b_\delta \geq n^{-1}y'Py + b_\delta'\Sigma_\delta b_\delta$. In case, $b_\delta' S b_\delta < n^{-1}y'Py + b_\delta'\Sigma_\delta b_\delta$, we take the value of R_δ^2 as 1.

So the goodness of fit statistic under measurement error model when Σ_δ is known is defined as

$$R_\delta^2 = \min\left(\frac{b_\delta' S b_\delta}{n^{-1}y'Py + b_\delta'\Sigma_\delta b_\delta}, 1\right) \tag{3.7}$$

for which $\text{plim}_{n\to\infty} R_\delta^2 = \theta$, see Cheng et al. (2014) for more details on the asymptotic properties of R_δ^2. Thus the goodness of fit statistic R_δ^2 can be used to judge the goodness of fit in the multiple linear measurement error model when the model is fitted using (3.5).

3.2.2 Goodness of Fit Statistics when Reliability matrix K_x is known

The information on the reliability ratio helps in obtaining a consistent estimator of the regression coefficient in measurement error model. The reliability ratio is defined as the ratio of the variances of true and observed values of explanatory variable. The reliability matrix is the multivariate generalization of reliability ratios and is defined as

$$K_x = \Sigma_x^{-1}(\Sigma_x - \Sigma_\delta).$$

When K_x is known, then the consistent estimator of β in measurement error model is given as

$$b_k = K_x^{-1} b, \tag{3.8}$$

see Cheng and Van Ness (1999) and Fuller (1987). The concept of reliability ratio is popular in Psychometrics and tools are available to estimate it. Using K_x and b_k, the goodness of fit statistic is given as

$$R_k^2 = \frac{b_k' S b_k}{b_k' S b_k + \{n^{-1}y'Py - b_k' S K_x b_k\}} = \frac{b_k' S b_k}{n^{-1}y'Py + b_k' S(I_p - K_x)b_k}; \quad 0 \leq R_k^2 \leq 1, \tag{3.9}$$

provided $b_k' S b_k \geq n^{-1}y'Py + b_k' S(I_p - K_x)b_k$. In case $b_k' S b_k < n^{-1}y'Py + b_k' S(I_p - K_x)b_k$, we take the value of R_k^2 as 1. So the goodness of fit statistic under the measurement error model and assumption of known reliability matrix can be defined as

$$R_k^2 = \min\left(\frac{b_k' S b_k}{n^{-1} y' P y + b_k' S (I_p - K_x) b_k}, 1\right) \tag{3.10}$$

and $\text{plim}_{n\to\infty} R_k^2 = \theta$. see Cheng et al. (2014) for more details on the asymptotic properties of R_k^2. Thus the proposed statistic R_k^2 measures the goodness of fit in the linear measurement error model when the model is fitted using (3.8).

3.2.3 Interpretation of R_δ^2 and R_k^2

Note that both $0 \le R_\delta^2 \le 1$ and $0 \le R_k^2 \le 1$. When all the explanatory variables in the model are not contributing in explaining the variation in the values of study variable, then ideally b_δ (or b_k) will be zero or close to zero. In turn, R_δ^2 (or R_k^2) will be zero or close to zero indicating that the model fit is poor. Similarly, the values of R_δ^2 (or R_k^2) close to 1 will indicate the perfect fitted model. Any other value of R_δ^2 (or R_k^2) between 0 and 1 will give an idea about the degree of goodness of fit in the model, just like the CoD in a usual multiple linear regression model. So R_δ^2 (or R_k^2) defines a goodness of fit statistic for the linear multiple measurement error model and can be used depending upon the form of available information.

Such goodness of fit statistics also helps a user to decide different issues. Suppose that both types of additional information, *viz.*, the known covariance matrix of measurement errors and the known reliability matrix are available. A question arises that which of the information will provide better estimator. Another question that can be answered is how to know which of the estimator should be used to get a good fitted model based on these estimators.

4. Goodness of Fit Statistics for Restricted Measurement Error Models

In several applications, some apriori information about the regression coefficients is available which can be represented in various forms like exact restrictions, stochastic restrictions or inequality restrictions, see Toutenburg (1982), Rao et al. (2008) for more details. Such information is available from different resources from outside the sample resources like similar type of studies conducted in the past, theoretical considerations, past experience of the experimenter, etc. If such information can be incorporated in the estimation procedures, then it improves the estimates of parameter. It is necessary that thus obtained estimators satisfy the restrictions on the regression coefficients. Such situations are common in economics and finance.

Suppose that such prior information can be expressed in the form of exact linear restrictions. When such constraints are incorporated in the least squares estimation procedures using the observations that are free from measurement errors, then the restricted regression estimator or the restricted least squares estimator is obtained, and it satisfies the exact restrictions. Such obtained estimator is unbiased, more efficient than the ordinary least squares estimator and consistent for the regression coefficients. When the observations contain the measurement errors, then the same restricted regression estimator also becomes biased and inconsistent for estimating the regression coefficients. Moreover, the estimators of regression coefficients obtained by using the additional information (known covariance matrix of the measurement errors involved in explanatory variables or the reliability matrix as in subsections (3.2.1) and (3.2.2)) are consistent under the multiple measurement error model but they do not satisfy the exact linear restrictions, and hence, are inappropriate for use in such situations. The estimators of regression coefficients that are consistent under the multiple measurement error model and satisfy the exact linear restrictions are obtained in Shalabh et al. (2007, 2009) which are based on the least squares approach and constrained minimization of the weighted error sum of squares under the given exact restrictions.

The question is how to judge the goodness of fit in such a situation where the measurement error models with given exact linear restrictions on the regression coefficients are fitted. The traditional R^2 statistic cannot be used as it is based on the inconsistent ordinary least squares estimator which does not satisfy even the exact restrictions. We now discuss the goodness of fit statistics for a situation when measurement errors are present in the data and available apriori information is utilized for the estimation of regression coefficient and fitting the models

4.1 The Exact Linear Restrictions and Model

Sometimes the prior information about the regression coefficients is available from some past experience, extraneous sources, similar kind of studies conducted in the past etc. There can also be situations when some restrictions are imposed on the regression coefficients. Such restrictions may arise from some theoretical considerations, need of analyst or the need of experimental conditions. For example, the Cobb-Douglas function in Economics requires the assumption of constant returns to scale. Similar constraints are used in the area of finance also. We assume that such prior information on the restrictions can be expressed in the form of exact linear restrictions as

$$H\beta = h, \tag{4.1}$$

which binds the $J(< p)$ restrictions. Here H is a known full row rank matrix of order $(J \times p)$ and h is a known vector of order $(J \times 1)$.

We continue to consider the ultrastructural form of the measurement error model (3.1)–(3.4).

The restricted least squares estimator (RLSE) of β when the measurement errors are absent as in the classical multiple linear regression model (2.1) is

$$b_H = b + S^{-1}H'(HS^{-1}H')^{-1}(h - Hb), \tag{4.2}$$

where $S = n^{-1}X'PX$ and b is the OLSE of β, see Rao et al. (2008, Chap. 5). The OLSE does not satisfy the restrictions, i.e., $Hb \neq h$ whereas RLSE satisfies the restriction, i.e., $Hb_H = h$. When measurement errors are present, then b_H is an inconsistent estimator of β and under the multiple ultrastructural model (3.1)–(3.4),

$$plim_{n\to\infty} b_H = [I_p - \sigma_\delta^2\Sigma^{-1}(I_p - H'H_\Sigma^{-1}H\Sigma_*^{-1})]\beta \neq \beta, \tag{4.3}$$

in general, where $\Sigma_* = \Sigma_\mu + \Sigma_\phi + \Sigma_\delta$ and $H_\Sigma = H\Sigma^{-1}H'$.

The consistent estimators of β in case of multiple measurement error models using two possible forms of additional information *viz.*, the known covariance matrix of measurement errors in explanatory variables (Σ_δ) and the known reliability matrix of explanatory variables (K_x) are given by b_δ (as in (3.5)) and b_k (as in (3.8)), respectively. But b_δ and b_k do not satisfy the restrictions, i.e., $Hb_\delta \neq h$ and $Hb_k \neq h$, respectively. So first we need such estimators which are consistent and satisfy the restrictions under the known Σ_δ and K_x. Such estimators of regression coefficients are proposed in Shalabh et al. (2007, 2009) and can be used to specify the goodness of fit statistics as discussed in the following subsections.

4.2 Goodness of Fit when Σ_δ is known

The three possible estimators of β under known Σ_δ which are consistent as well as satisfy the restrictions $H\beta = h$, obtained in Shalabh et al. (2007, 2009), are as follows:

$$b_\delta^{(1)} = b_\delta + S^{-1}H'H_s^{-1}(h - Hb_\delta) \tag{4.4}$$

$$b_\delta^{(2)} = [I_p - (I_p - S^{-1}H'H_s^{-1}H)S^{-1}\Sigma_\delta]^{-1}b_H \tag{4.5}$$

$$b_\delta^{(3)} = b_\delta + H'(HH')^{-1}(h - Hb_\delta), \tag{4.6}$$

where $H_s = HS^{-1}H'$, $plim_{n\to\infty} b_\delta^{(r)\prime} = \beta$ and $Hb_\delta^{(r)} = h, r = 1, 2, 3$ holds true.

We propose to develop the goodness of fit statistic based on the estimators (4.4)–(4.6) which are consistent and satisfy the exact linear restrictions $H\beta = h$. It is based on the consistent estimator of σ_ϵ^2 which is obtained from the following results: $plim_{n\to\infty}\left[n^{-1}y'Py - b_\delta^{(r)}(S - \Sigma_\delta)b_\delta^{(r)}\right] = \sigma_\epsilon^2$, and $plim_{n\to\infty}\left[b_\delta^{(r)\prime}Sb_\delta^{(r)}\right] = \beta'\Sigma\beta, \quad r = 1, 2, 3$. It may be noted that σ_ϵ^2 is variance, it is estimated by 0 in case the statistic $\{n^{-1}y'Py - b_\delta^{(r)}(S - \Sigma_\delta)b_\delta^{(r)}\}$ takes negative values.

The goodness of fit to the models that are fitted using $b_\delta^{(r)}$, $r = 1, 2, 3$ in (4.4)–(4.6) under the measurement error model using the apriori information $H\beta = h$ is given by

$$\widetilde{R^2}_\delta^{(r)} = \frac{b_\delta^{(r)\prime}Sb_\delta^{(r)}}{n^{-1}y'Py - b_\delta^{(r)\prime}(S - \Sigma_\delta)b_\delta^{(r)} + b_\delta^{(r)\prime}Sb_\delta^{(r)}} = \frac{b_\delta^{(r)\prime}Sb_\delta^{(r)}}{n^{-1}y'Py + b_\delta^{(r)\prime}\Sigma_\delta b_\delta^{(r)}}, \tag{4.7}$$

provided $n^{-1}y'Py > b_\delta^{(r)}(S - \Sigma_\delta)b_\delta^{(r)}$. When $n^{-1}y'Py \leq b_\delta^{(r)}(S - \Sigma_\delta)b_\delta^{(r)}$, $\widetilde{R^2}_\delta^{(r)} = 1$ for $r = 1, 2, 3$. Therefore, the modified $\widetilde{R^2}_\delta^{(r)}, r = 1, 2, 3$, is given by

$$\widetilde{R^2}_\delta^{(r)} = Min\left(\frac{b_\delta^{(r)\prime}Sb_\delta^{(r)}}{n^{-1}y'Py + b_\delta^{(r)\prime}\Sigma_\delta b_\delta^{(r)}}, 1\right), \quad 0 \leq \widetilde{R^2}_\delta^{(r)} \leq 1, \quad r = 1, 2, 3.$$

Note that $\widetilde{R^2}_\delta^{(r)}, r = 1, 2, 3$, are consistent estimator of θ, i.e., $plim_{n\to\infty}\widetilde{R^2}_\delta^{(r)} = \theta$ and can be used to measure the goodness of fit in restricted measurement error model.

4.3 Goodness of Fit when K_x is Known

The three possible estimators of β under known K_x which are consistent as well as satisfy the restrictions $H\beta = h$ are obtained in Shalabh et al. (2007, 2009) as follows:

$$b_k^{(1)} = b_k + S^{-1}H'H_s^{-1}(h - Hb_k) \tag{4.8}$$
$$b_k^{(2)} = [K_x + (I_p - K_x)H'(H(I_p - K_x)H')^{-1}H(I_p - K_x)]^{-1}b_H \tag{4.9}$$
$$b_k^{(3)} = b_k + H'(HH')^{-1}(h - Hb_k), \tag{4.10}$$

where $plim_{n\to\infty} b_k^{(r)\prime} = \beta$ and $Hb_k^{(r)} = h, r = 1,2,3$ holds true.

A consistent estimator of σ_ϵ^2 and $\beta'\Sigma\beta$ based on the use of $b_k^{(r)}$, $r = 1,2,3$ in (4.8)-(4.10) can be obtained from the following results: $plim_{n\to\infty}\left[n^{-1}y'Py - b_k^{(r)\prime}SK_x b_k^{(r)}\right] = \sigma_\epsilon^2$, $r = 1,2,3$ and $plim_{n\to\infty}\left[b_k^{(r)\prime}Sb_k^{(r)}\right] = \beta'\Sigma\beta$. When $n^{-1}y'Py \le b_k^{(r)\prime}SK_x b_k^{(r)}$, σ_ϵ^2 is estimated by 0.

The goodness of fit to the models that are fitted using $b_k^{(r)}$, $r = 1,2,3$ under the measurement error model using the apriori information $H\beta = h$ is

$$\widetilde{R^2}_k^{(r)} = \frac{b_k^{(r)\prime}Sb_k^{(r)}}{b_k^{(r)\prime}Sb_k^{(r)} + (n^{-1}y'Py - b_k^{(r)\prime}SK_x b_k^{(r)})} = \frac{b_k^{(r)\prime}Sb_k^{(r)}}{n^{-1}y'Py + b_k^{(r)\prime}S(I_p - K_x)b_k^{(r)}}, \tag{4.11}$$

provided $n^{-1}y'Py > b_k^{(r)\prime}SK_x b_k^{(r)}$. In case, $n^{-1}y'Py \le b_k^{(r)\prime}SK_x b_k^{(r)}$, then we take $\widetilde{R^2}_k^{(r)} = 1$, $r = 1,2,3$. Therefore, the modified $\widetilde{R^2}_k^{(r)}$ using K_x and $H\beta = h$ is given by

$$\widetilde{R^2}_k^{(r)} = Min\left(\frac{b_k^{(r)\prime}Sb_k^{(r)}}{n^{-1}y'Py + b_k^{(r)\prime}S(I_p - K_x)b_k^{(r)}},\ 1\right),\quad 0 \le \widetilde{R^2}_k^{(r)} \le 1,\quad r = 1,2,3.$$

As $\widetilde{R^2}_k^{(r)}$ is a consistent estimator of θ, so $\widetilde{R^2}_\delta^{(r)}$, $r = 1,2,3$, are consistent estimators of θ, i.e., $plim_{n\to\infty}\widetilde{R^2}_k^{(r)} = \theta$ for $r = 1,2,3$ and can be used to measure the goodness of fit in restricted measurement error model.

4.4 Interpretation of $\widetilde{R^2}_\delta^{(r)}$ and $\widetilde{R^2}_k^{(r)}$, $r = 1, 2, 3$

The interpretation of $\widetilde{R^2}_\delta^{(r)}$ and $\widetilde{R^2}_k^{(r)}$, $r = 1,2,3$ is similar to the interpretations of goodness of fit statistics described in subsections 2.1.1 and 3.2.3 in the classical regression model. The values of $\widetilde{R^2}_\delta^{(r)}$ (or $\widetilde{R^2}_k^{(r)}$, $r = 1,2,3$) equal to 0 and 1 indicate the worst and best fitted models, respectively. Any other value between 0 and 1 indicates the degree of goodness of fit provided by the respective estimators when used in the model. For example, suppose that in a given data set, $\widetilde{R^2}_\delta^{(2)} = 0.8$ and $\widetilde{R^2}_k^{(2)} = 0.5$, it indicates that use of K_x for obtaining the consistent estimator yields better model than using Σ_δ. Similarly, if $\widetilde{R^2}_\delta^{(2)} = 0.6$ and $\widetilde{R^2}_\delta^{(3)} = 0.8$, it indicates that the use of $b_\delta^{(3)}$ yields better fitted model than $b_\delta^{(2)}$.

5. Goodness of Fit for Generalized Shrinkage Estimation

The regression parameters in a multiple linear regression model (2.1) can be estimated by various estimation procedures. The ordinary least squares estimators (OLSEs) provide the estimator of regression coefficient having minimum variance in the class of best linear and unbiased estimators under the model (2.1). The shrinkage estimators are nonlinear and biased estimator of regression but they have smaller variability than the OLSE under some mild conditions. For example, the Stein rule family of estimators of regression coefficients provides biased and nonlinear estimators under (2.1) but have smaller variability than the OLSE when the model (2.1) has more than two explanatory variables, see Stein (1956) and James and Stein (1961). Similarly, another family of estimators based on shrinkage estimators is the family of double k-class estimators proposed by Ullah and Ullah (1978). It encompasses different type of estimators originating from various estimation methods as its special cases, see Wan and Chaturvedi (2001), Chaturvedi and Shalabh (2004) and Shalabh et al. (2012) for more details on the double k-class estimators. There are two characterizing scalars in the double k-class estimators. Different estimators can be obtained by choosing different values of these scalars. Various forms of feasible versions of double k-class estimators are discussed in Wan and Chaturvedi (2001) and Chaturvedi and Shalabh (2004) when the random errors in the linear regression model are non spherical and the covariance matrix of errors is unknown.

As discussed earlier, the CoD R^2 is based on the OLSE of regression coefficients, and it is also obtainable from analysis of variance in linear regression model. The double k-class estimators of regression coefficient are nonlinear and biased estimators of regression coefficient. So it is not possible to partition the total sum of squares in orthogonal components based on the double k-class estimators as in the analysis of variance. Hence the CoD is not a valid measure to judge the goodness of fit in such a case. Moreover, many other estimators which can be obtained as particular case from the double k-class estimators by substituting different choices of characterizing scalars does not partitions the total sum of squares into two orthogonal components. The R^2 will provide invalid conclusions if it is considered to measure the goodness of fit in such cases. The question is now how to judge the goodness of fit in the models which are fitted using the double k-class estimators. Such goodness of fit measures can be utilized to judge the goodness of fit of model obtained from various different estimation procedures which also arise as particular cases of double k-class estimator. How to know that which choice of scalars gives a better fitted model can also be answered using such goodness of fit statistic.

5.1 Model and the Estimators

Consider the following linear regression model between the $(n \times 1)$ vector y of n observations on the study variable and $(n \times p)$ matrix X of n observations on p explanatory variables:

$$y = \alpha \mathbf{e}_n + X\beta + w, \tag{5.1}$$

where α is the intercept term, $\mathbf{e}_n$ is the $(n \times 1)$ vector of elements unity $(1'\text{s})$, β is a $(p \times 1)$ vector of coefficients associated with them, and w is a $(n \times 1)$ vector of non-spherical disturbances. The disturbance vector w is assumed to follow a distribution with mean vector 0 and covariance matrix $\sigma^2\Omega^{-1}$.

Note that we are considering here a very general set up by considering the disturbances to be non-spherical with the two cases of having known and unknown covariance matrix. So this will provide a solution to measure the goodness of fit of a model under a variety of standard and non-standard conditions.

5.1.1 When Ω is known

Let $A = \Omega - \frac{1}{\mathbf{e}_n'\Omega\mathbf{e}_n}\Omega\mathbf{e}_n\mathbf{e}_n'\Omega$. The generalized least squares estimator (GLSE) to estimate β in is given by

$$\hat{\beta}_g = (X'AX)^{-1}X'Ay. \tag{5.2}$$

It is well known that the GLSE has lower variability than OLSE under the criterion of Löwner ordering and is the best linear unbiased estimator of β.

The family of generalized double k-class estimators (DKKE) characterized by two nonstochastic scalars $k_1 > 0$ and $0 < k_2 < 1$ which are based on the structure of double k-class estimator proposed by Ullah and Ullah (1978), (see also, Ullah and Ullah (1981)), is given as

$$\begin{aligned}\hat{\beta}_{kk} &= \left[1 - \left(\frac{k_1}{n-p+2}\right)\frac{(y - X\hat{\beta}_g)'A(y - X\hat{\beta}_g)}{y'Ay - k_2(y - X\hat{\beta}_g)'A(y - X\hat{\beta}_g)}\right]\hat{\beta}_g \\ &= \left[1 - \left(\frac{k_1}{n-p+2}\right)\frac{\xi}{\hat{\beta}_g'X'AX\hat{\beta}_g + (1 - k_2 v)}\right]\hat{\beta}_g,\end{aligned} \tag{5.3}$$

where $\xi = (y - X\hat{\beta}_g)'A(y - X\hat{\beta}_g)$. This is a very general class of estimator and various estimators which have been obtained by different estimation procedures separately in the literature can be obtained as a special case of DKKE. For example, GLSE can be obtained by substituting $k_1 = 0$; generalized Stein-rule estimator (GSRE) is obtainable by substituting $k_1 = p - 2$ and $k_2 = 1$; generalized minimum mean squared error estimator (GMMSE) is obtained with the choice $k_1 = \frac{1}{n-p}$ and $k_2 = 1 - k_1$; adjusted generalized minimum mean squared error estimator (AGMMSE) is obtainable by $k_1 = \frac{n-p+2}{n-p}$ and $k_2 = 1 - \frac{k_1}{n-p+2}$; generalized double k-class estimators (GKKCE) can be obtained by choosing $k_1 = \frac{(n-p+2)p}{n-p}$ and $k_2 = 1 - \frac{k_1}{n-p+2}$ etc. Cheng et al. (2019) discuss such measures of goodness of fit and derive their first and second order moments up to the first order of approximation.

5.1.2 When Ω is unknown

In case, the variance-covariance matrix $\sigma^2\Omega^{-1}$ is unknown, we assume that the elements of Ω are functions of a $(q \times 1)$ parameter vector ϑ that belongs to an open subset of q dimensional Euclidean space. We also assumed that a consistent estimator $\hat{\vartheta}$ of ϑ is available so that Ω is consistently estimated by $\hat{\Omega} = \Omega(\hat{\vartheta})$. So $\Omega \equiv \Omega(\vartheta)$ and $\hat{\Omega} \equiv \Omega(\hat{\vartheta})$. In such situation, a feasible version of GLSE of β is given by

$$\hat{\beta}_{fg} = (X'\hat{A}X)^{-1}X'\hat{A}y, \tag{5.4}$$

where $\hat{A}$ is obtained by replacing Ω by $\hat{\Omega}$ in A and then the family of feasible generalized double k-class estimators (FDKKE) is given as

$$\begin{aligned}\hat{\beta}_{fkk} &= \left[1 - \left(\frac{k_1}{n-p+2}\right) \frac{(y - X\hat{\beta}_{fg})'\hat{A}(y - X\hat{\beta}_{fg})}{y'\hat{A}y - k_2(y - X\hat{\beta}_{fg})'\hat{A}(y - X\hat{\beta}_{fg})}\right]\hat{\beta}_{fg} \\ &= \left[1 - \left(\frac{k_1}{n-p+2}\right) \frac{\hat{\xi}}{\hat{\beta}'_{fg}X'\hat{A}X\hat{\beta}_{fg} + (1-k_2)\hat{v}}\right]\hat{\beta}_{fg},\end{aligned} \tag{5.5}$$

where $\hat{\xi} = (y - X\hat{\beta}_{fg})'\hat{A}(y - X\hat{\beta}_{fg})$. Again, feasible versions of the estimators which were obtained by choosing different values of the characterizing scalars k_1 and k_2 in DKKE can also be obtained from FDKKE. Such estimators have been obtained in the literature by different researchers independently. For example, feasible generalized least squares estimator (FGLSE) is obtained by choosing $k_1 = 0$; feasible generalized Stein-rule estimator (FGSRE) is obtained by substituting $k_1 = p - 2$ and $k_2 = 1$; similarly feasible generalized minimum mean squared error estimator (FGMMSE) is obtainable by assigning $k_1 = \frac{1}{n-p}$ and $k_2 = 1 - k_1$ which was proposed by Farebrother (1975), see also Ohtani (1996), Ohtani (1997); adjusted feasible generalized minimum mean squared error estimator (AFGMMSE), proposed by Ohtani (1996), is obtained by $k_1 = \frac{n-p+2}{n-p}$ and $k_2 = 1 - \frac{k_1}{n-p+2}$; feasible generalized double k-class estimator (FGKKCE), proposed by Carter et al. (1993), can be obtained by substituting $k_1 = \frac{(n-p+2)p}{n-p}$ and $k_2 = 1 - \frac{k_1}{n-p+2}$.

So it is clear that by considering the family of double k-class estimator and its feasible version, we are essentially considering estimators arising from various estimation procedures in a unified way.

5.2 Goodness of Fit Statistics for GLS, DKK and FDKK Estimators

We consider the goodness of fit statistic for various type of estimators in the following subsections.

5.2.1 Goodness of Fit Statistics for GLSE

First we develop the goodness of fit for the GLSE of regression coefficient. Consider the form of CoD R^2 in the regression model (2.1) under the assumptions $V(v) = \sigma^2\Omega^{-1} = \Omega^*$, say which is a positive definite matrix. One can find a nonsingular matrix T such that $TT^{-1} = \Omega^*$. Pre-multiplying (2.1) by T^{-1} gives

$$T^{-1}Ay = T^{-1}AX\beta + T^{-1}Aw \tag{5.6}$$

or

$$y^* = X^*\beta + w^*, \tag{5.7}$$

where $y^* = T^{-1}Ay$, $X^* = T^{-1}AX$ and $w^* = T^{-1}Av$. Note that now $E(w^*) = 0$ and $V(w^*) = \sigma^2 I$ so that elements in w^* are identically and independently normally distributed. The goodness of fit statistic under (5.7) is given as

$$R_g^2 = \frac{y^{*\prime}X^*(X^{*\prime}X^*)^{-1}X^{*\prime}y^*}{y^{*\prime}y^*} = \frac{y'AX(X'AX)^{-1}X'Ay}{y'Ay} = \frac{\hat{\beta}'_g X'AX\hat{\beta}_g}{y'Ay} \quad , \quad 0 \leq R_g^2 \leq 1. \tag{5.8}$$

Assume that $plim_{n\to\infty}\frac{X'AX}{n} = \Sigma_{XX}$ (a positive definite matrix), and $plim_{n\to\infty}\frac{X'A\epsilon}{n} = 0$. Then $plim_{n\to\infty}R_g^2 = \theta$, where R_g^2 is a consistent but biased estimator of population multiple correlation coefficient θ.

Note that $0 \leq R_g^2 \leq 1$. When all the explanatory variables in the model are not contributing towards the study variable in explaining the variation in its values, then ideally $\hat{\beta}_g$ will be zero or close to zero. In turn, $R_g^2 = 0$ indicating that the model fit is poor. Similarly, $R_g^2 = 1$ will indicate the perfect fit. Any other value of R_g^2 between 0 and 1 will give an idea about the degree of goodness of fit in the model, similar to CoD in a usual multiple linear regression model. So R_g^2 in (5.8) defines a statistic which measures the goodness of fit in the linear regression model with a non-identity covariance matrix assuming A (or equivalently Ω) is known. This statistic can be used to judge the goodness of fit of model when GLSE is used to fit the model.

5.2.2 Goodness of Fit Statistics for DKKE

Next, the goodness of fit statistic based on the use of $\hat{\beta}_{kk}$ is given as

$$R_{kk}^2 = \frac{\hat{\beta}'_{kk}X'AX\hat{\beta}_{kk}}{y'Ay} \quad , \quad 0 \leq R_{kk}^2 \leq 1. \tag{5.9}$$

Assume that $plim_{n\to\infty}\frac{X'AX}{n} = \Sigma_{XX} > 0$, and $plim_{n\to\infty}\frac{X'A\epsilon}{n} = 0$. Then $plim_{n\to\infty}R^2_{kk} = \theta, \quad 0 \le \theta \le 1$, i.e., R^2_{kk} is a consistent but biased estimator of θ.

Note that $0 \le R^2_{kk} \le 1$, so this also has an interpretation like the CoD in a multiple linear regression model. For example, when $\hat{\beta}_{kk}$ is zero or close to zero, it indicates that all the elements in its vector are zero or close to zero meaning thereby that all the explanatory variables in the model are not capable of explaining the variation in the values of study variable. Hence, $R^2_{kk} = 0$ will indicate the poorest fit. Similarly, $R^2_{kk} = 1$ will indicate the best fit and any other value of R^2_{kk} between 0 and 1 will give an idea about the degree of goodness of fit in the model resulting by the use of any estimator from the family of DKKE. Thus this statistic can be used to judge the goodness of fit of a model based on DKKE $\hat{\beta}_{kk}$.

5.2.3 Goodness of Fit Statistics for FGLSE and FDKKE

Now we consider the case when A (or equivalently Ω) is unknown and is consistently estimated as $\hat{A}$ (or equivalently $\hat{\Omega}$) such that $plim_{n\to\infty}\hat{A} = A$ (or equivalently $plim_{n\to\infty}\hat{\Omega} = \Omega$). In such a case, the goodness of fit statistics based on FGLSE in (5.4) and FDKKE in (5.5) are proposed as

$$R^2_{fg} = \frac{\hat{\beta}'_{fg}X'\hat{A}X\hat{\beta}_{fg}}{y'\hat{A}y}, \quad 0 \le R^2_{fg} \le 1 \tag{5.10}$$

and

$$R^2_{fkk} = \frac{\hat{\beta}'_{fkk}X'\hat{A}X\hat{\beta}_{fkk}}{y'\hat{A}y}, \quad 0 \le R^2_{fkk} \le 1, \tag{5.11}$$

respectively.

Note that $plim_{n\to\infty}R^2_{fg} = \theta$ and $plim_{n\to\infty}R^2_{fkk} = \theta$, $0 \le \theta \le 1$. Thus both R^2_{fg} and R^2_{fkk} are the consistent but biased estimator of population multiple correlation coefficient θ. Both R^2_{fg} and R^2_{fkk} lie between 0 and 1. The interpretations of R^2_{fg} and R^2_{fkk} can be obtained just like the interpretations of R^2_g and R^2_{kk}, respectively. For example, $R^2_{fg} = 0$ and $R^2_{fkk} = 0$ indicate the poorest fit and $R^2_{fg} = 1$ and $R^2_{fkk} = 1$ indicate the best fit of the model using the estimators $\hat{\beta}_{fg}$ and $\hat{\beta}_{fkk}$, respectively.

6. AIC, BIC and Mallows' C_p Criteria

One approach to select a good model from a set of candidate models is to use Akaikes Information Criterion (AIC) due to Akaike (1973). It is based on quantifying the information and the Kullback-Leibler distance function as a measure of distance between the underlying true but unknown model f and any approximation to it denoted by $g(y|\lambda)$ which depends on the unknown parameters λ to be estimated. Then the Kullback-Leibler distance function is

$$I(f, g(y|\lambda)) = \int f(y)\cdot\log(f(y))dx - \int f(y)\cdot\log(g(y|\lambda))dy = E_f[\log(f(y))] - E_f[\log(g(y|\lambda))]. \tag{6.1}$$

Let $\hat{\lambda}(y)$ denote an maximum likelihood estimator of λ based on observations y. Then

$$E_y\left[I(f, g(\cdot|\hat{\lambda}(y)))\right] = C - E_yE_x\left[\log[g(x|\hat{\lambda}(y))]\right], \tag{6.2}$$

where C is a constant term. Akaike found a relationship between the second term of (6.2) and the log-likelihood function as

$$E_yE_x\left[\log[g(x|\hat{\lambda}(y))]\right] \approx L(\hat{\lambda}|y) - K_e \tag{6.3}$$

where $L(\hat{\lambda}|y)$ denotes the maximized log-likelihood for the model g and K_e is the the number of estimable parameters in the approximating model, see Burnham (2002) for proof. The AIC can further be expressed as

$$AIC = -2L(\hat{\lambda}|y) + 2K_e. \tag{6.4}$$

Burnham (2002) state that Akaike multiplied the bias–corrected log-likelihood by -2 due to the result that -2 times the logarithm of the ratio of two maximized likelihood values is asymptotically chi-squared under certain conditions and assumptions.

In case of a linear regression model, $SS_{error} = y'[I - X(X'X)^{-1}X']y$ and the AIC is given by

$$AIC = n \cdot \ln(SS_{error}) + 2K_e - n \cdot \ln(n). \tag{6.5}$$

The second terms in (6.4) and (6.5) are often termed as penalty term. The model with smallest AIC, which in turn is closest to the unknown true model, is said to be the best model under AIC criterion.

Another criterion for model selection is the Bayesian information criterion (BIC) developed by Schwarz (1978). The Bayesian information criterion (BIC) is

$$BIC = -2L(\hat{\theta}|y) + \ln n \cdot K_e, \tag{6.6}$$

which has a form similar to AIC. In case of a linear regression model, the Bayesian information criterion can also be written as

$$BIC_{LM} = n \cdot \ln(RSS) + \ln(n)\, K_e - n \cdot \ln(n). \tag{6.7}$$

Note that corresponding to AIC, the second term in (6.6) and (6.7) are called as penalty term. The model with smallest BIC is said to be the best model under BIC criterion.

Another method of model selection based on Mallows' C_p (see Mallows (1973)) aims in determining a good model which fits to the observed data well. Assume that the linear regression model

$$y = X^{(q)}\beta^{(q)} + u^{(q)}, \qquad u^{(q)} \sim N(0, \sigma^2 I) \tag{6.8}$$

is the true regression model with q explanatory variables. Consider a sub-model of (6.8) with $p < q$ explanatory variables

$$y = X^{(p)}\beta^{(p)} + u^{(p)}, \qquad u^{(p)} \sim N(0, \sigma^2 I) \tag{6.9}$$

as the model of interest. Assume that $u^{(q)}$ and $u^{(p)}$ are independent.

The model selection criterion C_p is defined as

$$C_p = \frac{\sum_{i=1}^{n}(y_t - \hat{y}^{(p)})^2}{\hat{\sigma}^2} - n + 2p = \frac{SS_{error}}{\hat{\sigma}^2} - n + 2p, \tag{6.10}$$

where σ^2 is estimated by $\hat{\sigma}^2$ and is based on the full model. A criterion to choose a good model is to choose a model with minimum C_p.

The asymptotic properties of these model selection criteria are also available, which enable us to compare the aforementioned methods when the sample size tends to infinity. The BIC is shown to be consistent (see Csiszar and Shields (2000)) while AIC and Mallows' C_p Criteria are proved inconsistent (see Nishii (1984); Rao and Wu (1989)). When the number of regressors does not increase with sample size, AIC and Mallows' C_p Criteria generally give inconsistent estimates of the true model (see Shibata (1984), Zhang (1993b)), mainly due to over-fitting. However, when the number of regressors and sample size simultaneously tend to infinity, the AIC exhibits consistency (see Yanagihara et al. (2015)).

Apart from the AIC, BIC and Mallows' C_p criteria, there are a few more variable selection/screening procedures available in the literature. For example, one may consider stepwise forward selection and stepwise backward elimination (see Miller (1984)), based on OLS methodology (see, Basu (2019)), Delete-1 cross-validation method (see Allen (1974) and Stone (1974)), the Delete-d cross validation method (see Geisser (1975), Burman (1989), Shao (1993), Zhang (1993a)). For various applications related to model selection, the readers may refer to Shalabh and Dhar (2020).

7. Goodness of Fit in Non-parametric Regression Model

7.1 Basics for Non-parametric Regression Model

Now we consider the goodness of fit measures in a non-parametric regression model. Unlike the parametric regression model, there is no parametric relationship among the explanatory variables and the study variable in a non-parametric regression. In other words, by estimating the regression function from observed data, we strive to find out the true nature of relationship between the study and the explanatory variables.

Let us define the regression model formally. Suppose that $((y_1, \mathbf{x}_1), \ldots, (y_n, \mathbf{x}_n))$ are n many observations, and the associated regression model is $y_i = m(\mathbf{x}_i) + \nu_i$ for all $i = 1, \ldots, n$, where y_i is the i-th observation of the study variable, $\mathbf{x}_i$ is the i-th observation of the d-dimensional ($d \geq 1$) explanatory variable, and ν_i is the error corresponds to i-th observation. Here $m : \mathbb{R}^d \to \mathbb{R}$ is the unknown non-parametric regression function. The more technical assumptions on m or the error random variables will be stated following the context of study.

In order to carry out any methodology using non-parametric regression model, one needs to estimate the non-parametric regression curve, and it is needless to mention that the "quality" of the estimated regression curve depends on various factors such as the smoothness of the curve. In this context, the basic fundamental idea of smoothness can be described as "If m is believed to be smooth, then the observations at $\mathbf{X}_i$ near $\mathbf{x}$ should contain information about the value of m at $\mathbf{x}$. Thus it should be possible to use something like a local average of the data near $\mathbf{x}$ to construct an estimator of $m(\mathbf{x})$" (see Eubank (1988, p. 7)). In the same spirit, one can reasonably estimate the regression curve $m(\mathbf{x})$ by a center of the collection of the response variables, whose corresponding covariate observations lying in the neighbourhood of $\mathbf{x}$. As a center of the observations of the response variables, a natural choice will be the mean or any other suitable measure of central tendency of those observations in the neighbourhood of $\mathbf{x}$. In other words, we may consider "local average", which is based on Y observations corresponding to co-variate observations in a small neighbourhood of $\mathbf{x}$ since if the covariate observations corresponding to Y observations are far away from $\mathbf{x}$, it will give the mean value, which is distant from the original value of $m(\mathbf{x})$. Such local averaging procedure can be considered as the basic of smoothing.

Let us now be formal about the smoothed estimator of $m(\mathbf{x})$, where $\mathbf{x}$ is a fixed point. Suppose that $\hat{m}_n(\mathbf{x})$ is a locally smoothed estimator of $m(\mathbf{x})$ at the fixed point $\mathbf{x}$, and $w_i(\mathbf{x})$ is the positive weight corresponding to i-th observation such that $\sum\limits_{i=1}^{n} w_i(\mathbf{x}) = 1$. Define the smoothed estimator $\hat{m}_n(\mathbf{x})$ as $\hat{m}_n(\mathbf{x}) = \sum\limits_{i=1}^{n} w_i(\mathbf{x})y_i$, i.e., a certain weighted average of the response variables, where weights $w_i(\mathbf{x})$ depend on the point of evaluation $\mathbf{x}$. Further, note that $\hat{m}_n(\mathbf{x})$ is the solution of a certain minimization problem. Note that

$$\hat{m}_n(\mathbf{x}) = arg\min_{\zeta} \sum_{i=1}^{n} (y_i - \zeta)^2 \omega_i(\mathbf{x}),$$

i.e., in other words, the locally smoothed estimator is nothing but the locally weighted least squares estimator. As we have observed, the smoothness of the estimator is controlled by the choices of $\omega_i(.)$, one now needs to address the possibility of the choices of $\omega_i(.)$, which is discussed in the following subsection.

7.2 Kernel Smoothing

A simple way to choose the weight function $w_i(.)$ by considering the kernel density estimator (see, e.g., Silverman (1986)) as the weight function, and in Statistics literature, it is well-known as kernel smoothing technique. The kernel is usually denoted by $k(.)$ satisfying the conditions $k(.) \geq 0$ and $\int k(v)dv = 1$, and one can consider $\omega_i(\mathbf{x}) = \frac{k(\frac{\mathbf{x}_i - \mathbf{x}}{h_n})}{\sum\limits_{i=1}^{n} k(\frac{\mathbf{x}_i - \mathbf{x}}{h_n})}$, where h_n is a sequence of bandwidth satisfying $h_n \to 0$ as $n \to \infty$ and some other technical conditions, which will be stated as required. Using the aforementioned $w_i(\mathbf{x})$, we have $\hat{m}_n(\mathbf{x}) = \frac{\sum\limits_{i=1}^{n} k(\frac{\mathbf{x}_i - \mathbf{x}}{h_n})y_i}{\sum\limits_{i=1}^{n} k(\frac{\mathbf{x}_i - \mathbf{x}}{h_n})}$, which is well-known Nadaraya-Watson estimator (see, e.g., Nadaraya (1964) and Watson (1964)). For $d = 1$, the following example can give us some insight how $\hat{m}_n(.)$ as $h_n \to 0$ and $h_n \to \infty$ as $n \to \infty$. Consider the following example from Härdle (1990).

Let $k(s) = 0.75(1 - s^2)1_{\{|s| \leq 1\}}$, where $1_A = 1$ if A is true, and $1_A = 0$ if A is not true. On the one hand, note that as $h_n \to 0$ as $n \to \infty$, we have $\hat{m}_n(x_i) \to \frac{k(0)y_i}{k(0)} = y_i$, i.e., as the bandwidth is getting smaller, the estimator $\hat{m}_n(.)$ converges to corresponding observation of the response variable. On the other hand, as $h_n \to \infty$ as $n \to \infty$, we have $\hat{m}_n(x) \to \frac{\sum\limits_{i=1}^{n} k(0)y_i}{\sum\limits_{i=1}^{n} k(0)} = \frac{1}{n}\sum\limits_{i=1}^{n} y_i$, i.e., the bandwidth is getting larger, the estimated curve will be flattened. Besides, apart from the Nadaraya-Watson estimator, for $d = 1$, there are a few more well-known estimators in the literature. For instance, Priestley-Chao estimator (see Priestley and Chao (1972)) and Gasser-Muller estimator (see Gasser and Muller (1984)) are two such examples. For Priestley-Chao estimator, $\omega_i(x) = n(x_i - x_{i-1})k(\frac{x - x_i}{h_n})$, where $x_0 = 0$, and for Gasser-Muller estimator, $\omega_i(x) = n\int_{s_{i-1}}^{s_i} k(\frac{x-u}{h_n})du$, where $x_{i-1} \leq s_{i-1} \leq x_i$. In the literature, there are many articles on the various large sample properties of the aforesaid estimators. For all those results, the readers are refer to Härdle (1990) and a few references therein.

7.3 Kernel Smoothing and Fitting Local Polynomial

We have already observed kernel smoothed estimator can be viewed as the solution of a certain minimization problem. In other words, one may consider it as the fitting a constant in the neighourhood of the point of evaluation. We now can extend this idea by fitting a polynomial in the neighbourhood of the point of evaluation. For instance, a simple algebra gives us for local linear fitting (considering $d = 1$ for notational simplicity):

$$(\hat{m}_n(x), h_n\hat{m}_n^{'}(x)) = arg \min_{\zeta_0,\zeta_1} \sum_{i=1}^{n} (y_i - \zeta_0 - \zeta_1(x_i - x)^2)^2 \omega_i(x),$$

where $f^{'}(.)$ denotes the derivative of f at the point $(.)$ for any f, and $\omega_i(x) = \frac{k(\frac{x_i - x}{h_n})}{\sum_{i=1}^{n} k(\frac{x_i - x}{h_n})}$. Note that local linear fitting gives us the estimator of the original regression function and its first derivative also whereas local constant fit only gives us the estimator of the original function. In the same spirit, one can extend this approach for local polynomial (of any finite order) fitting and able to find the estimator of l-th ($l \geq 0$) derivative of the estimator using l-th degree polynomial. However, it should be pointed out here that for a large l, the issue of variability of the estimators will arise, and this issue will be discussed in the subsequent section.

Let us consider a simple example, which will give us an insight how local polynomial fits a data. Consider the following example from Härdle (1990). Suppose that the model is a fixed design model, and the design points are equispaced, i.e., $x_i = \frac{i}{n}$. Here a local parabolic, i.e., local quadratic fitting is considered. We now consider a uniform kernel $k(s) = \frac{1}{2}1_{(|s|\leq 1)}$, and we then need to minimize

$$\frac{1}{n}\sum_{i=1}^{n} k(x - X_i)(Y_i - a_* - b_*(X_i - x)^2)^2$$

with respect to a and b. In the above expression, the linear term is not considered since it will be cancelled out due to the symmetry of the kernel k. hence, the normal equations will be

$$\frac{1}{n}\sum_{i=1}^{n} k(x - X_i)(Y_i - a_* - b_*(X_i - x)^2) = 0$$

and

$$\frac{1}{n}\sum_{i=1}^{n} k(x - X_i)(Y_i - a_* - b_*(X_i - x)^2)(X_i - x)^2 = 0.$$

Let us now denote $\hat{Y} = \frac{1}{n}\sum_{i=1}^{n} k(x - X_i)Y_i$, and note that using the idea of Riemann approximation,

$$\frac{1}{n}\sum_{i=1}^{n} k(x - X_i)(x - X_i)^2 - \int k(x-u)(x-u)^2 du = o_p(1).$$

Using the similar arguments on $\frac{1}{n}\sum_{i=1}^{n} k(x - X_i)(x - X_i)^4$, the normal equations become $\hat{Y} - a_* - \frac{h^2}{3}b_* = 0$ and $A - \frac{h^2}{3}a_* - \frac{h^4}{5}b_* = 0$, where $A = \frac{1}{n}\sum_{i=1}^{n} k(x - X_i)(x - X_i)^2 Y_i$.

Next, elementary algebra ensures that a satisfies the following equation:

$$3h^2\hat{Y} - 5A + \left(-3 + \frac{5}{3}\right) h^2 a_* = 0,$$

i.e., $\hat{a}_* = \frac{3}{4}\left[\frac{1}{n}\sum_{i=1}^{n} k(x - X_i)\left(3 - 5\left(\frac{x - X_i}{h}\right)^2\right) Y_i\right]$. Similarly, one can estimate b_* also. Note that one can write

$$\hat{a}_* = \hat{m}(x) = \frac{1}{n}\sum_{i=1}^{n} k^*(x - X_i)Y_i,$$

where $k^*(s) = \frac{3}{8}(3 - 5s^2)1_{(|s|\leq 1)}$. A technical point of interest is, the first three moments of k^* vanish. To be summarized, for this setting, k^* is a local parabolic fitting procedure.

It may be remarked that to estimate the derivative of the regression function, as we described here, one can definitely adopt the idea of local polynomial fitting. Besides, note that we can also directly take the derivative of the estimator of the original regression function. Although they sound similar; however they are in principle different. The local polynomial approach gives us the estimator of the derivative of the regression function whereas the later one gives us the derivative of the estimated regression function. In practice, they sometimes gives similar result when the original function is smooth enough but otherwise, they may not give the similar result.

7.4 Goodness of Fit in Non-parametric Regression Model

The issues related to the development of goodness of fit in the non-parametric setup, we need to move in a direction similar to the case in parametric setup. It may be noted that the CoD or R^2 in the multiple linear regression model (2.1) is formulated based on the partitioning of total sum of squares into two orthogonal components, *viz.*, the sum of squares due to regression and the sum of squares due to random errors which is not possible in the case of non-parametric setup. Such a criteria is not well-developed for a non-parametric regression model, to the best of our knowledge.

To define the goodness of fit statistic in the non-parametric regression model, one may use any estimator, e.g., well-known estimators like Nadaraya-Watson and Priestley-Chao estimators, to estimates the unknown regression function. Now, a relevant question is how to choose a better estimator, e.g., which estimator between Nadaraya-Watson and Priestley-Chao estimators perform better, based on proposed goodness of fit statistic and this issues will also be addressed subsequently.

We now first define CoD like statistic for a given data in non-parametric regression model. Let us consider the data $(y_1, x_1), \ldots, (y_n, x_n)$ associated with the regression model

$$y_i = m(x_i) + \nu_i,$$

where m is the unknown regression function, and ν_i is the error corresponding to i-th observation with variance $\sigma^2 < \infty$. Under this set up, one can define a goodness of fit statistic for a fixed z as

$$R^2_{n,NP}(z) = \frac{\hat{m}^2_n(z)}{\hat{\sigma}^2_n + \hat{m}^2_n(z)},$$

where $\hat{m}^2_n(z)$ is an appropriate estimator of $m^2(z)$, and $\hat{\sigma}^2_n$ is an appropriate estimator of the error variance, i.e., σ^2. One of the possible choice of $\hat{\sigma}^2_n$ can be

$$\hat{\sigma}^2_n = \frac{1}{n-1}\sum_{i=1}^{n}\{y_i - \hat{m}_n(x_i)\}^2,$$

and two well-known estimators of $m(z)$ are Nadaraya-Watson (NW) estimator and Priestley-Chao (PC) estimator, which are defined as the following.

$$\hat{m}_{n,NW}(z) = \frac{\sum_{i=1}^{n} k(\frac{x_i - z}{h_n})y_i}{\sum_{i=1}^{n} k(\frac{x_i - z}{h_n})} \quad \text{and} \quad \hat{m}_{n,PC}(z) = \frac{\sum_{i=1}^{n} k(\frac{x_i - z}{h_n})y_i}{nh_n}.$$

Note that $R^2_{n,NP}(z) \xrightarrow{p} 1$ when $\hat{\sigma}^2_n \xrightarrow{p} 0$ as $n \to \infty$, and $R^2_{n,NP}(z) \xrightarrow{p} 0$ when $\hat{\sigma}^2_n \xrightarrow{p} \infty$ as $n \to \infty$. In other words, the regression model fits the data well when the variability among the errors is less, and the regression model does not fit the data well when the variability among the errors is high. Overall, it is clear that the variance of the error random variable controls the performance of $R^2_{n,NP}(z)$.

7.4.1 Interpretation and Utility of $R^2_{n,NP}$

Now we address the utility of the proposed goodness of fit statistics in choosing an estimator for fitting the model and determining the model with better fit. First we address how to decide about an estimator. Suppose that $R^{NW,2}_{n,NP}(z)$ and $R^{PC,2}_{n,NP}(z)$ are based on $\hat{m}_{n,NW}(z)$ and $\hat{m}_{n,PC}(z)$, respectively, and now the question is which one of the estimator between Nadaraya-Watson and Priestley-Chao estimators can be used from an objective of goodness of model fitting. For a given data, if $R^{NW,2}_{n,NP}(z) \geq R^{PC,2}_{n,NP}(z)$ for all z, we can choose Nadaraya-Watson estimator over the Priestley-Chao estimator as this is providing a better fit to the model based on the given data. However, when $R^{NW,2}_{n,NP}(z)$ and $R^{PC,2}_{n,NP}(z)$ have overlapping between each other for various values of z, one may consider the goodness of fit criteria as $\sup_z R^2_n(z)$ or $\int R^2_n(z)dz$. For instance, one may check whether $\int_z R^{NW,2}_{n,NP}(z)dz$ is larger than $\int_z R^{PC,2}_{n,NP}(z)dz$ to prefer Nadaraya-Watson based estimator over Priestley-Chao based estimator.

Next utility is to select the model with better fit. A model with the higher value of $R^2_{n,NP}(z)$ for all z is preferable as this indicates the better model fit. In fact, $\int_z R^2_n(z)dz$ can also be used for the same purpose. For instance, suppose that the model involves five covariates (denote them as x_1, x_2, x_3, x_4 and x_5), and a is the value of $\int_z R^2_n(z)dz$ for that model. Now, suppose that two more covariates (denote them as x_6 and x_7) are added, and a^* is the value of $\int_z R_n^2(z)dz$ for the model involving all seven covariates (i.e., x_1, x_2, ..., x_7). At the end, if $a^* > a$, one may conclude that x_6 and x_7 are significant covariates.

Next use of $R^2_{n,NP}$ is in model selection using $\int_z R_n^2(z)dz$. Suppose that we have three models, namely, M_1, M_2 and M_3, and using $\int_z R_n^2(z)dz$, which model we will choose. Under such situation, one can compute $\int_z R_n^2(z)dz$ for all three models M_1, M_2 and M_3 and choose the model for which the value of $\int_z R_n^2(z)dz$ is maximum. This methodology can be applied for any finite number of models.

A new methodology related to variable screening/selection in multiple non-parametric regression will be discussed in the next section.

8. Variable Selection/screening in Multiple Non-parametric Regression

Over the past few decades, the search for efficient methods to determine the true regression model has become central goal of a large number of research problems in Statistics. As a result, the topic of variable screening has received valuable attention from the Statistics community. It is a well known fact that, the unnecessary explanatory variables add noise to the estimation of quantities of interest, resulting in reduction of degrees of freedom. Moreover, the collinearity may creep in the model, if there are a large number of predictors for the same response variable. In addition, the cost in terms of time and money can also be saved by not recording redundant explanatory variables has led to rapid growth in the use of model selection procedures, which are reminiscent of the variable screening procedures.

Several methods for variable screening are available in the literature of multiple linear regression model, which has already been discussed in Section 6. However, quite surprisingly, despite its wide range of applications, there are only a few research articles on the variable screening in multiple non-parametric regression model. Almost three decades ago, variable screening in non-parametric regression with continuous explanatory variables is studied by Zhang (1991), which is one of the earliest work in this direction. The article only addresses the estimation of number of significant variables and not the specific variables that are to be selected. The author used Nadaraya-Watson estimator for estimation of the regression function, and the estimation of number of the variables was based on the minimizing the mean squared prediction error obtained by cross validation technique. However, it was assumed that the explanatory variables are pre-ordered according to their importance. The variable screening in additive models has been explored to some extent in recent years, e.g., Meier et al. (2009), Ravikumar et al. (2009), and Bach (2008); however, none of them considered general non-parametric regression model. One of the most notable contribution towards this direction is Huang et al. (2010) in which they estimate the additive components using truncated series expansions with B-spline bases, and their selection criterion is based on the adaptive group Lasso. The variable selection in some generalized set ups has been explored by Li and Liang (2008), Meinshausen et al. (2009), and Storlie et al. (2011). However, the completely general non-parametric regression model is relatively unexplored to the best of our knowledge and the only work in such case is Zambom and Akritas (2014), that proposes an ANOVA-type test for significance of a particular variable.

We now propose an idea for variable screening in non-parametric regression model, which is capable of eliminating all redundant explanatory variables as the sample size grows to infinity. The method can be primarily based on local linear approximation. We first estimate the unknown regression function from the data using local linear approximation, and it gives us the least squares estimates of the partial derivatives of the regression function with respect to the particular variable at any point in the sample space. Technically speaking, the partial derivative of the true regression function with respect to the particular variable will be zero everywhere in the sample space if the corresponding explanatory variable is insignificant. Thus, we retain the explanatory variable in the model, if the estimate obtained from local linear approximation of the partial derivative of the regression function with respect to the corresponding component is above a certain threshold value everywhere in sample space. Otherwise, the respective variable will not be selected. We can investigate how good this methodology is in future.

9. Goodness of Fit in Time Series Models

Unlike the models studied in the earlier sections, there may exist situations where the errors in regression models are not independent or they are serially correlated. Such errors may be modelled using suitable time series models. A question arises that once a given set of data is available, different people may fit different time series models to match it with the real life phenomenon. So a couple of queries arise. For example, two different persons may fit autoregressive time series models of orders three and four to the same data, then how to know which model fitting is better. Similarly, how to know whether an autoregressive or a moving average model is better fitted to the same data set is another query, how to know the degree of fit to a given model etc. We now discuss the goodness of fit tests for time series models in this section. Let $\{Z_t\}$ be a time series, where the suffix t indicates that Z varies over time. While developing techniques for analyzing a time series $\{Z_t, t = 0, \pm 1, \pm 2, \ldots\}$, it is assumed that such a series is one of the realizations of a discrete parameter weakly stationary stochastic process. The time series is weakly stationary if $E(Z_t)$ is constant, $Var(Z_t) < \infty$ and for $h = \pm 1, \pm 2, \ldots$, $Cov(Z_t, Z_{t-h})$ is a function of h and free from t. For a weakly stationary time series $\{Z_t\}$, we define $E(Z_t) = m, Var(Z_t) = \sigma_z^2$ and the autocovariance function $Cov(Z_t, Z_{t-h}) = \gamma_h$ for $h = \pm 1, \pm 2, ...$, and the autocorrelation function $Corr(Z_t, Z_{t-h}) = \rho_h = \frac{\gamma_h}{\sigma_z^2}$. Further, $\rho_{-h} = \rho_h$. For details on the basic concepts of time series, methods of analysis and their applications, one may refer to Brockwell and Davis (1991) and Box et al. (1994).

9.1 AR, MA and ARMA Models

Let $\{a_t\}$ be a sequence of uncorrelated random variables with mean 0 and finite constant variance σ_a^2, which may be referred to as random shocks. A time series $\{Z_t, t = 0, \pm 1, \pm 2, \ldots\}$ is said to follow an autoregressive model of order p denoted as AR(p) if

$$Z_t = \phi_1 Z_{t-1} + \phi_2 Z_{t-2} + \ldots + \phi_p Z_{t-p} + a_t. \tag{9.1}$$

We assume that $\{Z_t, t = 0, \pm 1, \pm 2, \ldots\}$ defined by this model is a stationary time series. For a stationary AR(p) series, the autocovariance function $\gamma_k = \gamma_{-k}$ and satisfies the relation:

$$\gamma_k = \phi_1 \gamma_{k-1} + \phi_2 \gamma_{k-2} + \ldots + \phi_p \gamma_{k-p}, \quad k = 0, 1, 2, \ldots, \tag{9.2}$$

and $\sigma_z^2 = Var(Z_t) = \gamma_0 = \phi_1 \gamma_1 + \phi_2 \gamma_2 + \cdots + \phi_p \gamma_p + \sigma_a^2$. Hence, the autocorrelation function (ACF) ρ_k satisfies the Yule-Walker equations and helps in estimating the AR parameters

$$\rho_k = \phi_1 \rho_{k-1} + \phi_2 \rho_{k-2} + \cdots + \phi_p \rho_{k-p}, \quad k = 0, 1, 2, \ldots, p. \tag{9.3}$$

Note that the range of k in (9.2) and (9.3) are different because ACF can be computed for any value of for any $k = 0, 1, 2, \ldots$ in (9.2) whereas range of k in (9.3) corresponds to the ACF of an AR(p) in the form of Yule Walker equations.

Another time series model of practical utility is the finite moving average model in which, the observation Z_t at time t is expressed as a linear combination of the present and past shocks. A moving average model of order q, denoted as MA(q), is defined by

$$Z_t = a_t - \vartheta_1\, a_{t-1} - \vartheta_2\, a_{t-2} - \ldots - \vartheta_q\, a_{t-q}, \quad t = 0, \pm 1, \pm 2, \ldots \tag{9.4}$$

A model with p AR terms and q MA terms is called as an ARMA (p, q) model and is given as

$$Z_t - \phi_1 Z_{t-1} - \phi_2 Z_{t-2} - \ldots - \phi_p Z_{t-p} = a_t - \vartheta_1 a_{t-1} - \vartheta_2 a_{t-2} - \ldots - \vartheta_q a_{t-q} + \vartheta_0, \quad t = 0, \pm 1, \pm 2, \ldots \tag{9.5}$$

Assuming that the ARMA(p, q) process is stationary and invertible, we can express Z_t in the form of a moving average model of infinite order as

$$Z_t = \sum_{j=0}^{\infty} \psi_j a_{t-j} \text{ with } \psi_0 = 1, \tag{9.6}$$

where the coefficients $\{\psi_j, j = 1, 2, \ldots\}$ are suitable functions of the parameters $\phi_1, \phi_2, ..., \phi_p, \vartheta_0, \vartheta_1, \ldots, \vartheta_q$.

9.2 Goodness of Fit in ARMA Models

The goodness of fit of a model in time series can be tested using the residuals. The residuals are defined as the difference between the observed and fitted values of the variable. The residuals are expected to be approximately uncorrelated for a well fitted model. For example, the residuals in the class of ARMA(p, q) models can be computed after estimating the parameters as $\hat{a}_t$ for $t = 1, 2, \ldots$

$$\hat{a}_t = \hat{\vartheta}_1 \hat{a}_{t-1} + \hat{\vartheta}_2 \hat{a}_{t-2} + \ldots + \hat{\vartheta}_q \hat{a}_{t-q} - \hat{\vartheta}_0 + Z_t - \hat{\phi}_1 Z_{t-1} - \hat{\phi}_2 Z_{t-2} - \ldots - \hat{\phi}_p Z_{t-p}, \tag{9.7}$$

where $\hat{\phi}_1, \hat{\phi}_2, \ldots, \hat{\phi}_p, \hat{\vartheta}_0, \hat{\vartheta}_1, \ldots, \hat{\vartheta}_q$ are the estimates of $\phi_1, \phi_2, \ldots, \phi_p, \vartheta_0, \vartheta_1, \ldots, \vartheta_q$, respectively.

The portmanteau test measures the goodness of fit for ARMA(p, q) models and is based on the asymptotic distribution of the residual autocorrelations ($\hat{r}_k$). This test is considered as an overall benchmark of tests, assuming the same kind of role as the chi-square tests for goodness of fit in classical inference. The residual ACF is defined by

$$\hat{r}_k = \frac{\sum_{t=k+1}^{n} \hat{a}_t \hat{a}_{t-k}}{\sum_{t=1}^{n} \hat{a}_t^2}, \quad k = 1, 2, \ldots$$

If the model is a good fit for the given time series then the residual ACF will be negligible. The lack of correlation in residuals can be tested using the portmanteau test statistic (see, Box and Pierce (1970)) given as

$$Q_m = n \sum_{k=1}^{m} \hat{r}_k^2, \tag{9.8}$$

for a suitable m. Ljung and Box (1978) advocated the use of the modified statistic, termed as Ljung-Box statistic or the Ljung-Box-Pierce statistic given by

$$\tilde{Q}_m = n\,(n+2) \sum_{k=1}^{m} \frac{\hat{r}_k^2}{n-k}. \tag{9.9}$$

The motivation for the modification is from the fact that $var(\hat{r}_k) \cong (n-k)/\{n\,(n+2)\}$. That is, $\tilde{Q}_m$ is obtained by adjusting essentially each of the $\hat{r}_k$ in Q_m by its asymptotic variance. Li and McLeod (1981) suggested a modified statistic as

$$Q_m^* = Q_m + \frac{m(m+1)}{2n}.$$

Smaller value of the statistic Q_m^* indicates a better fit.

Even though, the structure of an ARMA time series model is similar to that of a regression model, the observations at different time points are not independent in the former one. In time series, the role of explanatory variable is taken over by the past values of the Z-series, which are serially correlated. It is clear from the definition of the model (9.5) that at any time point t, the values $Z_{t-1}, Z_{t-2}, \ldots, Z_{t-p}, \ldots$ and $a_{t-1}, a_{t-2}, \ldots, a_{t-q}, \ldots$ are known.

So, at time t, if the model parameters are specified, the only unknown quantity before observing Z_t is the error term, a_t, which has the theoretical variance σ_a^2. Moreover, the definition of the model implies that a_t is uncorrelated with past realizations of Z_t. The minimum mean square error (MMSE) forecast $\hat{Z}_{n-1}(1)$ of Z_n based on the realization up to the time $(n-1)$ for the model (9.5) is

$$\hat{Z}_{n-1}(1) = E(Z_n | Z_{n-1}, Z_{n-2}, \ldots) = \sum_{j=1}^{p} \phi_j Z_{n-j} - \sum_{k=1}^{q} \vartheta_k a_{n-k} + \vartheta_0, \tag{9.10}$$

where $\hat{Z}_{n-1}(1)$ denotes the one-step ahead forecast at time $(n-1)$, and the corresponding one-step ahead forecast error is $Z_n - \hat{Z}_{n-1}(1) = a_n$. The associated forecast error variance is σ_a^2, and the fraction of variation in Z_t which cannot be predicted at time t is σ_a^2/σ_z^2, where σ_z^2 and σ_a^2 are the theoretical variances of Z_t and a_t, respectively. So, the population proportion of variation that can be predicted based on the fitted model (9.5) is

$$\theta_T = 1 - \frac{\sigma_a^2}{\sigma_z^2}, \quad 0 \le \theta_T \le 1. \tag{9.11}$$

A goodness of fit test based on the sample version of θ_T for a stationary and invertible ARMA(p,q) model is given by

$$R_T^2 = 1 - \frac{\sum_{t=1}^{n} \hat{a}_t^2}{\sum_{t=1}^{n} (z_t - \bar{z})^2}, \quad 0 \le R_T^2 \le 1, \tag{9.12}$$

see Nelson (1976), where $\bar{z}$ is the sample mean of the observed time series. By the strong law of large numbers for stationary time series, it follows that $\text{plim}_{n\to\infty} R_T^2 = \theta_T$, see Hosking (1979). The value of R_T^2 close to 1 indicates that the model provides a better prediction of the future value whereas $R_T^2 = 0$ indicates the worst fitted model which cannot be used for prediction. The implications of R_T^2 may be suitably interpreted when $0 < R_T^2 < 1$. For example, if $R_T^2 = 0.75$ then it can be interpreted as if 75% of the variation in Z_t is captured by the model.

There are several methods of estimation in time series which depend on the model structures. For example, the Yule-Walker method of estimation is popular for autoregressive models whereas method of maximum likelihood is used for moving average models. For illustration, we obtain the expressions for θ_T for some of the selected time series models in terms of their parameters and then propose suitable expressions for goodness of fit statistic.

Using the random shock form (9.6) of a stationary and invertible ARMA(p,q) model, we get the variance of Z_t as

$$\sigma_z^2 = \sigma_a^2 \sum_{j=0}^{\infty} \psi_j^2 = \sigma_a^2\Big(1 + \sum_{j=1}^{\infty} \psi_j^2\Big)$$

and θ_T in (9.11) becomes

$$\theta_{Tpq} = \frac{\sum_{j=1}^{\infty} \psi_j^2}{1 + \sum_{j=1}^{\infty} \psi_j^2}. \tag{9.13}$$

Recall that the coefficients ψ_j are functions of the original AR and MA parameters. The θ_T in MA(q) process becomes

$$\theta_{Tq} = \frac{\sum_{j=1}^{q} \vartheta_j^2}{1 + \sum_{j=1}^{q} \vartheta_j^2}. \tag{9.14}$$

The AR(p) model defined by (9.1) has close resemblance with the linear regression model, except that the explanatory variables are the past values of the study variable, which are observable. Let θ_{Tp} be the theoretical measure for a stationary AR(p) model. Using (9.2), (9.3) and σ_z^2, we get

$$\theta_{Tp} = 1 - \frac{\sigma_a^2}{\sigma_z^2} = 1 - \phi_1\rho_1 - \phi_2\rho_2 - \cdots + \phi_p\rho_p = \boldsymbol{\rho_p}'\boldsymbol{\phi}_p, \text{ say,} \tag{9.15}$$

where $\boldsymbol{\rho}_p = (\rho_1, \rho_2, \ldots, \rho_p)'$ and $\boldsymbol{\phi}_p = (\phi_1, \phi_2, \ldots, \phi_p)'$. The Yule-Walker equations (9.3) is expressible as a system of linear equations as $\boldsymbol{\rho}_p = \boldsymbol{\Gamma}_p\boldsymbol{\phi}_p$, where $\boldsymbol{\Gamma}_p = ((\rho_{i-j})), i, j = 1, 2, \ldots, p$ is the matrix of correlations. This provides $\boldsymbol{\phi}_p = \boldsymbol{\Gamma}_p^{-1}\boldsymbol{\rho}_p$, and hence, $\theta_{Tp} = \boldsymbol{\rho}_p'\boldsymbol{\Gamma}_p^{-1}\boldsymbol{\rho}_p$, which is a function of ACF only. The linear relationship between the ACF and the autoregressive parameters helps in computing θ_{Tp}. Hence, the goodness of fit measure for a stationary AR(p) model is given by

$$R_{Tp}^2 = \hat{\boldsymbol{\rho}}_p'\hat{\boldsymbol{\Gamma}}_p^{-1}\hat{\boldsymbol{\rho}}_p, \tag{9.16}$$

where $\hat{\boldsymbol{\rho}}_p$ and $\hat{\boldsymbol{\Gamma}}_p$ are respectively the vector and matrix of estimated ACFs of the series $\{Z_t\}$.

Partial auto-correlation function (PACF) is another characteristic useful for model identification in time series. PACF of order k is defined as the correlation between Z_t and Z_{t-k}, adjusted for the intervening variables $Z_{t-1}, Z_{t-2}, \ldots, Z_{t-k-1}$. The PACF of order k of a time series can be computed by fitting AR models of orders k and then choosing its last coefficient for $k = 1, 2, \ldots$, see Brockwell and Davis (1991) and Box et al. (1994) for details on the relationships between ACF, PACF and their applications. A relation between the goodness of fit measure θ_{Tk} and PACF q_k based on an AR(k) model is given as $\theta_{T,k} = \theta_{T,k-1} + q_k^2(1 - \theta_{T,k-1})$, where $\theta_{T,k-1}$ is computed based on an $AR(k-1)$ model, see Lawrance (1979).

10. Model Selection Using Goodness of Fit Statistics and Conclusions

A model can be selected by choosing the suitable explanatory variable and goodness of fit statistics. For example, in case of multiple linear regression models, the CoD helps in this process. Let R_q^2 be the CoD based on q such variables which has a tendency as q increases. So proceed as follows: First choose an appropriate value of q, fit the model and obtain R_q^2. Now add explanatory one variable, again fit the model and obtain R_{q+1}^2, and needless to mention that $R_{q+1}^2 \geq R_q^2$. If $R_{q+1}^2 - R_q^2$ is small, then stop the process and choose the value of q; get the regression model based on this. In case, $R_{q+1}^2 - R_q^2$ is high, then keep on adding variables up to a point where an additional variable does not produce a large change in the value of R_q^2.

The same methodology can be easily extended to other type of parametric and non-parametric econometric models. Different measures of goodness of fit have been discussed in this chapter which can be suitably used in respective models. In general, the goodness of fit statistic can be obtained and the explanatory variables can be added one by one. The value of goodness of fit statistic can be evaluated after adding the observations on every explanatory variable and the process can be continued till the difference in the values of goodness of fit statistic statistic remains significant. For example, in the context of non-parametric modelling, $R_{n,NP}^{NW,2}(z)$ based on $\hat{m}_{n,NW}(z)$ can be computed as $R_{n,NP}^{NW,2}(z,q)$ based on q explanatory variables. An additional explanatory variable can be added and the difference $R_{n,NP}^{NW,2}(z,q+1) - R_{n,NP}^{NW,2}(z,q)$ is evaluated and the process of increasing q until this difference remains significant. Following the same lines, the model selection can also be done in case of any other model.

Acknowledgement

The authors gratefully acknowledges the support from the MATRICS project from Science and Engineering Research Board (SERB), Department of Science and Technology, Government of India.

References

Akaike, H. 1973. Information theory and an extension of the maximum likelihood principle. pp. 267–281. *In*: Proceeding of the Second International Symposium on Information Theory Budapest.

Allen, D. M. 1974. The relationship between variable selection and data agumentation and a method for prediction. Technometrics, 16: 125–127.

Bach, F. R. 2008. Consistency of the group lasso and multiple kernel learning. Journal of Machine Learning Research, 9: 1179–1225.

Basu, D. 2020. Bias of ols estimators due to exclusion of relevant variables and inclusion of irrelevant variables. Oxford Bulletin of Economics and Statistics, 82: 209–234.

Box, G. E. P. and Pierce. 1970. Distribution of the residual autocorrelations in autoregressive integrated moving average models. Journal of American Statistical Association, 65: 1509–1526.

Box, G. E. P., Jenkins, G. M. and Reinsel, G. C. 1994. Time Series Analysis, Forecasting and Control, 3rd edn. Pearson Education.

Brockwell, P. J. and Davis, R. A. 1991. Time Series: Theory and Methods, Second Edition, Springer.

Burman, P. 1989. A comparative study of ordinary cross-validation, *v*-fold cross validation and the repeated learning-testing methods. Biometrika, 76: 503–514.

Burnham, K. P. and Anderson, D. 2002. Model Selection and Multimodel Inference: A Practical Information-Theoretic Approach, Springer, New York.

Carter, R. A. L., Srivastava, V. K. and Chaturvedi, A. 1993. Selecting a double k-class estimator for regression coefficients. Statistics and Probability Letters, 18(5): 363–371.

Chaturvedi, A. and Shalabh. 2004. Risk and Pitman closeness properties of feasible generalized double k-class estimators in linear regression models with non-spherical disturbances under balanced loss function. Journal of Multivariate Analysis, 90(2): 229–256.

Cheng, C. -L. and Van Ness, J. W. 1999. Statistical Regression with Measurement Errors, London, Arnold.

Cheng, C. -L., Shalabh and Garg, G. 2014. Coefficient of determination for multiple measurement error models. Journal of Multivariate Analysis, 126: 137–152.

Cheng, C.-L., Shalabh and Chaturvedi, Anoop. 2019. Goodness of fit for generalized shrinkage estimation. Theory of Probability and Mathematical Statistics American Mathematical Society, 100: 177–197.

Cramer, J. S. 1987. Mean and variance of R^2 in small and moderate samples. Journal of Econometrics, 35: 253–266.

Csiszar, I. and Shields, P. C. 2000. The consistency of the bic markov order estimator. The Annals of Statistics, 28: 1601–1619.

Dolby, G. R. 1976. The ultrastructural relation: A synthesis of the functional and structural relations. Biometrika, 63: 39–50.

Eshima, N. and Tabata, M. 2010. Entropy coefficient of determination for generalized linear models. Computational Statistics and Data Analysis, 54(5): 1381–1389.

Eshima, N. and Tabata, M. 2011. Three predictive power measures for generalized linear models: The entropy coefficient of determination, the entropy correlation coefficient and the regression correlation coefficient. Computational Statistics and Data Analysis, 55(11): 3049–3058.

Eubank, R. 1988. Spline Smoothing and Nonparametric Regression, Dekker, New York.

Farebrother, R. W. 1975. The minimum mean square error linear estimator and ridge regression. Technometrics, 17: 127–128.

Fuller, W. A. 1987. Measurement Error Models, John Willey.

Gasser, T. and Muller, H. G. 1984. Estimating regression functions and their derivatives by the kernel method. Scandinavian Journal of Statistics, 11: 171–185.

Geisser, S. 1975. The predictive sample reuse method with applications. Journal of the American Statistical Association, 70: 320–328.

Hahn, G. J. 1973. The coefficient of determination exposed. Chemical Technology, 3: 609–614.

Härdle, W. 1990. Applied Nonparametric Regression, Cambridge University Press.

Hilliard, J. E. and Lloyd, W. P. 1980. Coefficient of determination in a simultaneous equation model: A pedagogic note. Journal of Business Research, 8(1): 1–6.

Hong, C. S., Ham, J. H. and Kim, H. I. 2005. Variable selection for logistic regression model using adjusted coefficients of determination. Korean Journal of Applied Statistics, 18(2): 435–443.

Hosking, J. R. M. 1979. The asymptotic distribution of R^2 for autoregressive-moving average time series models when parameters are estimated. Biometrika, 66: 156–157.

Hössjer, O. 2008. On the coefficient of determination for mixed regression models. Journal of Statistical Planning and Inference, 138(10): 3022–3038.

Huang, L. -S. and Chen, J. 2008. Analysis of variance, coefficient of determination and F-test for local polynomial regression. The Annals of Statistics, 36(5): 2085–2109.

Huang, J., Horowitz, J. L. and Wei, F. 2010. Variable selection in non-parametric additive models. The Annals of Statistics, 38: 2282–2313.

James, W. and Stein, C. 1961. Estimation with quadratic loss. pp. 361–379. *In*: Proceedings of the Fourth Berkeley Symposium on Mathematical Statistics and Probability, University of California Press, Berkeley.

Knight, J. L. 1980. The coefficient of determination and simultaneous equation systems. Journal of Econometrics, 14(2): 265–270.

Lawrance, A. J. 1979. Partial and multiple correlation for time series. American Statistician 33(3): 127–130.

Liao, J. G. and McGee, D. 2003. Adjusted coefficients of determination for logistic regression. The American Statistician, 57(3): 161–165.

Li, W. K. and Mcleod, A. I. 1981. Distribution of the residual autocorrelations in multivariate ARMA time series models. Journal of Royal Statistical Society B, 43: 231–239.

Lipsitz, S. R., Leong, T., Ibrahim, J. and Lipshultz, S. 2001. A partial correlation coefficient and coefficient of determination for multivariate normal repeated measures data. The Statistician, 50(1): 87–95.

Li, R. and Liang, H. 2008. Variable selection in semiparametric regression modeling. The Annals of Statistics, 36: 261–286.

Ljung, G. M. and Box, G. E. P. 1978. On measure of lack of fit in time series models. Biometrika, 65: 297–303.

Mallows, C. L. 1973. Some comments on c_p. Technometrics, 15: 661–675.

Marchand, E. 1997. On moments of beta mixtures, the noncentral beta distribution, and the coefficient of determination. Journal of Statistical Computation and Simulation, 59(2): 161–178.

Marchand, E. 2001. Point estimation of the coefficient of determination. Statistics and Decisions, 19(2): 137–154.

McKean, J. W. and Sievers, G. L. 1987. Coefficients of determination for least absolute deviation analysis. Statistics and Probability Letters, 5(1): 49–54.

Meier, L., Van de Geer, S. and Buhlmann, P. 2009. High-dimensional additive modeling. The Annals of Statistics, 37: 3779–3821.

Meinshausen, N., Meier, L. and Buhlmann, P. 2009. P-values for high-dimensional regression. Journal of the American Statistical Association, 104: 1671–1681.

Miller, A. J. 1984. Selection of subsets of regression variables. Journal of the Royal Statistical Society: Series A, 147: 389–410.

Nadaraya, E. A. 1964. On estimating regression function. Theory of Probability and its Application, 10: 186–190.

Nagelkerke, N. J. D. 1991. A note on a general definition of the coefficient of determination. Biometrika, 78(3): 691–692.

Nishii, R. 1984. Asymptotic properties of criteria for selection of variables in multiple regression. The Annals of Statistics, 12: 758–765.

Ohtani, K. 1994. The density functions of R^2 and $\bar{R}^2$, and their risk performance under asymmetric loss in misspecied linear regression models. Economic Modelling, 11(4): 463–471.

Ohtani, K. 1996. Exact small sample properties of an operational variant of the minimum mean squared error estimator. Communications in Statistics (Theory and Methods), 25: 1223–1231.

Ohtani, K. 1997. Minimum mean squared error estimation of each individual coefficients in a linear regression model. Journal of Statistical Planning and Inference, 62: 30–316.

Priestley, M. B. and Chao, M. T. 1972. Nonparametric function fitting. Journal of the Royal Statistical Society, Series B, 34: 385–392.

Ravikumar, P., Laerty, J., Liu, H. and Wasserman, L. 2009. Sparse additive models. Journal of the Royal Statistical Society: Series B, 71: 1009–1030.

Rao, C.R., Toutenburg, H., Shalabh and Heumann, C. 2008. Linear Models and Generalizations—Least Squares and Alternatives. Springer.

Rao, R. and Wu, Y. 1989. A strongly consistent procedure for model selection in a regression problem. Biometrika, 76: 369–374.

Renaud, O. and Victoria-Feser, M. -P. 2010. A robust coefficient of determination for regression. Journal of Statistical Planning and Inference, 140(7): 1852–1862.

Schneeweiss, H. 1991. Note on a linear model with errors in the variables and with trend. Statistical Papers, 32: 261–264.

Schwarz, G. 1978. Estimating the dimension of a model. Annals of Mathematical Statistics, 6: 461.

Shalabh and Dhar, S. S. 2020. Statistical modelling and variable selection in climate science. pp. 351–378. *In*: N. Roy, S. Roychoudhury, S. Nautiyal, S. K. Agarwal and S. Baksi (eds.). Socio-economic and Eco-biological Dimensions of Climate Change: Strategies for Resource Use, Conservation and Ecological Sustainability. Springer.

Shalabh, G. Garg and Heumann, C. 2012. Performance of double k-class estimators for coefficients in linear regression models with non spherical disturbances under asymmetric losses. Journal of Multivariate Analysis, 112: 35–47.

Shalabh, Garg, G. and Misra, N. 2007. Restricted regression estimation in measurement error models. Computational Statistics and Data Analysis, 52(2): 1149–1166.

Shalabh, Garg, G. and Misra, N. 2009. Use of prior information in the consistent estimation of regression coefficients in measurement error models. Journal of Multivariate Analysis, 100(7): 1498–1520.

Shao, J. 1993. Linear model selection by cross-validation. Journal of the American statistical Association, 88: 486–494.

Shibata, R. 1984. Approximate efficiency of a selection procedure for the number of regression variables. Biometrika, 71: 43–49.

Silverman, B. W. 1986. Density Estimation for Statistics and Data Analysis. Chapman and Hall, London.

Srivastava, A. K. and Shobhit. 2002. A family of estimators for the coefficient of determination in linear regression models. Journal of Applied Statistical Science, 11(2): 133–144.

Stein, C. 1956. Inadmissibility of the usual estimator for the mean of a multivariate normal distribution. pp. 197–206. *In*: Proceedings of the Third Berkeley Symposium on Mathematical Statistics and Probability, University of California Press, Berkeley.

Stone, M. 1974. Cross-validation and multinomial prediction. Biometrika, 61: 509–515.

Storlie, C. H., Bondell, H. D., Reich, B. J. and Zhang, H. L. 2011. Surface estimation, variable selection, and the non-parametric oracle property. Statistica Sinica, 21: 679–705.

Tanaka, J. S. and Huba, G. J. 1998. A general coefficient of determination for covariance structure models under arbitrary GLS estimation. British Journal of Mathematical and Statistical Psychology, 42(2): 233–239.

Tjur, T. 2009. Coefficients of determination in logistic regression models—a new proposal: The coefficient of discrimination. The American Statistician, 63(4): 366–372.

Toutenburg, H. 1982. Prior Information in Linear Models. John Wiley.

Ullah, A. and Ullah, S. 1978. Double k-class estimators of coefficients in linear regression. Econometrica, 46(3): 705–722.

Ullah, A. and Ullah, S. 1981. Errata: Double k-class estimators of coefficients in linear regression. Econometrica, 46(1978): 3: 705–722. Econometrica, 49, 2: 554.

van der Linde, A. and Tutz, G. 2008. On association in regression: the coefficient of determination revisited. Statistics, 42(1): 1–24.

Wan, Alan T. K. and Chaturvedi, A. 2001. Double k-class estimators in regression models with non-spherical disturbances. Journal of Multivariate Analysis, 79(2): 226–250.

Watson, G. S. 1964. Smooth regression analysis. Sankhya, 26: 359–372.

Yanagihara, H., Wakaki, H. and Fujikoshi, Y. 2015. A consistency property of the AIC for multivariate linear models when the dimension and the sample size are large. Electronic Journal of Statistics, 9: 869–897.

Zambom, A. Z. and Akritas, M. G. 2014. Nonparametric lack-of-fit testing and consistent variable selection. Statistica Sinica, 24: 1837–1858.

Zhang, P. 1991. Variable selection in nonparametric regression with continuous covariates. The Annals of Statistics, 19: 1869–1882.

Zhang, P. 1993a. Model selection via multifold cross validation. The Annals of Statistics, 21: 299–313.

Zhang, P. 1993b. On the convergence rate of model selection criteria. Communications in Statistics-Theory and Methods, 22: 2765–2775.

CHAPTER 6

Stochastic Models for Cancer Progression and its Optimal Programming for Control with Chemotherapy

Tirupathi Rao Padi

1. Introduction

Continuous, uninterrupted and unending cell division is usually referred to as carcinogenesis. Formulation of cancer-causing cells in a living body may be due to several reasons. The likely causes are exposure to hazardous environmental conditions, unhealthy food habits, unwarranted life styles, smoking, consumption of toxic beverages, among others. Natural mechanism of cell growth under normal conditions will happen whenever there is a requirement for cell construction due to the wear and tear processes. Normal cells have to undergo processes like cell division, growth, death/differentiation and transformation from one format to the other. Usually a normal/healthy cell will divide into two identical normal/healthy daughter cells, further the two daughter cells divide into subsequent progeny cells. The alleles of genes are used to regulate the mechanism of structured cell division. They play a vital role in protecting the living body from risks like over proliferation of specific cells above the required level and unnecessary invasion of similar cells from the original location to the invasive one. However, when the regulating mechanism disobeys the usual cell division process, the problem of cancer cell growth will come up. The three main factors like genetic materials or proteins or protein-encoding genes are responsible for regulating the cell's division, growth, death/differentiation and invasion to the other tissues. They are namely (i) *proto-oncogene;* responsible for enhancing the cell division, (ii) *tumor suppressor gene;* monitor the cell division or causes the cell death, (iii) *DNA repair gene;* protects the mutation causing gene by means of rectifying its unusual behavior. If mutation occurs in these genes it leads to loss of control of the complete cell cycle and leads to the development of cancer. Moreover, all kinds of cancers are having one of the features that disrupt the process of regulatory mechanism in the normal cell division. The process of abnormal cells growth is initiated with a simple mutation in the regulating genes. The operating characteristics and behavior patterns of cancerous cells are entirely different from those of normal cells. Needless to say that mathematical modelling is a suitable option for studying cancer growth. However, it has numerous limitations as model building is linked with many uncertain conditions. Hence, stochastic modeling is more rational in studying cancer spread and control.

The nature of a malignant tumor is defined with the type of cells in the organ where it is originally initiated. Carcinogenesis is a complex, random and multistep process consisting of initiation, promotion and progression of normal cells (Tan, 1989). Transforming normal cells into malignant cells in any organ is usually initiated with simple mutation in the gene of cells and aggravate with abnormal proliferation. The mutation process involved in normal stage cell to intermediate stage cells and intermediate stage to malignant stage cells are purely random in nature (Quinn, 1997). The cancer cell growth involves a series of molecular changes in the normal cells. While reaching the malignant stage, it may undergo one or more mutation (transformation of one state to another state) processes. The cancer cells have the capability to spread/migrate to neighboring locations of an organ or other part of the body through blood vessels which is termed as the metastasis. The migrated malignant cells form a new colony by means of further proliferation leading to formation of cancer tumors. In the genesis of cancer, the response variables such as cell division, differentiation/death, mutation and migration are subject to random variations. Studying the growth of abnormal cells through stochastic modelling is always a superior approach. Getting the Statistical measures from the observed stochastic possesses through the probability functions shall provide the most relevant picture for proper understanding of cancer cell growth.

There is established literature evidence in making use of deterministic and stochastic study models for carcinogenesis. Normally a cell division shall have to take several mutations to become a malignant cell. Many researchers have developed two/three stage growth stochastic models similar to birth-and-death and mutation processes. Pathophysiological behavior of cancerous cells is modeled with

Department of Statistics, Pondicherry University, Puducherry 605014, India.
Email: drtrpadi@gmail.com

suitable assumptions for quantifying functional relations in the contexts of primary and invasive states of growth. There is substantial evidence in mathematical modeling of cancer formation, growth and invasion using the birth and death processes. However, there is a significant lagging in applying the spread of cancer through migration processes. Hence this study is intended to contribute towards the research gap for making use of blended stochastic processes with a backdrop of bivariate and Trivariate birth-death—migration processes. Different states of cancer growth and its dynamics are modeled for exploring the joint probability distributions of the tandem arrival and departure processes. Development of optimization programming problems for regulating the cancer cell's growth with the controlling parameters is still an open problem for study. Optimal drug administration by assessing the varying health indicators through stochastic programming is another focused area where there is no significant evidence on reported research. Predicting the optimal decision parameters from the developed stochastic and optimization models is the core objective of this study. This study can be extended for healthcare management during the treatment of cancer with chemotherapy. Exploring the real time decision support systems from the developed stochastic optimization modeling is the motivating factor for selecting this study.

2. Stochastic Model for Cancer Cell Growth during Chemotherapy

Tirupathi Rao et al., 2011, 2012 have developed a bi-variate stochastic model for normal cell and mutant cell growth during drug administration and drug vacation. These models are for treatment dependent malignant tumor progression based on the assumption that the cell division of pre-malignant and malignant cells follow a Poisson distribution during and off the chemotherapy process. Madhavi et al., 2013 have developed a stochastic model for stage dependent cancer cell growth under the assumption of all possible cell divisions during drug administration and its vacation. Tirupathi Rao et al. (2013) have developed a Trivariate stochastic model for cancer cell growth progression under the assumptions of cell divisions from normal cells to premalignant cells and from premalignant cells to malignant cells during the presence and absence of chemotherapy. The focus of all the above-mentioned works are mostly based on the cancer cell processes within the organ in the region of formation of the cancer cell. They have neither discussed nor developed any prospective cancer growth model as result of migration from the region of formation to the region of destination. Hence, the development of cancer growth through migration of mutant cell or cancer cell deserves the attention of researchers for better understanding of cancer growth behavior/dynamics.

The above said gap of research is taken care of in this chapter. The proposed model will help in knowing the spontaneous and active growth of cells during drug administration and vacation periods. A linear function is defined to express the decision parameters in the dynamics of growth of any organ cell during drug administration and vacation periods using indicator variables. They will define the relationship between the decision parameters of cancer cell growth. This model can study the growth of a cancer cell as an overall phenomenon with drug administration and its vacation. In order to study the characteristics of the model, the first and second order moments were derived. This model will be well suited for carcinogenesis experimental trials. Data can be extracted for the said parameters on sample experimentation basis. However, in order to understand the model behavior and sensitivity analysis a numerical study is carried out with simulated data.

Mutation is a resulting change of cell division behavior due to violation of genetic instructions of cell division. Unusual, continuous and a never-ending cell division process is termed as cancer. This unwarranted cell proliferation within the membrane of a tissue/muscle leads to formation of tumors. However, the connectivity of each cell with the blood circulation system makes the flow of mutant/cancer causing cells to other regions of the body referred to as metastasis. Such cell differentiation, division and death will cause the abnormal structure and functioning of tissues in an organ. It is a threat to the normal and regulating functions of cell growth within the tissue as the initiated mutant cells will be converted to full-fledged accumulated malignant cells. These newly formatted malignant cells will further spread to the other parts of the body from the origin through the process of metastasis. This process will lead to form secondary tumors either in the neighboring places (within the organ where the mutation has occurred) or in distant locations (other location of the body away from the organ where the newly formulated cancer cell was originated). If the cell growth process is deterministic then mathematical models will provide appropriate and accurate results. Whereas for abnormal cell growth processes which are subject to non-deterministic, stochastic models will be suitable alternatives. Here, it is assumed that the behaviour of normal/mutant/malignant cells growth/loss/differentiation are fully stochastic. This process can be modeled with homogeneous birth, death, mutation and migration processes.

Notations & Postulates of the Model

β^*_{ijl} - Growth rate of ith staged cell in j^{th} staged tumor and l^{th} state of drug application

δ^*_{ijl} - Loss rate of ith staged cell in j^{th} staged tumor and l^{th} state of drug application

τ^*_{ijl} - Transformation rate of cell from i^{th} staged cell to $(i+1)^{th}$ staged cell, from the jth staged tumor to $(j+1)^{th}$ staged tumor in l^{th} state of drug application

Where,

$i = 1, 2$ and 3 are Normal cell, Malignant cell and Migrated Malignant cell respectively

$j = 1$ and 2 are Primary stage of tumor and Secondary stage of tumor respectively

$l = 0$ and 1 are drug vacation, drug administration respectively

a_r is an indicator variable representing the drug administration and vacation referred below.

$$a_r = \begin{cases} 1 & ;\text{Drug Administration period} \\ 0 & ;\text{Drug vacation period} \end{cases} \qquad \text{for } r = 1,\dots,9$$

The growth rate, migration/transformation (mutation) rate and loss rate of cells can be represented as $\beta_{ij}^* = \left[a_r\beta_{ijl} + (1-a_r)\beta_{ijl}\right]$; $\tau_{ij}^* = \left[a_r\tau_{ijl} + (1-a_r)\tau_{ijl}\right]$; $\delta_{ij}^* = \left[a_r\delta_{ijl} + (1-a_r)\delta_{ijl}\right]$ respectively.

The mechanisms involved in the cell divisions, death/ differentiation are purely stochastic during drug administration and vacation periods. Let us assume that the events occurred in non-overlapping intervals of time and are statistically independent in both the periods. The cell's growth, loss and migrations during the drug administration period and drug vacation period are also assumed to be stochastic and a model is developed based on the following postulates.

Let $\{N(t), t \geq 0\}$ be the process of normal cell mechanisms (growth/loss/transformation/migration) and $\{M(t), t \geq 0\}$ be the process of malignant cells mechanism (growth/loss/migration). Let $\{N(t), M(t); t \geq 0\}$ be a joint Bivariate stochastic process of individual stochastic processes of $\{N(t), t \geq 0\}$ and $\{M(t), t \geq 0\}$, the joint probability being $P\{[N(t), M(t)] = [n, m]\} = P_{n,m}(t)$ and, marginal probabilities with respect to the number of normal cells and number of malignant cells are $P\{N(t) = n\} = P_n(t)$ and $P\{M(t) = m\} = P_m(t)$,

Further,

$$P\{N(\Delta t) = u / N(t) = n\} = P_{nu};\ \text{for } u = n+1, n-1, n, n\pm 2$$

$$P\{M(\Delta t) = v / M(t) = m\} = P_{mv};\ \text{for } v = m+1, m-1, m, m\pm 2$$

$$P\left[\{(N(\Delta t), M(\Delta t)) = (u,v)\} / \{(N(t), M(t)) = (n,m)\}\right] = P_{nu,mv};\ \text{for } v = m+1, m-1, m, m\pm 2$$

If Δt be a very small interval of time.

Let us now define postulates of the process with respect to normal cells and malignant cells growth,

$$\begin{aligned} p_{n,u} &= P\{N(\Delta t) = u / N(t) = n\} \\ &= n\left(a_1\beta_{111} + (1-a_1)\beta_{110}\right)\Delta t + o(\Delta t) && ;u = n+1 \\ &= n\left(a_5\delta_{111} + (1-a_5)\delta_{110}\right)\Delta t + o(\Delta t) && ;u = n-1 \\ &= n\left(a_2\tau_{111} + (1-a_2)\tau_{111}\right)\Delta t + o(\Delta t) && ;u = n-1 \\ &= 1 - \left[n\begin{pmatrix} \left(a_1\beta_{111} + (1-a_1)\beta_{110}\right) \\ +\left(a_5\delta_{111} + (1-a_5)\delta_{110}\right) \\ +\left(a_2\tau_{111} + (1-a_2)\tau_{111}\right) \end{pmatrix}\Delta t + o(\Delta t)\right] && ;u = n \\ &= o(\Delta t)^2 && ;u = n\pm 2 \end{aligned} \tag{2.1}$$

For malignant cell growth processes,

$$\begin{aligned} p_{m,v} &= P\{M(\Delta t) = v / M(t) = m\} \\ &= m\left(a_3\beta_{211} + (1-a_3)\beta_{210}\right)\Delta t + o(\Delta t) && ;v = m+1 \\ &= m\left(a_6\delta_{211} + (1-a_6)\delta_{210}\right)\Delta t + o(\Delta t) && ;v = m-1 \\ &= m\left(a_7\tau_{211} + (1-a_7)\tau_{210}\right)\Delta t + o(\Delta t) && ;v = m-1 \\ &= \left(a_4\beta_{321} + (1-a_4)\beta_{320}\right)\Delta t + o(\Delta t) && ;v = 1 \\ &= \left(a_8\delta_{321} + (1-a_8)\delta_{320}\right)\Delta t + o(\Delta t) && ;v = 1 \\ &= \left(a_9\tau_{321} + (1-a_9)\tau_{321}\right)\Delta t + o(\Delta t) && ;v = 1 \\ &= 1 - \left[\left\{\begin{matrix} m\left[\left(a_3\beta_{211} + (1-a_3)\beta_{210}\right) + \left(a_6\delta_{211} + (1-a_6)\delta_{210}\right)\right. \\ \left. +\left(a_7\tau_{211} + (1-a_7)\tau_{210}\right)\right] + \left[\left(a_4\beta_{321} + (1-a_4)\beta_{320}\right)\right. \\ \left. +\left(a_8\delta_{321} + (1-a_8)\delta_{320}\right) + \left(a_9\tau_{321} + (1-a_9)\tau_{321}\right)\right] \end{matrix}\right\}\Delta t + o(\Delta t)\right] && ;v = m \\ &= o(\Delta t)^2 && ;v = m\pm 2 \end{aligned} \tag{2.2}$$

Considering the joint stochastic processes of normal cells and malignant cells dynamics during drug vacation period and drug administration period, we have

$$
\begin{aligned}
p_{nu,mv} &= P\{(N(\Delta t), M(\Delta t)) = (u,v) / (N(t), M(t)) = (n,m)\} \\
&= n\left(a_1\beta_{111} + (1-a_1)\beta_{110}\right)\Delta t + o(\Delta t) && ; u = n+1, v = m \\
&= n\left(a_5\delta_{111} + (1-a_5)\delta_{110}\right)\Delta t + o(\Delta t) && ; u = n-1, v = m \\
&= n\left(a_2\tau_{111} + (1-a_2)\tau_{111}\right)\Delta t + o(\Delta t) && ; u = n-1, v = m \\
&= m\left(a_3\beta_{211} + (1-a_3)\beta_{210}\right)\Delta t + o(\Delta t) && ; u = n, v = m+1 \\
&= m\left(a_6\delta_{211} + (1-a_6)\delta_{210}\right)\Delta t + o(\Delta t) && ; u = n, v = m-1 \\
&= m\left(a_7\tau_{211} + (1-a_7)\tau_{210}\right)\Delta t + o(\Delta t) && ; u = n, v = m-1 \\
&= \left(a_4\beta_{321} + (1-a_4)\beta_{320}\right)\Delta t + o(\Delta t) && ; u = n, v = 1 \\
&= \left(a_8\delta_{321} + (1-a_8)\delta_{320}\right)\Delta t + o(\Delta t) && ; u = n, v = 1 \\
&= \left(a_9\tau_{321} + (1-a_9)\tau_{320}\right)\Delta t + o(\Delta t) && ; u = n, v = 1 \\
&= 1 - \left[\left\{\begin{array}{l} n\left[\left(a_1\beta_{111} + (1-a_1)\beta_{110}\right) + \left(a_5\delta_{111} + (1-a_5)\delta_{110}\right)\right. \\ \left.+\left(a_2\tau_{111} + (1-a_2)\tau_{111}\right)\right] + m\left[\left(a_3\beta_{211} + (1-a_3)\beta_{210}\right)\right. \\ \left.+\left(a_6\delta_{211} + (1-a_6)\delta_{210}\right) + \left(a_7\tau_{211} + (1-a_7)\tau_{210}\right)\right] \\ +\left[\left(a_4\beta_{321} + (1-a_4)\beta_{320}\right) + \left(a_8\delta_{321} + (1-a_8)\delta_{320}\right)\right. \\ \left.+\left(a_9\tau_{321} + (1-a_9)\tau_{320}\right)\right] \end{array}\right\}\Delta t + o(\Delta t)\right] && ; u = n, v = m \\
&= o(\Delta t)^2 && ; u = n \pm 2, v = m \pm 2
\end{aligned}
\tag{2.3}
$$

Differential Equations & Probability Generating Functions of the Model:

Let $p_{n,m}(t+\Delta t)$ be the probability that the occurrence of any one possible event such as cell division, differentiation/death, transformation/ migration in an infinitesimal interval Δt, on the condition that there exist 'n' normal cells and 'm' malignant cells in the organ up to time 't' during the period of drug administration and drug vacation period.

$$
\begin{aligned}
p_{n,m}(t+\Delta t) = &\left\{1 - \left\{n\left[\left(a_1\beta_{111} + (1-a_1)\beta_{110}\right) + \left(a_2\tau_{111} + (1-a_2)\tau_{111}\right) + \left(a_5\delta_{111} + (1-a_5)\delta_{110}\right)\right]\right.\right. \\
&+ m\left[\left(a_3\beta_{211} + (1-a_3)\beta_{210}\right) + \left(a_6\delta_{211} + (1-a_6)\delta_{210}\right) + \left(a_7\tau_{211} + (1-a_7)\tau_{210}\right)\right] \\
&\left.\left.+ \left[\left(a_4\beta_{321} + (1-a_4)\beta_{320}\right) + \left(a_8\delta_{321} + (1-a_8)\delta_{320}\right) + \left(a_9\tau_{321} + (1-a_9)\tau_{320}\right)\right]\right\}\Delta t\right\} p_{n,m}(t) \\
&+ p_{n+1,m}(t)\left[(n-1)\left(a_1\beta_{111} + (1-a_1)\beta_{110}\right)\right] + p_{n+1,m-1}(t)\left[(n+1)\left(a_2\tau_{111} + (1-a_2)\tau_{111}\right)\right]\Delta t \\
&+ p_{n,m-1}(t)\left[(m-1)\left(a_3\beta_{211} + (1-a_3)\beta_{210}\right)\right] + p_{n+1,m}(t)\left[(n+1)\left(a_5\delta_{111} + (1-a_5)\delta_{110}\right)\right]\Delta t \\
&+ p_{n,m+1}(t)\left[(m+1)\left(a_6\delta_{211} + (1-a_6)\delta_{210}\right)\right] + p_{n,m+1}(t)\left[(m+1)\left(a_7\tau_{211} + (1-a_7)\tau_{210}\right)\right]\Delta t \\
&+ p_{n,m-1}(t)\left(a_4\beta_{321} + (1-a_4)\beta_{320}\right)\Delta t + p_{n,m+1}(t)\left(a_8\delta_{321} + (1-a_8)\delta_{320}\right)\Delta t \\
&+ p_{n,m+1}(t)\left(a_9\tau_{321} + (1-a_9)\tau_{320}\right)\Delta t + o(\Delta t^2) \qquad \text{for } n, m \geq 1
\end{aligned}
\tag{2.4}
$$

The initial conditions are

$$p_{N_0,M_0}(0) = 1,\ p_{i,j}(0) = 0 \quad \forall i \neq N_0; j \neq M_0 \tag{2.5}$$

The differential equations of the model when Δt tends to zero with the above equation are

$$\begin{aligned}
p'_{n,m}(t) = &-\Big\{n\big[\big(a_1\beta_{111}+(1-a_1)\beta_{110}\big)+\big(a_2\tau_{111}+(1-a_2)\tau_{111}\big)+\big(a_5\delta_{111}+(1-a_5)\delta_{110}\big)\big] \\
&+m\big[\big(a_3\beta_{211}+(1\text{-}a_3)\beta_{210}\big)+\big(a_6\delta_{211}+(1\text{-}a_6)\delta_{210}\big)+\big(a_7\tau_{211}+(1\text{-}a_7)\tau_{210}\big)\big] \\
&+\big[\big(a_4\beta_{321}+(1\text{-}a_4)\beta_{320}\big)+\big(a_8\delta_{321}+(1\text{-}a_8)\delta_{320}\big)+\big(a_9\tau_{321}+(1\text{-}a_9)\tau_{320}\big)\big]\Big\}p_{n,m}(t) \\
&+p_{n+1,m}(t)\big[(n-1)\big(a_1\beta_{111}+(1-a_1)\beta_{110}\big)\big]+p_{n+1,m-1}(t)\big[(n+1)\big(a_2\tau_{111}+(1-a_2)\tau_{111}\big)\big] \\
&+p_{n,m-1}(t)\big[(m-1)\big(a_3\beta_{211}+(1\text{-}a_3)\beta_{210}\big)\big]+p_{n+1,m}(t)\big[(n+1)\big(a_5\delta_{111}+(1-a_5)\delta_{110}\big)\big] \\
&+p_{n,m+1}(t)\big[(m+1)\big(a_6\delta_{211}+(1\text{-}a_6)\delta_{210}\big)\big]+p_{n,m+1}(t)\big[(m+1)\big(a_7\tau_{211}+(1\text{-}a_7)\tau_{210}\big)\big] \\
&+p_{n,m-1}(t)\big(a_4\beta_{321}+(1\text{-}a_4)\beta_{320}\big)+p_{n,m+1}(t)\big(a_8\delta_{321}+(1\text{-}a_8)\delta_{320}\big) \\
&+p_{n,m+1}(t)\big(a_9\tau_{321}+(1\text{-}a_9)\tau_{320}\big) \qquad \text{for } n,m\geq 1
\end{aligned} \tag{2.6}$$

$$\begin{aligned}
\frac{dp_{0,1}(t)}{dt} = &\text{-}\big[\big(a_3\beta_{211}+(1\text{-}a_3)\beta_{210}\big)+\big(a_4\beta_{321}+(1\text{-}a_4)\beta_{320}\big)+\big(a_6\delta_{211}+(1\text{-}a_6)\delta_{210}\big) \\
&+\big(a_8\delta_{321}+(1\text{-}a_8)\delta_{320}\big)+\big(a_9\tau_{321}+(1\text{-}a_9)\tau_{320}\big)\big]P_{0,1}(t)+\big(a_5\delta_{111}+(1-a_5)\delta_{110}\big)p_{1,1}(t) \\
&+\big(a_2\tau_{111}+(1-a_2)\tau_{111}\big)p_{1,0}(t)+\big\{2\big[\big(a_6\delta_{211}+(1\text{-}a_6)\delta_{210}\big)+\big(a_7\tau_{211}+(1\text{-}a_7)\tau_{210}\big)\big] \\
&+(a_8\mu_{321}+(1\text{-}a_8)\mu_{320})+\big(a_9\tau_{321}+(1\text{-}a_9)\tau_{320}\big)\big\}p_{0,2}(t)+\big(a_4\beta_{321}+(1\text{-}a_4)\beta_{320}\big)p_{0,0}(t)
\end{aligned} \tag{2.7}$$

$$\begin{aligned}
\frac{dp_{1,0}(t)}{dt} = &-\big\{\big(a_1\beta_{111}+(1-a_1)\beta_{110}\big)+\big(a_2\tau_{111}+(1-a_2)\tau_{111}\big)+\big(a_4\beta_{321}+(1\text{-}a_4)\beta_{320}\big) \\
&+\big(a_5\delta_{111}+(1-a_5)\delta_{110}\big)+\big(a_7\tau_{211}+(1\text{-}a_7)\tau_{210}\big)+\big(a_8\delta_{321}+(1\text{-}a_8)\delta_{320}\big) \\
&+\big(a_9\tau_{321}+(1\text{-}a_9)\tau_{320}\big)\big\}p_{1,0}(t)+2\big(a_5\delta_{111}+(1-a_5)\delta_{110}\big)p_{2,0}(t) \\
&+\big\{\big(a_5\delta_{111}+(1-a_5)\delta_{110}\big)+\big(a_7\tau_{211}+(1\text{-}a_7)\tau_{210}\big)+\big(a_8\delta_{321}+(1\text{-}a_8)\delta_{320}\big) \\
&+\big(a_9\tau_{321}+(1\text{-}a_9)\tau_{320}\big)\big\}p_{1,1}(t)
\end{aligned} \tag{2.8}$$

$$\begin{aligned}
\frac{dp_{0,0}(t)}{dt} = &-\big\{\big(a_4\beta_{321}+(1\text{-}a_4)\beta_{320}\big)+\big(a_8\delta_{321}+(1\text{-}a_8)\delta_{320}\big)+\big(a_9\tau_{321}+(1\text{-}a_9)\tau_{320}\big)\big\}p_{0,0}(t) \\
&+\big(a_5\delta_{111}+(1-a_5)\delta_{110}\big)p_{1,0}(t)+\big\{\big(a_5\delta_{111}+(1-a_5)\delta_{110}\big)+\big(a_7\tau_{211}+(1\text{-}a_7)\tau_{210}\big) \\
&+\big(a_8\delta_{321}+(1\text{-}a_8)\delta_{320}\big)+\big(a_9\tau_{321}+(1\text{-}a_9)\tau_{320}\big)\big\}p_{0,1}(t)
\end{aligned} \tag{2.9}$$

Let $P(x, y; t)$ be the probability generating function of random variable $X(t)$ and $Y(t)$ representing the number of normal cells and number of malignant cells in an organ during the drug administration and vacation period with probability function $P_{n,m}(t)$. Where, $P(x,y;t)=\sum_{m=0}^{\infty}\sum_{n=0}^{\infty}x^n y^m\, p_{n,m}(t))\ ;|x|<1,|y|<1$.

Multiplying the above differential Equations from (2.6) to (2.9) with $x^n y^m$ and summing over n, m, we get

$$\begin{aligned}
\frac{dP(x,y;t)}{dt} = &-\big[\big(a_1\beta_{111}+(1-a_1)\beta_{110}\big)+\big(a_2\tau_{111}+(1-a_2)\tau_{111}\big)+\big(a_5\delta_{111}+(1-a_5)\delta_{110}\big)\big]x\sum_{m=0}^{\infty}\sum_{n=0}^{\infty}nx^{n-1}y^m p_{n,m}(t) \\
&-\big[\big(a_3\beta_{211}+(1\text{-}a_3)\beta_{210}\big)+\big(a_6\delta_{211}+(1\text{-}a_6)\delta_{210}\big)+\big(a_7\tau_{211}+(1\text{-}a_7)\tau_{210}\big)\big]y\sum_{m=0}^{\infty}\sum_{n=0}^{\infty}mx^n y^{m-1}p_{n,m}(t) \\
&-\big[\big(a_4\beta_{321}+(1\text{-}a_4)\beta_{320}\big)+\big(a_8\delta_{321}+(1\text{-}a_8)\delta_{320}\big)+\big(a_9\tau_{321}+(1\text{-}a_9)\tau_{320}\big)\big]\sum_{m=0}^{\infty}\sum_{n=0}^{\infty}x^n y^m p_{n,m}(t) \\
&+\big(a_1\beta_{111}+(1-a_1)\beta_{110}\big)x^2\sum_{m=0}^{\infty}\sum_{n=0}^{\infty}(n-1)x^{n-2}y^m p_{n+1,m}(t) \\
&+\big(a_2\tau_{111}+(1-a_2)\tau_{111}\big)y\sum_{m=0}^{\infty}\sum_{n=0}^{\infty}(n+1)x^n y^{m-1}p_{n+1,m-1}(t) \\
&+\big(a_3\beta_{211}+(1\text{-}a_3)\beta_{210}\big)y^2\sum_{m=0}^{\infty}\sum_{n=0}^{\infty}(m-1)x^n y^{m-2}p_{n,m-1}(t)
\end{aligned}$$

$$+\left(a_5\delta_{111}+(1-a_5)\delta_{110}\right)\sum_{m=0}^{\infty}\sum_{n=0}^{\infty}(n+1)x^n y^m p_{n+1,m}(t)$$
$$+\left[\left(a_6\delta_{211}+(1\text{-}a_6)\delta_{210}\right)+\left(a_7\tau_{211}+(1\text{-}a_7)\tau_{210}\right)\right]\sum_{m=0}^{\infty}\sum_{n=0}^{\infty}(m+1)x^n y^m p_{n,m+1}(t)$$
$$+\left[\left(a_8\delta_{321}+(1\text{-}a_8)\delta_{320}\right)+\left(a_9\tau_{321}+(1\text{-}a_9)\tau_{320}\right)\right]\frac{1}{y}\sum_{m=0}^{\infty}\sum_{n=0}^{\infty}x^n y^{m+1} p_{n,m+1}(t)$$
$$+\left(a_4\beta_{321}+(1\text{-}a_4)\beta_{320}\right)y\sum_{m=0}^{\infty}\sum_{n=0}^{\infty}x^n y^{m-1} p_{n,m-1}(t) \tag{2.10}$$

where

$$\frac{dP(x,y;t)}{dx}=\sum_{m=0}^{\infty}\sum_{n=0}^{\infty}nx^{n-1}y^m p_{n,m}(t)=\sum_{m=0}^{\infty}\sum_{n=0}^{\infty}(n-1)x^{n-2}y^m p_{n-1,m}(t)$$
$$\frac{dP(x,y;t)}{dy}=\sum_{m=0}^{\infty}\sum_{n=0}^{\infty}mx^{n-1}y^m p_{n,m}(t)=\sum_{m=0}^{\infty}\sum_{n=0}^{\infty}(m-1)x^{n-1}y^{m-2} p_{n,m-1}(t)$$

On simplification, we obtain the differential equation of the form given below,

$$\frac{\partial}{\partial t}P(x,y;t)=\Big\{-\left[\left(a_1\beta_{111}+(1-a_1)\beta_{110}\right)+\left(a_2\tau_{111}+(1-a_2)\tau_{111}\right)+\left(a_5\delta_{111}+(1-a_5)\delta_{110}\right)\right]x$$
$$+\left(a_5\delta_{111}+(1-a_5)\delta_{110}\right)+\left(a_1\beta_{111}+(1-a_1)\beta_{110}\right)x^2+\left(a_2\tau_{111}+(1-a_2)\tau_{111}\right)y\Big\}\frac{\partial P(x,y;t)}{\partial x}$$
$$+\Big\{-\left[\left(a_3\beta_{211}+(1\text{-}a_3)\beta_{210}\right)+\left(a_6\delta_{211}+(1\text{-}a_6)\delta_{210}\right)+\left(a_7\tau_{211}+(1\text{-}a_7)\tau_{210}\right)\right]y \tag{2.11}$$
$$+\left(a_3\beta_{211}+(1\text{-}a_3)\beta_{210}\right)y^2+\left(a_6\delta_{211}+(1\text{-}a_6)\delta_{210}\right)+\left(a_7\tau_{211}+(1\text{-}a_7)\tau_{210}\right)\Big\}\frac{\partial P(x,y;t)}{\partial y}$$
$$+\Big\{-\left[\left(a_4\beta_{321}+(1\text{-}a_4)\beta_{320}\right)+\left(a_8\delta_{321}+(1\text{-}a_8)\delta_{320}\right)+\left(a_9\tau_{321}+(1\text{-}a_9)\tau_{320}\right)\right]$$
$$+\left[\left(a_8\delta_{321}+(1\text{-}a_8)\delta_{320}\right)+\left(a_9\tau_{321}+(1\text{-}a_9)\tau_{320}\right)\right]y^{-1}+\left(a_4\beta_{321}+(1\text{-}a_4)\beta_{320}\right)y\Big\}P(x,y;t)$$

We can obtain the characteristics of the model by using joint cumulant generating function of $p_{n,m}(t)$. Let $K(u, v\, t) = \log P(e^u, e^v, t)$ denote the cumulant generating function. By taking the Jacobian transformation of $x = e^u$ AND $y = e^v$ in the probability generating function, the joint cumulant generating function $K(u, v, t)$ of the PGF $P(x, y, t)$ is as the following expression

$$\frac{\partial}{\partial t}K(u,v;t)$$
$$=\Big\{-\left[\left(a_1\beta_{111}+(1-a_1)\beta_{110}\right)+\left(a_2\tau_{111}+(1-a_2)\tau_{111}\right)+\left(a_5\delta_{111}+(1-a_5)\delta_{110}\right)\right]$$
$$+\left(a_5\delta_{111}+(1-a_5)\delta_{110}\right)e^{-u}+\left(a_1\beta_{111}+(1-a_1)\beta_{110}\right)e^{u}+\left(a_2\tau_{111}+(1-a_2)\tau_{111}\right)e^{v-u}\Big\}\frac{\partial K(u,v;t)}{\partial u}$$
$$+\Big\{-\left[\left(a_3\beta_{211}+(1\text{-}a_3)\beta_{210}\right)+\left(a_6\delta_{211}+(1\text{-}a_6)\delta_{210}\right)+\left(a_7\tau_{211}+(1\text{-}a_7)\tau_{210}\right)\right] \tag{2.12}$$
$$+\left(a_3\beta_{211}+(1\text{-}a_3)\beta_{210}\right)e^{v}+\left[\left(a_6\delta_{211}+(1\text{-}a_6)\delta_{210}\right)+\left(a_7\tau_{211}+(1\text{-}a_7)\tau_{210}\right)\right]e^{-v}\Big\}\frac{\partial K(u,v;t)}{\partial v}$$
$$+\Big\{-\left[\left(a_4\beta_{321}+(1\text{-}a_4)\beta_{320}\right)+\left(a_8\delta_{321}+(1\text{-}a_8)\delta_{320}\right)+\left(a_9\tau_{321}+(1\text{-}a_9)\tau_{320}\right)\right]$$
$$+\left[\left(a_8\delta_{321}+(1\text{-}a_8)\delta_{320}\right)+\left(a_9\tau_{321}+(1\text{-}a_9)\tau_{320}\right)\right]e^{-v}+\left(a_4\beta_{321}+(1\text{-}a_4)\beta_{320}\right)e^{v}\Big\}K(u,v;t)$$

3. Statistical Measures & Moments

By making use of the cumulant generating function (CGF),

$$K(u,v;t)=u\,m^*_{1,0}(t)+v\,m^*_{0,1}(t)+\frac{1}{2}u^2\,m^*_{2,0}(t)+\frac{1}{2}v^2\,m^*_{0,2}(t)+uv\,m^*_{1,1}(t)+\cdots \tag{3.1}$$

Where $m_{i,j}(t)$ denotes the moments of order (i, j) of the normal cells, malignant cells in an organ at time t under the drug vacation period and drug administration period.

$m^*_{1,0}(t) = E[x(t)]$ - Expected number of normal cells in an organ at time't'

$m^*_{1,0}(t) = E[y(t)]$ - Expected number of malignant cells in an organ at time't'

$m^*_{2,0}(t) = Var[x(t)]$ - Variance of number of normal cells in an organ at time't'

$m^*_{2,0}(t) = Var[y(t)]$ - Variance of malignant cells in an organ at time't'.

$m^*_{1,1}(t) = Cov[x(t)y(t)]$ - Covariance between the number of normal cells and number of malignant cells in an organ at time't'.

On expanding the Equation (2.9) and applying the regulating conditions of cumulant generating function, the following linear differential equations are obtained.

$$\frac{d m^*_{1,0}(t)}{dt} = (\beta^*_{11} - \tau^*_{11} - \delta^*_{11}) m^*_{1,0}(t) \tag{3.2}$$

$$\frac{d m^*_{0,1}(t)}{dt} = \tau^*_{11} m^*_{1,0}(t) + (\beta^*_{21} - \delta^*_{21} - \tau^*_{21}) m^*_{0,1}(t) \tag{3.3}$$

$$\frac{d m_{2,0}(t)}{dt} = 2(\beta^*_{11} - \tau^*_{11} - \delta^*_{11}) m^*_{2,0}(t) + (\beta^*_{11} + \tau^*_{11} + \delta^*_{11}) m^*_{1,0}(t) \tag{3.4}$$

$$\begin{aligned}\frac{d m^*_{0,2}(t)}{dt} = {} & \tau^*_{11} m^*_{1,0}(t) + (\beta^*_{21} + \delta^*_{21} + \tau^*_{21}) m^*_{0,1}(t) + 2(\beta^*_{21} - \delta^*_{21} - \tau^*_{21}) m^*_{0,2}(t) \\ & + 2(\beta^*_{32} - \delta^*_{32} - \tau^*_{32}) m^*_{0,1}(t) + \tau^*_{11} m^*_{1,1}(t)\end{aligned} \tag{3.5}$$

$$\begin{aligned}\frac{d m^*_{1,1}(t)}{dt} = {} & (\beta^*_{11} - \tau^*_{11} - \delta^*_{11}) m^*_{1,1}(t) + 2\tau^*_{11} m^*_{2,0}(t) - \tau^*_{11} m^*_{1,0}(t) + (\beta^*_{21} - \delta^*_{21} - \tau^*_{21}) m^*_{1,1}(t) \\ & + (\beta^*_{32} - \delta^*_{32} - \tau^*_{32}) m^*_{1,0}(t)\end{aligned} \tag{3.6}$$

From (3.2) $\frac{d m^*_{1,0}(t)}{dt} = (\beta^*_{11} - \tau^*_{11} - \delta^*_{11}) m^*_{1,0}(t)$ with $m_{1,0}(0) = N_0$

On solving the above with the given initial condition, $m^*_{1,0}(t) = E[y(t)] = N_0 e^{A^*t}$ (3.7)

From (3.3) $\frac{d m^*_{0,1}(t)}{dt} = \tau^*_{11} m^*_{1,0}(t) + (\beta^*_{21} - \delta^*_{21} - \tau^*_{21}) m^*_{0,1}(t)$ with $m^*_{1,0}(0) = M_0$; Substituting $m_{0,1}(t)$ in the above equation and solving under the given initial condition,

$$m^*_{0,1}(t) = \frac{\tau^*_{11} N_0 e^{A^*t}}{A^* - B^*} + \left(M_0 - \frac{\tau^*_{11} N_0}{A^* - B^*}\right) e^{B^*t} \tag{3.8}$$

From (3.4) $\frac{d m^*_{2,0}(t)}{dt} = 2(\beta^*_{11} - \tau^*_{11} - \delta^*_{11}) m^*_{2,0}(t) + (\beta^*_{11} + \tau^*_{11} + \delta^*_{11}) m^*_{1,0}(t)$ with $m^*_{2,0}(0) = 0$. Substituting $m^*_{2,0}(t)$ and solving the above equation based on given initial condition,

$$m^*_{2,0}(t) = \frac{D^* N_0 e^{A^*t}}{A^*} \left(e^{A^*t} - 1\right) \tag{3.9}$$

From (3.5) $\frac{d m^*_{0,2}(t)}{dt} = \tau^*_{11} m^*_{1,0}(t) + (\beta^*_{21} + \delta^*_{21} + \tau^*_{21}) m^*_{0,1}(t) + 2(\beta^*_{21} - \delta^*_{21} - \tau^*_{21}) m^*_{0,2}(t)$ with

$$+ 2(\beta^*_{32} - \delta^*_{32} - \tau^*_{32}) m^*_{0,1}(t) + \tau^*_{11} m^*_{1,1}(t)$$

$m^*_{2,0}(0) = 0$. Substituting $m^*_{1,0}(t)$, $m^*_{0,1}(t)$, and solving the above equation with the given initial condition,

$$
\begin{aligned}
m^*_{0,2}(t) = {} & \frac{\tau^*_{11} N_0 e^{A^* t}}{\left(A^* - 2B^*\right)} + \left(F^* + 2E^*\right)\left\{\frac{\tau^*_{11} N_0 e^{A^* t}}{\left(A^* - 2B^*\right)\left(A - B^*\right)} - \left(M_0 - \frac{\tau^*_{11} N_0}{\left(A^* - B^*\right)}\right)\frac{e^{B^* t}}{B^*}\right\} \\
& + \tau^*_{11}\left\{\begin{array}{l} \dfrac{\tau^*_{11} N_0 D^*}{A^*}\left(\dfrac{e^{2A^* t}}{2\left(A^* - B^*\right)^2} + \dfrac{e^{A^* t}}{\left(A^* - 2B^*\right)B^*}\right) - \dfrac{\left(E^* - \tau^*_{11}\right) N_0 e^{A^* t}}{\left(A^* - 2B^*\right)B^*} \\ - \left(\dfrac{\tau^*_{11} N_0 D^* - \left(A^* - B^*\right)\left(E^* - \tau^*_{11}\right) N_0}{\left(A^* - B^*\right)B^*}\right)\dfrac{e^{\left(A^* + B^*\right)t}}{\left(A^* - B^*\right)} \end{array}\right\} \\
& - \left\{\begin{array}{l} \dfrac{\tau^*_{11} N_0}{\left(A^* - 2B^*\right)} + \left(F^* + 2E^*\right)\left(\dfrac{\tau^*_{11} N_0 - \left(A^* - 2B^*\right) M_0}{2\left(A^* - 2B^*\right)\left(A^* - B^*\right)^2 B^*}\right) \\ + \tau^{*\,2}_{11} N_0 \left(\dfrac{2\left(A^* - B^*\right)\left(\tau^*_{11} - E^*\right) + D^* \tau^*_{11}}{2\left(A^* - 2B^*\right)\left(A^* - B^*\right)^2}\right) \end{array}\right\} e^{2B^* t}
\end{aligned}
\tag{3.10}
$$

From (3.6)

$$
\begin{aligned}
\frac{d m^*_{1,1}(t)}{dt} = {} & \left(\beta^*_{11} - \tau^*_{11} - \delta^*_{11}\right) m^*_{1,1}(t) + 2\tau^*_{11} m^*_{2,0}(t) - \tau^*_{11} m^*_{1,0}(t) + \left(\beta^*_{21} - \delta^*_{21} - \tau^*_{21}\right) m^*_{1,1}(t) \\
& + \left(\beta^*_{32} - \delta^*_{32} - \tau^*_{32}\right) m^*_{1,0}(t)
\end{aligned}
$$

with $m^*_{1,1}(0) = 0$. Substituting $m^*_{1,0}(t)$ and $m^*_{2,0}(t)$; solving the above equation using the given initial condition,

$$
\begin{aligned}
m^*_{1,1}(t) = {} & \left\{\frac{(E^* - \tau^*_{11}) N_0 e^{-B^* t}}{-B^*} + \frac{\tau^*_{11} D^* N_0}{A^*}\left(\frac{e^{2A^* t}}{(A^* - B^*)} + \frac{e^{A^* t}}{B^*}\right)\right\} \\
& + \left\{\frac{\tau^*_{11} D^* N_0 - (A^* - B^*)(E^* - \tau^*_{11}) N_0}{(A^* - B^*) B^*}\right\} e^{(A^* + B^*)t}
\end{aligned}
$$

Where, N_0 & M_0 – are the initial number of normal and mutant cells in an organ

$$
\begin{array}{lll}
A^* = \beta^*_{11} - \tau^*_{11} - \delta^*_{11} & B^* = \beta^*_{21} - \delta^*_{21} - \tau^*_{21} & D^* = \beta^*_{11} + \tau^*_{11} + \delta^*_{11} \\
E^* = \beta^*_{32} - \delta^*_{32} - \tau^*_{32} & F^* = \beta^*_{21} + \delta^*_{21} + \tau^*_{21} & \\
\beta^*_{11} = a_1\beta_{111} + (1 - a_1)\beta_{110} & \tau^*_{11} = a_2\tau_{111} + (1 - a_2)\tau_{110} & \beta^*_{21} = a_3\beta_{211} + (1 - a_3)\beta_{210} \\
\beta^*_{32} = a_4\beta_{321} + (1 - a_4)\beta_{320} & \delta^*_{11} = a_5\delta_{111} + (1 - a_5)\delta_{110} & \delta^*_{21} = a_6\delta_{211} + (1 - a_6)\delta_{210} \\
\tau^*_{21} = a_7\tau_{211} + (1 - a_7)\tau_{210} & \delta^*_{32} = a_8\delta_{321} + (1 - a_8)\delta_{320} & \tau^*_{32} = a_9\tau_{321} + (1 - a_9)\tau_{320}
\end{array}
\tag{3.11}
$$

Hence, the characteristics of the model representing the period of drug administration and period of drug vacations are obtained as

- Average number of normal cells in an organ at time't' is $m_{1,0}(t)$;
- Average number of malignant cells in an organ at time't' is $m_{0,1}(t)$;
- Variance of number of normal cells in the organ $m_{2,0}(t)$;
- Variance of number of mutant cells in the organ at time't' is $m^*_{0,2}(t)$;
- Covariance of number of normal and mutant cells in an organ at time 't' is $m^*_{1,1}(t)$.

4. Sensitivity Analysis

In order carry out a better understanding of the developed models, a numerical illustration is considered. The variations in different moments on the number of cells in an organ during the periods of drug administration and vacation are observed by changing value of one parameter keeping the other parameters constant. The computed values of the characteristics of the model $m^*_{1,0}(t)$ $m^*_{0,1}(t)$ $m^*_{2,0}(t)$ $m^*_{0,2}(t)$ and $m^*_{1,1}(t)$ mentioned above from Equations (3.7) to (3.11) for the parameters are presented in the tables using the stimulated data sets for changing values of β_{111}, β_{110}, τ_{111}, τ_{110}, β_{211}, β_{210}, β_{321}, β_{320}, δ_{111}, δ_{110}, δ_{211}, δ_{210}, τ_{211}, τ_{210}, δ_{321}, δ_{320}, τ_{321}, τ_{320} N_0, M_0, t. The linear function defined using an indicator function a_r has taken values 0 and 1.

Sensitivity Analysis during drug vacation period

This deals with the assumption of $a_r = 0$, r =1,..,9. It has focused on the study of explaining the variation in different statistical measures of cell counts in an organ during the complete absence of a drug in the cancer affected body. The respective tables will give the variation in the statistical measures with respect to the changes in a single decision parameter when all other parameters are constant. Appendix IV (a) deals with fixed parameter values of $\beta_{110} = 1.0$, $\tau_{110} = 0.1$, $\beta_{210} = 4$, $\beta_{320} = 0.3$, $\delta_{110} = 0.5$, $\delta_{210} = 0.5$, $\tau_{210} = 0.3$, $\delta_{320} = 0.03$, $\tau_{320} = 0.01$, $N_0 = 200$, $M_0 = 500$, t = 2 and changing values of one among given parameters.

From the Appendix IV (a), it is observed that average and variance of number of normal cells is an increasing function of β_{110}, average number of malignant cells is an increasing function of β_{110}, variance of number of malignant cells is a decreasing function of β_{110}. The average & variance of numbers of normal cells are decreasing functions of τ_{110}; average and variance of number of malignant cells are increasing and decreasing functions of τ_{110} respectively. The average and variance of numbers of malignant cells are increasing functions of β_{210}. The variance of number of malignant cells is an increasing function of β_{320}. The average number of normal & malignant cells and variance of number of normal cells are decreasing functions of δ_{110}; variance of number of malignant cells is an increasing function of δ_{110}. The average & variance of number of malignant cells covariance between the normal and malignant cells are a decreasing function of δ_{210}. The average & variance numbers of malignant cells and covariance between cells are a decreasing function of τ_{210}. The variance of number of malignant cells and covariance between normal and malignant cells are decreasing functions of δ_{320}. The variance of number of malignant cells and covariance between normal and malignant cells are decreasing function of τ_{320}. The average & variance of number of normal & malignant cells along are increasing functions of time.

Sensitivity Analysis during drug administration period

This deals with the assumption of $a_r=1$, r=1,..,9. It has focused on the study of explaining the variation in different statistical measures of cell counts in an organ during complete presence of a drug in the cancer affected body. The respective tables will give the variation in statistical measures with respect to the changes in a single decision parameter when all other parameters are constant. Appendix IV (b) deals with changing values of one parameter and fixed value of remaining parameters of the set of parameters under study.

From the Appendix IV (b), it is observed that the average and variance of number of normal and malignant cells are increasing functions of β_{111}. The average number of normal cells and variance of number of malignant cells are decreasing functions of τ_{111}; average number of malignant cells and variance of number of normal cells are increasing functions of τ_{111}. The average & variance of number of malignant cells are increasing functions of β_{211}. The variance of number of malignant cells is an increasing function of β_{321}. The average and variance of number of normal and malignant cells is a decreasing function of δ_{111}. The average and variance of number of malignant cells are decreasing functions of δ_{211}. The average and variance of number of malignant cells are decreasing functions of τ_{211}. The variance of number of malignant cells is a decreasing function of δ_{321}. The variance of number of malignant cells is a decreasing function of τ_{321}. The average and variance of number of normal and malignant cells are an increasing function of N_0. The average and variance of number of malignant cells are an increasing function of M_0. The average number of normal & malignant cells and variance of number of normal & malignant cells are an increasing function of time.

Summary and Results

It is with regard to constructing the stochastic model and sensitivity analysis, a bivariate stochastic model is developed for the cancer cell growth in an organ based on homogeneous birth, death, mutation and migration processes under the period of drug administration and vacation by defining a linear function with an indicator variable. The differential difference equations are derived with the help of postulates possessing the model. The classical probability generating function approach is applied to represent the model in terms of partial differential equations for the random variables X(t) and Y(t) for the number of normal cells and number of malignant cells respectively during the drug administration period and drug vacation period. The first and second order moments are derived through cumulant generating functions. The numerical study was carried out to study the model behaviour for varying parameters during the period of drug administration and vacation. This study will help the medical practitioner to understand the behaviour of the cancer and dynamics in the cell growth over a period of time. Proper understanding about behaviour of cancer cell growth will give inputs to the medical practitioner regarding the condition of the patient. Treatment protocols can be designed with the developed model. Further, the statistical moments in this chapter were used for constructing the optimization programming problems in the following section.

5. Optimization Model for Cancer Control

The core objective of this section is to explore the optimal decision parameters which play a vital role in the processes of growth, death and invasion rates of cancer cells during drug vacation (absences of drug) and administration periods (presence of drug).

It is observed from the literature that several attempts have been made to predict the decision parameters in the abnormal cell dynamics (Tirupathi Rao, 2012). This study proposed the decision-making optimization models for minimizing the average number of malignant cells (or cancer cells), maximizing the average number of normal cells (or healthy cells), minimizing the variability (volatility) in the growth of normal cells, maximizing the variability in the growth of malignant (or cancer) cells, among other things. All the proposed programming problems shall have the common constraints such as average number of normal cells should be within threshold limits, average number of malignant cells should be lower than a specific danger limit, variance of the number of normal cells should be in a narrower range and variance of number of malignant cells should be in a wider range, are taken into account for the problem (Tirupathi Rao, 2012, 2013, 2014).

Optimal Programming for Cancer Control & Progression during Drug Administration

All the above-cited studies have considered the programming problems as a simple variable presentation and have solved each separately. They have considered the variables like average number normal and malignant cells, variance of normal and malignant cells, among others. However, in this study the researcher has addressed programming problems with a variable ratio between the normal and malignant cell growth. This approach will provide the relative behavior of the study variables. They can provide the comparative movements between the observed variables. While studying the growth or loss dynamics of cancer cells, desired situations like (i) the average number of healthy cells should always be greater than the average number of mutant or malignant cells and (ii) The variance of normal/healthy cells should always be less than the variance of mutant/malignant cells have to be taken in to consideration for better understanding. The phenomenon of programming problem formulation of programming has been carried out with two notions namely (i) in general context (assuming patient is not in treatment) and (ii) in drug administration context (the patient is in treatment). The objective functions are formulated with the relative measures (ratio between normal and mutant cells). First objective is to Minimize, where is the ratio between average number of normal cells to the average number of malignant cells and the second objective is to minimize the, where is the ratio between variance of normal cells to the variance of the malignant cells. Further it is assumed that $R_{E1} < 1$ and $R_{E2} > 1$. There are several other conditions which are formulated as constraints.

Optimal Programming for Treatment during Drug Administration and its Vacation

The statistical moments derived in the previous section are used to construct the optimization programming problem. Usually, growth rate of malignant cells will be more during drug absence period than the drug presence period. Similarly, the loss rate of malignant cells will be more during drug presence than the drug absence period. These stipulations are considered while formulating the proposed programming problem in this section. The health status is considered to be under control as long as the average number of normal cells is more than average number of malignant cells in the treatment of any cancer patient. In other words, the ratio of average normal to malignant cells shall be more than unity. However, the situation seems to be alarming when the ratio is less than unity. In order to study the cells dynamics in the regimen period and vacation periods of drug administration, a linear function is defined using a indicator variable for growth, death, migration, mutation, and transformation rates. The growth, migration/transformation and loss rate of cells can be represented as $\beta^*_{ij} = \left[a_r\beta_{ijl} + (1-a_r)\beta_{ijl}\right]$, $\delta^*_{ij} = \left[a_r\delta_{ijl} + (1-a_r)\delta_{ijl}\right]$ and $\tau^*_{ij} = \left[a_r\tau_{ijl} + (1-a_r)\tau_{ijl}\right]$ respectively. In this situation, we can define $R^*_{E_1}$ such that,

$$R^*_{E_1} = \frac{\text{Average number of normal cells}}{\text{Average number of malignant cells}}$$

From the derived results of Chapter –IV, it can be redefined as

$$R^*_{E_1} = \left\{\left[N_0\, e^{A^* t}\right] \Big/ \left[\frac{\tau^*_{12} N_0 e^{At}}{\left(A^* - B^*\right)} + \left(M_0 - \frac{\tau^*_{11} N_0}{\left(A^* - B^*\right)}\right) e^{B^* t}\right]\right\} \quad (5.1)$$

Thus, the objective is to maximize $R^*_{E_1}$.

The low fluctuations in the average number of normal cells in any organ is an indicator of consistency in health status. The lower the volatility in the sizes of normal and healthy cells is the greater the indication of the health of the patient. However, a contrary situation will be prevailing in the case of mutant or malignant cells. The increased volatility is desired in this case. Keeping such issues in mind the other program is formulated with the objective function ($R^*_{E_2}$) as the ratio between variance of normal cells to that of malignant cells.

$$R^*_{E_2} = \frac{\text{Variance of number of normal cells}}{\text{Variance of number of malignant cells}}$$

The results derived previously have defined the objective function as

$$R^*_{E_2} = \left\{\frac{D^* N_0 e^{A^* t}}{A^*}\left(e^{A^* t} - 1\right)\right\} \Big/ \left\{\frac{\tau^*_{11} N_0 e^{A^* t}}{\left(A^* - 2B^*\right)} + \left(F^* + 2E^*\right)\left\{\frac{\tau^*_{11} N_0 e^{A^* t}}{\left(A^* - 2B^*\right)\left(A - B^*\right)} - \left(M_0 - \frac{\tau^*_{11} N_0}{\left(A^* - B^*\right)}\right)\frac{e^{B^* t}}{B^*}\right\}\right.$$
$$+\tau^*_{11}\left\{\begin{array}{l}\dfrac{\tau^*_{11} N_0 D^*}{A^*}\left(\dfrac{e^{2A^* t}}{2\left(A^* - B^*\right)^2} + \dfrac{e^{A^* t}}{\left(A^* - 2B^*\right)B^*}\right) - \dfrac{\left(E^* - \tau^*_{11}\right) N_0 e^{A^* t}}{\left(A^* - 2B^*\right)B^*} \\ -\left(\dfrac{\tau^*_{11} N_0 D^* - \left(A^* - B^*\right)\left(E^* - \tau^*_{11}\right) N_0}{\left(A^* - B^*\right)B^*}\right)\dfrac{e^{\left(A^* + B^*\right)t}}{\left(A^* - B^*\right)}\end{array}\right\} - \left[\frac{\tau^*_{11} N_0}{\left(A^* - 2B^*\right)} + \left(F^* + 2E^*\right)\right. \tag{5.2}$$
$$\left.\left.\left(\frac{\tau^*_{11} N_0 - \left(A^* - 2B^*\right) M_0}{2\left(A^* - 2B^*\right)\left(A^* - B^*\right)^2 B^*}\right) + \tau^{*2}_{11} N_0\left(\frac{2\left(A^* - B^*\right)\left(\tau^*_{11} - E^*\right) + D^* \tau^*_{11}}{2\left(A^* - 2B^*\right)\left(A^* - B^*\right)^2}\right)\right] e^{2B^* t}\right\}$$

Thus, the objective is to minimize $R^*_{E_2}$.

For having a healthy status of the patient under treatment, maintaining the average number of normal cells to a value greater than the average number of malignant cells, and minimizing the fluctuations in the average number of normal cells is required. Hence, the growth of normal cells should be in a consistent environment. This implies that the programming problem shall have constraints like $R^*_{E_2} < 1$ and $R^*_{E_2} > 1$.

The presentation of the said constraints with derived results in the previous section are

$$\left[N_0\, e^{A^* t}\right] > \left[\frac{\tau^*_{12} N_0 e^{At}}{\left(A^* - B^*\right)} + \left(M_0 - \frac{\tau^*_{11} N_0}{\left(A^* - B^*\right)}\right) e^{B^* t}\right] \tag{5.3}$$

and

$$\left\{\frac{D^* N_0 e^{A^* t}}{A^*}\left(e^{A^* t} - 1\right)\right\} < \left\{\frac{\tau^*_{11} N_0 e^{A^* t}}{\left(A^* - 2B^*\right)} + \left(F^* + 2E^*\right)\left\{\frac{\tau^*_{11} N_0 e^{A^* t}}{\left(A^* - 2B^*\right)\left(A - B^*\right)} - \left(M_0 - \frac{\tau^*_{11} N_0}{\left(A^* - B^*\right)}\right)\frac{e^{B^* t}}{B^*}\right\}\right.$$
$$+\tau^*_{11}\left\{\begin{array}{l}\dfrac{\tau^*_{11} N_0 D^*}{A^*}\left(\dfrac{e^{2A^* t}}{2\left(A^* - B^*\right)^2} + \dfrac{e^{A^* t}}{\left(A^* - 2B^*\right)B^*}\right) - \dfrac{\left(E^* - \tau^*_{11}\right) N_0 e^{A^* t}}{\left(A^* - 2B^*\right)B^*} \\ -\left(\dfrac{\tau^*_{11} N_0 D^* - \left(A^* - B^*\right)\left(E^* - \tau^*_{11}\right) N_0}{\left(A^* - B^*\right)B^*}\right)\dfrac{e^{\left(A^* + B^*\right)t}}{\left(A^* - B^*\right)}\end{array}\right\} - \left[\frac{\tau^*_{11} N_0}{\left(A^* - 2B^*\right)} + \left(F^* + 2E^*\right)\right. \tag{5.4}$$
$$\left.\left.\left(\frac{\tau^*_{11} N_0 - \left(A^* - 2B^*\right) M_0}{2\left(A^* - 2B^*\right)\left(A^* - B^*\right)^2 B^*}\right) + \tau^{*2}_{11} N_0\left(\frac{2\left(A^* - B^*\right)\left(\tau^*_{11} - E^*\right) + D^* \tau^*_{11}}{2\left(A^* - 2B^*\right)\left(A^* - B^*\right)^2}\right)\right] e^{2B^* t}\right\}$$

There are some more additional constraints as below. The average size of normal cells should be within threshold limits. It implies a constraint,

$$L^*_n \le \left[N_0\, e^{A^* t}\right] \le U^*_n \tag{5.5}$$

Average number of malignant cells shall not be more than warning limits. Hence,

$$\left[\frac{\tau^*_{12} N_0 e^{At}}{\left(A^* - B^*\right)} + \left(M_0 - \frac{\tau^*_{11} N_0}{\left(A^* - B^*\right)}\right) e^{B^* t}\right] \le U^*_m \tag{5.6}$$

The consistency in average number of normal cells should be maintained at a certain level. More volatility in growth of normal cells leads to much complication in the functioning of an organ. Hence, the constraint with the above notion is

$$\left\{\frac{D^* N_0 e^{A^* t}}{A^*}\left(e^{A^* t}-1\right)\right\} \leq U_{vn}^* . \tag{5.7}$$

The consistency in the growth of malignant cells results in a threat to healthy functioning. So, higher volatility among malignant cells is always desired. Hence, the constraint with the above notion is

$$\begin{aligned}
&\left\{\frac{\tau_{11}^* N_0 e^{A^* t}}{\left(A^*-2B^*\right)}+\left(F^*+2E^*\right)\left\{\frac{\tau_{11}^* N_0 e^{A^* t}}{\left(A^*-2B^*\right)\left(A-B^*\right)}-\left(M_0-\frac{\tau_{11}^* N_0}{\left(A^*-B^*\right)}\right)\frac{e^{B^* t}}{B^*}\right\}\right. \\
&+\tau_{11}^*\left\{\begin{array}{l}\dfrac{\tau_{11}^* N_0 D^*}{A^*}\left(\dfrac{e^{2A^* t}}{2\left(A^*-B^*\right)^2}+\dfrac{e^{A^* t}}{\left(A^*-2B^*\right)B^*}\right)-\dfrac{\left(E^*-\tau_{11}^*\right)N_0 e^{A^* t}}{\left(A^*-2B^*\right)B^*} \\ -\left(\dfrac{\tau_{11}^* N_0 D^*-\left(A^*-B^*\right)\left(E^*-\tau_{11}^*\right)N_0}{\left(A^*-B^*\right)B^*}\right)\dfrac{e^{\left(A^*+B^*\right)t}}{\left(A^*-B^*\right)}\end{array}\right\} \\
&-\left[\frac{\tau_{11}^* N_0}{\left(A^*-2B^*\right)}+\left(F^*+2E^*\right)\left(\frac{\tau_{11}^* N_0-\left(A^*-2B^*\right)M_0}{2\left(A^*-2B^*\right)\left(A^*-B^*\right)^2 B^*}\right)\right. \\
&\left.\left.+\tau_{11}^{*\,2} N_0\left(\frac{2\left(A^*-B^*\right)\left(\tau_{11}^*-E^*\right)+D^*\tau_{11}^*}{2\left(A^*-2B^*\right)\left(A^*-B^*\right)^2}\right)\right] e^{2B^* t}\right\} \geq L_{vm}^*
\end{aligned} \tag{5.8}$$

Where,

$A^* = \beta_{11}^* - \tau_{11}^* - \delta_{11}^*$ $\quad B^* = \beta_{21}^* - \delta_{21}^* - \tau_{21}^*$ $\quad D^* = \beta_{11}^* + \tau_{11}^* + \delta_{11}^*$

$E^* = \beta_{32}^* - \delta_{32}^* - \tau_{32}^*$ $\quad F^* = \beta_{21}^* + \delta_{21}^* + \tau_{21}^*$ $\quad$ -

$\beta_{11}^* = a_1\beta_{111} + (1-a_1)\beta_{110}$ $\quad \tau_{11}^* = a_2\tau_{111} + (1-a_2)\tau_{110}$ $\quad \beta_{21}^* = a_3\beta_{211} + (1-a_3)\beta_{210}$

$\beta_{32}^* = a_4\beta_{321} + (1-a_4)\beta_{320}$ $\quad \delta_{11}^* = a_5\delta_{111} + (1-a_5)\delta_{110}$ $\quad \delta_{21}^* = a_6\delta_{211} + (1-a_6)\delta_{210}$

$\tau_{21}^* = a_7\tau_{211} + (1-a_7)\tau_{210}$ $\quad \delta_{32}^* = a_8\delta_{321} + (1-a_8)\delta_{320}$ $\quad \tau_{32}^* = a_9\tau_{321} + (1-a_9)\tau_{320}$

N_0 - Initial number of normal cells in an organ

M_0 - Initial number of malignant cells in an organ

L_n^*, U_n^* - Lower and upper threshold limits of the number of normal cells

U_m^* - Warning upper limit on the average number of malignant cells

U_{vn}^* - Upper allowable limit on the volatility of number of normal cells

L_{vm}^* - desired lower limit on the volatility of malignant cells

The non-negative decision parameters under study are $\beta_{11}^* \geq 0$, $\tau_{11}^* \geq 0$, $\delta_{11}^* \geq 0$, $\beta_{21}^* \geq 0$, $\tau_{21}^* \geq 0$, $\delta_{21}^* \geq 0$, $\beta_{32}^* \geq 0$, $\tau_{32}^* \geq 0$, $\delta_{32}^* \geq 0$.

6. Sensitivity Analysis with Optimization Models

This section consists of four sets of programming problems addressing the issues of (i) maximizing $R_{E_1}^*$ during drug vacation period; (ii) maximizing $R_{E_1}^*$ during drug administration period; (iii) minimizing $R_{E_2}^*$ during drug vacation period and (iv) minimizing $R_{E_2}^*$ during drug administration period under the well-defined subjective constraints.

Optimization problem of maximizing $R_{E_1}^*$ During Drug Vacation Period

The illustration has been presented here for aforementioned problems. The indicator function is defined for denoting the drug administration (presence of drug) and vacation periods (absence of drug). The results are obtained by solving the objective function in (5.1) and the set of constraints from (5.3) to (5.8) by considering the indicator variable as drug vacation period (absence of drug).

The optimal solutions pertaining to local maximization are obtained as the formulated programming problem is non-linear. The results are presented in terms of the values of decision parameter and corresponding objective function. While exploring the estimated values of parameters, it is considered for varying values of L_n^*, U_n^*, U_m^*, U_{vn}^*, L_{vm}^*, M_0, N_0, t, one at a time, when the remaining are fixed values of other target limits. From the appendix V(c), it is observed that the objectives function $R_{E_1}^*$ is an increasing function of N_0, L_n^*, U_n^*, and decreasing functions of M_0, t. β_{110} and δ_{110} are increasing functions of N_0. When increasing M_0: $\beta_{110} < \delta_{110}$, $\beta_{210} >$ (δ_{210} & τ_{210}) and $\beta_{210} > (\beta_{110}, \beta_{320})$; β_{110}, δ_{110} are decreasing functions of L_n^*. When increasing L_n^*: $\beta_{210} > (\beta_{110}, \beta_{320})$. When increasing U_n^*: β_{110}, δ_{110} and δ_{210} are decreasing functions of U_n^*. When increasing U_m^* : $\beta_{110} < \delta_{110}$ and invariant, $\beta_{110} < (\beta_{210}, \beta_{320})$. When increasing U_{vn}^*: $\beta_{110} < \delta_{110}$, $\beta_{110} < (\beta_{210}, \beta_{320})$. When increasing U_{vn}^*: $\beta_{110} < \delta_{110}$, $\delta_{210} > (\beta_{210}, \tau_{210})$, $\beta_{110} < (\beta_{210}, \beta_{320})$. When increasing t: $\beta_{110} < \delta_{110}$, $\beta_{210} < (\beta_{110}, \beta_{320})$ and $\beta_{210} < \delta_{210}$.

Optimization problem of Maximizing $R_{E_1}^*$ during drug administration period

An illustration has been presented here for aforementioned problems. The results are obtained by solving objective function in (5.1) with a set of constraints from (5.3) to (5.8) under consideration with the indicator variable in the drug administration period (presence of drug). Since the programming problem handled here is non-linear it has only a local optimum solution. The results of the sensitivity analysis of the problem are given in the tables from Annexure 3(a) & 3(b). The results are presented in terms of the values of the decision parameter and corresponding objective function. While exploring the estimated values of parameters, varying values of L_n^*, U_n^*, U_m^*, U_{vn}^*, L_{vm}^*, M_0, N_0, t one at a time is considered when the remaining are fixed values of other target limits.

From the appendix – V (d), it is observed that the objective function $R_{E_1}^*$ is an increasing function of N_0, U_m^*, U_{vn}^*, are decreasing function of M_0, L_n^* and t. When increasing N_0: $\beta_{111} < \delta_{111}$, $\beta_{211} > (\delta_{211}, \tau_{211})$, $\beta_{211} > (\beta_{111}, \beta_{321})$. When increasing M_0: $\beta_{111} < \delta_{111}$, $\beta_{211} > (\beta_{111}, \beta_{321})$. δ_{111} is a decreasing function of L_n^*. When increasing L_n^*: $\beta_{211} > (\delta_{211}, \tau_{211})$ and $\beta_{211} > \beta_{321}$. When increasing U_n^*: β_{111}, δ_{111} and $\beta_{211} > \delta_{211}$. When increasing U_m^* : $\beta_{111} < \delta_{111}$, $\beta_{211} > (\beta_{111}, \beta_{321})$ and $\beta_{211} > (\beta_{321}, \tau_{211})$. When increasing U_{vn}^*: $\beta_{111} < \delta_{111}$, $(\beta_{111}, \beta_{211}) > \beta_{321}$. When increasing L_{vm}^*: $\beta_{111} < \delta_{111}$, $\beta_{211} > (\beta_{321}, \tau_{211})$ and $\beta_{211} > (\beta_{111}, \beta_{321})$. When increasing t: $\beta_{111} < \delta_{111}$ and $\beta_{211} > (\beta_{111}, \beta_{321})$.

Optimization problem of minimizing $R_{E_2}^*$ during drug vacation period

The illustration given here is for discussing the situation of cancer growth related issues during the drug vacation period. The indicator variable is considered with drug vacation period (absence of drug). The results are obtained by solving objective function in (5.2) with the formulated constraints given from (5.3) to (5.8). This illustration has given local optimal results only as the formulated programming problem is nonlinear. The results are presented in terms of the values of the decision parameter and corresponding objective function. While exploring the estimated values of parameters, varying values of L_n^*, U_n^*, U_m^*, U_{vn}^*, L_{vm}^*, M_0, N_0, t is considered, when the remaining are fixed values of other target limits.

From the appendix –V (e), it is observed that the objective function $R_{E_2}^*$ is an increasing function of M_0, and decreasing function of U_{vn}^*. When increasing N_0: $\beta_{110} < (\delta_{110}, \tau_{110})$, $\beta_{210} > (\beta_{110}, \beta_{320})$ and $\beta_{210} > (\delta_{210}, \tau_{210})$. β_{110}, δ_{110} are decreasing functions of M_0. When increasing M_0 : $\beta_{110} < (\delta_{110}, \tau_{110})$, $(\beta_{110}, \beta_{210}) < \beta_{320}$ and $(\delta_{110}, \delta_{210}) > \delta_{320}$. β_{110} and τ_{210} are decreasing functions of L_n^*. When increasing L_n^* : $\beta_{110} < (\delta_{110}, \tau_{110})$, $(\beta_{110}, \beta_{210}) < \beta_{320}$ and $\tau_{210} > \tau_{320}$. When increasing U_m^* : $\beta_{110} < (\delta_{110}, \tau_{110})$ and $\beta_{210} > (\beta_{110}, \beta_{320})$. When increasing U_{vn}^*: $\beta_{110} < (\delta_{110}, \tau_{110})$, $(\beta_{110}, \beta_{210}) < \beta_{320}$, $(\beta_{210}, \delta_{210}) < \tau_{110}$, $\tau_{210} > \tau_{320}$ and δ_{210}, δ_{320}. When increasing L_{vm}^* : $\beta_{110} < (\delta_{110}, \tau_{110})$, $\beta_{110} < (\beta_{210}, \beta_{320})$, $\beta_{210} > \delta_{210}$. δ_{210} is a decreasing function of t. When increasing t : $\beta_{110} < (\delta_{110}, \tau_{110})$ and $\beta_{110} < (\beta_{210}, \beta_{320})$.

Optimization problem of minimizing $R_{E_2}^*$ during drug administration period

The given illustration here considers the minimization of $R_{E_2}^*$ during the drug administration period. The indicator variable has been functioned with the stipulation of drug presence. The results are obtained by solving the objective function in (5.2) along with developed constraints given in (5.3) through (5.8) Results are presented for local optimality only as the developed problem is nonlinear in nature. The results of the sensitivity analysis of the problem are given in the tables from Annexure 3(c) & 3(d). The results are presented in term of the values of decision parameter and corresponding objective function. While exploring the estimated values of parameters, varying values of L_n^*, U_n^*, U_m^*, U_{vn}^*, L_{vm}^*, M_0, N_0, t, one at a time when the remaining are fixed values of other target limits.

From the appendix V(f), it is observed that the objective function $R_{E_2}^*$ is an increasing function of M_0, and decreasing function of L_n^* and t. When increasing N_0: $\delta_{111} > (\beta_{111}, \tau_{111})$, $(\beta_{111}, \beta_{211}) < \beta_{321}$, $\tau_{211} < \tau_{321}$. δ_{111} is an increasing function of M_0. When increasing M_0: $\delta_{111} > (\beta_{111}, \tau_{111})$, $(\beta_{211}, \delta_{211}) < \tau_{211}$, $\beta_{211} < \beta_{321}$, $(\beta_{321}, \delta_{321}) < \tau_{321}$ and $(\beta_{111}, \beta_{211}) < \beta_{321}$. β_{111}, δ_{111} and τ_{111} are decreasing functions of L_n^*. When increasing L_n^* : $(\beta_{211}, \delta_{211}) < \tau_{211}$, $(\beta_{111}, \beta_{211}) < \beta_{321}$ and $\tau_{211} < \tau_{321}$. δ_{111} is a decreasing function of U_n^*. When increasing U_n^*: $\delta_{111} > (\beta_{111}, \tau_{111})$, $(\beta_{111}, \beta_{211}) < \beta_{321}$, and $\delta_{211} < \delta_{321}$. When increasing U_m^*: $\delta_{111} > (\beta_{111}, \tau_{111})$, $\beta_{111} < \beta_{211}$, $\beta_{211} < \beta_{321}$ and $\tau_{211} > \tau_{321}$. β_{111}, δ_{111} are increasing functions of U_{vn}^* and δ_{211}, τ_{111}, τ_{211} are decreasing functions of U_{vn}^*. When increasing U_{vn}^*: β_{111}, $< (\delta_{111}, \tau_{111})$ and $(\beta_{111}, \beta_{211}) < \beta_{321}$. δ_{111} is a decreasing function of t.

Summary of Optimal Drug Administration

This section deals with a set of non-linear optimization programming problems which are constructed with the derived mathematical relations for different statistical measures and there by defined relative efficiency such as average number of normal cells, average number of malignant cells, variance of number of normal cells and variance of number of malignant cells, etc. The decision parameters involved in the cell division dynamics such as growth rate of normal cells, death rate of normal cells, transformation rate (normal cells to malignant cells), growth rate of malignant cells, death rate of malignant cells, migration rate of malignant cells, growth rate of immigrant malignant cells, death rate of immigrant malignant cells and emigration rate of malignant cells are predicted and presented in a table. The discussions regarding the predicted parameters are also presented. The problems in this chapter are dealt with separately and solved. The scope of this programming can be extended with the goal programming problem approach for dealing the multi objective nonlinear programming problems.

References

Armitage, P. and Doll, R. 1957. A two stage theory of carcinogenesis in relation to the age distribution of human cancer. British. J. Cancer, 11: 161–169.

Armitage, P. and Doll, R. 1961. Stochastic models for carcinogenesis. Proc. 4th. Berkeley symposium on mathematical statistics and probability; 4: 19–38, University of California press. Berkeley, Losangles.

Bahrami, K. and Kim, M. 1975. Optimal control of multiplicative control system arising from cancer therapy. IEEE. Trans. Auton., Control., 20: 534–542.

Birkhead, B. G. and Gregory, W. M. 1984. A Mathematical model of the effect of drug resistance in cancer chemotherapy. Math. Biosci., 72: 59–69.

Birkhead, B. G. 1986. The transient solution of the linear birth death process with random spontaneous mutation. Math. Biosci., 82: 193–200.

Coldman, A. J. and Goldie, J. H. 1983. A model for the resistance of tumor cells to cancer chemotherapeutic agents. Math. Biosci., 65: 291–307.

Iversion, S. and Arley, N. 1950. On the mechanism of experimental carcinogenesis. Acta. Path. Micro. Biol. Scand., 27: 773–803.

Kendall, D. G. 1952. Les. Processus stochastiques de. Ann. Inst. Pioncari., 13: 43–108.

Kendall, D. G. 1960. Birth and death processes and the theory of carcinogenesis. Math. Biosci., 73: 103–107.

Madhavi et al. 2012. Stochastic modeling on stage dependent mutant cell growth. International Journal of Computational Science and Mathematics, 4(3): 187–196.

Madhavi et al. 2013. Optimal drug administration for cancer chemotherapy through stochastic programming. American Journal of Applied Mathematics and Mathematical Sciences, 2(1): 37–45.

Madhavi, K. 2015. Stochastic Modeling and Optimization Methods-Studies on Cancer Growth and Control.

Martin et al. 1990. A mathematical model of cancer chemotherapy with an optimal selection of parameters. Mathematical Biosciences, 99(2): 205–230. doi: 10.1016/0025-5564(90)90005-J.

Mayneord, W. V. 1932. On a law of growth of jensen's rat sarcoma. Am J Cancer, 16: 841–846.

Neyman, J. 1958. A stochastic model of carcinogenesis. Seminar talks at National Institute of Health.

Neyman, J. 1958. A two-step mutation theory of carcinogenesis. Mimeograph, US National Institute of Health.

Neyman, J. and Scott, E. L. 1967. Statistical aspect of the problem of carcinogenesis. Proceedings of Fifth Berkeley Symposium on Mathematical Statistics and Probability, 3: 745–776.

Quinn, D. W. 1997. The method of characteristics applied to a stochastic two-stage model of carcinogenesis. Mathematical and Computer Modelling, 25(7): 1–13. doi: 10.1016/S0895-7177(97)00044-7.

Serio, G. 1984. Two-stage stochastic model for carcinogenesis with time-dependent parameters. Statistics & Probability Letters, 2(March), 95–103.

Tan, W. Y. 1989. A stochastic model for the formation of metastatic foci at distant sites. Mathematical and Computer Modelling, 12(9): 1093–1102. doi:10.1016/0895-7177(89)90230-6.

Tirupathi et al. 2011. Multistage tumor growth through stochastic model with continuous time. Int. J. Enterprise Network Management, 4(4): 367–379.

Tirupathi Rao et al. 2010. Stochastic Models for Optimal Drug Administration in Cancer chemotherapy. International Journal of Engineering Science and Technology, 2(5): 859–865.

Tirupathi Rao et al. 2011. Two stage mutant cell growth through stochastic modeling. Global Journal of Mathematical Sciences: Theory and Practical, 3(2): 101–110.

Tirupathi Rao et al. 2012. Stochastic modeling for treatment dependent malignancy growth. International Journal of Advanced Scientific and Technical Research, 4(2): 607–618.

Tirupathi et al. 2013. Stochastic modeling for tumor growth within organ through bivariate birth, death and migration processes. Journal of International Academic Research for Multidisciplinary, 2(5): 205–217.

Tirupathi et al. 2013. Stochastic Modelling of tumor growth within organ during chemotherapy using bivariate birth, death and migration processes. IOSR Journal of Mathematics, 10(3): 01–08.

Tirupathi Rao et al. 2014. Stochastic modeling of inter & intra stage dependent cancer growth under chemotherapy through Trivariate Poisson Processes. Bulletin of Society for Mathematical Services and Standards, 3(2): 53–70.

Tirupathi Rao et al. 2014. Stochastic optimization models for cancer chemotherapy. British Journal of Applied Science and Technology, 4(28): 4097–4108.

Tirupathi Rao et al. 2014. Trivariate Poisson Processes for Modeling the stage dependent cancer cell growth. International Journal of Innovative Research in Science, Engineering and Technology, 5(5): 13007–13016.

Tirupathi Rao, P. and Srinivasa Rao, K. 2004. Stochastic model for cancer cell growth with spontaneous mutation and proliferation. International Journal of Management and Systems, 20(1): 85–93.

Annexure –IV (a)

Table for all statistical measures with varying values of one parameter when other parameters are fixed drug vacation period.

N_0	M_0	β_{110}	τ_{110}	β_{210}	β_{320}	δ_{110}	δ_{210}	τ_{210}	δ_{320}	τ_{320}	t	$m^*_{1,0}(t)$	$m^*_{0,1}(t)$	$m^*_{2,0}(t)$	$m^*_{0,2}(t)$	$m^*_{1,1}(t)$	r (n, m)
400	500	1.0	0.1	4.0	0.3	0.5	0.5	0.3	0.03	0.01	2.0	890.2164	253641.5	436.3986	7.12E+07	27683.67	0.1571
800	500	1.0	0.1	4.0	0.3	0.5	0.5	0.3	0.03	0.01	2.0	1780.4327	260908.6	872.7973	7.15E+07	55367.33	0.2216
1200	500	1.0	0.1	4.0	0.3	0.5	0.5	0.3	0.03	0.01	2.0	2670.6491	268175.6	1309.196	7.21E+07	83051	0.2704
1600	500	1.0	0.1	4.0	0.3	0.5	0.5	0.3	0.03	0.01	2.0	3560.8655	275442.6	1745.595	7.25E+07	110734.7	0.3113
2000	500	1.0	0.1	4.0	0.3	0.5	0.5	0.3	0.03	0.01	2.0	4451.0819	282709.6	2181.993	7.29E+07	138418.3	0.3470
200	**1000**	1.0	0.1	4.0	0.3	0.5	0.5	0.3	0.03	0.01	2.0	445.1082	496382.5	218.1993	1.42E+08	13841.83	0.0787
200	**1200**	1.0	0.1	4.0	0.3	0.5	0.5	0.3	0.03	0.01	2.0	445.1082	594932.4	218.1993	1.70E+08	13841.83	0.0719
200	**1300**	1.0	0.1	4.0	0.3	0.5	0.5	0.3	0.03	0.01	2.0	445.1082	644207.3	218.1993	1.84E+08	13841.83	0.0690
200	**1400**	1.0	0.1	4.0	0.3	0.5	0.5	0.3	0.03	0.01	2.0	445.1082	693482.2	218.1993	1.98E+08	13841.83	0.0665
200	**1500**	1.0	0.1	4.0	0.3	0.5	0.5	0.3	0.03	0.01	2.0	445.1082	742757.1	218.1993	2.12E+08	13841.83	0.0643
200	500	**1.2**	0.1	4.0	0.3	0.5	0.5	0.3	0.03	0.01	2.0	664.0234	305526.6	462.1836	1.01E+08	20794.15	0.0964
200	500	**1.4**	0.1	4.0	0.3	0.5	0.5	0.3	0.03	0.01	2.0	990.6065	305896.6	978.9749	1.00E+08	31269.26	0.0997
200	500	**1.6**	0.1	4.0	0.3	0.5	0.5	0.3	0.03	0.01	2.0	1477.8112	306326.7	2077.2	1.00E+08	47077.16	0.1031
200	500	**1.8**	0.1	4.0	0.3	0.5	0.5	0.3	0.03	0.01	2.0	2204.6353	306830.7	4419.49	1.00E+08	70978.65	0.1067
200	500	**2.0**	0.1	4.0	0.3	0.5	0.5	0.3	0.03	0.01	2.0	3288.9294	307427	9433.608	9.98E+07	107202.8	0.1105
200	500	1.0	**0.2**	4.0	0.3	0.5	0.5	0.3	0.03	0.01	2.0	364.4238	309198.7	339.5462	1.01E+08	5686.589	0.0307
200	500	1.0	**0.4**	4.0	0.3	0.5	0.5	0.3	0.03	0.01	2.0	244.2806	316422.5	411.0414	1.07E+08	–1976.5	–0.0094
200	500	1.0	**0.6**	4.0	0.3	0.5	0.5	0.3	0.03	0.01	2.0	163.7462	322778	373.995	8.85E+07	–3477.97	–0.0191
200	500	1.0	**0.7**	4.0	0.3	0.5	0.5	0.3	0.03	0.01	2.0	134.0640	325676.8	340.3263	9.25E+07	–3164.79	–0.0178
200	500	1.0	**0.8**	4.0	0.3	0.5	0.5	0.3	0.03	0.01	2.0	109.7623	328410.3	303.744	9.34E+07	–2530.78	–0.0150
200	500	1.0	0.1	**4.1**	0.3	0.5	0.5	0.3	0.03	0.01	2.0	445.1082	372601.9	218.1993	1.48E+08	16381.28	0.0910
200	500	1.0	0.1	**4.2**	0.3	0.5	0.5	0.3	0.03	0.01	2.0	445.1082	454894.5	218.1993	2.19E+08	19404.52	0.0888
200	500	1.0	0.1	**4.3**	0.3	0.5	0.5	0.3	0.03	0.01	2.0	445.1082	555377.3	218.1993	3.23E+08	23005.84	0.0866
200	500	1.0	0.1	**4.4**	0.3	0.5	0.5	0.3	0.03	0.01	2.0	445.1082	678072.9	218.1993	4.77E+08	27298.23	0.0846
200	500	1.0	0.1	**4.5**	0.3	0.5	0.5	0.3	0.03	0.01	2.0	445.1082	827893.8	218.1993	7.04E+08	32417.08	0.0827
200	500	1.0	0.1	4.0	**0.4**	0.5	0.5	0.3	0.03	0.01	2.0	445.1082	305205.5	218.1993	1.04E+08	22199.37	0.1470
200	500	1.0	0.1	4.0	**0.5**	0.5	0.5	0.3	0.03	0.01	2.0	445.1082	305205.5	218.1993	1.08E+08	30556.9	0.1989
200	500	1.0	0.1	4.0	**0.6**	0.5	0.5	0.3	0.03	0.01	2.0	445.1082	305205.5	218.1993	1.12E+08	38914.43	0.2490
200	500	1.0	0.1	4.0	**0.7**	0.5	0.5	0.3	0.03	0.01	2.0	445.1082	305205.5	218.1993	1.16E+08	47271.96	0.2975
200	500	1.0	0.1	4.0	**0.8**	0.5	0.5	0.3	0.03	0.01	2.0	445.1082	305205.5	218.1993	1.19E+08	55629.5	0.3446
200	500	1.0	0.1	4.0	0.3	**0.6**	0.5	0.3	0.03	0.01	2.0	364.4238	305060.6	169.7731	1.01E+08	11343.37	0.0867
200	500	1.0	0.1	4.0	0.3	**0.7**	0.5	0.3	0.03	0.01	2.0	298.3649	304924.9	132.0689	1.01E+08	9295.286	0.0806
200	500	1.0	0.1	4.0	0.3	**0.8**	0.5	0.3	0.03	0.01	2.0	244.2806	304797.5	102.7603	1.01E+08	7616.544	0.0748
200	500	1.0	0.1	4.0	0.3	**1.0**	0.5	0.3	0.03	0.01	2.0	163.7462	304565.1	62.3325	1.01E+08	5113.068	0.0645

N_0	M_0	β_{110}	τ_{110}	β_{210}	β_{320}	δ_{110}	δ_{210}	τ_{210}	δ_{320}	τ_{320}	t	$m^*_{1,0}(t)$	$m^*_{0,1}(t)$	$m^*_{2,0}(t)$	$m^*_{0,2}(t)$	$m^*_{1,1}(t)$	r (n, m)
200	500	1.0	0.1	4.0	0.3	**1.1**	0.5	0.3	0.03	0.01	2.0	134.0640	304458.8	48.618	1.01E+08	4189.035	0.0598
200	500	1.0	0.1	4.0	0.3	0.5	**0.6**	0.3	0.03	0.01	2.0	445.1082	250008	208.1993	7.10E+07	11707.44	0.0963
200	500	1.0	0.1	4.0	0.3	0.5	**0.7**	0.3	0.03	0.01	2.0	445.1082	204800.6	208.1993	5.00E+07	9912.341	0.0971
200	500	1.0	0.1	4.0	0.3	0.5	**0.8**	0.3	0.03	0.01	2.0	445.1082	167774.4	208.1993	3.53E+07	8401.583	0.0980
200	500	1.0	0.1	4.0	0.3	0.5	**0.9**	0.3	0.03	0.01	2.0	445.1082	137448.2	208.1993	2.49E+07	7129.236	0.0990
200	500	1.0	0.1	4.0	0.3	0.5	**1.0**	0.3	0.03	0.01	2.0	445.1082	112609.1	208.1993	1.76E+07	6056.891	0.1000
200	500	1.0	0.1	4.0	0.3	0.5	0.5	**0.4**	0.03	0.01	2.0	445.1082	250008	208.1993	7.10E+07	11707.44	0.0963
200	500	1.0	0.1	4.0	0.3	0.5	0.5	**0.5**	0.03	0.01	2.0	445.1082	204800.6	208.1993	5.00E+07	9912.341	0.0971
200	500	1.0	0.1	4.0	0.3	0.5	0.5	**0.6**	0.03	0.01	2.0	445.1082	167774.4	208.1993	3.53E+07	8401.583	0.0980
200	500	1.0	0.1	4.0	0.3	0.5	0.5	**0.7**	0.03	0.01	2.0	445.1082	137448.2	208.1993	2.49E+07	7129.236	0.0990
200	500	1.0	0.1	4.0	0.3	0.5	0.5	**0.8**	0.03	0.01	2.0	445.1082	112609.1	208.1993	1.76E+07	6056.891	0.1000
200	500	1.0	0.1	4.0	0.3	0.5	0.5	0.3	**0.04**	0.01	2.0	445.1082	250008	218.1993	7.07E+07	13006.08	0.1047
200	500	1.0	0.1	4.0	0.3	0.5	0.5	0.3	**0.06**	0.01	2.0	445.1082	250008	218.1993	7.02E+07	11334.57	0.0916
200	500	1.0	0.1	4.0	0.3	0.5	0.5	0.3	**0.08**	0.01	2.0	445.1082	250008	218.1993	6.97E+07	9663.066	0.0784
200	500	1.0	0.1	4.0	0.3	0.5	0.5	0.3	**0.10**	0.01	2.0	445.1082	250008	218.1993	6.92E+07	7991.56	0.0651
200	500	1.0	0.1	4.0	0.3	0.5	0.5	0.3	**0.12**	0.01	2.0	445.1082	250008	218.1993	6.86E+07	6320.053	0.0516
200	500	1.0	0.1	4.0	0.3	0.5	0.5	0.3	0.03	**0.02**	2.0	445.1082	250008	218.1993	7.07E+07	13006.08	0.1047
200	500	1.0	0.1	4.0	0.3	0.5	0.5	0.3	0.03	**0.03**	2.0	445.1082	250008	218.1993	7.05E+07	12170.33	0.0981
200	500	1.0	0.1	4.0	0.3	0.5	0.5	0.3	0.03	**0.04**	2.0	445.1082	250008	218.1993	7.02E+07	11334.57	0.0916
200	500	1.0	0.1	4.0	0.3	0.5	0.5	0.3	0.03	**0.05**	2.0	445.1082	250008	218.1993	6.99E+07	10498.82	0.0850
200	500	1.0	0.1	4.0	0.3	0.5	0.5	0.3	0.03	**0.06**	2.0	445.1082	250008	218.1993	6.97E+07	9663.066	0.0784
200	500	1.0	0.1	4.0	0.3	0.5	0.5	0.3	0.03	0.01	**3.0**	664.0234	5.55E+06	6.16E+02	3.50E+10	5.08E+05	0.1094
200	500	1.0	0.1	4.0	0.3	0.5	0.5	0.3	0.03	0.01	**3.2**	914.4451	1.03E+07	7.47E+02	1.21E+11	1.04E+06	0.1094
200	500	1.0	0.1	4.0	0.3	0.5	0.5	0.3	0.03	0.01	**3.4**	719.3280	1.92E+07	9.03E+02	4.18E+11	2.14E+06	0.1101
200	500	1.0	0.1	4.0	0.3	0.5	0.5	0.3	0.03	0.01	**3.6**	779.2387	3.57E+07	1.09E+03	1.44E+12	4.40E+06	0.1111
200	500	1.0	0.1	4.0	0.3	0.5	0.5	0.3	0.03	0.01	**3.8**	844.1392	6.63E+07	1.31E+03	4.99E+12	9.04E+06	0.1118

Annexure –IV (b)

Table for all statistical measures with varying values of one parameter when other parameters are fixed, during drug administration period

N_0	M_0	β_{111}	τ_{111}	β_{211}	β_{321}	δ_{111}	δ_{211}	τ_{211}	δ_{321}	τ_{321}	t	$m^*_{1,0}(t)$	$m^*_{0,1}(t)$	$m^*_{2,0}(t)$	$m^*_{0,2}(t)$	$m^*_{1,1}(t)$	r (n, m)
400	500	2.0	0.10	1.0	0.100	0.1	1.0	0.01	0.10	0.001	2.0	17526.42	580.28	8377.613	1302.274	–341.512	–0.10339
800	500	2.0	0.10	1.0	0.100	0.1	1.0	0.01	0.10	0.001	1.0	35052.83	670.46	16755.23	1292.301	–683.024	–0.14678
1200	500	2.0	0.10	1.0	0.100	0.1	1.0	0.01	0.10	0.001	1.0	52579.25	760.641	25132.84	1282.328	–1024.54	–0.18047
1600	500	2.0	0.10	1.0	0.100	0.1	1.0	0.01	0.10	0.001	1.0	70105.67	850.822	33510.45	1272.355	–1366.05	–0.2092
2000	500	2.0	0.10	1.0	0.100	0.1	1.0	0.01	0.10	0.001	1.0	87632.08	941.003	41888.07	1262.383	–1707.56	–0.23482
200	**600**	2.0	0.10	1.0	0.100	0.1	1.0	0.01	0.10	0.001	1.0	8763.208	633.2096	4188.807	1569.71	–170.756	–0.06659
200	**1000**	2.0	0.10	1.0	0.100	0.1	1.0	0.01	0.10	0.001	1.0	8763.208	1025.289	4188.807	2619.507	–170.756	–0.05155
200	**1200**	2.0	0.10	1.0	0.100	0.1	1.0	0.01	0.10	0.001	1.0	8763.208	1221.329	4188.807	3144.406	–170.756	–0.04705
200	**1600**	2.0	0.10	1.0	0.100	0.1	1.0	0.01	0.10	0.001	1.0	8763.208	1613.408	4188.807	4194.203	–170.756	–0.04074
200	**2000**	2.0	0.10	1.0	0.100	0.1	1.0	0.01	0.10	0.001	1.0	8763.208	2005.807	4188.807	5244.001	–170.756	–0.03643
200	500	**2.1**	0.10	1.0	0.100	0.1	1.0	0.01	0.10	0.001	1.0	10703.41	542.6362	6242.541	1301.313	–204.265	–0.07167
200	500	**2.3**	0.10	1.0	0.100	0.1	1.0	0.01	0.10	0.001	1.0	15967.61	561.7883	13853.15	1282.911	–288.293	–0.06838
200	500	**2.5**	0.10	1.0	0.100	0.1	1.0	0.01	0.10	0.001	1.0	23820.87	588.5361	30723.18	1251.576	–395.97	–0.06386
200	500	**2.7**	0.10	1.0	0.100	0.1	1.0	0.01	0.10	0.001	1.0	35536.56	626.0244	68120.26	1199.731	–519.642	–0.05748
200	500	**2.9**	0.10	1.0	0.100	0.1	1.0	0.01	0.10	0.001	1.0	53014.32	678.7361	151034.9	1115.636	–626.604	–0.04827
200	500	2.0	**0.02**	1.0	0.100	0.1	1.0	0.01	0.10	0.001	1.0	8589.685	578.921	8126.451	1217.76	–278.82	–0.08863
200	500	2.0	**0.03**	1.0	0.100	0.1	1.0	0.01	0.10	0.001	1.0	8419.598	621.3263	11824.18	1047.366	–345.146	–0.09808
200	500	2.0	**0.04**	1.0	0.100	0.1	1.0	0.01	0.10	0.001	1.0	8252.879	662.4381	15292.82	799.5986	–372.817	–0.10661
200	500	2.0	**0.05**	1.0	0.100	0.1	1.0	0.01	0.10	0.001	1.0	8089.461	702.2881	18542.73	477.3662	–364.757	–0.1226
200	500	2.0	**0.06**	1.0	0.100	0.1	1.0	0.01	0.10	0.001	1.0	7929.279	740.9071	21583.9	85.5232	–323.794	–0.23832
200	500	2.0	0.10	**0.8**	0.100	0.1	1.0	0.01	0.10	0.001	1.0	8763.208	369.6272	4188.807	848.6954	–138.915	–0.07368
200	500	2.0	0.10	**1.2**	0.100	0.1	1.0	0.01	0.10	0.001	1.0	8763.208	780.9702	4188.807	2253.43	–212.544	–0.06918
200	500	2.0	0.10	**1.4**	0.100	0.1	1.0	0.01	0.10	0.001	1.0	8763.208	1146.249	4188.807	3916.095	–267.83	–0.06613
200	500	2.0	0.10	**1.6**	0.100	0.1	1.0	0.01	0.10	0.001	1.0	8763.208	1689.59	4188.807	7081.028	–341.537	–0.06271
200	500	2.0	0.10	**1.8**	0.100	0.1	1.0	0.01	0.10	0.001	1.0	8763.208	2498.316	4188.807	13076.97	–440.502	–0.05952
200	500	2.0	0.10	1.0	**0.006**	0.1	1.0	0.01	0.10	0.001	1.0	8763.208	535.1897	4188.807	785.877	–1801.87	–0.99312
200	500	2.0	0.10	1.0	**0.008**	0.1	1.0	0.01	0.10	0.001	1.0	8763.208	535.1897	4188.807	796.9703	–1767.17	–0.96719
200	500	2.0	0.10	1.0	**0.010**	0.1	1.0	0.01	0.10	0.001	1.0	8763.208	535.1897	4188.807	808.0635	–1732.46	–0.94167
200	500	2.0	0.10	1.0	**0.012**	0.1	1.0	0.01	0.10	0.001	1.0	8763.208	535.1897	4188.807	819.1568	–1697.76	–0.91653
200	500	2.0	0.10	1.0	**0.014**	0.1	1.0	0.01	0.10	0.001	1.0	8763.208	535.1897	4188.807	830.2501	–1663.55	–0.89205
200	500	2.0	0.10	1.0	0.100	**0.2**	1.0	0.01	0.10	0.001	1.0	7174.708	528.8697	3089.154	1311.606	–140.858	–0.06998
200	500	2.0	0.10	1.0	0.100	**0.4**	1.0	0.01	0.10	0.001	1.0	4809.351	518.9325	1680.027	1316.927	–95.6889	–0.06433
200	500	2.0	0.10	1.0	0.100	**0.6**	1.0	0.01	0.10	0.001	1.0	3223.804	511.7262	915.2043	1319.26	–64.8721	–0.05904
200	500	2.0	0.10	1.0	0.100	**0.8**	1.0	0.01	0.10	0.001	1.0	2160.981	506.4738	500.3265	1319.802	–43.9019	–0.05403
200	500	2.0	0.10	1.0	0.100	**1.0**	1.0	0.01	0.10	0.001	1.0	1448.549	502.6244	274.9412	1319.265	–29.6655	–0.04926

N_0	M_0	β_{111}	τ_{111}	β_{211}	β_{321}	δ_{111}	δ_{211}	τ_{211}	δ_{321}	τ_{321}	t	$m^*_{1,0}(t)$	$m^*_{0,1}(t)$	$m^*_{2,0}(t)$	$m^*_{0,2}(t)$	$m^*_{1,1}(t)$	r (n, m)
200	500	2.0	0.10	1.0	0.100	0.1	**1.1**	0.01	0.10	0.001	1.0	8763.208	444.2729	4188.807	1170.627	–153.773	–0.06944
200	500	2.0	0.10	1.0	0.100	0.1	**1.2**	0.01	0.10	0.001	1.0	8763.208	369.6272	4188.807	1028.855	–138.915	–0.06692
200	500	2.0	0.10	1.0	0.100	0.1	**1.3**	0.01	0.10	0.001	1.0	8763.208	308.3159	4188.807	913.0742	–125.89	–0.06437
200	500	2.0	0.10	1.0	0.100	0.1	**1.4**	0.01	0.10	0.001	1.0	8763.208	257.9338	4188.807	818.5052	–114.445	–0.06181
200	500	2.0	0.10	1.0	0.100	0.1	**1.5**	0.01	0.10	0.001	1.0	8763.208	216.5103	4188.807	740.8735	–104.367	–0.05924
200	500	2.0	0.10	1.0	0.100	0.1	1.0	**0.30**	0.10	0.001	1.0	8763.208	313.9193	4188.807	923.6197	–127.118	–0.06463
200	500	2.0	0.10	1.0	0.100	0.1	1.0	**0.40**	0.10	0.001	1.0	8763.208	262.5393	4188.807	827.1362	–115.525	–0.06206
200	500	2.0	0.10	1.0	0.100	0.1	1.0	**0.50**	0.10	0.001	1.0	8763.208	220.298	4188.807	747.9785	–105.319	–0.0595
200	500	2.0	0.10	1.0	0.100	0.1	1.0	**0.60**	0.10	0.001	1.0	8763.208	185.5488	4188.807	682.6326	–96.3117	–0.05696
200	500	2.0	0.10	1.0	0.100	0.1	1.0	**0.70**	0.10	0.001	1.0	8763.208	156.9428	4188.807	628.3201	–88.3447	–0.05446
200	500	2.0	0.10	1.0	0.100	0.1	1.0	0.01	**0.02**	0.001	1.0	8763.208	535.1897	4188.807	1750.991	1217.429	0.449528
200	500	2.0	0.10	1.0	0.100	0.1	1.0	0.01	**0.04**	0.001	1.0	8763.208	535.1897	4188.807	1640.058	870.3828	0.332074
200	500	2.0	0.10	1.0	0.100	0.1	1.0	0.01	**0.06**	0.001	1.0	8763.208	535.1897	4188.807	1529.126	523.3347	0.206782
200	500	2.0	0.10	1.0	0.100	0.1	1.0	0.01	**0.08**	0.001	1.0	8763.208	535.1897	4188.807	1418.193	176.2902	0.072329
200	500	2.0	0.10	1.0	0.100	0.1	1.0	0.01	**0.12**	0.001	1.0	8763.208	535.1897	4188.807	1196.328	–517.802	–0.23131
200	500	2.0	0.10	1.0	0.100	0.1	1.0	0.01	0.10	**0.002**	1.0	8763.208	535.1897	4188.807	1301.714	–188.109	–0.08056
200	500	2.0	0.10	1.0	0.100	0.1	1.0	0.01	0.10	**0.004**	1.0	8763.208	535.1897	4188.807	1290.621	–222.813	–0.09583
200	500	2.0	0.10	1.0	0.100	0.1	1.0	0.01	0.10	**0.006**	1.0	8763.208	535.1897	4188.807	1279.527	–257.518	–0.11123
200	500	2.0	0.10	1.0	0.100	0.1	1.0	0.01	0.10	**0.008**	1.0	8763.208	535.1897	4188.807	1268.434	–292.222	–0.12678
200	500	2.0	0.10	1.0	0.100	0.1	1.0	0.01	0.10	**0.01**	1.0	8763.208	535.1897	4188.807	1257.341	–326.927	–0.14246
200	500	2.0	0.10	1.0	0.100	0.1	1.0	0.01	0.10	0.001	**2.5**	22546.1	605.292	28123.14	1623.921	–470.497	–0.06962
200	500	2.0	0.10	1.0	0.100	0.1	1.0	0.01	0.10	0.001	**3.0**	58006.91	789.5008	187176	1932.266	–919.701	–0.04836
200	500	2.0	0.10	1.0	0.100	0.1	1.0	0.01	0.10	0.001	**3.5**	149240.9	1267.265	1241607	2222.675	831.3823	0.015826
200	500	2.0	0.10	1.0	0.100	0.1	1.0	0.01	0.10	0.001	**4.0**	383969.1	2500.273	8225398	2467.901	26563.17	0.186439
200	500	2.0	0.10	1.0	0.100	0.1	1.0	0.01	0.10	0.001	**4.5**	987880.8	5676.365	54464291	2591.531	238355.3	0.63444

Appendix-3(a)

Table: Values of R_{E1}*, β_{11}, τ_{11}, δ_{11}, β_{21}, δ_{21}, τ_{21}, β_{32}, δ_{32}, τ_{32} for Varying values of one value of the following N_0, M_0, L_n, U_n, U_m, U_{vn}, L_{vm}, t when other parameters are constants (Drug Administration Period)

N_0	M_0	L_n	U_n	U_m	U_{vn}	L_{nm}	t	R_{E1}^*	β_{11}	β_{21}	δ_{11}	τ_{11}	δ_{21}	τ_{21}	β_{32}	δ_{32}	τ_{32}
900000	15000	15000	100000	20000	20000	2000	5.0	32.1683	0.0000	0.0000	1.0236	1.21E+06	6.09E+05	6.05E+05	6.09E+05	0.0000	0.0000
1000000	15000	15000	100000	20000	20000	2000	5.0	37.3954	0.0000	0.0000	1.0499	1.00E+06	5.09E+05	4.98E+05	5.09E+05	0.0000	0.0000
1100000	15000	15000	100000	20000	20000	2000	5.0	42.6861	0.2222	0.0000	1.2626	9.87E+06	4.98E+06	4.98E+06	5.00E+06	0.0000	0.0000
1500000	15000	15000	100000	20000	20000	2000	5.0	64.3578	0.0000	0.0000	1.8967	1.27E+06	6.43E+05	6.43E+05	6.43E+05	0.0000	0.0000
1600000	15000	15000	100000	20000	20000	2000	5.0	69.8791	0.2376	0.0000	2.1168	0.00E+00	7.38E-01	3.89E-01	2.12E+07	0.0000	0.0000
1000000	**10000**	15000	100000	20000	20000	2000	5.0	56.0931	0.2184	0.0000	1.2356	5.29E+06	2.71E+06	2.71E+06	2.70E+06	0.0000	0.0000
1000000	**12000**	15000	100000	20000	20000	2000	5.0	46.7442	0.2184	0.0000	1.2356	1.18E+06	5.98E+05	5.98E+05	5.97E+05	0.0000	0.0000
1000000	**13000**	15000	100000	20000	20000	2000	5.0	43.1485	0.0000	0.0000	1.0499	5.25E+06	5.24E+06	8.40E-01	1.44E+06	0.0000	0.0000
1000000	**16000**	15000	100000	20000	20000	2000	5.0	35.0582	0.2892	0.0000	1.2356	3.64E+06	2.10E+01	3.60E+06	3.64E+06	0.0000	0.0000
1000000	**18000**	15000	100000	20000	20000	2000	5.0	31.1628	0.2184	0.0000	1.2356	5.27E+06	2.66E+06	2.66E+06	2.67E+06	0.0000	0.0000
1000000	15000	**9000**	100000	20000	20000	2000	5.0	44.1474	0.0000	0.0000	1.1776	4.17E+06	2.10E+06	2.09E+06	2.10E+06	0.0000	0.0000
1000000	15000	**9500**	100000	20000	20000	2000	5.0	43.5187	0.0000	0.0000	1.1641	1.57E+06	8.00E+05	7.83E+05	8.00E+05	0.0000	0.0000
1000000	15000	**10000**	100000	20000	20000	2000	5.0	42.9052	0.0000	0.0000	1.1513	2.69E+06	1.35E+06	1.34E+06	1.35E+06	0.0000	0.0000
1000000	15000	**10500**	100000	20000	20000	2000	5.0	42.3057	0.0000	0.0000	1.1391	2.44E+06	1.24E+06	1.22E+06	1.24E+06	0.0000	0.0000
1000000	15000	**11000**	100000	20000	20000	2000	5.0	41.7191	0.0000	0.0000	1.1275	2.60E+06	1.31E+06	1.30E+06	1.31E+06	0.0000	0.0000
1000000	15000	15000	**19900**	20000	20000	2000	5.0	37.3954	0.1878	0.0000	1.2356	3.62E+06	2.02E-01	3.56E+06	3.62E+06	0.0000	0.0000
1000000	15000	15000	**20000**	20000	20000	2000	5.0	37.3954	0.1874	0.0000	1.2356	9.53E+06	4.82E+06	4.76E+06	4.88E+06	0.0000	0.0000
1000000	15000	15000	**20100**	20000	20000	2000	5.0	37.3954	0.1870	0.0000	1.2356	9.18E+05	4.62E+05	4.62E+05	4.62E+05	0.0000	0.0000
1000000	15000	15000	**20200**	20000	20000	2000	5.0	37.3954	0.1865	0.0000	1.2356	9.10E+05	4.64E+05	4.54E+05	4.64E+05	0.0000	0.0000
1000000	15000	15000	**20400**	20000	20000	2000	5.0	37.3954	0.1857	0.0000	1.2356	2.30E+06	1.16E+06	1.15E+06	1.16E+06	0.0000	0.0000
1000000	15000	15000	25000	**20100**	20000	2000	5.0	37.3954	0.0000	0.0000	1.0499	1.00E+06	5.09E+05	4.98E+05	5.09E+05	0.0000	0.0000
1000000	15000	15000	25000	**20800**	20000	2000	5.0	37.3954	0.2184	0.0000	1.2356	1.45E+06	7.34E+05	7.34E+05	7.34E+05	0.0000	0.0000
1000000	15000	15000	25000	**23000**	20000	2000	5.0	37.3954	0.2184	0.0000	1.2356	1.36E+06	6.93E+05	6.81E+05	6.93E+05	0.0000	0.0000
1000000	15000	15000	25000	**26000**	20000	2000	5.0	37.3954	0.0000	0.0000	1.0499	3.33E+06	1.70E+06	1.66E+06	1.70E+06	0.0000	0.0000
1000000	15000	15000	25000	**27000**	20000	2000	5.0	37.3954	0.0000	0.0000	1.0499	3.33E+06	1.70E+06	1.66E+06	1.70E+06	0.0000	0.0000
1000000	15000	15000	25000	20000	**16000**	2000	5.0	37.3954	0.0238	0.0000	1.0855	1.09E+06	1.09E+06	1.54E+00	1.09E+06	0.0000	0.0000
1000000	15000	15000	25000	20000	**17000**	2000	5.0	37.3954	0.0791	0.0000	1.1436	4.68E+06	6.70E+05	4.03E+06	5.26E+07	0.0000	0.0000
1000000	15000	15000	25000	20000	**22000**	2000	5.0	37.3954	0.0000	0.0000	1.0499	6.51E+06	3.30E+06	3.30E+06	3.29E+06	0.0000	0.0000
1000000	15000	15000	25000	20000	**25000**	2000	5.0	37.3954	0.3707	0.0000	1.4132	0.00E+00	9.52E-01	4.17E-02	5.60E+06	0.0000	0.0000
1000000	15000	15000	25000	20000	**26000**	2000	5.0	37.3954	0.4046	0.0000	1.4488	1.52E+06	1.52E+06	3.86E+00	1.53E+06	0.0000	0.0000
1000000	15000	15000	25000	20000	20000	**1700**	5.0	37.3954	0.0000	0.0000	1.0499	2.47E+06	2.24E-01	2.39E+06	8.66E+06	0.0000	0.0000
1000000	15000	15000	25000	20000	20000	**1800**	5.0	37.3954	0.1857	0.0000	1.2356	8.35E+05	4.23E+05	4.23E+05	4.23E+05	0.0000	0.0000
1000000	15000	15000	25000	20000	20000	**2100**	5.0	37.3954	0.2184	0.0000	1.2356	1.39E+06	7.07E+05	7.07E+05	7.07E+05	0.0000	0.0000
1000000	15000	15000	25000	20000	20000	**2200**	5.0	37.3954	0.0000	0.0000	1.0499	4.16E+06	2.14E+06	2.14E+06	2.13E+06	0.0000	0.0000

N_0	M_0	L_n	U_n	U_m	U_{vn}	L_{nm}	t	R_{E1}^*	β_{11}	β_{21}	δ_{11}	τ_{11}	δ_{21}	τ_{21}	β_{32}	δ_{32}	τ_{32}
1000000	15000	15000	25000	20000	20000	**2400**	5.0	37.3954	0.0000	0.0000	1.0499	5.43E+06	6.66E+04	5.37E+06	1.40E+06	0.0000	0.0000
1000000	15000	15000	25000	20000	20000	2000	**3.0**	49.2802	0.0274	0.0000	1.7772	1.09E+06	1.09E+06	3.48E-01	1.10E+06	0.0000	0.0000
1000000	15000	15000	25000	20000	20000	2000	**3.4**	46.9359	0.2730	0.0000	1.8170	4.95E+05	8.42E-01	4.95E+05	1.96E+06	0.0000	0.0000
1000000	15000	15000	25000	20000	20000	2000	**5.4**	34.9449	0.0000	0.0000	0.9722	3.46E+06	3.43E+06	8.91E-02	3.51E+06	0.0000	0.0000
1000000	15000	15000	25000	20000	20000	2000	**5.6**	33.7047	0.0000	0.0000	0.9374	3.23E+06	3.23E+06	9.70E-02	8.14E+05	0.0000	0.0000
1000000	15000	15000	25000	20000	20000	2000	**6.4**	28.604	0.1706	0.0000	0.9666	1.30E+06	1.30E+06	1.05E-02	1.30E+06	0.0000	0.0000

Appendix-3(b)

Table : Values of R_{EI}*, β_{11}, τ_{11}, δ_{11}, β_{21}, δ_{21}, τ_{21}, β_{32}, δ_{32}, τ_{32} for Varying values of one value of the following N_0, M_0, L_n, U_n, U_m, U_{vn}, L_{vm}, t when other parameters are constants (Vacation Period)

N_0	M_0	L_n	U_n	U_m	U_{vn}	L_{nm}	T	R_{EI}*	β_{11}	β_{21}	δ_{11}	τ_{11}	δ_{21}	τ_{21}	β_{32}	δ_{32}	τ_{32}
700000	25000	25000	200000	40000	40000	4000	2.0	36.9840	0.3483	0.0000	2.2698	9.98E+05	9.98E+05	1.36E-01	2.54E+05	0.0000	0.0000
750000	25000	25000	200000	40000	40000	4000	2.0	39.9791	0.0616	0.0000	2.0176	1.05E+06	5.23E+05	5.23E+05	5.23E+05	0.0000	0.0000
800000	25000	25000	200000	40000	40000	4000	2.0	42.9853	0.3567	0.0000	2.3450	8.33E+05	2.43E-01	8.33E+05	8.33E+05	0.0000	0.0000
850000	25000	25000	200000	40000	40000	4000	2.0	46.0015	0.3606	0.0000	2.3792	8.24E+05	8.24E+05	1.79E-01	8.24E+05	0.0000	0.0000
900000	25000	25000	200000	40000	40000	4000	2.0	49.0269	0.3643	0.0000	2.4115	1.18E+06	1.18E+06	3.09E-01	1.18E+06	0.0000	0.0000
1000000	**5000**	25000	200000	40000	40000	4000	2.0	165.3008	0.3713	0.0000	2.4712	4.52E+04	2.26E+04	2.26E+04	2.26E+04	0.0000	0.0000
1000000	**6000**	25000	200000	40000	40000	4000	2.0	137.7554	0.3713	0.0000	2.4712	3.40E+06	6.29E-01	3.40E+06	3.41E+06	0.0000	0.0000
1000000	**9000**	25000	200000	40000	40000	4000	2.0	91.8370	0.0575	0.0000	2.1574	2.64E+06	1.32E+06	1.32E+06	1.32E+06	0.0000	0.0000
1000000	**10000**	25000	200000	40000	40000	4000	2.0	82.6533	0.0546	0.0000	2.1545	2.45E+06	1.23E+06	1.23E+06	1.22E+06	0.0000	0.0000
1000000	**11000**	25000	200000	40000	40000	4000	2.0	75.1393	0.0575	0.0000	2.1574	2.50E+06	1.25E+06	1.25E+06	1.25E+06	0.0000	0.0000
1000000	25000	**8000**	200000	40000	40000	4000	2.0	58.2275	1.8350	0.0000	4.2491	0.00E+00	2.34E+00	6.19E-03	2.72E+07	0.0000	0.0000
1000000	25000	**9000**	200000	40000	40000	4000	2.0	57.7146	1.4631	0.0000	3.8184	6.45E+05	6.45E+05	1.19E-01	2.74E+07	0.0000	0.0000
1000000	25000	**10000**	200000	40000	40000	4000	2.0	57.2294	1.1746	0.0000	3.4771	4.22E+04	4.80E+03	3.74E+04	2.82E+07	0.0000	0.0000
1000000	25000	**11000**	200000	40000	40000	4000	2.0	56.7678	0.9453	0.0000	3.2002	0.00E+00	2.17E+00	6.19E-03	6.36E+07	0.0000	0.0000
1000000	25000	**12000**	200000	40000	40000	4000	2.0	56.3265	0.7595	0.0000	2.9710	9.18E+04	7.06E+03	8.47E+04	2.51E+07	0.0000	0.0000
1000000	25000	**13000**	200000	40000	40000	4000	2.0	55.9031	0.0000	0.0000	2.1714	5.57E+05	5.57E+05	1.19E-01	2.26E+07	0.0000	0.0000
1000000	25000	25000	**25000**	40000	40000	4000	2.0	11.4382	0.0231	0.0000	0.7163	0.00E+00	3.95E-01	5.68E-01	0.0000	73075	73071.9
1000000	25000	25000	**26000**	40000	40000	4000	2.0	11.5345	0.0133	0.0000	0.6868	0.00E+00	3.78E-01	5.70E-01	0.0000	73196.3	73195.2
1000000	25000	25000	**26500**	40000	40000	4000	2.0	11.5820	0.0089	0.0000	0.6729	0.00E+00	3.56E-01	5.84E-01	0.0000	65206.9	65207.9
1000000	25000	25000	**27000**	40000	40000	4000	2.0	11.6291	0.0048	0.0000	0.6595	0.00E+00	3.52E-01	5.81E-01	0.0000	71801.8	71801.2
1000000	25000	25000	**27500**	40000	40000	4000	2.0	11.6757	0.0010	0.0000	0.6465	0.00E+00	3.52E-01	5.74E-01	0.0000	71519.7	71519.1
1000000	25000	25000	200000	**15000**	40000	4000	2.0	55.1022	0.3713	0.0000	2.4712	3.49E+06	1.74E+06	1.74E+06	1.74E+06	0.0000	0.0000
1000000	25000	25000	200000	**14300**	40000	4000	2.0	55.1022	0.3713	0.0000	2.4712	1.26E+06	6.50E+05	6.11E+05	6.10E+05	0.0000	0.0000
1000000	25000	25000	200000	**14900**	40000	4000	2.0	55.1022	0.3713	0.0000	2.4712	1.07E+06	1.07E+06	2.20E-01	1.07E+06	0.0000	0.0000
1000000	25000	25000	200000	**15100**	40000	4000	2.0	55.1022	0.3713	0.0000	2.4712	6.13E+06	3.31E-03	6.13E+06	6.15E+06	0.0000	0.0000
1000000	25000	25000	200000	40000	**27000**	4000	2.0	55.1022	0.0616	0.0000	2.1614	1.28E+06	6.40E+05	6.39E+05	6.40E+05	0.0000	0.0000
1000000	25000	25000	200000	40000	**28000**	4000	2.0	55.1022	0.9398	0.0000	3.0396	8.84E+05	4.42E+05	4.42E+05	4.42E+05	0.0000	0.0000
1000000	25000	25000	200000	40000	**29000**	4000	2.0	55.1022	1.0108	0.0000	3.1107	8.84E+05	4.42E+05	4.42E+05	4.42E+05	0.0000	0.0000
1000000	25000	25000	200000	40000	**30000**	4000	2.0	55.1022	0.1427	0.0000	2.2426	5.92E+04	2.96E+04	2.96E+04	1.69E+06	0.0000	0.0000
1000000	25000	25000	200000	40000	**31000**	4000	2.0	55.1022	1.1530	0.0000	3.2528	1.30E+06	1.30E+06	3.62E-01	1.30E+06	0.0000	0.0000
1000000	25000	25000	200000	40000	40000	**2200**	2.0	55.1022	0.3713	0.0000	2.4712	6.52E+06	6.52E+06	1.19E-01	6.32E+06	0.0000	0.0000
1000000	25000	25000	200000	40000	40000	**2300**	2.0	55.1022	0.0000	0.0000	2.0999	2.87E+04	2.87E+04	1.19E-01	1.11E+07	0.0000	0.0000
1000000	25000	25000	200000	40000	40000	**2400**	2.0	55.1022	0.0000	0.0000	2.0999	1.15E+06	1.15E+06	1.23E-01	1.16E+06	0.0000	0.0000
1000000	25000	25000	200000	40000	40000	**2500**	2.0	55.1022	0.3713	0.0000	2.4712	1.03E+06	1.03E+06	1.23E-01	1.02E+06	0.0000	0.0000
1000000	25000	25000	200000	40000	40000	**2800**	2.0	55.1022	0.3713	0.0000	2.4712	2.81E+07	2.81E+07	1.19E-01	7.45E+06	0.0000	0.0000

N_0	M_0	$\mathbf{L_n}$	$\mathbf{U_n}$	$\mathbf{U_m}$	$\mathbf{U_{vn}}$	$\mathbf{L_{nm}}$	T	R_{EI}*	β_{11}	β_{21}	δ_{11}	τ_{11}	δ_{21}	τ_{21}	β_{32}	δ_{32}	τ_{32}
1000000	25000	25000	200000	40000	40000	4000	**2.3**	53.3602	0.3229	0.0000	2.1488	1.07E+06	2.13E-01	1.07E+06	1.07E+06	0.0000	0.0000
1000000	25000	25000	200000	40000	40000	4000	**2.5**	52.1969	0.2970	0.0000	1.9769	1.24E+06	6.19E+05	6.19E+05	6.19E+05	0.0000	0.0000
1000000	25000	25000	200000	40000	40000	4000	**2.8**	50.4484	0.0652	0.0000	1.5651	7.69E+05	6.11E+05	1.58E+05	6.11E+05	0.0000	0.0000
1000000	25000	25000	200000	40000	40000	4000	**3.0**	49.2802	0.0723	0.0000	1.4722	8.43E+05	4.22E+05	4.22E+05	4.22E+05	0.0000	0.0000
1000000	25000	25000	200000	40000	40000	4000	**3.8**	44.5790	0.0778	0.0000	1.1830	1.17E+06	5.83E+05	5.83E+05	5.83E+05	0.0000	0.0000

Appendix-3(c)

Table : Values of R_{E2}*, β_{11}, τ_{11}, δ_{11}, β_{21},δ_{21}, τ_{21}, β_{32}, δ_{32}, τ_{32} for Varying values of one value of the following N_0, M_0, L_n, U_n, U_m,U_{vn}, L_{vm}, t when other parameters are constants (Drug Administration Period)

N_0	M_0	L_n	U_n	U_m	U_{vn}	L_{nm}	t	R_{E2}*	β_{11}	β_{21}	δ_{11}	τ_{11}	δ_{21}	τ_{21}	β_{32}	δ_{32}	τ_{32}
400000	10000	15000	100000	20000	20000	2000	3.0	1.14E-12	0.2108	0.2493	1.0560	1.3344	1.3372	1.0916	1.3845	1.4105	1.4352
500000	10000	15000	100000	20000	20000	2000	3.0	5.67E-13	0.2189	0.3890	0.9987	1.3075	1.3358	1.1405	1.4852	1.3099	1.3346
800000	10000	15000	100000	20000	20000	2000	3.0	2.61E-13	0.0847	0.3653	0.9833	2.1476	1.3358	2.0758	1.6461	1.1490	1.1247
900000	10000	15000	100000	20000	20000	2000	3.0	1.59E-12	0.2100	0.4198	1.1433	1.3118	1.3358	1.3292	1.3855	1.4095	1.4342
1000000	10000	15000	100000	20000	20000	2000	3.0	1.80E-12	0.1999	0.4129	1.1703	1.3030	1.3358	1.3505	1.3856	1.4094	1.4341
1000000	**11000**	15000	100000	20000	20000	2000	3.0	1.83E-12	0.1997	0.4125	1.1705	1.3030	1.3358	1.3504	1.3856	1.4094	1.4341
1000000	**12000**	15000	100000	20000	20000	2000	3.0	1.84E-12	0.1995	0.4120	1.1707	1.3030	1.3358	1.3504	1.3856	1.4094	1.4341
1000000	**13000**	15000	100000	20000	20000	2000	3.0	1.85E-12	0.1994	0.4116	1.1709	1.3030	1.3358	1.3503	1.3856	1.4094	1.4341
1000000	**14000**	15000	100000	20000	20000	2000	3.0	1.88E-12	0.1992	0.4112	1.1711	1.3030	1.3358	1.3503	1.3856	1.4094	1.4341
1000000	**16000**	15000	100000	20000	20000	2000	3.0	1.89E-12	0.2115	0.4218	1.1772	1.3026	1.3358	1.3543	1.3852	1.4098	1.4345
1000000	10000	**10000**	100000	20000	20000	2000	3.0	2.03E-12	0.2577	0.4705	1.2240	1.2925	1.3523	1.3770	1.3899	1.4051	1.4298
1000000	10000	**13000**	100000	20000	20000	2000	3.0	2.03E-12	0.2468	0.4863	1.1965	1.3030	1.3358	1.4033	1.3856	1.4094	1.4341
1000000	10000	**16000**	100000	20000	20000	2000	3.0	1.85E-12	0.1626	0.3967	1.1363	1.3029	1.3358	1.3375	1.3855	1.4095	1.4342
1000000	10000	**17000**	100000	20000	20000	2000	3.0	1.66E-12	0.1161	0.3808	1.0904	1.3026	1.3358	1.3219	1.3852	1.4099	1.4345
1000000	10000	**18000**	100000	20000	20000	2000	3.0	1.49E-12	0.0526	0.3485	1.0350	1.3027	1.3358	1.2978	1.3852	1.4098	1.4344
1000000	10000	10000	**13000**	20000	20000	2000	3.0	4.11E-12	0.2450	0.4633	1.2293	1.2933	1.3581	1.3828	1.3891	1.4059	1.4306
1000000	10000	10000	**14000**	20000	20000	2000	3.0	2.00E-12	0.2557	0.4507	1.2282	1.2997	1.3498	1.3731	1.3874	1.4076	1.4323
1000000	10000	10000	**15000**	20000	20000	2000	3.0	1.86E-12	0.2659	0.4493	1.2226	1.3093	1.3421	1.3731	1.3893	1.4058	1.4305
1000000	10000	10000	**16000**	20000	20000	2000	3.0	1.89E-12	0.2647	0.4488	1.2215	1.3093	1.3421	1.3727	1.3893	1.4058	1.4305
1000000	10000	10000	**19000**	20000	20000	2000	3.0	3.57E-12	0.2647	0.4488	1.2215	1.3093	1.3421	1.3727	1.3893	1.4058	1.4305
1000000	10000	10000	100000	**5000**	20000	2000	3.0	1.55E-14	0.1960	0.1633	1.5677	2.2649	2.6124	1.1875	517.8064	0.8554	0.9828
1000000	10000	10000	100000	**6000**	20000	2000	3.0	1.65E-13	0.0045	0.1967	1.3429	0.8051	1.1764	1.1637	27.8915	1.0971	0.2705
1000000	10000	10000	100000	**8000**	20000	2000	3.0	1.08E-12	0.0135	0.1262	1.1914	1.3551	0.7743	1.8848	2.2967	0.7124	1.0352
1000000	10000	10000	100000	**9000**	20000	2000	3.0	1.08E-12	0.0312	0.2748	1.2914	0.8163	1.0020	1.3494	1.4769	0.2786	0.2835
1000000	10000	10000	100000	**10000**	20000	2000	3.0	2.08E-13	0.0198	0.3178	1.2271	4.2613	4.4370	1.3494	4.3662	0.1086	0.1135
1000000	10000	10000	100000	**14000**	20000	2000	3.0	7.17E-13	0.2371	0.4633	1.3088	0.8001	1.0171	1.3180	1.3906	0.2836	0.2886
1000000	10000	10000	100000	20000	**15000**	2000	3.0	1.98E-12	0.2241	0.5016	1.2066	1.2865	1.3729	1.3976	1.4060	1.3890	1.4137
1000000	10000	10000	100000	20000	**16000**	2000	3.0	2.07E-12	0.2313	0.4894	1.2077	1.2844	1.3627	1.3874	1.3939	1.4012	1.4259
1000000	10000	10000	100000	20000	**17000**	2000	3.0	2.05E-12	0.2363	0.4665	1.2178	1.2933	1.3583	1.3830	1.3894	1.4056	1.4303
1000000	10000	10000	100000	20000	**18000**	2000	3.0	2.07E-12	0.2473	0.4556	1.2251	1.2965	1.3526	1.3773	1.3869	1.4081	1.4328
1000000	10000	10000	100000	20000	**19000**	2000	3.0	1.97E-12	0.2574	0.4496	1.2273	1.3023	1.3490	1.3728	1.3877	1.4074	1.4320
1000000	10000	10000	100000	20000	20000	**18000**	3.0	4.17E-12	0.4688	0.5551	1.3897	1.2802	1.3935	1.3227	1.3762	1.4189	1.6483
1000000	10000	10000	100000	20000	20000	**20000**	3.0	3.81E-12	0.4071	0.6104	1.3317	1.0980	1.2920	1.3411	1.3814	1.4137	1.4383
1000000	10000	10000	100000	20000	20000	**21000**	3.0	2.25E-13	0.1564	0.4881	1.2034	4.4802	1.5643	4.4509	2.3499	0.4596	0.4701
1000000	10000	10000	100000	20000	20000	**22000**	3.0	2.24E-13	0.1553	0.4870	1.2033	4.5536	1.5643	4.5243	2.3400	0.4694	0.4800

N_0	M_0	L_n	U_n	U_m	U_{vn}	L_{nm}	t	R_{E2}*	β_{11}	β_{21}	δ_{11}	τ_{11}	δ_{21}	τ_{21}	β_{32}	δ_{32}	τ_{32}
1000000	10000	10000	100000	20000	20000	**23000**	3.0	2.26E-14	0.3547	0.6115	1.2782	63.4283	1.8054	63.1579	1.5786	1.2164	1.2411
1000000	10000	10000	100000	20000	20000	2000	**2.5**	2.18E-13	0.2000	0.5917	1.4503	4.5465	1.5645	4.8240	4.6329	0.8940	0.8978
1000000	10000	10000	100000	20000	20000	2000	**3.4**	2.23E-13	0.1980	0.2527	1.1734	1.6698	1.3116	1.5864	5.4597	0.8551	0.8798
1000000	10000	10000	100000	20000	20000	2000	**3.8**	1.53E-13	0.0945	0.0109	1.1514	0.8654	0.6326	0.2327	0.3241	1.6480	1.6529
1000000	10000	10000	100000	20000	20000	2000	**4.0**	2.26E-23	0.0379	0.0984	0.9312	2.2591	1.0861	2.1647	0.3449	0.8422	1.0230
1000000	10000	10000	100000	20000	20000	2000	**4.8**	9.80E-24	0.0482	0.2260	0.6561	3.3709	1.4835	2.7213	0.3501	0.2129	0.8665

Appendix-3(d)

Table : Values of R_{E2}*, β_{11}, τ_{11}, δ_{11}, β_{21}, δ_{21}, τ_{21}, β_{32}, δ_{32}, τ_{32} for Varying values of one value of the following N_0, M_0, L_n, U_n, U_m, U_{vn}, L_{vm}, t when other parameters are constants (Drug vacation period)

N_0	M_0	L_n	U_n	U_m	U_{vn}	L_{nm}	T	R_{E2}*	β_{11}	β_{21}	δ_{11}	τ_{11}	δ_{21}	τ_{21}	β_{32}	δ_{32}	τ_{32}
100000	20000	25000	200000	40000	40000	4000	3.0	1.28E-13	0.10975	0.28007	0.46059	1.37058	1.00075	1.00075	1.00074	0.99926	0.99926
300000	20000	25000	200000	40000	40000	4000	3.0	8.04E-24	0.15077	0.37113	0.75378	1.72874	1.33910	1.36378	1.40164	1.42221	1.76575
400000	20000	25000	200000	40000	40000	4000	3.0	8.32E-14	0.18622	0.42415	0.84744	1.36755	1.36248	1.09045	4.04545	1.39667	1.42136
600000	20000	25000	200000	40000	40000	4000	3.0	2.84E-13	0.00000	0.31627	0.80575	1.84173	1.37454	1.58921	1.42562	1.38177	1.40646
900000	20000	25000	200000	40000	40000	4000	3.0	2.92E-13	0.00000	0.32464	0.93922	1.58933	1.41425	1.43894	1.47423	1.37335	0.41677
1000000	**11000**	25000	200000	40000	40000	4000	3.0	2.41E-12	0.18621	0.29323	1.27133	1.32902	1.34135	1.36602	1.39192	1.41547	1.44016
1000000	**12000**	25000	200000	40000	40000	4000	3.0	2.42E-12	0.18620	0.29324	1.27132	1.32902	1.34135	1.36602	1.39192	1.41547	1.44016
1000000	**14000**	25000	200000	40000	40000	4000	3.0	2.44E-12	0.18620	0.29325	1.27131	1.32902	1.34135	1.36602	1.39192	1.41546	1.44015
1000000	**16000**	25000	200000	40000	40000	4000	3.0	2.47E-12	0.18620	0.29326	1.27129	1.32902	1.34135	1.36602	1.39193	1.41546	1.44015
1000000	**18000**	25000	200000	40000	40000	4000	3.0	2.50E-12	0.18620	0.29327	1.27128	1.32902	1.34135	1.36602	1.39193	1.41546	1.44015
1000000	**24000**	25000	200000	40000	40000	4000	3.0	2.57E-12	0.18621	0.29330	1.27127	1.32892	1.34130	1.36598	1.39193	1.41561	1.44015
1000000	20000	**16000**	200000	40000	40000	4000	3.0	2.28E-12	0.16648	0.29272	1.24474	1.33132	1.33881	1.36349	1.39169	1.41570	1.44039
1000000	20000	**17000**	200000	40000	40000	4000	3.0	2.06E-12	0.12830	0.29212	1.19221	1.33610	1.33372	1.35840	1.39138	1.41601	1.44070
1000000	20000	**18000**	200000	40000	40000	4000	3.0	5.45E-11	0.06943	0.35799	1.04233	1.34780	1.33023	1.34845	1.39135	1.41604	1.57612
1000000	20000	**19000**	200000	40000	40000	4000	3.0	1.66E-12	0.03701	0.34195	1.01081	1.35007	1.33260	1.33322	1.39131	1.41608	1.55700
1000000	20000	**20000**	200000	40000	40000	4000	3.0	1.42E-12	0.00000	0.29233	1.01168	1.33990	1.33074	1.31317	1.39135	1.41604	1.44073
1000000	20000	25000	**15500**	40000	40000	4000	3.0	2.58E-13	0.21573	0.42677	1.17793	1.74898	1.55663	1.58132	3.78418	1.21249	1.23718
1000000	20000	25000	**15750**	40000	40000	4000	3.0	6.17E-13	0.20080	0.41998	1.16449	1.74886	1.55392	1.57861	3.73562	1.22033	1.24502
1000000	20000	25000	**16500**	40000	40000	4000	3.0	5.93E-14	0.14346	0.42148	1.10686	12.47859	1.74432	12.11915	4.24540	1.07232	1.08528
1000000	20000	25000	**17000**	40000	40000	4000	3.0	1.70E-09	0.00000	0.01421	1.38569	1.89847	0.13200	1.76647	0.46211	1.44938	1.46790
1000000	20000	25000	**17500**	40000	40000	4000	3.0	1.92E-13	0.00000	0.01402	1.36542	3.76969	1.87250	1.89719	5.20203	0.82074	0.84543
1000000	20000	25000	**90000**	40000	40000	4000	3.0	2.40E-12	0.18621	0.29323	1.27133	1.32902	1.34135	1.36602	1.39192	1.41547	1.44016
1000000	20000	25000	200000	**10000**	40000	4000	3.0	5.63E-13	0.09929	0.15560	1.28767	2.74342	2.28740	1.80000	1.82921	0.59846	1.05999
1000000	20000	25000	200000	**12000**	40000	4000	3.0	2.57E-13	0.21278	0.25380	1.34690	6.09295	1.33075	6.15012	1.60417	0.91273	0.92013
1000000	20000	25000	200000	**13000**	40000	4000	3.0	3.96E-11	0.00261	0.00398	1.29714	0.68243	0.41975	0.26268	0.51561	2.29179	1.32201
1000000	20000	25000	200000	**15000**	40000	4000	3.0	2.36E-12	0.18363	0.30824	1.25279	1.30879	1.33076	1.35543	1.39135	1.41604	1.44073
1000000	20000	25000	200000	**16000**	40000	4000	3.0	2.46E-12	0.19053	0.29206	1.27841	1.32426	1.33825	1.36594	1.39135	1.41604	1.44073
1000000	20000	25000	200000	**17000**	40000	4000	3.0	3.49E-12	0.18621	0.29317	1.27139	1.32902	1.34135	1.36602	1.39192	1.41547	1.44016
1000000	20000	25000	200000	40000	**16000**	4000	3.0	4.58E-12	0.00000	0.38110	0.99182	1.51032	1.42928	1.45397	0.65592	1.32923	1.35392
1000000	20000	25000	200000	40000	**18500**	4000	3.0	2.35E-12	0.14267	0.29416	1.23636	1.32619	1.34468	1.36936	1.39243	1.41496	1.43965
1000000	20000	25000	200000	40000	**19000**	4000	3.0	2.48E-12	0.18145	0.29662	1.27796	1.32528	1.34686	1.37154	1.39369	1.41370	1.43839
1000000	20000	25000	200000	40000	**19500**	4000	3.0	2.47E-12	0.18502	0.29368	1.27741	1.32662	1.34401	1.36868	1.39218	1.41521	1.43990
1000000	20000	25000	200000	40000	**20000**	4000	3.0	2.40E-12	0.18621	0.29323	1.27133	1.32902	1.34135	1.36602	1.39192	1.41547	1.44016
1000000	20000	25000	200000	40000	**21000**	4000	3.0	2.36E-12	0.18853	0.29232	1.25971	1.33366	1.33624	1.36092	1.39147	1.41592	1.44061
1000000	20000	25000	200000	40000	40000	**11500**	3.0	2.40E-12	0.18621	0.29323	1.27133	1.32902	1.34135	1.36602	1.39192	1.41547	1.44016

N_0	M_0	L_n	U_n	U_m	U_{vn}	L_{nm}	T	R_{E2}*	β_{11}	β_{21}	δ_{11}	τ_{11}	δ_{21}	τ_{21}	β_{32}	δ_{32}	τ_{32}
1000000	20000	25000	200000	40000	40000	**12500**	3.0	3.02E-13	0.00000	0.44110	0.95879	1.49336	1.36359	1.52967	2.87962	1.22993	1.25462
1000000	20000	25000	200000	40000	40000	**13000**	3.0	2.03E-13	0.04722	0.39220	1.01606	1.99353	1.36359	1.99098	2.12748	0.67991	0.31540
1000000	20000	25000	200000	40000	40000	**14000**	3.0	3.13E-13	0.16449	0.44111	1.12328	1.81143	1.36359	1.84774	1.92475	0.88264	0.35843
1000000	20000	25000	200000	40000	40000	**14500**	3.0	9.26E-14	0.20982	0.41474	1.18196	8.49868	4.30962	5.57595	5.59806	2.01671	2.06609
1000000	20000	25000	200000	40000	40000	4000	**2.0**	2.45E-13	0.00000	0.66167	1.43819	1.18336	2.00524	1.27797	4.04410	0.47539	0.48280
1000000	20000	25000	200000	40000	40000	4000	**2.4**	2.44E-13	0.08033	0.53637	1.29383	3.68143	3.90038	1.53093	1.97249	0.91645	1.03945
1000000	20000	25000	200000	40000	40000	4000	**3.2**	2.51E-12	0.21550	0.28858	1.23933	1.35103	1.31938	1.34406	1.39197	1.41542	1.44011
1000000	20000	25000	200000	40000	40000	4000	**4.0**	3.67E-23	0.01978	0.25226	0.75045	1.84748	1.36798	1.46242	1.42788	1.37950	1.67864
1000000	20000	25000	200000	40000	40000	4000	**4.2**	2.14E-13	0.00000	0.31508	0.68485	3.06915	1.35357	2.71551	1.44206	1.36533	1.11618

CHAPTER 7

A New Unrelated Question Model with Two Questions Per Card

Tonghui Xu, Stephen A Sedory* and *Sarjinder Singh**

1. Proposed New Model

Consider a population Ω in which two types of people are living, a few bearing a special characteristic are considered as members of a sensitive group A and the rest are considered as members of a non-sensitive group A^c, such that $\Omega = A \cup A^c$. In addition, let there be another characteristic unrelated to the sensitive characteristic prevailing in the same population. Again, the people bearing the unrelated characteristic are considered as members of a non-sensitive group Y, and rest of the people are members of the group Y^c. Such a population setup is shown in the Venn diagram in Figure 1.

Assume π to be the true proportion of the people belonging to the sensitive group A; π_y be the true known proportion of people belonging to unrelated group Y; Presume, A and Y are independent, $\pi\pi_y$ be the proportion of the people belonging to the both sensitive and unrelated groups $A \cap Y$. Indicate that $(A \cup Y)^c \neq \Phi$ and $(A \cup Y)^c \cup (A \cup Y) = \Omega$.

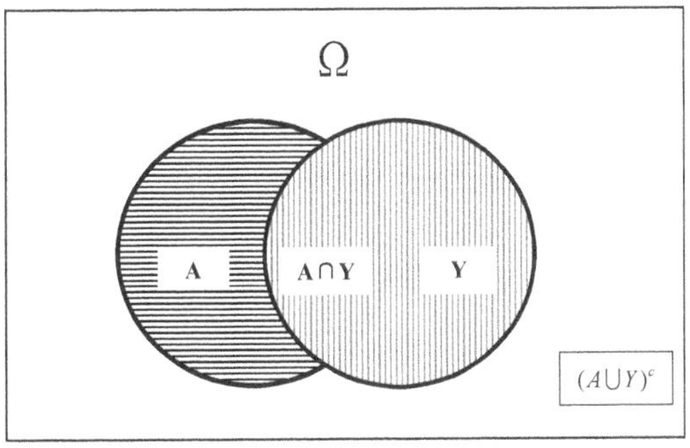

Figure 1: Population of interest.

Assume a simple random sample with replacement (SRSWR) and n respondents are selected from the population Ω. From the selected respondents, each respondent is asked to draw one card from a deck that consists of two different types of cards, each card has two questions printed in order. The respondent is instructed to reply in the same order as the question is being asked on the card drawn by him/her.

Let P be the proportion of the cards bearing the two-questions:

(1) "Are you a member of the sensitive group A?"

(2) "Are you a member of the unrelated group Y?"

Let $(1 - P)$ be the proportion of the cards bearing the two-questions:

(1) "Are you a member of non-sensitive group A^c?"

(2) "Are you a member of unrelated group Y^c?"

A pictorial representation of such a randomization device is shown in Figure 2.

Department of Mathematics, Texas A&M University – Kingsville, Kingsville, Texas, 78363, USA.
* Corresponding authors: tonghui.xu@students.edu; sarjinder.singh@tamuk.edu

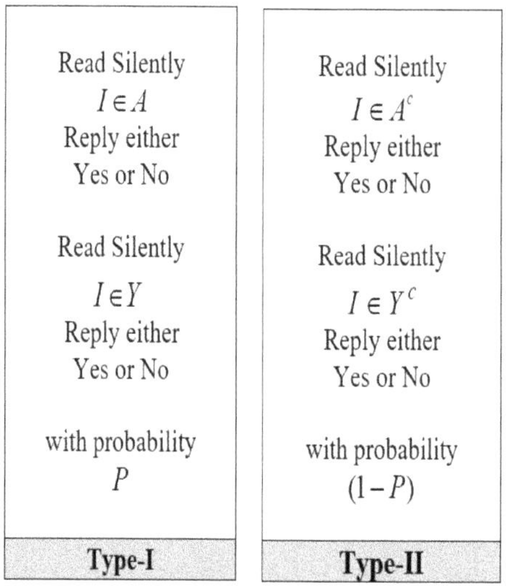

Figure 2: Two types of cards, each with two questions.

Scenario I: in the first situation, assume the respondent originates from sensitive group A and unrelated group Y; If he/she draws the card with the statements $I \in A$ and $I \in Y$ with probability P, then his/her response will be "(Yes, Yes)". Presume, in the second situation, the respondent belongs to the non-sensitive group A^c and unrelated group Y^c; if he/she draws the card with the statements $I \in A^c$ and $I \in Y^c$ with probability $(1 - P)$, then his/her response will also be "(Yes, Yes)". The response "(Yes, Yes)" can come from both types of respondents either belonging to group A and unrelated group Y or originates from non-sensitive group A^c and unrelated group Y^c. The respondent is protected.
Thus, the probability of getting "(Yes, Yes)" response is given by:

$$\begin{aligned} P(\text{Yes,Yes}) = \alpha_{11} &= P\pi\pi_y + (1-P)(1-\pi)(1-\pi_y) \\ &= \pi(P+\pi_y-1) + (1-P)(1-\pi_y) \end{aligned} \tag{1.1}$$

Scenario II: in the first situation, presume the respondent only belongs to sensitive group A but not to unrelated group Y; If he/she draws the card with the statement $I \in A$ and $I \in Y$ with probability P, then his/her response will be "(Yes, No)". Assume, in the second situation, the respondent belongs to the non-sensitive group A^c and to the unrelated group Y; if he/she draws the card with the statements $I \in A^c$ and $I \in Y^c$ with probability $(1 - P)$, then his/her responses will also be "(Yes, No)". The response "(Yes, No)" can come from both types of respondents belonging to group A and not belonging to unrelated group Y or stay in the non-sensitive group A^c and stay in unrelated group Y. The respondent is protected.
Thus, the probability of getting "(Yes, No)" response is given by:

$$\begin{aligned} P(\text{Yes,No}) = \alpha_{10} &= P\pi(1-\pi_y) + (1-P)(1-\pi)\pi_y \\ &= (P-\pi_y)\pi + \pi_y(1-P) \end{aligned} \tag{1.2}$$

Scenario III: in the first situation, assume the respondent is not in the sensitive group A and is only in unrelated group Y; If he/she draws the card with the statement $I \in A$ and $I \in Y$ with probability P, then his/her response will be "(No, Yes)". Presume, in the second situation, the respondent belongs to the sensitive group A and not to the unrelated group Y; if he/she draws the card with the statements $I \in A^c$ and $I \in Y^c$ with probability $(1 - P)$, then his/her response will also be "(No, Yes)". The response "(No, Yes)" can come from both types of respondents either belonging to group A^c and group Y or belonging to group A and unrelated group Y^c. The respondent is protected.
Thus, the probability of getting "(No, Yes)" response is given by:

$$\begin{aligned} P(\text{No,Yes}) = \alpha_{01} &= P(1-\pi)\pi_y + (1-P)\pi(1-\pi_y) \\ &= (1-P-\pi_y)\pi + P\pi_y \end{aligned} \tag{1.3}$$

Scenario IV: in the first situation, assume the respondent does not reside in both the sensitive group A and in the unrelated group Y; If he/she draws the card containing the statements $I \in A$ and $I \in Y$ with probability P, then his/her response will be "(No, No)". Presume, in the second situation, the respondent resides in both the sensitive group A and in the unrelated group Y; if he/she draws the card containing the statements $I \in A^c$ and $I \in Y^c$ with probability $(1- P)$, then his/her responses will also be "(No, No)". The response "(No, No)" can come from both types of respondents pertaining to both groups A^c and group Y^c or both in sensitive group A and unrelated group Y. The respondent is protected.

Thus, the probability of getting "(No, No)" response is given by:

$$P(\text{No,No}) = \alpha_{00} = P(1-\pi)(1-\pi_y) + (1-P)\pi\pi_y$$
$$= (\pi_y - P)\pi + P(1-\pi_y) \tag{1.4}$$

Note that $\alpha_{11} + \alpha_{10} + \alpha_{01} + \alpha_{00} = 1$.

2. Proposed Unbiased Estimator and Variance with new Model

The responses from the n respondents can be classified into a 2×2 contingency table as shown in Table 1.

Table 1: Observed responses.

Responses	Yes	No
Yes	n_{11}	n_{10}
No	n_{01}	n_{00}

Note that: $n_{11} + n_{10} + n_{01} + n_{00} = n$.

The purpose is to estimate the unknown proportion π of the persons in the population belonging to the sensitive group A.

Let $\hat{\alpha}_{11} = n_{11}/n$, $\hat{\alpha}_{10} = n_{10}/n$, $\hat{\alpha}_{01} = n_{01}/n$, and $\hat{\alpha}_{00} = n_{00}/n$ be the observed proportions of "(Yes, Yes)", "(Yes, No)", "(No, Yes)" and "(No, No)" responses. Following Odumade and Singh (2009), we consider a least squared distance between the observed proportions and the true proportions as:

$$D = \frac{1}{2}\sum_{i=0}^{1}\sum_{j=0}^{1}\left(\alpha_{ij} - \hat{\alpha}_{ij}\right)^2$$

$$= \frac{1}{2}\left[\alpha_{11} - \hat{\alpha}_{11}\right]^2 + \frac{1}{2}\left[\alpha_{10} - \hat{\alpha}_{10}\right]^2 + \frac{1}{2}\left[\alpha_{01} - \hat{\alpha}_{01}\right]^2 + \frac{1}{2}\left[\alpha_{00} - \hat{\alpha}_{00}\right]^2 \tag{2.1}$$

$$= \frac{1}{2}\left[(P + \pi_y - 1)\pi + (1-P)(1-\pi_y) - \hat{\alpha}_{11}\right]^2 + \frac{1}{2}\left[(P - \pi_y)\pi + \pi_y(1-P) - \hat{\alpha}_{10}\right]^2$$

$$+ \frac{1}{2}\left[(1 - P - \pi_y)\pi + P\pi_y - \hat{\alpha}_{01}\right]^2 + \frac{1}{2}\left[(\pi_y - P)\pi + P(1-\pi_y) - \hat{\alpha}_{00}\right]^2$$

We decide to find π such that the squared distance D is minimum. Thus, in order to find π we set:

$$\frac{\partial D}{\partial \pi} = 0$$

we have

$$[(P + \pi_y - 1)\pi + (1-P)(1-\pi_y) - \hat{\alpha}_{11}](P + \pi_y - 1) + [(P - \pi_y)\pi + \pi_y(1-P) - \hat{\alpha}_{10}](P - \pi_y)$$
$$+ [(1 - P - \pi_y)\pi + P\pi_y - \hat{\alpha}_{01}](1 - P - \pi_y) + [(\pi_y - P)\pi + P(1-\pi_y) - \hat{\alpha}_{00}](\pi_y - P) = 0$$

or

$$2\pi\left[(P + \pi_y - 1)^2 + (P - \pi_y)^2\right] - (P + \pi_y - 1)^2 - (P - \pi_y)^2 = (P + \pi_y - 1)(\hat{\alpha}_{11} - \hat{\alpha}_{01}) + (P - \pi_y)(\hat{\alpha}_{10} - \hat{\alpha}_{00})$$

or

$$\pi = \frac{(P + \pi_y - 1)(\hat{\alpha}_{11} - \hat{\alpha}_{01}) + (P - \pi_y)(\hat{\alpha}_{10} - \hat{\alpha}_{00}) + (P + \pi_y - 1)^2 + (P - \pi_y)^2}{2\left[(P + \pi_y - 1)^2 + (P - \pi_y)^2\right]}$$

or

$$\pi = \frac{1}{2} + \frac{(P + \pi_y - 1)(\hat{\alpha}_{11} - \hat{\alpha}_{01}) + (P - \pi_y)(\hat{\alpha}_{10} - \hat{\alpha}_{00})}{2\left[(P + \pi_y - 1)^2 + (P - \pi_y)^2\right]} \tag{2.2}$$

From (2.2), by the method of moments, we have the following theorem:

Theorem 2.1: An unbiased estimator of the population proportion π is given by:

$$\hat{\pi}_{xs} = \frac{1}{2} + \frac{(P+\pi_y-1)(\hat{\alpha}_{11}-\hat{\alpha}_{01}) + (P-\pi_y)(\hat{\alpha}_{10}-\hat{\alpha}_{00})}{2\left[(P+\pi_y-1)^2+(P-\pi_y)^2\right]} \tag{2.3}$$

Proof: Note that:

$$E(\hat{\alpha}_{11}) = E\left(\frac{n_{11}}{n}\right) = \alpha_{11};\ E(\hat{\alpha}_{10}) = E\left(\frac{n_{10}}{n}\right) = \alpha_{10};\ E(\hat{\alpha}_{01}) = E\left(\frac{n_{01}}{n}\right) = \alpha_{01}$$

and, $E(\hat{\alpha}_{00}) = E\left(\frac{n_{00}}{n}\right) = \alpha_{00}$.

Now taking the expected values on both sides of (2.3), we have:

$$E(\hat{\pi}_{xs}) = \frac{1}{2} + \frac{(P+\pi_y-1)(E(\hat{\alpha}_{11})-E(\hat{\alpha}_{01})) + (P-\pi_y)(E(\hat{\alpha}_{10})-E(\hat{\alpha}_{00}))}{2\left[(P+\pi_y-1)^2+(P-\pi_y)^2\right]}$$

$$= \frac{1}{2} + \frac{(P+\pi_y-1)(\alpha_{11}-\alpha_{01}) + (P-\pi_y)(\alpha_{10}-\alpha_{00})}{2\left[(P+\pi_y-1)^2+(P-\pi_y)^2\right]}$$

$$= \frac{1}{2} + \frac{(P+\pi_y-1)\left[2(P+\pi_y-1)\pi-(P+\pi_y-1)\right] + (P-\pi_y)\left[2(P-\pi_y)\pi-(P-\pi_y)\right]}{2\left[(P+\pi_y-1)^2+(P-\pi_y)^2\right]}$$

$$= \frac{1}{2} + \frac{2\pi\left[(P+\pi_y-1)^2+(P-\pi_y)^2\right] - (P+\pi_y-1)^2-(P-\pi_y)^2}{2\left[(P+\pi_y-1)^2+(P-\pi_y)^2\right]}$$

$$= \frac{1}{2} + \pi - \frac{1}{2} = \pi$$

which proves the theorem.

Now from the concept of multinomial distribution, we have:

$$V(\hat{\alpha}_{11}) = \frac{\alpha_{11}(1-\alpha_{11})}{n},\quad Cov(\hat{\alpha}_{11},\hat{\alpha}_{10}) = \frac{-\alpha_{11}\alpha_{10}}{n},\quad Cov(\hat{\alpha}_{11},\hat{\alpha}_{01}) = \frac{-\alpha_{11}\alpha_{01}}{n}$$

$$Cov(\hat{\alpha}_{11},\hat{\alpha}_{00}) = \frac{-\alpha_{11}\alpha_{00}}{n},\ V(\hat{\alpha}_{10}) = \frac{\alpha_{10}(1-\alpha_{10})}{n},\quad Cov(\hat{\alpha}_{10},\hat{\alpha}_{01}) = \frac{-\alpha_{10}\alpha_{01}}{n}$$

$$Cov(\hat{\alpha}_{10},\hat{\alpha}_{00}) = \frac{-\alpha_{10}\alpha_{00}}{n},\ V(\hat{\alpha}_{01}) = \frac{\alpha_{01}(1-\alpha_{01})}{n},\quad Cov(\hat{\alpha}_{01},\hat{\alpha}_{00}) = \frac{-\alpha_{01}\alpha_{00}}{n}$$

and

$$V(\hat{\alpha}_{00}) = \frac{\alpha_{00}(1-\alpha_{00})}{n}$$

Now we have the following theorem:

Theorem 2.2: The variance of the estimator $\hat{\pi}_{xs}$ is given by:

$$V(\hat{\pi}_{xs}) = \frac{(P+\pi_y-1)^2\{P\pi_y+(1-P)(1-\pi_y)\}+(P-\pi_y)^2\{\pi_y(1-P)+P(1-\pi_y)\}}{4n\left[(P+\pi_y-1)^2+(P-\pi_y)^2\right]^2} - \frac{(2\pi-1)^2}{4n} \tag{2.4}$$

Proof: We have

$$V(\hat{\pi}_{xs}) = \frac{1}{4\left[(P+\pi_y-1)^2+(P-\pi_y)^2\right]^2}\Big[(P+\pi_y-1)^2\cdot V(\hat{\alpha}_{11}-\hat{\alpha}_{01})+(P-\pi_y)^2\cdot V(\hat{\alpha}_{10}-\hat{\alpha}_{00})$$

$$+2(P+\pi_y-1)(P-\pi_y)\cdot Cov(\hat{\alpha}_{11}-\hat{\alpha}_{01},\hat{\alpha}_{10}-\hat{\alpha}_{00})\Big]$$

$$= \frac{1}{4\left[(P+\pi_y-1)^2+(P-\pi_y)^2\right]^2}\Big[(P+\pi_y-1)^2\{V(\hat{\alpha}_{11})+V(\hat{\alpha}_{01})-2Cov(\hat{\alpha}_{11},\hat{\alpha}_{01})\}$$

$$+(P-\pi_y)^2\{V(\hat{\alpha}_{10})+V(\hat{\alpha}_{00})-2Cov(\hat{\alpha}_{10},\hat{\alpha}_{00})\}$$

$$+2(P+\pi_y-1)(P-\pi_y)\{Cov(\hat{\alpha}_{11},\hat{\alpha}_{10})-Cov(\hat{\alpha}_{01},\hat{\alpha}_{10})$$

$$-Cov(\hat{\alpha}_{11},\hat{\alpha}_{00})+Cov(\hat{\alpha}_{01},\hat{\alpha}_{00})\}\Big]$$

$$= \frac{1}{4\left[(P+\pi_y-1)^2+(P-\pi_y)^2\right]^2}\left[(P+\pi_y-1)^2\left\{\frac{\alpha_{11}(1-\alpha_{11})}{n}+\frac{\alpha_{01}(1-\alpha_{01})}{n}+2\frac{\alpha_{11}\alpha_{01}}{n}\right\}\right.$$

$$+(P-\pi_y)^2\left\{\frac{\alpha_{10}(1-\alpha_{10})}{n}+\frac{\alpha_{00}(1-\alpha_{00})}{n}+2\frac{\alpha_{10}\alpha_{00}}{n}\right\}$$

$$\left.+2(P+\pi_y-1)(P-\pi_y)\left\{-\frac{\alpha_{11}\alpha_{10}}{n}+\frac{\alpha_{01}\alpha_{10}}{n}+\frac{\alpha_{11}\alpha_{00}}{n}-\frac{\alpha_{01}\alpha_{00}}{n}\right\}\right]$$

$$= \frac{1}{4n\left[(P+\pi_y-1)^2+(P-\pi_y)^2\right]^2}\Big[(P+\pi_y-1)^2\{\alpha_{11}-\alpha_{11}^2+\alpha_{01}-\alpha_{01}^2+2\alpha_{11}\alpha_{01}\}$$

$$+(P-\pi_y)^2\{\alpha_{10}-\alpha_{10}^2+\alpha_{00}-\alpha_{00}^2+2\alpha_{10}\alpha_{00}\}$$

$$+2(P+\pi_y-1)(P-\pi_y)\{-\alpha_{11}\alpha_{10}+\alpha_{01}\alpha_{10}+\alpha_{11}\alpha_{00}-\alpha_{01}\alpha_{00}\}\Big]$$

$$= \frac{1}{4n\left[(P+\pi_y-1)^2+(P-\pi_y)^2\right]^2}\Big[(P+\pi_y-1)^2\{(\alpha_{11}+\alpha_{01})-(\alpha_{11}-\alpha_{01})^2\}+(P-\pi_y)^2\{(\alpha_{10}+\alpha_{00})-(\alpha_{10}-\alpha_{00})^2\}$$

$$+2(P+\pi_y-1)(P-\pi_y)\{\alpha_{10}(\alpha_{01}-\alpha_{11})+\alpha_{00}(\alpha_{11}-\alpha_{01})\}\Big]$$

$$= \frac{1}{4n\left[(P+\pi_y-1)^2+(P-\pi_y)^2\right]^2}\Big[(P+\pi_y-1)^2\{(\alpha_{11}+\alpha_{01})-(\alpha_{11}-\alpha_{01})^2\}+(P-\pi_y)^2\{(\alpha_{10}+\alpha_{00})-(\alpha_{10}-\alpha_{00})^2\}$$

$$-2(P+\pi_y-1)(P-\pi_y)\{(\alpha_{11}-\alpha_{01})(\alpha_{10}-\alpha_{00})\}\Big]$$

$$=\frac{1}{4n\left[(P+\pi_y-1)^2+(P-\pi_y)^2\right]^2}\Big[(P+\pi_y-1)^2\{(1-P)(1-\pi_y)+P\pi_y-(2\pi-1)^2(P+\pi_y-1)^2\}$$

$$+(P-\pi_y)^2\{\pi_y(1-P)+P(1-\pi_y)-(2\pi-1)^2(P-\pi_y)^2\}$$

$$-2(P+\pi_y-1)(P-\pi_y)(2\pi-1)(P+\pi_y-1)(2\pi-1)(P-\pi_y)\Big]$$

$$=\frac{1}{4n\left[(P+\pi_y-1)^2+(P-\pi_y)^2\right]^2}\Big[(P+\pi_y-1)^2\{(1-P-\pi_y+P\pi_y)+P\pi_y-(2\pi-1)^2(P+\pi_y-1)^2\}$$

$$+(P-\pi_y)^2\{\pi_y-\pi_yP+P-P\pi_y-(2\pi-1)^2(P-\pi_y)^2\}-2(P+\pi_y-1)^2(P-\pi_y)^2(2\pi-1)^2\Big]$$

$$=\frac{1}{4n\left[(P+\pi_y-1)^2+(P-\pi_y)^2\right]^2}\Big[(P+\pi_y-1)^2\{P\pi_y+(1-P)(1-\pi_y)\}-(2\pi-1)^2(P+\pi_y-1)^4$$

$$+(P-\pi_y)^2\{\pi_y(1-P)+P(1-\pi_y)\}-(2\pi-1)^2(P-\pi_y)^4-2(P+\pi_y-1)^2(P-\pi_y)^2(2\pi-1)^2\Big]$$

$$=\frac{1}{4n\left[(P+\pi_y-1)^2+(P-\pi_y)^2\right]^2}\Big[(P+\pi_y-1)^2\{P\pi_y+(1-P)(1-\pi_y)\}+(P-\pi_y)^2\{\pi_y(1-P)+P(1-\pi_y)\}$$

$$-(2\pi-1)^2\{(P+\pi_y-1)^4+(P-\pi_y)^4+2(P+\pi_y-1)^2(P-\pi_y)^2\}\Big]$$

$$=\frac{1}{4n\left[(P+\pi_y-1)^2+(P-\pi_y)^2\right]^2}\Big[(P+\pi_y-1)^2\{P\pi_y+(1-P)(1-\pi_y)\}+(P-\pi_y)^2\{\pi_y(1-P)+P(1-\pi_y)\}$$

$$-(2\pi-1)^2\{(P+\pi_y-1)^2+(P-\pi_y)^2\}^2\Big]$$

$$=\frac{(P+\pi_y-1)^2\{P\pi_y+(1-P)(1-\pi_y)\}+(P-\pi_y)^2\{\pi_y(1-P)+P(1-\pi_y)\}}{4n\left[(P+\pi_y-1)^2+(P-\pi_y)^2\right]^2}-\frac{(2\pi-1)^2}{4n}$$

which proves the theorem.

Now we have the following theorem:

Theorem 2.3: An unbiased estimator of the variance of $\hat{\pi}_{xs}$ is given by:

$$\hat{v}(\hat{\pi}_{xs})=\frac{(P+\pi_y-1)^2\{P\pi_y+(1-P)(1-\pi_y)\}+(P-\pi_y)^2\{\pi_y(1-P)+P(1-\pi_y)\}}{4(n-1)\left[(P+\pi_y-1)^2+(P-\pi_y)^2\right]^2}-\frac{(2\hat{\pi}_{xs}-1)^2}{4(n-1)} \tag{2.5}$$

Proof: Taking expected on both sides of (2.5) we have:

$$E[\hat{v}(\hat{\pi}_{xs})]=\frac{1}{4(n-1)}\left[\frac{(P+\pi_y-1)^2\{P\pi_y+(1-P)(1-\pi_y)\}+(P-\pi_y)^2\{\pi_y(1-P)+P(1-\pi_y)\}}{\{(P+\pi_y-1)^2+(P-\pi_y)^2\}^2}-E(2\hat{\pi}_{xs}-1)^2\right]$$

$$=\frac{n}{(n-1)}\left[\frac{(P+\pi_y-1)^2\{P\pi_y+(1-P)(1-\pi_y)\}+(P-\pi_y)^2\{\pi_y(1-P)+P(1-\pi_y)\}}{4n\{(P+\pi_y-1)^2+(P-\pi_y)^2\}^2}\right.$$

$$
\left. - \frac{4\{V(\hat{\pi}_{xs}) + \pi^2\} + 1 - 4\pi}{4n} \right]
$$

$$
= \frac{n}{(n-1)} \left[\frac{(P+\pi_y-1)^2\{P\pi_y+(1-P)(1-\pi_y)\}+(P-\pi_y)^2\{\pi_y(1-P)+P(1-\pi_y)\}}{4n\{(P+\pi_y-1)^2+(P-\pi_y)^2\}^2} \right.
$$

$$
\left. - \frac{4\pi^2+1-4\pi}{4n} - \frac{4V(\hat{\pi}_{xs})}{4n} \right]
$$

$$
= \frac{n}{(n-1)} \left[\frac{(P+\pi_y-1)^2\{P\pi_y+(1-P)(1-\pi_y)\}+(P-\pi_y)^2\{\pi_y(1-P)+P(1-\pi_y)\}}{4n\{(P+\pi_y-1)^2+(P-\pi_y)^2\}^2} \right.
$$

$$
\left. - \frac{(2\pi-1)^2}{4n} - \frac{V(\hat{\pi}_{xs})}{n} \right]
$$

$$
= \frac{n}{(n-1)} \left[V(\hat{\pi}_{xs}) - \frac{V(\hat{\pi}_{xs})}{n} \right] = V(\hat{\pi}_{xs})
$$

which proves the theorem.
Now we have the following corollary:

Corollary 2.1: If $P = \pi_y = P_0$ (say), then the variance of the proposed estimator $\hat{\pi}_{xs}$ in (2.4) becomes:

$$
V(\hat{\pi}_{xs})_{P=\pi_y=P_0} = \frac{\pi(1-\pi)}{n} + \frac{P_0(1-P_0)}{2n(2P_0-1)^2} = V(\hat{\pi}_w)_{q=2} \tag{2.6}
$$

which means if the proportion of unrelated characteristics π_y is same as the proportion P of cards bearing the first two-question, the proposed model has the variance twice that from the use of the Warner (1965) model.

Proof: On setting $\pi_y = P = P_0$ in (2.4), we have:

$$
V(\hat{\pi}_{xs})_{P=\pi_y=P_0} = \frac{(2P_0-1)^2\left[P_0^2+(1-P_0)^2\right]}{4n\{(2P_0-1)^2\}^2} - \frac{(2\pi-1)^2}{4n}
$$

$$
= \frac{P_0^2+(1-P_0)^2}{4n(2P_0-1)^2} - \frac{(2\pi-1)^2}{4n}
$$

$$
= \frac{(2P_0-1)^2+2P_0(1-P_0)}{4n(2P_0-1)^2} - \frac{(2\pi-1)^2}{4n}
$$

$$
= \frac{1}{4n} - \frac{(2\pi-1)^2}{4n} + \frac{P_0(1-P_0)}{2n(2P_0-1)^2}
$$

$$
= \frac{1-(2\pi-1)^2}{4n} + \frac{P_0(1-P_0)}{2n(2P_0-1)^2}
$$

$$
= \frac{4\pi(1-\pi)}{4n} + \frac{P_0(1-P_0)}{2n(2P_0-1)^2}
$$

which proves the corollary.

In section 3, we compare the new proposed model with the Warner (1965) and the Greenberg et al. (1969) models in order to calculate the relative efficiency.

3. Relative Efficiency in Different Models

Now, we compare the proposed new estimator with the Warner (1965) estimator and the Greenberg et al. (1969) estimator in order to figure out which model is more efficient. In Appendix A, we develop SAS codes to compute the relative efficiency.
The relative efficiency of the new model with respect to the Warner (1965) model is given by:

$$RE(1) = \frac{V(\hat{\pi}_w)_{q=2}}{V(\hat{\pi}_{xs})} \times 100\%$$

$$= \frac{4\left[\pi(1-\pi) + \dfrac{P_0(1-P_0)}{2(2P_0-1)^2}\right] \times 100\%}{\dfrac{(P+\pi_y-1)^2\{P\pi_y+(1-P)(1-\pi_y)\} + (P-\pi_y)^2\{\pi_y(1-P)+P(1-\pi_y)\}}{\left[(P+\pi_y-1)^2+(P-\pi_y)^2\right]^2} - (2\pi-1)^2} \quad (2.7)$$

The relative efficiency of the new model respect to the Greenberg et al. (1969) model is given by:

$$RE(2) = \frac{V(\hat{\pi}_g)}{V(\hat{\pi}_{xs})} \times 100\%$$

$$= \frac{4\left[\dfrac{(P_1\pi+(1-P_1)\pi_y)(1-P_1\pi-(1-P_1)\pi_y)}{P_1^2}\right] \times 100\%}{\dfrac{(P+\pi_y-1)^2\{P\pi_y+(1-P)(1-\pi_y)\} + (P-\pi_y)^2\{\pi_y(1-P)+P(1-\pi_y)\}}{\left[(P+\pi_y-1)^2+(P-\pi_y)^2\right]^2} - (2\pi-1)^2} \quad (2.8)$$

where P_0 and P_1 are the device parameters in the Warner (1965) and Greenberg et al. (1969) models.

4. Discussion of Results

In Table 2, we computed the relative efficiency values RE(1) and RE(2), for different values of π between 0.05 and 0.5 with a skip of 0.05 for $P = P_0 = P_1 = 0.7$. We simulated three values of π_y between 0.85 and 0.95 with a skip of 0.05, such that, the proposed estimator is more efficient than the Warner (1965) and the Greenberg et al. (1969) models. We provided the SAS code in APPENDIX A. Based on the simulation study; we suggested the better choice of π_y close to 1. We observed that for $\pi = 0.05$, the average value of RE(1) is 397.80% with STD of 157.30%, with minimum value of 256.00%, and maximum value of 567.10%; the average of RE(2) is 246.10% with STD of 103.70%, with minimum value of 152.90%, and maximum value of 357.80%. As the value of the π increases to 0.5 by 0.05, the average value of RE(1) is 232.70% with STD of 43.90%, with minimum value of 189.00%, and

Table 2: Descriptive statistics of RE(1) and RE(2) for $P = P_0 = P_1 = 0.7$, and $\pi_y \in [0.85, 0.95]$.

		RE(1)				RE(2)			
π	Freq	Mean	STD	MIN	MAX	Mean	STD	MIN	MAX
0.05	3	397.80	157.30	256.00	567.10	246.10	103.70	152.90	357.80
0.10	3	334.70	107.10	235.10	447.90	206.20	70.20	141.10	280.50
0.15	3	298.40	81.90	220.80	384.00	182.50	52.90	132.50	237.90
0.25	3	259.50	58.20	203.40	319.50	155.10	35.80	120.60	191.90
0.30	3	248.70	52.20	198.00	302.20	146.20	31.00	116.10	178.00
0.35	3	241.20	48.20	194.30	290.60	139.00	27.50	112.20	167.20
0.40	3	236.40	45.70	191.80	283.10	133.00	24.90	108.60	158.40
0.45	3	233.60	44.30	190.30	278.80	127.80	22.90	105.30	151.10
0.50	3	232.70	43.90	189.80	277.50	123.20	21.30	102.20	144.80

maximum value of 277.50%; the average of RE(2) is 123.20% with STD of 21.30%, with minimum value of 102.20%, and maximum value of 144.80%. It is worth pointing out that the proposed model shows more efficient result when π is close to zero which is more practical in case sensitive variables.

Following the Greenberg et al. (1969) model, the optimal choice of π_y is close to π. Thus, we simulated the choice of device parameter P such that the proposed model can perform better than the Greenberg et al. (1969) model under their optimal conditions.

In Table 3, we choose the value of $\pi_y = \pi - 0.05$ for π between 0.1 and 0.5 with a skip of 0.05 and simulate a practical value choice of $P = P_0 = P_1 = 0.3$, such that the proposed estimator is more efficient than the Warner (1965) and the Greenberg et al. (1969) models. We noted that for $\pi = 0.1$, a choice of $\pi_y =$ 0.05 will yield RE(2) value of 405.32% and that of RE(1) value 447.92%. As the value of π increased to 0.35 by 0.05, a choice of $\pi_y = 0.3$, the value of RE(2) is 271.29% and RE(1) is 100%. Thus a good guess of π_y below the true value of π by 0.05, leads to efficient results.

Table 3: RE(1) and RE(2) values under the 1st optimal condition of Greenberg et al. (1969).

π	π_y	**RE(1)**	**RE(2)**
0.10	0.05	447.92	405.32
0.15	0.10	290.28	418.83
0.20	0.15	210.70	395.16
0.25	0.20	160.78	357.35
0.30	0.25	125.77	314.22
0.35	0.30	100.00	271.29

In Table 4, we change the optimal criterion to choose π_y such that, $\pi_y = \pi + 0.05$. We noted that if $\pi = 0.1$, a choice of $\pi_y = 0.15$ will yield RE(2) value 408.80% and RE(1) is 235.12%. Comparing with Table 3 we noted that there is lose in relative efficiency value, if one chosen value of π_y is higher than the value of π in its neighborhood.

Table 4: RE(1) and RE(2) values under the 2nd optimal condition of Greenberg *et al.* (1969).

π	π_y	**RE(1)**	**RE(2)**
0.05	0.10	370.40	454.83
0.10	0.15	235.12	408.80
0.15	0.20	168.63	360.45
0.20	0.25	127.79	312.73
0.25	0.30	100.00	268.35

However, when the value of π is increased to 0.25 with a skip of 0.05, the value of RE(2) becomes 268.35% and RE(1) becomes 100% which are slightly higher than the corresponding values in Table 3. In Table 4, for a value of $\pi = 0.05$, a choice of $\pi_y = 0.1$ will yield RE(2) value of 454.83% and RE(1) value of 370.40%.

In conclusion, the proposed estimator can be used more efficiently than the Warner (1965) and the Greenberg et al. (1969) models by making a choice of device parameters. The original version of this chapter is available in Xu (2019).

References

Greenberg, B. G., Abul-Ela, A. L. A., Simmons, W. R. and Horuitz, D. G. 1969. The unrelated question randomized response model: Theoretical framework, J. Am. Statist. Assoc., 64: 520–539.

Odumade, O. and Singh, S. 2009. Efficient use of two decks of cards in randomized response sampling. Communications in Statistics-Theory and Methods, 38: 439–446.

Warner, S. L. 1965. Randomized response: a survey technique for eliminating evasive answer bias. J. Amer. Statist. Assoc., 60: 63–69.

Xu, T. 2019. On improved randomized response strategies in survey sampling. Unpublished MS Thesis submitted to the Department of Mathematics, Texas A&M University-Kingsville, Kingsville, TX.

APPENDIX A.

PROGRAM 1: SAS CODE FOR RESULT IN TABLE 1.1

```
data data1;
n =1;
p0 = 0.7;
p1 = 0.7;
p = 0.7;
dopie =0.05 to0.5by0.05;
dopiey = 0.85to0.95by0.05;
*variance of the new model;
a1 = (p+piey-1)**2*(p*piey+(1-p)*(1-piey));
a2 = (p-piey)**2*(piey*(1-p)+p*(1-piey));
a3 = (2*pie-1)**2/(4*n);
a4 = 4*n*((p+piey-1)**2+(p-piey)**2)**2;
var_new = ((a1+a2)/a4)-a3;
*variance of warner;
var_w = (pie*(1-pie)+(p0*(1-p0))/(2*(2*p0-1)**2))/n;
*Greenberg;
th_g = p1*pie +(1-p1)*piey;
var_g = (th_g*(1-th_g))/(n*p1**2);
*relative efficiency;
re1 = (var_w)*100/(var_new);
re2 = (var_g)*100/(var_new);
output;
end;
end;
data data2;
set data1;
keep n p0 p1 p pie piey re1 re2;
if re1 ge100 and re2 ge100;
procprintdata = data2;
run;
```

CHAPTER 8

Hybrid of Simple Model and a New Unrelated Question Model for Two Sensitive Characteristics

Renhua Zheng,* Stephen A Sedory* and *Sarjinder Singh*

1. Simple Model for two Sensitive Characteristics

Consider a situation where the population Ω of interest consists of four types of persons; that is, some people belong to either sensitive group A, or sensitive group B, or both $A \cap B$, or none of these. In Figure 1 a Venn diagram has been used to display such a population. In this population of interest, let π_A and π_B be the proportions of people possessing the sensitive attribute A and B, and π_{AB} be the proportion of the people possessing both the sensitive attributes $A \cap B$. Note that $(A \cup B)^C$ may or may not be an empty set.

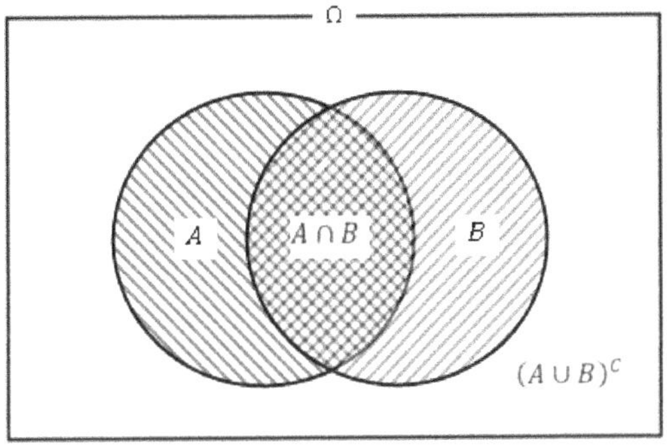

Figure 1: Population of interest.

In the Simple Model developed by Lee, Sedory and Singh (2013), each person selected in a sample, taken using SRSWR of n persons, is asked to experience in order two decks of cards, as shown in Figure 2.

The respondent is asked to keep the cards taken from the first and the second deck in order and match his/her status with the statements written on the drawn cards. Each respondent will obviously fall into one of the four categories with a response of: (Yes, Yes); (Yes, No); (No, Yes), or (No, No).

Following Lee, Sedory and Singh (2013), the probability of observing a (*Yes, Yes*) response is given by:

$$\theta_{11} = (2P-1)(2T-1)\pi_{AB} + (2P-1)(1-T)\pi_A + (1-P)(2T-1)\pi_B + (1-P)(1-T) \tag{1.1}$$

The probability of getting a (*Yes, No*) response is given by:

$$\theta_{10} = -(2P-1)(2T-1)\pi_{AB} + (2P-1)T\pi_A - (1-P)(2T-1)\pi_B + (1-P)T \tag{1.2}$$

$I \in A$ with probability P
$I \in A^c$ with probability $(1-P)$
Deck-I

$I \in B$ with probability T
$I \in B^c$ with probability $(1-T)$
Deck-II

Figure 2: Simple Model: Two decks of cards.

Department of Mathematics, Texas A&M University-Kingsville, Kingsville, TX 78363, USA.
* Corresponding authors: renhua.zheng@students.tamuk.edu; sarjinder.singh@tamuk.edu

The probability of getting a (*No, Yes*) response is given by:

$$\theta_{01} = -(2P-1)(2T-1)\pi_{AB} - (2P-1)(1-T)\pi_A + P(2T-1)\pi_B + P(1-T) \tag{1.3}$$

and the probability of getting a (*No, No*) response is given by:

$$\theta_{00} = (2P-1)(2T-1)\pi_{AB} - T(2P-1)\pi_A - P(2T-1)\pi_B + PT \tag{1.4}$$

Assuming $\hat{\theta}_{11} = n_{11}/n$, $\hat{\theta}_{10} = n_{10}/n$, $\hat{\theta}_{01} = n_{01}/n$ and $\hat{\theta}_{00} = n_{00}/n$ be the observed proportions of (*Yes, Yes*), (*Yes, No*), (*No, Yes*) and (*No, No*) responses, Lee et al. (2013) considered minimizing the distance function between the observed proportions and the true proportions as:

$$D = \frac{1}{2}\sum_{i=0}^{1}\sum_{j=0}^{1}(\theta_{ij} - \hat{\theta}_{ij})^2 \tag{1.5}$$

On setting $\frac{\partial D}{\partial \pi_{AB}} = 0$, $\frac{\partial D}{\partial \pi_A} = 0$, and $\frac{\partial D}{\partial \pi_B} = 0$, they obtained the following unbiased estimators of π_A, π_B and π_{AB} as:

$$\hat{\pi}_A = \frac{\hat{\theta}_{11} + \hat{\theta}_{10} - \hat{\theta}_{01} - \hat{\theta}_{00} + (2P-1)}{2(2P-1)} \tag{1.6}$$

$$\hat{\pi}_B = \frac{\hat{\theta}_{11} - \hat{\theta}_{10} + \hat{\theta}_{01} - \hat{\theta}_{00} + (2T-1)}{2(2T-1)} \tag{1.7}$$

and

$$\hat{\pi}_{AB} = \frac{(P+T)\hat{\theta}_{11} + (T-P)\hat{\theta}_{10} + (P-T)\hat{\theta}_{01} + (2-P-T)\hat{\theta}_{00} - T(1-P) - P(1-T)}{2(2P-1)(2T-1)} \tag{1.8}$$

for $T \neq 0.5$ and $P \neq 0.5$.

The variance of the estimator $\hat{\pi}_A$ is given by:

$$V(\hat{\pi}_A) = \frac{\pi_A(1-\pi_A)}{n} + \frac{P(1-P)}{n(2P-1)^2} \tag{1.9}$$

The variance of the estimator $\hat{\pi}_B$ is given by:

$$V(\hat{\pi}_B) = \frac{\pi_B(1-\pi_B)}{n} + \frac{T(1-T)}{n(2T-1)^2} \tag{1.10}$$

and the variance of the estimator $\hat{\pi}_{AB}$ is given by:

$$V(\hat{\pi}_{AB}) = \frac{\pi_{AB}(1-\pi_{AB})}{n} + \frac{(2P-1)^2T(1-T)\pi_A + P(1-P)(2T-1)^2\pi_B + PT(1-P)(1-T)}{n(2P-1)^2(2T-1)^2} \tag{1.11}$$

The covariance between $\hat{\pi}_A$ and $\hat{\pi}_B$ is given by:

$$Cov(\hat{\pi}_A, \hat{\pi}_B) = \frac{\pi_{AB} - \pi_A\pi_B}{n} \tag{1.12}$$

The covariance between $\hat{\pi}_A$ and $\hat{\theta}_{00}$ is given by:

$$Cov(\hat{\pi}_A, \hat{\theta}_{00}) = \frac{-\theta_{00}}{n(2P-1)}\left[(2P-1)\pi_A + (1-P)\right] \tag{1.13}$$

The covariance between $\hat{\pi}_B$ and $\hat{\theta}_{00}$ is given by:

$$Cov(\hat{\pi}_B, \hat{\theta}_{00}) = \frac{-\theta_{00}}{n(2T-1)}\left[(2T-1)\pi_B + (1-T)\right] \tag{1.14}$$

The covariance between $\hat{\pi}_{AB}$ and $\hat{\pi}_A$ is given by:

$$Cov(\hat{\pi}_{AB}, \hat{\pi}_A) = \frac{\pi_{AB}(1-\pi_A)}{n} + \frac{P(1-P)\pi_B}{n(2P-1)^2} \tag{1.15}$$

The covariance between $\hat{\pi}_{AB}$ and $\hat{\pi}_B$ is given by:

$$Cov(\hat{\pi}_{AB}, \hat{\pi}_B) = \frac{\pi_{AB}(1-\pi_B)}{n} + \frac{T(1-T)\pi_A}{n(2T-1)^2} \tag{1.16}$$

Researchers are often interested in estimating the population proportion of individuals who belong to a sensitive group *A* and possess the second sensitive characteristic of group *B*. As a result, we are interested in estimating the population proportion defined as:

$$\pi_{A|B} = \frac{\pi_{AB}}{\pi_B} \tag{1.17}$$

Lee et al. (2013) consider a natural estimator of the conditional proportion $\pi_{A|B}$ as;

$$\hat{\pi}_{A|B} = \frac{\hat{\pi}_{AB}}{\hat{\pi}_B} \tag{1.18}$$

The bias, to the first order of approximation, in the estimator $\hat{\pi}_{A|B}$ is given by:

$$B(\hat{\pi}_{A|B}) = \pi_{A|B}\left[\frac{V(\hat{\pi}_B)}{\pi_B^2} - \frac{Cov(\hat{\pi}_{AB}, \hat{\pi}_B)}{\pi_{AB}\pi_B}\right] \tag{1.19}$$

The mean squared error, to the first order of approximation, of the estimator $\hat{\pi}_{A|B}$ is given by:

$$MSE(\hat{\pi}_{A|B}) = \pi_{A|B}^2\left[\frac{V(\hat{\pi}_{AB})}{\pi_{AB}^2} + \frac{V(\hat{\pi}_B)}{\pi_B^2} - \frac{2Cov(\hat{\pi}_B, \hat{\pi}_{AB})}{\pi_B\pi_{AB}}\right] \tag{1.20}$$

Researchers may also be interested in the problem of estimating the proportion of those persons possessing only the characteristic *A* and not the characteristic *B* in the population defined as:

$$\pi_{A-B} = \pi_A - \pi_{AB} \tag{1.21}$$

A natural unbiased estimator of π_{A-B} is given by:

$$\hat{\pi}_{A-B} = \hat{\pi}_A - \hat{\pi}_{AB} \tag{1.22}$$

The variance of the estimator of $\hat{\pi}_{A-B}$ is given by:

$$V(\hat{\pi}_{A-B}) = V(\hat{\pi}_A) + V(\hat{\pi}_{AB}) - 2Cov(\hat{\pi}_A, \hat{\pi}_{AB}) \tag{1.23}$$

The proportion of those persons possessing at least one of the two characteristics *A* or *B* in the population is defined as:

$$\pi_{A\cup B} = \pi_A + \pi_B - \pi_{AB} \tag{1.24}$$

A natural unbiased estimator of $\pi_{A\cup B}$ is given by:

$$\hat{\pi}_{A\cup B} = \hat{\pi}_A + \hat{\pi}_B - \hat{\pi}_{AB} \tag{1.25}$$

The variance of the estimator of $\hat{\pi}_{A\cup B}$ is given by:

$$V(\hat{\pi}_{A\cup B}) = V(\hat{\pi}_A) + V(\hat{\pi}_B) + V(\hat{\pi}_{AB}) + 2Cov(\hat{\pi}_A, \hat{\pi}_B) - 2Cov(\hat{\pi}_A, \hat{\pi}_{AB}) - 2Cov(\hat{\pi}_B, \hat{\pi}_{AB}) \tag{1.26}$$

Very often, researchers are found to be interested in the problem of estimating the relative risk of a respondent belonging to characteristic *B* given that the respondent belongs to characteristic *A*.
For example, the relative risk of attribute B given that the attribute A occurred is defined as;

$$RR(B|A) = \frac{\pi_{AB}(1-\pi_A)}{\pi_A(\pi_B - \pi_{AB})} \tag{1.27}$$

An obvious estimator of the RR is given by:

$$\hat{RR}(B|A) = \frac{\hat{\pi}_{AB}(1-\hat{\pi}_A)}{\hat{\pi}_A(\hat{\pi}_B - \hat{\pi}_{AB})} \tag{1.28}$$

The bias in the estimator $\hat{RR}(B|A)$, to the first order of approximation, is given by;

$$B\left(\hat{RR}(B\mid A)\right)=RR\left[\begin{array}{l}\dfrac{V(\hat{\pi}_A)}{\pi_A^2(1-\pi_A)}+\dfrac{V(\hat{\pi}_B)}{(\pi_B-\pi_{AB})^2}+\dfrac{\pi_B V(\hat{\pi}_{AB})}{\pi_{AB}(\pi_B-\pi_{AB})^2}+\dfrac{Cov(\hat{\pi}_A,\hat{\pi}_B)}{\pi_A(1-\pi_A)(\pi_B-\pi_{AB})}\\ -\dfrac{\pi_B Cov(\hat{\pi}_A,\hat{\pi}_{AB})}{\pi_A\pi_{AB}(1-\pi_A)(\pi_B-\pi_{AB})}-\dfrac{(\pi_B+\pi_{AB})Cov(\hat{\pi}_B,\hat{\pi}_{AB})}{\pi_{AB}(\pi_B-\pi_{AB})^2}\end{array}\right] \quad (1.29)$$

The mean squared error of the estimator $\hat{RR}(B|A)$ to the first order of approximation is given by;

$$MSE\left(\hat{RR}(B\mid A)\right)=RR^2\left[\begin{array}{l}\dfrac{\pi_B^2 V(\hat{\pi}_{AB})}{\pi_{AB}^2(\pi_B-\pi_{AB})^2}+\dfrac{V(\hat{\pi}_A)}{\pi_A^2(1-\pi_A)^2}+\dfrac{V(\hat{\pi}_B)}{(\pi_B-\pi_{AB})^2}\\ -\dfrac{2\pi_B Cov(\hat{\pi}_A,\hat{\pi}_{AB})}{\pi_A\pi_{AB}(1-\pi_A)(\pi_B-\pi_{AB})}-\dfrac{2\pi_B Cov(\hat{\pi}_B,\hat{\pi}_{AB})}{\pi_{AB}(\pi_B-\pi_{AB})^2}+\dfrac{2Cov(\hat{\pi}_A,\hat{\pi}_B)}{\pi_A(1-\pi_A)(\pi_B-\pi_{AB})}\end{array}\right] \quad (1.30)$$

Lee et al. (2013) have also considered the estimation of the correlation coefficient between the two sensitive attributes A and B defined as:

$$\rho_{AB}=\frac{\pi_{AB}-\pi_A\pi_B}{\sqrt{\pi_A(1-\pi_A)}\sqrt{\pi_B(1-\pi_B)}} \quad (1.31)$$

Then they considered a natural estimator of the correlation coefficient ρ_{AB} as:

$$\hat{\rho}_{AB}=\frac{\hat{\pi}_{AB}-\hat{\pi}_A\hat{\pi}_B}{\sqrt{\hat{\pi}_A(1-\hat{\pi}_A)}\sqrt{\hat{\pi}_B(1-\hat{\pi}_B)}} \quad (1.32)$$

The bias in the estimator $\hat{\rho}_{AB}$, to the first order of approximation, is:

$$B(\hat{\rho}_{AB})=\rho_{AB}\left[F_4\frac{V(\hat{\pi}_A)}{\pi_A^2}+F_5\frac{V(\hat{\pi}_B)}{\pi_B^2}+F_6\frac{Cov(\hat{\pi}_A,\hat{\pi}_B)}{\pi_A\pi_B}-F_7\frac{Cov(\hat{\pi}_A,\hat{\pi}_{AB})}{\pi_A\pi_{AB}}-F_8\frac{Cov(\hat{\pi}_B,\hat{\pi}_{AB})}{\pi_B\pi_{AB}}\right] \quad (1.33)$$

The mean squared error of the estimator $\hat{\rho}_{AB}$, to the first order of approximation, is given by;

$$MSE(\hat{\rho}_{AB})=\rho_{AB}^2\left[\begin{array}{l}F_1^2\dfrac{V(\hat{\pi}_{AB})}{\pi_{AB}^2}+F_2^2\dfrac{V(\hat{\pi}_A)}{\pi_A^2}+F_3^2\dfrac{V(\hat{\pi}_B)}{\pi_B^2}-\dfrac{2F_1F_2Cov(\hat{\pi}_A,\hat{\pi}_{AB})}{\pi_A\pi_{AB}}\\ -\dfrac{2F_1F_3Cov(\hat{\pi}_B,\hat{\pi}_{AB})}{\pi_B\pi_{AB}}+\dfrac{2F_2F_3Cov(\hat{\pi}_A,\hat{\pi}_B)}{\pi_A\pi_B}\end{array}\right] \quad (1.34)$$

where

$$F_1=\frac{\pi_{AB}}{\pi_{AB}-\pi_A\pi_B}, \quad (1.35)$$

$$F_2=\frac{\pi_A\pi_B}{\pi_{AB}-\pi_A\pi_B}+\frac{(1-2\pi_A)}{2(1-\pi_A)}, \quad (1.36)$$

$$F_3=\frac{\pi_A\pi_B}{\pi_{AB}-\pi_A\pi_B}+\frac{(1-2\pi_B)}{2(1-\pi_B)}, \quad (1.37)$$

$$F_4=\frac{\pi_A\pi_B(1-2\pi_A)}{2(1-\pi_A)(\pi_{AB}-\pi_A\pi_B)}+\frac{\pi_A}{2(1-\pi_A)}+\frac{3(1-2\pi_A)^2}{8(1-\pi_A)^2}, \quad (1.38)$$

$$F_5=\frac{\pi_A\pi_B(1-2\pi_B)}{2(1-\pi_B)(\pi_{AB}-\pi_A\pi_B)}+\frac{\pi_B}{2(1-\pi_B)}+\frac{3(1-2\pi_B)^2}{8(1-\pi_B)^2}, \quad (1.39)$$

$$F_6=\frac{1-2\pi_A-2\pi_B+4\pi_A\pi_B}{4(1-\pi_A)(1-\pi_B)}-\frac{\pi_A\pi_B}{\pi_{AB}-\pi_A\pi_B}+\frac{\pi_A\pi_B}{2(\pi_{AB}-\pi_A\pi_B)}\left[\frac{1-2\pi_B}{1-\pi_B}+\frac{1-2\pi_A}{1-\pi_A}\right] \quad (1.40)$$

$$F_7 = \frac{\pi_{AB}(1-2\pi_A)}{2(1-\pi_A)(\pi_{AB}-\pi_A\pi_B)}, \tag{1.41}$$

and

$$F_8 = \frac{\pi_{AB}(1-2\pi_B)}{2(1-\pi_B)(\pi_{AB}-\pi_A\pi_B)} \tag{1.42}$$

The problem of estimating the difference between two proportions is defined as:

$$\pi_d = \pi_A - \pi_B \tag{1.43}$$

An unbiased estimator of π_d is given by:

$$\hat{\pi}_d = \hat{\pi}_A - \hat{\pi}_B \tag{1.44}$$

The variance of the difference estimator $\hat{\pi}_d$ is given by;

$$V(\hat{\pi}_d) = V(\hat{\pi}_A) + V(\hat{\pi}_B) - 2Cov(\hat{\pi}_A, \hat{\pi}_B) \tag{1.45}$$

We introduce a new paired unrelated question model in the next section and discuss its complications while implementing it in practice.

2. Paired Unrelated Question Model for two Sensitive Questions: A Challenge

In the proposed model, assume we introduce the simultaneous use of two unrelated questions randomized response model consisting of an ordered pair of deck of cards as seen in Figure 3. It leads to a new challenge to be resolved while making use of it in practice.

Deck-I	Deck-II
$I \in A$ with probability P_1	$I \in B$ with probability T_1
$I \in Y$ with probability $(1-P_1)$	$I \in Y$ with probability $(1-T_1)$
Deck-I	Deck-II

Figure 3: Unrelated Question Model: Two decks of cards.

Every participant in the sample is asked to experience the ordered pair of decks of cards as shown in Figure 3. Now if the individual respondent is belonging to both sensitive groups A and B, and he/she draws the first card from the Deck-I with the statement $I \in A$, and the second card from Deck-II with the statement $I \in B$, then the probability of his/her response (Yes, Yes) will be represented as $P_1T_1\pi_{AB}$. Now if a selected respondent draws the first card from Deck-I with a statement $I \in A$ and the second card from Deck-II with statement $I \in Y$, and if he/she is a member of both unrelated groups A and Y, then the probability of his/her response (Yes, Yes) is represented as $P_1(1-T_1)\pi_y.\pi_A$ Now if a selected respondent draws the first card with a statement $I \in Y$ and the second card with a statement $I \in B$, and he/she is a member of both unrelated groups B and Y, then the probability of his/her response (Yes, Yes) will be represented as $(1-P_1)T_1\pi_y\pi_B$. Last, if a selected respondent draws the first card with statement $I \in Y$ from the first deck and the second card from the second deck with statement $I \in Y$, and the selected respondent is a member of the unrelated group Y, then his/her response will be (Yes, Yes) with probability $(1-P_1)(1-T_1)\pi_y$. Note that the response (Yes, Yes) can come from four different types of respondents who either belong to both sensitive groups A and B; or belong to the first sensitive group A and the unrelated group Y; or belong to the unrelated group Y and the second sensitive group B, or belong to only unrelated group Y. The overall probability of observing (Yes, Yes) response is given by:

$$P(Yes, Yes) = \theta_{11}^u = P_1T_1\pi_{AB} + P_1(1-T_1)\pi_A\pi_y + (1-P_1)T_1\pi_y\pi_B + (1-P_1)(1-T_1)\pi_y \tag{2.1}$$

Now if a respondent belongs to the sensitive group A and not to B, and he/she draws the first card with the statement $I \in A$ from the first deck and second card with the statement $I \in B$ from the second deck, then his/her response will be (Yes, No) with probability $P_1T_1(\pi_A - \pi_{AB})$. Now if a selected respondent draws the first card with statement $I \in A$ from the first deck and second card with statement $I \in Y$ from the second deck, and if he/she is a member of the group A but not a member of unrelated group Y, then his/her response will be (Yes, No) with probability $P_1(1-T_1)\pi_A(1-\pi_y)$. Now if a selected respondent draws first card with statement $I \in Y$ and the second card with statement $I \in B$, and he/she is a member of unrelated group Y and not a member of the sensitive group B, then his/her response will be (Yes, No) with probability $(1-P_1)T_1\pi_y(1-\pi_B)$. Last, if a selected respondent draws the first card with statement $I \in Y$ from the first deck and the second card with statement $I \in Y$ from the second deck, then his/her response will be (Yes, No) with probability 0. Note that the response (Yes, No) can come from only three different types of respondents who either belong to the sensitive group A and not to group B; or belongs to the first sensitive group A and not to the unrelated group Y; or belongs to

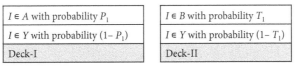

the unrelated group Y and not to the second sensitive group B. Thus the respondents reporting (Yes, No) are at the risk of losing their privacy that they may belong to the sensitive groups A but not to B.
The overall probability of observing (Yes, No) response is given by:

$$\theta_{10}^{u} = -P_1T_1\pi_{AB} + P_1\{T_1 + (1-T_1)(1-\pi_y)\}\pi_A - (1-P_1)T_1\pi_y\pi_B + (1-P_1)T_1\pi_y \quad (2.2)$$

Now if a respondent belongs to the sensitive group B and not to A, and he/she draws the first card with the statement $I \in A$ from the first deck and second card with the statement $I \in B$ from the second deck, then his/her response will be (No, Yes) with probability $P_1T_1(\pi_B - \pi_{AB})$. Now if a selected respondent draws first card with statement $I \in A$ from the first deck and second card with statement $I \in Y$ from the second deck, and if he/she is not a member of the group A and is a member of unrelated group Y, then his/her response will be (No, Yes) with probability $P_1(1 - T_1)\pi_y(1 - \pi_A)$. Now if a selected respondent draws first card with statement $I \in Y$ and the second card with statement $I \in B$, and he/she is not a member of unrelated group Y and is a member of the sensitive group B, then his/her response will be (No, Yes) with probability $(1 - P_1)T_1\pi_B(1 - \pi_y)$. Last, if a selected respondent draws first card with statement $I \in Y$ from the first deck and the second card with statement $I \in Y$ from the second deck, then his/her response will be (No, Yes) with probability 0. Note that the response (No, Yes) can come only from three different types of respondents who either belong to sensitive group B and not A; or does not belong to the first sensitive group A and belongs to unrelated group Y; or does not belong to unrelated group Y and belongs to the sensitive group B. Thus the respondents reporting (No, Yes) are at the risk of losing their privacy that they may belong to the sensitive group B but not to A.
The overall probability of observing (No, Yes) response is given by:

$$\theta_{01}^{u} = -P_1T_1\pi_{AB} + T_1\{P_1 + (1-P_1)(1-\pi_y)\}\pi_B - P_1(1-T_1)\pi_y\pi_A + (1-T_1)P_1\pi_y \quad (2.3)$$

Now if a respondent does not belong to both sensitive groups A and B, and he/she draws the first card with the statement $I \in A$ from the first deck and second card with the statement $I \in B$ from the second deck, then his/her response will be (No, No) with probability $P_1T_1(1 - \pi_A - \pi_B + \pi_{AB})$. Now if a selected respondent draws the first card with statement $I \in A$ from the first deck and second card with statement $I \in Y$ from the second deck, and if he/she is neither a member of groups A nor a member of Y, then his/her response will be (No, No) with probability $P_1(1 - T_1)(1 - \pi_A)(1 - \pi_y)$. Now if a selected respondent draws the first card with statement $I \in Y$ and the second card with statement $I \in B$, and he/she is a not a member of both groups B and Y, then his/her response will be (No, No) with probability $(1 - P_1)T_1(1 - \pi_y)(1 - \pi_B)$. Last, if a selected respondent draws the first card with statement $I \in Y$ from the first deck and the second card with statement $I \in Y$ from the second deck, and he/she is not a member of the group Y, then his/her response will be (No, No) with probability $(1 - P_1)(1 - T_1)(1 - \pi_y)$. Note that the response (No, No) can come four different types of respondents who do not belong to either of the two sensitive groups A and B; or do not belong to the first sensitive group A and do not belong to the unrelated group Y; or do not belong to the unrelated group Y and do not belong to sensitive group B, or do not belong to the unrelated group Y.
The overall probability of observing (No, No) response is given by:

$$\theta_{00}^{u} = P_1T_1\pi_{AB} - P_1\{T_1 + (1-T_1)(1-\pi_y)\}\pi_A - T_1\{P_1 + (1-P_1)(1-\pi_y)\}\pi_B + P_1T_1\pi_y + 1 - \pi_y \quad (2.4)$$

Note that:

$$\sum_{i=0}^{1}\sum_{j=0}^{1}\theta_{ij}^{u} = 1 \quad (2.5)$$

Caution: It should be noted that a (Yes, No) or (No, Yes) response could not have come from an individual who drew an "$I \in Y$" card from both decks. He/she must be addressing at least one of the statements "$I \in A$" or "$I \in B$". While no conclusion can be drawn as to their membership in either of the sensitive groups, and it may result in cooperation being lost. Thus, the use of unrelated question model due to Horvitz, Shah, and Simmons (1967) two-times is quite different than the use of the Warner (1965) model as in the Simple Model. This allows us to think for alternatives to make the use of paired unrelated questions model functional when estimating the two sensitive characteristics. In other words, "A Necessity is the Mother of an Invention." Following Johnson, Sedory and Singh (2019), and Yennum, Sedory and Singh (2019), in the following sections, we develop a new hybrid randomized response two stage technique, which we label the Decision Maker Deck Models I, abbreviated as DMDM-I.

3. Decision Maker Deck Model-I for two Sensitive Characteristics

In the proposed decision maker deck model-I, we use a two-stage question model where the added new stage helps the respondents to make a decision about the choice of model to be used at the second stage while responding to the sensitive questions. As both the respondent and the interviewer are uncertain about the final model being used by a respondent, and hence increases cooperation between both. A pictorial presentation of such a hybrid two-stage model is given in Figure 4.

Use Simple Model	With probability w
Use paired unrelated question model	With probability $(1 - w)$

Figure 4: Decision Maker Deck.

In other words, each selected respondent in the sample is directed to use a Decision Maker Deck, consisting of two types of cards bearing the statements "Use Simple Model" and "Use paired unrelated question model" with known relative frequencies w and $(1-w)$, respectively. Thus, in the proposed DMDM-I, the probability of responding (Yes, Yes) is given by;

$$\begin{aligned}\Psi_{11} &= w\theta_{11} + (1-w)\theta_{11}^{u}\\ &= w[(2P-1)(2T-1)\pi_{AB} + (2P-1)(1-T)\pi_A + (1-P)(2T-1)\pi_B + (1-P)(1-T)]\\ &\quad + (1-w)[P_1T_1\pi_{AB} + P_1(1-T_1)\pi_Y\pi_A + (1-P_1)T_1\pi_Y\pi_B + (1-P_1)(1-T_1)\pi_Y]\\ &= C_1\pi_{AB} + C_2\pi_A + C_3\pi_B + H_1 \end{aligned} \tag{3.1}$$

The probability of responding (*Yes, No*) is given by:

$$\begin{aligned}\Psi_{10} &= w\theta_{10} + (1-w)\theta_{10}^{u}\\ &= w[-(2P-1)(2T-1)\pi_{AB} + (2P-1)T\pi_A - (1-P)(2T-1)\pi_B + (1-P)T]\\ &\quad + (1-w)[-P_1T_1\pi_{AB} + P_1\{T_1 + (1-T_1)(1-\pi_Y)\}\pi_A - (1-P_1)T_1\pi_Y\pi_B + (1-P_1)T_1\pi_Y]\\ &= -[w(2P-1)(2T-1) + (1-w)P_1T_1]\pi_{AB} + [w(2P-1)T + (1-w)P_1\{T_1 + (1-T_1)(1-\pi_Y)\}]\pi_A\\ &\quad - [w(1-P)(2T-1) + (1-w)(1-P_1)T_1\pi_Y]\pi_B + (1-P)Tw + (1-w)(1-P_1)T_1\pi_Y\\ &= -C_1\pi_{AB} + C_4\pi_A - C_3\pi_B + H_2 \end{aligned} \tag{3.2}$$

The probability of responding (*No, Yes*) is given by:

$$\begin{aligned}\Psi_{01} &= w\theta_{01} + (1-w)\theta_{01}^{u}\\ &= w[-(2P-1)(2T-1)\pi_{AB} - (2P-1)(1-T)\pi_A + P(2T-1)\pi_B + (1-T)P]\\ &\quad + (1-w)[-P_1T_1\pi_{AB} + T_1\{P_1 + (1-P_1)(1-\pi_Y)\}\pi_B - P_1(1-T_1)\pi_Y\pi_A + (1-T_1)P_1\pi_Y]\\ &= -[w(2P-1)(2T-1) + (1-w)P_1T_1]\pi_{AB} + [w(2P-1)T + (1-w)P_1\{T_1 + (1-T_1)(1-\pi_Y)\}]\pi_A\\ &\quad - [w(1-P)(2T-1) + (1-w)(1-P_1)T_1\pi_Y]\pi_B + (1-T)Pw + (1-w)(1-T_1)P_1\pi_Y\\ &= -C_1\pi_{AB} - C_2\pi_A + C_5\pi_B + H_3 \end{aligned} \tag{3.3}$$

and the probability of responding (*No, No*) is given by:

$$\begin{aligned}\Psi_{00} &= w\theta_{00} + (1-w)\theta_{00}^{u}\\ &= w[(2P-1)(2T-1)\pi_{AB} - T(2P-1)\pi_A - P(2T-1)\pi_B + PT]\\ &\quad + (1-w)[P_1T_1\pi_{AB} - P_1\{T_1 + (1-\pi_Y)(1-T_1)\}]\pi_A - T_1\{P_1 + (1-P_1)(1-\pi_Y)\}\pi_B + P_1T_1\pi_Y + 1 - \pi_Y]\\ &= [w(2P-1)(2T-1) + (1-w)P_1T_1]\pi_{AB} - [wT(2P-1) + (1-w)P_1\{T_1 + (1-T_1)(1-\pi_Y)\}]\pi_A\\ &\quad - [wP(2T-1) + (1-w)T_1\{P_1 + (1-P_1)(1-\pi_Y)\}]\pi_B + wPT + (1-w)(P_1T_1\pi_Y + 1 - \pi_Y)\\ &= C_1\pi_{AB} - C_4\pi_A - C_5\pi_B + H_4 \end{aligned} \tag{3.4}$$

where

$$C_1 = w(2P-1)(2T-1) + (1-w)P_1T_1 \tag{3.5}$$

$$C_2 = w(2P-1)(1-T) + (1-w)P_1(1-T_1)\pi_Y \tag{3.6}$$

$$C_3 = w(1-P)(2T-1) + (1-w)(1-P_1)T_1\pi_Y \tag{3.7}$$

$$C_4 = w(2P-1)T + (1-w)P_1\{T_1 + (1-T_1)(1-\pi_Y)\} \tag{3.8}$$

$$C_5 = wP(2T-1) + (1-w)T_1\{P_1 + (1-P_1)(1-\pi_Y)\} \tag{3.9}$$

$$H_1 = w(1-P)(1-T) + (1-w)(1-P_1)(1-T_1)\pi_Y \tag{3.10}$$

$$H_2 = w(1-P)T + (1-w)(1-P_1)T_1\pi_Y \tag{3.11}$$

$$H_3 = w(1-T)P + (1-w)(1-T_1)P_1\pi_Y \tag{3.12}$$

and

$$H_4 = wPT + (1-w)(P_1T_1\pi_Y + 1 - \pi_Y) \tag{3.13}$$

Again note that

$$\sum_{i=0}^{1}\sum_{j=0}^{1} \Psi_{ij} = 1 \tag{3.14}$$

Assume we selected a simple random sample with replacement (SRSWR) of n persons from the population of interest Ω. By making the use of DMDM-I, the observed responses are classified into a 2x2 contingency Figure 5.

Observed:	Response-II	
Response-I	Yes	No
Yes	n_{11}	n_{10}
No	n_{01}	n_{00}

Figure 5: Observed responses.

In other words, out of the sample of n persons, there are n_{11} persons who responded (*Yes, Yes*); n_{10} persons responded (*Yes, No*); n_{01} persons responded (*No, Yes*); and n_{00} persons responded (*No, No*). Let $\hat{\Psi}_{11} = \frac{n_{11}}{n}$, $\hat{\Psi}_{10} = \frac{n_{10}}{n}$, $\hat{\Psi}_{01} = \frac{n_{01}}{n}$ and $\hat{\Psi}_{00} = \frac{n_{00}}{n}$ be the unbiased estimators of Ψ_{11}, Ψ_{10}, Ψ_{01} and Ψ_{00}, respectively. Thus n_{ij} (i=0,1; j=0,1) follows multinomial distribution with parameters n and Ψ_{ij}. The probability mass function for n_{ij} is given by:

$$P(n_{ij}) = \binom{n}{n_{11}, n_{10}, n_{01}, n_{00}} \Psi_{11}^{n_{11}} \Psi_{10}^{n_{10}} \Psi_{01}^{n_{01}} \Psi_{00}^{n_{00}} \tag{3.15}$$

Following Odumade and Singh (2009) we considered a squared distance between the observed proportion and the true proportions as:

$$\begin{aligned} D &= \frac{1}{2}\sum_{i=0}^{1}\sum_{j=0}^{1}\left(\Psi_{ij} - \hat{\Psi}_{ij}\right)^2 \\ &= \frac{1}{2}\left(\Psi_{11} - \hat{\Psi}_{11}\right)^2 + \frac{1}{2}\left(\Psi_{10} - \hat{\Psi}_{10}\right)^2 + \frac{1}{2}\left(\Psi_{01} - \hat{\Psi}_{01}\right)^2 + \frac{1}{2}\left(\Psi_{00} - \hat{\Psi}_{00}\right)^2 \end{aligned} \tag{3.16}$$

or equivalently

$$\begin{aligned} D &= \frac{1}{2}\left(C_1\pi_{AB} + C_2\pi_A + C_3\pi_B + H_1 - \hat{\Psi}_{11}\right)^2 + \frac{1}{2}\left(-C_1\pi_{AB} + C_4\pi_A - C_3\pi_B + H_2 - \hat{\Psi}_{10}\right)^2 \\ &+ \frac{1}{2}\left(-C_1\pi_{AB} - C_2\pi_A + C_5\pi_B + H_3 - \hat{\Psi}_{01}\right)^2 + \frac{1}{2}\left(C_1\pi_{AB} - C_4\pi_A - C_5\pi_B + H_4 - \hat{\Psi}_{00}\right)^2 \end{aligned} \tag{3.17}$$

On setting:

$$\frac{\partial D}{\partial \pi_{AB}} = 0 \tag{3.18}$$

we get

$$(C_1\pi_{AB} + C_2\pi_A + C_3\pi_B + H_1 - \hat{\Psi}_{11})(C_1) + (-C_1\pi_{AB} + C_4\pi_A - C_3\pi_B + H_2 - \hat{\Psi}_{10})(-C_1)$$
$$+ (-C_1\pi_{AB} - C_2\pi_A + C_5\pi_B + H_3 - \hat{\Psi}_{01})(-C_1) + (C_1\pi_{AB} - C_4\pi_A - C_5\pi_B + H_4 - \hat{\Psi}_{00})(C_1) = 0$$

or

$$C_1\pi_{AB} + C_2\pi_A + C_3\pi_B + H_1 - \hat{\Psi}_{11} + C_1\pi_{AB} - C_4\pi_A + C_3\pi_B - H_2 + \hat{\Psi}_{10}$$
$$+ C_1\pi_{AB} + C_2\pi_A - C_5\pi_B - H_3 + \hat{\Psi}_{01} + C_1\pi_{AB} - C_4\pi_A - C_5\pi_B + H_4 - \hat{\Psi}_{00} = 0$$

or

$$4C_1\pi_{AB} + 2(C_2 - C_4)\pi_A + 2(C_3 - C_5)\pi_B = \hat{\Psi}_{11} - \hat{\Psi}_{10} - \hat{\Psi}_{01} + \hat{\Psi}_{00} - H_1 + H_2 + H_3 - H_4$$

or

$$4C_1\pi_{AB} + 2(C_2 - C_4)\pi_A + 2(C_3 - C_5)\pi_B = D_1 \quad (3.19)$$

On setting

$$\frac{\partial D}{\partial \pi_A} = 0 \quad (3.20)$$

we get

$$(C_1\pi_{AB} + C_2\pi_A + C_3\pi_B + H_1 - \hat{\Psi}_{11})(C_2) + (-C_1\pi_{AB} + C_4\pi_A - C_3\pi_B + H_2 - \hat{\Psi}_{10})(C_4)$$
$$+ (-C_1\pi_{AB} - C_2\pi_A + C_5\pi_B + H_3 - \hat{\Psi}_{01})(-C_2) + (C_1\pi_{AB} - C_4\pi_A - C_5\pi_B + H_4 - \hat{\Psi}_{00})(-C_4) = 0$$

or

$$C_1C_2\pi_{AB} + C_2^2\pi_A + C_2C_3\pi_B + C_2H_1 - C_2\hat{\Psi}_{11} - C_1C_4\pi_{AB} + C_4^2\pi_A - C_3C_4\pi_B + C_4H_2 - C_4\hat{\Psi}_{10}$$
$$+ C_1C_2\pi_{AB} + C_2^2\pi_A - C_2C_5\pi_B - C_2H_3 + C_2\hat{\Psi}_{01} - C_1C_4\pi_{AB} + C_4^2\pi_A + C_4C_5\pi_B - C_4H_4 + C_4\hat{\Psi}_{00} = 0$$

or

$$2C_1(C_2 - C_4)\pi_{AB} + 2(C_2^2 + C_4^2)\pi_A + (C_2 - C_4)(C_3 - C_5)\pi_B = C_2(\hat{\Psi}_{11} - \hat{\Psi}_{01}) + C_4(\hat{\Psi}_{10} - \hat{\Psi}_{00}) - C_2(H_1 - H_3) - C_4(H_2 - H_4)$$

or

$$2C_1(C_2 - C_4)\pi_{AB} + 2(C_2^2 + C_4^2)\pi_A + (C_2 - C_4)(C_3 - C_5)\pi_B = D_2 \quad (3.21)$$

and, on setting

$$\frac{\partial D}{\partial \pi_B} = 0 \quad (3.22)$$

we get

$$(C_1\pi_{AB} + C_2\pi_A + C_3\pi_B + H_1 - \hat{\Psi}_{11})(C_3) + (-C_1\pi_{AB} + C_4\pi_A - C_3\pi_B + H_2 - \hat{\Psi}_{10})(-C_3)$$
$$+ (-C_1\pi_{AB} - C_2\pi_A + C_5\pi_B + H_3 - \hat{\Psi}_{01})(C_5) + (C_1\pi_{AB} - C_4\pi_A - C_5\pi_B + H_4 - \hat{\Psi}_{00})(-C_5) = 0$$

or

$$C_1C_3\pi_{AB} + C_2C_3\pi_A + C_3^2\pi_B + C_3H_1 - C_3\hat{\Psi}_{11} + C_1C_3\pi_{AB} - C_3C_4\pi_A + C_3^2\pi_B - C_3H_2 + C_4\hat{\Psi}_{10}$$
$$- C_1C_5\pi_{AB} - C_2C_5\pi_A + C_5^2\pi_B + C_5H_3 - C_5\hat{\Psi}_{01} - C_1C_5\pi_{AB} + C_4C_5\pi_A + C_5^2\pi_B - C_5H_4 + C_5\hat{\Psi}_{00} = 0$$

or

$$2C_1(C_3 - C_5)\pi_{AB} + (C_2 - C_4)(C_3 - C_5)\pi_A + 2(C_3^2 + C_5^2)\pi_B$$
$$= C_3(\hat{\Psi}_{11} - \hat{\Psi}_{10}) + C_5(\hat{\Psi}_{01} - \hat{\Psi}_{00}) - C_3(H_1 - H_2) - C_5(H_3 - H_4)$$

or

$$2C_1(C_3 - C_5)\pi_{AB} + (C_2 - C_4)(C_3 - C_5)\pi_A + 2(C_3^2 + C_5^2)\pi_B = D_3 \quad (3.23)$$

where

$$D_1 = \hat{\Psi}_{11} - \hat{\Psi}_{10} - \hat{\Psi}_{01} + \hat{\Psi}_{00} - H_1 + H_2 + H_3 - H_4 \tag{3.24}$$

$$D_2 = C_2(\hat{\Psi}_{11} - \hat{\Psi}_{01}) + C_4(\hat{\Psi}_{10} - \hat{\Psi}_{00}) - C_2(H_1 - H_3) - C_4(H_2 - H_4) \tag{3.25}$$

and

$$D_3 = C_3(\hat{\Psi}_{11} - \hat{\Psi}_{10}) + C_5(\hat{\Psi}_{01} - \hat{\Psi}_{00}) - C_3(H_1 - H_2) - C_5(H_3 - H_4) \tag{3.26}$$

The system of equations in $\frac{\partial D}{\partial \pi_{AB}} = 0$, $\frac{\partial D}{\partial \pi_A} = 0$ and $\frac{\partial D}{\partial \pi_B} = 0$ can be written as:

$$\begin{bmatrix} 4C_1 & 2(C_2 - C_4) & 2(C_3 - C_5) \\ 2C_1(C_2 - C_4) & 2(C_2^2 + C_4^2) & (C_2 - C_4)(C_3 - C_5) \\ 2C_1(C_3 - C_5) & (C_2 - C_4)(C_3 - C_5) & 2(C_3^2 + C_5^2) \end{bmatrix} \begin{bmatrix} \pi_{AB} \\ \pi_A \\ \pi_B \end{bmatrix} = \begin{bmatrix} D_1 \\ D_2 \\ D_3 \end{bmatrix}$$

We apply the Cramer's rule as follows:

$$\Delta = \begin{vmatrix} 4C_1 & 2(C_2 - C_4) & 2(C_3 - C_5) \\ 2C_1(C_2 - C_4) & 2(C_2^2 + C_4^2) & (C_2 - C_4)(C_3 - C_5) \\ 2C_1(C_3 - C_5) & (C_2 - C_4)(C_3 - C_5) & 2(C_3^2 + C_5^2) \end{vmatrix}$$

$$\begin{aligned} &= 16C_1(C_2^2 + C_4^2)(C_3^2 + C_5^2) + 8C_1(C_2 - C_4)^2(C_3 - C_5)^2 - 8C_1(C_3 - C_5)^2(C_2^2 + C_4^2) \\ &\quad - 4C_1(C_2 - C_4)^2(C_3 - C_5)^2 - 8C_1(C_2 - C_4)^2(C_3^2 + C_5^2) \end{aligned} \tag{3.27}$$

$$= 4C_1[2(C_3^2 + C_5^2) - (C_3 - C_5)^2][2(C_2^2 + C_4^2) - (C_2 - C_4)^2]$$

$$\Delta_{AB} = \begin{vmatrix} D_1 & 2(C_2 - C_4) & 2(C_3 - C_5) \\ D_2 & 2(C_2^2 + C_4^2) & (C_2 - C_4)(C_3 - C_5) \\ D_3 & (C_2 - C_4)(C_3 - C_5) & 2(C_3^2 + C_5^2) \end{vmatrix} \tag{3.28}$$

$$\begin{aligned} &= 4D_1(C_2^2 + C_4^2)(C_3^2 + C_5^2) + 2D_2(C_2 - C_4)(C_3 - C_5)^2 + 2D_3(C_2 - C_4)^2(C_3 - C_5) \\ &\quad - 4D_3(C_2^2 + C_4^2)(C_3 - C_5) - 4D_2(C_2 - C_4)(C_3^2 + C_5^2) - D_1(C_2 - C_4)^2(C_3 - C_5)^2 \end{aligned}$$

$$\begin{aligned} &= D_1[4(C_2^2 + C_4^2)(C_3^2 + C_5^2) - (C_2 - C_4)^2(C_3 - C_5)^2] \\ &\quad + D_2[2(C_2 - C_4)(C_3 - C_5)^2 - 4(C_2 - C_4)(C_3^2 + C_5^2)] \\ &\quad + D_3[2(C_2 - C_4)^2(C_3 - C_5) - 4(C_2^2 + C_4^2)(C_3 - C_5)] \end{aligned} \tag{3.29}$$

$$= D_1K_1 + D_2K_2 + D_3K_3$$

$$\Delta_A = \begin{vmatrix} 4C_1 & D_1 & 2(C_3 - C_5) \\ 2C_1(C_2 - C_4) & D_2 & (C_2 - C_4)(C_3 - C_5) \\ 2C_1(C_3 - C_5) & D_3 & 2(C_3^2 + C_5^2) \end{vmatrix} \tag{3.30}$$

$$\begin{aligned} &= D_1[2C_1(C_2 - C_4)(C_3 - C_5)^2] + D_2[8C_1(C_3^2 - C_5^2)] + D_3[4C_1(C_2 - C_4)(C_3 - C_5)] \\ &\quad - D_1[4C_1(C_3^2 + C_5^2)(C_2 - C_4)] - D_2[4C_1(C_3 - C_5)^2] - D_3[4C_1(C_2 - C_4)(C_3 - C_5)] \\ &= D_1\left[2C_1(C_2 - C_4)\{(C_3 - C_5)^2 - 2(C_3^2 + C_5^2)\}\right] + D_2\left[4C_1\{2(C_3^2 + C_5^2) - (C_3 - C_5)^2\}\right] \\ &= D_1L_1 + D_2L_2 \end{aligned} \tag{3.31}$$

$$\Delta_B = \begin{vmatrix} 4C_1 & 2(C_2 - C_4) & D_1 \\ 2C_1(C_2 - C_4) & 2(C_2^2 + C_4^2) & D_2 \\ 2C_1(C_3 - C_5) & (C_2 - C_4)(C_3 - C_5) & D_3 \end{vmatrix} \tag{3.32}$$

$$
\begin{aligned}
&= D_1[2C_1(C_2-C_4)^2(C_3-C_5)] + D_2[4C_1(C_2-C_4)(C_3-C_5)] + D_3[8C_1(C_2^2+C_4^2)] \\
&\quad - D_1[4C_1(C_2^2+C_4^2)(C_3-C_5)] - D_2[4C_1(C_2-C_4)(C_3-C_5)] - D_3[4C_1(C_2-C_4)^2] \\
&= D_1\left[2C_1(C_3-C_5)\left\{(C_2-C_4)^2 - 2(C_2^2+C_4^2)\right\}\right] + D_3\left[4C_1\left\{2(C_2^2+C_4^2)-(C_2-C_4)^2\right\}\right] \\
&= D_1M_1 + D_3M_2
\end{aligned}
\tag{3.33}
$$

where

$$K_1 = 4(C_2^2+C_4^2)(C_3^2+C_5^2) - (C_2-C_4)^2(C_3-C_5)^2 \tag{3.34}$$

$$K_2 = 2(C_2-C_4)(C_3-C_5)^2 - 4(C_2-C_4)(C_3^2+C_5^2) \tag{3.35}$$

$$K_3 = 2(C_2-C_4)^2(C_3-C_5) - 4(C_2^2+C_4^2)(C_3-C_5) \tag{3.36}$$

$$L_1 = 2C_1(C_2-C_4)[(C_3-C_5)^2 - 2(C_3^2+C_5^2)] \tag{3.37}$$

$$L_2 = 4C_1[2(C_3^2+C_5^2) - (C_3-C_5)^2] \tag{3.38}$$

$$M_1 = 2C_1(C_3-C_5)\left[(C_2-C_4)^2 - 2(C_2^2+C_4^2)\right] \tag{3.39}$$

and

$$M_2 = 4C_1[2(C_2^2+C_4^2) - (C_2-C_4)^2] \tag{3.40}$$

We have the following theorems based on the method of moments.

Theorem 3.1. An unbiased estimator of π_A is given by:

$$\hat{\pi}_A^{Amy} = \frac{1}{\Delta}(\hat{\Psi}_{11}W_{11}^a - \hat{\Psi}_{10}W_{10}^a - \hat{\Psi}_{01}W_{01}^a + \hat{\Psi}_{00}W_{00}^a + W_c^a) \tag{3.41}$$

where

$$W_{11}^a = L_1 + C_2L_2 \tag{3.42}$$

$$W_{10}^a = L_1 - C_4L_2 \tag{3.43}$$

$$W_{01}^a = L_1 + C_2L_2 \tag{3.44}$$

$$W_{00}^a = L_1 - C_4L_2 \tag{3.45}$$

and

$$W_c^a = (-H_1+H_2+H_3-H_4)L_1 - [C_2(H_1-H_3)+C_4(H_2-H_4)]L_2 \tag{3.46}$$

Proof. Applying the Cramer's rule and by the method of moments, an estimator of π_A is given by:

$$
\begin{aligned}
\hat{\pi}_A^{Amy} &= \frac{\Delta_A}{\Delta} = \frac{D_1L_1 + D_2L_2}{\Delta} \\
&= \frac{1}{\Delta}\left[\begin{aligned}&(\hat{\Psi}_{11} - \hat{\Psi}_{10} - \hat{\Psi}_{01} + \hat{\Psi}_{00} - H_1 + H_2 + H_3 - H_4)L_1 \\ &+ \left\{C_2(\hat{\Psi}_{11} - \hat{\Psi}_{01}) + C_4(\hat{\Psi}_{10} - \hat{\Psi}_{00}) - C_2(H_1-H_3) - C_4(H_2-H_4)\right\}L_2\end{aligned}\right] \\
&= \frac{1}{\Delta}\left[\begin{aligned}&\hat{\Psi}_{11}(L_1+C_2L_2) - \hat{\Psi}_{10}(L_1-C_4L_2) - \hat{\Psi}_{01}(L_1+C_2L_2) + \hat{\Psi}_{00}(L_1-C_4L_2) \\ &+ (-H_1+H_2+H_3-H_4)L_1 - \{C_2(H_1-H_3)+C_4(H_2-H_4)\}L_2\end{aligned}\right] \\
&= \frac{1}{\Delta}(\hat{\Psi}_{11}W_{11}^a - \hat{\Psi}_{10}W_{10}^a - \hat{\Psi}_{01}W_{01}^a + \hat{\Psi}_{00}W_{00}^a + W_c^a)
\end{aligned}
\tag{3.47}
$$

Taking the expected value of both sides on Equation (3.41), we have

$$E\left(\hat{\pi}_A^{Amy}\right)= E\left[\frac{1}{\Delta}(\hat{\Psi}_{11}W_{11}^a - \hat{\Psi}_{10}W_{10}^a - \hat{\Psi}_{01}W_{01}^a + \hat{\Psi}_{00}W_{00}^a + W_c^a)\right]$$

$$= \frac{1}{\Delta}\left[E(\hat{\Psi}_{11})W_{11}^a - E(\hat{\Psi}_{10})W_{10}^a - E(\hat{\Psi}_{01})W_{01}^a + E(\hat{\Psi}_{00})W_{00}^a + W_c^a\right] \tag{3.48}$$

$$= \frac{1}{\Delta}(\Psi_{11}W_{11}^a - \Psi_{10}W_{10}^a - \Psi_{01}W_{01}^a + \Psi_{00}W_{00}^a + W_c^a)$$

$$= \pi_A$$

which proves the theorem.

Theorem 3.2. An unbiased estimator of π_B is given by:

$$\hat{\pi}_B^{Amy} = \frac{1}{\Delta}(\hat{\Psi}_{11}W_{11}^b - \hat{\Psi}_{10}W_{10}^b - \hat{\Psi}_{01}W_{01}^b + \hat{\Psi}_{00}W_{00}^b + W_c^b) \tag{3.49}$$

where

$$W_{11}^b = M_1 + C_3M_2 \tag{3.50}$$

$$W_{10}^b = M_1 + C_3M_2 \tag{3.51}$$

$$W_{01}^b = M_1 - C_5M_2 \tag{3.52}$$

$$W_{00}^b = M_1 - C_5M_2 \tag{3.53}$$

and

$$W_c^b = (-H_1 + H_2 + H_3 - H_4)M_1 - [C_3(H_1 - H_2) + C_5(H_3 - H_4)]M_2 \tag{3.54}$$

Proof. The Cramer's rule is applied and based on the method of moments, an estimator of π_B is given by:

$$\begin{aligned}\hat{\pi}_B^{Amy} &= \frac{\Delta_B}{\Delta} = \frac{D_1M_1 + D_3M_2}{\Delta}\\ &= \frac{1}{\Delta}\Big[(\hat{\Psi}_{11} - \hat{\Psi}_{10} - \hat{\Psi}_{01} + \hat{\Psi}_{00} - H_1 + H_2 + H_3 - H_4)M_1\\ &\quad + \{C_3(\hat{\Psi}_{11} - \hat{\Psi}_{10}) + C_5(\hat{\Psi}_{01} - \hat{\Psi}_{00}) - C_3(H_1 - H_2) - C_5(H_3 - H_4)\}M_2\Big]\\ &= \frac{1}{\Delta}\begin{bmatrix}\hat{\Psi}_{11}(M_1 + C_3M_2) - \hat{\Psi}_{10}(M_1 + C_3M_2) - \hat{\Psi}_{01}(M_1 - C_5M_2) + \hat{\Psi}_{00}(M_1 - C_5M_2)\\ + (-H_1 + H_2 + H_3 - H_4)M_1 - \{C_3(H_1 - H_2) + C_5(H_3 - H_4)\}M_2\end{bmatrix}\\ &= \frac{1}{\Delta}(\hat{\Psi}_{11}W_{11}^b - \hat{\Psi}_{10}W_{10}^b - \hat{\Psi}_{01}W_{01}^b + \hat{\Psi}_{00}W_{00}^b + W_c^b)\end{aligned} \tag{3.55}$$

Taking the expected value of both sides on Equation (3.49), we have

$$E\left(\hat{\pi}_B^{Amy}\right) = \frac{1}{\Delta}E(\hat{\Psi}_{11}W_{11}^b - \hat{\Psi}_{10}W_{10}^b - \hat{\Psi}_{01}W_{01}^b + \hat{\Psi}_{00}W_{00}^b + W_c^b)$$

$$= \frac{1}{\Delta}\left[E(\hat{\Psi}_{11})W_{11}^b - E(\hat{\Psi}_{10})W_{10}^b - E(\hat{\Psi}_{01})W_{01}^b + E(\hat{\Psi}_{00})W_{00}^b + W_c^b\right] \tag{3.56}$$

$$= \frac{1}{\Delta}(\Psi_{11}W_{11}^b - \Psi_{10}W_{10}^b - \Psi_{01}W_{01}^b + \Psi_{00}W_{00}^b + W_c^b)$$

$$= \pi_B$$

which proves the theorem.

Theorem 3.3. An unbiased estimator of π_{AB} is given by

$$\hat{\pi}_{AB}^{Amy} = \frac{1}{\Delta}(\hat{\Psi}_{11}W_{11}^{ab} - \hat{\Psi}_{10}W_{10}^{ab} - \hat{\Psi}_{01}W_{01}^{ab} + \hat{\Psi}_{00}W_{00}^{ab} + W_c^{ab}) \tag{3.57}$$

where

$$W_{11}^{ab} = K_1 + C_2K_2 + C_3K_3 \tag{3.58}$$

$$W_{10}^{ab} = K_1 - C_4K_2 + C_3K_3 \tag{3.59}$$

$$W_{01}^{ab} = K_1 + C_2K_2 - C_5K_3 \tag{3.60}$$

$$W_{00}^{ab} = K_1 - C_4K_2 - C_5K_3 \tag{3.61}$$

And

$$W_c^{ab} = (-H_1 + H_2 + H_3 - H_4)K_1 - [C_2(H_1 - H_3) + C_4(H_2 - H_4)]K_2 - [C_3(H_1 - H_2) + C_5(H_3 - H_4)]K_3 \tag{3.62}$$

Proof. Applying the Cramer's rule and by the method of moments, we have

$$\hat{\pi}_{AB}^{Amy} = \frac{\Delta_{AB}}{\Delta} = \frac{D_1K_1 + D_2K_2 + D_3K_3}{\Delta}$$

$$= \frac{1}{\Delta}\begin{bmatrix}(\hat{\Psi}_{11} - \hat{\Psi}_{10} - \hat{\Psi}_{01} + \hat{\Psi}_{00} - H_1 + H_2 + H_3 - H_4)K_1 \\ + \{C_2(\hat{\Psi}_{11} - \hat{\Psi}_{01}) + C_4(\hat{\Psi}_{10} - \hat{\Psi}_{00}) - C_2(H_1 - H_3) - C_4(H_2 - H_4)\}K_2 \\ + \{C_3(\hat{\Psi}_{11} - \hat{\Psi}_{10}) + C_5(\hat{\Psi}_{01} - \hat{\Psi}_{00}) - C_3(H_1 - H_2) - C_5(H_3 - H_4)\}K_3\end{bmatrix}$$

$$= \frac{1}{\Delta}\begin{bmatrix}\hat{\Psi}_{11}(K_1 + C_2K_2 + C_3K_3) - \hat{\Psi}_{10}(K_1 - C_4K_2 + C_3K_3) - \hat{\Psi}_{01}(K_1 + C_2K_2 - C_5K_3) \\ + \hat{\Psi}_{00}(K_1 - C_4K_2 - C_5K_3) + (-H_1 + H_2 + H_3 - H_4)K_1 - \{C_2(H_1 - H_3) + C_4(H_2 - H_4)\}K_2 \\ - \{C_3(H_1 - H_2) + C_5(H_3 - H_4)\}K_3\end{bmatrix} \tag{3.63}$$

$$= \frac{1}{\Delta}(\hat{\Psi}_{11}W_{11}^{ab} - \hat{\Psi}_{10}W_{10}^{ab} - \hat{\Psi}_{01}W_{01}^{ab} + \hat{\Psi}_{00}W_{00}^{ab} + W_c^{ab})$$

Taking the expected value on both sides of (3.57) we have

$$E\left(\hat{\pi}_{AB}^{Amy}\right) = \frac{1}{\Delta}E\left[\hat{\Psi}_{11}W_{11}^{ab} - \hat{\Psi}_{10}W_{10}^{ab} - \hat{\Psi}_{01}W_{01}^{ab} + \hat{\Psi}_{00}W_{00}^{ab} + W_c^{ab}\right]$$

$$= \frac{1}{\Delta}\left[E(\hat{\Psi}_{11})W_{11}^{ab} - E(\hat{\Psi}_{10})W_{10}^{ab} - E(\hat{\Psi}_{01})W_{01}^{ab} + E(\hat{\Psi}_{00})W_{00}^{ab} + W_c^{ab}\right] \tag{3.64}$$

$$= \frac{1}{\Delta}\left[\Psi_{11}W_{11}^{ab} - \Psi_{10}W_{10}^{ab} - \Psi_{01}W_{01}^{ab} + \Psi_{00}W_{00}^{ab} + W_c^{ab}\right]$$

$$= \pi_{AB}$$

which proves the theorem.

Theorem 3.3.4. The variance of the unbiased estimator $\hat{\pi}_A^{Amy}$ of π_A is given by

$$V\left(\hat{\pi}_A^{Amy}\right) = \frac{1}{n\Delta^2}\begin{bmatrix}(W_{11}^a)^2\Psi_{11}(1-\Psi_{11}) + (W_{10}^a)^2\Psi_{10}(1-\Psi_{10}) + (W_{01}^a)^2\Psi_{01}(1-\Psi_{01}) + (W_{00}^a)^2\Psi_{00}(1-\Psi_{00}) \\ + 2W_{11}^aW_{10}^a\Psi_{11}\Psi_{10} + 2W_{11}^aW_{01}^a\Psi_{11}\Psi_{01} - 2W_{11}^aW_{00}^a\Psi_{11}\Psi_{00} \\ - 2W_{10}^aW_{01}^a\Psi_{10}\Psi_{01} + 2W_{10}^aW_{00}^a\Psi_{10}\Psi_{00} + 2W_{01}^aW_{00}^a\Psi_{01}\Psi_{00}\end{bmatrix} \tag{3.65}$$

Proof. The variance of the estimator $\hat{\pi}_A^{Amy}$ is given by:

$$V\left(\hat{\pi}_A^{Amy}\right)=V\left[\frac{1}{\Delta}(\hat{\Psi}_{11}W_{11}^a-\hat{\Psi}_{10}W_{10}^a-\hat{\Psi}_{01}W_{01}^a+\hat{\Psi}_{00}W_{00}^a+W_c^a)\right]$$

$$=\frac{1}{\Delta^2}\begin{bmatrix}(W_{11}^a)^2V(\hat{\Psi}_{11})+(W_{10}^a)^2V(\hat{\Psi}_{10})+(W_{01}^a)^2V(\hat{\Psi}_{01})+(W_{00}^a)^2V(\hat{\Psi}_{00})\\-2W_{11}^aW_{10}^a\operatorname{cov}(\hat{\Psi}_{11},\hat{\Psi}_{10})-2W_{11}^aW_{01}^a\operatorname{cov}(\hat{\Psi}_{11},\hat{\Psi}_{01})+2W_{11}^aW_{00}^a\operatorname{cov}(\hat{\Psi}_{11},\hat{\Psi}_{00})\\+2W_{10}^aW_{01}^a\operatorname{cov}(\hat{\Psi}_{10},\hat{\Psi}_{01})-2W_{10}^aW_{00}^a\operatorname{cov}(\hat{\Psi}_{10},\hat{\Psi}_{00})-2W_{01}^aW_{00}^a\operatorname{cov}(\hat{\Psi}_{01},\hat{\Psi}_{00})\end{bmatrix}$$

$$=\frac{1}{n\Delta^2}\begin{bmatrix}(W_{11}^a)^2\Psi_{11}(1-\Psi_{11})+(W_{10}^a)^2\Psi_{10}(1-\Psi_{10})+(W_{01}^a)^2\Psi_{01}(1-\Psi_{01})+(W_{00}^a)^2\Psi_{00}(1-\Psi_{00})\\+2W_{11}^aW_{10}^a\Psi_{11}\Psi_{10}+2W_{11}^aW_{01}^a\Psi_{11}\Psi_{01}-2W_{11}^aW_{00}^a\Psi_{11}\Psi_{00}\\-2W_{10}^aW_{01}^a\Psi_{10}\Psi_{01}+2W_{10}^aW_{00}^a\Psi_{10}\Psi_{00}+2W_{01}^aW_{00}^a\Psi_{01}\Psi_{00}\end{bmatrix}$$

which proves the theorem.

Theorem 3.5. The variance of the unbiased estimator $\hat{\pi}_B^{Amy}$ of π_B is given by:

$$V\left(\hat{\pi}_B^{Amy}\right)=\frac{1}{n\Delta^2}\begin{bmatrix}(W_{11}^b)^2\Psi_{11}(1-\Psi_{11})+(W_{10}^b)^2\Psi_{10}(1-\Psi_{10})+(W_{01}^b)^2\Psi_{01}(1-\Psi_{01})+(W_{00}^b)^2\Psi_{00}(1-\Psi_{00})\\+2W_{11}^bW_{10}^b\Psi_{11}\Psi_{10}+2W_{11}^bW_{01}^b\Psi_{11}\Psi_{01}-2W_{11}^bW_{00}^b\Psi_{11}\Psi_{00}\\-2W_{10}^bW_{01}^b\Psi_{10}\Psi_{01}+2W_{10}^bW_{00}^b\Psi_{10}\Psi_{00}+2W_{01}^bW_{00}^b\Psi_{01}\Psi_{00}\end{bmatrix} \tag{3.66}$$

Proof. The variance of the estimator $\hat{\pi}_B^{Amy}$ is given by

$$V(\hat{\pi}_B^{Amy})=V\left[\frac{1}{\Delta}(\hat{\Psi}_{11}W_{11}^b-\hat{\Psi}_{10}W_{10}^b-\hat{\Psi}_{01}W_{01}^b+\hat{\Psi}_{00}W_{00}^b+W_c^b)\right]$$

$$=\frac{1}{\Delta^2}\begin{bmatrix}(W_{11}^b)^2V(\hat{\Psi}_{11})+(W_{10}^b)^2V(\hat{\Psi}_{10})+(W_{01}^b)^2V(\hat{\Psi}_{01})+(W_{00}^b)^2V(\hat{\Psi}_{00})\\-2W_{11}^bW_{10}^b\operatorname{cov}(\hat{\Psi}_{11},\hat{\Psi}_{10})-2W_{11}^bW_{01}^b\operatorname{cov}(\hat{\Psi}_{11},\hat{\Psi}_{01})+2W_{11}^bW_{00}^b\operatorname{cov}(\hat{\Psi}_{11},\hat{\Psi}_{00})\\+2W_{10}^bW_{01}^b\operatorname{cov}(\hat{\Psi}_{10},\hat{\Psi}_{01})-2W_{10}^bW_{00}^b\operatorname{cov}(\hat{\Psi}_{10},\hat{\Psi}_{00})-2W_{01}^bW_{00}^b\operatorname{cov}(\hat{\Psi}_{01},\hat{\Psi}_{00})\end{bmatrix}$$

$$=\frac{1}{n\Delta^2}\begin{bmatrix}(W_{11}^b)^2\Psi_{11}(1-\Psi_{11})+(W_{10}^b)^2\Psi_{10}(1-\Psi_{10})+(W_{01}^b)^2\Psi_{01}(1-\Psi_{01})+(W_{00}^b)^2\Psi_{00}(1-\Psi_{00})\\+2W_{11}^bW_{10}^b\Psi_{11}\Psi_{10}+2W_{11}^bW_{01}^b\Psi_{11}\Psi_{01}-2W_{11}^bW_{00}^b\Psi_{11}\Psi_{00}\\-2W_{10}^bW_{01}^b\Psi_{10}\Psi_{01}+2W_{10}^bW_{00}^b\Psi_{10}\Psi_{00}+2W_{01}^bW_{00}^b\Psi_{01}\Psi_{00}\end{bmatrix}$$

which proves the theorem.

Theorem 3.6. The variance of the unbiased estimator $\hat{\pi}_{AB}^{Amy}$ of π_{AB} is given by

$$V(\hat{\pi}_{AB}^{Amy})=\frac{1}{n\Delta^2}\begin{bmatrix}(W_{11}^{ab})^2\Psi_{11}(1-\Psi_{11})+(W_{10}^{ab})^2\Psi_{10}(1-\Psi_{10})+(W_{01}^{ab})^2\Psi_{01}(1-\Psi_{01})\\+(W_{00}^{ab})^2\Psi_{00}(1-\Psi_{00})+2W_{11}^{ab}W_{10}^{ab}\Psi_{11}\Psi_{10}+2W_{11}^{ab}W_{01}^{ab}\Psi_{11}\Psi_{01}-2W_{11}^{ab}W_{00}^{ab}\Psi_{11}\Psi_{00}\\-2W_{10}^{ab}W_{01}^{ab}\Psi_{10}\Psi_{01}+2W_{10}^{ab}W_{00}^{ab}\Psi_{10}\Psi_{00}+2W_{01}^{ab}W_{00}^{ab}\Psi_{01}\Psi_{00}\end{bmatrix} \tag{3.67}$$

Proof. The variance of the estimator $\hat{\pi}_{AB}^{Amy}$ is given by:

$$V(\hat{\pi}_{AB}^{Amy})=V\left[\frac{1}{\Delta}(\hat{\Psi}_{11}W_{11}^{ab}-\hat{\Psi}_{10}W_{10}^{ab}-\hat{\Psi}_{01}W_{01}^{ab}+\hat{\Psi}_{00}W_{00}^{ab}+W_c^{ab})\right]$$

$$=\frac{1}{\Delta^2}\begin{bmatrix}(W_{11}^{ab})^2V(\hat{\Psi}_{11})+(W_{10}^{ab})^2V(\hat{\Psi}_{10})+(W_{01}^{ab})^2V(\hat{\Psi}_{01})+(W_{00}^{ab})^2V(\hat{\Psi}_{00})\\-2W_{11}^{ab}W_{10}^{ab}\operatorname{cov}(\hat{\Psi}_{11},\hat{\Psi}_{10})-2W_{11}^{ab}W_{01}^{ab}\operatorname{cov}(\hat{\Psi}_{11},\hat{\Psi}_{01})+2W_{11}^{ab}W_{00}^{ab}\operatorname{cov}(\hat{\Psi}_{11},\hat{\Psi}_{00})\\+2W_{10}^{ab}W_{01}^{ab}\operatorname{cov}(\hat{\Psi}_{10},\hat{\Psi}_{01})-2W_{10}^{ab}W_{00}^{ab}\operatorname{cov}(\hat{\Psi}_{10},\hat{\Psi}_{00})-2W_{01}^{ab}W_{00}^{ab}\operatorname{cov}(\hat{\Psi}_{01},\hat{\Psi}_{00})\end{bmatrix}$$

$$=\frac{1}{n\Delta^2}\begin{bmatrix}(W_{11}^{ab})^2\Psi_{11}(1-\Psi_{11})+(W_{10}^{ab})^2\Psi_{10}(1-\Psi_{10})+(W_{01}^{ab})^2\Psi_{01}(1-\Psi_{01})+(W_{00}^{ab})^2\Psi_{00}(1-\Psi_{00})\\+2W_{11}^{ab}W_{10}^{ab}\Psi_{11}\Psi_{10}+2W_{11}^{ab}W_{01}^{ab}\Psi_{11}\Psi_{01}-2W_{11}^{ab}W_{00}^{ab}\Psi_{11}\Psi_{00}\\-2W_{10}^{ab}W_{01}^{ab}\Psi_{10}\Psi_{01}+2W_{10}^{ab}W_{00}^{ab}\Psi_{10}\Psi_{00}+2W_{01}^{ab}W_{00}^{ab}\Psi_{01}\Psi_{00}\end{bmatrix}$$

which proves the theorem.

Lemma 3.1. An estimator of the variance of the unbiased estimator $\hat{\pi}_A^{Amy}$ of π_A is given by

$$\hat{V}\left(\hat{\pi}_A^{Amy}\right)=\frac{1}{(n-1)\Delta^2}\begin{bmatrix}(W_{11}^a)^2\hat{\Psi}_{11}(1-\hat{\Psi}_{11})+(W_{10}^a)^2\hat{\Psi}_{10}(1-\hat{\Psi}_{10})+(W_{01}^a)^2\hat{\Psi}_{01}(1-\hat{\Psi}_{01})\\+(W_{00}^a)^2\hat{\Psi}_{00}(1-\hat{\Psi}_{00})+2W_{11}^aW_{10}^a\hat{\Psi}_{11}\hat{\Psi}_{10}+2W_{11}^aW_{01}^a\hat{\Psi}_{11}\hat{\Psi}_{01}-2W_{11}^aW_{00}^a\hat{\Psi}_{11}\hat{\Psi}_{00}\\-2W_{10}^aW_{01}^a\hat{\Psi}_{10}\hat{\Psi}_{01}+2W_{10}^aW_{00}^a\hat{\Psi}_{10}\hat{\Psi}_{00}+2W_{01}^aW_{00}^a\hat{\Psi}_{01}\hat{\Psi}_{00}\end{bmatrix} \tag{3.68}$$

Lemma 3.2. An estimator of the variance of the unbiased estimator $\hat{\pi}_B^{Amy}$ of π_B is given by

$$\hat{V}\left(\hat{\pi}_B^{Amy}\right)=\frac{1}{(n-1)\Delta^2}\begin{bmatrix}(W_{11}^b)^2\hat{\Psi}_{11}(1-\hat{\Psi}_{11})+(W_{10}^b)^2\hat{\Psi}_{10}(1-\hat{\Psi}_{10})+(W_{01}^b)^2\hat{\Psi}_{01}(1-\hat{\Psi}_{01})\\+(W_{00}^b)^2\hat{\Psi}_{00}(1-\hat{\Psi}_{00})+2W_{11}^bW_{10}^b\hat{\Psi}_{11}\hat{\Psi}_{10}+2W_{11}^bW_{01}^b\hat{\Psi}_{11}\hat{\Psi}_{01}-2W_{11}^bW_{00}^b\hat{\Psi}_{11}\hat{\Psi}_{00}\\-2W_{10}^bW_{01}^b\hat{\Psi}_{10}\hat{\Psi}_{01}+2W_{10}^bW_{00}^b\hat{\Psi}_{10}\hat{\Psi}_{00}+2W_{01}^bW_{00}^b\hat{\Psi}_{01}\hat{\Psi}_{00}\end{bmatrix} \tag{3.69}$$

Lemma 3.3. An estimator of the variance of the unbiased estimator $\hat{\pi}_{AB}^{Amy}$ of π_{AB} is given by

$$\hat{V}(\hat{\pi}_{AB}^{Amy})=\frac{1}{(n-1)\Delta^2}\begin{bmatrix}(W_{11}^{ab})^2\hat{\Psi}_{11}(1-\hat{\Psi}_{11})+(W_{10}^{ab})^2\hat{\Psi}_{10}(1-\hat{\Psi}_{10})+(W_{01}^{ab})^2\hat{\Psi}_{01}(1-\hat{\Psi}_{01})\\+(W_{00}^{ab})^2\hat{\Psi}_{00}(1-\hat{\Psi}_{00})+2W_{11}^{ab}W_{10}^{ab}\hat{\Psi}_{11}\hat{\Psi}_{10}+2W_{11}^{ab}W_{01}^{ab}\hat{\Psi}_{11}\hat{\Psi}_{01}-2W_{11}^{ab}W_{00}^{ab}\hat{\Psi}_{11}\hat{\Psi}_{00}\\-2W_{10}^{ab}W_{01}^{ab}\hat{\Psi}_{10}\hat{\Psi}_{01}+2W_{10}^{ab}W_{00}^{ab}\hat{\Psi}_{10}\hat{\Psi}_{00}+2W_{01}^{ab}W_{00}^{ab}\hat{\Psi}_{01}\hat{\Psi}_{00}\end{bmatrix} \tag{3.70}$$

We investigate the percent relative efficiency of the proposed estimators in the next section with respect to the three estimators of Lee et al. (2013) for the simple model.

4. Relative Efficiency

The percent relative efficiency of the proposed estimators $\hat{\pi}_A^{Amy}$, $\hat{\pi}_B^{Amy}$ and $\hat{\pi}_{AB}^{Amy}$ with respect to the three estimators $\hat{\pi}_A$, $\hat{\pi}_B$, and $\hat{\pi}_{AB}$, respectively, is given by:

$$RE(1)=\frac{V(\hat{\pi}_A)}{V(\hat{\pi}_A^{Amy})}\times 100\% \tag{4.1}$$

$$RE(2)=\frac{V(\hat{\pi}_B)}{V(\hat{\pi}_B^{Amy})}\times 100\% \tag{4.2}$$

and

$$RE(3)=\frac{V(\hat{\pi}_{AB})}{V(\hat{\pi}_{AB}^{Amy})}\times 100\% \tag{4.3}$$

We wrote SAS Codes to compute the percent relative efficiency values RE(1), RE(2) and RE(3) for various choices of parameters. We observed that the proposed estimators are almost always more efficient that their competitors for any choice of w between 0 and 1. For $w = 1$, the proposed estimators reduce to the three estimators of Lee et al. (2013). For $w = 0$ the proposed estimators show maximum relative efficiency values, but this choice is not recommended as the respondents reporting (Yes, No) or (No, Yes) are at the risk of losing their privacy towards at least one sensitive variable either A or B. Following Lanke (1975, 1976), the randomized response models should be compared based on some protection criterion. Thus, we defined a new protection criterion for the proposed model with respect to the Lee et al. (2013) as follows.

In the proposed model: the conditional probabilities of a respondent belonging to the sensitive group A, given that the respondent replied (Yes, Yes), (Yes, No), (No, Yes) or (No, No) are, respectively, given by:

$$P(A\mid YY)=\frac{\pi_A(1-w)[P_1(1-T_1)\pi_y+P_1T_1\pi_B+(1-P_1)\pi_yT_1\pi_B+(1-P_1)(1-T_1)\pi_y]+\pi_Aw[PT\pi_B+(1-T)P(1-\pi_B)]}{\psi_{11}} \tag{4.4}$$

$$P(A\mid YN)=\frac{\pi_A(1-w)[P_1(1-T_1)(1-\pi_y)+P_1T_1(1-\pi_B)+(1-P_1)\pi_yT_1(1-\pi_B)]+\pi_Aw[PT(1-\pi_B)+(1-T)P\pi_B]}{\psi_{10}} \tag{4.5}$$

$$P(A\mid NY)=\frac{\pi_A(1-w)[(1-P_1)(1-\pi_y)T_1\pi_B]+\pi_Aw[(1-P)T\pi_B+(1-P)(1-T)(1-\pi_B)]}{\psi_{01}} \tag{4.6}$$

and

$$P(A\mid NN)=\frac{\pi_A(1-w)[(1-P_1)(1-\pi_y)(1-T_1)+(1-P_1)(1-\pi_y)(1-\pi_B)T_1]+\pi_Aw[(1-P)T(1-\pi_B)+(1-P)(1-T)\pi_B]}{\psi_{00}} \tag{4.7}$$

Then the least protection of a respondent belonging to the group A is computed as:

$$PROT(A)_{Amy} = Max\left[P(A\,|\,YY), P(A\,|\,YN),\; P(A\,|\,NY), P(A\,|\,NN)\right] \tag{4.8}$$

In the proposed model: the conditional probabilities of a respondent belonging to the sensitive characteristic B, given that the respondent replied (Yes, Yes), (Yes, No), (No, Yes) or (No, No) are respectively given by:

$$P(B\,|\,YY) = \frac{\pi_B(1-w)[(1-P_1)T_1\pi_y + P_1T_1\pi_A + (1-T_1)\pi_y P_1\pi_A + (1-P_1)(1-T_1)\pi_y] + \pi_B w[PT\pi_A + (1-P)T(1-\pi_A)]}{\psi_{11}} \tag{4.9}$$

$$P(B\,|\,YN) = \frac{\pi_B(1-w)P_1(1-T_1)(1-\pi_y) + \pi_B w[P(1-T)\pi_A + (1-P)(1-T)(1-\pi_A)]}{\psi_{10}} \tag{4.10}$$

$$P(B\,|\,NY) = \frac{\pi_B(1-w)[P_1T_1(1-\pi_A) + (1-P_1)T_1(1-\pi_y) + (1-T_1)\pi_y P_1(1-\pi_A)] + \pi_B w[PT(1-\pi_A) + (1-P)T\pi_A]}{\psi_{01}} \tag{4.11}$$

and

$$P(B\,|\,NN) = \frac{\pi_B(1-w)[P_1(1-T_1)(1-\pi_A)(1-\pi_y) + (1-P_1)(1-T_1)(1-\pi_y)] + \pi_B w[P(1-T)(1-\pi_A) + (1-P)(1-T)\pi_A]}{\psi_{00}} \tag{4.12}$$

Then the least protection of a respondent belonging to the characteristic B is computed as:

$$PROT(B)_{Amy} = Max\left[P(B\,|\,YY), P(B\,|\,YN),\; P(B\,|\,NY), P(B\,|\,NN)\right] \tag{4.13}$$

In the proposed model: the conditional probabilities of a respondent belonging to both the sensitive attribute A and attribute B, given that the participant response (Yes, Yes), (Yes, No), (No, Yes) or (No, No) are, respectively, given by:

$$P(AB\,|\,YY) = \frac{[(1-w)\{P_1T_1 + \pi_y(1-P_1T_1)\} + wPT]\pi_{AB}}{\psi_{11}} \tag{4.14}$$

$$P(AB\,|\,YN) = \frac{[(1-w)P_1(1-T_1)(1-\pi_y) + wP(1-T)]\pi_{AB}}{\psi_{10}} \tag{4.15}$$

$$P(AB\,|\,NY) = \frac{[(1-w)(1-P_1)T_1(1-\pi_y) + w(1-P)T]\pi_{AB}}{\psi_{01}} \tag{4.16}$$

and

$$P(AB\,|\,NN) = \frac{[(1-w)(1-P_1)(1-T_1)(1-\pi_y) + w(1-P)(1-T)]\pi_{AB}}{\psi_{00}} \tag{4.17}$$

Then the least protection of a respondent belonging to the characteristic A and B is computed as:

$$PROT(AB)_{Amy} = Max\left[P(AB\,|\,YY), P(AB\,|\,YN),\; P(AB\,|\,NY), P(AB\,|\,NN)\right] \tag{4.18}$$

In the Lee et al. (2013)'s simple model: the conditional probabilities of a respondent to belonging to the sensitive group A, given that the respondent replied (Yes, Yes), (Yes, No), (No, Yes) or (No, No) are, respectively, given by:

$$P(A\,|\,YY)_{Lee} = \frac{\pi_A[PT\pi_B + (1-T)P(1-\pi_B)]}{\theta_{11}} \tag{4.19}$$

$$P(A\,|\,YN)_{Lee} = \frac{\pi_A[PT(1-\pi_B) + (1-T)P\pi_B]}{\theta_{10}} \tag{4.20}$$

$$P(A\,|\,NY)_{Lee} = \frac{\pi_A[(1-P)T\pi_B + (1-P)(1-T)(1-\pi_B)]}{\theta_{01}} \tag{4.21}$$

and

$$P(A\,|\,NN)_{Lee} = \frac{\pi_A[(1-P)T(1-\pi_B) + (1-P)(1-T)\pi_B]}{\theta_{00}} \tag{4.22}$$

Then the least protection of a respondent belonging to the group A is computed as:

$$PROT(A)_{Lee} = Max\left[P(A|YY)_{Lee}, P(A|YN)_{Lee}, \ P(A|NY)_{Lee}, P(A|NN)_{Lee}\right] \tag{4.23}$$

In the Lee et al.'s simple model: the conditional probabilities of a respondent of belonging to the sensitive group B, given that the respondent replied (Yes, Yes), (Yes, No), (No, Yes) or (No, No) are, respectively, given by

$$P(B|YY)_{Lee} = \frac{\pi_B[PT\pi_A + (1-P)T(1-\pi_A)]}{\theta_{11}} \tag{4.24}$$

$$P(B|YN)_{Lee} = \frac{\pi_B[P(1-T)\pi_A + (1-P)(1-T)(1-\pi_A)]}{\theta_{10}} \tag{4.25}$$

$$P(B|NY)_{Lee} = \frac{\pi_B[PT(1-\pi_A) + (1-P)T\pi_A]}{\theta_{01}} \tag{4.26}$$

and

$$P(B|NN)_{Lee} = \frac{\pi_B[P(1-T)(1-\pi_A) + (1-P)(1-T)\pi_A]}{\theta_{00}} \tag{4.27}$$

Then the least protection of a respondent belonging to the group B is computed as:

$$PROT(B)_{Lee} = Max\left[P(B|YY)_{Lee}, P(B|YN)_{Lee}, \ P(B|NY)_{Lee}, P(B|NN)_{Lee}\right] \tag{4.28}$$

In the Lee et al.'s simple model: the conditional probabilities of a respondent belonging to both the sensitive groups A and B, given that the respondent replied (Yes, Yes), (Yes, No), (No, Yes) or (No, No) are, respectively, given by:

$$P(AB|YY)_{Lee} = \frac{PT\pi_{AB}}{\theta_{11}} \tag{4.29}$$

$$P(AB|YN)_{Lee} = \frac{P(1-T)\pi_{AB}}{\theta_{10}} \tag{4.30}$$

$$P(AB|NY)_{Lee} = \frac{(1-P)T\pi_{AB}}{\theta_{01}} \tag{4.31}$$

and

$$P(AB|NN)_{Lee} = \frac{(1-P)(1-T)\pi_{AB}}{\theta_{00}} \tag{4.32}$$

Then the least protection of a respondent belonging to the group A and B is computed as:

$$PROT(AB)_{Lee} = Max\left[P(AB|YY)_{Lee}, P(AB|YN)_{Lee}, \ P(AB|NY)_{Lee}, P(AB|NN)_{Lee}\right] \tag{4.33}$$

The percent relative protection of the proposed model with respect to the Lee et al. simple model for three types of respondents possessing membership in group A, B or A and B are, respectively,
defined as:

$$RP(1) = \frac{PROT\ (A)_{Lee}}{PROT\ (A)_{Amy}} \times 100\% \tag{4.34}$$

$$RP(2) = \frac{PROT\ (B)_{Lee}}{PROT\ (B)_{Amy}} \times 100\% \tag{4.35}$$

and

$$RP(3) = \frac{PROT\ (AB)_{Lee}}{PROT\ (AB)_{Amy}} \times 100\% \tag{4.36}$$

We also examined the relative protection of the respondents by using the SAS Macro provided in Appendix A. Again, we found a lot of simulated situations where the proposed randomized response model is both more protective and more efficient than the simple model of Lee et al. (2013) when estimating the three proportions viz. π_A, π_B and π_{AB}. The behavior of the simulated results is very similar to the situation where one is comparing the Greenberg et al. (1969)'s unrelated model with the Warner (1965) model based on the criteria of both efficiency and protection. Useful choices of the device parameters P, T and π_y depend on the parameters of interest π_A, π_B and π_{AB}. Although a broad choice of such parameters does not exist, there are many situations that have been simulated by the SAS Codes given in Appendix A such that the proposed model gives more efficient and protective estimates of all three parameters π_A, π_B and π_{AB}. For illustration purposes, a set of such simulated results for $P = T = 0.7$, that includes many choices of the other parameters π_y, P_1, T_1 and W are listed in Table B1. in Appendix B. The values of w considered are between 0.1 and 0.9 with a step size of 0.1; the values of π_y are between 0.9 and 1.0 with a step size of 0.05; the values of P_1 are 0.4 and 0.5; the values of P_2 are also 0.4 and 0.5; the values of π_{AB} considered between 0.05 to 0.20 with a step size of 0.05; the values of π_A and π_B considered between 0.1 and 0.35 with a step size of 0.05 with an additional restrictions that $\pi_{AB} < \pi_A$ and $\pi_{AB} < \pi_B$. There are 1604 results; in Table B1. of Appendix B which show percent relative efficiencies and relative protection values greater than 100%. Here we providing a summary of such detailed results for a various choice of parameters. Table 1 provides a summary of the values of RE(1), RE(2), RE(3), RP(1), RP(2), and RP(3) for different choice of w.

Table 1: Summary of the values of RE(1), RE(2), RE(3), RP(1), RP(2) and RP(3) for different choices of w.

Variable	w	Freq	Mean	StDev	Minimum	Maximum
RE(1)	0.1	526	135.35	20.07	100.31	162.85
	0.3	394	126.43	13.80	100.56	142.29
	0.5	280	118.72	7.72	100.00	126.41
	0.7	202	111.56	1.79	107.79	113.95
	0.9	202	103.45	0.558	102.20	104.27
RE(2)	0.1	526	135.35	20.07	100.31	162.85
	0.3	394	126.43	13.80	100.56	142.29
	0.5	280	118.72	7.72	100.00	126.41
	0.7	202	111.56	1.79	107.79	113.95
	0.9	202	103.45	0.558	102.20	104.27
RE(3)	0.1	526	181.73	33.88	116.61	260.44
	0.3	394	162.35	22.62	108.16	208.55
	0.5	280	144.74	13.30	119.40	167.67
	0.7	202	127.66	3.01	120.00	135.31
	0.9	202	107.97	0.892	105.53	110.11
RP(1)	0.1	526	109.81	6.82	100.04	125.71
	0.3	394	106.78	5.00	100.03	121.31
	0.5	280	104.08	3.08	100.02	114.83
	0.7	202	102.00	1.39	100.01	105.47
	0.9	202	100.76	0.594	100.00	102.21
RP(2)	0.1	526	109.81	6.82	100.04	125.71
	0.3	394	106.78	5.00	100.03	121.31
	0.5	280	104.08	3.08	100.02	114.83
	0.7	202	102.00	1.39	100.01	105.47
	0.9	202	100.76	0.594	100.00	102.21
RP(3)	0.1	526	125.23	8.08	112.32	148.69
	0.3	394	119.42	5.69	110.73	139.32
	0.5	280	113.73	3.40	108.72	121.59
	0.7	202	108.25	1.25	106.06	111.90
	0.9	202	103.24	0.500	102.36	104.72

If $w = 0.1$ then there are 526 useful choices of the other parameters, which have an average value of RE(1) of 135.35% with a standard deviation of 20.07%, a minimum value of 100.31% and a maximum value of 162.85%; the average value of RE(2) is also 135.35% with a standard deviation of 20.07%, a minimum value of 100.31% and a maximum value of 162.85%. The underlying reason is clear, because we consider the values of π_A and π_B between 0.1 and 0.35 with a step size of 0.05 in the simulation study. RE(3) has an average value of 181.73% with a standard deviation of 33.88%, a minimum value of 116.61%, and a maximum value of 260.44%. RP(1) (or RP(2)) has an average value of 109.81% with a standard deviation of 6.82%, a minimum value of 100.04%, and a maximum value of 125.71%. RP(3) has an average value of 125.23% with a standard deviation of 8.08%, a minimum value of 112.32%, and maximum value of 148.69%. As the value of w increases to 0.9 then the number of situations where the proposed model shows efficient results reduces to 202. Thus in practice, the choice of w close to 0.1 is suggested so that maximum gain in efficiencies and protection are achieved. A graphical representation of such results is also shown in Figure 6.

In Figure 6 the notations RE_PIA, RE_PIB, RE_PIAB stand for RE(1), RE(2) and RE(3) respectively, and the notations RPA, RPB and RPAB stand for RP(1), RP(2) nad RP(3) respectively. For demonstration purposes, we extract a few results from the detailed results in Appendix B as cited in Table 2. For $\pi_A = 0.1$, $\pi_B = 0.1$ and $\pi_{AB} = 0.05$, RE(1) and RE(2) have an average value of 117.23% with a standard deviation of 11.90%, minimum value of 102.20%, and a maximum value of 135.15%, RE(3) has an average value of 143.17% with a standard deviation of 30.06% a minimum value of 105.53%, and a maximum value of 188.10%. RP(1) and RP(2) have an average value of 103.40% with a standard deviation of 3.13% a minimum value of 100.18%, and a maximum value of 110.64%, while RP(3) has an average value of 114.79% with a standard deviation of 6.84%, a minimum value of 103.93%, and a maximum value of 124.19%. Thus, the proposed model is found to be quite helpful in estimating such tiny proportions of sensitive attributes and their overlaps.

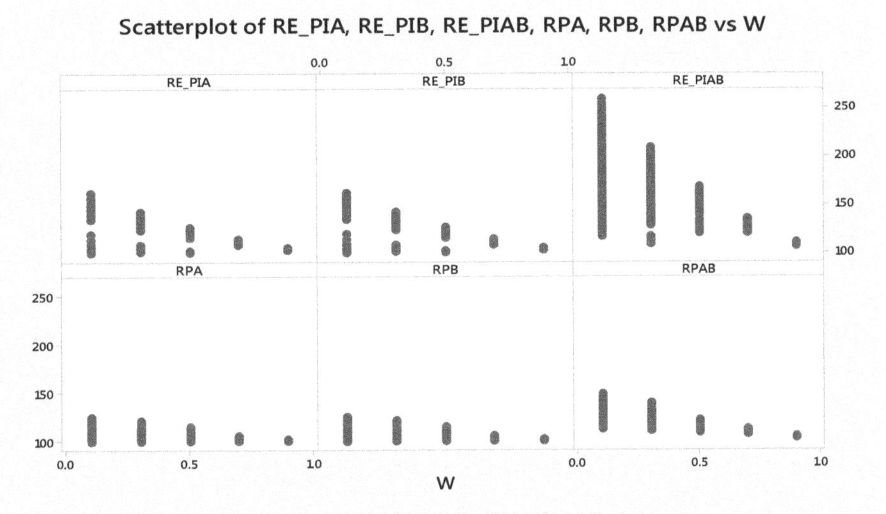

Figure 6: RE(1), RE(2), RE(3), RP(1), RP(2) and RP(3) values for different choice of w.

Table 2: Summary of the values of RE(1), RE(2), RE(3), RP(1), RP(2) and RP(3) for different choices of π_A, π_B and π_{AB}.

Results for $\pi_A = 0.10$ and $\pi_B = 0.10$

Variable	π_{AB}	Freq	Mean	StDev	Minimum	Maximum
RE(1)	0.05	15	117.23	11.90	102.20	135.15
RE(2)	0.05	15	117.23	11.90	102.20	135.15
RE(3)	0.05	15	143.17	30.06	105.53	188.10
RP(1)	0.05	15	103.40	3.13	100.18	110.64
RP(2)	0.05	15	103.40	3.13	100.18	110.64
RP(3)	0.05	15	114.79	6.84	103.93	124.19

Results for $\pi_A = 0.10$ and $\pi_B = 0.15$

Variable	π_{AB}	Freq	Mean	StDev	Minimum	Maximum
RE(1)	0.05	15	117.23	11.90	102.20	135.15
RE(2)	0.05	15	119.28	13.28	102.59	139.90
RE(3)	0.05	15	145.37	31.63	105.90	193.21
RP(1)	0.05	15	103.32	3.15	100.16	110.63
RP(2)	0.05	15	103.07	2.80	100.17	109.54
RP(3)	0.05	15	114.11	6.54	103.74	123.15

Likewise, the results for $\pi_A = 0.1$, $\pi_B = 0.15$ and $\pi_{AB} = 0.05$ can be seen from Table 2. In addition, results for any such combination of the three proportions π_A, π_B and π_{AB} can be extracted, if needed, from the detailed results give in Table B1. in Appendix B. Table 3 helps to make an appropriate choice of the proportion of unrelated question π_y.

Table 3: RE(1), RE(2), RE(3), RP(1), RP(2) and RP(3) for different values of π_y.

Variable	π_y	Freq	Mean	StDev	Minimum	Maximum
RE(1)	0.90	280	122.52	16.52	100.00	154.93
	0.95	582	122.98	17.46	100.31	158.45
	1.00	742	123.72	18.12	100.88	162.85
RE(2)	0.90	280	122.52	16.52	100.00	154.93
	0.95	582	122.98	17.46	100.31	158.45
	1.00	742	123.72	18.12	100.88	162.85
RE(3)	0.90	280	151.31	33.17	106.43	228.31
	0.95	582	153.96	33.44	105.98	243.13
	1.00	742	155.94	35.19	105.53	260.44
RP(1)	0.90	280	103.76	4.75	100.00	120.87
	0.95	582	105.07	5.44	100.07	123.05
	1.00	742	107.45	6.16	100.21	125.71
RP(2)	0.90	280	103.76	4.75	100.00	120.87
	0.95	582	105.07	5.44	100.07	123.05
	1.00	742	107.45	6.16	100.21	125.71
RP(3)	0.90	280	113.76	7.54	102.36	139.66
	0.95	582	116.39	9.10	102.41	143.97
	1.00	742	118.45	10.16	102.40	148.69

For $\pi_y = 1$ there are combinations of parameter *freq* = 742 where the proposed estimators are more efficient and more protective than the simple model of Lee et al. (2013), and it gives us a very truthful message that in the proposed model "Forced Yes" response could be used instead of trying to find an appropriate unrelated characteristic in a population under study. This finding through simulation study enhances the usefulness of the proposed model in practice. Behavior of the values of RE(1), RE(2), RE(3), RP(1), RP(2) and RP(3) versus π_y are shown in Figure 7. In Figure 7 the notation Y stands for π_y.

Table 4 provides a summary of results for different choices of P_1 and T_1. For $P_1 = 0.4$ and $T_1 = 0.4$, the average value of RE(1) (or RE(2)) is 112.08% with a standard deviation of 5.55%, and a minimum value of 100.88% and a maximum value of 119.67%; the value of RE(3) is 124.60% with a standard deviation of 8.75%, a minimum value of 108.16% and a maximum value of 144.46%.

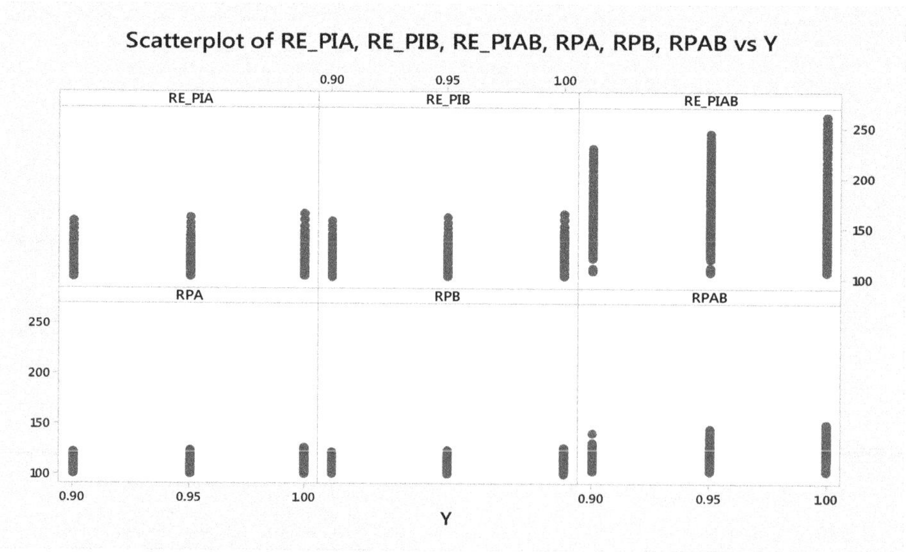

Figure 7: RE(1), RE(2), RE(3), RP(1), RP(2), RP(3) versus π_y.

Table 4: RE(1), RE(2), RE(3), RP(1), RP(2) and RP(3) for different P_1 and T_1.

Results for $P_1 = 0.4$

Variable	T_1	Freq	Mean	StDev	Minimum	Maximum
RE(1)	0.4	58	112.08	5.55	100.88	119.67
	0.5	268	106.33	5.52	100.00	119.67
RE(2)	0.4	58	112.08	5.55	100.88	119.67
	0.5	268	140.63	12.32	114.99	162.85
RE(3)	0.4	58	124.60	8.75	108.16	144.46
	0.5	268	151.06	17.04	119.40	194.52
RP(1)	0.4	58	118.59	3.20	111.21	125.71
	0.5	268	115.11	4.19	106.24	125.31
RP(2)	0.4	58	118.59	3.20	111.21	125.71
	0.5	268	104.97	3.06	100.09	111.15
RP(3)	0.4	58	140.55	4.32	131.49	148.69
	0.5	268	125.34	4.65	115.10	136.94

Results for $P_1 = 0.5$

Variable	T_1	Freq	Mean	StDev	Minimum	Maximum
RE(1)	0.4	268	140.63	12.32	114.99	162.85
	0.5	1010	123.76	16.70	102.20	162.85
RE(2)	0.4	268	106.33	5.52	100.00	119.67
	0.5	1010	123.76	16.70	102.20	162.85
RE(3)	0.4	268	151.06	17.04	119.40	194.52
	0.5	1010	157.91	40.43	105.53	260.44
RP(1)	0.4	268	104.97	3.06	100.09	111.15
	0.5	1010	103.04	2.58	100.00	111.04
RP(2)	0.4	268	115.11	4.19	106.24	125.31
	0.5	1010	103.04	2.58	100.00	111.04
RP(3)	0.4	268	125.34	4.65	115.10	136.94
	0.5	1010	111.04	5.20	102.36	124.19

The average value of RP(1) (or RP(2)) is 118.59% with a standard deviation of 3.20%, with a minimum value of 111.21% and a maximum value of 125.71%, the average value of RP(3) is 140.55% with a standard deviation of 4.32%, with a minimum of 131.49% and a maximum of 148.69%. In the same way the other results from Table 4 can be interpreted.

We consider the problem of estimating other parameters in the next section such as conditional proportion, proportion of membership in at least one group, difference between two proportions, relative risk, and correlation between two sensitive characteristics.

5. Estimation of Other Parameters

For estimating the other parameters such as $\pi_{A|B}$, π_{A-B}, $\pi_{A\cup B}$, $RR(B|A)$, and ρ_{AB}, we need the following lemmas.

Lemma 5.1. The covariance between $Cov(\hat{\pi}_A^{Amy}, \hat{\pi}_B^{Amy})$ is given by

$$\begin{aligned} Cov\left(\hat{\pi}_A^{Amy}, \hat{\pi}_B^{Amy}\right) = \frac{1}{n\Delta^2}\Big[& W_{11}^a W_{11}^b \Psi_{11}(1-\Psi_{11}) + W_{11}^a W_{10}^b \Psi_{11}\Psi_{10} + W_{11}^a W_{01}^b \Psi_{11}\Psi_{01} - W_{11}^a W_{00}^b \Psi_{11}\Psi_{00} \\ & + W_{10}^a W_{11}^b \Psi_{10}\Psi_{11} + W_{10}^a W_{10}^b \Psi_{10}(1-\Psi_{10}) - W_{10}^a W_{01}^b \Psi_{10}\Psi_{01} + W_{10}^a W_{00}^b \Psi_{10}\Psi_{00} \\ & + W_{01}^a W_{11}^b \Psi_{01}\Psi_{11} - W_{01}^a W_{10}^b \Psi_{01}\Psi_{10} + W_{01}^a W_{01}^b \Psi_{01}(1-\Psi_{01}) + W_{01}^a W_{00}^b \Psi_{01}\Psi_{00} \\ & - W_{00}^a W_{11}^b \Psi_{00}\Psi_{11} + W_{00}^a W_{10}^b \Psi_{00}\Psi_{10} + W_{00}^a W_{01}^b \Psi_{00}\Psi_{01} + W_{00}^a W_{00}^b \Psi_{00}(1-\Psi_{00})\Big] \end{aligned} \quad (5.1)$$

Proof. We have

$$Cov(\hat{\pi}_A^{Amy}, \hat{\pi}_B^{Amy})$$

$$= Cov\left\{\frac{1}{\Delta}(\hat{\Psi}_{11}W_{11}^a - \hat{\Psi}_{10}W_{10}^a - \hat{\Psi}_{01}W_{01}^a + \hat{\Psi}_{00}W_{00}^a + W_c^a), \frac{1}{\Delta}(\hat{\Psi}_{11}W_{11}^b - \hat{\Psi}_{10}W_{10}^b - \hat{\Psi}_{01}W_{01}^b + \hat{\Psi}_{00}W_{00}^b + W_c^b)\right\}$$

$$= \frac{1}{\Delta^2}Cov\left[\hat{\Psi}_{11}W_{11}^a - \hat{\Psi}_{10}W_{10}^a - \hat{\Psi}_{01}W_{01}^a + \hat{\Psi}_{00}W_{00}^a + W_c^a, \hat{\Psi}_{11}W_{11}^b - \hat{\Psi}_{10}W_{10}^b - \hat{\Psi}_{01}W_{01}^b + \hat{\Psi}_{00}W_{00}^b + W_c^b\right]$$

$$\begin{aligned}
= \frac{1}{\Delta^2}\Big[& W_{11}^aW_{11}^bCov(\hat{\Psi}_{11},\hat{\Psi}_{11}) - W_{11}^aW_{10}^bCov(\hat{\Psi}_{11},\hat{\Psi}_{10}) - W_{11}^aW_{01}^bCov(\hat{\Psi}_{11},\hat{\Psi}_{01}) + W_{11}^aW_{00}^bCov(\hat{\Psi}_{11},\hat{\Psi}_{00}) \\
& - W_{10}^aW_{11}^bCov(\hat{\Psi}_{10},\hat{\Psi}_{11}) + W_{10}^aW_{10}^bCov(\hat{\Psi}_{10},\hat{\Psi}_{10}) + W_{10}^aW_{01}^bCov(\hat{\Psi}_{10},\hat{\Psi}_{01}) - W_{10}^aW_{00}^bCov(\hat{\Psi}_{10},\hat{\Psi}_{00}) \\
& - W_{01}^aW_{11}^bCov(\hat{\Psi}_{01},\hat{\Psi}_{11}) + W_{01}^aW_{10}^bCov(\hat{\Psi}_{01},\hat{\Psi}_{10}) + W_{01}^aW_{01}^bCov(\hat{\Psi}_{01},\hat{\Psi}_{01}) - W_{01}^aW_{00}^bCov(\hat{\Psi}_{01},\hat{\Psi}_{00}) \\
& + W_{00}^aW_{11}^bCov(\hat{\Psi}_{00},\hat{\Psi}_{11}) - W_{00}^aW_{10}^bCov(\hat{\Psi}_{00},\hat{\Psi}_{10}) - W_{00}^aW_{01}^bCov(\hat{\Psi}_{00},\hat{\Psi}_{01}) + W_{00}^aW_{00}^bCov(\hat{\Psi}_{00},\hat{\Psi}_{00})\Big]
\end{aligned}$$

$$\begin{aligned}
= \frac{1}{n\Delta^2}\Big[& W_{11}^aW_{11}^b\Psi_{11}(1-\Psi_{11}) + W_{11}^aW_{10}^b\Psi_{11}\Psi_{10} + W_{11}^aW_{01}^b\Psi_{11}\Psi_{01} - W_{11}^aW_{00}^b\Psi_{11}\Psi_{00} \\
& + W_{10}^aW_{11}^b\Psi_{10}\Psi_{11} + W_{10}^aW_{10}^b\Psi_{10}(1-\Psi_{10}) - W_{10}^aW_{01}^b\Psi_{10}\Psi_{01} + W_{10}^aW_{00}^b\Psi_{10}\Psi_{00} \\
& + W_{01}^aW_{11}^b\Psi_{01}\Psi_{11} - W_{01}^aW_{10}^b\Psi_{01}\Psi_{10} + W_{01}^aW_{01}^b\Psi_{01}(1-\Psi_{01}) + W_{01}^aW_{00}^b\Psi_{01}\Psi_{00} \\
& - W_{00}^aW_{11}^b\Psi_{00}\Psi_{11} + W_{00}^aW_{10}^b\Psi_{00}\Psi_{10} + W_{00}^aW_{01}^b\Psi_{00}\Psi_{01} + W_{00}^aW_{00}^b\Psi_{00}(1-\Psi_{00})\Big]
\end{aligned}$$

which proves the lemma.

Lemma 5.2. The covariance between $Cov(\hat{\pi}_A^{Amy}, \hat{\pi}_{AB}^{Amy})$ is given by

$$\begin{aligned}
Cov\left(\hat{\pi}_A^{Amy}, \hat{\pi}_{AB}^{Amy}\right) = \frac{1}{n\Delta^2}\Big[& W_{11}^aW_{11}^{ab}\Psi_{11}(1-\Psi_{11}) + W_{11}^aW_{10}^{ab}\Psi_{11}\Psi_{10} + W_{11}^aW_{01}^{ab}\Psi_{11}\Psi_{01} - W_{11}^aW_{00}^{ab}\Psi_{11}\Psi_{00} \\
& + W_{10}^aW_{11}^{ab}\Psi_{10}\Psi_{11} + W_{10}^aW_{10}^{ab}\Psi_{10}(1-\Psi_{10}) - W_{10}^aW_{01}^{ab}\Psi_{10}\Psi_{01} + W_{10}^aW_{00}^{ab}\Psi_{10}\Psi_{00} \\
& + W_{01}^aW_{11}^{ab}\Psi_{01}\Psi_{11} - W_{01}^aW_{10}^{ab}\Psi_{01}\Psi_{10} + W_{01}^aW_{01}^{ab}\Psi_{01}(1-\Psi_{01}) + W_{01}^aW_{00}^{ab}\Psi_{01}\Psi_{00} \\
& - W_{00}^aW_{11}^{ab}\Psi_{00}\Psi_{11} + W_{00}^aW_{10}^{ab}\Psi_{00}\Psi_{10} + W_{00}^aW_{01}^{ab}\Psi_{00}\Psi_{01} + W_{00}^aW_{00}^{ab}\Psi_{00}(1-\Psi_{00})\Big]
\end{aligned} \tag{5.2}$$

Proof. we have

$$Cov(\hat{\pi}_A^{Amy}, \hat{\pi}_{AB}^{Amy})$$

$$= Cov\left\{\frac{1}{\Delta}(\hat{\Psi}_{11}W_{11}^a - \hat{\Psi}_{10}W_{10}^a - \hat{\Psi}_{01}W_{01}^a + \hat{\Psi}_{00}W_{00}^a + W_c^a), \frac{1}{\Delta}(\hat{\Psi}_{11}W_{11}^{ab} - \hat{\Psi}_{10}W_{10}^{ab} - \hat{\Psi}_{01}W_{01}^{ab} + \hat{\Psi}_{00}W_{00}^{ab} + W_c^{ab})\right\}$$

$$= \frac{1}{\Delta^2}Cov\left[\hat{\Psi}_{11}W_{11}^a - \hat{\Psi}_{10}W_{10}^a - \hat{\Psi}_{01}W_{01}^a + \hat{\Psi}_{00}W_{00}^a + W_c^a, \hat{\Psi}_{11}W_{11}^{ab} - \hat{\Psi}_{10}W_{10}^{ab} - \hat{\Psi}_{01}W_{01}^{ab} + \hat{\Psi}_{00}W_{00}^{ab} + W_c^{ab}\right]$$

$$= \frac{1}{\Delta^2}\Big[W_{11}^aW_{11}^{ab}Cov(\hat{\Psi}_{11},\hat{\Psi}_{11}) - W_{11}^aW_{10}^{ab}Cov(\hat{\Psi}_{11},\hat{\Psi}_{10}) - W_{11}^aW_{01}^{ab}Cov(\hat{\Psi}_{11},\hat{\Psi}_{01}) + W_{11}^aW_{00}^{ab}Cov(\hat{\Psi}_{11},\hat{\Psi}_{00})$$

$$\begin{aligned}
= \frac{1}{\Delta^2}\Big[& W_{11}^aW_{11}^{ab}Cov(\hat{\Psi}_{11},\hat{\Psi}_{11}) - W_{11}^aW_{10}^{ab}Cov(\hat{\Psi}_{11},\hat{\Psi}_{10}) - W_{11}^aW_{01}^{ab}Cov(\hat{\Psi}_{11},\hat{\Psi}_{01}) + W_{11}^aW_{00}^{ab}Cov(\hat{\Psi}_{11},\hat{\Psi}_{00}) \\
& - W_{10}^aW_{11}^{ab}Cov(\hat{\Psi}_{10},\hat{\Psi}_{11}) + W_{10}^aW_{10}^{ab}Cov(\hat{\Psi}_{10},\hat{\Psi}_{10}) + W_{10}^aW_{01}^{ab}Cov(\hat{\Psi}_{10},\hat{\Psi}_{01}) - W_{10}^aW_{00}^{ab}Cov(\hat{\Psi}_{10},\hat{\Psi}_{00}) \\
& - W_{01}^aW_{11}^{ab}Cov(\hat{\Psi}_{01},\hat{\Psi}_{11}) + W_{01}^aW_{10}^{ab}Cov(\hat{\Psi}_{01},\hat{\Psi}_{10}) + W_{01}^aW_{01}^{ab}Cov(\hat{\Psi}_{01},\hat{\Psi}_{01}) - W_{01}^aW_{00}^{ab}Cov(\hat{\Psi}_{01},\hat{\Psi}_{00}) \\
& + W_{00}^aW_{11}^{ab}Cov(\hat{\Psi}_{00},\hat{\Psi}_{11}) - W_{00}^aW_{10}^{ab}Cov(\hat{\Psi}_{00},\hat{\Psi}_{10}) - W_{00}^aW_{01}^{ab}Cov(\hat{\Psi}_{00},\hat{\Psi}_{01}) + W_{00}^aW_{00}^{ab}Cov(\hat{\Psi}_{00},\hat{\Psi}_{00})\Big]
\end{aligned}$$

$$= \frac{1}{n\Delta^2}\Big[W_{11}^{a}W_{11}^{ab}\Psi_{11}(1-\Psi_{11}) + W_{11}^{a}W_{10}^{ab}\Psi_{11}\Psi_{10} + W_{11}^{a}W_{01}^{ab}\Psi_{11}\Psi_{01} - W_{11}^{a}W_{00}^{ab}\Psi_{11}\Psi_{00}$$
$$+ W_{10}^{a}W_{11}^{ab}\Psi_{10}\Psi_{11} + W_{10}^{a}W_{10}^{ab}\Psi_{10}(1-\Psi_{10}) - W_{10}^{a}W_{01}^{ab}\Psi_{10}\Psi_{01} + W_{10}^{a}W_{00}^{ab}\Psi_{10}\Psi_{00}$$
$$+ W_{01}^{a}W_{11}^{ab}\Psi_{01}\Psi_{11} - W_{01}^{a}W_{10}^{ab}\Psi_{01}\Psi_{10} + W_{01}^{a}W_{01}^{ab}\Psi_{01}(1-\Psi_{01}) + W_{01}^{a}W_{00}^{ab}\Psi_{01}\Psi_{00}$$
$$- W_{00}^{a}W_{11}^{ab}\Psi_{00}\Psi_{11} + W_{00}^{a}W_{10}^{ab}\Psi_{00}\Psi_{10} + W_{00}^{a}W_{01}^{ab}\Psi_{00}\Psi_{01} + W_{00}^{a}W_{00}^{ab}\Psi_{00}(1-\Psi_{00})\Big]$$

which proves the lemma.

Lemma 5.3. The covariance between $Cov(\hat{\pi}_{AB}^{Amy}, \hat{\pi}_{B}^{Amy})$ is given by

$$Cov\left(\hat{\pi}_{AB}^{Amy}, \hat{\pi}_{B}^{Amy}\right) = \frac{1}{n\Delta^2}\Big[W_{11}^{ab}W_{11}^{b}\Psi_{11}(1-\Psi_{11}) + W_{11}^{ab}W_{10}^{b}\Psi_{11}\Psi_{10} + W_{11}^{ab}W_{01}^{b}\Psi_{11}\Psi_{01} - W_{11}^{ab}W_{00}^{b}\Psi_{11}\Psi_{00}$$
$$+ W_{10}^{ab}W_{11}^{b}\Psi_{10}\Psi_{11} + W_{10}^{ab}W_{10}^{b}\Psi_{10}(1-\Psi_{10}) - W_{10}^{ab}W_{01}^{b}\Psi_{10}\Psi_{01} + W_{10}^{ab}W_{00}^{b}\Psi_{10}\Psi_{00} \tag{5.3}$$
$$+ W_{01}^{ab}W_{11}^{b}\Psi_{01}\Psi_{11} - W_{01}^{ab}W_{10}^{b}\Psi_{01}\Psi_{10} + W_{01}^{ab}W_{01}^{b}\Psi_{01}(1-\Psi_{01}) + W_{01}^{ab}W_{00}^{b}\Psi_{01}\Psi_{00}$$
$$- W_{00}^{ab}W_{11}^{b}\Psi_{00}\Psi_{11} + W_{00}^{ab}W_{10}^{b}\Psi_{00}\Psi_{10} + W_{00}^{ab}W_{01}^{b}\Psi_{00}\Psi_{01} + W_{00}^{ab}W_{00}^{b}\Psi_{00}(1-\Psi_{00})\Big]$$

Proof. we have

$$Cov(\hat{\pi}_{AB}^{Amy}, \hat{\pi}_{B}^{Amy})$$
$$= Cov\left\{\frac{1}{\Delta}(\hat{\Psi}_{11}W_{11}^{ab} - \hat{\Psi}_{10}W_{10}^{ab} - \hat{\Psi}_{01}W_{01}^{ab} + \hat{\Psi}_{00}W_{00}^{ab} + W_{c}^{ab}), \frac{1}{\Delta}(\hat{\Psi}_{11}W_{11}^{b} - \hat{\Psi}_{10}W_{10}^{b} - \hat{\Psi}_{01}W_{01}^{b} + \hat{\Psi}_{00}W_{00}^{b} + W_{c}^{b})\right\}$$
$$= \frac{1}{\Delta^2}Cov\left[\hat{\Psi}_{11}W_{11}^{ab} - \hat{\Psi}_{10}W_{10}^{ab} - \hat{\Psi}_{01}W_{01}^{ab} + \hat{\Psi}_{00}W_{00}^{ab} + W_{c}^{ab}, \hat{\Psi}_{11}W_{11}^{b} - \hat{\Psi}_{10}W_{10}^{b} - \hat{\Psi}_{01}W_{01}^{b} + \hat{\Psi}_{00}W_{00}^{b} + W_{c}^{b}\right]$$

$$= \frac{1}{\Delta^2}\Big[W_{11}^{ab}W_{11}^{b}Cov(\hat{\Psi}_{11}, \hat{\Psi}_{11}) - W_{11}^{ab}W_{10}^{b}Cov(\hat{\Psi}_{11}, \hat{\Psi}_{10}) - W_{11}^{ab}W_{01}^{b}Cov(\hat{\Psi}_{11}, \hat{\Psi}_{01}) + W_{11}^{ab}W_{00}^{b}Cov(\hat{\Psi}_{11}, \hat{\Psi}_{00})$$
$$- W_{10}^{ab}W_{11}^{b}Cov(\hat{\Psi}_{10}, \hat{\Psi}_{11}) + W_{10}^{ab}W_{10}^{b}Cov(\hat{\Psi}_{10}, \hat{\Psi}_{10}) + W_{10}^{ab}W_{01}^{b}Cov(\hat{\Psi}_{10}, \hat{\Psi}_{01}) - W_{10}^{ab}W_{00}^{b}Cov(\hat{\Psi}_{10}, \hat{\Psi}_{00})$$
$$- W_{01}^{ab}W_{11}^{b}Cov(\hat{\Psi}_{01}, \hat{\Psi}_{11}) + W_{01}^{ab}W_{10}^{b}Cov(\hat{\Psi}_{01}, \hat{\Psi}_{10}) + W_{01}^{ab}W_{01}^{b}Cov(\hat{\Psi}_{01}, \hat{\Psi}_{01}) - W_{01}^{ab}W_{00}^{b}Cov(\hat{\Psi}_{01}, \hat{\Psi}_{00})$$
$$+ W_{00}^{ab}W_{11}^{b}Cov(\hat{\Psi}_{00}, \hat{\Psi}_{11}) - W_{00}^{ab}W_{10}^{b}Cov(\hat{\Psi}_{00}, \hat{\Psi}_{10}) - W_{00}^{ab}W_{01}^{b}Cov(\hat{\Psi}_{00}, \hat{\Psi}_{01}) + W_{00}^{ab}W_{00}^{b}Cov(\hat{\Psi}_{00}, \hat{\Psi}_{00})\Big]$$
$$= \frac{1}{n\Delta^2}\Big[W_{11}^{ab}W_{11}^{b}\Psi_{11}(1-\Psi_{11}) + W_{11}^{ab}W_{10}^{b}\Psi_{11}\Psi_{10} + W_{11}^{ab}W_{01}^{b}\Psi_{11}\Psi_{01} - W_{11}^{ab}W_{00}^{b}\Psi_{11}\Psi_{00}$$
$$+ W_{10}^{ab}W_{11}^{b}\Psi_{10}\Psi_{11} + W_{10}^{ab}W_{10}^{b}\Psi_{10}(1-\Psi_{10}) - W_{10}^{ab}W_{01}^{b}\Psi_{10}\Psi_{01} + W_{10}^{ab}W_{00}^{b}\Psi_{10}\Psi_{00}$$
$$+ W_{01}^{ab}W_{11}^{b}\Psi_{01}\Psi_{11} - W_{01}^{ab}W_{10}^{b}\Psi_{01}\Psi_{10} + W_{01}^{ab}W_{01}^{b}\Psi_{01}(1-\Psi_{01}) + W_{01}^{ab}W_{00}^{b}\Psi_{01}\Psi_{00}$$
$$- W_{00}^{ab}W_{11}^{b}\Psi_{00}\Psi_{11} + W_{00}^{ab}W_{10}^{b}\Psi_{00}\Psi_{10} + W_{00}^{ab}W_{01}^{b}\Psi_{00}\Psi_{01} + W_{00}^{ab}W_{00}^{b}\Psi_{00}(1-\Psi_{00})\Big]$$

which proves the lemma.
Now we have the following corollaries.

Corollary 5.1. A new natural estimator of the parameter $\pi_{A|B}$ is given by:

$$\hat{\pi}_{A|B}^{Amy} = \frac{\hat{\pi}_{AB}^{Amy}}{\hat{\pi}_{B}^{Amy}} \tag{5.4}$$

The bias, to the first order of approximation, in the new estimator $\hat{\pi}_{A|B}^{Amy}$ is given by

$$B\left(\hat{\pi}_{A|B}^{Amy}\right) = \pi_{A|B}\left[\frac{V(\hat{\pi}_{B}^{Amy})}{\pi_B^2} - \frac{Cov(\hat{\pi}_{AB}^{Amy}, \hat{\pi}_{B}^{Amy})}{\pi_{AB}\pi_B}\right] \tag{5.5}$$

and, the mean squared error, to the first order of approximation, of the new estimator $\hat{\pi}_{A|B}^{Amy}$ is given by

$$MSE\left(\hat{\pi}_{A|B}^{Amy}\right) = \pi_{A|B}^2\left[\frac{V(\hat{\pi}_{AB}^{Amy})}{\pi_{AB}^2} + \frac{V(\hat{\pi}_{B}^{Amy})}{\pi_B^2} - \frac{2Cov(\hat{\pi}_{B}^{Amy}, \hat{\pi}_{AB}^{Amy})}{\pi_B\pi_{AB}}\right] \tag{5.6}$$

Corollary 5.2. A natural new unbiased estimator of the parameter π_{A-B} is given by:

$$\hat{\pi}_{A-B}^{Amy} = \hat{\pi}_{A}^{Amy} - \hat{\pi}_{AB}^{Amy} \tag{5.7}$$

The variance of the new estimator of $\hat{\pi}_{A-B}^{Amy}$ is given by;

$$V\left(\hat{\pi}_{A-B}^{Amy}\right) = V\left(\hat{\pi}_{A}^{Amy}\right) + V\left(\hat{\pi}_{AB}^{Amy}\right) - 2Cov\left(\hat{\pi}_{A}^{Amy}, \hat{\pi}_{AB}^{Amy}\right) \tag{5.8}$$

Corollary 5.3. A natural new unbiased estimator of the parameter $\pi_{A\cup B}$ is given by

$$\hat{\pi}_{A\cup B}^{Amy} = \hat{\pi}_{A}^{Amy} + \hat{\pi}_{B}^{Amy} - \hat{\pi}_{AB}^{Amy} \tag{5.9}$$

The variance of the new estimator of $\hat{\pi}_{A\cup B}^{Amy}$ is given by;

$$V\left(\hat{\pi}_{A\cup B}^{Amy}\right) = V\left(\hat{\pi}_{A}^{Amy}\right) + V\left(\hat{\pi}_{B}^{Amy}\right) + V\left(\hat{\pi}_{AB}^{Amy}\right) + 2Cov\left(\hat{\pi}_{A}^{Amy}, \hat{\pi}_{B}^{Amy}\right) - 2Cov\left(\hat{\pi}_{A}^{Amy}, \hat{\pi}_{AB}^{Amy}\right) - 2Cov\left(\hat{\pi}_{B}^{Amy}, \hat{\pi}_{AB}^{Amy}\right) \tag{5.10}$$

Corollary 5.4. A natural new estimator of the parameter $RR(B|A)$ is given by

$$\hat{RR}(B|A)_{Amy} = \frac{\hat{\pi}_{AB}^{Amy}\left(1 - \hat{\pi}_{A}^{Amy}\right)}{\hat{\pi}_{A}^{Amy}\left(\hat{\pi}_{B}^{Amy} - \hat{\pi}_{AB}^{Amy}\right)} \tag{5.11}$$

The bias in the new estimator $\hat{RR}(B|A)_{Amy}$, to the first order of approximation, is given by

$$\begin{aligned} B\left(\hat{RR}(B|A)_{Amy}\right) = RR\Bigg[& \frac{V\left(\hat{\pi}_{A}^{Amy}\right)}{\pi_A^2(1-\pi_A)} + \frac{V\left(\hat{\pi}_{B}^{Amy}\right)}{(\pi_B - \pi_{AB})^2} + \frac{\pi_B V\left(\hat{\pi}_{AB}^{Amy}\right)}{\pi_{AB}(\pi_B - \pi_{AB})^2} + \frac{Cov\left(\hat{\pi}_{A}^{Amy}, \hat{\pi}_{B}^{Amy}\right)}{\pi_A(1-\pi_A)(\pi_B - \pi_{AB})} \\ & - \frac{\pi_B Cov\left(\hat{\pi}_{A}^{Amy}, \hat{\pi}_{AB}^{Amy}\right)}{\pi_A \pi_{AB}(1-\pi_A)(\pi_B - \pi_{AB})} - \frac{(\pi_B + \pi_{AB}) Cov\left(\hat{\pi}_{B}^{Amy}, \hat{\pi}_{AB}^{Amy}\right)}{\pi_{AB}(\pi_B - \pi_{AB})^2}\Bigg] \end{aligned} \tag{5.12}$$

The mean squared error of the new estimator $\hat{RR}(B|A)_{Amy}$, to the first order of approximation, is given by

$$\begin{aligned} & MSE\left(\hat{RR}(B|A)_{Amy}\right) \\ & = RR^2\Bigg[\frac{\pi_B^2 V\left(\hat{\pi}_{AB}^{Amy}\right)}{\pi_{AB}^2(\pi_B - \pi_{AB})^2} + \frac{V\left(\hat{\pi}_{A}^{Amy}\right)}{\pi_A^2(1-\pi_A)^2} + \frac{V\left(\hat{\pi}_{B}^{Amy}\right)}{(\pi_B - \pi_{AB})^2} - \frac{2\pi_B Cov\left(\hat{\pi}_{A}^{Amy}, \hat{\pi}_{AB}^{Amy}\right)}{\pi_A \pi_{AB}(1-\pi_A)(\pi_B - \pi_{AB})} \\ & \quad - \frac{2\pi_B Cov\left(\hat{\pi}_{B}^{Amy}, \hat{\pi}_{AB}^{Amy}\right)}{\pi_{AB}(\pi_B - \pi_{AB})^2} + \frac{2Cov\left(\hat{\pi}_{A}^{Amy}, \hat{\pi}_{B}^{Amy}\right)}{\pi_A(1-\pi_A)(\pi_B - \pi_{AB})}\Bigg] \end{aligned} \tag{5.13}$$

Corollary 5.5. A natural new estimator of the parameter ρ_{AB} is given by

$$\hat{\rho}_{AB}^{Amy} = \frac{\hat{\pi}_{AB}^{Amy} - \hat{\pi}_{A}^{Amy}\hat{\pi}_{B}^{Amy}}{\sqrt{\hat{\pi}_{A}^{Amy}(1-\hat{\pi}_{A}^{Amy})}\sqrt{\hat{\pi}_{B}^{Amy}(1-\hat{\pi}_{B}^{Amy})}} \tag{5.14}$$

The bias in the new estimator $\hat{\rho}_{AB}^{Amy}$, to the first order of approximation, is:

$$B\left(\hat{\rho}_{AB}^{Amy}\right) = \rho_{AB}\left[F_4 \frac{V\left(\hat{\pi}_{A}^{Amy}\right)}{\pi_A^2} + F_5 \frac{V\left(\hat{\pi}_{B}^{Amy}\right)}{\pi_B^2} + F_6 \frac{Cov\left(\hat{\pi}_{A}^{Amy}, \hat{\pi}_{B}^{Amy}\right)}{\pi_A \pi_B} - F_7 \frac{Cov\left(\hat{\pi}_{A}^{Amy}, \hat{\pi}_{AB}^{Amy}\right)}{\pi_A \pi_{AB}} - F_8 \frac{Cov\left(\hat{\pi}_{B}^{Amy}, \hat{\pi}_{AB}^{Amy}\right)}{\pi_B \pi_{AB}}\right] \tag{5.15}$$

The mean squared error of the estimator $\hat{\rho}_{AB}^{Amy}$, to the first order of approximation, is given by

$$\begin{aligned} MSE(\hat{\rho}_{AB}^{Amy}) = \rho_{AB}^2\Bigg[& F_1^2 \frac{V\left(\hat{\pi}_{AB}^{Amy}\right)}{\pi_{AB}^2} + F_2^2 \frac{V\left(\hat{\pi}_{A}^{Amy}\right)}{\pi_A^2} + F_3^2 \frac{V\left(\hat{\pi}_{B}^{Amy}\right)}{\pi_B^2} - \frac{2F_1F_2 Cov\left(\hat{\pi}_{A}^{Amy}, \hat{\pi}_{AB}^{Amy}\right)}{\pi_A \pi_{AB}} \\ & - \frac{2F_1F_3 Cov\left(\hat{\pi}_{B}^{Amy}, \hat{\pi}_{AB}^{Amy}\right)}{\pi_B \pi_{AB}} + \frac{2F_2F_3 Cov\left(\hat{\pi}_{A}^{Amy}, \hat{\pi}_{B}^{Amy}\right)}{\pi_A \pi_B}\Bigg] \end{aligned} \tag{5.16}$$

where F_j, $j = 1,2,3,4,5,6,7,8$ are same as defined earlier.

Corollary 5.6. A natural new estimator of the parameter $\pi_d = \pi_A - \pi_B$ is given by

$$\hat{\pi}_d^{Amy} = \hat{\pi}_A^{Amy} - \hat{\pi}_B^{Amy} \tag{5.17}$$

The variance of the difference estimator $\hat{\pi}_d^{Amy}$ is given by

$$V(\hat{\pi}_d^{Amy}) = V(\hat{\pi}_A^{Amy}) + V(\hat{\pi}_B^{Amy}) - 2Cov(\hat{\pi}_A^{Amy}, \hat{\pi}_B^{Amy}) \tag{5.18}$$

We investigate the percent relative efficiency of the proposed estimators with respect to the estimators of other parameters of Lee et al. (2013) for the simple model in the next section.

6. Relative Efficiency of Estimators of Other Parameters

The percent relative efficiencies of the proposed estimators $\hat{\pi}_{A|B}^{Amy}$, $\hat{\pi}_{A-B}^{Amy}$, $\hat{\pi}_{A\cup B}^{Amy}$, $\hat{RR}(B|A)_{Amy}$, $\hat{\rho}_{AB}^{Amy}$ and $\hat{\pi}_d^{Amy}$ with respect to the Lee et al. (2013) estimators $\hat{\pi}_{A|B}$, $\hat{\pi}_{A-B}$, $\hat{\pi}_{A\cup B}$, $\hat{RR}(B|A)$, $\hat{\rho}_{AB}$ and $\hat{\pi}_d$ are, respectively, defined as

$$RE(4) = \frac{V(\hat{\pi}_{A|B})}{V(\hat{\pi}_{A|B}^{Amy})} \times 100\% \tag{6.1}$$

$$RE(5) = \frac{V(\hat{\pi}_{A-B})}{V(\hat{\pi}_{A-B}^{Amy})} \times 100\% \tag{6.2}$$

$$RE(6) = \frac{V(\hat{\pi}_{A\cup B})}{V(\hat{\pi}_{A\cup B}^{Amy})} \times 100\% \tag{6.3}$$

$$RE(7) = \frac{MSE(\hat{RR}(B\,|\,A))}{MSE(\hat{RR}(B\,|\,A))_{Amy}} \times 100\% \tag{6.4}$$

$$RE(8) = \frac{MSE(\hat{\rho}_{AB})}{MSE(\hat{\rho}_{AB}^{Amy})} \times 100\% \tag{6.5}$$

and

$$RE(9) = \frac{V(\hat{\pi}_d)}{V(\hat{\pi}_d^{Amy})} \times 100\% \tag{6.6}$$

The same SAS Codes provided in the Appendix A also give the values of percent relative efficiencies RE(4), RE(5), RE(6), RE(7), RE(8), and RE(9). The detailed results are given in Appendix C in Table C1. For the same protection level, and same choice of other parameters in the study, the summary statistics of the percent relative efficiencies RE(4) to RE(9) are given in Table 5. The values of RE(4), RE(5), RE(6), RE(7), RE(8) and RE(9) are higher than 100% indicating that the proposed unrelated question model performs efficiently for estimating all the parameters considered here. For $w = 0.1$ in Table 5 there are 526 such cases, the average value of RE(4) is 178.27% with a standard deviation of 34.44%, a minimum value of 115.11% and a maximum value of 256.70%; the average value of RE(5) is 162.29% with a standard deviation of 28.48%, a minimum value of 109.16% and a maximum value of 233.69%; the average value of RE(6) is 144.26% with a standard deviation of 24.4%, a minimum value of 100.02% and a maximum value of 234.96%; the average value of RE(7) is 176.12% with a standard deviation of 35.79%, a minimum value of 111.34% and a maximum value of 266.77%; the average value of RE(8) is 179.25% with a standard deviation of 36.50%, a minimum value of 113.67% and a maximum value of 270.65%; and the average value of RE(9) is 142.05% with a standard deviation of 14.50%, a minimum value of 112.56% and a maximum value of 172.64%.

In the same way, 202 cases for the value of $w = 0.9$ are where all the proposed estimators are more efficient than the corresponding estimators of Lee et al. (2013). A pictorial presentation of such results is given in Figure 8.

Again, for demonstration purposes, we extract a few results from the detailed results in Appendix C as cited in Table 6. For $\pi_A = 0.1$, $\pi_B = 0.1$ and $\pi_{AB} = 0.05$, the average value of RE(4) is 139.01% with a standard deviation of 26.67%, a minimum value of 105.17% and a maximum value of 180.67%; the average value of RE(5) is 130.89% with a standard deviation of 20.70%, a minimum value of 104.24% and a maximum value of 163.75%; the average value of RE(6) is 120.87% with a standard deviation of 14.94%, a minimum value of 102.35% and a maximum value of 145.53%; the average value of RE(7) is 136.32% with a standard deviation of 24.92%, a minimum value of 104.75% and a maximum value of 175.07%; the average value of RE(8) is 139.78% with a standard deviation of 27.43%, a minimum value of 105.17% and a maximum value of 182.33%; and the average value of RE(9) is 124.24% with a standard deviation of 14.92%, a minimum value of 104.13% and a maximum value of 150.37%; Likewise the results for

Table 5: RE(4) to RE(9) values for different values of w.

Variable	w	Freq	Mean	StDev	Minimum	Maximum
RE(4)	0.1	526	178.27	34.44	115.11	256.70
	0.3	394	159.81	22.63	108.02	203.57
	0.5	280	143.53	13.04	118.70	164.18
	0.7	202	127.22	3.11	117.87	133.71
	0.9	202	108.00	0.910	105.17	109.92
RE(5)	0.1	526	162.29	28.48	109.16	233.69
	0.3	394	147.58	18.68	107.16	185.11
	0.5	280	135.28	10.78	111.16	151.93
	0.7	202	122.61	3.03	114.10	128.13
	0.9	202	106.91	0.915	104.24	108.61
RE(6)	0.1	526	144.26	24.94	100.02	234.96
	0.3	394	131.09	16.14	100.31	175.82
	0.5	280	122.58	9.48	104.47	142.74
	0.7	202	114.87	2.73	108.28	121.66
	0.9	202	104.55	0.859	102.35	106.69
RE(7)	0.1	526	176.12	35.79	111.34	266.77
	0.3	394	157.50	23.25	108.10	205.71
	0.5	280	142.00	13.32	117.68	163.90
	0.7	202	126.50	3.46	116.28	133.02
	0.9	202	107.86	0.985	104.75	109.78
RE(8)	0.1	526	179.25	36.50	113.67	270.65
	0.3	394	159.71	23.65	107.72	205.90
	0.5	280	143.49	13.51	118.32	163.39
	0.7	202	127.38	3.11	117.98	133.12
	0.9	202	108.08	0.923	105.17	109.80
RE(9)	0.1	526	142.05	14.50	112.56	172.64
	0.3	394	135.34	9.57	113.33	153.69
	0.5	280	127.30	5.32	115.28	136.48
	0.7	202	117.48	1.62	112.54	120.83
	0.9	202	105.63	0.482	104.13	106.61

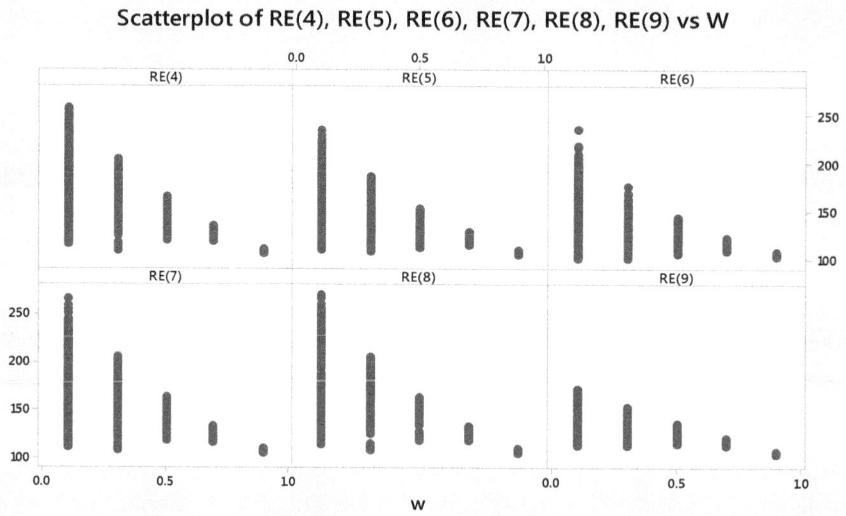

Figure 8: RE(4) to RE(9) for different values of w.

Table 6: Summary of the values of RE(4), RE(5), RE(6), RE(7), RE(8) and RE(9) for different choices of π_A, π_B and π_{AB}.

Results for $\pi_A = 0.10$ and $\pi_B = 0.10$

Variable	π_{AB}	Freq	Mean	StDev	Minimum	Maximum
RE(4)	0.05	15	139.01	26.67	105.17	180.67
RE(5)	0.05	15	130.89	20.70	104.24	163.75
RE(6)	0.05	15	120.87	14.94	102.35	145.53
RE(7)	0.05	15	136.32	24.92	104.75	175.07
RE(8)	0.05	15	139.78	27.43	105.17	182.33
RE(9)	0.05	15	124.24	14.92	104.13	150.37

Results for $\pi_A = 0.10$ and $\pi_B = 0.15$

Variable	π_{AB}	Freq	Mean	StDev	Minimum	Maximum
RE(4)	0.05	15	143.48	30.03	105.75	188.84
RE(5)	0.05	15	133.28	22.30	104.66	168.24
RE(6)	0.05	15	122.66	16.22	102.66	150.16
RE(7)	0.05	15	140.47	27.93	105.35	182.93
RE(8)	0.05	15	143.24	29.92	105.69	188.74
RE(9)	0.05	15	125.30	15.57	104.34	152.14

$\pi_A = 0.1$, $\pi_B = 0.15$ and $\pi_{AB} = 0.05$ can be seen from Table 6. In addition, results for any such combination of the three proportions π_A, π_B and π_{AB} can be extracted, if needed, from the detailed results give in Appendix C in Table C1. Thus, the proposed model is found to be quite helpful in estimating such tiny proportions of sensitive attributes and their overlaps for estimating all the parameters considered.

Table 7 helps to make a choice of the proportion of unrelated question π_y. Again for $\pi_y = 1$ there are combinations of *freq* = 742 of the parameters where the proposed estimators of the others considered are more efficient and more protective than the simple model of Lee et al. (2013), and it again ensures that in the proposed model "Forced Yes" response could be used instead of finding any unrelated characteristic in a population under study.

This finding through simulation study enhances the usefulness of the proposed model in practice. A behavior of the values of RE(4), RE(5), RE(6), RE(7), RE(8) and RE(9) versus π_y are shown in Figure 9. In Figure 9 the notation Y stands for π_y.

Table 7: RE(4), RE(5), RE(6), RE(7), RE(8) and RE(9) for different values of π_y.

Variable	π_y	Freq	Mean	StDev	Minimum	Maximum
RE(4)	0.90	280	152.54	34.05	106.47	235.39
	0.95	582	152.28	32.78	105.81	246.07
	1.00	742	152.42	33.83	105.17	256.70
RE(5)	0.90	280	143.02	27.21	105.65	212.63
	0.95	582	141.70	26.29	104.94	220.76
	1.00	742	141.83	27.37	104.24	233.69
RE(6)	0.90	280	129.34	20.25	100.02	193.32
	0.95	582	127.45	20.35	100.51	209.04
	1.00	742	129.09	23.11	100.31	234.96
RE(7)	0.90	280	152.80	34.46	106.22	240.51
	0.95	582	150.39	32.56	105.48	250.82
	1.00	742	150.25	33.99	104.75	266.77
RE(8)	0.90	280	153.34	34.90	106.54	239.76
	0.95	582	152.34	33.62	105.85	251.78
	1.00	742	152.77	35.25	105.17	270.65
RE(9)	0.90	280	132.30	17.59	105.09	169.95
	0.95	582	130.96	15.82	104.61	171.24
	1.00	742	128.70	14.75	104.13	172.64

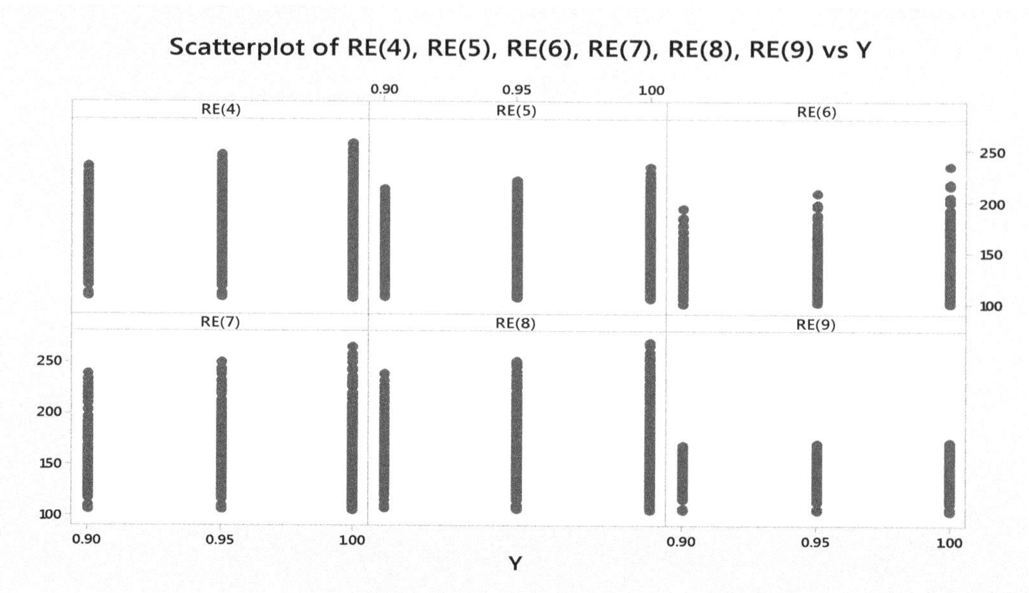

Figure 9: RE(4), RE(5), RE(6), RE(7), RE(8), and RE(9) versus π_y.

Table 8 provides a summary of results for different choices of P_1 and T_1. For $P_1 = 0.4$ and $T_1 = 0.4$, the average value of RE(4) is 122.63% with a standard deviation of 7.76%, a minimum value of 108.2% and a maximum value of 140.06%; the average value of

Table 8: RE(4), RE(5), RE(6), RE(7), RE(8) and RE(9) for different P_1 and T_1.

Results for $\pi_A = 0.10$ and $\pi_B = 0.10$

Variable	T_1	Freq	Mean	StDev	Minimum	Maximum
RE(4)	0.4	58	122.63	7.76	108.02	140.06
	0.5	268	147.92	16.18	118.70	186.98
RE(5)	0.4	58	118.21	7.09	107.16	139.34
	0.5	268	132.50	12.70	111.16	178.37
RE(6)	0.4	58	109.38	8.50	100.02	138.47
	0.5	268	123.54	13.22	104.47	180.36
RE(7)	0.4	58	120.68	7.93	108.10	140.87
	0.5	268	146.27	16.60	117.89	193.80
RE(8)	0.4	58	121.42	8.12	107.72	143.90
	0.5	268	147.28	16.86	118.32	197.05
RE(9)	0.4	58	118.81	4.16	112.56	127.26
	0.5	268	129.83	6.84	115.28	145.32

Results for $\pi_A = 0.10$ and $\pi_B = 0.15$

Variable	T_1	Freq	Mean	StDev	Minimum	Maximum
RE(4)	0.4	268	148.03	16.39	119.24	192.23
	0.5	1010	156.45	39.47	105.17	256.70
RE(5)	0.4	268	144.06	14.77	116.58	182.44
	0.5	1010	145.33	31.33	104.24	233.69
RE(6)	0.4	268	123.54	13.22	104.47	180.36
	0.5	1010	132.29	24.50	102.35	234.96
RE(7)	0.4	268	144.64	16.86	117.68	193.57
	0.5	1010	155.28	39.37	104.75	266.77
RE(8)	0.4	268	147.28	16.86	118.32	197.05
	0.5	1010	157.39	40.65	105.17	270.65
RE(9)	0.4	268	129.83	6.84	115.28	145.32
	0.5	1010	130.97	18.94	104.13	172.64

RE(5) is 118.21% with a standard deviation of 7.09%, a minimum value of 107.16% and a maximum value of 139.34%; the average value of RE(6) is 109.38% with a standard deviation of 8.50%, a minimum value of 100.02% and a maximum value of 138.47%.

The average value of RE(7) is 120.68% with a standard deviation of 7.93%, a minimum value of 108.10% and a maximum value of 140.87%;the average value of RE(8) is 121.42% with a standard deviation of 8.12%, a minimum value of 107.72% and a maximum value of 143.90%; and the average value of RE(9) is 118.81% with a standard deviation of 4.16%, a minimum value of 112.56% and a maximum value of 127.26%. Likewise, other results from Table 8 can be interpreted. It could be worth highlighting that the choice of $P_1 = 0.5$ and $T_1 = 0.5$ would be a natural preference by the participants. In the next section, we develop the Cramer-Rao lower bounds of variance-covariance for verifying the results.

7. Cramer-Rao Lower Bounds of Variance-Covariances

Theorem 7.1. The maximum likelihood estimates $\hat{\pi}_{AB}^{(mle)}$, $\hat{\pi}_{A}^{(mle)}$, and $\hat{\pi}_{B}^{(mle)}$ of π_{AB}, π_A, and π_B are a solution to the three non-linear Equations given as:

$$\frac{\hat{\Psi}_{11}}{\Psi_{11}} - \frac{\hat{\Psi}_{10}}{\Psi_{10}} - \frac{\hat{\Psi}_{01}}{\Psi_{01}} + \frac{\hat{\Psi}_{00}}{\Psi_{00}} = 0 \tag{7.1}$$

$$\frac{\hat{\Psi}_{11}}{\Psi_{11}}C_2 + \frac{\hat{\Psi}_{10}}{\Psi_{10}}C_4 - \frac{\hat{\Psi}_{01}}{\Psi_{01}}C_2 - \frac{\hat{\Psi}_{00}}{\Psi_{00}}C_4 = 0 \tag{7.2}$$

and

$$\frac{\hat{\Psi}_{11}}{\Psi_{11}}C_3 - \frac{\hat{\Psi}_{10}}{\Psi_{10}}C_3 + \frac{\hat{\Psi}_{01}}{\Psi_{01}}C_5 - \frac{\hat{\Psi}_{00}}{\Psi_{00}}C_5 = 0 \tag{7.3}$$

Proof. In case of the proposed model, consider the likelihood function given by:

$$L = \binom{n}{n_{11}\ n_{10}\ n_{01}\ n_{00}} (\Psi_{11})^{n_{11}} (\Psi_{10})^{n_{10}} (\Psi_{01})^{n_{01}} (\Psi_{00})^{n_{00}} \tag{7.4}$$

Taking log on both sides, we get:

$$\begin{aligned} \log L &= \log C + n_{11}\log\Psi_{11} + n_{10}\log\Psi_{10} + n_{01}\log\Psi_{01} + n_{00}\log\Psi_{00} \\ &= \log(C) + n\left[\begin{aligned} &\hat{\Psi}_{11}\log\{C_1\pi_{AB} + C_2\pi_A + C_3\pi_B + H_1\} + \hat{\Psi}_{10}\log\{-C_1\pi_{AB} + C_4\pi_A - C_3\pi_B + H_2\} \\ &+ \hat{\Psi}_{01}\log\{-C_1\pi_{AB} - C_2\pi_A + C_5\pi_B + H_3\} + \hat{\Psi}_{00}\log\{C_1\pi_{AB} - C_4\pi_A - C_5\pi_B + H_4\} \end{aligned}\right] \end{aligned} \tag{7.5}$$

On taking the first order partial derivates of log L with respect to π_{AB}, π_A and π_B respectively, we have

$$\frac{\partial \log L}{\partial \pi_{AB}} = nC_1\left[\frac{\hat{\Psi}_{11}}{\Psi_{11}} - \frac{\hat{\Psi}_{10}}{\Psi_{10}} - \frac{\hat{\Psi}_{01}}{\Psi_{01}} + \frac{\hat{\Psi}_{00}}{\Psi_{00}}\right] \tag{7.6}$$

$$\frac{\partial \log L}{\partial \pi_{A}} = n\left[\frac{\hat{\Psi}_{11}}{\Psi_{11}}C_2 + \frac{\hat{\Psi}_{10}}{\Psi_{10}}C_4 - \frac{\hat{\Psi}_{01}}{\Psi_{01}}C_2 - \frac{\hat{\Psi}_{00}}{\Psi_{00}}C_4\right] \tag{7.7}$$

and

$$\frac{\partial \log L}{\partial \pi_{B}} = n\left[\frac{\hat{\Psi}_{11}}{\Psi_{11}}C_3 - \frac{\hat{\Psi}_{10}}{\Psi_{10}}C_3 + \frac{\hat{\Psi}_{01}}{\Psi_{01}}C_5 - \frac{\hat{\Psi}_{00}}{\Psi_{00}}C_5\right] \tag{7.8}$$

On setting $\frac{\partial \log L}{\partial \pi_{AB}} = 0$, $\frac{\partial \log L}{\partial \pi_{A}} = 0$ and $\frac{\partial \log L}{\partial \pi_{B}} = 0$, we have the theorem.

Theorem 7.2. The lower bounds of variance-covariance of the maximum likelihood estimates $\hat{\pi}_{AB}^{mle}$, $\hat{\pi}_{A}^{mle}$ and $\hat{\pi}_{B}^{mle}$ of π_{AB}, π_A, and π_B for the simple model are, respectively, given by:

$$V\begin{bmatrix}\hat{\pi}_{AB}^{mle}\\ \hat{\pi}_{A}^{mle}\\ \hat{\pi}_{B}^{mle}\end{bmatrix} \geq \frac{1}{\Delta}\begin{bmatrix}a_{11} & a_{12} & a_{13}\\ a_{21} & a_{22} & a_{23}\\ a_{31} & a_{32} & a_{33}\end{bmatrix} \tag{7.9}$$

where the inequality sign ($\geq$) has its usual meaning, and Δ and a_{ij} are given by

$$\Delta = n(gb^2 - 2bcf + dc^2 + af^2 - adg)$$

$$a_{11} = f^2 - gd, \qquad a_{22} = c^2 - ag, \qquad a_{33} = b^2 - ad,$$

$$a_{12} = a_{21} = bg - cf, \qquad a_{13} = a_{31} = cd - bf, \text{ and} \qquad a_{23} = a_{32} = af - bc$$

Proof. By the definition of the Cramer-Rao lower bound of variance covariance, we have

$$V\begin{bmatrix}\hat{\pi}_{AB}^{(mle)}\\ \hat{\pi}_{A}^{(mle)}\\ \hat{\pi}_{B}^{(mle)}\end{bmatrix} \geq -\begin{bmatrix}E(\frac{\partial^2 \log L}{\partial \pi_{AB}^2}) & E(\frac{\partial^2 \log L}{\partial \pi_A \partial \pi_{AB}}) & E(\frac{\partial^2 \log L}{\partial \pi_B \partial \pi_{AB}})\\ E(\frac{\partial^2 \log L}{\partial \pi_A \partial \pi_{AB}}) & E(\frac{\partial^2 \log L}{\partial \pi_A^2}) & E(\frac{\partial^2 \log L}{\partial \pi_B \partial \pi_A})\\ E(\frac{\partial^2 \log L}{\partial \pi_B \partial \pi_{AB}}) & E(\frac{\partial^2 \log L}{\partial \pi_B \partial \pi_A}) & E(\frac{\partial^2 \log L}{\partial \pi_B^2})\end{bmatrix}^{-1} = \begin{bmatrix}a & b & c\\ b & d & f\\ c & f & g\end{bmatrix}^{-1},$$

where

$$a = -E\left[\frac{\partial^2 \log L}{\partial \pi_{AB}^2}\right] = nC_1^2\left[\frac{1}{\Psi_{11}} + \frac{1}{\Psi_{10}} + \frac{1}{\Psi_{01}} + \frac{1}{\Psi_{00}}\right],$$

$$b = -E\left[\frac{\partial^2 \log L}{\partial \pi_A \partial \pi_{AB}}\right] = nC_1\left[\frac{C_2}{\Psi_{11}} - \frac{C_4}{\Psi_{10}} + \frac{C_2}{\Psi_{01}} - \frac{C_4}{\Psi_{00}}\right],$$

$$c = -E\left[\frac{\partial^2 \log L}{\partial \pi_B \partial \pi_{AB}}\right] = nC_1\left[\frac{C_3}{\Psi_{11}} + \frac{C_3}{\Psi_{10}} - \frac{C_5}{\Psi_{01}} - \frac{C_5}{\Psi_{00}}\right],$$

$$d = -E\left[\frac{\partial^2 \log L}{\partial \pi_A^2}\right] = n\left[\frac{C_2^2}{\Psi_{11}} + \frac{C_4^2}{\Psi_{10}} + \frac{C_2^2}{\Psi_{01}} + \frac{C_4^2}{\Psi_{00}}\right],$$

$$f = -E\left[\frac{\partial^2 \log L}{\partial \pi_B \partial \pi_A}\right] = n\left[\frac{C_2C_3}{\Psi_{11}} - \frac{C_3C_4}{\Psi_{10}} - \frac{C_2C_5}{\Psi_{01}} + \frac{C_4C_5}{\Psi_{00}}\right],$$

and

$$g = -E\left[\frac{\partial^2 \log L}{\partial \pi_B^2}\right] = n\left[\frac{C_3^2}{\Psi_{11}} + \frac{C_3^2}{\Psi_{10}} + \frac{C_5^2}{\Psi_{01}} + \frac{C_5^2}{\Psi_{00}}\right]$$

which proves the theorem.

8. SAS Codes Output for the Cramer-Rao Lower Bounds

We also included the Cramer-Rao lower bounds of variance and covariance matrix in the SAS Codes. We abbreviate for our convenience $V(\hat{\pi}_A^{Amy}) = V_1$(say), $V(\hat{\pi}_B^{Amy}) = V_2$, $V(\hat{\pi}_{AB}^{Amy}) = V_3$, $Cov(\hat{\pi}_A^{Amy}, \hat{\pi}_B^{Amy}) = C_{12}$, $Cov(\hat{\pi}_A^{Amy}, \hat{\pi}_B^{Amy}) = C_{13}$, and $Cov(\hat{\pi}_B^{Amy}, \hat{\pi}_{AB}^{Amy}) = C_{23}$ for the closed form of the estimators $\hat{\pi}_A^{Amy}$, $\hat{\pi}_B^{Amy}$ and $\hat{\pi}_{AB}^{Amy}$ and V_1^{LB}, V_2^{LB}, V_3^{LB}, C_{12}^{LB}, C_{13}^{LB} and C_{23}^{LB} as the corresponding Cramer-Rao lower-

Table 9: The variance and covariance from the closed forms of the estimators.

Obs	V_3	V_1	V_2	C_{12}	C_{13}	C_{23}
1	2.281	1.410	1.410	0.142	0.433	0.433
2	1.634	1.478	1.027	0.131	0.325	0.248
3	1.655	1.478	1.013	0.119	0.403	0.238
4	1.675	1.478	0.994	0.107	0.481	0.228
5	1.608	1.447	1.037	0.133	0.237	0.308

Table 10: The Cramer-Rao lower bounds of variance-covariance.

Obs	V_3^{LB}	V_1^{LB}	V_2^{LB}	C_{12}^{LB}	C_{13}^{LB}	C_{23}^{LB}
1	2.281	1.410	1.410	0.142	0.433	0.433
2	1.634	1.478	1.027	0.131	0.325	0.248
3	1.655	1.478	1.013	0.119	0.403	0.238
4	1.675	1.478	0.994	0.107	0.481	0.228
5	1.608	1.447	1.037	0.133	0.237	0.308

bounds entries from the variance co-variance matrix. We provide the first five outcomes from the SAS Codes in Table 9 and Table 10. It is found that the corresponding Cramer-Rao lower bounds of variance-covariance matrix obtained from the maximum-likelihood estimates match with the variance-covariance expressions of the closed forms of the estimators.

One can see that the Cramer-Rao lower bounds of variance-covariance are quite helpful in checking the derivations of the variance-covariance of the closed forms of the estimators. For $w = 1$ the Cramer-Rao lower bounds of variance-covariance match with those obtained by Lee, Sedory and Singh (2016).

9. Survey data Application

Electronic cigarettes (e-cigarettes) have been more and more popular in the world. Lots of researchers are interested in whether the use of e-cigarette is associated with the conventional cigarette. We considered the new designed methodology by using a new unrelated question randomized response technique to investigate whether the use of e-cigarette and conventional cigarette are related. In the study design, every respondent was explained how to use the proposed device which consists of five decks of cards in five colors: green, first-pair (blue and white), and second pair (yellow and pink). The procedure was as follows: every respondent was requested to draw one card from the green deck to decide if he/she had to use the first-pair or the second-pair of decks. In the first pair, the blue-deck was consisting of cards bearing two sets of questions: 14 cards with the statement, "Over the past year, "Have you ever used a Conventional Cigarette?" and 6 cards with the statement, "Over the past year, and Have you never used Conventional Cigarette?". The white-deck was consisting of cards bearing two-types of questions: 14 cards with the statement, "Over the past year, Have you ever used E-cigarette?", and 6 cards with the statement, "Over the past year, Have you never used E-Cigarette?". In the second pair of decks, the yellow-deck was consisting of cards bearing two statements: 14 cards with the statement, "Over the past year, have you ever used Conventional Cigarette?" and 6 cards with the statement "Report Yes". The pink-deck was consisting of cards bearing two statements: 14 cards with the statement, "Over the past year, have you ever used E-Cigarette?" and 6 cards with the statement "Report Yes". During three days of data collection, a total of 121 students participated in the survey. Using the proposed DMDM-I model, and the black box estimates of the true proportion of the required parameters are given in Table 11.

Table 11: Real survey results.

DMDM-I Estimates			Black Box Estimates		
$\hat{\pi}_A^{Amy}$	$\hat{\pi}_B^{Amy}$	$\hat{\pi}_{AB}^{Amy}$	$\hat{\pi}_A^{b}$	$\hat{\pi}_B^{b}$	$\hat{\pi}_{AB}^{b}$
0.206	0.191	0.030	0.198	0.231	0.091

From Table 11, the DMDM-I estimate of the proportion of conventional cigarette is 0.206, that of e-cigarette users is 0.191, and that of both e-cigarette and conventional cigarette users is 0.03. The black box estimate of the proportion of conventional cigarettes is 0.198, that of e-cigarette users is 0.231, and that of both e-cigarette and conventional cigarette users is 0.091. We calculated the Z scores to compare the two types of methods used for estimating the proportions as follows:

$$Z_{AB} = \frac{\hat{\pi}_{AB}^{Amy} - \hat{\pi}_{AB}^{bb}}{\sqrt{\hat{v}(\hat{\pi}_{AB}^{Amy}) + \hat{v}(\hat{\pi}_{AB}^{bb})}} = -\ 0.802 \tag{9.1}$$

$$Z_A = \frac{\hat{\pi}_A^{Amy} - \hat{\pi}_A^{bb}}{\sqrt{\hat{v}(\hat{\pi}_A^{Amy}) + \hat{v}(\hat{\pi}_A^{bb})}} = 0.088 \tag{9.2}$$

$$Z_B = \frac{\hat{\pi}_B^{Amy} - \hat{\pi}_B^{bb}}{\sqrt{\hat{v}(\hat{\pi}_B^{Amy}) + \hat{v}(\hat{\pi}_B^{bb})}} = -0.446 \tag{9.3}$$

The absolute values of the computed Z scores are less than 1.96 indicating that there is no significant difference between the proposed DMDM-I and the black box estimates. We conclude that the proposed DMDM-I method can be used in real practice. The original version of this work can be had from Zheng(2019).

Acknowledgements

Thanks are due to the IRB Committee Members, Research Compliance, Office of the Research & Sponsored Programs, Texas A&M University-Kingsville for the permission to collect data from the students.

References

Greenberg, B. G., Abul-Ela, A.-L. A., Simmons, W. R. and Horvitz, D. G. 1969. The unrelated question randomized response model: Theoretical framework. Journal of the American Statistical Association, 64(326): 520–539.

Johnson, M. L., Sedory, S. A. and Singh, S. 2019. Alternative methods to make efficient use of two decks of cards in randomized response sampling. Sociological Methods and Research, 48(1): 62–91.

Lanke, J. 1975. On the choice of the unrelated question in Simmons' version of randomized response. Journal of the American Statistical Association, 70(349): 80–83.

Lanke, J. 1976. On the degree of protection in randomized interviews. International Statistical Review/Revue Internationale de Statistique, 197–203.

Lee, C.-S., Sedory, S. A. and Singh, S. 2013. Estimating at least seven measures of qualitative variables from a single sample using randomized response technique. Statistics & Probability Letters, 83(1): 399–409.

Lee, C. S., Sedory, S. A. and Singh, S. 2016. Cramer-Rao lower bounds of variance for estimating two proportions and their overlap by using two-decks of cards. Handbook of Statistics, 34: ELSEVIER, PP. 353–385.

Odumade, O. and Singh, S. 2009. Efficient use of two decks of cards in randomized response sampling. Communications in Statistics—Theory and Methods, 38(4): 439–446.

Warner, S. L. 1965. Randomized response: A survey technique for eliminating evasive answer bias. Journal of the American Statistical Association, 60(309): 63–69.

Yennum, N., Sedory, S. A. and Singh, S. 2019. Improved strategy to collect sensitive data by using geometric distribution as a randomization device. Communications In Statistics-Theory and Methods, 48(23): 5777–5795.

Zheng, R. 2019. On unrelated question randomized response techniques. Unpublished MS thesis submitted to the Department of Mathematics, Texas A&M University-Kingsville, Kingsville, TX.

Appendix A

SAS CODES USED IN THE SIMULATION STUDY

```
*SAS Codes used in the simulation study
DATA DATA1;
DO W = 0.1 TO 0.9 BY 0.2;
DO Y = 0.90 TO 1.0 BY 0.05;
*DO P = 0.7 TO 0.7 BY 0.1;
*DO T = 0.7 TO 0.7 BY 0.1;
P=0.7;
T=0.7;
DO P1 = 0.4 TO 0.5 BY 0.1;
DO T1 = 0.4 TO 0.5 BY 0.1;
DO PIAB = 0.05 TO 0.2 BY 0.05;
DO PIA = 0.05 TO 0.35 BY 0.05;
DO PIB = 0.05 TO 0.35 BY 0.05;
TH11_L=(2*P-1)*(2*T-1)*PIAB+(2*P-1)*(1-T)*PIA+(1-P)*(2*T-1)*PIB+(1-P)*(1-T);
TH10_L=-(2*P-1)*(2*T-1)*PIAB+(2*P-1)*T*PIA-(1-P)*(2*T-1)*PIB+(1-P)*T;
TH01_L=-(2*P-1)*(2*T-1)*PIAB-(2*P-1)*(1-T)*PIA+P*(2*T-1)*PIB+P*(1-T);
TH00_L=(2*P-1)*(2*T-1)*PIAB-T*(2*P-1)*PIA-P*(2*T-1)*PIB+P*T;
SUMTH_L=TH11_L+TH10_L+TH01_L+TH00_L;
PABYY_L=P*T*PIAB/TH11_L;
PABYN_L=P*(1-T)*PIAB/TH10_L;
PABNY_L=(1-P)*T*PIAB/TH01_L;
PABNN_L=(1-P)*(1-T)*PIAB/TH00_L;
PROT_AB_L=MAX(PABYY_L, PABYN_L, PABNY_L, PABNN_L);
PAYY_L=(P*T*PIB+(1-T)*P*(1-PIB))*PIA/TH11_L;
PAYN_L=(P*T*(1-PIB)+P*(1-T)*PIB)*PIA/TH10_L;
PANY_L=((1-P)*T*PIB+(1-P)*(1-T)*(1-PIB))*PIA/TH01_L;
PANN_L=((1-P)*T*(1-PIB)+(1-P)*(1-T)*PIB)*PIA/TH00_L;
PROT_A_L=MAX(PAYY_L, PAYN_L, PANY_L, PANN_L);
PBYY_L=(P*T*PIA+(1-P)*T*(1-PIA))*PIB/TH11_L;
PBYN_L=(P*(1-T)*PIA+(1-P)*(1-PIA)*(1-T))*PIB/TH10_L;
PBNY_L=(P*(1-PIA)*T+(1-P)*PIA*T)*PIB/TH01_L;
PBNN_L=(P*(1-PIA)*(1-T)+(1-P)*PIA*(1-T))*PIB/TH00_L;
PROT_B_L=MAX(PBYY_L, PBYN_L, PBNY_L, PBNN_L);
VAR_PIAB_LEE =PIAB*(1-PIAB)+( (2*P-1)**2*T*(1-T)*PIA + P*(1-P)*(2*T-1)**2*PIB+P*T*(1-P)*(1-T))/
((2*P-1)**2*(2*T-1)**2);
VAR_PIA_LEE =PIA*(1-PIA)+P*(1-P)/(2*P-1)**2;
VAR_PIB_LEE =PIB*(1-PIB)+T*(1-T)/(2*T-1)**2;
CPIA_PIB_LEE=(PIAB-PIA*PIB);
CPIA_PIAB_LEE=PIAB*(1-PIA)+P*(1-P)*PIB/(2*P-1)**2;
CPIAB_PIB_LEE=PIAB*(1-PIB)+T*(1-T)*PIA/(2*T-1)**2;
F1=PIAB/(PIAB-PIA*PIB);
F2=PIA*PIB/(PIAB-PIA*PIB)+(1-2*PIA)/(2*(1-PIA));
F3=PIA*PIB/(PIAB-PIA*PIB)+(1-2*PIB)/(2*(1-PIB));
C1 = W*(2*P-1)*(2*T-1)+(1-W)*P1*T1;
C2 = W*(2*P-1)*(1-T)+(1-W)*P1*(1-T1)*Y;
C3 = W*(1-P)*(2*T-1)+(1-W)*(1-P1)*T1*Y;
C4 = W*(2*P-1)*T+(1-W)*P1*(T1+(1-T1)*(1-Y));
C5 = W*P*(2*T-1)+(1-W)*T1*(P1+(1-P1)*(1-Y));
H1 = (1-P)*(1-T)*W+(1-W)*(1-P1)*(1-T1)*Y;
H2 = (1-P)*T*W+(1-W)*(1-P1)*T1*Y;
H3 = P*(1-T)*W+(1-W)*(1-T1)*P1*Y;
H4 = W*P*T + (1-W)*(P1*T1*Y + 1-Y);
```

```
SH11 = C1*PIAB+C2*PIA+C3*PIB+H1;
SH10 = -C1*PIAB+C4*PIA-C3*PIB+H2;
SH01 = -C1*PIAB -C2*PIA+C5*PIB+H3;
SH00 = C1*PIAB-C4*PIA-C5*PIB+H4;
SUMSH = SH11+SH10+SH01+SH00;
PABYY_AMY=(W*P*T+(1-W)*(P1*T1+Y*(1-P1*T1)))*PIAB/SH11;
PABYN_AMY=(W*P*(1-T)+(1-W)*P1*(1-T1)*(1-Y))*PIAB/SH10;
PABNY_AMY=(W*(1-P)*T+(1-W)*(1-P1)*(1-Y)*T1)*PIAB/SH01;
PABNN_AMY=(W*(1-P)*(1-T)+(1-W)*(1-P1)*(1-T1)*(1-Y))*PIAB/SH00;
PROT_AB_AMY=MAX(PABYY_AMY, PABYN_AMY, PABNY_AMY, PABNN_AMY);
RPAB=(PROT_AB_L*100)/PROT_AB_AMY;
PAYY_AMY=(W*(P*T*PIB+(1-T)*P*(1-PIB))+(1-W)*(P1*(1-T1)*Y+P1*T1*PIB+(1-P1)*Y*T1*PIB+(1-P1)*(1-
T1)*Y))*PIA/SH11;
PAYN_AMY=(W*(P*T*(1-PIB)+P*(1-T)*PIB)+(1-W)*(P1*(1-T1)*(1-Y)+P1*T1*(1-PIB)+(1-P1)*Y*T1*(1-PIB)))*PIA/
SH10;
PANY_AMY=(W*((1-P)*T*PIB+(1-P)*(1-T)*(1-PIB))+(1-W)*((1-P1)*(1-Y)*PIB))*PIA/SH01;
PANN_AMY=(W*((1-P)*T*(1-PIB)+(1-P)*(1-T)*PIB)+(1-W)*((1-P1)*(1-Y)*(1-T1)+(1-P1)*(1-Y)*(1-PIB)*T1))*PIA/SH00;
PROT_A_AMY=MAX(PAYY_AMY, PAYN_AMY, PANY_AMY, PANN_AMY);
RPA=(PROT_A_L*100)/PROT_A_AMY;
PBYY_AMY=(W*(P*T*PIA+(1-P)*T*(1-PIA))+(1-W)*((1-P1)*T1*Y+P1*T1*PIA+(1-T1)*Y*P1*PIA+(1-P1)*(1-
T1)*Y))*PIB/SH11;
PBYN_AMY=(W*(P*PIA*(1-T)+(1-P)*(1-PIA)*(1-T))+(1-W)*(P1*PIA*(1-T1)*(1-Y)))*PIB/SH10;
PBNY_AMY=(W*(P*(1-PIA)*T+(1-P)*PIA*T)+(1-W)*(P1*(1-PIA)*T1+(1-P1)*T1*(1-Y)+(1-T1)*Y*P1*(1-PIA)))*PIB/
SH01;
PBNN_AMY=(W*(P*(1-PIA)*(1-T)+(1-P)*PIA*(1-T))+(1-W)*(P1*(1-PIA)*(1-T1)*(1-Y)+(1-P1)*(1-T1)*(1-Y)))*PIB/SH00;
PROT_B_AMY=MAX(PBYY_AMY, PBYN_AMY, PBNY_AMY, PBNN_AMY);
RPB=(PROT_B_L*100)/PROT_B_AMY;
K1 = 4*(C2**2+C4**2)*(C3**2+C5**2)-(C2-C4)**2*(C3-C5)**2;
K2 = 2*(C2-C4)*(C3-C5)**2-4*(C2-C4)*(C3**2+C5**2);
K3 = 2*(C2-C4)**2*(C3-C5)-4*(C2**2+C4**2)*(C3-C5);
L1 = 2*C1*(C2-C4)*((C3-C5)**2-2*(C3**2+C5**2));
L2 = 4*C1*(2*(C3**2+C5**2)-(C3-C5)**2);
M1 = 2*C1*(C3-C5)*((C2-C4)**2-2*(C2**2+C4**2));
M2 = 4*C1*(2*(C2**2+C4**2)-(C2-C4)**2);
W11AB = K1+C2*K2+C3*K3;
W10AB = K1-C4*K2+C3*K3;
W01AB = K1+C2*K2-C5*K3;
W00AB = K1-C4*K2-C5*K3;
TERM1 = W11AB**2*SH11*(1-SH11)+W10AB**2*SH10*(1-SH10)+ W01AB**2 *SH01*(1-SH01) +
W00AB**2*SH00*(1-SH00);
TERM2 = 2*W11AB*W10AB*SH11*SH10+2*W11AB*W01AB*SH11*SH01-2*W11AB*W00AB*SH11*SH00
-2*W10AB*W01AB*SH10*SH01+ 2*W10AB*W00AB*SH10*SH00 +2*W01AB*W00AB*SH01*SH00;
W11A = L1+C2*L2;
W10A = L1-C4*L2;
W01A = L1+C2*L2;
W00A = L1-C4*L2;
TERM3 = W11A**2*SH11*(1-SH11)+W10A**2*SH10*(1-SH10)+ W01A**2*SH01*(1-SH01) + W00A**2*SH00*(1-
SH00);
TERM4 = 2*W11A*W10A*SH11*SH10+2*W11A*W01A*SH11*SH01-2*W11A* W00A*SH11*SH00
-2*W10A*W01A*SH10*SH01 +2*W10A*W00A*SH10*SH00 +2*W01A*W00A*SH01*SH00;
```

```
W11B = M1+C3*M2;
W10B = M1+C3*M2;
W01B = M1-C5*M2;
W00B = M1-C5*M2;
TERM5 = W11B**2*SH11*(1-SH11)+W10B**2*SH10*(1-SH10) +W01B**2*SH01*(1-SH01)+ W00B**2*SH00*(1-SH00);
TERM6 = 2*W11B*W10B*SH11*SH10+2*W11B*W01B*SH11*SH01-2*W11B*W00B*SH11*SH00-
2*W10B*W01B*SH10*SH01 +2*W10B*W00B*SH10*SH00+2*W01B*W00B*SH01*SH00;

DELTA1=4*C1*(2*(C3**2+C5**2)-(C3-C5)**2)*(2*(C2**2+C4**2)-(C2-C4)**2);
CPIA_PIB=(W11A*W11B*SH11*(1-SH11)+W11A*W10B*SH11*SH10 +W11A*W01B*SH11*SH01
-W11A*W00B*SH11*SH00
 +W10A*W11B*SH10*SH11 +W10A*W10B*SH10*(1-SH10)-W10A*W01B*SH10*SH01 +W10A*W00B*SH10*SH00
 +W01A*W11B*SH01*SH11 -W01A*W10B*SH01*SH10 +W01A*W01B*SH01*(1-SH01)+W01A*W00B*SH01*SH00
 -W00A*W11B*SH00*SH11 +W00A*W10B*SH00*SH10 +W00A*W01B*SH00*SH01 +W00A*W00B*SH00*(1-SH00))/
DELTA1**2;
CPIA_PIAB=( W11A*W11AB*SH11*(1-SH11) +W11A*W10AB*SH11*SH10
 +W11A*W01AB*SH11*SH01-W11A*W00AB*SH11*SH00
 +W10A*W11AB*SH10*SH11 +W10A*W10AB*SH10*(1-SH10)
 -W10A*W01AB*SH10*SH01+W10A*W00AB*SH10*SH00
 +W01A*W11AB*SH01*SH11-W01A*W10AB*SH01*SH10
 +W01A*W01AB*SH01*(1-SH01)+W01A*W00AB*SH01*SH00
 -W00A*W11AB*SH00*SH11+W00A*W10AB*SH00*SH10
 +W00A*W01AB*SH00*SH01 +W00A*W00AB*SH00*(1-SH00))/DELTA1**2;
CPIAB_PIB=(W11AB*W11B*SH11*(1-SH11) +W11AB*W10B*SH11*SH10
 +W11AB*W01B*SH11*SH01-W11AB*W00B*SH11*SH00
 +W10AB*W11B*SH10*SH11 +W10AB*W10B*SH10*(1-SH10)
 -W10AB*W01B*SH10*SH01 +W10AB*W00B*SH10*SH00
 +W01AB*W11B*SH01*SH11-W01AB*W10B*SH01*SH10
 +W01AB*W01B*SH01*(1-SH01) +W01AB*W00B*SH01*SH00
 -W00AB*W11B*SH00*SH11 +W00AB*W10B*SH00*SH10
 +W00AB*W01B*SH00*SH01+W00AB*W00B*SH00*(1-SH00))/DELTA1**2;
VAR_PIAB = (TERM1+TERM2)/DELTA1**2;
VAR_PIA = (TERM3+TERM4)/DELTA1**2;
VAR_PIB = (TERM5+TERM6)/DELTA1**2;
RE_PIAB = VAR_PIAB_LEE*100/VAR_PIAB;
RE_PIA = VAR_PIA_LEE*100/VAR_PIA;
RE_PIB = VAR_PIB_LEE*100/VAR_PIB;
RPB=(PROT_B_L*100)/PROT_B_AMY;
RPA=(PROT_A_L*100)/PROT_A_AMY;
RPAB=(PROT_AB_L*100)/PROT_AB_AMY;
MSE_PIA_given_B_LEE =VAR_PIAB_LEE/(PIAB**2)+VAR_PIB_LEE/(PIB**2)
-2*CPIAB_PIB_LEE/(PIB*PIAB);
MSE_PIA_given_B=VAR_PIAB/(PIAB**2)+VAR_PIB/(PIB**2)
-2*CPIAB_PIB/(PIB*PIAB);
VAR_PIA_minus_B_LEE=VAR_PIA_LEE+VAR_PIAB_LEE-2*CPIA_PIAB_LEE;
VAR_PIA_minus_B=VAR_PIA+VAR_PIAB-2*CPIA_PIAB;
VAR_PI_AUB_LEE=VAR_PIA_LEE+VAR_PIB_LEE+VAR_PIAB_LEE+2*CPIA_PIB_LEE-2*CPIA_PIAB_LEE-2*CPIAB_
PIB_LEE;
VAR_PI_AUB = VAR_PIA+VAR_PIB+VAR_PIAB+2*CPIA_PIB-2*CPIA_PIAB
-2*CPIAB_PIB;
```

```
MSE_RR_B_given_A_LEE=(PIB**2)*VAR_PIAB_LEE/((PIAB**2)*(PIB-PIAB)**2)
 +VAR_PIA_LEE/((PIA**2)*(1-PIA)**2)
 +VAR_PIB_LEE/(PIB-PIAB)**2
 -2*PIB*CPIA_PIAB_LEE/(PIA*PIAB*(1-PIA)*(PIB-PIAB))
 -2*PIB*CPIAB_PIB_LEE/(PIAB*(PIB-PIAB)**2)
 +2*CPIA_PIB_LEE/(PIA*(1-PIA)*(PIB-PIAB));
MSE_RR_B_given_A=(PIB**2)*VAR_PIAB/(PIAB**2*(PIB-PIAB)**2)
 + VAR_PIA/(PIA**2*(1-PIA)**2)
 +VAR_PIB/(PIB-PIAB)**2
 -2*PIB*CPIA_PIAB/ (PIA*PIAB*(1-PIA)*(PIB-PIAB))
 -2*PIB*CPIAB_PIB/(PIAB*(PIB-PIAB)**2)
 +2*CPIA_PIB/(PIA*(1-PIA)*(PIB-PIAB));
MSE_RHO_AB_LEE=(F1**2)*VAR_PIAB_LEE/(PIAB**2)
+(F2**2)*VAR_PIA_LEE/PIA**2 +(F3**2)*VAR_PIB_LEE/PIB**2
-2*F1*F2*CPIA_PIAB_LEE/(PIA*PIAB)-2*F1*F3*CPIAB_PIB_LEE/(PIB*PIAB)
+2*F2*F3*CPIA_PIB_LEE/(PIA*PIB);
MSE_RHO_AB=(F1**2)*VAR_PIAB/PIAB**2 + F2**2*VAR_PIA/PIA**2 +F3**2*VAR_PIB/PIB**2 -2*F1*F2*CPIA_
PIAB/(PIA*PIAB)
 -2*F1*F3*CPIAB_PIB/(PIB*PIAB)+2*F2*F3*CPIA_PIB/(PIA*PIB);

VAR_PI_d_LEE=VAR_PIA_LEE+VAR_PIB_LEE-2*CPIA_PIB_LEE;
VAR_PI_d=VAR_PIA+VAR_PIB-2*CPIA_PIB;
RE_PIA_given_B = MSE_PIA_given_B_LEE*100/MSE_PIA_given_B;
RE_PIA_minus_B = VAR_PIA_minus_B_LEE*100/VAR_PIA_minus_B;
RE_PI_AUB=VAR_PI_AUB_LEE*100/VAR_PI_AUB;
RE_RR_B_given_A = MSE_RR_B_given_A_LEE*100/MSE_RR_B_given_A;
RE_RHO_AB = MSE_RHO_AB_LEE*100/MSE_RHO_AB;
RE_PI_d =VAR_PI_d_LEE*100/VAR_PI_d;
*******CRAMER –RAO LOWER BOUNDS;
a_Amy=C1**2*(1/SH11+1/SH10+1/SH01+1/SH00);
b_Amy=C1*(C2/SH11-C4/SH10+C2/SH01-C4/SH00);
c_Amy=C1*(C3/SH11+C3/SH10-C5/SH01-C5/SH00);
d_Amy=C2**2/SH11+C4**2/SH10+C2**2/SH01+C4**2/SH00;
f_Amy=C2*C3/SH11-C3*C4/SH10-C2*C5/SH01+C4*C5/SH00;
g_Amy=C3**2/SH11+C3**2/SH10+C5**2/SH01+C5**2/SH00;
DELTA_AMY=g_Amy*b_Amy**2-2*b_Amy*c_Amy*f_Amy+ d_Amy*c_Amy**2+a_Amy*f_Amy**2-a_Amy*d_Amy*g_
Amy;
a11_Amy=f_Amy**2-d_Amy*g_Amy;
a22_Amy=c_Amy**2-a_Amy*g_Amy;
a33_Amy=b_Amy**2-a_Amy*d_Amy;
a12_Amy=b_Amy*g_Amy-c_Amy*f_Amy;
a13_Amy=c_Amy*d_Amy-b_Amy*f_Amy;
a23_Amy= a_Amy*f_Amy-b_Amy*c_Amy;
crpiab=a11_Amy/DELTA_AMY;
crpia=a22_Amy/DELTA_AMY;

crpib=a33_Amy/DELTA_AMY;
cr_piabpia=a12_Amy/DELTA_AMY;
cr_piabpib=a13_Amy/DELTA_AMY;
cr_piapib=a23_Amy/DELTA_AMY;
OUTPUT;
```

```
END;
END;
END;
END;
END;
END;
END;
DATA DATA2;
SET DATA1;
KEEP P T P1 T1 PIAB PIA PIB Y W RE_PIA RE_PIB RE_PIAB RPA RPB RPAB
RE_PIA_given_B RE_PIA_minus_B RE_PI_AUB RE_RR_B_given_A RE_RHO_AB RE_PI_d;
IF RE_PIAB GT 100;
IF RE_PIA GT 100;
IF RE_PIB GT 100;
IF RPAB GT 100;
IF RPA GT 100;
IF RPB GT 100;
IF RE_RHO_AB GT 100;
IF RE_PI_AUB GT 100;
IF RE_PI_d GT 100;
IF PIAB LT PIA AND PIAB LT PIB;
DATA DATA3;
SET DATA2;
KEEP P T P1 T1 PIAB PIA PIB Y W RE_PIA RE_PIB RE_PIAB RPA RPB RPAB;
DATA DATA4;
SET DATA2;
KEEP P T W Y P1 T1 PIAB PIA PIB RE_PIA_given_B RE_PIA_minus_B RE_PI_AUB RE_RR_B_given_A RE_RHO_AB
RE_PI_d;
PROC PRINT DATA=DATA3;
VAR P T W Y P1 T1 PIAB PIA PIB RE_PIA RE_PIB RE_PIAB RPA RPB RPAB;
PROC PRINT DATA=DATA4;
VAR P T W Y P1 T1 PIAB PIA PIB RE_PIA_given_B RE_PIA_minus_B RE_PI_AUB RE_RR_B_given_A RE_RHO_AB
RE_PI_d;
DATA DATA5;
SET DATA1;
KEEP P T W Y P1 T1 PIAB PIA PIB crpiab VAR_PIAB crpia VAR_PIA crpib VAR_PIB cr_piabpia CPIA_PIAB cr_piabpib
CPIAB_PIB cr_piapib CPIA_PIB;
IF RE_PIAB GT 100;
IF RE_PIA GT 100;
IF RE_PIB GT 100;
IF RPAB GT 100;
IF RPA GT 100;
IF RPB GT 100;
IF RE_RHO_AB GT 100;

IF RE_PI_AUB GT 100;
IF RE_PI_d GT 100;
IF PIAB LT PIA AND PIAB LT PIB;
PROC PRINT DATA=DATA5;
VAR P T W Y P1 T1 PIAB PIA PIB crpiab VAR_PIAB crpia VAR_PIA crpib VAR_PIB cr_piabpia CPIA_PIAB cr_piabpib
cr_piapib CPIA_PIB CPIAB_PIB;
RUN;
```

Appendix B

Table B1: Detailted results for different choice of parameters.

Obs	w	π_y	P_1	T_1	π_{AB}	π_A	π_B	RE(1)	RE(2)	RE(3)	RP(1)	RP(2)	RP(3)
1	0.1	0.9	0.4	0.4	0.05	0.35	0.35	109.2	109.2	117.9	120.7	120.7	139.7
2	0.1	0.9	0.4	0.5	0.05	0.25	0.25	101.5	146.0	148.5	116.0	102.6	130.5
3	0.1	0.9	0.4	0.5	0.05	0.25	0.30	101.5	150.3	150.6	118.4	105.0	130.1
4	0.1	0.9	0.4	0.5	0.05	0.25	0.35	101.5	154.9	152.7	120.8	107.2	129.7
5	0.1	0.9	0.4	0.5	0.05	0.30	0.20	105.2	142.0	150.9	114.1	101.8	129.1
TO GET COMPLETE TABLE EXECUTE THE SAS CODE													
1600	0.9	1	1	1	0.20	0.30	0.30	103.7	103.7	109.7	100.2	100.2	102.5
1601	0.9	1	1	1	0.20	0.30	0.35	103.7	104.1	109.9	100.2	100.2	102.5
1602	0.9	1	1	1	0.20	0.35	0.25	104.1	103.3	109.7	100.2	100.2	102.5
1603	0.9	1	1	1	0.20	0.35	0.30	104.1	103.7	109.9	100.2	100.2	102.5
1604	0.9	1	1	1	0.20	0.35	0.35	104.1	104.1	110.1	100.3	100.3	102.4

Appendix C

Table C1: Detailed results for different choice of parameters.

Obs	w	π_y	P_1	T_1	π_{AB}	π_A	π_B	RE(4)	RE(5)	RE(6)	RE(7)	RE(8)	RE(9)
1	0.1	0.9	0.4	0.4	0.05	0.35	0.35	118.0	114.8	100.0	117.5	115.7	127.1
2	0.1	0.9	0.4	0.5	0.05	0.25	0.25	148.5	129.8	121.0	148.5	148.1	134.8
3	0.1	0.9	0.4	0.5	0.05	0.25	0.30	150.5	134.5	125.0	152.3	151.6	136.3
4	0.1	0.9	0.4	0.5	0.05	0.25	0.35	152.5	139.8	129.4	156.2	155.6	137.9
5	0.1	0.9	0.4	0.5	0.05	0.30	0.20	151.6	130.0	120.6	150.4	150.0	135.9
TO GET COMPLETE TABLE EXECUTE THE SAS CODES													
1600	0.9	1	0.5	0.5	0.20	0.30	0.30	109.0	107.8	104.3	108.2	108.9	106.3
1601	0.9	1	0.5	0.5	0.20	0.30	0.35	109.4	108.3	104.7	108.7	109.4	106.5
1602	0.9	1	0.5	0.5	0.20	0.35	0.25	108.8	107.5	104.3	108.3	108.9	106.3
1603	0.9	1	0.5	0.5	0.20	0.35	0.30	109.5	108.0	104.7	108.8	109.4	106.5
1604	0.9	1	0.5	0.5	0.20	0.35	0.35	109.9	108.5	105.1	109.3	109.8	106.6

CHAPTER 9

Hybrid of Crossed Model and a New Unrelated Question Model for Two Sensitive Characteristics

Renhua Zheng,* Stephen A Sedory* and *Sarjinder Singh*

1. Crossed Model for two Sensitive Characteristics

Consider a situation where the population Ω of interest consists of four types of persons; that is, some people possess either sensitive attribute A, or sensitive attribute B, or both $A \cap B$, or none of these. In Figure 1 a Venn diagram has been used to display such a population.

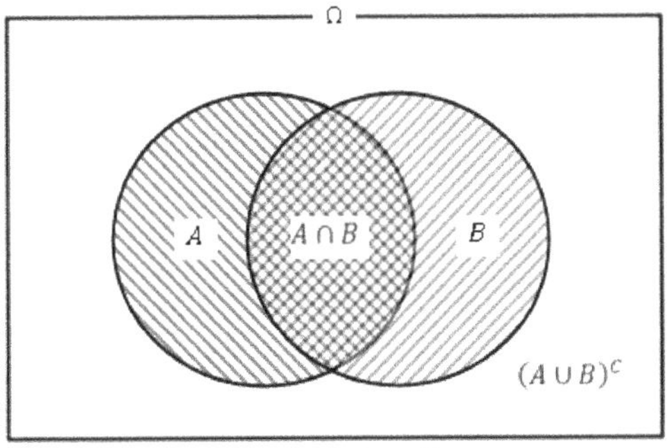

Figure 1: Population of interest.

In this population of interest, let π_A, π_B and π_{AB} be the proportions of people possessing the sensitive attribute A, the sensitive attribute B and proportion of the people possessing the both sensitive attributes $A \cap B$. Note that $(A \cup B)^c$ may or may not be an empty set. A new variation in the field of randomized response sampling by proposing a crossed model is shown in Figure 2.

$I \in A$ with probability P
$I \in B^c$ with probability $(1-P)$
Deck-I

$I \in B$ with probability T
$I \in A^c$ with probability $(1-T)$
Deck-II

Figure 2: Crossed Model: Two decks of cards.

The data collection procedure is the same as discussed for the case of the simple model, then the probability of responding (Yes, Yes) is given by:

$$\theta_{11}^* = \pi_{AB}\{PT + (1-P)(1-T)\} - \pi_A(1-P)(1-T) - \pi_B(1-P)(1-T) + (1-P)(1-T) \quad (1.1)$$

The probability of getting a (Yes, No) response in the crossed model is shown as:

$$\theta_{10}^* = -\pi_{AB}\{PT + (1-P)(1-T)\} - \pi_A\{(1-P)T - 1\} - \pi_B(1-P)T + (1-P)T \quad (1.2)$$

Thus the probability of getting a (No, Yes) response in the crossed model is described as:

$$\theta_{01}^* = -\pi_{AB}\{PT + (1-P)(1-T)\} - \pi_A P(1-T) - \pi_B\{P(1-T) - 1\} + P(1-T) \quad (1.3)$$

Department of Mathematics, Texas A&M University-Kingsville, Kingsville, TX 78363, USA.
* Corresponding authors: renhua.zheng@students.tamuk.edu; sarjinder.singh@tamuk.edu

The probability of getting a (No, No) response is shown as:

$$\theta_{00}^{*} = \pi_{AB}\{PT + (1-P)(1-T)\} - \pi_A PT - \pi_B PT + PT \tag{1.4}$$

Assume $\hat{\theta}_{11}^{*} = \frac{n_{11}^{*}}{n}$, $\hat{\theta}_{10}^{*} = \frac{n_{10}^{*}}{n}$, $\hat{\theta}_{01}^{*} = \frac{n_{01}^{*}}{n}$, and $\hat{\theta}_{00}^{*} = \frac{n_{00}^{*}}{n}$ to be the observed proportions of (Yes, Yes), (Yes, No), (No, Yes) and (No, No) responses. Lee, Sedory and Singh (2013) consider the distance between the observed proportions and the true proportions as:

$$D^{*} = \frac{1}{2}\sum_{i=0}^{1}\sum_{j=0}^{1}(\theta_{ij}^{*} - \hat{\theta}_{ij}^{*})^2 \tag{1.5}$$

On setting

$$\frac{\partial D^{*}}{\partial \pi_A} = 0 \tag{1.6}$$

$$\frac{\partial D^{*}}{\partial \pi_B} = 0 \tag{1.7}$$

and

$$\frac{\partial D^{*}}{\partial \pi_{AB}} = 0 \tag{1.8}$$

They obtained unbiased estimators of π_A, π_B and π_{AB} respectively given by:

$$\hat{\pi}_A^{*} = \frac{1}{2} + \frac{(T-P+1)\left(\hat{\theta}_{11}^{*} - \hat{\theta}_{00}^{*}\right) + (P+T-1)\left(\hat{\theta}_{10}^{*} - \hat{\theta}_{01}^{*}\right)}{2(P+T-1)} \tag{1.9}$$

$$\hat{\pi}_B^{*} = \frac{1}{2} + \frac{(P-T+1)\left(\hat{\theta}_{11}^{*} - \hat{\theta}_{00}^{*}\right) + (P+T-1)\left(\hat{\theta}_{01}^{*} - \hat{\theta}_{10}^{*}\right)}{2(P+T-1)} \tag{1.10}$$

and

$$\hat{\pi}_{AB}^{*} = \frac{PT\hat{\theta}_{11}^{*} - (1-P)(1-T)\hat{\theta}_{00}^{*}}{\{PT + (1-P)(1-T)\}(P+T-1)} \tag{1.11}$$

for $P + T \neq 1$.

The variance of the estimator $\hat{\pi}_A^{*}$ is given by:

$$V(\hat{\pi}_A^{*}) = \frac{\pi_A(1-\pi_A)}{n} + \frac{(1-P)\left[T\{PT+(1-P)(1-T)\}(1-\pi_A-\pi_B+2\pi_{AB})\right]}{n(P+T-1)^2} \tag{1.12}$$

The variance of the estimator $\hat{\pi}_B^{*}$ is given by

$$V(\hat{\pi}_B^{*}) = \frac{\pi_B(1-\pi_B)}{n} + \frac{(1-T)\left[P\{PT+(1-P)(1-T)\}(1-\pi_A-\pi_B+2\pi_{AB})\right]}{n(P+T-1)^2} \tag{1.13}$$

The variance of the estimator $\hat{\pi}_{AB}^{*}$ is given by

$$V\left(\hat{\pi}_{AB}^{*}\right) = \frac{\pi_{AB}(1-\pi_{AB})}{n} + \frac{\left[\begin{array}{l}\pi_{AB}\left\{P^2T^2 + (1-P)^2(1-T)^2 - \{PT+(1-P)(1-T)\}(P+T-1)^2\right\} \\ +PT(1-P)(1-T)(1-\pi_A-\pi_B)\end{array}\right]}{n\{PT+(1-P)(1-T)\}(P+T-1)^2} \tag{1.14}$$

The covariance between $\hat{\pi}_A^{*}$ and $\hat{\theta}_{11}^{*}$ is given by

$$Cov\left(\hat{\pi}_A^{*}, \hat{\theta}_{11}^{*}\right) = \frac{\theta_{11}^{*}}{n(P+T-1)}\left[T - (P+T-1)\pi_A\right] \tag{1.15}$$

The covariance between $\hat{\pi}_A^*$ and $\hat{\theta}_{00}^*$ is given by

$$Cov\left(\hat{\pi}_A^*, \hat{\theta}_{00}^*\right) = \frac{-\theta_{00}^*}{n(P+T-1)}\left[(P+T-1)\pi_A + (1-P)\right] \tag{1.16}$$

The covariance between $\hat{\pi}_{AB}^*$ and $\hat{\pi}_A^*$ is given by

$$Cov\left(\hat{\pi}_{AB}^*, \hat{\pi}_A^*\right) = \frac{\pi_{AB}(1-\pi_A)}{n} + \frac{\pi_{AB}T(1-P)(P-T+1)}{n(P+T-1)^2} + \frac{PT(1-P)(1-T)(T-P+1)(1-\pi_A-\pi_B)}{n\{PT+(1-P)(1-T)\}(P+T-1)^2} \tag{1.17}$$

The covariance between $\hat{\pi}_{AB}^*$ and $\hat{\pi}_B^*$ is given by

$$Cov\left(\hat{\pi}_{AB}^*, \hat{\pi}_B^*\right) = \frac{\pi_{AB}(1-\pi_B)}{n} + \frac{\pi_{AB}P(1-T)(T-P+1)}{n(P+T-1)^2} + \frac{PT(1-P)(1-T)(P-T+1)(1-\pi_A-\pi_B)}{n\{PT+(1-P)(1-T)\}(P+T-1)^2} \tag{1.18}$$

The covariance between $\hat{\pi}_A^*$ and $\hat{\pi}_B^*$ is given by

$$Cov\left(\hat{\pi}_A^*, \hat{\pi}_B^*\right) = -\frac{(\pi_A\pi_B - \pi_{AB})}{n} - \frac{2PT(1-T)(1-P)(\pi_A + \pi_B - 2\pi_{AB} - 1)}{n(P+T-1)^2} \tag{1.19}$$

A natural estimator of the conditional proportion $\pi_{A|B}$ is given by:

$$\hat{\pi}_{A|B}^* = \frac{\hat{\pi}_{AB}^*}{\hat{\pi}_B^*} \tag{1.20}$$

The bias, to the first order of approximation, in the estimator $\hat{\pi}_{A|B}^*$ is given by

$$B\left(\hat{\pi}_{A|B}^*\right) = \pi_{A|B}\left[\frac{V(\hat{\pi}_B^*)}{\pi_B^2} - \frac{Cov(\hat{\pi}_{AB}^*, \hat{\pi}_B^*)}{\pi_{AB}\pi_B}\right] \tag{1.21}$$

The mean squared error, to the first order of approximation, of the estimator $\hat{\pi}_{A|B}^*$ is given by

$$MSE\left(\hat{\pi}_{A|B}^*\right) = \pi_{A|B}^2\left[\frac{V(\hat{\pi}_{AB}^*)}{\pi_{AB}^2} + \frac{V(\hat{\pi}_B^*)}{\pi_B^2} - \frac{2Cov(\hat{\pi}_B^*, \hat{\pi}_{AB}^*)}{\pi_B\pi_{AB}}\right] \tag{1.22}$$

An unbiased estimator of π_{A-B} is given by

$$\hat{\pi}_{A-B}^* = \hat{\pi}_A^* - \hat{\pi}_{AB}^* \tag{1.23}$$

The variance of the estimator of $\hat{\pi}_{A-B}^*$ is given by

$$V\left(\hat{\pi}_{A-B}^*\right) = V\left(\hat{\pi}_A^*\right) + V\left(\hat{\pi}_{AB}^*\right) - 2Cov\left(\hat{\pi}_A^*, \hat{\pi}_{AB}^*\right) \tag{1.24}$$

An unbiased estimator of $\pi_{A\cup B}$ is given by

$$\hat{\pi}_{A\cup B}^* = \hat{\pi}_A^* + \hat{\pi}_B^* - \hat{\pi}_{AB}^* \tag{1.25}$$

The variance of the estimator of $\hat{\pi}_{A\cup B}^*$ is given by;

$$V\left(\hat{\pi}_{A\cup B}^*\right) = V\left(\hat{\pi}_A^*\right) + V\left(\hat{\pi}_B^*\right) + V\left(\hat{\pi}_{AB}^*\right) + 2Cov\left(\hat{\pi}_A^*, \hat{\pi}_B^*\right) - 2Cov\left(\hat{\pi}_A^*, \hat{\pi}_{AB}^*\right) - 2Cov\left(\hat{\pi}_B^*, \hat{\pi}_{AB}^*\right) \tag{1.26}$$

A natural estimator of the RR is given by:

$$\hat{RR}^*(B|A) = \frac{\hat{\pi}_{AB}^*\left(1-\hat{\pi}_A^*\right)}{\hat{\pi}_A^*\left(\hat{\pi}_B^* - \hat{\pi}_{AB}^*\right)} \tag{1.27}$$

The bias in the estimator $\hat{RR}^*(B|A)$, to the first order of approximation, is given by:

$$B\left(\hat{RR}^*(B\mid A)\right) = RR\left[\frac{V(\hat{\pi}_A^*)}{\pi_A^2(1-\pi_A)} + \frac{V(\hat{\pi}_B^*)}{(\pi_B-\pi_{AB})^2} + \frac{\pi_B V(\hat{\pi}_{AB}^*)}{\pi_{AB}(\pi_B-\pi_{AB})^2} + \frac{Cov(\hat{\pi}_A^*,\hat{\pi}_B^*)}{\pi_A(1-\pi_A)(\pi_B-\pi_{AB})} - \frac{\pi_B Cov(\hat{\pi}_A^*,\hat{\pi}_{AB}^*)}{\pi_A\pi_{AB}(1-\pi_A)(\pi_B-\pi_{AB})} - \frac{(\pi_B+\pi_{AB})Cov(\hat{\pi}_B^*,\hat{\pi}_{AB}^*)}{\pi_{AB}(\pi_B-\pi_{AB})^2}\right] \tag{1.28}$$

The mean squared error of the estimator $\hat{RR}^*(B|A)$ to the first order of approximation, is given by:

$$MSE\left(\hat{RR}^*(B\mid A)\right) = RR^2\left[\frac{\pi_B^2 V(\hat{\pi}_{AB}^*)}{\pi_{AB}^2(\pi_B-\pi_{AB})^2} + \frac{V(\hat{\pi}_A^*)}{\pi_A^2(1-\pi_A)^2} + \frac{V(\hat{\pi}_B^*)}{(\pi_B-\pi_{AB})^2} - \frac{2\pi_B Cov(\hat{\pi}_A^*,\hat{\pi}_{AB}^*)}{\pi_A\pi_{AB}(1-\pi_A)(\pi_B-\pi_{AB})} - \frac{2\pi_B Cov(\hat{\pi}_B^*,\hat{\pi}_{AB}^*)}{\pi_{AB}(\pi_B-\pi_{AB})^2} + \frac{2Cov(\hat{\pi}_A^*,\hat{\pi}_B^*)}{\pi_A(1-\pi_A)(\pi_B-\pi_{AB})}\right] \tag{1.29}$$

A usual estimator of the correlation coefficient ρ_{AB} is given by:

$$\hat{\rho}_{AB}^* = \frac{\hat{\pi}_{AB}^* - \hat{\pi}_A^*\hat{\pi}_B^*}{\sqrt{\hat{\pi}_A^*(1-\hat{\pi}_A^*)}\sqrt{\hat{\pi}_B^*(1-\hat{\pi}_B^*)}} \tag{1.30}$$

The bias in the estimator $\hat{\rho}_{AB}^*$, to the first order of approximation, is:

$$B(\hat{\rho}_{AB}^*) = \rho_{AB}\left[F_4\frac{V(\hat{\pi}_A^*)}{\pi_A^2} + F_5\frac{V(\hat{\pi}_B^*)}{\pi_B^2} + F_6\frac{Cov(\hat{\pi}_A^*,\hat{\pi}_B^*)}{\pi_A\pi_B} - F_7\frac{Cov(\hat{\pi}_A^*,\hat{\pi}_{AB}^*)}{\pi_A\pi_{AB}} - F_8\frac{Cov(\hat{\pi}_B^*,\hat{\pi}_{AB}^*)}{\pi_B\pi_{AB}}\right] \tag{1.31}$$

The mean squared error of the estimator $\hat{\rho}_{AB}^*$, to the first order of approximation, is given by

$$MSE(\hat{\rho}_{AB}^*) = \rho_{AB}^2\left[F_1^2\frac{V(\hat{\pi}_{AB}^*)}{\pi_{AB}^2} + F_2^2\frac{V(\hat{\pi}_A^*)}{\pi_A^2} + F_3^2\frac{V(\hat{\pi}_B^*)}{\pi_B^2} - \frac{2F_1F_2Cov(\hat{\pi}_A^*,\hat{\pi}_{AB}^*)}{\pi_A\pi_{AB}} - \frac{2F_1F_3Cov(\hat{\pi}_B^*,\hat{\pi}_{AB}^*)}{\pi_B\pi_{AB}} + \frac{2F_2F_3Cov(\hat{\pi}_A^*,\hat{\pi}_B^*)}{\pi_A\pi_B}\right] \tag{1.32}$$

where

$$F_1 = \frac{\pi_{AB}}{\pi_{AB}-\pi_A\pi_B}, \tag{1.33}$$

$$F_2 = \frac{\pi_A\pi_B}{\pi_{AB}-\pi_A\pi_B} + \frac{(1-2\pi_A)}{2(1-\pi_A)}, \tag{1.34}$$

$$F_3 = \frac{\pi_A\pi_B}{\pi_{AB}-\pi_A\pi_B} + \frac{(1-2\pi_B)}{2(1-\pi_B)}, \tag{1.35}$$

$$F_4 = \frac{\pi_A\pi_B(1-2\pi_A)}{2(1-\pi_A)(\pi_{AB}-\pi_A\pi_B)} + \frac{\pi_A}{2(1-\pi_A)} + \frac{3(1-2\pi_A)^2}{8(1-\pi_A)^2}, \tag{1.36}$$

$$F_5 = \frac{\pi_A\pi_B(1-2\pi_B)}{2(1-\pi_B)(\pi_{AB}-\pi_A\pi_B)} + \frac{\pi_B}{2(1-\pi_B)} + \frac{3(1-2\pi_B)^2}{8(1-\pi_B)^2}, \tag{1.37}$$

$$F_6 = \frac{1-2\pi_A-2\pi_B+4\pi_A\pi_B}{4(1-\pi_A)(1-\pi_B)} - \frac{\pi_A\pi_B}{\pi_{AB}-\pi_A\pi_B} + \frac{\pi_A\pi_B}{2(\pi_{AB}-\pi_A\pi_B)}\left\{\frac{(1-2\pi_A)}{(1-\pi_A)} + \frac{(1-2\pi_B)}{(1-\pi_B)}\right\}, \tag{1.38}$$

$$F_7 = \frac{\pi_{AB}\left(1-2\pi_A\right)}{2\left(1-\pi_A\right)\left(\pi_{AB}-\pi_A\pi_B\right)}, \tag{1.39}$$

and

$$F_8 = \frac{\pi_{AB}\left(1-2\pi_B\right)}{2\left(1-\pi_B\right)\left(\pi_{AB}-\pi_A\pi_B\right)} \tag{1.40}$$

An unbiased estimator of π_d is defined as:

$$\hat{\pi}_d^* = \hat{\pi}_A^* - \hat{\pi}_B^* \tag{1.41}$$

The variance of the estimator $\hat{\pi}_d^*$ is given by

$$V(\hat{\pi}_d^*) = V(\hat{\pi}_A^*) + V(\hat{\pi}_B^*) - 2Cov(\hat{\pi}_A^*, \hat{\pi}_B^*) \tag{1.42}$$

We introduced a new unrelated question randomized response model in the following section.

2. Paired Unrelated Question Model for two Sensitive Questions: A Challenge

In the proposed model, assume we introduce the simultaneous use of an unrelated question in a randomized response model consisting of an ordered pair of decks of cards as shown in Figure 3. It leads to a new challenge to be resolved while making its use in practice.

$I \in A$ with probability P_1	$I \in B$ with probability T_1
$I \in Y$ with probability $(1- P_1)$	$I \in Y$ with probability $(1- T_1)$
Deck-I	Deck-II

Figure 3: Unrelated Question Model: Two decks of cards.

Every participant in the sample pool is requested to experience the ordered pair of decks of cards in Figure 3. Now if a respondent belongs to both sensitive characteristic A and B, and he (or she) draws the first card from Deck-I with the statement $I \in A$, and the second card from the Deck-II with the statement $I \in B$. As a result, his/her response will be (Yes, Yes) with probability $P_1T_1\pi_{AB}$. Now if a selected respondent draws the first card with statement $I \in A$ from the first deck and second card with the statement $I \in Y$ from the second deck, and if he (she) is a member of both unrelated groups A and Y, then his/her response will be (Yes, Yes) with probability $P_1(1 - T_1)\pi_y\pi_A$. Now if a selected respondent draws the first card with statement $I \in Y$ and the second card with statement $I \in B$, and he/she is a member of both unrelated groups B and Y, then his/her response will be (Yes, Yes) with probability $(1 - P_1)T_1\pi_y\pi_B$. Last, if a selected respondent draws the first card from Deck-I with statement $I \in Y$ and the second card from Deck-II with statement $I \in Y$, and the selected respondent is a member of the unrelated group Y, then his/her response will be (Yes, Yes) with probability $(1 - P_1)(1 - T_1)\pi_y$. Note that the response (Yes, Yes) can come from four different types of respondents who either belong to both sensitive groups A and B; or belong to the first sensitive group A and the unrelated group Y; or belong to the unrelated group Y and the second sensitive group B, or belong to only unrelated group Y. The overall probability of observing (Yes, Yes) response is given by:

$$P(Yes, Yes) = \theta_{11}^u = P_1T_1\pi_{AB} + P_1(1-T_1)\pi_A\pi_y + (1-P_1)T_1\pi_y\pi_B + (1-P_1)(1-T_1)\pi_y$$

or

$$\theta_{11}^u = P_1T_1\pi_{AB} + P_1(1-T_1)\pi_y\pi_A + (1-P_1)T_1\pi_y\pi_B + (1-P_1)(1-T_1)\pi_y \tag{2.1}$$

Now if a respondent belongs to the sensitive group A and not to group B, and he (or she) draws the first card from the first deck with the statement $I \in A$, and the second card from the Deck-II with the statement $I \in B$, then his/her response (Yes, No) has probability $P_1T_1(\pi_A - \pi_{AB})$. Now if a selected respondent draws a first card with statement $I \in A$ from the first deck and second card with statement $I \in Y$ from the second deck, and if he/she is a member of the group A but not in the unrelated group Y, then his/her response will be (Yes, No) with probability $P_1(1 - T_1)\pi_A(1 - \pi_y)$. Now if a selected respondent draws the first card with statement $I \in Y$, and the second card with statement $I \in B$, and he (or she) is a member of unrelated group Y and not belonging to the sensitive attribute B, then his/her response will be (Yes, No) with probability $(1 - P_1)T_1\pi_y(1 - \pi_B)$. Last, if a selected respondent draws a first card with statement $I \in Y$ from the first deck and the second card with statement $I \in Y$ from the second deck, then his/her response will be (Yes, No) with probability 0. Note that the response (Yes, No) can come from only three different types of respondents who either belong to the sensitive groups A and not to B; or belong to the first sensitive group A and not to the unrelated group Y; or belonging to unrelated group Y and not to the second sensitive group B. Thus the respondents reporting (Yes, No) are at the risk of losing their privacy that they may belong to the sensitive group A but not to B. The overall probability of observing (Yes, No) response is given by:

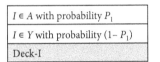

$$P(Yes, No) = \theta_{10}^u = P_1T_1(\pi_A - \pi_{AB}) + P_1(1-T_1)\pi_A(1-\pi_y) + (1-P_1)T_1\pi_y(1-\pi_B) + (1-P_1)(1-T_1)(0)$$

or

$$\theta_{10}^u = -P_1T_1\pi_{AB} + P_1\{T_1 + (1-T_1)(1-\pi_y)\}\pi_A - (1-P_1)T_1\pi_y\pi_B + (1-P_1)T_1\pi_y \tag{2.2}$$

Now if a respondent belongs to the sensitive group B and not to A, and he/she draws the first card with the statement $I \in A$ from the first deck and second card with the statement $I \in B$ from the second deck, then his/her response will be (No, Yes) with probability $P_1T_1(\pi_B - \pi_{AB})$. Now if a selected respondent draws a first card with statement $I \in A$ from the first deck and second card with statement $I \in Y$ from the second deck, and if he/she is not a member of the group A and is a member of unrelated group Y, then his/her response will be (No, Yes) with probability $P_1(1 - T_1)\pi_y(1 - \pi_A)$. Now if a selected respondent draws a first card with statement $I \in Y$ and the second card with statement $I \in B$, and he/she is a not a member of unrelated group Y and is a member of the sensitive group B, then his/her response will be (No, Yes) with probability $(1 - P_1)T_1\pi_B(1 - \pi_y)$. Last, if a selected respondent draws first card with statement $I \in Y$ from the first deck and the second card with statement $I \in Y$ from the second deck, then his/her response will be (No, Yes) with probability 0. Note that the response (No, Yes) can come only from three different types of respondents who either belong to the sensitive group B and not to A; or does not belong to the sensitive group A and belongs to unrelated group Y; or does not belong to unrelated group Y and belongs to the sensitive group B. Thus the respondents reporting (No, Yes) are at the risk of losing their privacy that they may belong to the sensitive group B but not to A. The overall probability of observing (No, Yes) response is given by:

$$P(No, Yes) = \theta_{01}^u = P_1T_1(\pi_B - \pi_{AB}) + P_1(1-T_1)\pi_y(1-\pi_A) + (1-P_1)T_1\pi_B(1-\pi_y) + (1-P_1)(1-T_1)(0)$$

or

$$\theta_{01}^u = -P_1T_1\pi_{AB} + T_1\{P_1 + (1-P_1)(1-\pi_y)\}\pi_B - P_1(1-T_1)\pi_y\pi_A + (1-T_1)P_1\pi_y \tag{2.3}$$

Now if a respondent does not belong to both sensitive groups A and B, and he/she draws the first card with the statement $I \in A$ from the first deck and second card with the statement $I \in B$ from the second deck, then his/her response will be (No, No) with probability $P_1T_1(1 - \pi_A - \pi_B + \pi_{AB})$. Now if a selected respondent draws first card with statement $I \in A$ from the first deck and second card with statement $I \in Y$ from the second deck, and if he/she is neither a member of group A nor a member of Y, then his/her response will be (No, No) with probability $P_1(1 - T_1)(1 - \pi_A)(1 - \pi_y)$. Now if a selected respondent draws first card with statement $I \in Y$ and the second card with statement $I \in B$, and he/she is a not a member of both groups B and Y, then his/her response will be (No, No) with probability $(1 - P_1)T_1(1 - \pi_y)(1 - \pi_B)$. Last, if a selected respondent draws first card with statement $I \in Y$ from the first deck and the second card with statement $I \in Y$ from the second deck, and he/she is a member of the group Y^c, then his/her response will be (No, No) with probability $(1 - P_1)(1 - T_1)(1 - \pi_y)$. Note that the response (No, No) can come from four different types of respondents who do not belong to either of the two sensitive groups A and B; or does not belong to the first sensitive group A and nor to the unrelated group Y; or does not belong to the unrelated group Y and does not belong to sensitive group B, or does not belong to the unrelated group Y. The overall probability of observing (No, No) response is given by:

$$P(No, No) = \theta_{00}^u = P_1T_1(1-\pi_A-\pi_B+\pi_{AB}) + P_1(1-T_1)(1-\pi_A)(1-\pi_y) + (1-P_1)T_1(1-\pi_y)(1-\pi_B)$$
$$+(1-P_1)T_1(1-\pi_B)\pi_y + (1-P_1)(1-T_1)(1-\pi_y)$$

or

$$\theta_{00}^u = P_1T_1\pi_{AB} - P_1\{T_1 + (1-T_1)(1-\pi_y)\}\pi_A - T_1\{P_1 + (1-P_1)(1-\pi_y)\}\pi_B + P_1T_1\pi_y + 1 - \pi_y \tag{2.4}$$

Note that:

$$\sum_{i=0}^{1}\sum_{j=0}^{1}\theta_{ij}^u = 1 \tag{2.5}$$

Again it should be noted that a (Yes, No) or (No, Yes) response could not have come from an individual who drew a card with "$I \in Y$" from both decks. He/she must be addressing at least one of the statements "$I \in A$" or "$I \in B$". While no conclusion can be drawn as to their membership in either of the sensitive groups, and it may result in cooperation being lost. Thus the use of Simmons' (Horvitz, Shah and Simmons, 1967) unrelated question model twice for collecting data on two sensitive characteristics is quite different than the use of Warner (1965) model. This makes us think of an alternative method to make it functional while estimating the two sensitive characteristics.

On the lines of Johnson, Sedory and Singh (2019), and Yennum, Sedory and Singh (2019), in the following sections, we develop a new two stage model, which we name as Decision Maker Deck Model -II and abbreviate it to DMDM-II.

3. Decision Maker Deck Model-II for two Sensitive Characteristics

In the proposed decision maker deck model-II, we use a two-stage model where the added new stage helps the respondent to make a decision about the choice of model to be used at the second stage while responding to the sensitive questions. As both the respondent and the interviewer are uncertain about the final model being used by a respondent, and hence it increases cooperation between both. A pictorial presentation of such a two-stage model is given in Figure 4.

Use crossed Model	With probability w
Use paired unrelated question model	With probability $(1 - w)$

Figure 4: Decision Maker Deck.

In other words, each respondent selected in the sample is directed to use a Decision Maker Deck (DMD), consisting of two types of cards bearing the statements "Use Crossed Model" and "Use paired unrelated question model" with known relative frequencies w and $(1 - w)$, respectively.

Thus in the proposed DMDM-II, the probability of (Yes, Yes) response is given by:

$$\begin{aligned}\Psi_{11}^* &= w\theta_{11}^* + (1-w)\theta_{11}^u \\ &= w[(PT+(1-P)(1-T))\pi_{AB} - (1-P)(1-T)\pi_A - (1-P)(1-T)\pi_B + (1-P)(1-T)] \\ &\quad + (1-w)[P_1T_1\pi_{AB} + P_1(1-T_1)\pi_Y\pi_A + (1-P_1)T_1\pi_Y\pi_B + (1-P_1)(1-T_1)\pi_Y] \\ &= [w(PT+(1-P)(1-T)) + (1-w)P_1T_1]\pi_{AB} + [-w(1-P)(1-T) + (1-w)P_1(1-T_1)\pi_Y]\pi_A \\ &\quad + [-w(1-P)(1-T) + (1-w)(1-P_1)T_1\pi_Y]\pi_B + w(1-P)(1-T) + (1-w)(1-P_1)(1-T_1)\pi_Y \\ &= C_1^*\pi_{AB} + C_2^*\pi_A + C_3^*\pi_B + H_1^*\end{aligned} \tag{3.1}$$

the probability of getting a (*Yes, No*) response is given by:

$$\begin{aligned}\Psi_{10}^* &= w\theta_{10}^* + (1-w)\theta_{10}^u \\ &= w[-(PT+(1-P)(1-T))\pi_{AB} - ((1-P)T-1)\pi_A - (1-P)T\pi_B + (1-P)T] \\ &\quad + (1-w)\left[-P_1T_1\pi_{AB} + P_1\{T_1 + (1-T_1)(1-\pi_Y)\}\pi_A - (1-P_1)T_1\pi_Y\pi_B + (1-P_1)T_1\pi_Y\right] \\ &= -[w(PT+(1-P)(1-T)) + (1-w)P_1T_1]\pi_{AB} \\ &\quad -[w\{(1-P)T-1\} - (1-w)P_1\{T_1 + (1-T_1)(1-\pi_Y)\}]\pi_A \\ &\quad -[w(1-P)T + (1-w)(1-P_1)T_1\pi_Y]\pi_B + (1-P)Tw + (1-w)(1-P_1)T_1\pi_Y \\ &= -C_1^*\pi_{AB} - C_4^*\pi_A - C_5^*\pi_B + H_2^*\end{aligned} \tag{3.2}$$

the probability of getting a (*No, Yes*) response is given by:

$$\begin{aligned}\Psi_{01}^* &= w\theta_{01}^* + (1-w)\theta_{01}^u \\ &= w[-(PT+(1-P)(1-T))\pi_{AB} - P(1-T)\pi_A - \{P(1-T)-1\}\pi_B + (1-T)P] \\ &\quad + (1-w)[-P_1T_1\pi_{AB} + T_1\{P_1 + (1-P_1)(1-\pi_Y)\}\pi_B - P_1(1-T_1)\pi_Y\pi_A + (1-T_1)P_1\pi_Y] \\ &= -[w(PT+(1-P)(1-T)) + (1-w)P_1T_1]\pi_{AB} - [wP(1-T) + (1-w)P_1(1-T_1)\pi_Y]\pi_A \\ &\quad -[w\{P(1-T)-1\} - (1-w)T_1\{P_1 + (1-P_1)(1-\pi_Y)\}]\pi_B + w(1-T)P + (1-w)(1-T_1)P_1\pi_Y \\ &= -C_1^*\pi_{AB} - C_6^*\pi_A - C_7^*\pi_B + H_3^*\end{aligned} \tag{3.3}$$

and the probability of getting a (*No, No*) response is given by:

$$\begin{aligned}\Psi^*_{00} &= w\theta^*_{00} + (1-w)\theta^u_{00} \\ &= w[(PT + (1-P)(1-T))\pi_{AB} - PT\pi_A - PT\pi_B + PT] \\ &\quad +(1-w)[P_1T_1\pi_{AB} - P_1\{T_1 + (1-T_1)(1-\pi_Y)\}\pi_A - T_1\{P_1 + (1-P_1)(1-\pi_Y)\}\pi_B + P_1T_1\pi_Y + 1 - \pi_Y] \\ &= [w\{PT + (1-P)(1-T)\} + (1-w)P_1T_1]\pi_{AB} - [wPT + (1-w)P_1\{T_1 + (1-T_1)(1-\pi_Y)\}]\pi_A \\ &\quad -[wPT + (1-w)T_1\{P_1 + (1-P_1)(1-\pi_Y)\}]\pi_B + wPT + (1-w)(P_1T_1\pi_Y + 1 - \pi_Y) \\ &= C^*_1\pi_{AB} - C^*_8\pi_A - C^*_9\pi_B + H^*_4\end{aligned} \tag{3.4}$$

where

$$C^*_1 = w\{PT + (1-P)(1-T)\} + (1-w)P_1T_1 \tag{3.5}$$

$$C^*_2 = -w(1-P)(1-T) + (1-w)P_1(1-T_1)\pi_Y \tag{3.6}$$

$$C^*_3 = -w(1-P)(1-T) + (1-w)(1-P_1)T_1\pi_Y \tag{3.7}$$

$$C^*_4 = w\{(1-P)T - 1\} - (1-w)P_1\{T_1 + (1-T_1)(1-\pi_Y)\} \tag{3.8}$$

$$C^*_5 = wT(1-P) + (1-w)T_1(1-P_1)\pi_Y \tag{3.9}$$

$$C^*_6 = wP(1-T) + (1-w)P_1(1-T_1)\pi_Y \tag{3.10}$$

$$C^*_7 = w\{P(1-T) - 1\} - (1-w)T_1\{P_1 + (1-P_1)(1-\pi_Y)\} \tag{3.11}$$

$$C^*_8 = wPT + (1-w)P_1\{T_1 + (1-T_1)(1-\pi_Y)\} \tag{3.12}$$

$$C^*_9 = wPT + (1-w)T_1\{P_1 + (1-P_1)(1-\pi_Y)\} \tag{3.13}$$

$$H^*_1 = w(1-P)(1-T) + (1-w)(1-P_1)(1-T_1)\pi_Y \tag{3.14}$$

$$H^*_2 = w(1-P)T + (1-w)(1-P_1)T_1\pi_Y \tag{3.15}$$

$$H^*_3 = wP(1-T) + (1-w)P_1(1-T_1)\pi_Y \tag{3.16}$$

and

$$H^*_4 = wPT + (1-w)(P_1T_1\pi_Y + 1 - \pi_Y) \tag{3.17}$$

Again note that:

$$\sum_{i=0}^{1}\sum_{j=0}^{1}\Psi^*_{ij} = 1 \tag{3.18}$$

Assume we selected a simple random sample with replacement (SRSWR) of n persons from the population of interest Ω. By making the use of DMDM-II, the observed responses are classified into a 2x2 contingency table in Figure 5.

In other words, out of the sample of n persons, there are n_{11} persons who responded (*Yes, Yes*); n_{10} persons responded (*Yes, No*); n_{01} persons responded (*No, Yes*); and n_{00} persons responded (*No, No*). Let $\hat{\Psi}^*_{11} = \frac{n_{11}}{n}$, $\hat{\Psi}^*_{10} = \frac{n_{10}}{n}$, $\hat{\Psi}^*_{01} = \frac{n_{01}}{n}$ and $\hat{\Psi}^*_{00} = \frac{n_{00}}{n}$ be the unbiased

Observed:	Response-II	
Response-I	Yes	No
Yes	n_{11}	n_{10}
No	n_{01}	n_{00}

Figure 5: Observed Responses.

estimators of Ψ^*_{11}, Ψ^*_{10}, Ψ^*_{01} and Ψ^*_{00}, respectively. Thus n_{ij} (i=0,1; j=0,1) follows a multinomial distribution with parameters n and Ψ^*_{ij}. The probability mass function for n_{ij} is given by:

$$P(n_{ij}) = \binom{n}{n_{11}, n_{10}, n_{01}, n_{00}} (\Psi^*_{11})^{n_{11}} (\Psi^*_{10})^{n_{10}} (\Psi^*_{01})^{n_{01}} (\Psi^*_{00})^{n_{00}} \tag{3.19}$$

Following Odumade and Singh (2009), we consider a squared distance between the true proportion and the observed proportions as:

$$\begin{aligned} D^* &= \frac{1}{2}\sum_{i=0}^{1}\sum_{j=0}^{1}\left(\Psi^*_{ij} - \hat{\Psi}^*_{ij}\right)^2 \\ &= \frac{1}{2}\left(\Psi^*_{11} - \hat{\Psi}^*_{11}\right)^2 + \frac{1}{2}\left(\Psi^*_{10} - \hat{\Psi}^*_{10}\right)^2 + \frac{1}{2}\left(\Psi^*_{01} - \hat{\Psi}^*_{01}\right)^2 + \frac{1}{2}\left(\Psi^*_{00} - \hat{\Psi}^*_{00}\right)^2 \\ &= \frac{1}{2}\left(C^*_1\pi_{AB} + C^*_2\pi_A + C^*_3\pi_B + H^*_1 - \hat{\Psi}^*_{11}\right)^2 + \frac{1}{2}\left(-C^*_1\pi_{AB} - C^*_4\pi_A - C^*_5\pi_B + H^*_2 - \hat{\Psi}^*_{10}\right)^2 \\ &+ \frac{1}{2}\left(-C^*_1\pi_{AB} - C^*_6\pi_A - C^*_7\pi_B + H^*_3 - \hat{\Psi}^*_{01}\right)^2 + \frac{1}{2}\left(C^*_1\pi_{AB} - C^*_8\pi_A - C^*_9\pi_B + H^*_4 - \hat{\Psi}^*_{00}\right)^2 \end{aligned} \tag{3.20}$$

On setting:

$$\frac{\partial D^*}{\partial \pi_{AB}} = 0 \tag{3.21}$$

we get

$$\begin{aligned} &(C^*_1\pi_{AB} + C^*_2\pi_A + C^*_3\pi_B + H^*_1 - \hat{\Psi}^*_{11})(C^*_1) + (-C^*_1\pi_{AB} - C^*_4\pi_A - C^*_5\pi_B + H^*_2 - \hat{\Psi}^*_{10})(-C^*_1) \\ &+(-C^*_1\pi_{AB} - C^*_6\pi_A - C^*_7\pi_B + H^*_3 - \hat{\Psi}^*_{01})(-C^*_1) + (C^*_1\pi_{AB} - C^*_8\pi_A - C^*_9\pi_B + H^*_4 - \hat{\Psi}^*_{00})(C^*_1) = 0 \end{aligned}$$

or

$$\begin{aligned} &C^*_1\pi_{AB} + C^*_2\pi_A + C^*_3\pi_B + H^*_1 - \hat{\Psi}^*_{11} + C^*_1\pi_{AB} + C^*_4\pi_A + C^*_5\pi_B - H^*_2 + \hat{\Psi}^*_{10} \\ &+C^*_1\pi_{AB} + C^*_6\pi_A + C^*_7\pi_B - H^*_3 + \hat{\Psi}^*_{01} + C^*_1\pi_{AB} - C^*_8\pi_A - C^*_9\pi_B + H^*_4 - \hat{\Psi}^*_{00} = 0 \end{aligned}$$

or

$$\begin{aligned} &4C^*_1\pi_{AB} + (C^*_2 + C^*_4 + C^*_6 - C^*_8)\pi_A + (C^*_3 + C^*_5 + C^*_7 - C^*_9)\pi_B \\ &= \hat{\Psi}^*_{11} - \hat{\Psi}^*_{10} - \hat{\Psi}^*_{01} + \hat{\Psi}^*_{00} - H^*_1 + H^*_2 + H^*_3 - H^*_4 \end{aligned}$$

or

$$4C^*_1\pi_{AB} + (C^*_2 + C^*_4 + C^*_6 - C^*_8)\pi_A + (C^*_3 + C^*_5 + C^*_7 - C^*_9)\pi_B = D^*_1 \tag{3.22}$$

On setting:

$$\frac{\partial D^*}{\partial \pi_A} = 0 \tag{3.23}$$

we get

$$\begin{aligned} &(C^*_1\pi_{AB} + C^*_2\pi_A + C^*_3\pi_B + H^*_1 - \hat{\Psi}^*_{11})(C^*_2) + (-C^*_1\pi_{AB} - C^*_4\pi_A - C^*_5\pi_B + H^*_2 - \hat{\Psi}^*_{10})(-C^*_4) \\ &+(-C^*_1\pi_{AB} - C^*_6\pi_A - C^*_7\pi_B + H^*_3 - \hat{\Psi}^*_{01})(-C^*_6) + (C^*_1\pi_{AB} - C^*_8\pi_A - C^*_9\pi_B + H^*_4 - \hat{\Psi}^*_{00})(-C^*_8) = 0 \end{aligned}$$

or

$$\begin{aligned} &\pi_{AB}\left(C^*_1C^*_2 + C^*_1C^*_4 + C^*_1C^*_6 - C^*_1C^*_8\right) + \pi_A\left[(C^*_2)^2 + (C^*_4)^2 + (C^*_6)^2 + (C^*_8)^2\right] + \pi_B\left(C^*_2C^*_3 + C^*_4C^*_5 + C^*_6C^*_7 + C^*_8C^*_9\right) \\ &+C^*_2H^*_1 - C^*_4H^*_2 - C^*_6H^*_3 - C^*_8H^*_4 - C^*_2\hat{\Psi}^*_{11} + C^*_4\hat{\Psi}^*_{10} + C^*_6\hat{\Psi}^*_{01} + C^*_8\hat{\Psi}^*_{00} = 0 \end{aligned}$$

or

$$\begin{aligned} &\pi_{AB}C^*_1\left(C^*_2 + C^*_4 + C^*_6 - C^*_8\right) + \pi_A\left[(C^*_2)^2 + (C^*_4)^2 + (C^*_6)^2 + (C^*_8)^2\right] + \pi_B\left(C^*_2C^*_3 + C^*_4C^*_5 + C^*_6C^*_7 + C^*_8C^*_9\right) \\ &= C^*_4H^*_2 + C^*_6H^*_3 + C^*_8H^*_4 - C^*_2H^*_1 + C^*_2\hat{\Psi}^*_{11} - C^*_4\hat{\Psi}^*_{10} - C^*_6\hat{\Psi}^*_{01} - C^*_8\hat{\Psi}^*_{00} \end{aligned}$$

or

$$\pi_{AB}C^*_1\left(C^*_2 + C^*_4 + C^*_6 - C^*_8\right) + \pi_A\left[(C^*_2)^2 + (C^*_4)^2 + (C^*_6)^2 + (C^*_8)^2\right] + \pi_B\left(C^*_2C^*_3 + C^*_4C^*_5 + C^*_6C^*_7 + C^*_8C^*_9\right) = D^*_2 \tag{3.24}$$

and, on setting

$$\frac{\partial D^*}{\partial \pi_B} = 0 \tag{3.25}$$

we get

$$\begin{aligned}&(C_1^*\pi_{AB} + C_2^*\pi_A + C_3^*\pi_B + H_1^* - \hat{\Psi}_{11}^*)(C_3^*) + (-C_1^*\pi_{AB} - C_4^*\pi_A - C_5^*\pi_B + H_2^* - \hat{\Psi}_{10}^*)(-C_5^*)\\ &+(-C_1^*\pi_{AB} - C_6^*\pi_A - C_7^*\pi_B + H_3^* - \hat{\Psi}_{01}^*)(-C_7^*) + (C_1^*\pi_{AB} - C_8^*\pi_A - C_9^*\pi_B + H_4^* - \hat{\Psi}_{00}^*)(-C_9^*) = 0\end{aligned}$$

or

$$\begin{aligned}&\pi_{AB}\left(C_1^*C_3^* + C_1^*C_5^* + C_1^*C_7^* - C_1^*C_9^*\right) + \pi_A(C_2^*C_3^* + C_4^*C_5^* + C_6^*C_7^* + C_8^*C_9^*) + \pi_B\left[(C_3^*)^2 + (C_5^*)^2 + (C_7^*)^2 + (C_9^*)^2\right]\\ &+(C_3^*H_1^* - C_5^*H_2^* - C_7^*H_3^* - C_9^*H_4^*) - C_3^*\hat{\Psi}_{11}^* + C_5^*\hat{\Psi}_{10}^* + C_7^*\hat{\Psi}_{01}^* + C_9^*\hat{\Psi}_{00}^* = 0\end{aligned}$$

or

$$\begin{aligned}&\pi_{AB}C_1^*\left(C_3^* + C_5^* + C_7^* - C_9^*\right) + \pi_A(C_2^*C_3^* + C_4^*C_5^* + C_6^*C_7^* + C_8^*C_9^*) + \pi_B\left[(C_3^*)^2 + (C_5^*)^2 + (C_7^*)^2 + (C_9^*)^2\right]\\ &= (C_5^*H_2^* + C_7^*H_3^* + C_9^*H_4^* - C_3^*H_1^*) + C_3^*\hat{\Psi}_{11}^* - C_5^*\hat{\Psi}_{10}^* - C_7^*\hat{\Psi}_{01}^* - C_9^*\hat{\Psi}_{00}^*\end{aligned}$$

or

$$\pi_{AB}C_1^*\left(C_3^* + C_5^* + C_7^* - C_9^*\right) + \pi_A(C_2^*C_3^* + C_4^*C_5^* + C_6^*C_7^* + C_8^*C_9^*) + \pi_B\left[(C_3^*)^2 + (C_5^*)^2 + (C_7^*)^2 + (C_9^*)^2\right] = D_3^* \tag{3.26}$$

where

$$D_1^* = \hat{\Psi}_{11}^* - \hat{\Psi}_{10}^* - \hat{\Psi}_{01}^* + \hat{\Psi}_{00}^* - H_1^* + H_2^* + H_3^* - H_4^* \tag{3.27}$$

$$D_2^* = C_4^*H_2^* + C_6^*H_3^* + C_8^*H_4^* - C_2^*H_1^* + C_2^*\hat{\Psi}_{11}^* - C_4^*\hat{\Psi}_{10}^* - C_6^*\hat{\Psi}_{01}^* - C_8^*\hat{\Psi}_{00}^* \tag{3.28}$$

and

$$D_3^* = (C_5^*H_2^* + C_7^*H_3^* + C_9^*H_4^* - C_3^*H_1^*) + C_3^*\hat{\Psi}_{11}^* - C_5^*\hat{\Psi}_{10}^* - C_7^*\hat{\Psi}_{01}^* - C_9^*\hat{\Psi}_{00}^* \tag{3.29}$$

The system of equations in $\frac{\partial D^*}{\partial \pi_{AB}} = 0$, $\frac{\partial D^*}{\partial \pi_A} = 0$ and $\frac{\partial D^*}{\partial \pi_B} = 0$ can be written as:

$$\begin{bmatrix} 4C_1^* & \left(C_2^* + C_4^* + C_6^* - C_8^*\right) & \left(C_3^* + C_5^* + C_7^* - C_9^*\right)\\ C_1^*\left(C_2^* + C_4^* + C_6^* - C_8^*\right) & (C_2^*)^2 + (C_4^*)^2 + (C_6^*)^2 + (C_8^*)^2 & (C_2^*C_3^* + C_4^*C_5^* + C_6^*C_7^* + C_8^*C_9^*)\\ C_1^*\left(C_3^* + C_5^* + C_7^* - C_9^*\right) & (C_2^*C_3^* + C_4^*C_5^* + C_6^*C_7^* + C_8^*C_9^*) & (C_3^*)^2 + (C_5^*)^2 + (C_7^*)^2 + (C_9^*)^2 \end{bmatrix}\begin{bmatrix}\pi_{AB}\\ \pi_A\\ \pi_B\end{bmatrix} = \begin{bmatrix}D_1^*\\ D_2^*\\ D_3^*\end{bmatrix}$$

We apply the Cramer's rule as follows:

$$\Delta^* = \begin{vmatrix} 4C_1^* & \left(C_2^* + C_4^* + C_6^* - C_8^*\right) & \left(C_3^* + C_5^* + C_7^* - C_9^*\right)\\ C_1^*\left(C_2^* + C_4^* + C_6^* - C_8^*\right) & (C_2^*)^2 + (C_4^*)^2 + (C_6^*)^2 + (C_8^*)^2 & (C_2^*C_3^* + C_4^*C_5^* + C_6^*C_7^* + C_8^*C_9^*)\\ C_1^*\left(C_3^* + C_5^* + C_7^* - C_9^*\right) & (C_2^*C_3^* + C_4^*C_5^* + C_6^*C_7^* + C_8^*C_9^*) & (C_3^*)^2 + (C_5^*)^2 + (C_7^*)^2 + (C_9^*)^2 \end{vmatrix}$$

or

$$\begin{aligned}\Delta^* &= 4C_1^*[(C_2^*)^2 + (C_4^*)^2 + (C_6^*)^2 + (C_8^*)^2]\left[(C_3^*)^2 + (C_5^*)^2 + (C_7^*)^2 + (C_9^*)^2\right]\\ &+ 2C_1^*\left(C_3^* + C_5^* + C_7^* - C_9^*\right)\left(C_2^* + C_4^* + C_6^* - C_8^*\right)(C_2^*C_3^* + C_4^*C_5^* + C_6^*C_7^* + C_8^*C_9^*)\\ &- C_1^*\left(C_3^* + C_5^* + C_7^* - C_9^*\right)^2\left[(C_2^*)^2 + (C_4^*)^2 + (C_6^*)^2 + (C_8^*)^2\right]\\ &- 4C_1^*(C_2^*C_3^* + C_4^*C_5^* + C_6^*C_7^* + C_8^*C_9^*)^2\\ &- C_1^*\left(C_2^* + C_4^* + C_6^* - C_8^*\right)^2\left[\left(C_3^*\right)^2 + \left(C_5^*\right)^2 + \left(C_7^*\right)^2 + \left(C_9^*\right)^2\right]\end{aligned} \tag{3.30}$$

$$\Delta^*_{AB} = \begin{vmatrix} D_1^* & \left(C_2^* + C_4^* + C_6^* - C_8^*\right) & \left(C_3^* + C_5^* + C_7^* - C_9^*\right) \\ D_2^* & (C_2^*)^2 + (C_4^*)^2 + (C_6^*)^2 + (C_8^*)^2 & (C_2^*C_3^* + C_4^*C_5^* + C_6^*C_7^* + C_8^*C_9^*) \\ D_3^* & (C_2^*C_3^* + C_4^*C_5^* + C_6^*C_7^* + C_8^*C_9^*) & (C_3^*)^2 + (C_5^*)^2 + (C_7^*)^2 + (C_9^*)^2 \end{vmatrix}$$

or

$$\begin{aligned} \Delta^*_{AB} = {} & D_1^*\left[(C_2^*)^2 + (C_4^*)^2 + (C_6^*)^2 + (C_8^*)^2\right]\left[(C_3^*)^2 + (C_5^*)^2 + (C_7^*)^2 + (C_9^*)^2\right] \\ & + D_2^*(C_2^*C_3^* + C_4^*C_5^* + C_6^*C_7^* + C_8^*C_9^*)\left(C_3^* + C_5^* + C_7^* - C_9^*\right) \\ & + D_3^*(C_2^*C_3^* + C_4^*C_5^* + C_6^*C_7^* + C_8^*C_9^*)\left(C_2^* + C_4^* + C_6^* - C_8^*\right) \\ & - D_3^*\left(C_3^* + C_5^* + C_7^* - C_9^*\right)\left[(C_2^*)^2 + (C_4^*)^2 + (C_6^*)^2 + (C_8^*)^2\right] \\ & - D_2^*\left(C_2^* + C_4^* + C_6^* - C_8^*\right)\left[(C_3^*)^2 + (C_5^*)^2 + (C_7^*)^2 + (C_9^*)^2\right] \\ & - D_1^*(C_2^*C_3^* + C_4^*C_5^* + C_6^*C_7^* + C_8^*C_9^*)^2 \end{aligned} \tag{3.31}$$

or

$$\begin{aligned} \Delta^*_{AB} = {} & D_1^*\left[\left\{(C_2^*)^2 + (C_4^*)^2 + (C_6^*)^2 + (C_8^*)^2\right\}\left\{(C_3^*)^2 + (C_5^*)^2 + (C_7^*)^2 + (C_9^*)^2\right\} - (C_2^*C_3^* + C_4^*C_5^* + C_6^*C_7^* + C_8^*C_9^*)^2\right] \\ & + D_2^*\left[(C_2^*C_3^* + C_4^*C_5^* + C_6^*C_7^* + C_8^*C_9^*)\left(C_3^* + C_5^* + C_7^* - C_9^*\right) - \left(C_2^* + C_4^* + C_6^* - C_8^*\right)\left\{(C_3^*)^2 + (C_5^*)^2 + (C_7^*)^2 + (C_9^*)^2\right\}\right] \\ & + D_3^*\left[(C_2^*C_3^* + C_4^*C_5^* + C_6^*C_7^* + C_8^*C_9^*)\left(C_2^* + C_4^* + C_6^* - C_8^*\right) - \left(C_3^* + C_5^* + C_7^* - C_9^*\right)\left\{(C_2^*)^2 + (C_4^*)^2 + (C_6^*)^2 + (C_8^*)^2\right\}\right] \end{aligned}$$

or

$$\Delta^*_{AB} = K_1^*D_1^* + K_2^*D_2^* + K_3^*D_3^* \tag{3.32}$$

$$\Delta^*_{A} = \begin{vmatrix} 4C_1^* & D_1^* & \left(C_3^* + C_5^* + C_7^* - C_9^*\right) \\ C_1^*\left(C_2^* + C_4^* + C_6^* - C_8^*\right) & D_2^* & (C_2^*C_3^* + C_4^*C_5^* + C_6^*C_7^* + C_8^*C_9^*) \\ C_1^*\left(C_3^* + C_5^* + C_7^* - C_9^*\right) & D_3^* & (C_3^*)^2 + (C_5^*)^2 + (C_7^*)^2 + (C_9^*)^2 \end{vmatrix}$$

or

$$\begin{aligned} \Delta^*_{A} = {} & D_1^*C_1^*(C_2^*C_3^* + C_4^*C_5^* + C_6^*C_7^* + C_8^*C_9^*)\left(C_3^* + C_5^* + C_7^* - C_9^*\right) \\ & + 4D_2^*C_1^*\left[(C_3^*)^2 + (C_5^*)^2 + (C_7^*)^2 + (C_9^*)^2\right] \\ & + D_3^*C_1^*(C_2^* + C_4^* + C_6^* - C_8^*)\left(C_3^* + C_5^* + C_7^* - C_9^*\right) \\ & - D_1^*C_1^*\left(C_2^* + C_4^* + C_6^* - C_8^*\right)\left[(C_3^*)^2 + (C_5^*)^2 + (C_7^*)^2 + (C_9^*)^2\right] \\ & - D_2^*C_1^*\left(C_3^* + C_5^* + C_7^* - C_9^*\right)^2 \\ & - 4D_3^*C_1^*(C_2^*C_3^* + C_4^*C_5^* + C_6^*C_7^* + C_8^*C_9^*) \end{aligned} \tag{3.33}$$

or

$$\begin{aligned} \Delta^*_{A} = {} & D_1^*C_1^*\begin{bmatrix} (C_2^*C_3^* + C_4^*C_5^* + C_6^*C_7^* + C_8^*C_9^*)\left(C_3^* + C_5^* + C_7^* - C_9^*\right) \\ -\left(C_2^* + C_4^* + C_6^* - C_8^*\right)\left\{(C_3^*)^2 + (C_5^*)^2 + (C_7^*)^2 + (C_9^*)^2\right\} \end{bmatrix} \\ & + D_2^*C_1^*\left[4\left\{(C_3^*)^2 + (C_5^*)^2 + (C_7^*)^2 + (C_9^*)^2\right\} - \left(C_3^* + C_5^* + C_7^* - C_9^*\right)^2\right] \\ & + D_3^*C_1^*\begin{bmatrix} (C_2^*C_3^* + C_4^*C_5^* + C_6^*C_7^* + C_8^*C_9^*)(C_3^* + C_5^* + C_7^* - C_9^*) \\ -(C_2^* + C_4^* + C_6^* + C_8^*)\left\{(C_3^*)^2 + (C_5^*)^2 + (C_7^*)^2 + (C_9^*)^2\right\} \end{bmatrix} \end{aligned}$$

or

$$\Delta^*_{A} = L_1^*D_1^* + L_2^*D_2^* + L_3^*D_3^* \tag{3.34}$$

$$\Delta_B^* = \begin{bmatrix} 4C_1^* & \left(C_2^* + C_4^* + C_6^* - C_8^*\right) & D_1^* \\ C_1^*\left(C_2^* + C_4^* + C_6^* - C_8^*\right) & (C_2^*)^2 + (C_4^*)^2 + (C_6^*)^2 + (C_8^*)^2 & D_2^* \\ C_1^*\left(C_3^* + C_5^* + C_7^* - C_9^*\right) & (C_2^*C_3^* + C_4^*C_5^* + C_6^*C_7^* + C_8^*C_9^*) & D_3^* \end{bmatrix}$$

or

$$\begin{aligned} \Delta_B^* &= 4D_3^*C_1^*\left[\left(C_2^*\right)^2 + \left(C_4^*\right)^2 + \left(C_6^*\right)^2 + \left(C_8^*\right)^2\right] \\ &+ D_2^*C_1^*(C_2^* + C_4^* + C_6^* - C_8^*)^2 \\ &+ D_1^*C_1^*(C_2^* + C_4^* + C_6^* - C_8^*)(C_2^*C_3^* + C_4^*C_5^* + C_6^*C_7^* + C_8^*C_9^*) \\ &- D_1^*C_1^*\left(C_3^* + C_5^* + C_7^* - C_9^*\right)\left[(C_2^*)^2 + (C_4^*)^2 + (C_6^*)^2 + (C_8^*)^2\right] \\ &- 4D_2^*C_1^*(C_2^*C_3^* + C_4^*C_5^* + C_6^*C_7^* + C_8^*C_9^*) \\ &- D_3^*C_1^*\left(C_2^* + C_4^* + C_6^* - C_8^*\right)^2 \end{aligned} \tag{3.35}$$

or

$$\begin{aligned} \Delta_B^* = D_1^*C_1^*\Big[&(C_2^*C_3^* + C_4^*C_5^* + C_6^*C_7^* + C_8^*C_9^*)\left(C_2^* + C_4^* + C_6^* - C_8^*\right) - \left(C_3^* + C_5^* + C_7^* - C_9^*\right) \\ &\left\{(C_2^*)^2 + (C_4^*)^2 + (C_6^*)^2 + (C_8^*)^2\right\}\Big] \\ &+ D_2^*C_1^*\left[\left(C_2^* + C_4^* + C_6^* - C_8^*\right)\left(C_3^* + C_5^* + C_7^* - C_9^*\right) - 4\left(C_2^*C_3^* + C_4^*C_5^* + C_6^*C_7^* + C_8^*C_9^*\right)\right] \\ &+ D_3^*C_1^*\left[4\left\{(C_2^*)^2 + (C_4^*)^2 + (C_6^*)^2 + (C_8^*)^2\right\} - (C_2^* + C_4^* + C_6^* + C_8^*)^2\right] \end{aligned}$$

or

$$\Delta_B^* = M_1^*D_1^* + M_2^*D_2^* + M_3^*D_3^* \tag{3.36}$$

where

$$\begin{aligned} K_1^* = &\left\{(C_2^*)^2 + (C_4^*)^2 + (C_6^*)^2 + (C_8^*)^2\right\}\left\{(C_3^*)^2 + (C_5^*)^2 + (C_7^*)^2 + (C_9^*)^2\right\} \\ &- (C_2^*C_3^* + C_4^*C_5^* + C_6^*C_7^* + C_8^*C_9^*)^2 \end{aligned} \tag{3.37}$$

$$\begin{aligned} K_2^* = &(C_2^*C_3^* + C_4^*C_5^* + C_6^*C_7^* + C_8^*C_9^*)\left(C_3^* + C_5^* + C_7^* - C_9^*\right) \\ &- \left(C_2^* + C_4^* + C_6^* - C_8^*\right)\left\{(C_3^*)^2 + (C_5^*)^2 + (C_7^*)^2 + (C_9^*)^2\right\} \end{aligned} \tag{3.38}$$

$$\begin{aligned} K_3^* = &(C_2^*C_3^* + C_4^*C_5^* + C_6^*C_7^* + C_8^*C_9^*)\left(C_2^* + C_4^* + C_6^* - C_8^*\right) \\ &- \left(C_3^* + C_5^* + C_7^* - C_9^*\right)\left\{(C_2^*)^2 + (C_4^*)^2 + (C_6^*)^2 + (C_8^*)^2\right\} \end{aligned} \tag{3.39}$$

$$L_1^* = C_1^*\begin{bmatrix} (C_2^*C_3^* + C_4^*C_5^* + C_6^*C_7^* + C_8^*C_9^*)\left(C_3^* + C_5^* + C_7^* - C_9^*\right) \\ -\left(C_2^* + C_4^* + C_6^* - C_8^*\right)\left\{(C_3^*)^2 + (C_5^*)^2 + (C_7^*)^2 + (C_9^*)^2\right\} \end{bmatrix} \tag{3.40}$$

$$L_2^* = C_1^*\left[4\left\{(C_3^*)^2 + (C_5^*)^2 + (C_7^*)^2 + (C_9^*)^2\right\} - \left(C_3^* + C_5^* + C_7^* - C_9^*\right)^2\right] \tag{3.41}$$

$$L_3^* = C_1^*\left[(C_2^* + C_4^* + C_6^* - C_8^*)(C_3^* + C_5^* + C_7^* - C_9^*) - 4(C_2^*C_3^* + C_4^*C_5^* + C_6^*C_7^* + C_8^*C_9^*)\right] \tag{3.42}$$

$$M_1^* = C_1^*\left[(C_2^*C_3^* + C_4^*C_5^* + C_6^*C_7^* + C_8^*C_9^*)\left(C_2^* + C_4^* + C_6^* - C_8^*\right) - \left(C_3^* + C_5^* + C_7^* - C_9^*\right)\left\{(C_2^*)^2 + (C_4^*)^2 + (C_6^*)^2 + (C_8^*)^2\right\}\right] \tag{3.43}$$

$$M_2^* = C_1^*\left[\left(C_2^* + C_4^* + C_6^* - C_8^*\right)\left(C_3^* + C_5^* + C_7^* - C_9^*\right) - 4\left(C_2^*C_3^* + C_4^*C_5^* + C_6^*C_7^* + C_8^*C_9^*\right)\right] \tag{3.44}$$

and

$$M_3^* = C_1^*\left[4\left\{(C_2^*)^2 + (C_4^*)^2 + (C_6^*)^2 + (C_8^*)^2\right\} - (C_2^* + C_4^* + C_6^* + C_8^*)^2\right] \tag{3.45}$$

Theorem 3.1. An unbiased estimator of π_A is given by

$$\hat{\pi}_A^{Amy^*} = \frac{1}{\Delta^*}(\hat{\Psi}_{11}^* W_{11}^{a^*} - \hat{\Psi}_{10}^* W_{10}^{a^*} - \hat{\Psi}_{01}^* W_{01}^{a^*} + \hat{\Psi}_{00}^* W_{00}^{a^*} + W_c^{a^*}) \tag{3.46}$$

where

$$W_{11}^{a^*} = L_1^* + C_2^* L_2^* + C_3^* L_3^* \tag{3.47}$$

$$W_{10}^{a^*} = L_1^* + C_4^* L_2^* + C_5^* L_3^* \tag{3.48}$$

$$W_{01}^{a^*} = L_1^* + C_6^* L_2^* + C_7^* L_3^* \tag{3.49}$$

$$W_{00}^{a^*} = L_1^* - C_8^* L_2^* - C_9^* L_3^* \tag{3.50}$$

and

$$W_c^{a^*} = (H_2^* + H_3^* - H_1^* - H_4^*)L_1^* + (C_4^* H_2^* + C_6^* H_3^* + C_8^* H_4^* - C_2^* H_1^*)L_2^* + (C_5^* H_2^* + C_7^* H_3^* + C_9^* H_4^* - C_3^* H_1^*)L_3^* \tag{3.51}$$

Proof. Applying the Cramer's rule and by the method of moments, an estimator of π_A is given by

$$\hat{\pi}_A^{Amy^*} = \frac{\Delta_A^*}{\Delta^*} = \frac{L_1^* D_1^* + L_2^* D_2^* + L_3^* D_3^*}{\Delta^*}$$

$$= \frac{1}{\Delta^*}\begin{bmatrix} (H_2^* + H_3^* - H_4^* - H_1^* + \hat{\Psi}_{11}^* + \hat{\Psi}_{00}^* - \hat{\Psi}_{10}^* - \hat{\Psi}_{01}^*)L_1^* \\ +\{(C_4^* H_2^* + C_6^* H_3^* + C_8^* H_4^* - C_2^* H_1^*) + (C_2^* \hat{\Psi}_{11}^* - C_4^* \hat{\Psi}_{10}^* - C_6^* \hat{\Psi}_{01}^* - C_8^* \hat{\Psi}_{00}^*)\} L_2^* \\ +\{(C_5^* H_2^* + C_7^* H_3^* + C_9^* H_4^* - C_3^* H_1^*) + (C_3^* \hat{\Psi}_{11}^* - C_5^* \hat{\Psi}_{10}^* - C_7^* \hat{\Psi}_{01}^* - C_9^* \hat{\Psi}_{00}^*)\} L_3^* \end{bmatrix} \tag{3.52}$$

$$= \frac{1}{\Delta^*}(\hat{\Psi}_{11}^* W_{11}^{a^*} - \hat{\Psi}_{10}^* W_{10}^{a^*} - \hat{\Psi}_{01}^* W_{01}^{a^*} + \hat{\Psi}_{00}^* W_{00}^{a^*} + W_c^{a^*})$$

Taking expected value on both sides of (3.52), we have:

$$E(\hat{\pi}_A^{Amy^*}) = E\left[\frac{1}{\Delta^*}(\hat{\Psi}_{11}^* W_{11}^{a^*} - \hat{\Psi}_{10}^* W_{10}^{a^*} - \hat{\Psi}_{01}^* W_{01}^{a^*} + \hat{\Psi}_{00}^* W_{00}^{a^*} + W_c^{a^*})\right]$$

$$= \frac{1}{\Delta^*}\left[E(\hat{\Psi}_{11}^*)W_{11}^{a^*} - E(\hat{\Psi}_{10}^*)W_{10}^{a^*} - E(\hat{\Psi}_{01}^*)W_{01}^{a^*} + E(\hat{\Psi}_{00}^*)W_{00}^{a^*} + W_c^{a^*}\right] \tag{3.53}$$

$$= \frac{1}{\Delta^*}(\Psi_{11}^* W_{11}^{a^*} - \Psi_{10}^* W_{10}^{a^*} - \Psi_{01}^* W_{01}^{a^*} + \Psi_{00}^* W_{00}^{a^*} + W_c^{a^*})$$

$$= \pi_A$$

which proves the theorem.

Theorem 3.2. An unbiased estimator of π_B is given by

$$\hat{\pi}_B^{Amy^*} = \frac{1}{\Delta^*}(\hat{\Psi}_{11}^* W_{11}^{b^*} - \hat{\Psi}_{10}^* W_{10}^{b^*} - \hat{\Psi}_{01}^* W_{01}^{b^*} + \hat{\Psi}_{00}^* W_{00}^{b^*} + W_c^{b^*}) \tag{5.54}$$

where

$$W_{11}^{b^*} = M_1^* + C_2^* M_2^* + C_3^* M_3^* \tag{5.55}$$

$$W_{10}^{b^*} = M_1^* + C_4^* M_2^* + C_5^* M_3^* \tag{5.56}$$

$$W_{01}^{b^*} = M_1^* + C_6^* M_2^* + C_7^* M_3^* \tag{5.57}$$

$$W_{00}^{b^*} = M_1^* - C_8^* M_2^* - C_9^* M_3^* \tag{5.58}$$

and

$$W_c^{b^*} = (H_2^* + H_3^* - H_1^* - H_4^*)M_1^* + (C_4^*H_2^* + C_6^*H_3^* + C_8^*H_4^* - C_2^*H_1^*)M_2^* + (C_5^*H_2^* + C_7^*H_3^* + C_9^*H_4^* - C_3^*H_1^*)M_3^* \quad (5.59)$$

Proof. Applying the Cramer's rule and by the method of moments, an estimator of π_B is given by

$$\begin{aligned} \hat{\pi}_B^{Amy^*} &= \frac{\Delta_B^*}{\Delta^*} = \frac{M_1^*D_1^* + M_2^*D_2^* + M_3^*D_3^*}{\Delta^*} \\ &= \frac{1}{\Delta^*}\begin{bmatrix} (H_2^* + H_3^* - H_4^* - H_1^* + \hat{\Psi}_{11}^* + \hat{\Psi}_{00}^* - \hat{\Psi}_{10}^* - \hat{\Psi}_{01}^*)M_1^* \\ +\left\{(C_4^*H_2^* + C_6^*H_3^* + C_8^*H_4^* - C_2^*H_1^*) + (C_2^*\hat{\Psi}_{11}^* - C_4^*\hat{\Psi}_{10}^* - C_6^*\hat{\Psi}_{01}^* - C_8^*\hat{\Psi}_{00}^*)\right\} M_2^* \\ +\left\{(C_5^*H_2^* + C_7^*H_3^* + C_9^*H_4^* - C_3^*H_1^*) + (C_3^*\hat{\Psi}_{11}^* - C_5^*\hat{\Psi}_{10}^* - C_7^*\hat{\Psi}_{01}^* - C_9^*\hat{\Psi}_{00}^*)\right\} M_3^* \end{bmatrix} \\ &= \frac{1}{\Delta^*}(\hat{\Psi}_{11}^*W_{11}^{b^*} - \hat{\Psi}_{10}^*W_{10}^{b^*} - \hat{\Psi}_{01}^*W_{01}^{b^*} + \hat{\Psi}_{00}^*W_{00}^{b^*} + W_c^{b^*}) \end{aligned} \quad (3.60)$$

Taking the expected value on both sides of (3.60), we have

$$\begin{aligned} E\left(\hat{\pi}_B^{Amy^*}\right) &= E\left[\frac{1}{\Delta^*}(\hat{\Psi}_{11}^*W_{11}^{b^*} - \hat{\Psi}_{10}^*W_{10}^{b^*} - \hat{\Psi}_{01}^*W_{01}^{b^*} + \hat{\Psi}_{00}^*W_{00}^{b^*} + W_c^{b^*})\right] \\ &= \frac{1}{\Delta^*}\left[E(\hat{\Psi}_{11}^*)W_{11}^{b^*} - E(\hat{\Psi}_{10}^*)W_{10}^{b^*} - E(\hat{\Psi}_{01}^*)W_{01}^{b^*} + E(\hat{\Psi}_{00}^*)W_{00}^{b^*} + W_c^{b^*}\right] \\ &= \frac{1}{\Delta^*}(\Psi_{11}^*W_{11}^{b^*} - \Psi_{10}^*W_{10}^{b^*} - \Psi_{01}^*W_{01}^{b^*} + \Psi_{00}^*W_{00}^{b^*} + W_c^{b^*}) \\ &= \pi_B \end{aligned} \quad (3.61)$$

which proves the theorem.

Theorem 3.3. An unbiased estimator of π_{AB} is given by

$$\hat{\pi}_{AB}^{Amy^*} = \frac{1}{\Delta^*}(\hat{\Psi}_{11}^*W_{11}^{ab^*} - \hat{\Psi}_{10}^*W_{10}^{ab^*} - \hat{\Psi}_{01}^*W_{01}^{ab^*} + \hat{\Psi}_{00}^*W_{00}^{ab^*} + W_c^{ab^*}) \quad (3.62)$$

where

$$\begin{aligned} W_{11}^{ab^*} &= K_1^* + C_2^*K_2^* + C_3^*K_3^* \\ W_{10}^{ab^*} &= K_1^* + C_4^*K_2^* + C_5^*K_3^* \\ W_{01}^{ab^*} &= K_1^* + C_6^*K_2^* + C_7^*K_3^* \\ W_{00}^{ab^*} &= K_1^* - C_8^*K_2^* - C_9^*K_3^* \end{aligned} \quad (3.63)$$

and

$$W_c^{ab^*} = (H_2^* + H_3^* - H_1^* - H_4^*)K_1^* + (C_4^*H_2^* + C_6^*H_3^* + C_8^*H_4^* - C_2^*H_1^*)K_2^* + (C_5^*H_2^* + C_7^*H_3^* + C_9^*H_4^* - C_3^*H_1^*)K_3^* \quad (3.64)$$

Proof. Applying the Cramer's rule and by the method of moments, an estimator of π_{AB} is given by

$$\begin{aligned} \hat{\pi}_{AB}^{Amy^*} &= \frac{\Delta_{AB}^*}{\Delta^*} = \frac{K_1^*D_1^* + K_2^*D_2^* + K_3^*D_3^*}{\Delta^*} \\ &= \frac{1}{\Delta^*}\Big[(H_2^* + H_3^* - H_4^* - H_1^* + \hat{\Psi}_{11}^* + \hat{\Psi}_{00}^* - \hat{\Psi}_{10}^* - \hat{\Psi}_{01}^*)K_1^* \\ &\quad +\left\{(C_4^*H_2^* + C_6^*H_3^* + C_8^*H_4^* - C_2^*H_1^*) + (C_2^*\hat{\Psi}_{11}^* - C_4^*\hat{\Psi}_{10}^* - C_6^*\hat{\Psi}_{01}^* - C_8^*\hat{\Psi}_{00}^*)\right\} K_2^* \\ &\quad +\left\{(C_5^*H_2^* + C_7^*H_3^* + C_9^*H_4^* - C_3^*H_1^*) + (C_3^*\hat{\Psi}_{11}^* - C_5^*\hat{\Psi}_{10}^* - C_7^*\hat{\Psi}_{01}^* - C_9^*\hat{\Psi}_{00}^*)\right\} K_3^*\Big] \\ &= \frac{1}{\Delta^*}\Big[\hat{\Psi}_{11}^*(K_1^* + C_6^*K_2^* + C_7^*K_3^*) - \hat{\Psi}_{10}^*(K_1^* - C_8^*K_2^* - C_9^*K_3^*) \\ &\quad - \hat{\Psi}_{01}^*(K_1^* + C_6^*K_2^* + C_7^*K_3^*) + \hat{\Psi}_{00}^*(K_1^* - C_8^*K_2^* - C_9^*K_3^*) + (H_2^* + H_3^* - H_1^* - H_4^*)K_1^* \\ &\quad + (C_4^*H_2^* + C_6^*H_3^* + C_8^*H_4^* - C_2^*H_1^*)K_2^* + (C_5^*H_2^* + C_7^*H_3^* + C_9^*H_4^* - C_3^*H_1^*)K_3^*\Big] \\ &= \frac{1}{\Delta^*}(\hat{\Psi}_{11}^*W_{11}^{ab^*} - \hat{\Psi}_{10}^*W_{10}^{ab^*} - \hat{\Psi}_{01}^*W_{01}^{ab^*} + \hat{\Psi}_{00}^*W_{00}^{ab^*} + W_c^{ab^*}) \end{aligned} \quad (3.65)$$

Taking the expected value on both sides of (3.65), we have

$$E\left(\hat{\pi}_{AB}^{Amy^*}\right) = E\left[\frac{1}{\Delta^*}(\hat{\Psi}_{11}^* W_{11}^{ab^*} - \hat{\Psi}_{10}^* W_{10}^{ab^*} - \hat{\Psi}_{01}^* W_{01}^{ab^*} + \hat{\Psi}_{00}^* W_{00}^{ab^*} + W_c^{ab^*})\right]$$
$$= \frac{1}{\Delta^*}\left[E\left(\hat{\Psi}_{11}^*\right) W_{11}^{ab^*} - E\left(\hat{\Psi}_{10}^*\right) W_{10}^{ab^*} - E\left(\hat{\Psi}_{01}^*\right) W_{01}^{ab^*} + E\left(\hat{\Psi}_{00}^*\right) W_{00}^{ab^*} + W_c^{ab^*}\right] \quad (3.66)$$
$$= \frac{1}{\Delta^*}(\Psi_{11}^* W_{11}^{ab^*} - \Psi_{10}^* W_{10}^{ab^*} - \Psi_{01}^* W_{01}^{ab^*} + \Psi_{00}^* W_{00}^{ab^*} + W_c^{ab^*})$$
$$= \pi_{AB}$$

which proves the theorem.

Theorem 3.4. The variance of the unbiased estimator of $\hat{\pi}_A^{Amy^*}$ is given by

$$V\left(\hat{\pi}_A^{Amy^*}\right) = \frac{1}{n\left(\Delta^*\right)^2}\Big[(W_{11}^{a^*})^2 \Psi_{11}^*(1-\Psi_{11}^*) + (W_{10}^{a^*})^2 \Psi_{10}^*(1-\Psi_{10}^*) + (W_{01}^{a^*})^2 \Psi_{01}^*(1-\Psi_{01}^*) + (W_{00}^{a^*})^2 \Psi_{00}^*(1-\Psi_{00}^*)$$
$$+ 2W_{11}^{a^*} W_{10}^{a^*} \Psi_{11}^* \Psi_{10}^* + 2W_{11}^{a^*} W_{01}^{a^*} \Psi_{11}^* \Psi_{01}^* - 2W_{11}^{a^*} W_{00}^{a^*} \Psi_{11}^* \Psi_{00}^* \quad (3.67)$$
$$-2W_{10}^{a^*} W_{01}^{a^*} \Psi_{10}^* \Psi_{01}^* + 2W_{10}^{a^*} W_{00}^{a^*} \Psi_{10}^* \Psi_{00}^* + 2W_{01}^{a^*} W_{00}^{a^*} \Psi_{01}^* \Psi_{00}^*\Big]$$

Proof. The variance of the estimator $\hat{\pi}_A^{Amy^*}$ is given by

$$V\left(\hat{\pi}_A^{Amy^*}\right) = V\left[\frac{1}{\Delta^*}(\hat{\Psi}_{11}^* W_{11}^{a^*} - \hat{\Psi}_{10}^* W_{10}^{a^*} - \hat{\Psi}_{01}^* W_{01}^{a^*} + \hat{\Psi}_{00}^* W_{00}^{a^*} + W_c^{a^*})\right]$$
$$= \frac{1}{\left(\Delta^*\right)^2}\Big[(W_{11}^{a^*})^2 V(\hat{\Psi}_{11}^*) + (W_{10}^{a^*})^2 V(\hat{\Psi}_{10}^*) + (W_{01}^{a^*})^2 V(\hat{\Psi}_{01}^*) + (W_{00}^{a^*})^2 V(\hat{\Psi}_{00}^*)$$
$$-2W_{11}^{a^*} W_{10}^{a^*} Cov\left(\hat{\Psi}_{11}^*, \hat{\Psi}_{10}^*\right) - 2W_{11}^{a^*} W_{01}^{a^*} Cov\left(\hat{\Psi}_{11}^*, \hat{\Psi}_{01}^*\right) + 2W_{11}^{a^*} W_{00}^{a^*} Cov\left(\hat{\Psi}_{11}^*, \hat{\Psi}_{00}^*\right) \quad (5.68)$$
$$+2W_{10}^{a^*} W_{01}^{a^*} Cov\left(\hat{\Psi}_{10}^*, \hat{\Psi}_{01}^*\right) - 2W_{10}^{a^*} W_{00}^{a^*} Cov\left(\hat{\Psi}_{10}^*, \hat{\Psi}_{00}^*\right) - 2W_{01}^{a^*} W_{00}^{a^*} Cov\left(\hat{\Psi}_{01}^*, \hat{\Psi}_{00}^*\right)\Big]$$
$$= \frac{1}{n\left(\Delta^*\right)^2}\Big[(W_{11}^{a^*})^2 \Psi_{11}^*(1-\Psi_{11}^*) + (W_{10}^{a^*})^2 \Psi_{10}^*(1-\Psi_{10}^*) + (W_{01}^{a^*})^2 \Psi_{01}^*(1-\Psi_{01}^*) + (W_{00}^{a^*})^2 \Psi_{00}^*(1-\Psi_{00}^*)$$
$$+ 2W_{11}^{a^*} W_{10}^{a^*} \Psi_{11}^* \Psi_{10}^* + 2W_{11}^{a^*} W_{01}^{a^*} \Psi_{11}^* \Psi_{01}^* - 2W_{11}^{a^*} W_{00}^{a^*} \Psi_{11}^* \Psi_{00}^*$$
$$- 2W_{10}^{a^*} W_{01}^{a^*} \Psi_{10}^* \Psi_{01}^* + 2W_{10}^{a^*} W_{00}^{a^*} \Psi_{10}^* \Psi_{00}^* + 2W_{01}^{a^*} W_{00}^{a^*} \Psi_{01}^* \Psi_{00}^*\Big]$$

which proves the theorem.

Theorem 3.5. The variance of the unbiased estimator of $\hat{\pi}_B^{Amy^*}$ is given by

$$V\left(\hat{\pi}_B^{Amy^*}\right) = \frac{1}{n\left(\Delta^*\right)^2}\begin{bmatrix}(W_{11}^{b^*})^2 \Psi_{11}^*(1-\Psi_{11}^*) + (W_{10}^{b^*})^2 \Psi_{10}^*(1-\Psi_{10}^*) + (W_{01}^{b^*})^2 \Psi_{01}^*(1-\Psi_{01}^*) + (W_{00}^{b^*})^2 \Psi_{00}^*(1-\Psi_{00}^*) \\ +2W_{11}^{b^*} W_{10}^{b^*} \Psi_{11}^* \Psi_{10}^* + 2W_{11}^{b^*} W_{01}^{b^*} \Psi_{11}^* \Psi_{01}^* - 2W_{11}^{b^*} W_{00}^{b^*} \Psi_{11}^* \Psi_{00}^* \\ -2W_{10}^{b^*} W_{01}^{b^*} \Psi_{10}^* \Psi_{01}^* + 2W_{10}^{b^*} W_{00}^{b^*} \Psi_{10}^* \Psi_{00}^* + 2W_{01}^{b^*} W_{00}^{b^*} \Psi_{01}^* \Psi_{00}^*\end{bmatrix} \quad (3.69)$$

Proof. The variance of the estimator $\hat{\pi}_B^{Amy^*}$ is given by

$$V\left(\hat{\pi}_B^{Amy^*}\right) = V\left[\frac{1}{\Delta^*}(\hat{\Psi}_{11}^* W_{11}^{b^*} - \hat{\Psi}_{10}^* W_{10}^{b^*} - \hat{\Psi}_{01}^* W_{01}^{b^*} + \hat{\Psi}_{00}^* W_{00}^{b^*} + W_c^{b^*})\right]$$
$$= \frac{1}{\left(\Delta^*\right)^2}\Big[(W_{11}^{b^*})^2 V(\hat{\Psi}_{11}^*) + (W_{10}^{b^*})^2 V(\hat{\Psi}_{10}^*) + (W_{01}^{b^*})^2 V(\hat{\Psi}_{01}^*) + (W_{00}^{b^*})^2 V(\hat{\Psi}_{00}^*)$$
$$-2W_{11}^{b^*} W_{10}^{b^*} Cov\left(\hat{\Psi}_{11}^*, \hat{\Psi}_{10}^*\right) - 2W_{11}^{b^*} W_{01}^{b^*} Cov\left(\hat{\Psi}_{11}^*, \hat{\Psi}_{01}^*\right) + 2W_{11}^{b^*} W_{00}^{b^*} Cov\left(\hat{\Psi}_{11}^*, \hat{\Psi}_{00}^*\right)$$
$$+2W_{10}^{b^*} W_{01}^{b^*} Cov\left(\hat{\Psi}_{10}^*, \hat{\Psi}_{01}^*\right) - 2W_{10}^{b^*} W_{00}^{b^*} Cov\left(\hat{\Psi}_{10}^*, \hat{\Psi}_{00}^*\right) - 2W_{01}^{b^*} W_{00}^{b^*} Cov\left(\hat{\Psi}_{01}^*, \hat{\Psi}_{00}^*\right)\Big]$$
$$= \frac{1}{n\left(\Delta^*\right)^2}\Big[(W_{11}^{b^*})^2 \Psi_{11}^*(1-\Psi_{11}^*) + (W_{10}^{b^*})^2 \Psi_{10}^*(1-\Psi_{10}^*) + (W_{01}^{b^*})^2 \Psi_{01}^*(1-\Psi_{01}^*) + (W_{00}^{b^*})^2 \Psi_{00}^*(1-\Psi_{00}^*)$$

$$+2W_{11}^{b^*}W_{10}^{b^*}\Psi_{11}^*\Psi_{10}^*+2W_{11}^{b^*}W_{01}^{b^*}\Psi_{11}^*\Psi_{01}^*-2W_{11}^{b^*}W_{00}^{b^*}\Psi_{11}^*\Psi_{00}^*$$
$$\left.-2W_{10}^{b^*}W_{01}^{b^*}\Psi_{10}^*\Psi_{01}^*+2W_{10}^{b^*}W_{00}^{b^*}\Psi_{10}^*\Psi_{00}^*+2W_{01}^{b^*}W_{00}^{b^*}\Psi_{01}^*\Psi_{00}^*\right]$$

which proves the theorem.

Theorem 3.6. The variance of the unbiased estimator $\hat{\pi}_{AB}^{Amy^*}$ is given by

$$V\left(\hat{\pi}_{AB}^{Amy^*}\right)=\frac{1}{n\left(\Delta^*\right)^2}\left[(W_{11}^{ab^*})^2\Psi_{11}^*(1-\Psi_{11}^*)+(W_{10}^{ab^*})^2\Psi_{10}^*(1-\Psi_{10}^*)+(W_{01}^{ab^*})^2\Psi_{01}^*(1-\Psi_{01}^*)+(W_{00}^{ab^*})^2\Psi_{00}^*(1-\Psi_{00}^*)\right.$$
$$+2W_{11}^{ab^*}W_{10}^{ab^*}\Psi_{11}^*\Psi_{10}^*+2W_{11}^{ab^*}W_{01}^{ab^*}\Psi_{11}^*\Psi_{01}^*-2W_{11}^{ab^*}W_{00}^{ab^*}\Psi_{11}^*\Psi_{00}^* \tag{3.70}$$
$$\left.-2W_{10}^{ab^*}W_{01}^{ab^*}\Psi_{10}^*\Psi_{01}^*+2W_{10}^{ab^*}W_{00}^{ab^*}\Psi_{10}^*\Psi_{00}^*+2W_{01}^{ab^*}W_{00}^{ab^*}\Psi_{01}^*\Psi_{00}^*\right]$$

Proof. The variance of the estimator $\hat{\pi}_{AB}^{Amy^*}$ is given by

$$V\left(\hat{\pi}_{AB}^{Amy^*}\right)=V\left[\frac{1}{\Delta^*}(\hat{\Psi}_{11}^*W_{11}^{ab^*}-\hat{\Psi}_{10}^*W_{10}^{ab^*}-\hat{\Psi}_{01}^*W_{01}^{ab^*}+\hat{\Psi}_{00}^*W_{00}^{ab^*}+W_c^{ab^*})\right]$$
$$=\frac{1}{\left(\Delta^*\right)^2}\left[(W_{11}^{ab^*})^2V(\hat{\Psi}_{11}^*)+(W_{10}^{ab^*})^2V(\hat{\Psi}_{10}^*)+(W_{01}^{ab^*})^2V(\hat{\Psi}_{01}^*)+(W_{00}^{ab^*})^2V(\hat{\Psi}_{00}^*)\right.$$
$$-2W_{11}^{ab^*}W_{10}^{ab^*}Cov\left(\hat{\Psi}_{11}^*,\hat{\Psi}_{10}^*\right)-2W_{11}^{ab^*}W_{01}^{ab^*}Cov\left(\hat{\Psi}_{11}^*,\hat{\Psi}_{01}^*\right)+2W_{11}^{ab^*}W_{00}^{ab^*}Cov\left(\hat{\Psi}_{11}^*,\hat{\Psi}_{00}^*\right)$$
$$\left.+2W_{10}^{ab^*}W_{01}^{ab^*}Cov\left(\hat{\Psi}_{10}^*,\hat{\Psi}_{01}^*\right)-2W_{10}^{ab^*}W_{00}^{ab^*}Cov\left(\hat{\Psi}_{10}^*,\hat{\Psi}_{00}^*\right)-2W_{01}^{ab^*}W_{00}^{ab^*}Cov\left(\hat{\Psi}_{01}^*,\hat{\Psi}_{00}^*\right)\right]$$
$$=\frac{1}{n\left(\Delta^*\right)^2}\left[(W_{11}^{ab^*})^2\Psi_{11}^*(1-\Psi_{11}^*)+(W_{10}^{ab^*})^2\Psi_{10}^*(1-\Psi_{10}^*)+(W_{01}^{ab^*})^2\Psi_{01}^*(1-\Psi_{01}^*)+(W_{00}^{ab^*})^2\Psi_{00}^*(1-\Psi_{00}^*)\right.$$
$$+2W_{11}^{ab^*}W_{10}^{ab^*}\Psi_{11}^*\Psi_{10}^*+2W_{11}^{ab^*}W_{01}^{ab^*}\Psi_{11}^*\Psi_{01}^*-2W_{11}^{ab^*}W_{00}^{ab^*}\Psi_{11}^*\Psi_{00}^*$$
$$\left.-2W_{10}^{ab^*}W_{01}^{ab^*}\Psi_{10}^*\Psi_{01}^*+2W_{10}^{ab^*}W_{00}^{ab^*}\Psi_{10}^*\Psi_{00}^*+2W_{01}^{ab^*}W_{00}^{ab^*}\Psi_{01}^*\Psi_{00}^*\right]$$

which proves the theorem.
The following lemmas are trivial:

Lemma 3.1. An estimator of the variance of the unbiased estimator $\hat{\pi}_{A}^{Amy^*}$ is given by

$$\hat{V}\left(\hat{\pi}_{A}^{Amy^*}\right)=\frac{1}{(n-1)\left(\Delta^*\right)^2}\left[(W_{11}^{a^*})^2\hat{\Psi}_{11}^*(1-\hat{\Psi}_{11}^*)+(W_{10}^{a^*})^2\hat{\Psi}_{10}^*(1-\hat{\Psi}_{10}^*)+(W_{01}^{a^*})^2\hat{\Psi}_{01}^*(1-\hat{\Psi}_{01}^*)+(W_{00}^{a^*})^2\hat{\Psi}_{00}^*(1-\hat{\Psi}_{00}^*)\right.$$
$$+2W_{11}^{a^*}W_{10}^{a^*}\hat{\Psi}_{11}^*\hat{\Psi}_{10}^*+2W_{11}^{a^*}W_{01}^{a^*}\hat{\Psi}_{11}^*\hat{\Psi}_{01}^*-2W_{11}^{a^*}W_{00}^{a^*}\hat{\Psi}_{11}^*\hat{\Psi}_{00}^* \tag{3.71}$$
$$\left.-2W_{10}^{a^*}W_{01}^{a^*}\hat{\Psi}_{10}^*\hat{\Psi}_{01}^*+2W_{10}^{a^*}W_{00}^{a^*}\hat{\Psi}_{10}^*\hat{\Psi}_{00}^*+2W_{01}^{a^*}W_{00}^{a^*}\hat{\Psi}_{01}^*\hat{\Psi}_{00}^*\right]$$

Lemma 3.2. An estimator of the variance of the unbiased estimator $\hat{\pi}_{B}^{Amy^*}$ is given by

$$\hat{V}\left(\hat{\pi}_{B}^{Amy^*}\right)=\frac{1}{(n-1)\left(\Delta^*\right)^2}\left[(W_{11}^{b^*})^2\hat{\Psi}_{11}^*(1-\hat{\Psi}_{11}^*)+(W_{10}^{b^*})^2\hat{\Psi}_{10}^*(1-\hat{\Psi}_{10}^*)+(W_{01}^{b^*})^2\hat{\Psi}_{01}^*(1-\hat{\Psi}_{01}^*)+(W_{00}^{b^*})^2\hat{\Psi}_{00}^*(1-\hat{\Psi}_{00}^*)\right.$$
$$+2W_{11}^{b^*}W_{10}^{b^*}\hat{\Psi}_{11}^*\hat{\Psi}_{10}^*+2W_{11}^{b^*}W_{01}^{b^*}\hat{\Psi}_{11}^*\hat{\Psi}_{01}^*-2W_{11}^{b^*}W_{00}^{b^*}\hat{\Psi}_{11}^*\hat{\Psi}_{00}^* \tag{3.72}$$
$$\left.-2W_{10}^{b^*}W_{01}^{b^*}\hat{\Psi}_{10}^*\hat{\Psi}_{01}^*+2W_{10}^{b^*}W_{00}^{b^*}\hat{\Psi}_{10}^*\hat{\Psi}_{00}^*+2W_{01}^{b^*}W_{00}^{b^*}\hat{\Psi}_{01}^*\hat{\Psi}_{00}^*\right]$$

Lemma 3.3. An estimator of the variance of the unbiased estimator $\hat{\pi}_{AB}^{Amy^*}$ is given by

$$\hat{V}\left(\hat{\pi}_{AB}^{Amy^*}\right)=\frac{1}{(n-1)\left(\Delta^*\right)^2}\left[(W_{11}^{ab^*})^2\hat{\Psi}_{11}^*(1-\hat{\Psi}_{11}^*)+(W_{10}^{ab^*})^2\hat{\Psi}_{10}^*(1-\hat{\Psi}_{10}^*)+(W_{01}^{ab^*})^2\hat{\Psi}_{01}^*(1-\hat{\Psi}_{01}^*)+(W_{00}^{ab^*})^2\hat{\Psi}_{00}^*(1-\hat{\Psi}_{00}^*)\right.$$
$$+2W_{11}^{ab^*}W_{10}^{ab^*}\hat{\Psi}_{11}^*\hat{\Psi}_{10}^*+2W_{11}^{ab^*}W_{01}^{ab^*}\hat{\Psi}_{11}^*\hat{\Psi}_{01}^*-2W_{11}^{ab^*}W_{00}^{ab^*}\hat{\Psi}_{11}^*\hat{\Psi}_{00}^* \tag{3.73}$$
$$\left.-2W_{10}^{ab^*}W_{01}^{ab^*}\hat{\Psi}_{10}^*\hat{\Psi}_{01}^*+2W_{10}^{ab^*}W_{00}^{ab^*}\hat{\Psi}_{10}^*\hat{\Psi}_{00}^*+2W_{01}^{ab^*}W_{00}^{ab^*}\hat{\Psi}_{01}^*\hat{\Psi}_{00}^*\right]$$

We investigate the percent relative efficiency of the designed estimators with respect to the three estimators of Lee *et al's*. (2013) crossed model in the following section.

4. Relative Efficiency

The percent relative efficiency of the proposed estimators $\hat{\pi}_A^{Amy*}$, $\hat{\pi}_B^{Amy*}$ and $\hat{\pi}_{AB}^{Amy*}$ with respect to the three estimators $\hat{\pi}^*_A$, $\hat{\pi}^*_B$, and $\hat{\pi}^*_{AB}$, respectively, is defined as:

$$RE(1) = \frac{V(\hat{\pi}^*_A)}{V(\hat{\pi}_A^{Amy^*})} \times 100\% \tag{4.1}$$

$$RE(2) = \frac{V(\hat{\pi}^*_B)}{V(\hat{\pi}_B^{Amy^*})} \times 100\% \tag{4.2}$$

and

$$RE(3) = \frac{V(\hat{\pi}^*_{AB})}{V(\hat{\pi}_{AB}^{Amy^*})} \times 100\% \tag{4.3}$$

We wrote SAS codes to compute the percent relative efficiency values RE(1), RE(2) and RE(3) for different choices of parameters. We observed that the proposed estimators are most of the times more efficient than their competitors for any choice of w between 0.1 and 0.9 with a skip of 0.2. For $w = 1$, the proposed estimators reduce to the three estimators of Lee et al. (2013). For $w = 0$ the proposed estimators show maximum relative efficiency values, but this choice is not recommended as the respondents reporting (Yes, No) or (No, Yes) are at the risk of losing their privacy towards at least one sensitive group either A or B. Following Lanke (1975, 1976), the randomized response models should be compared based on some protection criterion. Thus, we define a new protection criterion for the proposed model with respect to the Lee et al. (2013) as follows.

In the proposed model: the conditional probabilities of a respondent to belong to the sensitive group A given that the respondent replied (Yes, Yes), (Yes, No), (No, Yes) or (No, No) are, respectively, given by:

$$P(A \mid YY) = \frac{\pi_A\left[wPT\pi_B + (1-w)\left\{P_1(1-T_1)\pi_y + P_1T_1\pi_B + (1-P_1)\pi_yT_1\pi_B + (1-P_1)(1-T_1)\pi_y\right\}\right]}{\Psi^*_{11}} \tag{4.4}$$

$$P(A \mid YN) = \frac{\pi_A\begin{bmatrix} w\{PT(1-\pi_B) + P(1-T) + (1-P)(1-\pi_B\} \\ +(1-w)\left\{P_1(1-T_1)(1-\pi_y) + P_1T_1(1-\pi_B) + (1-P_1)\pi_yT_1(1-\pi_B)\right\}\end{bmatrix}}{\Psi^*_{10}} \tag{4.5}$$

$$P(A \mid NY) = \frac{\pi_A\left[w(1-P)T\pi_B + (1-w)(1-P_1)(1-\pi_y)\pi_B\right]}{\Psi^*_{01}} \tag{4.6}$$

and

$$P(A \mid NN) = \frac{\pi_A\left[w(1-P)(1-T)\pi_B + (1-w)\left\{(1-P_1)(1-\pi_y)(1-T_1) + (1-P_1)(1-\pi_y)(1-\pi_B)T_1\right\}\right]}{\Psi^*_{00}} \tag{4.7}$$

Then the least protection of a respondent belonging to the group A is computed as:

$$PROT(A)^*_{Amy} = Max\left[P(A \mid YY), P(A \mid YN),\ P(A \mid NY), P(A \mid NN)\right] \tag{4.8}$$

In the following proposed model, we consider the conditional probabilities of a respondent belonging to the sensitive group B given that the respondent replied (Yes, Yes), (Yes, No), (No, Yes) or (No, No) are, separately, shown as:

$$P(B \mid YY) = \frac{\pi_B\left[wPT\pi_A + (1-w)\left\{P_1(1-T_1)\pi_y + P_1T_1\pi_B + (1-P_1)\pi_yT_1\pi_B + (1-P_1)(1-T_1)\pi_y\right\}\right]}{\Psi^*_{11}} \tag{4.9}$$

$$P(B \mid YN) = \frac{\pi_B\left[w\pi_AP(1-T) + (1-w)P_1(1-T_1)(1-\pi_y)\right]}{\Psi^*_{10}} \tag{4.10}$$

$$P(B \mid NY) = \frac{\pi_B \begin{bmatrix} w\{P(1-\pi_A)+(1-P)T+(1-P)(1-T)(1-\pi_A)\} \\ +(1-w)\{P_1(1-\pi_A)T_1+(1-P_1)T_1(1-\pi_y)+(1-T_1)\pi_y P_1(1-\pi_A)\} \end{bmatrix}}{\Psi^*_{01}} \tag{4.11}$$

and

$$P(B \mid NN) = \frac{\pi_B\left[w(1-P)(1-T)\pi_A+(1-w)\{P_1(1-\pi_A)(1-\pi_y)(1-T_1)+(1-P_1)(1-T_1)(1-\pi_y)\}\right]}{\Psi^*_{00}} \tag{4.12}$$

Then the least protection of a respondent belonging to the group B is computed as:

$$PROT\,(B)^*_{Amy} = Max\left[P(B \mid YY), P(B \mid YN), \; P(B \mid NY), P(B \mid NN)\right] \tag{4.13}$$

In this proposed model, we have the conditional probabilities of a respondent belonging to both sensitive characteristic A and B, given that the respondent replied (Yes, Yes), (Yes, No), (No, Yes) or (No, No) are, respectively, shown as:

$$P(AB \mid YY) = \frac{[wPT+(1-w)\{P_1T_1+\pi_y(1-P_1T_1)\}]\pi_{AB}}{\Psi^*_{11}} \tag{4.14}$$

$$P(AB \mid YN) = \frac{[wP(1-T)+(1-w)P_1(1-T_1)(1-\pi_y)]\pi_{AB}}{\Psi^*_{10}} \tag{4.15}$$

$$P(AB \mid NY) = \frac{[w(1-P)T+(1-w)(1-P_1)T_1(1-\pi_y)]\pi_{AB}}{\Psi^*_{01}} \tag{4.16}$$

and

$$P(AB \mid NN\;) = \frac{[w(1-P)(1-T)+(1-w)(1-P_1)(1-T_1)(1-\pi_y)]\pi_{AB}}{\Psi^*_{00}} \tag{4.17}$$

Then the least protection of a respondent belonging to the group A and B is computed as:

$$PROT(AB)^*_{Amy} = Max\;\left[P(AB \mid YY), P(AB \mid YN), \; P(AB \mid NY), P(AB \mid NN)\right] \tag{4.18}$$

In the Lee et al. (2013) crossed model: we considered the conditional probability of a respondent belonging to the sensitive attribute A given that the respondent replied (Yes, Yes), (Yes, No), (No, Yes) or (No, No) are, respectively, shown as:

$$P(A \mid YY)^*_{Lee} = \frac{\pi_A PT\pi_B}{\theta^*_{11}} \tag{4.19}$$

$$P(A \mid YN)^*_{Lee} = \frac{\pi_A[PT(1-\pi_B)+(1-T)P+(1-P)(1-\pi_B)]}{\theta^*_{10}} \tag{4.20}$$

$$P(A \mid NY)^*_{Lee} = \frac{\pi_A(1-P)T\pi_B}{\theta^*_{01}} \tag{4.21}$$

and

$$P(A \mid NN)^*_{Lee} = \frac{\pi_A(1-P)(1-T)\pi_B}{\theta^*_{00}} \tag{4.22}$$

Then the least protection of a respondent belonging to the group A is computed as:

$$PROT(A)^*_{Lee} = Max\left[P(A \mid YY)^*_{Lee}, P(A \mid YN)^*_{Lee}, \; P(A \mid NY)^*_{Lee}, P(A \mid NN)^*_{Lee}\right] \tag{4.23}$$

In the Lee et al. (2013) crossed model, we considered the conditional probabilities of a respondent belonging to the sensitive attribute B given that the respondent replied (Yes, Yes), (Yes, No), (No, Yes) or (No, No) are, respectively, shown as:

$$P(B \mid YY)^*_{Lee} = \frac{\pi_B PT\pi_A}{\theta^*_{11}} \tag{4.24}$$

$$P(B \mid YN)^*_{Lee} = \frac{\pi_B P(1-T)\pi_A}{\theta^*_{10}} \tag{4.25}$$

$$P(B \mid NY)^*_{Lee} = \frac{\pi_B[P(1-\pi_A)+(1-P)T+(1-P)(1-T)(1-\pi_A)]}{\theta^*_{01}} \tag{4.26}$$

and

$$P(B \mid NN)^*_{Lee} = \frac{\pi_B(1-P)(1-T)\pi_A}{\theta^*_{00}} \tag{4.27}$$

Then the least protection of a respondent belonging to the group B is computed as:

$$PROT(B)^*_{Lee} = Max\left[P(B \mid YY)^*_{Lee}, P(B \mid YN)^*_{Lee}, P(B \mid NY)^*_{Lee}, P(B \mid NN)^*_{Lee}\right] \tag{4.28}$$

In the Lee et al. (2013) crossed model, we considered the conditional probability of a respondent belonging to both of the sensitive groups A and B given that the respondent replied (Yes, Yes), (Yes, No), (No, Yes) or (No, No) is, respectively, given by

$$P(AB \mid YY)^*_{Lee} = \frac{PT\pi_{AB}}{\theta^*_{11}} \tag{4.29}$$

$$P(AB \mid YN)^*_{Lee} = \frac{P(1-T)\pi_{AB}}{\theta^*_{10}} \tag{4.30}$$

$$P(AB \mid NY)^*_{Lee} = \frac{(1-P)T\pi_{AB}}{\theta^*_{01}} \tag{4.31}$$

and

$$P(AB \mid NN)^*_{Lee} = \frac{(1-P)(1-T)\pi_{AB}}{\theta^*_{00}} \tag{4.32}$$

Then the least protection of a respondent that having both of characteristics A and B is computed as:

$$PROT(AB)^*_{Lee} = Max\left[P(AB \mid YY)^*_{Lee}, P(AB \mid YN)^*_{Lee}, P(AB \mid NY)^*_{Lee}, P(AB \mid NN)^*_{Lee}\right] \tag{4.33}$$

The percent relative protection of the proposed model with respect to the cross model proposed by the Lee et al. (2103) for three types of respondents possessing membership in groups A, B or $A \cap B$ are, respectively, defined as:

$$RP(1) = \frac{PROT\ (A)^*_{Lee}}{PROT\ (A)^*_{Amy}} \times 100\% \tag{4.34}$$

$$RP(2) = \frac{PROT\ (B)^*_{Lee}}{PROT\ (B)^*_{Amy}} \times 100\% \tag{4.35}$$

and

$$RP(3) = \frac{PROT\ (AB)^*_{Lee}}{PROT\ (AB)^*_{Amy}} \times 100\% \tag{4.36}$$

We also examined the relative protection of the respondents by using the SAS codes provided in Appendix A. Again we found a lot of simulated situations where the suggested randomized response model is both more protective and efficient than that of the crossed model of Lee et al. (2013) when estimating the three proportions *viz.* π_A, π_B and π_{AB}. The behavior of the simulated results is very similar to a situation where one is comparing the Greenberg et al. (1969) model with the Warner (1965) model based on criteria of both efficiency and protection. Useful choices of the device parameters P, T and π_y depend on the parameters of interest π_A, π_B and π_{AB}.

Although broad choice of such parameters does not exist, there are many situations that have been simulated by the SAS code given in Appendix A such that the proposed model leads to an efficient and protective model for all the three parameters π_A, π_B and π_{AB}. For an illustration purpose, a set of such simulated results for $P = T = 0.7$, that includes many choices of the other parameters π_y, P_1, T_1 and W are listed in Table B1 in Appendix B. The values of W considered are between 0.1 and 0.9 with a step size of 0.2; the values of π_y are between 0.9 and 1.0 with a step size of 0.05; the values of P_1 are 0.6 and 0.7; the values of T_1 are also 0.6 and 0.7; the values of π_{AB} are considered one between 0.05 to 0.20 with a step size of 0.05; the values of π_A and π_B are considered ones between 0.05 and 0.35 with a step size of 0.05 with an additional restriction that $\pi_{AB} < \pi_A$ and $\pi_{AB} < \pi_B$. There are 2943 results; in Table B1. in Appendix B which show percent relative efficiencies and relative protection values greater than 100%. Here we provide a summary of such detailed results for various choices of parameters.

Table 1: Summary of the values of RE(1), RE(2), RE(3), RP(1), RP(2) and RP(3) for *w*.

Variable	*w*	Freq	Mean	StDev	Minimum	Maximum
RE(1)	0.1	237	134.41	12.87	102.45	161.30
	0.3	520	128.77	13.21	100.33	152.56
	0.5	731	119.96	9.97	100.02	138.73
	0.7	731	113.14	6.01	100.44	124.20
	0.9	724	104.73	2.01	100.20	108.33
RE(2)	0.1	237	138.56	13.22	102.45	163.64
	0.3	520	129.09	13.71	100.33	155.58
	0.5	731	121.14	9.88	100.02	140.64
	0.7	731	113.85	5.91	100.69	124.52
	0.9	724	104.96	1.96	100.34	108.34
RE(3)	0.1	237	139.29	20.89	102.05	184.86
	0.3	520	130.40	15.63	100.47	168.28
	0.5	731	122.18	12.16	100.06	150.08
	0.7	731	114.84	7.82	100.46	130.63
	0.9	724	105.35	2.77	100.01	110.84
RP(1)	0.1	237	116.69	8.37	105.15	140.85
	0.3	520	114.53	8.52	103.19	144.17
	0.5	731	111.97	7.28	102.04	136.67
	0.7	731	106.59	4.03	101.11	120.24
	0.9	724	102.03	1.25	100.34	106.33
RP(2)	0.1	237	113.85	6.48	104.67	138.81
	0.3	520	113.56	7.74	102.95	142.40
	0.5	731	110.30	5.69	101.89	127.97
	0.7	731	105.65	3.13	101.02	115.44
	0.9	724	101.73	0.971	100.31	104.76
RP(3)	0.1	237	120.27	15.80	100.28	161.52
	0.3	520	117.51	12.37	100.13	153.61
	0.5	731	114.82	9.68	100.10	141.72
	0.7	731	110.22	6.65	100.07	129.00
	0.9	724	103.98	2.60	100.03	111.49

Table 1 provides the values of RE(1), RE(2), RE(3), RP(1), RP(2), and RP(3) for different choices of *w*. If $w = 0.1$ then there are 237 useful choices of the other parameters, which have an average value of RE(1) of 134.41% with a standard deviation of 12.87%, a minimum value of 102.45% and a maximum value of 161.30%; RE(2) has an average value of 138.56% with a standard deviation of 13.22%, a minimum value of 102.45% and a maximum value of 163.64%. RE(3) has an average value of 139.29% with a standard deviation of 20.89%, a minimum value of 102.05% and a maximum value of 184.86%. RP(1) has an average value of 116.69% with a standard deviation of 8.37%, a minimum value of 105.15% and a maximum value of 140.85%. RE(2) has an average value of 113.85% with a standard deviation of 6.48%, a minimum value of 104.67% and maximum value of 138.81%. RP(3) has an average value of 120.27% with a standard deviation of 15.80%, a minimum value of 100.28% and maximum value of 161.52%. As the value of *w* increases to 0.9 then there are 724 useful outcomes where the proposed model shows efficient results. Thus, in practice the choice of *w* close to 0.9 is suggested so that maximum gain in efficiencies and protection are achieved. A graphical representation of such results is also shown in Figure 5.

For demonstration purposes, we detailed a part of the results in Appendix B as cited in Table 2. For $\pi_A = 0.1$, $\pi_B = 0.1$ and $\pi_{AB} = 0.05$ the average value of RE(1) is 108.97% with a standard deviation of 4.16%, a minimum value of 102.97% and a maximum value of 114.31%. RE(2) has an average value of 117.13% with a standard deviation of 14.02%, a minimum value of 102.97% and maximum value of 114.31%. RE(3) has an average value of 119.27% with a standard deviation of 12.73%, a minimum value of

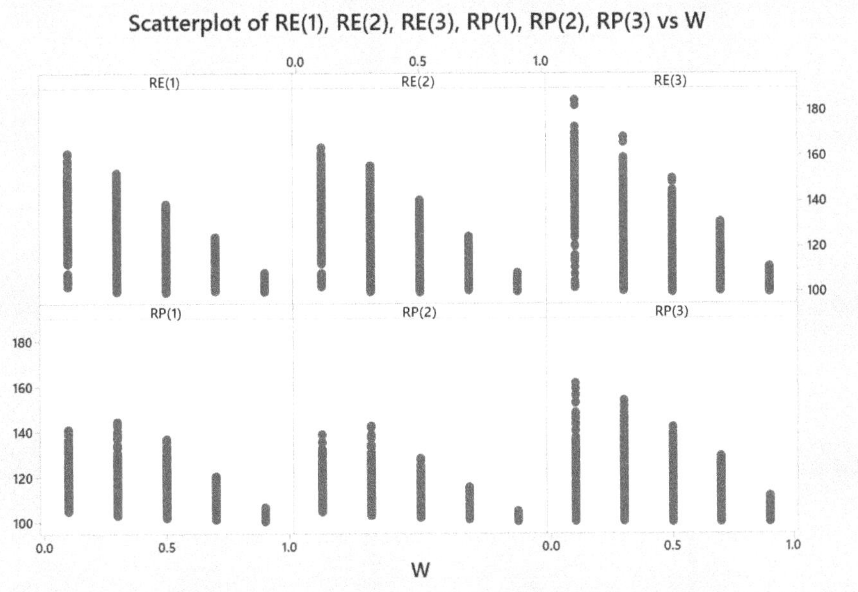

Figure 5: RE(1), RE(2), RE(3), RP(1), RP(2) and RP(3) values for different choice of w.

Table 2: Summary of the values of RE(1), RE(2), RE(3), RP(1), RP(2) and RP(3) for different choices of π_A, π_B and π_{AB}.

Results for $\pi_A = 0.10$ and $\pi_B = 0.10$

Variable	π_{AB}	Freq	Mean	StDev	Minimum	Maximum
RE(1)	0.05	16	108.97	4.16	102.97	114.31
RE(2)	0.05	16	117.13	14.02	103.47	150.00
RE(3)	0.05	16	119.27	12.73	105.11	143.59
RP(1)	0.05	16	115.10	10.22	103.24	137.59
RP(2)	0.05	16	108.25	5.89	101.23	120.08
RP(3)	0.05	16	104.88	4.98	100.07	113.99

Results for $\pi_A = 0.10$ and $\pi_B = 0.15$

Variable	π_{AB}	Freq	Mean	StDev	Minimum	Maximum
RE(1)	0.05	15	117.23	11.90	102.20	135.15
RE(2)	0.05	15	119.28	13.28	102.59	139.90
RE(3)	0.05	15	145.37	31.63	105.90	193.21
RP(1)	0.05	15	103.32	3.15	100.16	110.63
RP(2)	0.05	15	103.07	2.80	100.17	109.54
RP(3)	0.05	15	114.11	6.54	103.74	123.15

105.11% and a maximum value of 143.59%. The average value of RP(1) is found to be 115.10% with a standard deviation of 10.22%, a minimum value of 103.24% and a maximum value of 137.59%. RP(3) has an average value of 104.88% with a standard deviation of 4.98%, a minimum value of 100.07% and a maximum value of 113.99%. Thus, the proposed model is found to be quite helpful in estimating such tiny proportions of sensitive attributes and their overlaps.

Likewise, the results for $\pi_A = 0.1$, $\pi_B = 0.15$ and $\pi_{AB} = 0.05$ can be seen from Table 2. In addition, results for any such combination of the three proportions π_A, π_B and π_{AB} can be extracted, if needed, from the detailed results give in Table B1 in Appendix B.

Table 3 helps to make a choice on appropriate proportion of unrelated question π_y. For $\pi_y = 1$ there are combinations of the parameters of $freq = 991$ where the proposed estimators are more efficient and more protective than that of the crossed model of Lee et al. (2013), and it gives us a very truthful message that in the proposed model "Forced Yes" response could be used instead trying to find an appropriate unrelated characteristic in a population under study.

Table 3: Summary of the values of RE(1), RE(2), RE(3), RP(1), RP(2) and RP(3) for π_y.

Variable	π_y	Freq	Mean	StDev	Minimum	Maximum
RE(1)	0.90	966	117.15	13.02	100.33	154.42
	0.95	986	117.37	13.15	100.02	160.58
	1.00	991	117.20	13.01	100.19	161.30
RE(2)	0.90	966	118.06	13.63	100.33	163.64
	0.95	986	118.33	13.76	100.02	163.61
	1.00	991	118.07	13.47	100.51	161.30
RE(3)	0.90	966	118.15	14.62	100.02	167.56
	0.95	986	119.31	15.74	100.13	182.09
	1.00	991	119.66	16.10	100.01	184.86
RP(1)	0.90	966	107.76	7.08	100.34	140.00
	0.95	986	109.06	7.88	100.49	144.17
	1.00	991	110.21	8.36	100.65	142.97
RP(2)	0.90	966	106.70	5.98	100.31	134.31
	0.95	986	107.91	6.65	100.47	138.36
	1.00	991	109.06	7.26	100.63	142.40
RP(3)	0.90	966	111.20	9.98	100.06	156.12
	0.95	986	111.83	10.56	100.05	158.88
	1.00	991	112.73	11.16	100.03	161.52

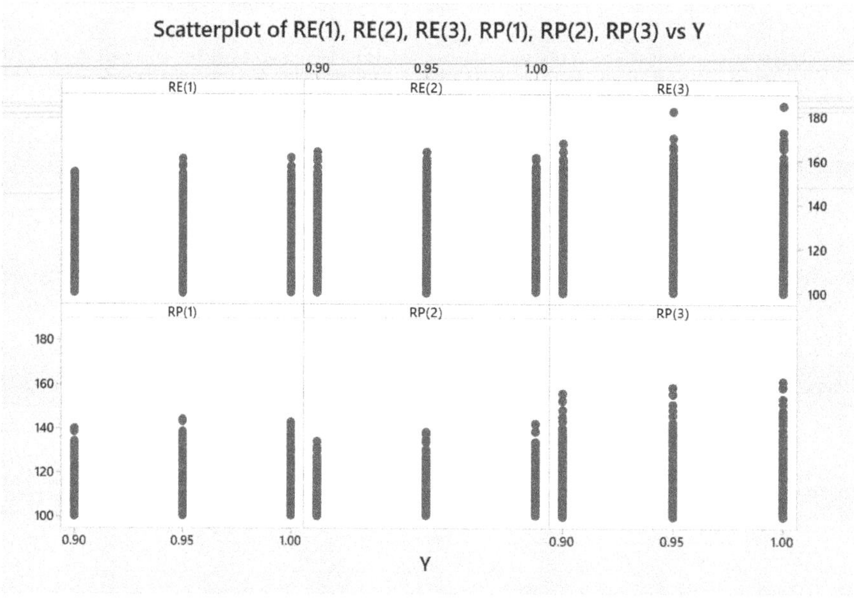

Figure 6: RE(1), RE(2), RE(3), RP(1), RP(2), RP(3) versus π_y.

These findings through a simulation study enhance the usefulness of the proposed model in practice. A behavior of the values of RE(1), RE(2), RE(3), RP(1), RP(2) and RP(3) versus π_y are shown in Figure 6. In Figure 6 the notation Y stands for π_y.

Table 4 provides a summary of results for different choices of P_1 and T_1. For $P_1 = 0.6$ and $T_1 = 0.6$, the average value of RE(1) is 107.55% with a standard deviation of 4.66%, and a minimum value of 100.19% and a maximum value of 120.27%; the average value of RE(2) is 108.67% with a standard deviation of 5.20%, and a minimum value of 100.96% and a maximum value of 122.76%; the average value of RE(3) is 109.46% with a standard deviation of 7.83%, a minimum value of 100.02% and a maximum value of 138.88%. The average value of RP(1) is 110.10% with a standard deviation of 7.16%, with a minimum value of 101.36% and a

Table 4: RE(1), RE(2), RE(3), RP(1), RP(2) and RP(3) for different P_1 and T_1.

Results for $P_1 = 0.4$

Variable	T_1	Freq	Mean	StDev	Minimum	Maximum
RE(1)	0.6	495	107.55	4.66	100.19	120.27
	0.7	804	108.92	5.61	100.02	126.92
RE(2)	0.6	495	108.67	5.20	100.96	122.76
	0.7	804	124.84	14.35	104.05	163.64
RE(3)	0.6	495	109.46	7.83	100.02	138.88
	0.7	804	118.18	13.57	100.01	165.38
RP(1)	0.6	495	110.10	7.16	101.36	133.29
	0.7	804	113.85	9.70	101.51	144.17
RP(2)	0.6	495	108.40	5.66	101.35	125.21
	0.7	804	104.79	4.08	100.31	124.17
RP(3)	0.6	495	110.74	7.61	100.07	131.36
	0.7	804	111.36	9.56	100.06	146.60

Results for $P_1 = 0.5$

Variable	T_1	Freq	Mean	StDev	Minimum	Maximum
RE(1)	0.6	766	123.66	13.18	104.05	158.22
	0.7	878	124.71	13.08	103.54	161.30
RE(2)	0.6	766	109.34	5.91	100.02	132.16
	0.7	878	125.07	13.41	103.54	161.30
RE(3)	0.6	766	117.69	13.04	100.01	169.78
	0.7	878	126.43	18.63	100.13	184.86
RP(1)	0.6	766	105.34	4.62	100.34	126.55
	0.7	878	107.20	6.03	100.44	140.50
RP(2)	0.6	766	112.26	8.53	101.51	142.40
	0.7	878	106.65	5.26	100.44	131.67
RP(3)	0.6	766	111.52	9.77	100.03	146.60
	0.7	878	113.46	13.19	100.05	161.52

maximum value of 133.29%, the average value of RP(2) is 108.40% with a standard deviation of 5.66%, with a minimum of 101.35% and a maximum of 125.21%. The average value of RP(3) is 110.74% with a standard deviation of 7.61%, with a minimum of 100.07% and a maximum of 131.36%. In the same way the other results from Table 4 can be interpreted. We consider the problem of estimation of other parameters such as conditional proportions in the next section, proportions of membership in at least one group, difference between two proportions, relative risk, and correlation between two sensitive characteristics.

5. Estimation of Other Parameters

For estimating the other parameters such as $\pi_{A|B}$, π_{A-B}, $\pi_{A\cup B}$, $RR(B|A)$ and ρ_{AB}. We need the following lemmas.

Lemma 5.1. The covariance between $Cov(\hat{\pi}_A^{Amy*}, \hat{\pi}_B^{Amy*})$ is given by

$$\begin{aligned} Cov(\hat{\pi}_A^{Amy^*}, \hat{\pi}_B^{Amy^*}) &= \frac{1}{n(\Delta^*)^2}\Big[W_{11}^{a^*}W_{11}^{b^*}\Psi_{11}^*\left(1-\Psi_{11}^*\right)+W_{11}^{a^*}W_{10}^{b^*}\Psi_{11}^*\Psi_{10}^*+W_{11}^{a^*}W_{01}^{b^*}\Psi_{11}^*\Psi_{01}^*-W_{11}^{a^*}W_{00}^{b^*}\Psi_{11}^*\Psi_{00}^* \\ &+W_{10}^{a^*}W_{11}^{b^*}\Psi_{10}^*\Psi_{11}^*+W_{10}^{a^*}W_{10}^{b^*}\Psi_{10}^*\left(1-\Psi_{10}^*\right)-W_{10}^{a^*}W_{01}^{b^*}\Psi_{10}^*\Psi_{01}^*+W_{10}^{a^*}W_{00}^{b^*}\Psi_{10}^*\Psi_{00}^* \\ &+W_{01}^{a^*}W_{11}^{b^*}\Psi_{01}^*\Psi_{11}^*-W_{01}^{a^*}W_{10}^{b^*}\Psi_{01}^*\Psi_{10}^*+W_{01}^{a^*}W_{01}^{b^*}\Psi_{01}^*(1-\Psi_{01}^*)+W_{01}^{a^*}W_{00}^{b^*}\Psi_{01}^*\Psi_{00}^* \\ &-W_{00}^{a^*}W_{11}^{b^*}\Psi_{00}^*\Psi_{11}^*+W_{00}^{a^*}W_{10}^{b^*}\Psi_{00}^*\Psi_{10}^*+W_{00}^{a^*}W_{01}^{b^*}\Psi_{00}^*\Psi_{01}^*+W_{00}^{a^*}W_{00}^{b^*}\Psi_{00}^*\left(1-\Psi_{00}^*\right)\Big] \end{aligned} \quad (5.1)$$

Proof. We have

$$
\begin{aligned}
&Cov(\hat{\pi}_A^{Amy^*}, \hat{\pi}_B^{Amy^*}) \\
&= Cov\left\{\frac{1}{\Delta^*}(\hat{\Psi}^*_{11}W_{11}^{a^*} - \hat{\Psi}^*_{10}W_{10}^{a^*} - \hat{\Psi}^*_{01}W_{01}^{a^*} + \hat{\Psi}^*_{00}W_{00}^{a^*} + W_c^{a^*}), \frac{1}{\Delta^*}(\hat{\Psi}^*_{11}W_{11}^{b^*} - \hat{\Psi}^*_{10}W_{10}^{b^*} - \hat{\Psi}^*_{01}W_{01}^{b^*} + \hat{\Psi}^*_{00}W_{00}^{b^*} + W_c^{b^*})\right\} \\
&= \frac{1}{(\Delta^*)^2} Cov\left\{\hat{\Psi}^*_{11}W_{11}^{a^*} - \hat{\Psi}^*_{10}W_{10}^{a^*} - \hat{\Psi}^*_{01}W_{01}^{a^*} + \hat{\Psi}^*_{00}W_{00}^{a^*} + W_c^{a^*}, \hat{\Psi}^*_{11}W_{11}^{b^*} - \hat{\Psi}^*_{10}W_{10}^{b^*} - \hat{\Psi}^*_{01}W_{01}^{b^*} + \hat{\Psi}^*_{00}W_{00}^{b^*} + W_c^{b^*}\right\} \\
&= \frac{1}{(\Delta^*)^2}\Big[W_{11}^{a^*}W_{11}^{b^*}Cov(\hat{\Psi}^*_{11}, \hat{\Psi}^*_{11}) - W_{11}^{a^*}W_{10}^{b^*}Cov(\hat{\Psi}^*_{11}, \hat{\Psi}^*_{10}) - W_{11}^{a^*}W_{01}^{b^*}Cov(\hat{\Psi}^*_{11}, \hat{\Psi}^*_{01}) + W_{11}^{a^*}W_{00}^{b^*}Cov(\hat{\Psi}^*_{11}, \hat{\Psi}^*_{00}) \\
&\quad - W_{10}^{a^*}W_{11}^{b^*}Cov(\hat{\Psi}^*_{10}, \hat{\Psi}^*_{11}) + W_{10}^{a^*}W_{10}^{b^*}Cov(\hat{\Psi}^*_{10}, \hat{\Psi}^*_{10}) + W_{10}^{a^*}W_{01}^{b^*}Cov(\hat{\Psi}^*_{10}, \hat{\Psi}^*_{01}) - W_{10}^{a^*}W_{00}^{b^*}Cov(\hat{\Psi}^*_{10}, \hat{\Psi}^*_{00}) \\
&\quad - W_{01}^{a^*}W_{11}^{b^*}Cov(\hat{\Psi}^*_{01}, \hat{\Psi}^*_{11}) + W_{01}^{a^*}W_{10}^{b^*}Cov(\hat{\Psi}^*_{01}, \hat{\Psi}^*_{10}) + W_{01}^{a^*}W_{01}^{b^*}Cov(\hat{\Psi}^*_{01}, \hat{\Psi}^*_{01}) - W_{01}^{a^*}W_{00}^{b^*}Cov(\hat{\Psi}^*_{01}, \hat{\Psi}^*_{00}) \\
&\quad + W_{00}^{a^*}W_{11}^{b^*}Cov(\hat{\Psi}^*_{00}, \hat{\Psi}^*_{11}) - W_{00}^{a^*}W_{10}^{b^*}Cov(\hat{\Psi}^*_{00}, \hat{\Psi}^*_{10}) - W_{00}^{a^*}W_{01}^{b^*}Cov(\hat{\Psi}^*_{00}, \hat{\Psi}^*_{01}) + W_{00}^{a^*}W_{00}^{b^*}Cov(\hat{\Psi}^*_{00}, \hat{\Psi}^*_{00})\Big] \\
&= \frac{1}{n(\Delta^*)^2}\Big[W_{11}^{a^*}W_{11}^{b^*}\Psi^*_{11}(1-\Psi^*_{11}) + W_{11}^{a^*}W_{10}^{b^*}\Psi^*_{11}\Psi^*_{10} + W_{11}^{a^*}W_{01}^{b^*}\Psi^*_{11}\Psi^*_{01} - W_{11}^{a^*}W_{00}^{b^*}\Psi^*_{11}\Psi^*_{00} \\
&\quad + W_{10}^{a^*}W_{11}^{b^*}\Psi^*_{10}\Psi^*_{11} + W_{10}^{a^*}W_{10}^{b^*}\Psi^*_{10}(1-\Psi^*_{10}) - W_{10}^{a^*}W_{01}^{b^*}\Psi^*_{10}\Psi^*_{01} + W_{10}^{a^*}W_{00}^{b^*}\Psi^*_{10}\Psi^*_{00} \\
&\quad + W_{01}^{a^*}W_{11}^{b^*}\Psi^*_{01}\Psi^*_{11} - W_{01}^{a^*}W_{10}^{b^*}\Psi^*_{01}\Psi^*_{10} + W_{01}^{a^*}W_{01}^{b^*}\Psi^*_{01}(1-\Psi^*_{01}) + W_{01}^{a^*}W_{00}^{b^*}\Psi^*_{01}\Psi^*_{00} \\
&\quad - W_{00}^{a^*}W_{11}^{b^*}\Psi^*_{00}\Psi^*_{11} + W_{00}^{a^*}W_{10}^{b^*}\Psi^*_{00}\Psi^*_{10} + W_{00}^{a^*}W_{01}^{b^*}\Psi^*_{00}\Psi^*_{01} + W_{00}^{a^*}W_{00}^{b^*}\Psi^*_{00}(1-\Psi^*_{00})\Big]
\end{aligned}
$$

which proves the lemma.

Lemma 5.2. The covariance between $Cov(\hat{\pi}_{AB}^{Amy^*}, \hat{\pi}_A^{Amy^*})$ is given by:

$$
\begin{aligned}
&Cov(\hat{\pi}_{AB}^{Amy^*}, \hat{\pi}_A^{Amy^*}) \\
&= \frac{1}{n(\Delta^*)^2}\Big[W_{11}^{ab^*}W_{11}^{a^*}\Psi^*_{11}(1-\Psi^*_{11}) + W_{11}^{ab^*}W_{10}^{a^*}\Psi^*_{11}\Psi^*_{10} + W_{11}^{ab^*}W_{01}^{a^*}\Psi^*_{11}\Psi^*_{01} - W_{11}^{ab^*}W_{00}^{a^*}\Psi^*_{11}\Psi^*_{00} \\
&\quad + W_{10}^{ab^*}W_{11}^{a^*}\Psi^*_{10}\Psi^*_{11} + W_{10}^{ab^*}W_{10}^{a^*}\Psi^*_{10}(1-\Psi^*_{10}) - W_{10}^{ab^*}W_{01}^{a^*}\Psi^*_{10}\Psi^*_{01} + W_{10}^{ab^*}W_{00}^{a^*}\Psi^*_{10}\Psi^*_{00} \\
&\quad + W_{01}^{ab^*}W_{11}^{a^*}\Psi^*_{01}\Psi^*_{11} - W_{01}^{ab^*}W_{10}^{a^*}\Psi^*_{01}\Psi^*_{10} + W_{01}^{ab^*}W_{01}^{a^*}\Psi^*_{01}(1-\Psi^*_{01}) + W_{01}^{ab^*}W_{00}^{a^*}\Psi^*_{01}\Psi^*_{00} \\
&\quad - W_{00}^{ab^*}W_{11}^{a^*}\Psi^*_{00}\Psi^*_{11} + W_{00}^{ab^*}W_{10}^{a^*}\Psi^*_{00}\Psi^*_{10} + W_{00}^{ab^*}W_{01}^{a^*}\Psi^*_{00}\Psi^*_{01} + W_{00}^{ab^*}W_{00}^{a^*}\Psi^*_{00}(1-\Psi^*_{00})\Big]
\end{aligned}
\tag{5.2}
$$

Proof. We have

$$
\begin{aligned}
&Cov\,(\hat{\pi}_{AB}^{Amy}, \hat{\pi}_A^{Amy}) \\
&= Cov\left\{\frac{1}{\Delta^*}(\hat{\Psi}^*_{11}W_{11}^{ab^*} - \hat{\Psi}^*_{10}W_{10}^{ab^*} - \hat{\Psi}^*_{01}W_{01}^{ab^*} + \hat{\Psi}^*_{00}W_{00}^{ab^*} + W_c^{ab^*}), \frac{1}{\Delta^*}(\hat{\Psi}^*_{11}W_{11}^{a^*} - \hat{\Psi}^*_{10}W_{10}^{a^*} - \hat{\Psi}^*_{01}W_{01}^{a^*} + \hat{\Psi}^*_{00}W_{00}^{a^*} + W_c^{a^*})\right\} \\
&= \frac{1}{(\Delta^*)^2} Cov\left\{\hat{\Psi}^*_{11}W_{11}^{ab^*} - \hat{\Psi}^*_{10}W_{10}^{ab^*} - \hat{\Psi}^*_{01}W_{01}^{ab^*} + \hat{\Psi}^*_{00}W_{00}^{ab^*} + W_c^{ab^*}, \hat{\Psi}^*_{11}W_{11}^{a^*} - \hat{\Psi}^*_{10}W_{10}^{a^*} - \hat{\Psi}^*_{01}W_{01}^{a^*} + \hat{\Psi}^*_{00}W_{00}^{a^*} + W_c^{a^*}\right\} \\
&= \frac{1}{(\Delta^*)^2}\Big[W_{11}^{ab^*}W_{11}^{a^*}Cov(\hat{\Psi}^*_{11}, \hat{\Psi}^*_{11}) - W_{11}^{ab^*}W_{10}^{a^*}Cov(\hat{\Psi}^*_{11}, \hat{\Psi}^*_{10}) - W_{11}^{ab^*}W_{01}^{a^*}Cov(\hat{\Psi}^*_{11}, \hat{\Psi}^*_{01}) + W_{11}^{ab^*}W_{00}^{a^*}Cov(\hat{\Psi}^*_{11}, \hat{\Psi}^*_{00}) \\
&\quad - W_{10}^{ab^*}W_{11}^{a^*}Cov(\hat{\Psi}^*_{10}, \hat{\Psi}^*_{11}) + W_{10}^{ab^*}W_{10}^{a^*}Cov(\hat{\Psi}^*_{10}, \hat{\Psi}^*_{10}) + W_{10}^{ab^*}W_{01}^{a^*}Cov(\hat{\Psi}^*_{10}, \hat{\Psi}^*_{01}) - W_{10}^{ab^*}W_{00}^{a^*}Cov(\hat{\Psi}^*_{10}, \hat{\Psi}^*_{00}) \\
&\quad - W_{01}^{ab^*}W_{11}^{a^*}Cov(\hat{\Psi}^*_{01}, \hat{\Psi}^*_{11}) + W_{01}^{ab^*}W_{10}^{a^*}Cov(\hat{\Psi}^*_{01}, \hat{\Psi}^*_{10}) + W_{01}^{ab^*}W_{01}^{a^*}Cov(\hat{\Psi}^*_{01}, \hat{\Psi}^*_{01}) - W_{01}^{ab^*}W_{00}^{a^*}Cov(\hat{\Psi}^*_{01}, \hat{\Psi}^*_{00}) \\
&\quad + W_{00}^{ab^*}W_{11}^{a^*}Cov(\hat{\Psi}^*_{00}, \hat{\Psi}^*_{11}) - W_{00}^{ab^*}W_{10}^{a^*}Cov(\hat{\Psi}^*_{00}, \hat{\Psi}^*_{10}) - W_{00}^{ab^*}W_{01}^{a^*}Cov(\hat{\Psi}^*_{00}, \hat{\Psi}^*_{01}) + W_{00}^{ab^*}W_{00}^{a^*}Cov(\hat{\Psi}^*_{00}, \hat{\Psi}^*_{00}) \\
&= \frac{1}{n(\Delta^*)^2}\Big[W_{11}^{ab^*}W_{11}^{a^*}\Psi^*_{11}(1-\Psi^*_{11}) + W_{11}^{ab^*}W_{10}^{a^*}\Psi^*_{11}\Psi^*_{10} + W_{11}^{ab^*}W_{01}^{a^*}\Psi^*_{11}\Psi^*_{01} - W_{11}^{ab^*}W_{00}^{a^*}\Psi^*_{11}\Psi^*_{00} \\
&\quad + W_{10}^{ab^*}W_{11}^{a^*}\Psi^*_{10}\Psi^*_{11} + W_{10}^{ab^*}W_{10}^{a^*}\Psi^*_{10}(1-\Psi^*_{10}) - W_{10}^{ab^*}W_{01}^{a^*}\Psi^*_{10}\Psi^*_{01} + W_{10}^{ab^*}W_{00}^{a^*}\Psi^*_{10}\Psi^*_{00} \\
&\quad + W_{01}^{ab^*}W_{11}^{a^*}\Psi^*_{01}\Psi^*_{11} - W_{01}^{ab^*}W_{10}^{a^*}\Psi^*_{01}\Psi^*_{10} + W_{01}^{ab^*}W_{01}^{a^*}\Psi^*_{01}(1-\Psi^*_{01}) + W_{01}^{ab^*}W_{00}^{a^*}\Psi^*_{01}\Psi^*_{00} \\
&\quad - W_{00}^{ab^*}W_{11}^{a^*}\Psi^*_{00}\Psi^*_{11} + W_{00}^{ab^*}W_{10}^{a^*}\Psi^*_{00}\Psi^*_{10} + W_{00}^{ab^*}W_{01}^{a^*}\Psi^*_{00}\Psi^*_{01} + W_{00}^{ab^*}W_{00}^{a^*}\Psi^*_{00}(1-\Psi^*_{00})\Big]
\end{aligned}
$$

which proves the lemma.

Lemma 5.3. The covariance between $Cov(\hat{\pi}_{AB}^{Amy*}, \hat{\pi}_{B}^{Amy*})$ is given by

$$
\begin{aligned}
&Cov(\hat{\pi}_{AB}^{Amy^*}, \hat{\pi}_{B}^{Amy^*}) \\
&= \frac{1}{n(\Delta^*)^2}\Big[W_{11}^{ab^*}W_{11}^{b^*}\Psi_{11}^*(1-\Psi_{11}^*) + W_{11}^{ab^*}W_{10}^{b^*}\Psi_{11}^*\Psi_{10}^* + W_{11}^{ab^*}W_{01}^{b^*}\Psi_{11}^*\Psi_{01}^* - W_{11}^{ab^*}W_{00}^{b^*}\Psi_{11}^*\Psi_{00}^* \\
&\quad + W_{10}^{ab^*}W_{11}^{b^*}\Psi_{10}^*\Psi_{11}^* + W_{10}^{ab^*}W_{10}^{b^*}\Psi_{10}^*(1-\Psi_{10}^*) - W_{10}^{ab^*}W_{01}^{b^*}\Psi_{10}^*\Psi_{01}^* + W_{10}^{ab^*}W_{00}^{b^*}\Psi_{10}^*\Psi_{00}^* \\
&\quad + W_{01}^{ab^*}W_{11}^{b^*}\Psi_{01}^*\Psi_{11}^* - W_{01}^{ab^*}W_{10}^{b^*}\Psi_{01}^*\Psi_{10}^* + W_{01}^{ab^*}W_{01}^{b^*}\Psi_{01}^*(1-\Psi_{01}^*) + W_{01}^{ab^*}W_{00}^{b^*}\Psi_{01}^*\Psi_{00}^* \\
&\quad - W_{00}^{ab^*}W_{11}^{b^*}\Psi_{00}^*\Psi_{11}^* + W_{00}^{ab^*}W_{10}^{b^*}\Psi_{00}^*\Psi_{10}^* + W_{00}^{ab^*}W_{01}^{b^*}\Psi_{00}^*\Psi_{01}^* + W_{00}^{ab^*}W_{00}^{b^*}\Psi_{00}^*(1-\Psi_{00}^*)\Big]
\end{aligned}
\tag{5.3}
$$

Proof. We have

$$
\begin{aligned}
&Cov(\hat{\pi}_{AB}^{Amy^*}, \hat{\pi}_{B}^{Amy^*}) \\
&= Cov\left\{\frac{1}{\Delta^*}(\hat{\Psi}_{11}^*W_{11}^{ab^*} - \hat{\Psi}_{10}^*W_{10}^{ab^*} - \hat{\Psi}_{01}^*W_{01}^{ab^*} + \hat{\Psi}_{00}^*W_{00}^{ab^*} + W_c^{ab^*}), \frac{1}{\Delta^*}(\hat{\Psi}_{11}^*W_{11}^{b^*} - \hat{\Psi}_{10}^*W_{10}^{b^*} - \hat{\Psi}_{01}^*W_{01}^{b^*} + \hat{\Psi}_{00}^*W_{00}^{b^*} + W_c^{b^*})\right\} \\
&= \frac{1}{(\Delta^*)^2}Cov\left\{\hat{\Psi}_{11}^*W_{11}^{ab^*} - \hat{\Psi}_{10}^*W_{10}^{ab^*} - \hat{\Psi}_{01}^*W_{01}^{ab^*} + \hat{\Psi}_{00}^*W_{00}^{ab^*} + W_c^{ab^*}, \hat{\Psi}_{11}^*W_{11}^{b^*} - \hat{\Psi}_{10}^*W_{10}^{b^*} - \hat{\Psi}_{01}^*W_{01}^{b^*} + \hat{\Psi}_{00}^*W_{00}^{b^*} + W_c^{b^*}\right\} \\
&= \frac{1}{(\Delta^*)^2}\Big[W_{11}^{ab^*}W_{11}^{b^*}Cov(\hat{\Psi}_{11}^*, \hat{\Psi}_{11}^*) - W_{11}^{ab^*}W_{10}^{b^*}Cov(\hat{\Psi}_{11}^*, \hat{\Psi}_{10}^*) - W_{11}^{ab^*}W_{01}^{b^*}Cov(\hat{\Psi}_{11}^*, \hat{\Psi}_{01}^*) + W_{11}^{ab^*}W_{00}^{b^*}Cov(\hat{\Psi}_{11}^*, \hat{\Psi}_{00}^*) \\
&\quad - W_{10}^{ab^*}W_{11}^{b^*}Cov(\hat{\Psi}_{10}^*, \hat{\Psi}_{11}^*) + W_{10}^{ab^*}W_{10}^{b^*}Cov(\hat{\Psi}_{10}^*, \hat{\Psi}_{10}^*) + W_{10}^{ab^*}W_{01}^{b^*}Cov(\hat{\Psi}_{10}^*, \hat{\Psi}_{01}^*) - W_{10}^{ab^*}W_{00}^{b^*}Cov(\hat{\Psi}_{10}^*, \hat{\Psi}_{00}^*) \\
&\quad - W_{01}^{ab^*}W_{11}^{b^*}Cov(\hat{\Psi}_{01}^*, \hat{\Psi}_{11}^*) + W_{01}^{ab^*}W_{10}^{b^*}Cov(\hat{\Psi}_{01}^*, \hat{\Psi}_{10}^*) + W_{01}^{ab^*}W_{01}^{b^*}Cov(\hat{\Psi}_{01}^*, \hat{\Psi}_{01}^*) - W_{01}^{ab^*}W_{00}^{b^*}Cov(\hat{\Psi}_{01}^*, \hat{\Psi}_{00}^*) \\
&\quad + W_{00}^{ab^*}W_{11}^{b^*}Cov(\hat{\Psi}_{00}^*, \hat{\Psi}_{11}^*) - W_{00}^{ab^*}W_{10}^{b^*}Cov(\hat{\Psi}_{00}^*, \hat{\Psi}_{10}^*) - W_{00}^{ab^*}W_{01}^{b^*}Cov(\hat{\Psi}_{00}^*, \hat{\Psi}_{01}^*) + W_{00}^{ab^*}W_{00}^{b^*}Cov(\hat{\Psi}_{00}^*, \hat{\Psi}_{00}^*)\Big] \\
&= \frac{1}{n(\Delta^*)^2}\Big[W_{11}^{ab^*}W_{11}^{b^*}\Psi_{11}^*(1-\Psi_{11}^*) + W_{11}^{ab^*}W_{10}^{b^*}\Psi_{11}^*\Psi_{10}^* + W_{11}^{ab^*}W_{01}^{b^*}\Psi_{11}^*\Psi_{01}^* - W_{11}^{ab^*}W_{00}^{b^*}\Psi_{11}^*\Psi_{00}^* \\
&\quad + W_{10}^{ab^*}W_{11}^{b^*}\Psi_{10}^*\Psi_{11}^* + W_{10}^{ab^*}W_{10}^{b^*}\Psi_{10}^*(1-\Psi_{10}^*) - W_{10}^{ab^*}W_{01}^{b^*}\Psi_{10}^*\Psi_{01}^* + W_{10}^{ab^*}W_{00}^{b^*}\Psi_{10}^*\Psi_{00}^* \\
&\quad + W_{01}^{ab^*}W_{11}^{b^*}\Psi_{01}^*\Psi_{11}^* - W_{01}^{ab^*}W_{10}^{b^*}\Psi_{01}^*\Psi_{10}^* + W_{01}^{ab^*}W_{01}^{b^*}\Psi_{01}^*(1-\Psi_{01}^*) + W_{01}^{ab^*}W_{00}^{b^*}\Psi_{01}^*\Psi_{00}^* \\
&\quad - W_{00}^{ab^*}W_{11}^{b^*}\Psi_{00}^*\Psi_{11}^* + W_{00}^{ab^*}W_{10}^{b^*}\Psi_{00}^*\Psi_{10}^* + W_{00}^{ab^*}W_{01}^{b^*}\Psi_{00}^*\Psi_{01}^* + W_{00}^{ab^*}W_{00}^{b^*}\Psi_{00}^*(1-\Psi_{00}^*)\Big]
\end{aligned}
$$

which proves the lemma. Now we have the following corollaries.

Corollary 5.1. A new natural estimator of the parameter $\pi_{A|B}$ is given by

$$\hat{\pi}_{A|B}^{Amy^*} = \frac{\hat{\pi}_{AB}^{Amy^*}}{\hat{\pi}_{B}^{Amy^*}} \tag{5.4}$$

The bias, to the first order of approximation, in the new estimator $\hat{\pi}_{A|B}^{Amy*}$ is given by

$$B\left(\hat{\pi}_{A|B}^{Amy^*}\right) = \pi_{A|B}\left[\frac{V(\hat{\pi}_{B}^{Amy^*})}{\pi_B^2} - \frac{Cov(\hat{\pi}_{AB}^{Amy^*}, \hat{\pi}_{B}^{Amy^*})}{\pi_{AB}\pi_B}\right] \tag{5.5}$$

and, the mean squared error, to the first order of approximation, of the new estimator $\hat{\pi}_{A|B}^{Amy}$ is given by

$$MSE\left(\hat{\pi}_{A|B}^{Amy^*}\right) = \pi_{A|B}^2\left[\frac{V(\hat{\pi}_{AB}^{Amy^*})}{\pi_{AB}^2} + \frac{V(\hat{\pi}_{B}^{Amy^*})}{\pi_B^2} - \frac{2Cov(\hat{\pi}_{B}^{Amy^*}, \hat{\pi}_{AB}^{Amy^*})}{\pi_B\pi_{AB}}\right] \tag{5.6}$$

Corollary 5.2. A natural new unbiased estimator of the parameter π_{A-B} is given by

$$\hat{\pi}_{A-B}^{Amy^*} = \hat{\pi}_{A}^{Amy^*} - \hat{\pi}_{AB}^{Amy^*} \tag{5.7}$$

The variance of the new estimator of $\hat{\pi}_{A-B}^{Amy*}$ is given by;

$$V\left(\hat{\pi}_{A-B}^{Amy^*}\right) = V\left(\hat{\pi}_{A}^{Amy^*}\right) + V\left(\hat{\pi}_{AB}^{Amy^*}\right) - 2Cov\left(\hat{\pi}_{A}^{Amy^*}, \hat{\pi}_{AB}^{Amy^*}\right) \tag{5.8}$$

Corollary 5.3. A natural new unbiased estimator of the parameter $\pi_{A\cup B}$ is given by

$$\hat{\pi}_{A\cup B}^{Amy^*} = \hat{\pi}_{A}^{Amy^*} + \hat{\pi}_{B}^{Amy^*} - \hat{\pi}_{AB}^{Amy^*} \tag{5.9}$$

The variance of the new estimator of $\hat{\pi}_{A\cup B}^{Amy}$ is given by;

$$\begin{aligned} V\left(\hat{\pi}_{A\cup B}^{Amy^*}\right) &= V\left(\hat{\pi}_{A}^{Amy^*}\right) + V\left(\hat{\pi}_{B}^{Amy^*}\right) + V\left(\hat{\pi}_{AB}^{Amy^*}\right) \\ &+2Cov\left(\hat{\pi}_{A}^{Amy^*}, \hat{\pi}_{B}^{Amy^*}\right) - 2Cov\left(\hat{\pi}_{A}^{Amy^*}, \hat{\pi}_{AB}^{Amy^*}\right) - 2Cov\left(\hat{\pi}_{B}^{Amy^*}, \hat{\pi}_{AB}^{Amy^*}\right) \end{aligned} \tag{5.10}$$

Corollary 5.4. A natural new estimator of the parameter $RR(B|A)$ is given by

$$\hat{RR}\left(B|A\right)_{Amy}^{*} = \frac{\hat{\pi}_{AB}^{Amy^*}\left(1-\hat{\pi}_{A}^{Amy^*}\right)}{\hat{\pi}_{A}^{Amy^*}\left(\hat{\pi}_{B}^{Amy^*} - \hat{\pi}_{AB}^{Amy^*}\right)} \tag{5.11}$$

The bias in the new estimator $\hat{RR}(B|A)_{Amy}$, to the first order of approximation, is given by

$$\begin{aligned} B\left(\hat{RR}(B\mid A)_{Amy}^{*}\right) = RR\Bigg[&\frac{V\left(\hat{\pi}_{A}^{Amy^*}\right)}{\pi_A^2\left(1-\pi_A\right)} + \frac{V\left(\hat{\pi}_{B}^{Amy^*}\right)}{\left(\pi_B-\pi_{AB}\right)^2} + \frac{\pi_B V\left(\hat{\pi}_{AB}^{Amy^*}\right)}{\pi_{AB}\left(\pi_B-\pi_{AB}\right)^2} + \frac{Cov\left(\hat{\pi}_{A}^{Amy^*}, \hat{\pi}_{B}^{Amy^*}\right)}{\pi_A\left(1-\pi_A\right)\left(\pi_B-\pi_{AB}\right)} \\ &- \frac{\pi_B Cov\left(\hat{\pi}_{A}^{Amy^*}, \hat{\pi}_{AB}^{Amy^*}\right)}{\pi_A\pi_{AB}\left(1-\pi_A\right)\left(\pi_B-\pi_{AB}\right)} - \frac{\left(\pi_B+\pi_{AB}\right)Cov\left(\hat{\pi}_{B}^{Amy^*}, \hat{\pi}_{AB}^{Amy^*}\right)}{\pi_{AB}\left(\pi_B-\pi_{AB}\right)^2}\Bigg] \end{aligned} \tag{5.12}$$

The mean squared error of the new estimator $\hat{RR}(B|A)_{Amy}^{*}$, to the first order of approximation, is given by

$$\begin{aligned} &MSE\left(\hat{RR}(B\mid A)_{Amy}^{*}\right) \\ &= RR^2\Bigg[\frac{\pi_B^2 V\left(\hat{\pi}_{AB}^{Amy^*}\right)}{\pi_{AB}^2\left(\pi_B-\pi_{AB}\right)^2} + \frac{V\left(\hat{\pi}_{A}^{Amy^*}\right)}{\pi_A^2\left(1-\pi_A\right)^2} + \frac{V\left(\hat{\pi}_{B}^{Amy^*}\right)}{\left(\pi_B-\pi_{AB}\right)^2} - \frac{2\pi_B Cov\left(\hat{\pi}_{A}^{Amy^*}, \hat{\pi}_{AB}^{Amy^*}\right)}{\pi_A\pi_{AB}\left(1-\pi_A\right)\left(\pi_B-\pi_{AB}\right)} \\ &- \frac{2\pi_B Cov\left(\hat{\pi}_{B}^{Amy^*}, \hat{\pi}_{AB}^{Amy^*}\right)}{\pi_{AB}\left(\pi_B-\pi_{AB}\right)^2} + \frac{2Cov\left(\hat{\pi}_{A}^{Amy^*}, \hat{\pi}_{B}^{Amy^*}\right)}{\pi_A\left(1-\pi_A\right)\left(\pi_B-\pi_{AB}\right)}\Bigg] \end{aligned} \tag{5.13}$$

Corollary 5.5. A natural new estimator of the parameter ρ_{AB} is given by

$$\hat{\rho}_{AB}^{Amy^*} = \frac{\hat{\pi}_{AB}^{Amy^*} - \hat{\pi}_{A}^{Amy^*}\hat{\pi}_{B}^{Amy^*}}{\sqrt{\hat{\pi}_{A}^{Amy^*}(1-\hat{\pi}_{A}^{Amy^*})}\sqrt{\hat{\pi}_{B}^{Amy^*}(1-\hat{\pi}_{B}^{Amy^*})}} \tag{5.14}$$

The bias in the new estimator $\hat{\rho}_{AB}^{Amy^*}$, to the first order of approximation, is:

$$\begin{aligned} &B\left(\hat{\rho}_{AB}^{Amy^*}\right) \\ &= \rho_{AB}\left[F_4\frac{V\left(\hat{\pi}_{A}^{Amy^*}\right)}{\pi_A^2} + F_5\frac{V\left(\hat{\pi}_{B}^{Amy^*}\right)}{\pi_B^2} + F_6\frac{Cov\left(\hat{\pi}_{A}^{Amy^*}, \hat{\pi}_{B}^{Amy^*}\right)}{\pi_A\pi_B} - F_7\frac{Cov\left(\hat{\pi}_{A}^{Amy^*}, \hat{\pi}_{AB}^{Amy^*}\right)}{\pi_A\pi_{AB}} - F_8\frac{Cov\left(\hat{\pi}_{B}^{Amy^*}, \hat{\pi}_{AB}^{Amy^*}\right)}{\pi_B\pi_{AB}}\right] \end{aligned} \tag{5.15}$$

The mean squared error of the estimator $\hat{\rho}_{AB}^{Amy^*}$, to the first order of approximation, is given by

$$MSE(\hat{\rho}_{AB}^{Amy^*}) = \rho_{AB}^2\left[\begin{aligned} &F_1^2\frac{V\left(\hat{\pi}_{AB}^{Amy^*}\right)}{\pi_{AB}^2} + F_2^2\frac{V\left(\hat{\pi}_{A}^{Amy^*}\right)}{\pi_A^2} + F_3^2\frac{V\left(\hat{\pi}_{B}^{Amy^*}\right)}{\pi_B^2} \\ &- \frac{2F_1F_2Cov\left(\hat{\pi}_{A}^{Amy^*}, \hat{\pi}_{AB}^{Amy^*}\right)}{\pi_A\pi_{AB}} - \frac{2F_1F_3Cov\left(\hat{\pi}_{B}^{Amy^*}, \hat{\pi}_{AB}^{Amy^*}\right)}{\pi_B\pi_{AB}} + \frac{2F_2F_3Cov\left(\hat{\pi}_{A}^{Amy^*}, \hat{\pi}_{B}^{Amy^*}\right)}{\pi_A\pi_B} \end{aligned}\right] \tag{5.16}$$

where

$$F_1 = \frac{\pi_{AB}}{\pi_{AB} - \pi_A\pi_B}, \tag{5.17}$$

$$F_2 = \frac{\pi_A \pi_B}{\pi_{AB} - \pi_A \pi_B} + \frac{(1-2\pi_A)}{2(1-\pi_A)}, \tag{5.18}$$

$$F_3 = \frac{\pi_A \pi_B}{\pi_{AB} - \pi_A \pi_B} + \frac{(1-2\pi_B)}{2(1-\pi_B)}, \tag{5.19}$$

$$F_4 = \frac{\pi_A \pi_B (1-2\pi_A)}{2(1-\pi_A)(\pi_{AB} - \pi_A \pi_B)} + \frac{\pi_A}{2(1-\pi_A)} + \frac{3(1-2\pi_A)^2}{8(1-\pi_A)^2}, \tag{5.20}$$

$$F_5 = \frac{\pi_A \pi_B (1-2\pi_B)}{2(1-\pi_B)(\pi_{AB} - \pi_A \pi_B)} + \frac{\pi_B}{2(1-\pi_B)} + \frac{3(1-2\pi_B)^2}{8(1-\pi_B)^2}, \tag{5.21}$$

$$F_6 = \frac{1-2\pi_A - 2\pi_B + 4\pi_A\pi_B}{4(1-\pi_A)(1-\pi_B)} - \frac{\pi_A \pi_B}{\pi_{AB} - \pi_A \pi_B} + \frac{\pi_A \pi_B}{2(\pi_{AB} - \pi_A \pi_B)}\left(\frac{1-2\pi_A}{1-\pi_A} + \frac{1-2\pi_B}{1-\pi_B}\right), \tag{5.22}$$

$$F_7 = \frac{\pi_{AB}(1-2\pi_A)}{2(1-\pi_A)(\pi_{AB} - \pi_A \pi_B)}, \tag{5.23}$$

and

$$F_8 = \frac{\pi_{AB}(1-2\pi_B)}{2(1-\pi_B)(\pi_{AB} - \pi_A \pi_B)} \tag{5.24}$$

Corollary 5.6. A natural new estimator of the parameter $\pi_d = \pi_A - \pi_B$ is given by

$$\hat{\pi}_d^{Amy^*} = \hat{\pi}_A^{Amy^*} - \hat{\pi}_B^{Amy^*} \tag{5.25}$$

The variance of the difference estimator $\hat{\pi}_d^{Amy*}$ is given by

$$V(\hat{\pi}_d^{Amy^*}) = V(\hat{\pi}_A^{Amy^*}) + V(\hat{\pi}_B^{Amy^*}) - 2Cov(\hat{\pi}_A^{Amy^*}, \hat{\pi}_B^{Amy^*}) \tag{5.26}$$

In the next section, we investigate the percent relative efficiency of the proposed estimators with respect to the estimators of other parameters of Lee et al. (2013) for the cross model.

6. Relative Efficiency of Estimators of Other Parameters

The percent relative efficiencies of the proposed estimators $\hat{\pi}_{A|B}^{Amy*}$, $\hat{\pi}_{A-B}^{Amy*}$, $\hat{\pi}_{A\cup B}^{Amy*}$, $\hat{RR}(B|A)_{Amy}^*$, $\hat{\rho}_{AB}^{Amy*}$ and $\hat{\pi}_d^{Amy*}$ with respect to the Lee et al. (2013) estimators $\hat{\pi}_{A|B}^*$, $\hat{\pi}_{A-B}^*$, $\hat{\pi}_{A\cup B}^*$, $\hat{RR}(B|A)^*$, $\hat{\rho}_{AB}^*$ and $\hat{\pi}_d^*$ are respectively defined as

$$RE(4) = \frac{V(\hat{\pi}_{A|B}^*)}{V(\hat{\pi}_{A|B}^{Amy^*})} \times 100\% \tag{6.1}$$

$$RE(5) = \frac{V(\hat{\pi}_{A-B}^*)}{V(\hat{\pi}_{A-B}^{Amy^*})} \times 100\% \tag{6.2}$$

$$RE(6) = \frac{V(\hat{\pi}_{A\cup B}^*)}{V(\hat{\pi}_{A\cup B}^{Amy^*})} \times 100\% \tag{6.3}$$

$$RE(7) = \frac{MSE(R\hat{R}(B \mid A)^*)}{MSE(R\hat{R}(B \mid A))_{Amy}^*} \times 100\% \tag{6.4}$$

$$RE(8) = \frac{MSE(\hat{\rho}_{AB}^*)}{MSE(\hat{\rho}_{AB}^{Amy^*})} \times 100\% \tag{6.5}$$

and

$$RE(9) = \frac{V(\hat{\pi}_d^*)}{V(\hat{\pi}_d^{Amy^*})} \times 100\% \tag{6.6}$$

The same SAS Codes provided in the Appendix A also give the values of percent relative efficiencies RE(4), RE(5), RE(6), RE(7), RE(8), and RE(9). The detailed results are given in Appendix C in Table C1. For the same protection level, and same choice of other parameters in the study, the summary statistics of the percent relative efficiencies RE(4) to RE(9) are given in Table 5.

The values of RE(4), RE(5), RE(6), RE(7), RE(8) and RE(9) are higher than 100% indicating that the proposed unrelated question model performs efficiently for estimating all the parameters considered here. For $w = 0.1$ in Table 5 there are 237 such cases, the average value of RE(4) is 83.57% with a standard deviation of 13.60%, a minimum value of 55.80% and a maximum value of 119.39%; the average value of RE(5) is 59.24% with a standard deviation of 6.08%, a minimum value of 44.19% and a maximum value of 69.60%; the average value of RE(6) is 111.80% with a standard deviation of 8.60%, a minimum value of 93.07% and a maximum value of 122.79%; the average value of RE(7) is 59.42% with a standard deviation of 5.40%, a minimum value of 50.04% and a maximum value of 70.52%; the average value of RE(8) is 63.00% with a standard deviation of 7.18%, a minimum value of 50.32% and a maximum value of 80.90%; and the average value of RE(9) is 65.51% with a standard deviation of 4.66%, a minimum value of 55.00% and a maximum value of 74.30%. A pictorial presentation of such results is given in Figure 7.

In the same way for the value of $w = 0.9$ there are 724 such cases where all the proposed estimators are more efficient than the corresponding estimators of Lee et al. (2013).

Again, for demonstration purposes, we described a few results from the detailed results in Appendix C as cited in Table 6. For $\pi_A = 0.1$, $\pi_B = 0.1$ and $\pi_{AB} = 0.05$, the average value of RE(4) is 83.72% with a standard deviation of 11.02%, a minimum value of 69.56% and a maximum value of 96.21%; the average value of RE(5) is 72.38% with a standard deviation of 14.13%, a minimum value of

Table 5: RE(4) to RE(9) values for different values of w.

Variable	w	Freq	Mean	StDev	Minimum	Maximum
RE(4)	0.1	237	83.57	13.60	55.80	119.39
	0.3	520	81.97	12.87	57.34	122.67
	0.5	731	84.85	11.97	61.14	122.74
	0.7	731	91.75	8.44	74.63	118.14
	0.9	724	97.60	3.35	90.50	107.63
RE(5)	0.1	237	59.24	6.08	44.19	69.60
	0.3	520	62.03	5.97	48.11	74.81
	0.5	731	67.70	6.24	52.93	80.64
	0.7	731	78.60	4.16	68.17	87.34
	0.9	724	91.99	1.57	87.80	95.32
RE(6)	0.1	237	111.80	8.60	93.07	122.79
	0.3	520	105.62	7.38	97.40	120.54
	0.5	731	102.87	7.08	89.90	116.86
	0.7	731	103.06	4.29	95.04	111.53
	0.9	724	101.51	1.46	98.73	104.35
RE(7)	0.1	237	59.42	5.40	50.04	70.52
	0.3	520	59.56	6.63	50.05	77.17
	0.5	731	64.92	6.98	50.42	84.38
	0.7	731	76.31	5.01	66.34	91.72
	0.9	724	90.91	2.15	86.37	98.17
RE(8)	0.1	237	63.00	7.18	50.32	80.90
	0.3	520	64.21	7.17	50.08	87.95
	0.5	731	69.43	7.84	51.77	94.72
	0.7	731	80.21	5.90	66.92	100.02
	0.9	724	92.82	2.64	86.93	101.61
RE(9)	0.1	237	65.51	4.66	55.00	74.30
	0.3	520	69.41	4.54	60.72	80.00
	0.5	731	75.86	4.53	64.19	85.71
	0.7	731	85.30	2.87	77.56	91.43
	0.9	724	95.04	1.01	92.20	97.14

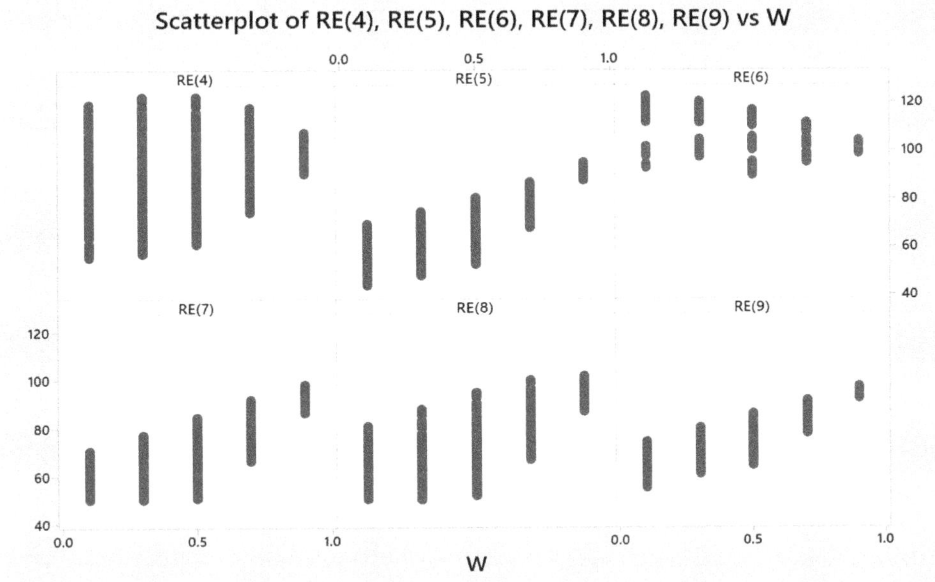

Figure 7: RE(4) to RE(9) for different values of w.

55.44% and a maximum value of 89.74%; the average value of RE(6) is 95.39% with a standard deviation of 3.60%, a minimum value of 89.90% and a maximum value of 99.52%; the average value of RE(7) is 68.87% with a standard deviation of 15.71%, a minimum value of 50.41% and a maximum value of 88.20%; the average value of RE(8) is 80.17% with a standard deviation of 12.57%, a minimum value of 64.42% and a maximum value of 94.71%; and the average value of RE(9) is 79.22% with a standard deviation of 11.59%, a minimum value of 64.68% and a maximum value of 93.10%. Likewise, the results for $\pi_A = 0.1$, $\pi_B = 0.15$ and $\pi_{AB} = 0.05$ can be seen from Table 6. In addition, results for any such combination of the three proportions π_A, π_B and π_{Ab} can be extracted, if needed, from the detailed results give in Appendix C in Table C1.

Thus, the proposed model is found to be quite helpful in estimating such tiny proportions of sensitive attributes and their overlaps for estimating all the parameters considered. Table 7 helps one to make a choice of the proportion of unrelated question π_y. Again for $\pi_y = 1$ there are combinations of parameters of *freq* = 991 where the proposed estimators of the other parameters being considered are more efficient and more protective than the crossed model of Lee et al. (2013), and it again ensures that in the proposed model "Forced Yes" response could be used instead of having to find an appropriate unrelated characteristic in a population under study.

Table 6: Summary of the values of RE(4), RE(5), RE(6), RE(7), RE(8) and RE(9) for different choices of π_A, π_B and π_{AB}.

Results for $\pi_A = 0.10$ and $\pi_B = 0.10$

Variable	π_{AB}	Freq	Mean	StDev	Minimum	Maximum
RE(4)	0.05	9	83.72	11.02	69.56	96.21
RE(5)	0.05	9	72.38	14.13	55.44	89.74
RE(6)	0.05	9	95.39	3.60	89.90	99.52
RE(7)	0.05	9	68.87	15.71	50.41	88.20
RE(8)	0.05	9	80.17	12.57	64.42	94.71
RE(9)	0.05	9	79.22	11.59	64.68	93.10

Results for $\pi_A = 0.10$ and $\pi_B = 0.15$

Variable	π_{AB}	Freq	Mean	StDev	Minimum	Maximum
RE(4)	0.05	16	95.02	6.98	83.08	102.89
RE(5)	0.05	16	71.47	14.32	48.79	89.89
RE(6)	0.05	16	97.877	3.935	90.252	102.532
RE(7)	0.05	16	70.77	14.93	50.20	91.28
RE(8)	0.05	16	83.76	11.41	67.22	98.92
RE(9)	0.05	16	79.76	11.28	61.33	94.12

Table 7: RE(4), RE(5), RE(6), RE(7), RE(8) and RE(9) for different values of π_y.

Variable	π_A	*freq*	Mean	StDev	Minimum	Maximum
RE(4)	0.90	966	89.433	11.641	58.241	122.735
	0.95	986	89.070	11.651	56.430	122.047
	1.00	991	88.775	11.640	55.798	121.438
RE(5)	0.90	966	75.237	12.583	44.895	95.321
	0.95	986	74.627	12.724	44.189	95.129
	1.00	991	74.243	12.739	44.280	94.906
RE(6)	0.90	966	104.04	6.30	91.76	122.79
	0.95	986	103.81	6.39	91.00	122.34
	1.00	991	103.53	6.39	89.90	121.83
RE(7)	0.90	966	73.228	13.187	50.041	98.165
	0.95	986	72.699	13.208	50.050	97.865
	1.00	991	72.338	13.189	50.101	97.571
RE(8)	0.90	966	76.775	12.699	50.879	101.610
	0.95	986	76.409	12.721	50.078	101.384
	1.00	991	76.095	12.718	50.471	101.164
RE(9)	0.90	966	81.552	10.494	56.540	97.142
	0.95	986	80.891	10.771	55.067	97.035
	1.00	991	80.419	10.953	54.999	96.929

This finding through a simulation study enhances the usefulness of the proposed model in practice. A behavior of the values of RE(4), RE(5), RE(6), RE(7), RE(8) and RE(9) versus π_y are shown in Figure 8. In Figure 8 the notation Y stands for π_y.

Table 8 provides a summary of results for different choices of P_1 and T_1. For $P_1 = 0.6$ and $T_1 = 0.6$, the average value of RE(4) is 83.96% with a standard deviation of 10.38%, a minimum value of 61.14% and a maximum value of 100.67%; the average value of RE(5) is 74.52% with a standard deviation of 12.44%, a minimum value of 52.93% and a maximum value of 91.83%; the average value of RE(6) is 96.43% with a standard deviation of 2.83%, a minimum value of 89.90% and a maximum value of 100.13%; the average value of RE(7) is 71.60% with a standard deviation of 13.30%, a minimum value of 50.42% and a maximum value of 90.49%; the average value of RE(8) is 74.63% with a standard deviation of 12.68%, a minimum value of 51.77% and a maximum

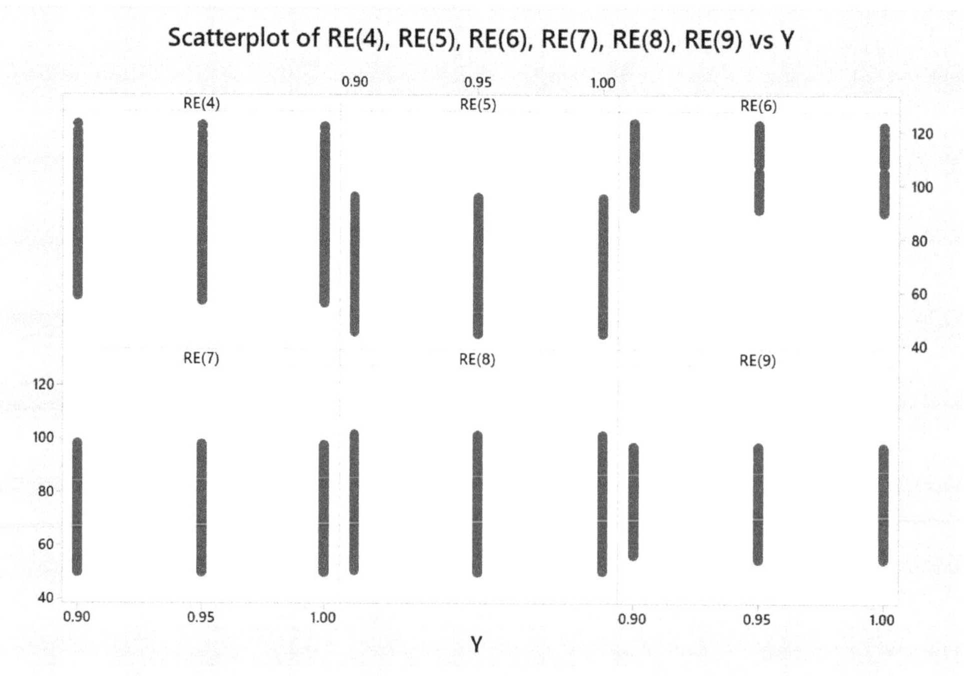

Figure 8: RE(4), RE(5), RE(6), RE(7), RE(8), and RE(9) versus π_y.

value of 94.84%; and the average value of RE(9) is 81.79% with a standard deviation of 9.74%, a minimum value of 64.19% and a maximum value of 95.18%. Likewise, other results from Table 8 can be interpreted.

From this simulation study we conclude that if one makes a new model more efficient and more protective than the crossed model for estimating the three parameters π_A, π_B and π_{AB}, then it may or may not produce efficient estimators of other parameters such as $\pi_{A|B}$, π_{A-B}, $\pi_{A\cup B}$, $RR(B|A)$ and ρ_{AB}.
It the next section, we develop the Cramer-Rao lower bounds of variances-covariance for verifying the results.

Table 8: RE(1), RE(2), RE(3), RP(1), RP(2) and RP(3) for different P_1 and T_1.

Results for $P_1 = 0.6$

Variable	T_1	Freq	Mean	StDev	Minimum	Maximum
RE(4)	0.6	495	83.963	10.381	61.135	100.666
	0.7	804	85.621	11.053	55.798	107.654
RE(5)	0.6	495	74.524	12.441	52.932	91.830
	0.7	804	71.431	14.229	44.189	93.113
RE(6)	0.6	495	96.431	2.833	89.902	100.131
	0.7	804	101.74	1.78	93.07	105.94
RP(7)	0.6	495	71.597	13.298	50.415	90.486
	0.7	804	71.257	13.608	50.101	94.038
RP(8)	0.6	495	74.626	12.679	51.772	94.839
	0.7	804	74.909	13.313	50.078	98.916
RP(9)	0.6	495	81.790	9.741	64.192	95.179
	0.7	804	80.008	11.364	54.999	96.061

Results for $P_1 = 0.7$

Variable	T_1	Freq	Mean	StDev	Minimum	Maximum
RE(4)	0.6	766	88.637	11.624	57.341	112.883
	0.7	878	95.553	9.821	65.911	122.735
RE(5)	0.6	766	76.035	11.867	50.418	93.971
	0.7	878	76.621	11.372	53.435	95.321
RE(6)	0.6	766	101.69	1.79	93.07	105.83
	0.7	878	111.65	5.03	103.01	122.79
RP(7)	0.6	766	71.204	13.754	50.041	93.989
	0.7	878	76.120	11.562	54.347	98.165
RP(8)	0.6	766	75.622	12.862	50.882	98.916
	0.7	878	79.522	11.439	53.739	101.610
RP(9)	0.6	766	80.393	11.081	54.999	96.061
	0.7	878	81.821	10.325	60.136	97.142

7. Cramer-Rao Lower Bounds of Variance-Covariance

Theorem 7.1. The maximum likelihood estimates $\hat{\pi}_{AB}^{mle*}$, $\hat{\pi}_{A}^{mle*}$, and $\hat{\pi}_{B}^{mle*}$ of π_{AB}, π_A, and π_B are a solution to the three non-linear equations given as:

$$\frac{\hat{\Psi}^*_{11}}{\Psi^*_{11}} - \frac{\hat{\Psi}^*_{10}}{\Psi^*_{10}} - \frac{\hat{\Psi}^*_{01}}{\Psi^*_{01}} + \frac{\hat{\Psi}^*_{00}}{\Psi^*_{00}} = 0 \tag{7.1}$$

$$\frac{\hat{\Psi}^*_{11}}{\Psi^*_{11}} C_2^* + \frac{\hat{\Psi}^*_{10}}{\Psi^*_{10}} C_4^* - \frac{\hat{\Psi}^*_{01}}{\Psi^*_{01}} C_6^* - \frac{\hat{\Psi}^*_{00}}{\Psi^*_{00}} C_8^* = 0, \tag{7.2}$$

and

$$\frac{\hat{\Psi}^*_{11}}{\Psi^*_{11}}C^*_3 - \frac{\hat{\Psi}^*_{10}}{\Psi^*_{10}}C^*_5 + \frac{\hat{\Psi}^*_{01}}{\Psi^*_{01}}C^*_7 - \frac{\hat{\Psi}^*_{00}}{\Psi^*_{00}}C^*_9 = 0 \tag{7.3}$$

Proof. In case of the proposed model, consider the likelihood function given by:

$$L = \binom{n}{n_{11}\ n_{10}\ n_{01}\ n_{00}} \left(\Psi^*_{11}\right)^{n_{11}} \left(\Psi^*_{10}\right)^{n_{10}} \left(\Psi^*_{01}\right)^{n_{01}} \left(\Psi^*_{00}\right)^{n_{00}} \tag{7.4}$$

Taking log on both sides of it, we get:

$$\log L = \log C + n_{11}\log\Psi^*_{11} + n_{10}\log\Psi^*_{10} + n_{01}\log\Psi^*_{01} + n_{00}\log\Psi^*_{00}$$

$$= \log(C) + n\begin{bmatrix} \hat{\Psi}^*_{11}\log\left\{C^*_1\pi_{AB} + C^*_2\pi_A + C^*_3\pi_B + H^*_1\right\} + \hat{\Psi}^*_{10}\log\left\{-C^*_1\pi_{AB} - C^*_4\pi_A - C^*_5\pi_B + H^*_2\right\} \\ +\hat{\Psi}^*_{01}\log\left\{-C^*_1\pi_{AB} - C^*_6\pi_A - C^*_7\pi_B + H^*_3\right\} + \hat{\Psi}^*_{00}\log\left\{C^*_1\pi_{AB} - C^*_8\pi_A - C^*_9\pi_B + H^*_4\right\} \end{bmatrix} \tag{7.5}$$

On taking the first order partial derivates of log L with respect to π_{AB}, π_A and π_B respectively, we have

$$\frac{\partial \log L}{\partial \pi_{AB}} = nC^*_1\left[\frac{\hat{\Psi}^*_{11}}{\Psi^*_{11}} - \frac{\hat{\Psi}^*_{10}}{\Psi^*_{10}} - \frac{\hat{\Psi}^*_{01}}{\Psi^*_{01}} + \frac{\hat{\Psi}^*_{00}}{\Psi^*_{00}}\right] \tag{7.6}$$

$$\frac{\partial \log L}{\partial \pi_{A}} = n\left[\frac{\hat{\Psi}^*_{11}}{\Psi^*_{11}}C^*_2 - \frac{\hat{\Psi}^*_{10}}{\Psi^*_{10}}C^*_4 - \frac{\hat{\Psi}^*_{01}}{\Psi^*_{01}}C^*_6 - \frac{\hat{\Psi}^*_{00}}{\Psi^*_{00}}C^*_8\right] \tag{7.7}$$

and

$$\frac{\partial \log L}{\partial \pi_{B}} = n\left[\frac{\hat{\Psi}^*_{11}}{\Psi^*_{11}}C^*_3 - \frac{\hat{\Psi}^*_{10}}{\Psi^*_{10}}C^*_5 - \frac{\hat{\Psi}^*_{01}}{\Psi^*_{01}}C^*_7 - \frac{\hat{\Psi}^*_{00}}{\Psi^*_{00}}C^*_9\right] \tag{7.8}$$

On setting $\frac{\partial \log L}{\partial \pi_{AB}} = 0$, $\frac{\partial \log L}{\partial \pi_{A}} = 0$ and $\frac{\partial \log L}{\partial \pi_{B}} = 0$, we have the theorem.

Theorem 7.2. The lower bounds of variance-covariance of the maximum likelihood estimates $\hat{\pi}^{mle^*}_{AB}$, $\hat{\pi}^{mle^*}_{A}$ and $\hat{\pi}^{mle^*}_{B}$ of π_{AB}, π_A, and π_B for the proposed model are, respectively, given by:

$$V\begin{bmatrix} \hat{\pi}^{mle^*}_{AB} \\ \hat{\pi}^{mle^*}_{A} \\ \hat{\pi}^{mle^*}_{B} \end{bmatrix} \geq \frac{1}{\Delta}\begin{bmatrix} a_{11} & a_{12} & a_{13} \\ a_{21} & a_{22} & a_{23} \\ a_{31} & a_{32} & a_{33} \end{bmatrix} \tag{7.9}$$

where the inequality sign ($\geq$) has its usual meaning, and Δ and a_{ij} are given by

$$\Delta = n(gb^2 - 2bcf + dc^2 + af^2 - adg)$$

$$a_{11} = f^2 - gd\,,$$

$$a_{22} = c^2 - ag\,,$$

$$a_{33} = b^2 - ad\,,$$

$$a_{12} = a_{21} = bg - cf\,,$$

$$a_{13} = a_{31} = cd - bf$$

and

$$a_{23} = a_{32} = af - bc\,.$$

Proof. By the definition of Cramer-Rao lower bound of variance, we have

$$V\begin{bmatrix}\hat{\pi}_{AB}^{mle^*}\\ \hat{\pi}_{A}^{mle^*}\\ \hat{\pi}_{B}^{mle^*}\end{bmatrix} \geq -\begin{bmatrix} E(\frac{\partial^2 \log L}{\partial \pi_{AB}^2}) & E(\frac{\partial^2 \log L}{\partial \pi_A \partial \pi_{AB}}) & E(\frac{\partial^2 \log L}{\partial \pi_B \partial \pi_{AB}}) \\ E(\frac{\partial^2 \log L}{\partial \pi_A \partial \pi_{AB}}) & E(\frac{\partial^2 \log L}{\partial \pi_A^2}) & E(\frac{\partial^2 \log L}{\partial \pi_B \partial \pi_A}) \\ E(\frac{\partial^2 \log L}{\partial \pi_B \partial \pi_{AB}}) & E(\frac{\partial^2 \log L}{\partial \pi_B \partial \pi_A}) & E(\frac{\partial^2 \log L}{\partial \pi_B^2}) \end{bmatrix}^{-1} = \begin{bmatrix} a & b & c \\ b & d & f \\ c & f & g \end{bmatrix}^{-1}, \tag{7.10}$$

where

$$a = -E\left[\frac{\partial^2 \log L}{\partial \pi_{AB}^2}\right] = n(C_1^*)^2\left[\frac{1}{\Psi_{11}^*}+\frac{1}{\Psi_{10}^*}+\frac{1}{\Psi_{01}^*}+\frac{1}{\Psi_{00}^*}\right], \tag{7.11}$$

$$b = -E\left[\frac{\partial^2 \log L}{\partial \pi_A \partial \pi_{AB}}\right] = nC_1^*\left[\frac{C_2^*}{\Psi_{11}^*}-\frac{C_4^*}{\Psi_{10}^*}-\frac{C_6^*}{\Psi_{01}^*}-\frac{C_8^*}{\Psi_{00}^*}\right], \tag{7.12}$$

$$c = -E\left[\frac{\partial^2 \log L}{\partial \pi_B \partial \pi_{AB}}\right] = nC_1^*\left[\frac{C_3^*}{\Psi_{11}^*}+\frac{C_5^*}{\Psi_{10}^*}+\frac{C_7^*}{\Psi_{01}^*}-\frac{C_9^*}{\Psi_{00}^*}\right], \tag{7.13}$$

$$d = -E\left[\frac{\partial^2 \log L}{\partial \pi_A^2}\right] = n\left[\frac{(C_2^*)^2}{\Psi_{11}^*}+\frac{(C_4^*)^2}{\Psi_{10}^*}+\frac{(C_6^*)^2}{\Psi_{01}^*}+\frac{(C_8^*)^2}{\Psi_{00}^*}\right], \tag{7.14}$$

$$f = -E\left[\frac{\partial^2 \log L}{\partial \pi_B \partial \pi_A}\right] = n\left[\frac{C_2^*C_3^*}{\Psi_{11}^*}+\frac{C_4^*C_5^*}{\Psi_{10}^*}+\frac{C_6^*C_7^*}{\Psi_{01}^*}+\frac{C_8^*C_9^*}{\Psi_{00}^*}\right], \tag{7.15}$$

and

$$g = -E\left[\frac{\partial^2 \log L}{\partial \pi_B^2}\right] = n\left[\frac{(C_3^*)^2}{\Psi_{11}^*}+\frac{(C_5^*)^2}{\Psi_{10}^*}+\frac{(C_7^*)^2}{\Psi_{01}^*}+\frac{(C_9^*)^2}{\Psi_{00}^*}\right], \tag{7.16}$$

which proves the theorem.

8. SAS Cods Output for the Cramer-Rao Lower Bounds

We also included the Cramer-Rao lower bounds of variance and covariance matrices in the SAS Codes. We abbreviate for our convenience $V(\hat{\pi}_A^{Amy^*}) = V_1^*$, $V(\hat{\pi}_B^{Amy^*}) = V_2^*$, $V(\hat{\pi}_{AB}^{Amy^*}) = V_3^*$, $Cov(\hat{\pi}_A^{Amy^*}, \hat{\pi}_B^{Amy^*}) = C_{12}^*$, $Cov(\hat{\pi}_A^{Amy^*}, \hat{\pi}_{AB}^{Amy^*}) = C_{13}^*$ and $Cov(\hat{\pi}_B^{Amy^*}, \hat{\pi}_{AB}^{Amy^*}) = C_{23}^*$ for the closed form of the estimators $\hat{\pi}_A^{Amy^*}$, $\hat{\pi}_B^{Amy^*}$ and $\hat{\pi}_{AB}^{Amy^*}$ and $V_1^{LB^*}$, $V_2^{LB^*}$, $V_3^{LB^*}$, $C_{12}^{LB^*}$, $C_{13}^{LB^*}$ and $C_{23}^{LB^*}$ as the corresponding Cramer-Rao lower-bounds entries from the variance co-variance matrix. We provide the first five outcomes from the SAS Codes in Table 9 and Table 10. It is found that the corresponding Cramer-Rao lower bounds of variance-covariance matrix obtained from the maximum-likelihood estimates match with the variance-covariance expressions of the closed forms of the estimators.

Table 9: The variance and covariance from the closed forms of the estimators.

Obs	V_3^*	V_1^*	V_2^*	C_{12}^*	C_{13}^*	C_{23}^*
1	0.05	0.20	0.15	0.082	0.140	0.137
2	0.05	0.20	0.20	0.071	0.169	0.134
3	0.05	0.25	0.15	0.073	0.136	0.155
4	0.05	0.30	0.15	0.065	0.132	0.173
5	0.15	0.30	0.20	0.151	0.188	0.222

Table 10: The Cramer-Rao lower bounds of variance-covariance.

Obs	$V_3^{LB^*}$	$V_1^{LB^*}$	$V_2^{LB^*}$	$C_{12}^{LB^*}$	$C_{13}^{LB^*}$	$C_{23}^{LB^*}$
1	0.05	0.20	0.15	0.082	0.140	0.137
2	0.05	0.20	0.20	0.071	0.169	0.134
3	0.05	0.25	0.15	0.073	0.136	0.155
4	0.05	0.30	0.15	0.065	0.132	0.173
5	0.15	0.30	0.20	0.151	0.188	0.222

One can see that the Cramer-Rao lower bounds of variance-covariance are quite helpful in checking the derivations of the variance-covariance of the closed forms of the estimators. If $w = 1$ the Cramer-Rao lower bounds of variance-covariance match with those obtained by Lee, Sedory and Singh (2016).

9. Survey Data Application

Electronic cigarettes (e-cigarettes) have been more and more popular in the world. Lots of researchers are interested in whether the use of e-cigarettes is associated with the conventional cigarette. We considered the new designed methodology by using a new unrelated question randomized response technique to investigate whether the use of e-cigarette and conventional cigarette are related. In the study design, every respondent was explained how to use the designed device which consists of five decks of cards in five colors: green, first-pair (blue and white), and second pair (yellow and pink). The procedure was as follows: every respondent was requested to draw one card from the green deck to decide if he/she had to use the first pair or the second pair of decks. In the first pair, the blue-deck was consisting of cards bearing two sets of questions: 14 cards with the statement, "Over the past year, Have you ever used Conventional Cigarette?" and 6 cards with the statement, "Over the past year, Have you never used E-Cigarette?". The white-deck was consisting of cards bearing two-types of questions: 14 cards with the statement, "Over the past year, Have you ever used E-cigarette?", and 6 cards with the statement, "Over the past year, Have you never used Conventional Cigarette?". In the second pair, the yellow-deck was consisting of cards bearing two statements: 14 cards with the statement, "Over the past year, Have you ever used Conventional Cigarette?" and 6 cards with the statement "Report Yes". The pink-deck was consisting of cards bearing two statements: 14 cards with the statement, "Over the past year, Have you ever used an E-Cigarette?" and 6 cards with the statement "Report Yes". During three days of data collection, a total of 121 students participated in this survey. Using the proposed DMDM-II model, and the black box estimates of the true proportion of the required parameters are given in Table 11.

Table 11: Real survey results.

DMDM-II Estimates			Black Box Estimates		
$\hat{\pi}_A^{Amy*}$	$\hat{\pi}_B^{Amy*}$	$\hat{\pi}_{AB}^{Amy*}$	$\hat{\pi}_A^{bb}$	$\hat{\pi}_B^{bb}$	$\hat{\pi}_{AB}^{bb}$
0.209	0.218	0.108	0.198	0.231	0.091

From Table 11, the DMDM-II estimate of the proportion of conventional cigarettes is 0.209, that of e-cigarette users is 0.218, and that of both e-cigarette and conventional cigarette users is 0.108.

The black box estimates of the proportion of conventional cigarettes is 0.198, that of e-cigarette users is 0.231, and that of both e-cigarette and conventional cigarette users is 0.091.

We calculated the Z scores to compare the two types of methods used for estimating the proportions as follows:

$$Z_{AB} = \frac{\hat{\pi}_{AB}^{Amy^*} - \hat{\pi}_{AB}^{bb}}{\sqrt{\hat{v}(\hat{\pi}_{AB}^{Amy^*}) + \hat{v}(\hat{\pi}_{AB}^{bb})}} = 0.284 \tag{9.1}$$

$$Z_{A} = \frac{\hat{\pi}_{A}^{Amy^*} - \hat{\pi}_{A}^{bb}}{\sqrt{\hat{v}(\hat{\pi}_{A}^{Amy^*}) + \hat{v}(\hat{\pi}_{A}^{bb})}} = 0.131 \tag{9.2}$$

$$Z_{B} = \frac{\hat{\pi}_{B}^{Amy^*} - \hat{\pi}_{B}^{bb}}{\sqrt{\hat{v}(\hat{\pi}_{B}^{Amy^*}) + \hat{v}(\hat{\pi}_{B}^{bb})}} = -0.167 \tag{9.3}$$

The absolute values of the computed Z scores are less than 1.96 indicating that there is no significant difference between the proposed DMDM-II and the black box estimates. It is concluded that the proposed DMDM-II method can be utilized in real practice. The original version of this work can be had from Zheng (2019).

Acknowledgements

Thanks are due to the IRB Committee Members, Research Compliance, Office of the Research & Sponsored Programs, Texas A&M University-Kingsville for the permission to collect data from the students.

References

Greenberg, B. G., Abul-Ela, A.-L. A., Simmons, W. R. and Horvitz, D. G. 1969. The unrelated question randomized response model: Theoretical framework. Journal of the American Statistical Association, 64(326): 520–539.

Johnson, M. L., Sedory, S. A. and Singh, S. 2019. Alternative methods to make efficient use of two decks of cards in randomized response sampling. Sociological Methods and Research, 48(1): 62–91.

Lanke, J. 1975. On the choice of the unrelated question in Simmons' version of randomized response. Journal of the American Statistical Association, 70(349): 80–83.

Lanke, J. 1976. On the degree of protection in randomized interviews. International Statistical Review/Revue Internationale de Statistique, 197–203.
Lee, C.-S., Sedory, S. A. and Singh, S. 2013. Estimating at least seven measures of qualitative variables from a single sample using randomized response technique. Statistics & Probability Letters, 83(1): 399–409.
Lee, C. S., Sedory, S. A. and Singh, S. 2016. Cramer-Rao lower bounds of variance for estimating two proportions and their overlap by using two-decks of cards. Handbook of Statistics, 34, ELSEVIER, PP. 353–385.
Odumade, O. and Singh, S. 2009. Efficient use of two decks of cards in randomized response sampling. Communications in Statistics—Theory and Methods, 38(4): 439–446.
Warner, S. L. 1965. Randomized response: A survey technique for eliminating evasive answer bias. Journal of the American Statistical Association, 60(309): 63–69.
Yennum, N., Sedory, S. A. and Singh, S. 2019. Improved strategy to collect sensitive data by using geometric distribution as a randomization device. Communications In Statistics-Theory and Methods, 48(23): 5777–5795.
Zheng, R. 2019. On unrelated question randomized response techniques. Unpublished MS thesis submitted to the Department of Mathematics, Texas A&M University-Kingsville, Kingsville, TX.

APPENDIX A

SAS CODES USED IN THE SIMULATION STUDY

```
*SAS Codes used in the simulation study
DATA DATA1;
DO W = 0.1 TO 0.9 BY 0.2;
DO Y = 0.9 TO 1.0 BY 0.05;
*DO P = 0.6 TO 0.9 BY 0.1;
*DO T = 0.6 TO 0.9 BY 0.1;
P=0.7;
T=0.7;
DO P1 = 0.6 TO 0.7 BY 0.1;
DO T1 = 0.6 TO 0.7 BY 0.1;
DO PIAB = 0.05 TO 0.2 BY 0.05;
DO PIA = 0.05 TO 0.35 BY 0.05;
DO PIB = 0.05 TO 0.35 BY 0.05;
VAR_PIA_LS=PIA*(1-PIA)+(1-P)*(T*(P*T+(1-P)*(1-T))*(1-PIA-PIB+2*PIAB))/(P+T-1)**2;
VAR_PIB_LS=PIB*(1-PIB)+(1-T)*(P*(P*T+(1-P)*(1-T))*(1-PIA-PIB+2*PIAB))/(P+T-1)**2;
TERM1=P**2*T**2+(1-P)**2*(1-T)**2-(P*T+(1-P)*(1-T))*(P+T-1)**2;
TERM2=P*T*(1-P)*(1-T)*(1-PIA-PIB);
TERM3=(P*T+(1-P)*(1-T))*(P+T-1)**2;
VAR_PIAB_LS=PIAB*(1-PIAB)+(PIAB*TERM1+TERM2)/TERM3;
CPIAB_PIA_LS=PIAB*(1-PIA)+PIAB*T*(1-P)*(P-T+1)/(P+T-1)**2+P*T*(1-P)*(1-T)*(T-P+1)*(1-PIA-PIB)/((P*T+(1-
P)*(1-T))*(P+T-1)**2);
CPIAB_PIB_LS=PIAB*(1-PIB)+PIAB*P*(1-T)*(T-P+1)/(P+T-1)**2+P*T*(1-P)*(1-T)*(P-T+1)*(1-PIA-PIB)/((P*T+(1-P)*(1-
T))*(P+T-1)**2);
CPIA_PIB_LS =-(PIA*PIB-PIAB)-2*(1-T)*(1-P)*P*T*(PIA+PIB-2*PIAB-1)/(P+T-1)**2;
TH11= PIAB*(P*T+(1-P)*(1-T))-PIA*(1-P)*(1-T)-PIB*(1-P)*(1-T)+(1-P)*(1-T);
TH10=-PIAB*(P*T+(1-P)*(1-T))-PIA*((1-P)*T-1)-PIB*(1-P)*T+(1-P)*T;
TH01=-PIAB*(P*T+(1-P)*(1-T))-PIA*P*(1-T)-PIB*(P*(1-T)-1)+P*(1-T);
TH00= PIAB*(P*T+(1-P)*(1-T))-PIA*P*T-PIB*P*T+P*T;
SUMTH=TH11+TH10+TH01+TH00;
PABYY_LC=PIAB*(P*T)/TH11;
PABYN_LC=PIAB*P*(1-T)/TH10;
PABNY_LC=PIAB*(1-P)*T/TH01;
PABNN_LC=PIAB*(1-P)*(1-T)/TH00;
PAYY_LC=PIA*(P*T*PIB)/TH11;
PAYN_LC=PIA*(P*T*(1-PIB)+P*(1-T)+(1-P)*(1-PIB))/TH10;
PANY_LC=PIA*(1-P)*PIB*T/TH01;
PANN_LC=PIA*((1-P)*PIB*(1-T))/TH00;
PBYY_LC=PIB*P*T*PIA/TH11;
PBYN_LC=PIB*PIA*P*(1-T)/TH10;
PBNY_LC=PIB*(P*(1-PIA)+(1-P)*T+(1-P)*(1-T)*(1-PIA))/TH01;
```

```
PBNN_LC=PIB*(1-P)*(1-T)*PIA/TH00;
PROT_AB_LC=MAX(PABYY_LC, PABYN_LC, PABNY_LC, PABNN_LC);
PROT_A_LC=MAX(PAYY_LC, PAYN_LC, PANY_LC, PANN_LC);
PROT_B_LC=MAX(PBYY_LC, PBYN_LC, PBNY_LC, PBNN_LC);
MSE_PIA_given_PIB_LS=VAR_PIAB_LS/PIAB**2+VAR_PIB_LS/PIB**2-2*CPIAB_PIB_LS/(PIB*PIAB);
VAR_PIA_minus_PIB_LS=VAR_PIA_LS+VAR_PIAB_LS-2*CPIAB_PIA_LS;
VAR_PIAUB_LS=VAR_PIA_LS+VAR_PIB_LS+VAR_PIAB_LS+2*CPIA_PIB_LS-2*CPIAB_PIA_LS-2*CPIAB_PIB_LS;
MSE_RR_B_A_LS=(PIB**2*VAR_PIAB_LS)/(PIAB**2*(PIB-PIAB)**2)
 +VAR_PIA_LS/(PIA**2*(1-PIA)**2)
 +VAR_PIB_LS/(PIB-PIAB)**2
-2*PIB*CPIAB_PIA_LS/(PIA*PIAB*(1-PIA)*(PIB-PIAB))
 -2*PIB*CPIAB_PIB_LS/(PIAB*(PIB-PIAB)**2)
 +2*CPIA_PIB_LS/(PIA*(1-PIA)*(PIB-PIAB));
F1=PIAB/(PIAB-PIA*PIB);
F2=PIA*PIB/(PIAB-PIA*PIB)+(1-2*PIA)/(2*(1-PIA));
F3=PIA*PIB/(PIAB-PIA*PIB)+(1-2*PIB)/(2*(1-PIB));
MSE_RHOAB_LS=F1**2*VAR_PIAB_LS/PIAB**2
 +F2**2*VAR_PIA_LS/PIA**2
 +F3**2*VAR_PIB_LS/PIB**2
 -2*F1*F2*CPIAB_PIA_LS/(PIA*PIAB)
 -2*F1*F3*CPIAB_PIB_LS/(PIB*PIAB)
 +2*F2*F3*CPIA_PIB_LS/(PIA*PIB);
VAR_PI_d_LS=VAR_PIA_LS+VAR_PIB_LS-2*CPIA_PIB_LS;
CS1 = W*(P*T+(1-P)*(1-T))+(1-W)*P1*T1;
CS2 =-W*(1-P)*(1-T)+(1-W)*P1*(1-T1)*Y;
CS3 =-W*(1-P)*(1-T)+(1-W)*(1-P1)*T1*Y;
CS4 = W*((1-P)*T-1)-(1-W)*P1*(T1+(1-T1)*(1-Y));
CS5 = W*(1-P)*T+(1-W)*T1*(1-P1)*Y;
CS6 = W*P*(1-T)+(1-W)*P1*(1-T1)*Y;
CS7 = W*(P*(1-T)-1)-(1-W)*T1*(P1+(1-P1)*(1-Y));
CS8 = W*P*T+(1-W)*P1*(T1+(1-T1)*(1-Y));
CS9 = W*P*T+(1-W)*T1*(P1+(1-P1)*(1-Y));
HS1 = W*(1-P)*(1-T)+(1-W)*(1-P1)*(1-T1)*Y;
HS2 = W*(1-P)*T+(1-W)*(1-P1)*T1*Y;
HS3 = W*P*(1-T)+(1-W)*(1-T1)*P1*Y;
HS4 = W*P*T + (1-W)*(P1*T1*Y + 1-Y);
SHS11 = CS1*PIAB+CS2*PIA+CS3*PIB+HS1;
SHS10 = -CS1*PIAB-CS4*PIA-CS5*PIB+HS2;
SHS01 = -CS1*PIAB-CS6*PIA-CS7*PIB+HS3;
SHS00 = CS1*PIAB-CS8*PIA-CS9*PIB+HS4;
SUMSH = SHS11+SHS10+SHS01+SHS00;
KS1 = (CS2**2+CS4**2+CS6**2+CS8**2)*(CS3**2+CS5**2+CS7**2+CS9**2)
 -(CS2*CS3+CS4*CS5+CS6*CS7+CS8*CS9)**2;
KS2 = (CS2*CS3+CS4*CS5+CS6*CS7+CS8*CS9)*(CS3+CS5+CS7-CS9)
 -(CS2+CS4+CS6-CS8)*(CS3**2+CS5**2+CS7**2+CS9**2);
KS3 = (CS2*CS3+CS4*CS5+CS6*CS7+CS8*CS9)*(CS2+CS4+CS6-CS8)
 -(CS3+CS5+CS7-CS9)*(CS2**2+CS4**2+CS6**2+CS8**2);
LS1 = CS1*(CS2*CS3+CS4*CS5+CS6*CS7+CS8*CS9)*(CS3+CS5+CS7-CS9)
 -CS1*(CS2+CS4+CS6-CS8)*(CS3**2+CS5**2+CS7**2+CS9**2);
LS2 = 4*CS1*(CS3**2+CS5**2+CS7**2+CS9**2)-CS1*(CS3+CS5+CS7-CS9)**2;
```

```
LS3 = CS1*(CS2+CS4+CS6-CS8)*(CS3+CS5+CS7-CS9)
 -4*CS1*(CS2*CS3+CS4*CS5+CS6*CS7+CS8*CS9);
MS1 = CS1*(CS2+CS4+CS6-CS8)*(CS2*CS3+CS4*CS5+CS6*CS7+CS8*CS9)
 -CS1*(CS3+CS5+CS7-CS9)*(CS2**2+CS4**2+CS6**2+CS8**2);
MS2=CS1*(CS2+CS4+CS6-CS8)*(CS3+CS5+CS7-CS9)-4*CS1*(CS2*CS3+CS4*CS5+CS6*CS7+CS8*CS9);
MS3 = 4*CS1*(CS2**2+CS4**2+CS6**2+CS8**2)-CS1*(CS2+CS4+CS6-CS8)**2;
WS11AB = KS1+CS2*KS2+CS3*KS3;
WS10AB = KS1+CS4*KS2+CS5*KS3;
WS01AB = KS1+CS6*KS2+CS7*KS3;
WS00AB = KS1-CS8*KS2-CS9*KS3;
TERMS1 = WS11AB**2*SHS11*(1-SHS11)+WS10AB**2*SHS10*(1-SHS10)+ WS01AB**2*SHS01*(1-SHS01) +
WS00AB**2*SHS00*(1-SHS00);
TERMS2 = 2*WS11AB*WS10AB*SHS11*SHS10+2*WS11AB*WS01AB*SHS11*SHS01-
2*WS11AB*WS00AB*SHS11*SHS00;
TERMS3 =-2*WS10AB*WS01AB*SHS10*SHS01+2*WS10AB*WS00AB*SHS10*SHS00+2*WS01AB*WS00AB*SHS01*
SHS00;
WS11A = LS1+CS2*LS2+CS3*LS3;
WS10A = LS1+CS4*LS2+CS5*LS3;
WS01A = LS1+CS6*LS2+CS7*LS3;
WS00A = LS1-CS8*LS2-CS9*LS3;
TERMS4 = WS11A**2*SHS11*(1-SHS11)+WS10A**2*SHS10*(1-SHS10)+ WS01A**2*SHS01*(1-SHS01) +
WS00A**2*SHS00*(1-SHS00);
TERMS5 = 2*WS11A*WS10A*SHS11*SHS10+2*WS11A*WS01A*SHS11*SHS01-2*WS11A*WS00A*SHS11*SHS00;
TERMS6 =-2*WS10A*WS01A*SHS10*SHS01+2*WS10A*WS00A*SHS10*SHS00+2*WS01A*WS00A*SHS01*SHS00;
WS11B = MS1+CS2*MS2+CS3*MS3;
WS10B = MS1+CS4*MS2+CS5*MS3;
WS01B = MS1+CS6*MS2+CS7*MS3;
WS00B = MS1-CS8*MS2-CS9*MS3;
TERMS7 = WS11B**2*SHS11*(1-SHS11)+WS10B**2*SHS10*(1-SHS10)+ WS01B**2*SHS01*(1-
SHS01)+WS00B**2*SHS00*(1-SHS00);
TERMS8 = 2*WS11B*WS10B*SHS11*SHS10+2*WS11B*WS01B*SHS11*SHS01-2*WS11B*WS00B*SHS11*SHS00;
TERMS9 =-2*WS10B*WS01B*SHS10*SHS01+2*WS10B*WS00B*SHS10*SHS00+2*WS01B*WS00B*SHS01*SHS00;
DELTS1=4*CS1*(CS2**2+CS4**2+CS6**2+CS8**2)*(CS3**2+CS5**2+CS7**2+CS9**2);
DELTS2=2*CS1*(CS2+CS4+CS6-CS8)*(CS3+CS5+CS7-CS9)*(CS2*CS3+CS4*CS5+CS6*CS7+CS8*CS9);
DELTS3=-CS1*(CS3+CS5+CS7-CS9)**2*(CS2**2+CS4**2+CS6**2+CS8**2)-4*CS1*(CS2*CS3+CS4*CS5+CS6*CS7+CS
8*CS9)**2;
DELTS4=-CS1*(CS2+CS4+CS6-CS8)**2*(CS3**2+CS5**2+CS7**2+CS9**2);
DELTS=DELTS1+DELTS2+DELTS3+DELTS4;
VARS_PIAB = (TERMS1+TERMS2+TERMS3)/DELTS**2;
VARS_PIA = (TERMS4+TERMS5+TERMS6)/DELTS**2;
VARS_PIB = (TERMS7+TERMS8+TERMS9)/DELTS**2;
CPIAB_PIA=(WS11AB*WS11A*SHS11*(1-SHS11)+WS11AB*WS10A*SHS11*SHS10 +WS11AB*WS01A*SHS11*SHS01
-WS11AB*WS00A*SHS11*SHS00
 +WS10AB*WS11A*SHS10*SHS11 +WS10AB*WS10A*SHS10*(1-SHS10)-WS10AB*WS01A*SHS10*SHS01
+WS10AB*WS00A*SHS10*SHS00
 +WS01AB*WS11A*SHS01*SHS11 -WS01AB*WS10A*SHS01*SHS10 +WS01AB*WS01A*SHS01*(1-
SHS01)+WS01AB*WS00A*SHS01*SHS00
 -WS00AB*WS11A*SHS00*SHS11 +WS00AB*WS10A*SHS00*SHS10 +WS00AB*WS01A*SHS00*SHS01
+WS00AB*WS00A*SHS00*(1-SHS00))/DELTS**2;
```

```
CPIAB_PIB=(WS11AB*WS11B*SHS11*(1-SHS11)+WS11AB*WS10B*SHS11*SHS10 +WS11AB*WS01B*SHS11*SHS01
-WS11AB*WS00B*SHS11*SHS00
 +WS10AB*WS11B*SHS10*SHS11 +WS10AB*WS10B*SHS10*(1-SHS10)-WS10AB*WS01B*SHS10*SHS01
+WS10AB*WS00B*SHS10*SHS00
 +WS01AB*WS11B*SHS01*SHS11 -WS01AB*WS10B*SHS01*SHS10 +WS01AB*WS01B*SHS01*(1-
SHS01)+WS01AB*WS00B*SHS01*SHS00
 -WS00AB*WS11B*SHS00*SHS11 +WS00AB*WS10B*SHS00*SHS10 +WS00AB*WS01B*SHS00*SHS01
+WS00AB*WS00B*SHS00*(1-SHS00))/DELTS**2;
CPIA_PIB=(WS11A*WS11B*SHS11*(1-SHS11) +WS11A*WS10B*SHS11*SHS10 +WS11A *WS01B*SHS11*SHS01
-WS11A*WS00B*SHS11*SHS00
 +WS10A*WS11B*SHS10*SHS11 +WS10A*WS10B*SHS10*(1-SHS10) -WS10A *WS01B*SHS10*SHS01
+WS10A*WS00B*SHS10*SHS00
 +WS01A*WS11B*SHS01*SHS11 -WS01A*WS10B*SHS01*SHS10 +WS01A *WS01B*SHS01*(1-
SHS01)+WS01A*WS00B*SHS01*SHS00
 -WS00A*WS11B*SHS00*SHS11 +WS00A*WS10B*SHS00*SHS10 +WS00A *WS01B*SHS00*SHS01
+WS00A*WS00B*SHS00*(1-SHS00))/DELTS**2;
a_Amy=CS1**2*(1/SHS11+1/SHS10+1/SHS01+1/SHS00);
b_Amy=CS1*(CS2/SHS11+CS4/SHS10+CS6/SHS01-CS8/SHS00);
c_Amy=CS1*(CS3/SHS11+CS5/SHS10+CS7/SHS01-CS9/SHS00);
d_Amy=CS2**2/SHS11+CS4**2/SHS10+CS6**2/SHS01+CS8**2/SHS00;
f_Amy=CS2*CS3/SHS11+CS4*CS5/SHS10+CS6*CS7/SHS01+CS8*CS9/SHS00;
g_Amy=CS3**2/SHS11+CS5**2/SHS10+CS7**2/SHS01+CS9**2/SHS00;
DELTA_AMY=g_Amy*b_Amy**2-2*b_Amy*c_Amy*f_Amy+ d_Amy*c_Amy**2+a_Amy*f_Amy**2-a_Amy*d_Amy*g_
Amy;
a11_Amy=f_Amy**2-d_Amy*g_Amy;
a22_Amy=c_Amy**2-a_Amy*g_Amy;
a33_Amy=b_Amy**2-a_Amy*d_Amy;
a12_Amy=b_Amy*g_Amy-c_Amy*f_Amy;
a13_Amy=c_Amy*d_Amy-b_Amy*f_Amy;
a23_Amy=a_Amy*f_Amy-b_Amy*c_Amy;
crpiab=a11_Amy/DELTA_AMY;
crpia=a22_Amy/DELTA_AMY;
crpib=a33_Amy/DELTA_AMY;
cr_piabpia=a12_Amy/DELTA_AMY;
cr_piabpib=a13_Amy/DELTA_AMY;
cr_piapib=a23_Amy/DELTA_AMY;
PABYY_AMY_C=PIAB*(W*P*T+(1-W)*(P1*T1+Y*(1-P1*T1)))/SHS11;
PABYN_AMY_C=PIAB*(W*P*(1-T)+(1-W)*(P1*(1-T1)*(1-Y)))/SHS10;
PABNY_AMY_C=PIAB*(W*(1-P)*T+(1-W)*(1-P1)*(1-Y)*T1)/SHS01;
PABNN_AMY_C=PIAB*(W*(1-T)*(1-P)+(1-W)*(1-P1)*(1-T1)*(1-Y))/SHS00;
PAYY_AMY_C=PIA*(W*P*T*PIB+(1-W)*(P1*(1-T1)*Y+P1*T1*PIB+(1-P1)*Y*T1*PIB+(1-P1)*(1-T1)*Y))/SHS11;
PAYN_AMY_C=PIA*(W*(P*T*(1-PIB)+P*(1-T)+(1-P)*(1-PIB))+(1-W)*(P1*(1-T1)*(1-Y)+P1*T1*(1-PIB)+(1-
P1)*Y*T1*(1-PIB)))/SHS10;
PANY_AMY_C=PIA*(W*(1-P)*PIB*T+(1-W)*(1-P1)*(1-Y)*PIB)/SHS01;
PANN_AMY_C=PIA*(W*(1-P)*PIB*(1-T)+(1-W)*((1-P1)*(1-Y)*(1-T1)+(1-P1)*(1-Y)*(1-PIB)*T1))/SHS00;
PBYY_AMY_C=PIB*(W*(P*T*PIA)+(1-W)*((1-P1)*T1*Y+P1*T1*PIA+(1-T1)*Y*P1*PIA+(1-P1)*(1-T1)*Y))/SHS11;
PBYN_AMY_C=PIB*(W*P*(1-T)*PIA+(1-W)*P1*PIA*(1-T1)*(1-Y))/SHS10;
PBNY_AMY_C=PIB*(W*(P*(1-PIA)+(1-P)*T+(1-P)*(1-T)*(1-PIA))+(1-W)*(P1*(1-PIA)*T1+(1-P1)*T1*(1-Y)+(1-
T1)*Y*P1*(1-PIA)))/SHS01;
PBNN_AMY_C=PIB*(W*(1-P)*(1-T)*PIA+(1-W)*(P1*(1-PIA)*(1-T1)*(1-Y)+(1-P1)*(1-T1)*(1-Y)))/SHS00;
```

```
PROT_AB_AMY_C=MAX(PABYY_AMY_C, PABYN_AMY_C,PABNY_AMY_C,PABNN_AMY_C);
PROT_A_AMY_C =MAX(PAYY_AMY_C, PAYN_AMY_C, PANY_AMY_C, PANN_AMY_C);
PROT_B_AMY_C =MAX(PBYY_AMY_C, PBYN_AMY_C, PBNY_AMY_C, PBNN_AMY_C);
MSE_PIA_given_PIB = VARS_PIAB/PIAB**2+VARS_PIB/PIB**2-2*CPIAB_PIB/(PIB*PIAB);
VAR_PIA_minus_PIB=VARS_PIA+VARS_PIAB-2*CPIAB_PIA;
VAR_PIAUB=VARS_PIA+VARS_PIB+VARS_PIAB+2*CPIA_PIB-2*CPIAB_PIA-2*CPIAB_PIB;
MSE_RR_B_A=(PIB**2*VARS_PIAB)/(PIAB**2*(PIB-PIAB)**2)
 +VARS_PIA/(PIA**2*(1-PIA)**2)
 +VARS_PIB/(PIB-PIAB)**2
-2*PIB*CPIAB_PIA/(PIA*PIAB*(1-PIA)*(PIB-PIAB))
 -2*PIB*CPIAB_PIB/(PIAB*(PIB-PIAB)**2)
 +2*CPIA_PIB/(PIA*(1-PIA)*(PIB-PIAB));
F1=PIAB/(PIAB-PIA*PIB);
F2=PIA*PIB/(PIAB-PIA*PIB)+(1-2*PIA)/(2*(1-PIA));
F3=PIA*PIB/(PIAB-PIA*PIB)+(1-2*PIB)/(2*(1-PIB));
MSE_RHOAB=F1**2*VARS_PIAB/PIAB**2
 +F2**2*VARS_PIA/PIA**2
 +F3**2*VARS_PIB/PIB**2
 -2*F1*F2*CPIAB_PIA/(PIA*PIAB)
 -2*F1*F3*CPIAB_PIB/(PIB*PIAB)
 +2*F2*F3*CPIA_PIB/(PIA*PIB);
VAR_PI_d_LS=VAR_PIA_LS+VAR_PIB_LS-2*CPIA_PIB_LS;
VAR_PI_d=VARS_PIA+VARS_PIB-2*CPIA_PIB;
RE_VAR_PIAB=VAR_PIAB_LS*100/VARS_PIAB;
RE_VAR_PIA=VAR_PIA_LS*100/VARS_PIA;
RE_VAR_PIB=VAR_PIB_LS*100/VARS_PIB;
RE_PROT_AB=PROT_AB_LC*100/PROT_AB_AMY_C;
RE_PROT_A=PROT_A_LC*100/PROT_A_AMY_C;
RE_PROT_B=PROT_B_LC*100/PROT_B_AMY_C;
RE_MSE_PIA_given_B=MSE_PIA_given_PIB_LS*100/MSE_PIA_given_PIB;
RE_VAR_PIA_minus_B=VAR_PIA_minus_PIB_LS*100/VAR_PIA_minus_PIB;
RE_VAR_PIAUB=VAR_PIAUB_LS*100/VAR_PIAUB;
RE_RR_B_A=MSE_RR_B_A_LS*100/MSE_RR_B_A;
RE_MSE_RHOAB=MSE_RHOAB_LS*100/MSE_RHOAB;
RE_VAR_PI_d=VAR_PI_d_LS*100/VAR_PI_d;
OUTPUT;
END;
END;
END;
END;
END;
END;
END;
DATA DATA2;
SET DATA1;
KEEP P T P1 T1 PIAB PIA PIB Y W RE_VAR_PIAB RE_VAR_PIA RE_VAR_PIB RE_PROT_AB RE_PROT_A RE_
PROT_B
 RE_MSE_PIA_given_B RE_VAR_PIA_minus_B RE_VAR_PIAUB RE_RR_B_A RE_MSE_RHOAB RE_VAR_PI_d;
IF RE_VAR_PIAB GT 100;
IF RE_VAR_PIA GT 100;
```

```
IF RE_VAR_PIB GT 100;
IF RE_PROT_AB GT 100;
IF RE_PROT_A GT 100;
IF RE_PROT_B GT 100;
IF RE_MSE_RHOAB GT 50;
IF RE_VAR_PIAUB GT 50;
IF RE_VAR_PI_d GT 50;
IF RE_RR_B_A GT 50;
IF PIAB LE PIA;
IF PIAB LE PIB;
PROC PRINT DATA=DATA2;
VAR P T P1 T1 PIAB PIA PIB Y W RE_VAR_PIAB RE_VAR_PIA RE_VAR_PIB RE_PROT_AB RE_PROT_A RE_PROT_B;
RUN;
DATA DATA3;
SET DATA1;
IF RE_VAR_PIAB GT 100;
IF RE_VAR_PIA GT 100;
IF RE_VAR_PIB GT 100;
IF RE_PROT_AB GT 100;
IF RE_PROT_A GT 100;
IF RE_PROT_B GT 100;
IF RE_MSE_RHOAB GT 50;
IF RE_VAR_PIAUB GT 50;
IF RE_VAR_PI_d GT 50;
IF RE_RR_B_A GT 50;
IF PIAB LE PIA;
IF PIAB LE PIB;
PROC PRINT DATA=DATA3;
VAR P T P1 T1 PIAB PIA PIB Y W RE_MSE_PIA_given_B RE_VAR_PIA_minus_B RE_VAR_PIAUB RE_RR_B_A RE_
MSE_RHOAB RE_VAR_PI_d;
RUN;
DATA DATA4;
SET DATA1;
KEEP P T W Y P1 T1 PIAB PIA PIB crpiab VARS_PIAB crpia VARS_PIA crpib VARS_PIB
 cr_piabpia CPIAB_PIA cr_piabpib CPIAB_PIB cr_piapib CPIA_PIB;
IF RE_VAR_PIAB GT 100;
IF RE_VAR_PIA GT 100;
IF RE_VAR_PIB GT 100;
IF RE_PROT_AB GT 100;
IF RE_PROT_A GT 100;
IF RE_PROT_B GT 100;
IF RE_MSE_RHOAB GT 50;
IF RE_VAR_PIAUB GT 50;
IF RE_VAR_PI_d GT 50;
IF RE_RR_B_A GT 50;
IF PIAB LE PIA;
IF PIAB LE PIB;
PROC PRINT DATA=DATA4;
VAR P T W Y P1 T1 PIAB PIA PIB crpiab VARS_PIAB crpia VARS_PIA crpib VARS_PIB
 cr_piabpia CPIAB_PIA cr_piabpib CPIAB_PIB cr_piapib CPIA_PIB;
RUN;
```

APPENDIX B

Table B1: Detailed results for different choice of parameters.

Obs	w	π_y	P_1	T_1	π_{AB}	π_A	π_B	RE(1)	RE(2)	RE(3)	RP(1)	RP(2)	RP(3)
1	0.1	0.9	0.6	0.7	0.05	0.20	0.15	108.0	144.7	120.4	131.4	113.8	117.4
2	0.1	0.9	0.6	0.7	0.05	0.20	0.20	102.6	139.3	107.9	131.5	111.6	130.9
3	0.1	0.9	0.6	0.7	0.05	0.25	0.15	106.3	137.3	111.0	126.2	114.5	128.3
4	0.1	0.9	0.6	0.7	0.05	0.30	0.15	104.5	129.8	102.0	122.0	115.3	140.4
5	0.1	0.9	0.6	0.7	0.15	0.30	0.20	119.9	152.8	145.8	124.5	110.5	106.0
6	0.1	0.9	0.6	0.7	0.15	0.35	0.20	118.4	145.7	137.1	120.3	111.0	112.7
7	0.1	0.9	0.6	0.7	0.15	0.35	0.25	113.0	140.9	125.6	120.3	109.4	121.7
8	0.1	0.9	0.6	0.7	0.20	0.25	0.30	121.1	163.6	155.7	132.4	105.6	102.3
9	0.1	0.9	0.6	0.7	0.20	0.25	0.35	115.8	159.8	144.0	132.4	104.7	109.2
10	0.1	0.9	0.6	0.7	0.20	0.30	0.25	124.8	161.3	159.5	126.0	107.1	100.9
To Get Complete Table Execute The Sas Code													
2934	0.9	1	0.7	0.7	0.15	0.35	0.20	107.9	107.0	109.2	100.8	101.5	101.3
2935	0.9	1	0.7	0.7	0.15	0.35	0.25	107.1	106.4	107.5	100.9	101.3	102.8
2936	0.9	1	0.7	0.7	0.15	0.35	0.30	106.2	105.8	105.7	100.9	101.1	104.4
2937	0.9	1	0.7	0.7	0.15	0.35	0.35	105.2	105.2	103.8	100.9	100.9	106.2
2938	0.9	1	0.7	0.7	0.20	0.25	0.35	107.8	108.3	110.3	101.1	100.7	100.2
2939	0.9	1	0.7	0.7	0.20	0.30	0.30	108.0	108.0	110.3	100.9	100.9	100.2
2940	0.9	1	0.7	0.7	0.20	0.30	0.35	107.2	107.6	108.9	101.0	100.8	101.4
2941	0.9	1	0.7	0.7	0.20	0.35	0.25	108.3	107.8	110.3	100.7	101.1	100.2
2942	0.9	1	0.7	0.7	0.20	0.35	0.30	107.6	107.2	108.9	100.8	101.0	101.4
2943	0.9	1	0.7	0.7	0.20	0.35	0.35	106.7	106.7	107.3	100.8	100.8	102.7

APPENDIX C

Table C1: Detailed results for different choice of parameters.

Obs	w	π_y	P_1	T_1	π_{AB}	π_A	π_B	RE(4)	RE(5)	RE(6)	RE(7)	RE(8)	RE(9)
1	0.1	0.9	0.6	0.7	0.05	0.20	0.15	73.10	44.90	94.58	51.22	60.28	56.54
2	0.1	0.9	0.6	0.7	0.05	0.20	0.20	75.45	45.30	94.06	52.31	56.43	57.65
3	0.1	0.9	0.6	0.7	0.05	0.25	0.15	69.98	46.13	93.87	51.97	56.98	57.69
4	0.1	0.9	0.6	0.7	0.05	0.30	0.15	66.81	46.98	93.07	51.49	53.21	58.61
5	0.1	0.9	0.6	0.7	0.15	0.30	0.20	58.24	48.77	98.72	52.37	52.89	59.54
6	0.1	0.9	0.6	0.7	0.15	0.35	0.20	58.62	49.96	98.11	52.76	53.43	60.56
7	0.1	0.9	0.6	0.7	0.15	0.35	0.25	63.78	50.65	97.65	50.21	52.93	62.04
8	0.1	0.9	0.6	0.7	0.20	0.25	0.30	65.22	48.08	101.48	53.07	51.30	60.14
9	0.1	0.9	0.6	0.7	0.20	0.25	0.35	69.38	48.82	101.20	51.49	51.49	60.85
10	0.1	0.9	0.6	0.7	0.20	0.30	0.25	60.75	49.58	101.27	55.79	53.11	60.73
To Get Complete Table Execute The Sas Codes													
2934	0.9	1	0.7	0.7	0.15	0.35	0.20	96.08	94.09	103.78	93.43	95.06	96.30
2935	0.9	1	0.7	0.7	0.15	0.35	0.25	97.91	94.15	103.71	92.86	94.55	96.52
2936	0.9	1	0.7	0.7	0.15	0.35	0.30	98.34	94.22	103.65	92.26	93.50	96.68
2937	0.9	1	0.7	0.7	0.15	0.35	0.35	97.85	94.29	103.61	91.60	92.28	96.81
2938	0.9	1	0.7	0.7	0.20	0.25	0.35	100.10	94.86	104.04	93.59	95.31	96.63
2939	0.9	1	0.7	0.7	0.20	0.30	0.30	98.55	94.80	104.04	93.74	95.45	96.69
2940	0.9	1	0.7	0.7	0.20	0.30	0.35	99.48	94.91	103.99	93.43	95.05	96.80
2941	0.9	1	0.7	0.7	0.20	0.35	0.25	96.51	94.69	104.04	94.21	95.31	96.63
2942	0.9	1	0.7	0.7	0.20	0.35	0.30	98.16	94.78	103.99	93.72	95.05	96.80
2943	0.9	1	0.7	0.7	0.20	0.35	0.35	98.83	94.87	103.95	93.29	94.43	96.93

CHAPTER 10

Modified Regression Type Estimator by Ingeniously Utilizing Probabilities for more Efficient Results in Randomized Response Sampling

Roberto Arias,* Stephen A Sedory* and *Sarjinder Singh*

1. Introduction

Survey sampling statisticians have long dealt with the difficulties of estimating the true population proportion of those individuals belonging to a group defined by a sensitive characteristic. One popular solution, first proposed by Warner (1965), is the implementation of a randomization device which protects the privacy of those individuals being surveyed. The idea, first proposed by Warner (1965), instructs the respondents, while keeping to themselves, to make use of a randomization device, such as a deck of cards. The respondent will select a card from the deck. Each card in this deck will either have the statement "I belong to group A" or "I do not belong to group A" with proportion P_0 and $(1 - P_0)$ respectively. After selecting a card, the respondents will read it to themselves and only tell the interviewer 'yes' or 'no' if the statement on the drawn card matches their status. By letting π represent the true population proportion of those individuals belonging to group A, the Warner (1965) model gives the following estimator of true proportion π of those individuals who belong to group A, for a given P_0, as:

$$\hat{\pi}_{w(P_0)} = \frac{\hat{\theta}_w - (1 - P_0)}{2P_0 - 1}, \; P_0 \neq 0.5 \tag{1.1}$$

where $\hat{\theta}_w = n_w/n$ is the observed proportion of 'yes' replies out of n respondents selected from the population utilizing a simple random sample with replacement sampling (SRSWR) scheme, and n_w be the observed number of 'yes' replies received by the interviewer. Then, the above estimator is unbiased and provides the following variance for two trials per respondent, for a given P_0, as:

$$V(\hat{\pi}_w)_{P_0} = \frac{\pi(1-\pi)}{n} + \frac{P_0(1-P_0)}{2n(2P_0-1)^2} \tag{1.2}$$

Since the Warner (1965) model randomized response model only requires the respondent to deal with a single randomization device, such as a deck of cards, we also consider the case where it is performed twice, independently. We do this because the randomization devices that have been proposed, and which will be discussed, after the Warner (1965) model makes use of two randomization devices, such as two decks of cards. These devices that use a second device gain efficiency and also improve protection for the respondents participating. Thus, while still using Warner's (1965) model, consider the case in which the interviewer receives 0, 1, or 2 'yes' replies based on using two independent randomizing devices with parameters, P and T. Then by letting π represent the true population proportion of people belonging to group A, the probability mass function (p.m.f) of the i-th Z_i reply is obtained as

$$Z_i = \begin{cases} 0 & \text{with prob. } \pi(1-P)(1-T) + (1-\pi)PT \\ 1 & \text{with prob. } \pi\left[P(1-T) + T(1-P)\right] + (1-\pi)[P(1-T) + T(1-P)] \\ 2 & \text{with prob. } \pi PT + (1-\pi)(1-P)(1-T) \end{cases} \tag{1.3}$$

Department of Mathematics, Texas A&M University-Kingsville, Kingsville, TX 78363.
* Corresponding authors: roberto.arias@tamuk.edu; sarjinder.singh@tamuk.edu

From (1.3), the expected value of Z_i is given as:

$$\begin{aligned} E(Z_i) &= (0)\left[\pi(1-P)(1-T)+(1-\pi)PT\right] \\ &+ (1)\left[\pi\{P(1-T)+T(1-P)\}+(1-\pi)\{P(1-T)+T(1-P)\}\right] \\ &+ (2)[\pi PT+(1-\pi)(1-P)(1-T)] \end{aligned} \tag{1.4}$$

$$= 2(P+T-1)\pi + P(1-T)+T(1-P)+2(1-P)(1-T)$$

The variance of the response Z_i is given by

$$V(Z_i) = E(Z_i^2) - \{E(Z_i)\}^2 \tag{1.5}$$

where

$$\begin{aligned} E(Z_i^2) &= (0)^2\,[\pi\,(1-P)(1-T)(1-\pi)PT] \\ &+ (1)^2[\pi\{P(1-T)+T(1-P)\}+(1-\pi)\{P(1-T)+T(1-P)\}] \\ &+ (2)^2\left[\pi PT+(1-\pi)(1-P)(1-T)\right] \end{aligned} \tag{1.6}$$

By plugging (1.4) and (1.6) into (1.5), the variance of Z_i is given as

$$V(Z_i) = 4(P+T-1)^2\pi(1-\pi)+P(1-P)+T(1-T) \tag{1.7}$$

From (1.4), an unbiased estimator of π is given by

$$\hat{\pi}_{W(PT)} = \frac{\frac{1}{n}\sum_{i=1}^{n} Z_i - [P(1-T)+T(1-P)+2(1-P)(1-T)]}{2(P+T-1)} \tag{1.8}$$

The variance of the Warner (1965) estimator $\hat{\pi}_{W(PT)}$ for two trials per respondent with two independent devices with parameters, P and T, is given as:

$$V(\hat{\pi}_W)_{PT} = \frac{\pi(1-\pi)}{n} + \frac{P(1-P)+T(1-T)}{4n(P+T-1)^2} \tag{1.9}$$

Clearly, if we let $P = T$, this model reduces back to the original Warner (1965) model with two independent trials per respondent as in (1.2).

Mangat and Singh (1990) improved on the Warner (1965) model by proposing two stage randomized response by making use of two decks of cards. In the Mangat and Singh (1990) model, each respondent is asked to use two randomized devices as R_1 and R_2. The device R_1 consists of two outcomes, "Are you a member of group *A*", with relative frequency T_0 and "Go to the second randomization device R_2" with relative frequency $(1 - T_0)$. The second randomization device R_2 is the same as the Warner (1965) randomization device. Similar to the Warner (1965) model, by letting π represent the true population proportion of those individuals belonging to group *A* and considering n respondents, with n_{ms} being the number of 'yes' replies received by the interviewer, respondents are selected from the population utilizing a SRSWR. Then, the following estimator of true proportion is derived for the Mangat and Singh (1990) model:

$$\hat{\pi}_{ms} = \frac{\hat{\theta}_{ms} - (1-T_0)(1-P_0)}{(2P_0-1)+2T_0(1-P_0)} \tag{1.10}$$

where $\hat{\theta}_{ms} = n_{ms}/n$ is the proportion of the observed 'yes' answers in the SRSWR. The estimator in (1.10), provided by Mangat and Singh (1990) is unbiased and has the variance:

$$V(\hat{\pi}_{ms}) = \frac{\pi(1-\pi)}{n} + \frac{(1-P_0)(1-T_0)\{1-(1-P_0)(1-T_0)\}}{n\{(2P_0-1)+2T_0(1-P_0)\}^2} \tag{1.11}$$

Another such idea, proposed by Kuk (1990), instructs the respondents, while keeping to themselves, to make use of two decks of cards. Suppose the respondent belongs to group *A* defined by a sensitive characteristic, then he/she will select a card from the 1st deck. Each card in this deck will either have the statement "I belong to group *A* " or "I do not belong to group *A*" with proportion θ_1 and $(1 - \theta_1)$ respectively. If the respondents belongs to the non-sensitive group A^c, then they will select a card from the second deck. In this deck, each card will bear the complements of the two statements as in the first deck, with proportion θ_2 and $(1 - \theta_2)$ respectively. After selecting a card, the respondents will read it to themselves and only tell the interviewer 'yes' or 'no'. By letting π represent the true population proportion of those individuals belonging to group A, Kuk's (1990) model gives the probability of obtaining a 'yes' from a respondent:

$$P(Yes) = \theta_{kuk} = \theta_1\pi + (1-\pi)\theta_2 \tag{1.12}$$

Additionally, consider that n respondents are selected from the population utilizing a SRSWR. Then, let n_{kuk} be the number of 'yes' replies received by the interviewer. The number of 'yes' replies follows a Binomial distribution with parameters n and $\theta_{kuk} = \theta_1\pi + \theta_2(1-\pi)$. If each individual being interviewed is requested to give $k \geq 1$ replies, then Kuk's (1990) model gives the variance:

$$V(\hat{\pi}_{kuk}) = \frac{\theta_{kuk}(1-\theta_{kuk})}{nk(\theta_1-\theta_2)^2} + \frac{\pi(1-\pi)}{n}\left(1-\frac{1}{k}\right) \tag{1.13}$$

The model proposed by Kuk (1990) serves as a special case of many suggested randomized response models such as those proposed by Warner (1965), and Mangat and Singh (1990).

While all these models posed as improvements upon one another, a recent paper has shown a more efficient way than all of them. Odumade and Singh (2009) proposed a more efficient use of two decks of cards as a randomization device. Similarl to Warner (1965), and the other proposed models, by letting π represent the true population proportion of those individuals belonging to group A and considering n respondents, with n_{11}, n_{10}, n_{01} and n_{00} being the number of (yes, yes), (yes, no), (no, yes) and (no, no) respective replies received by the interviewer. The respondents are selected from the population utilizing a SRSWR. Then, the estimator of true proportion derived from the Odumade and Singh (2009) model is given as

$$\hat{\pi}_{os} = \frac{(P+T-1)(\hat{\theta}_{11}-\hat{\theta}_{00}) + (P-T)(\hat{\theta}_{10}-\hat{\theta}_{01})}{2[(P+T-1)^2+(P-T)^2]} + \frac{1}{2} \tag{1.14}$$

where $\hat{\theta}_{ij} = n_{ij}/n$, $i = 0,1; j = 0,1$ are the observed proportions of (yes, yes), (yes, no), (no, yes) and (no, no) replies. The estimator $\hat{\pi}_{os}$, provided by Odumade and Singh (2009) is unbiased and has the variance:

$$V(\hat{\pi}_{os}) = \frac{(P+T-1)^2[PT+(1-P)(1-T)] + (P-T)^2[T(1-P)+P(1-T)]}{4n[(P+T-1)^2+(P-T)^2]^2} - \frac{(2\pi-1)^2}{4n} \tag{1.15}$$

In the sections that follow, we will modify the above Odumade and Singh (2009) estimator for proportion in such a way that it will give us more efficient results.

2. New Regression Type Randomized Response Estimator Based on the Addition of the Probabilities θ_{10} and θ_{01}

We became curious as to what types of methods we could use to produce a new and efficient randomized response model. We knew that the Odumade and Singh (2009) randomized response model proved to be more efficient than that of Warner (1965), Mangat and Singh (1990) and Kuk (1990) models. Naturally, this means that the Odumade and Singh (2009) estimator is what we wish to dethrone. To start, we considered the same randomized response model that Odumade and Singh (2009) did as shown below in Figure 1:

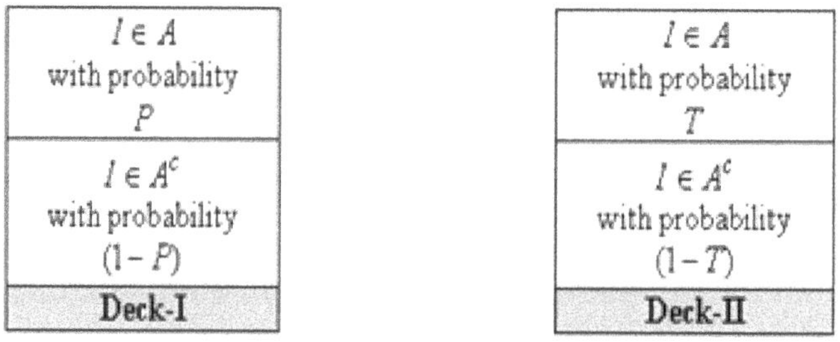

Figure 1: Two deck randomized response model.

Following them, this randomized response model produces the following probabilities:

$$P(\text{Yes, Yes}) = \theta_{11} = (P+T-1)\pi + (1-P)(1-T) \tag{2.1}$$

$$P(\text{Yes, No}) = \theta_{10} = (P-T)\pi + T(1-P) \tag{2.2}$$

$$P(\text{No, Yes}) = \theta_{01} = (T-P)\pi + P(1-T) \tag{2.3}$$

$$P(\text{No}, \text{No}) = \theta_{00} = (1 - P - T)\pi + PT \tag{2.4}$$

Thus instead of minimizing the distance function, we used the following idea to produce the following theorems.

Theorem 2.1. An unbiased estimator of the proportion, π, of individuals who belong to the sensitive group is given by

$$\hat{\pi}_{REG} = \hat{\pi}_{os} + \beta\left[\hat{\theta}_{10} + \hat{\theta}_{01} - (P + T - 2PT)\right] \tag{2.5}$$

where β is a constant to be determined.

Proof. On taking the expected value of $\hat{\pi}_{REG}$ on both sides, we have

$$\begin{aligned} E(\hat{\pi}_{REG}) &= E\left[\hat{\pi}_{os} + \beta\{\hat{\theta}_{10} + \hat{\theta}_{01} - (P + T - 2PT)\}\right] \\ &= E(\hat{\pi}_{os}) + \beta\{E(\hat{\theta}_{10}) + E(\hat{\theta}_{01}) - (P + T - 2PT)\} \end{aligned} \tag{2.6}$$

At this point, it is important to note that we consider the following as did Odumade and Singh (2009):

$$E(\hat{\theta}_{ij}) = \theta_{ij} \tag{2.7}$$

Thus, we get

$$\begin{aligned} E(\hat{\pi}_{REG}) &= \pi + \beta\{\theta_{10} + \theta_{01} - (P + T - 2PT)\} \\ &= \pi + \beta\left[(P - T)\pi + T(1 - P) + (T - P)\pi + P(1 - T) - (P + T - 2PT)\right] \\ &= \pi + \beta\left[T(1 - P) + P(1 - T) - (P + T - 2PT)\right] \\ &= \pi \end{aligned} \tag{2.8}$$

which proves the theorem.

Theorem 2.2. The minimum variance of $\hat{\pi}_{REG}$ is given by

$$\min.V(\hat{\pi}_{REG}) = V(\hat{\pi}_{os}) - \frac{(2\pi - 1)^2[(P + T - 1)^2(2PT - P - T) + (P - T)^2\{1 - (P + T - 2PT)\}]^2}{4n[(P + T - 1)^2 + (P - T)^2]^2(P + T - 2PT)[1 - (P + T - 2PT)]} \tag{2.9}$$

Proof. The variance of $\hat{\pi}_{REG}$ is given as

$$\begin{aligned} V(\hat{\pi}_{REG}) &= V\left[\hat{\pi}_{os} + \beta\{\hat{\theta}_{10} + \hat{\theta}_{01} - (P + T - 2PT)\}\right] \\ &= V(\hat{\pi}_{os}) + \beta^2 V(\hat{\theta}_{10} + \hat{\theta}_{01}) + 2\beta Cov\{\hat{\pi}_{os}, \hat{\theta}_{10} + \hat{\theta}_{01}\} \end{aligned} \tag{2.10}$$

Naturally, the variance of $\hat{\pi}_{REG}$ is minimum when

$$\frac{\partial V(\hat{\pi}_{REG})}{\partial \beta} = 0$$

we get

$$\beta = -\frac{Cov(\hat{\pi}_{os}, \hat{\theta}_{10} + \hat{\theta}_{01})}{V(\hat{\theta}_{10} + \hat{\theta}_{01})} \tag{2.11}$$

Now, by plugging in (2.11) to (2.10), we get minimum variance as

$$\begin{aligned} \min.V(\hat{\pi}_{REG}) &= V(\hat{\pi}_{os}) + \left\{-\frac{Cov(\hat{\pi}_{os}, \hat{\theta}_{10} + \hat{\theta}_{01})}{V(\hat{\theta}_{10} + \hat{\theta}_{01})}\right\}^2 V(\hat{\theta}_{10} + \hat{\theta}_{01}) \\ &\quad + 2\left\{-\frac{Cov(\hat{\pi}_{os}, \hat{\theta}_{10} + \hat{\theta}_{01})}{V(\hat{\theta}_{10} + \hat{\theta}_{01})}\right\} Cov(\hat{\pi}_{os}, \hat{\theta}_{10} + \hat{\theta}_{01}) \\ &= V(\hat{\pi}_{os}) + \frac{\{Cov(\hat{\pi}_{os}, \hat{\theta}_{10} + \hat{\theta}_{01})\}^2}{V(\hat{\theta}_{10} + \hat{\theta}_{01})} - 2\frac{\{Cov(\hat{\pi}_{os}, \hat{\theta}_{10} + \hat{\theta}_{01})\}^2}{V(\hat{\theta}_{10} + \hat{\theta}_{01})} \\ &= V(\hat{\pi}_{os}) - \frac{\{Cov(\hat{\pi}_{os}, \hat{\theta}_{10} + \hat{\theta}_{01})\}^2}{V(\hat{\theta}_{10} + \hat{\theta}_{01})} \end{aligned} \tag{2.12}$$

Since we know that

$$V(\hat{\pi}_{os}) = \frac{(P+T-1)^2[PT+(1-P)(1-T)]+(P-T)^2[T(1-P)+P(1-T)]}{4n[(P+T-1)^2+(P-T)^2]^2} - \frac{(2\pi-1)^2}{4n} \tag{2.13}$$

Then, we can derive the following

$$V(\hat{\theta}_{10}+\hat{\theta}_{01}) = V(\hat{\theta}_{10}) + V\left(\hat{\theta}_{01}\right) + 2Cov(\hat{\theta}_{10},\hat{\theta}_{01}) \tag{2.14}$$

Note that we consider the following as did the Odumade and Singh (2009) model:

$$V(\hat{\theta}_{ij}) = \frac{\theta_{ij}(1-\theta_{ij})}{n} \tag{2.15}$$

and

$$Cov(\hat{\theta}_{ij},\hat{\theta}_{i^*j^*}) = -\frac{\theta_{ij}\theta_{i^*j^*}}{n} \tag{2.16}$$

By solving (2.14), we now get

$$\begin{aligned} V(\hat{\theta}_{10}+\hat{\theta}_{01}) &= \frac{1}{n}\left[\theta_{10}(1-\theta_{10}) + \theta_{01}(1-\theta_{10}) - 2\theta_{10}\theta_{01}\right] \\ &= \frac{(\theta_{10}+\theta_{01})[1-(\theta_{10}+\theta_{01})]}{n} \\ &= \frac{(P+T-2PT)[1-(P+T-2PT)]}{n} \end{aligned} \tag{2.17}$$

Next, we derive the following:

$$\begin{aligned} Cov\left(\hat{\pi}_{os},\hat{\theta}_{10}+\hat{\theta}_{01}\right) &= Cov\left[\frac{(P+T-1)(\hat{\theta}_{11}-\hat{\theta}_{00})+(P-T)(\hat{\theta}_{10}-\hat{\theta}_{01})}{2\{(P+T-1)^2+(P-T)^2\}}+\frac{1}{2},\hat{\theta}_{10}+\hat{\theta}_{01}\right] \\ &= \frac{1}{2[(P+T-1)^2+(P-T)^2]}\left[(P+T-1)Cov\left(\hat{\theta}_{11}-\hat{\theta}_{00},\hat{\theta}_{10}+\hat{\theta}_{01}\right)+(P-T)Cov\left(\hat{\theta}_{10}-\hat{\theta}_{01},\hat{\theta}_{10}+\hat{\theta}_{01}\right)\right] \\ &= \frac{1}{2[(P+T-1)^2+(P-T)^2]}\Big[(P+T-1)\{Cov(\hat{\theta}_{11},\hat{\theta}_{10})+Cov(\hat{\theta}_{11},\hat{\theta}_{01})-Cov(\hat{\theta}_{00},\hat{\theta}_{10})-Cov(\hat{\theta}_{00},\hat{\theta}_{01})\} \\ &\qquad +(P-T)\{Cov(\hat{\theta}_{10},\hat{\theta}_{10})+Cov(\hat{\theta}_{10},\hat{\theta}_{01})-Cov(\hat{\theta}_{01},\hat{\theta}_{10})-Cov(\hat{\theta}_{01},\hat{\theta}_{01})\}\Big] \\ &= \frac{1}{2n[(P+T-1)^2+(P-T)^2]}\Big[(P+T-1)\{-\theta_{11}\theta_{10}-\theta_{11}\theta_{01}+\theta_{00}\theta_{10}+\theta_{00}\theta_{01}\} \\ &\qquad +(P-T)\{\theta_{10}(1-\theta_{10})-\theta_{10}\theta_{01}+\theta_{01}\theta_{10}-\theta_{01}(1-\theta_{01})\}\Big] \\ &= \frac{1}{2n[(P+T-1)^2+(P-T)^2]}\left[(P+T-1)(\theta_{10}+\theta_{01})(\theta_{00}-\theta_{11})+(P-T)\{(\theta_{10}-\theta_{01})-(\theta_{10}^2-\theta_{01}^2)\}\right] \\ &= \frac{1}{2n[(P+T-1)^2+(P-T)^2]}\Big[(P+T-1)(\theta_{10}+\theta_{01})(\theta_{00}-\theta_{11}) \\ &\qquad +(P-T)\{(\theta_{10}-\theta_{01})-(\theta_{10}-\theta_{01})(\theta_{10}+\theta_{01})\}\Big] \\ &= \frac{1}{2n[(P+T-1)^2+(P-T)^2]}\Big[(P+T-1)(\theta_{10}+\theta_{01})(\theta_{00}-\theta_{11}) \\ &\qquad +(P-T)(\theta_{10}-\theta_{01})\{1-(\theta_{10}+\theta_{01})\}\Big] \\ &= \frac{1}{2n[(P+T-1)^2+(P-T)^2]}\Big[(P+T-1)(P+T-2PT)(\theta_{00}-\theta_{11}) \\ &\qquad +(P-T)(\theta_{10}-\theta_{01})\{1-(P+T-2PT)\}\Big] \end{aligned} \tag{2.18}$$

where,

$$\theta_{00}-\theta_{11}=(2\pi-1)(1-P-T) \tag{2.19}$$

and

$$\theta_{10}-\theta_{01}=(2\pi-1)(P-T) \tag{2.20}$$

Now by plugging (2.19) and (2.20) back into (2.18), we get

$$Cov\left(\hat{\pi}_{os},\hat{\theta}_{10}+\hat{\theta}_{01}\right)=\frac{1}{2n[(P+T-1)^2+(P-T)^2]}[(P+T-1)(P+T-2PT)(2\pi-1)(1-P-T) \\ +(P-T)(2\pi-1)(P-T)\{1-(P+T-2PT)\}] \tag{2.21}$$

$$=\frac{(2\pi-1)}{2n[(P+T-1)^2+(P-T)^2]}\left[(P+T-1)^2(2PT-P-T)+(P-T)^2\{1-(P+T-2PT)\}\right]$$

Note that we have a derivation for $V(\hat{\theta}_{10}+\hat{\theta}_{01})$ in (2.17) and $Cov(\hat{\pi}_{os}+\hat{\theta}_{10}+\hat{\theta}_{01})$ in (2.21), so we can substitute in (2.12) and solve for minimum variance as:

$$\min.V(\hat{\pi}_{REG})=V(\hat{\pi}_{os})-\frac{\left[\dfrac{(2\pi-1)[(P+T-1)^2(2PT-P-T)+(P-T)^2\{1-(P+T-2PT)\}]}{2n[(P+T-1)^2+(P-T)^2]}\right]^2}{\dfrac{(P+T-2PT)\{1-(P+T-2PT)\}}{n}} \tag{2.22}$$

$$=V(\hat{\pi}_{os})-\frac{(2\pi-1)^2[(P+T-1)^2(2PT-P-T)+(P-T)^2\{1-(P+T-2PT)\}]^2}{4n[(P+T-1)^2+(P-T)^2]^2(P+T-2PT)\{1-(P+T-2PT)\}}$$

which proves the theorem.

Theorem 2.3. The optimum value of β is given as

$$\beta=-\frac{(2\pi-1)[(P+T-1)^2(2PT-P-T)+(P-T)^2\{1-(P+T-2PT)\}}{2[(P+T-1)^2+(P-T)^2](P+T-2PT)\{1-(P+T-2PT)\}} \tag{2.23}$$

Proof. From (2.11), (2.17) and (2.21) the optimum value of β is given by

$$\beta=-\frac{Cov(\hat{\pi}_{OS},\hat{\theta}_{10}+\hat{\theta}_{01})}{V(\hat{\theta}_{10}+\hat{\theta}_{01})}$$

$$=-\frac{\dfrac{(2\pi-1)[(P+T-1)^2(2PT-P-T)+(P-T)^2\{1-(P+T-2PT)\}]}{2n[(P+T-1)^2+(P-T)^2]}}{\dfrac{(P+T-2PT)\{1-(P+T-2PT)\}}{n}} \tag{2.24}$$

$$=-\frac{(2\pi-1)[(P+T-1)^2(2PT-P-T)+(P-T)^2\{1-(P+T-2PT)\}}{2[(P+T-1)^2+(P-T)^2](P+T-2PT)\{1-(P+T-2PT)\}}$$

which proves the theorem.

Corollary 2.1. An estimator of $V(\hat{\pi}_{REG})$ is given by

$$v(\hat{\pi}_{REG})=v(\hat{\pi}_{os})-\frac{(2\hat{\pi}_{OS}-1)^2[(P+T-1)^2(2PT-P-T)+(P-T)^2\{1-(P+T-2PT)\}]^2}{4(n-1)[(P+T-1)^2+(P-T)^2]^2(P+T-2PT)\{1-(P+T-2PT)\}} \tag{2.25}$$

3. Efficiency Comparisons

In order to show that our new estimator is better than the randomized response estimators derived for two trials by Warner (1965), Mangat and Singh (1990), Kuk (1990), and Odumade and Singh (2009), we must compare using relative efficiencies. With respect to both cases suggested via the Warner (1965) model, the relative efficiency criterions are given by

$$RE(w)_{P_0} = \frac{V(\hat{\pi}_w)_{P_0}}{V(\hat{\pi}_{REG})} \times 100\% \tag{3.1}$$

and

$$RE(w)_{PT} = \frac{V(\hat{\pi}_w)_{PT}}{V(\hat{\pi}_{REG})} \times 100\% \tag{3.2}$$

Similarly, the relative efficiency criterions with respect to Mangat and Singh (1990), Kuk (1990) and Odumade and Singh (2009) are given by

$$RE(ms) = \frac{V(\hat{\pi}_{ms})}{V(\hat{\pi}_{REG})} \times 100\% \tag{3.3}$$

$$RE(kuk) = \frac{V(\hat{\pi}_{kuk})}{V(\hat{\pi}_{REG})} \times 100\% \tag{3.4}$$

and

$$RE(os) = \frac{V(\hat{\pi}_{os})}{V(\hat{\pi}_{REG})} \times 100\% \tag{3.5}$$

Usually, we compare protection criterions as proposed by Lanke (1975, 1976) and Leysieffer and Warner (1976). However, since we are using the same randomized response model as Odumade and Singh (2009), we do not gain or lose any protection.

We executed a code in SAS (Program 1 in Appendix B) to test the efficiency of the proposed model with respect to Warner (1965), Mangat and Singh (1990), Kuk (1990), and Odumade and Singh (2009). For Warner (1965) shown in (1.2), Mangat and Singh (1990), we set $P = P_0$ and $T = T_0$. For Kuk (1990), we set $\theta_1 = P$, $\theta_2 = T$ and $k = 2$. For Odumade and Singh (2009), Warner (1965) shown in (1.9), and the suggested model, we allowed SAS to run a loop for values of P and T. So P will range $0.5 < P < 0.70$ and T will range $0.15 < T < 0.30$. Below is a summarized view of results, while the full outcome of results can be found in Appendix-A. The summary of results was conducted as follows; while keeping the value of π constant, we found the mean, median, standard deviation, maximum, and minimum of the found relative efficiencies of P and T.

In Table 1 *freq* stands for the number of times the proposed estimator is more efficient than all the competitors considered out of the 16 possible combinations for P and T. As one can clearly see from Table 1, the relative efficiency of the suggested model with respect to the models of the competitors is much better. As the value for π ranges from $0.05 < \pi < 0.50$ with a step of .05, the proposed estimator performs much better than the models proposed by Warner (1965) for both cases, Mangat and Singh (1990), Kuk (1990), and Odumade and Singh (2009) without any loss in protection with respect to the Odumade and Singh (2009) model. Now in order to visualize this concept, we can use the Figures 2, 3, 4, 5 and 6 shown below to see the relative efficiency. In these figures, we will put, into visuals, the relative efficiencies over the unknown proportion π.

From Figures 2, 3, 4, 5 and 6, one can clearly see that when compared with each of the randomized response models produced by Warner (1965) for two trials, Mangat and Singh (1990), Kuk (1990), and Odumade and Singh (2009), the suggested model performs better. In each respective case, it is easy to see that each of the values for relative efficiency, with respect to the other models, will remain above 100%. Naturally, proper values for P and T must be chosen in order for the model to perform at its peak!

In order to take the visuals further and get another idea of how the suggested model is performing with respect to the other models provided by Warner (1965), Mangat and Singh (1990), Kuk (1990), and Odumade and Singh (2009), 3D models were obtained in Figures 7, 8, 9, 10 and 11. Below are the 3D scatter plots of relative efficiencies versus the two choices of P and T.

From Figures 7, 8, 9, 10 and 11, of the 3D scatter plots, we can clearly see that for each loop that was performed, there exist optimal values of P and T that can be used to have more accuracy. For all of the relative efficiencies compared in the 3D plots, one can clearly see that there exist many values that are above the 100 percent mark. These graphs and summary of results all show that the suggested estimator can perform much better than all of the other estimators provided by Warner (1965), Mangat and Singh (1990), Kuk (1990), and Odumade and Singh (2009).

Table 1: Summarized results of relative efficiencies.

RE(W)$_{P0}$						
π	*freq*	*Mean*	*Med*	*StDev*	*Max*	*Min*
0.05	14	999	511	1136	4164	176
0.10	14	898	451	1008	3637	159
0.15	14	828	411	919	3274	148
0.20	14	777	384	854	3015	140
0.25	13	776	381	827	2826	135
0.30	13	747	370	792	2689	131
0.35	13	726	362	766	2591	128
0.40	13	712	357	749	2526	126
0.45	13	704	354	739	2489	125
0.50	13	701	353	736	2476	124
RE(W)$_{PT}$						
π	*freq*	*Mean*	*Med*	*StDev*	*Max*	*Min*
0.05	14	1702	846	2519	9966	355
0.10	14	1463	738	2136	8504	330
0.15	14	1308	667	1883	7532	306
0.20	14	1200	619	1708	6853	289
0.25	13	1138	577	1647	6369	277
0.30	13	1081	545	1554	6021	267
0.35	13	1041	523	1489	5776	261
0.40	13	1014	508	1446	5613	257
0.45	13	998	500	1421	5519	254
0.50	13	993	497	1413	5489	253
RE(ms)						
π	*freq*	*Mean*	*Med*	*StDev*	*Max*	*Min*
0.05	14	422.3	297.8	361.4	1449.6	106.5
0.10	14	375.3	274.9	314.2	1274.2	104.1
0.15	14	344.0	258.8	282.2	1153.4	102.2
0.20	14	322.0	247.2	259.6	1067.0	100.7
0.25	13	322.2	249.7	245.5	1004.3	121.8
0.30	13	310.0	236.9	233.6	958.7	118.5
0.35	13	301.4	228.1	225.2	926.3	116.2
0.40	13	295.6	222.4	219.5	904.5	114.6
0.45	13	292.3	219.2	216.3	892.1	113.7
0.50	13	291.2	218.1	215.2	888.0	113.4
RE(kuk)						
π	*freq*	*Mean*	*Med*	*StDev*	*Max*	*Min*
0.05	14	134.63	131.7	10.55	155.76	119.64
0.10	14	129.89	124.58	12.38	153.47	117.19
0.15	14	127.01	121.8	13.91	152.46	113.09
0.20	14	125.25	120.12	15.11	152.34	108.89
0.25	13	125.13	120.23	16.41	152.90	105.95
0.30	13	124.76	119.84	17.31	154.04	103.91
0.35	13	124.88	119.83	18.11	155.70	102.57
0.40	13	125.45	120.16	18.86	157.89	101.78
0.45	13	126.44	120.81	19.6	160.63	101.49
0.50	13	127.87	121.83	20.34	163.97	101.65
RE(os)						
π	*freq*	*Mean*	*Med*	*StDev*	*Max*	*Min*
0.05	14	116.27	111.34	12.31	137.04	102.99
0.10	14	110.88	107.94	7.78	122.76	102.07
0.15	14	107.37	105.64	5.08	115.01	101.42
0.20	14	104.93	103.94	3.31	109.84	100.96
0.25	13	103.22	102.62	2.18	106.27	100.62
0.30	13	101.95	101.62	1.3	103.78	100.38
0.35	13	101.05	100.89	0.69	102.04	100.20
0.40	13	100.45	100.39	0.29	100.88	100.09
0.45	13	100.11	100.10	0.07	100.22	100.02
0.50	13	100	100.00	0.00	100.00	100.00

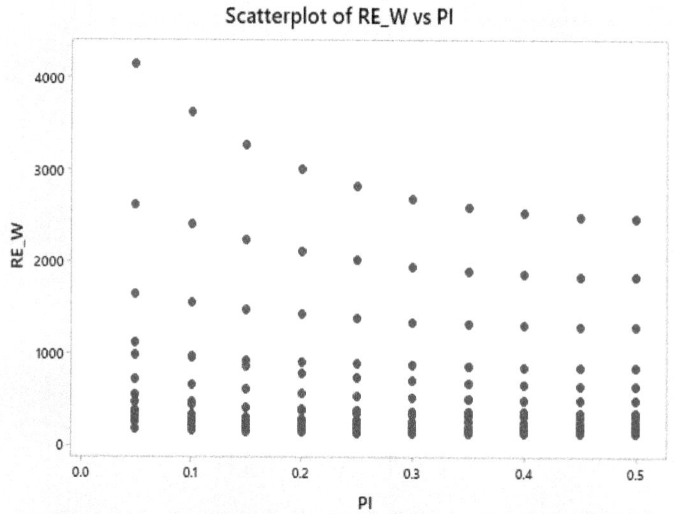

Figure 2: RE with respect to Warner (1965) as shown in (3.1).

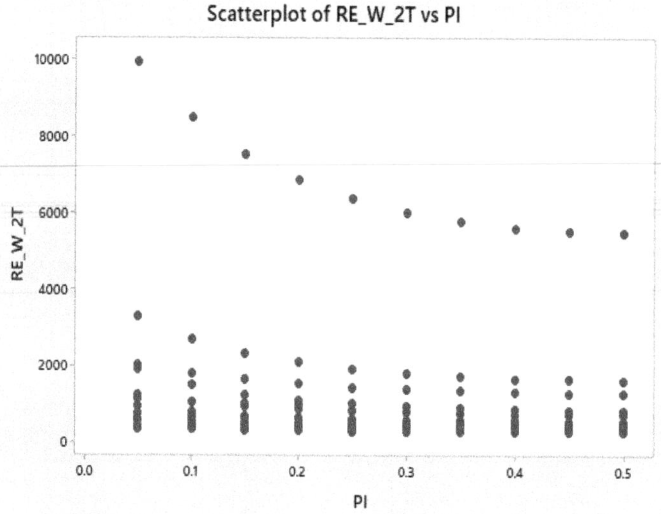

Figure 3: RE with respect to Warner (1965) as shown in (3.2).

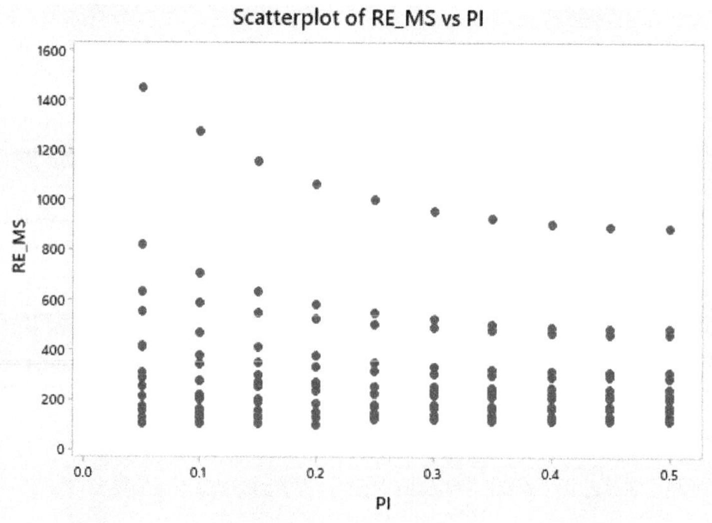

Figure 4: RE with respect to Mangat and Singh (1990).

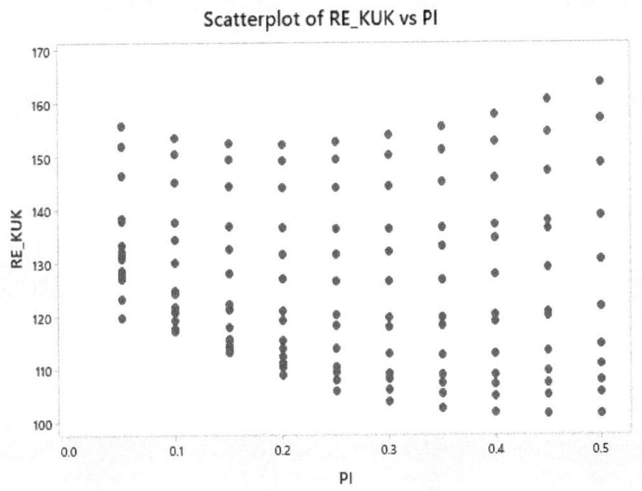

Figure 5: RE with respect to Kuk (1990).

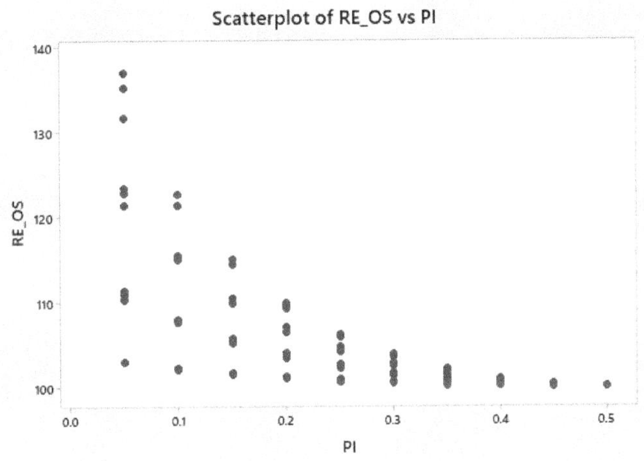

Figure 6: RE with respect to Odumade and Singh (1990).

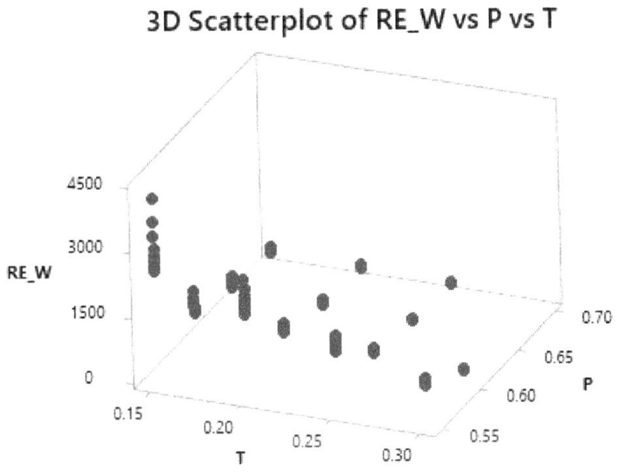

Figure 7: 3D scatter plot of $RE(W)P_0$ vs. P vs T.

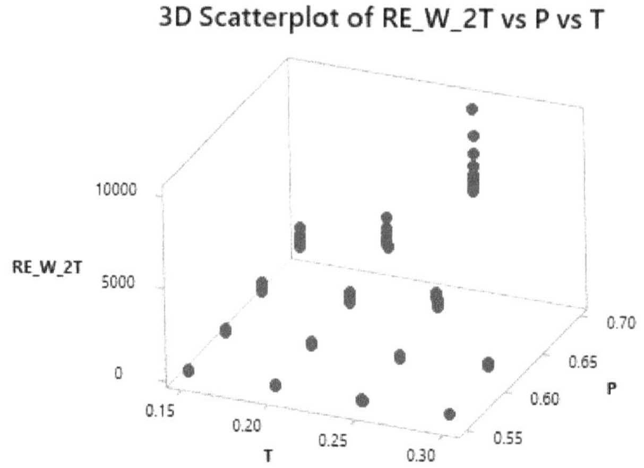

Figure 8: 3D scatter plot of $RE(W)P_T$ vs. P vs T.

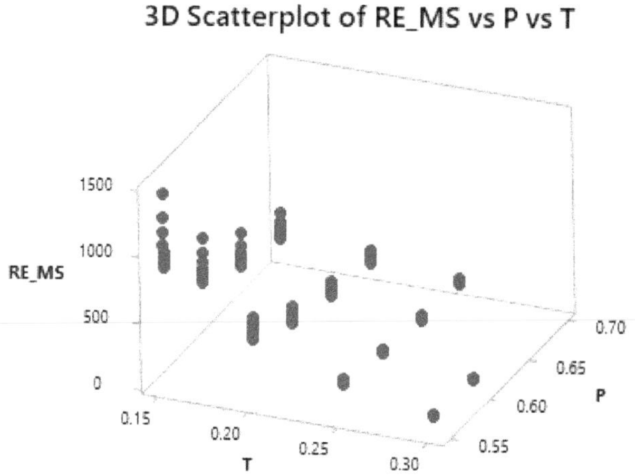

Figure 9: 3D scatter plot of *RE*(*ms*) vs. P vs T.

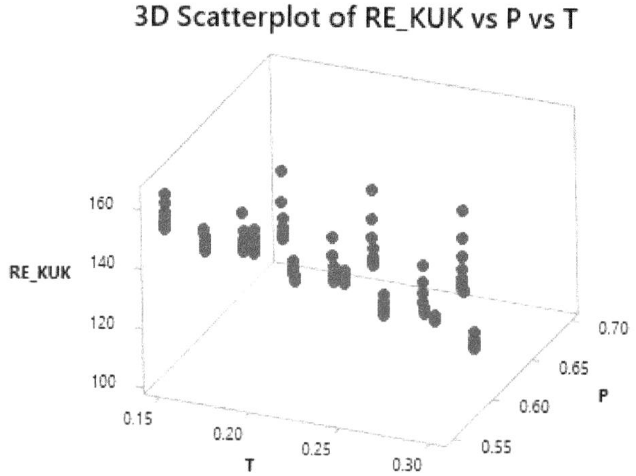

Figure 10: 3D scatter plot of *RE*(*kuk*) vs. P vs T.

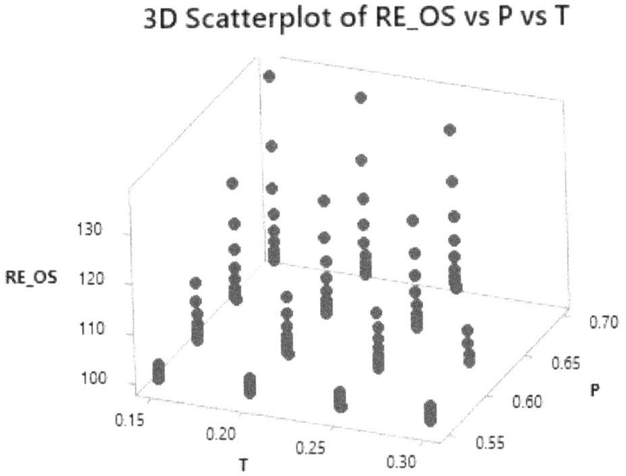

Figure 11: 3D scatter plot of *RE*(*os*) vs. P vs T.

4. Practical Proposed Estimator

Note that the estimator given for $\hat{\pi}_{REG}$ in (2.5) is not a practical estimator when working with real world data values as the value of β depends on the unknown parameter, π. Now, we have the following theorem.

Theorem 4.1. We suggest an unbiased estimator of β as

$$\hat{\beta} = -\frac{(2\hat{\pi}_{os}-1)[(P+T-1)^2(2PT-P-T)+(P-T)^2\{1-(P+T-2PT)\}}{2[(P+T-1)^2+(P-T)^2](P+T-2PT)\{1-(P+T-2PT)\}} \tag{4.1}$$

Proof. It is easy to verify that $E(\hat{\beta}) = \beta$.
By utilizing the result from Theorem 4.1, we suggest a new practical estimator of π as

$$\hat{\pi}^*_{REG} = \hat{\pi}_{os} + \hat{\beta}\left[\hat{\theta}_{10} + \hat{\theta}_{01} - (P+T-2PT)\right] \tag{4.2}$$

In order to find bias and mean squared error of (4.2) we define the following

$$\varepsilon_0 = \frac{\hat{\pi}_{os}}{\pi} - 1, \varepsilon_1 = \frac{\hat{\theta}_{10}+\hat{\theta}_{01}}{\theta_{10}+\theta_{01}} - 1 \text{ and } \varepsilon_2 = \frac{\hat{\beta}}{\beta} - 1$$

such that

$$E(\varepsilon_0) = E(\varepsilon_1) = E(\varepsilon_2) = 0$$

$$E(\varepsilon_0^2) = \frac{V(\hat{\pi}_{os})}{\pi^2};\quad E(\varepsilon_1^2) = \frac{V(\hat{\theta}_{10}+\hat{\theta}_{01})}{(\theta_{10}+\theta_{01})^2};\quad E(\varepsilon_2^2) = \frac{V(\hat{\beta})}{\beta^2};\quad E(\varepsilon_0\varepsilon_1) = \frac{Cov(\hat{\pi}_{os}, \hat{\theta}_{10}+\hat{\theta}_{01})}{\pi(\theta_{10}+\theta_{01})};$$

$$E(\varepsilon_0\varepsilon_2) = \frac{Cov(\hat{\pi}_{os}, \hat{\beta})}{\pi\beta} \text{ and } E(\varepsilon_1\varepsilon_2) = \frac{Cov(\hat{\beta}, \hat{\theta}_{10}+\hat{\theta}_{01})}{\beta(\theta_{10}+\theta_{01})}.$$

The estimator $\hat{\pi}^*_{REG}$ in terms of ε_0, ε_1 and ε_2 can be written as:

$$\begin{aligned}\hat{\pi}^*_{REG} &= \pi(1+\varepsilon_0) + \beta(1+\varepsilon_2)[(\theta_{10}+\theta_{01})(1+\varepsilon_1) - (P+T-2PT)] \\ &= \pi + \pi\varepsilon_0 + \beta(1+\varepsilon_2)(P+T-2PT)\varepsilon_1 \\ &= \pi + \pi\varepsilon_0 + \beta(P+T-2PT)\varepsilon_1 + \beta(P+T-2PT)\varepsilon_1\varepsilon_2\end{aligned} \tag{4.3}$$

By taking expected value of (4.3) on both sides, we obtain

$$\begin{aligned}E\left(\hat{\pi}^*_{REG}\right) &= E\left[\pi + \pi\varepsilon_0 + \beta(P+T-2PT)\varepsilon_1 + \beta(P+T-2PT)\varepsilon_1\varepsilon_2\right] \\ &= \pi + \pi E(\varepsilon_0) + \beta(P+T-2PT)E(\varepsilon_1) + \beta(P+T-2PT)E(\varepsilon_1\varepsilon_2) \\ &= \pi + 0 + 0 + \beta(P+T-2PT)\left[\frac{Cov(\hat{\theta}_{10}+\hat{\theta}_{01},\hat{\beta})}{\beta(\theta_{10}+\theta_{01})}\right] \\ &= \pi + Cov(\hat{\theta}_{10}+\hat{\theta}_{01},\hat{\beta})\end{aligned} \tag{4.4}$$

Naturally, the bias of the estimator $\hat{\pi}^*_{REG}$ is given by

$$\begin{aligned}B(\hat{\pi}^*_{REG}) &= E(\hat{\pi}^*_{REG}) - \pi = Cov(\hat{\theta}_{10}+\hat{\theta}_{01},\hat{\beta}) \\ &= Cov\left[\hat{\theta}_{10}+\hat{\theta}_{01}, -\frac{(2\hat{\pi}_{OS}-1)[(P+T-1)^2(2PT-P-T)+(P-T)^2\{1-(P+T-2PT)\}]}{2[(P+T-1)^2+(P-T)^2]^2(P+T-2PT)[1-(P+T-2PT)]}\right] \\ &= -\frac{[(P+T-1)^2(2PT-P-T)+(P-T)^2\{1-(P+T-2PT)\}]}{[(P+T-1)^2+(P-T)^2]^2(P+T-2PT)[1-(P+T-2PT)]}Cov\left[\hat{\theta}_{10}+\hat{\theta}_{01},\hat{\pi}_{OS}\right] \\ &= -\frac{(2\pi-1)[(P+T-1)^2(2PT-P-T)+(P-T)^2\{1-(P+T-2PT)\}]^2}{2n[(P+T-1)^2+(P-T)^2]^2(P+T-2PT)[1-(P+T-2PT)]}\end{aligned} \tag{4.5}$$

Note that as $n \to \infty$, then $B(\hat{\pi}^*_{REG}) \to 0$. Thus $\hat{\pi}^*_{REG}$ is a consistent estimator of π.
The mean squared error, to the first order of approximation, of the estimator $\hat{\pi}^*_{REG}$ is given by

$$\begin{aligned}MSE(\hat{\pi}^*_{REG}) &= E\left[\hat{\pi}^*_{REG} - \pi\right]^2 \\ &= E\left[\pi + \pi\varepsilon_0 + \beta(P+T-2PT)\varepsilon_1 + \beta(P+T-2PT)\varepsilon_1\varepsilon_2 - \pi\right]^2 \\ &= E\left[\pi\varepsilon_0 + \beta(P+T-2PT)\varepsilon_1 + \beta(P+T-2PT)\varepsilon_1\varepsilon_2\right]^2 \\ &\approx E\left[\pi\varepsilon_0 + \beta(P+T-2PT)\varepsilon_1\right]^2 \\ &= E\left[\pi^2\varepsilon_0^2 + \beta^2(P+T-2PT)^2\varepsilon_1^2 + 2\pi\beta(P+T-2PT)\varepsilon_0\varepsilon_1\right] \\ &= \pi^2E(\varepsilon_0^2) + \beta^2(P+T-2PT)^2E(\varepsilon_1^2) + 2\pi\beta(P+T-2PT)E(\varepsilon_0\varepsilon_1) \\ &= \pi^2\left[\frac{V(\hat{\pi}_{os})}{\pi^2}\right] + \beta^2(P+T-2PT)^2\left[\frac{V(\hat{\theta}_{10}+\hat{\theta}_{01})}{(\theta_{10}+\theta_{01})^2}\right] \\ &\quad + 2\pi\beta(P+T-2PT)\left[\frac{Cov(\hat{\pi}_{os},\hat{\theta}_{10}+\hat{\theta}_{01})}{\pi(\theta_{10}+\theta_{01})}\right] \\ &= V(\hat{\pi}_{os}) + \beta^2V(\hat{\theta}_{10}+\hat{\theta}_{01}) + 2\beta Cov(\hat{\pi}_{OS},\hat{\theta}_{10}+\hat{\theta}_{01}) \\ &= V(\hat{\pi}_{REG})\end{aligned} \tag{4.6}$$

which shows that the estimator $\hat{\pi}^*_{REG}$ is a more practical approach when studying real world data values.

5. Simulation Study

Now, while utilizing SAS (Program 2 in Appendix B), a simulation study, close to a real survey, was conducted. We wanted to see how the suggestedestimatorwouldcomparewiththeestimatoritwasbasedoff,OdumadeandSingh(2009).Whileweleto0.05≤π≤0.50andset$p$=0.7and $T = 0.7$, we first determined the true probabilities of θ_{11}, θ_{10}, θ_{01} and θ_{00}. Then, utilizing SAS, we created a sample size of $n = 50$ replies by utilizing the call function *Multinominal* (ntrials, ns, prob) where "ntrials" represents the number of trials in each simulation, "ns" is the sample size and "prob" represents the probabilities. We then used the function $NITR = 1{,}000{,}000$ which means each trial had 1,000,000 iterations. Naturally, we computed the variances of $\hat{\pi}_{OS}$ and mean squared error of $\hat{\pi}^*_{REG}$ as follows:

$$V_{sim}(\hat{\pi}_{OS}) = \frac{1}{NITR}\sum_{i=1}^{NITR}\left(\hat{\pi}_{OS(i)} - \pi\right)^2 \tag{5.1}$$

and

$$MSE_{sim}(\hat{\pi}^*_{REG}) = \frac{1}{NITR}\sum_{i=1}^{NITR}\left(\hat{\pi}^*_{REG(i)} - \pi\right)^2 \qquad (5.2)$$

The relative efficiency of $\hat{\pi}^*_{REG}$ with respect to $\hat{\pi}_{OS}$ can be found from (5.1) and (5.2) as:

$$RE_{sim} = \frac{V_{sim}(\hat{\pi}_{OS})}{MSE_{sim}(\hat{\pi}^*_{REG})} \times 100\% \qquad (5.3)$$

where RE_{sim} is the relative efficiency of $\hat{\pi}^*_{REG}$ with respect to $\hat{\pi}_{os}$.
We also determined the relative bias of $\hat{\pi}^*_{REG}$ which was computed empirically as:

$$RB_{sim} = \frac{\frac{1}{NITR}\sum_{i=1}^{NITR}\left(\hat{\pi}^*_{REG(i)} - \pi\right)}{\pi} \times 100\% \qquad (5.4)$$

A total of 10 simulations were run for $0.05 \le \pi \le 0.50$ with a step of .05. Each individual case considered 1,000,000 different trials. In each trial a sample size of 50 individuals was used in order to produce accurate results. For each study we determined the relative efficiency of the suggested estimator with respect to the Odumade and Singh (2009) estimator. The results are given below in Table 2.

Table 2: Results of SAS simulation study (n = 50 and n = 75).

P	T	π	n	RB_{sim}	RE_{sim}	n	RB_{sim}	RE_{sim}
0.7	0.7	0.05	50	12.783	126.702	75	8.48008	126.547
0.7	0.7	0.10	50	5.6283	119.175	75	3.72554	118.870
0.7	0.7	0.15	50	3.2693	113.8	75	2.16119	113.434
0.7	0.7	0.2	50	2.0837	109.888	75	1.37404	109.452
0.7	0.7	0.25	50	1.3689	106.985	75	0.90098	106.533
0.7	0.7	0.30	50	0.8955	104.854	75	0.59086	104.375
0.7	0.7	0.35	50	0.5916	103.359	75	0.38780	102.865
0.7	0.7	0.40	50	0.3309	102.334	75	0.22324	101.836
0.7	0.7	0.45	50	0.1351	101.738	75	0.08740	101.237
0.7	0.7	0.50	50	–0.0239	101.538	75	-0.01821	101.027

Clearly from the Table 2, the results of the simulation studies run on SAS show that the suggested estimator will perform better than that of Odumade and Singh (2009). The added β coefficient forces the suggested estimator to always be more efficient, especially when the optimum choice is derived from minimum variance of the estimator. In terms of the relative bias, the results of the simulation study show that the bias of the suggested estimator is not significantly higher or lower than 0. Note that Cochran (1963) claims that if the relative bias is less than 10% everywhere, then the estimator is practical. In Table 2, when $\pi = 0.05$, this is not the case, so we ran another simulation with $n = 75$. The results are also shown in Table 2.

The results continue to remain efficient, but now the relative bias of the estimators is less than 10% everywhere which satisfies Cochran (1963).

6. Conclusion

We knew that the model produced by Odumade and Singh (2009) could perform better than that of Warner (1965), Mangat and Singh (1990), and Kuk (1990). However, the natural question posed was, how do we beat the Odumade and Singh (2009) model? The use of the addition of probabilities pulled from the utilized randomized response model proved to be a genius . The addition of those probabilities in the form of a regression type estimator forced the new model to become more efficient. As one can clearly see from the proofs, tables, and figures, the idea to utilize a regression type estimator was the way to go. The suggested estimator became more efficient than all the other estimators we were testing. Naturally, additional research is required to see if we can further improve an estimator in the field of randomized response sampling. The original version of this chapter can be had from Arias (2019).

7. Future Study

No doubt this chapter opens a new window to study many new possibilities to develop new estimators in randomized response sampling, we list here two more options which could be explored in the same setup considered in this chapter.

(a) An unbiased estimator of $B(\hat{\pi}^{*}_{REG})$ is given by

$$\hat{B}\left(\hat{\pi}^{*}_{REG}\right) = -\frac{(2\hat{\pi}_{OS}-1)[(P+T-1)^2(2PT-P-T)+(P-T)^2\{1-(P+T-2PT)\}]^2}{2n[(P+T-1)^2+(P-T)^2]^2(P+T-2PT)[1-(P+T-2PT)]} \tag{7.1}$$

Thus a new unbiased practical regression type estimator given by

$$\hat{\pi}^{**}_{REG} = \hat{\pi}_{os} + \hat{\beta}\left[\hat{\theta}_{10} + \hat{\theta}_{01} - (P+T-2PT)\right] - \hat{B}\left(\hat{\pi}^{*}_{REG}\right) \tag{7.2}$$

would be a natural estimator of further interest.

(b) Note that the use of the sum of the probabilities θ_{11} and θ_{00} results in a new estimator and can also be investigated.

References

Arias, R. 2019. New methods for efficient results using randomized response sampling. Unpublished MS Thesis submitted to the Department of Mathematics, Texas A&M University-Kingsville, TX.
Cochran, W. G. 1963. Sampling Techniques, 2nd Edition, John Wiley and Sons, New York.
Kuk, A. Y. C. 1990. Asking sensitive questions indirectly. Biometrika, 77(2): 436–438.
Lanke, J. 1975. On the choice of the unrelated question in Simons version of randomized response. J. Amer. Stat. Assoc., 70: 80–83.
Lanke, J. 1976. On the degree of protection in randomized interviews. Int. Stat. Rev., 44: 197–203.
Leysieffer, F. W. and Warner, S. L. 1976. Respondent jeopardy and optimal designs in randomized response models. J. Amer. Stat. Assoc., 71: 649–656.
Mangat, N. S and Singh, R. 1990. An alternative randomized response procedure. Biometrika, 77(2): 439–442.
Odumade, Oluseun and Singh, Sarjinder. 2009. Efficient Use of Two Decks of Cards in Randomized Response Sampling, Communications in Statistics - Theory and Methods, 38: 4, 439–446.
Warner, S. L. 1965. Randomized response: a survey technique for eliminating evasive answer bias. J. Amer. Stat. Assoc., 60: 63–69.

APPENDIX A. Complete SAS Output

Table A1: Detailed results of SAS code.

P	T	PI	RE_W_P0	RE_MS	RE_KUK	RE_OS	RE_W_PT
0.55	0.15	0.05	4164.07	1449.57	155.765	103.008	365.09
0.55	0.15	0.1	3637.19	1274.24	153.465	102.069	330.21
0.55	0.15	0.15	3274.01	1153.38	152.457	101.422	306.17
.	.	.	.	.	.	.	.
	Execute the SASA codes to get full Table						
.	.	.	.	.	.	.	.
0.7	0.25	0.35	127.68	116.15	102.568	102.037	5775.98
0.7	0.25	0.4	125.8	114.59	101.785	100.88	5612.9
0.7	0.25	0.45	124.71	113.7	101.49	100.216	5519.44
0.7	0.25	0.5	124.36	113.41	101.648	100	5488.98

APPENDIX B. SAS Codes

PROGRAM 1: SAS code used to obtain Table A1 and Table 1

```
DATA DATA1;
DO P=.55 TO .70 BY .05;
DO T=.15 TO .30 BY .05;
DO PI=.05 TO .5 BY .05;
TH1=P;
TH2=T;
K=2;
P0=P;
T0=T;
WARNER1=PI*(1-PI);
WARNER2=P0*(1-P0);
WARNER3=2*(2*P0-1)**2;
V_W=WARNER1+(WARNER2/WARNER3);
TH_KUK=TH1*PI+TH2*(1-PI);
V_KUK=TH_KUK*(1-TH_KUK)/(K*(TH1-TH2)**2)+PI*(1-PI)*(1-(1/K));
MS1=PI*(1-PI);
MS2=(1-T0)*(1-P0)*(1-(1-T0)*(1-P0));
MS3=((2*P0-1)+2*T0*(1-P0))**2;
V_MS=MS1+(MS2/MS3);
OS1=((P+T-1)**2)*(P*T+(1-P)*(1-T));
OS2=((P-T)**2)*(T*(1-P)+P*(1-T));
OS3=4*(((P+T-1)**2)+((P-T)**2))**2;
OS4=((2*PI-1)**2)/4;
V_OS=((OS1+OS2)/OS3)-OS4;
REG2_1=(2*PI-1)**2;
REG2_2=((P+T-1)**2)*(2*P*T-P-T);
REG2_3=((P-T)**2)*(1-(P+T-2*P*T));
REG2_4=4*(((P+T-1)**2)+((P-T)**2))**2;
REG2_5=(P+T-2*P*T)*(1-(P+T-2*P*T));
REG2_ALL=(REG2_1*(REG2_2+REG2_3)**2)/(REG2_4*REG2_5);
V_REG2=V_OS-REG2_ALL;
V_NEW=PI*(1-PI)+(P*(1-P)+T*(1-T))/(4*(P+T-1)**2);
RE_W=(V_W/V_REG2)*100;
RE_MS=(V_MS/V_REG2)*100;
RE_KUK=(V_KUK/V_REG2)*100;
RE_OS=(V_OS/V_REG2)*100;
RE_W_2T=(V_NEW/V_REG2)*100;
KEEP PI P T RE_W RE_MS RE_KUK RE_OS RE_W_2T;
KEEP PI P T V_W V_OS V_REG2 V_MS V_KUK V_NEW;
OUTPUT;
END;
END;
END;
DATA DATA2;
SET DATA1;
IF RE_W GE 100;
IF RE_KUK GE 100;
IF RE_MS GE 100;
IF RE_OS GE 100;
IF RE_W_2T GE 100;
PROC PRINT DATA=DATA2;
RUN;
```

PROGRAM 2: SAS code used to obtain Table 2

```
%MACRO ROBERTO(JJJ,NS_IN, P_IN, T_IN,PI_IN,);
DATA DATA0;
P = &P_IN;
T = &T_IN;
PI = &PI_IN;
NS = &NS_IN;
TH11 = (P+T-1)*PI+(1-P)*(1-T);
TH10 = (P-T)*PI+T*(1-P);
TH01 = (T-P)*PI+P*(1-T);
TH00 = (1-P-T)*PI+P*T;
*NS = 50;
%let ntrials = 1000000;
proc iml;
use DATA0;
 read all var _ALL_;
close DATA0;
prob = TH11||TH10||TH01||TH00;
call randseed(1303);
x = RandMultinomial(&ntrials,ns,prob);
run;
*print x;
run;
Create DATA1 from x;
append from x;
RUN;
*PROC PRINT DATA=DATA1;
RUN;
DATA DATA2;
IF _N_ = 1 THEN SET DATA0;
SET DATA1;
TH11E = COL1/NS;
TH10E = COL2/NS;
TH01E = COL3/NS;
TH00E = COL4/NS;
TERM1_OS = (P+T-1)*(TH11E-TH00E)+(P-T)*(TH10E-TH01E);
TERM2_OS = 2*((P+T-1)**2 +(P-T)**2);
PI_OS = TERM1_OS/TERM2_OS + 0.5;
TERM1 = (P+T-1)**2*(2*P*T-P-T)+(P-T)**2*(1-(P+T-2*P*T));
TERM2 = 2*(((P+T-1)**2+(P-T)**2)*(P+T-2*P*T)*(1-(P+T-2*P*T)));
BETAE =-(2*PI_OS-1)*TERM1/TERM2;
PI_ROB = PI_OS + BETAE*(TH10E+TH01E - (P+T-2*P*T));
VAR_OS = (PI_OS-PI)**2;
VAR_ROB = (PI_ROB-PI)**2;
DATA DATA3;
SET DATA2;
KEEP PI PI_OS PI_ROB VAR_OS VAR_ROB;
*proc print data = data3;
RUN;
PROC MEANS DATA = DATA3 NOPRINT;
VAR VAR_OS VAR_ROB PI_ROB ;
OUTPUT OUT = DATA4 MEAN=VAR_OS_SIM VAR_ROB_SIM MEAN_PI_ROB;
```

```
DATA DATA5&JJJ;
SET DATA4;
P=&P_IN;
T=&T_IN;
PI=&PI_IN;
NS =&NS_IN;
RB_ROB = (MEAN_PI_ROB-PI)*100/PI;
RE_ROB = VAR_OS_SIM*100/VAR_ROB_SIM;
KEEP P T PI NS RE_ROB RB_ROB RB_OS;
%MEND ROBERTO;
%ROBERTO(1,50, 0.70, 0.70, 0.05);
%ROBERTO(2,50, 0.70, 0.70, 0.10);
%ROBERTO(3,50, 0.70, 0.70, 0.15);
%ROBERTO(4,50, 0.70, 0.70, 0.20);
%ROBERTO(5,50, 0.70, 0.70, 0.25);
%ROBERTO(6,50, 0.70, 0.70, 0.30);
%ROBERTO(7,50, 0.70, 0.70, 0.35);
%ROBERTO(8,50, 0.70, 0.70, 0.40);
%ROBERTO(9,50, 0.70, 0.70, 0.45);
%ROBERTO(10,50, 0.70, 0.70, 0.50);
DATA FINAL;
SET DATA51 DATA52 DATA53 DATA54 DATA55 DATA56 DATA57 DATA58 DATA59 DATA510;
PROC PRINT DATA = FINAL;
RUN;
```

CHAPTER 11

Ratio and Regression Type Estimators for a New Measure of Coefficient of Dispersion Relative to the Empirical Mode

Christin Variathu Eappen,* Stephen A Sedory* and *Sarjinder Singh*

1. Introduction

The problems of estimation of population mean and variance of a study variable, say Y, have been addressed well in the field of survey sampling. Not much attention has been paid towards measures of dispersion. One of the important measures of dispersion is the well-known parameter called the coefficient of variation which is defined as

$$C_y = \frac{\sigma_y}{\bar{Y}} \tag{1.1}$$

where $\bar{Y} = \frac{1}{N}\sum_{i=1}^{N} Y_i$ and $\sigma_y^2 = \frac{1}{N}\sum_{i=1}^{N}(Y_i - \bar{Y})^2$ have their usual meaning. Clearly the coefficient of variation C_y measures the variation σ_y in the study variable relative to the population mean $\bar{Y}$ that is relative to the center of a data set for the study variable.

Bonett (2006) considered the problem of estimating the population coefficient of dispersion (or coefficient quartile deviation), which is defined as

$$CQV_{y(old)} = \frac{(Q_{3y} - Q_{1y})/2}{(Q_{3y} + Q_{1y})/2} \tag{1.2}$$

where Q_{1y} and Q_{3y} are the first and third quartiles of the population. Bonett and Seier (2006) studies the properties of the estimator of coefficient of dispersion for non-normal distributions. Ambati et al. (2018) further explored the possibility of developing new ratio and regression type estimators for estimating the coefficient of quartile deviation in the presence of auxiliary information. It will be worth pointing out that the coefficient of variation C_y is defined relative to the mean value of the variable, and the coefficient of dispersion $CQV_{y(old)}$ is defined relative to the average values of the first and third quartiles. Thus the value of coefficient of dispersion $CQV_{y(old)}$ depends only on the first and the third quartiles of a data set. This allows us to ask whether one can develop a new measure of coefficient of dispersion which would give importance to some other measures as well.
Recently Eappen et al. (2020) proposed a new coefficient of dispersion for the study variable as

$$CQV_y = \frac{Q_{3y} - Q_{1y}}{2Q_{2y}} \tag{1.3}$$

which clearly depends on the three quartiles Q_{1y}, Q_{2y} and Q_{3y}, and hence is a more informative parameter than the old one.

Further if $\bar{Y} = Q_{2y} = \frac{Q_{3y} + Q_{1y}}{2}$, then the Eappen et al. (2020) measure of dispersion reduces to the traditional measure of coefficient of dispersion studied by Bonett (2006).

The problem of estimation of mode of a study variable has importance while running any type of business for any type of product such as shoes, notebooks and pens. The application of estimation of mode in several skewed variables such as income, drugs, abortions, cancer, AIDS has also been found playing an important role in other allied fields such as medical science or sociology. Doodson (1917) was the first to define an empirical relationship between the three measures of center, namely, mean, median and mode as:

$$\text{Mode} = 3\ \text{Median} - 2\ \text{Mean} \tag{1.4}$$

Department of Mathematics, Texas A&M University-Kingsville, Kingsville, TX 78363, USA.
* Corresponding authors: christin.variathu_eappen@students.tamuk.edu; sarjinder.singh@tamuk.edu

The problem of estimation of empirical mode using auxiliary information has been explained by Sedory and Singh (2014) while developing ratio and regression type estimators. However, in diverse disciplines such as sociology, demography, business, economics, psychology, engineering and medicine, there could be situations where the researchers would be interested to find the variation in a data set relative to the empirical mode. To our knowledge there is no such measure of variation relative to the empirical mode available in literature, thus, following Eappen et al. (2020), we define a new measure of coefficient of dispersion relative to the empirical mode as

$$CQV_{y(m)} = \frac{Q_{3y} - Q_{1y}}{6Q_{2y} - 4\bar{Y}} \tag{1.5}$$

Note that if $\bar{Y} = Q_{2y}$, then the new measure of coefficient of dispersion reduces to the new measure of Eappen et al. (2020).

Thus the proposed new measure of coefficient of dispersion relative to the empirical mode is more of a general case than the other two measures of the coefficient of dispersion.

We define a naïve estimator of the coefficient of dispersion relative to empirical mode as:

$$C\hat{Q}V_{y(m)} = \frac{\left(\hat{Q}_{3y} - \hat{Q}_{1y}\right)}{6\hat{Q}_{2y} - 4\bar{y}} \tag{1.6}$$

We have the following theorem,

Theorem 1.1: The bias, to the first order of approximation, in the proposed naïve estimator $C\hat{Q}V_{y(m)}$ is given by

$$B\left[C\hat{Q}V_{y(m)}\right] = \left(\frac{1-f}{n}\right)CQV_{y(m)}\left[\frac{1}{\left(6Q_{2y} - 4\bar{Y}\right)^2}\left\{\frac{9}{\left\{f_y\left(Q_{2y}\right)\right\}^2} + 16S_y^2 + 48S_{YQ_{2y}}\right\}\right.$$

$$\left. - \frac{1}{\left(Q_{3y} - Q_{1y}\right)\left(6Q_{2y} - 4\bar{Y}\right)}\left\{\frac{3}{4f_y\left(Q_{2y}\right)f_y\left(Q_{3y}\right)} - \frac{3}{4f_y\left(Q_{1y}\right)f_y\left(Q_{2y}\right)} + 4S_{YQ_{3y}} - 4S_{YQ_{1y}}\right\}\right] \tag{1.7}$$

Proof: The estimator $C\hat{O}V_{y(m)}$ may be written as

$$\begin{aligned}
C\hat{Q}V_{y(m)} &= \frac{\hat{Q}_{3y} - \hat{Q}_{1y}}{6\hat{Q}_{2y} - 4\bar{y}} = \frac{Q_{3y}\left(1+\varepsilon_{3y}\right) - Q_{1y}\left(1+\varepsilon_{1y}\right)}{6Q_{2y}\left(1+\varepsilon_{2y}\right) - 4\bar{Y}\left(1+\eta_y\right)} \\
&= \frac{Q_{3y} + Q_{3y}\varepsilon_{3y} - Q_{1y} - Q_{1y}\varepsilon_{1y}}{6Q_{2y} + 6Q_{2y}\varepsilon_{2y} - 4\bar{Y} - 4\bar{Y}\eta_y} = \frac{\left(Q_{3y} - Q_{1y}\right) + Q_{3y}\varepsilon_{3y} - Q_{1y}\varepsilon_{1y}}{6Q_{2y} - 4\bar{Y} + 6Q_{2y}\varepsilon_{2y} - 4\bar{Y}\eta_y} \\
&= \frac{\left(Q_{3y} - Q_{1y}\right)\left[1 + \frac{Q_{3y}\varepsilon_{3y} - Q_{1y}\varepsilon_{1y}}{Q_{3y} - Q_{1y}}\right]}{\left(6Q_{2y} - 4\bar{Y}\right)\left[1 + \frac{6Q_{2y}\varepsilon_{2y} - 4\bar{Y}\eta_y}{6Q_{2y} - 4\bar{Y}}\right]} \\
&= CQV_{y(m)}\left[1 + \frac{Q_{3y}\varepsilon_{3y} - Q_{1y}\varepsilon_{1y}}{Q_{3y} - Q_{1y}}\right]\left[1 + \frac{6Q_{2y}\varepsilon_{2y} - 4\bar{Y}\eta_y}{6Q_{2y} - 4\bar{Y}}\right]^{-1}
\end{aligned} \tag{1.8}$$

Let $\left|\varepsilon_{2y}\right| < 1$, $\left|\eta_y\right| < 1$ and $\left|\frac{6Q_{2y}\varepsilon_{2y} - 4\bar{Y}\eta_y}{6Q_{2y} - 4\bar{Y}}\right| < 1$, then by the binomial expansion of (1.8), to the first order of approximation, we have

$$\hat{CQV}_{y_{(m)}} \approx CQV_{y(m)}\left[1+\frac{Q_{3y}\varepsilon_{3y}-Q_{1y}\varepsilon_{1y}}{Q_{3y}-Q_{1y}}\right]\left[1-\frac{6Q_{2y}\varepsilon_{2y}-4\bar{Y}\eta_y}{6Q_{2y}-4\bar{Y}}+\frac{\left(6Q_{2y}\varepsilon_{2y}-4\bar{Y}\eta_y\right)^2}{\left(6Q_{2y}-4\bar{Y}\right)^2}+\dots\right]$$

$$\approx CQV_{y(m)}\left[1+\frac{Q_{3y}\varepsilon_{3y}-Q_{1y}\varepsilon_{1y}}{Q_{3y}-Q_{1y}}-\frac{6Q_{2y}\varepsilon_{2y}-4\bar{Y}\eta_y}{6Q_{2y}-4\bar{Y}}+\frac{\left(6Q_{2y}\varepsilon_{2y}-4\bar{Y}\eta_y\right)^2}{\left(6Q_{2y}-4\bar{Y}\right)^2}\right.$$

$$\left.-\frac{\left(Q_{3y}\varepsilon_{3y}-Q_{1y}\varepsilon_{1y}\right)\left(6Q_{2y}\varepsilon_{2y}-4\bar{Y}\eta_y\right)}{\left(Q_{3y}-Q_{1y}\right)\left(6Q_{2y}-4\bar{Y}\right)}+\dots\right] \tag{1.9}$$

$$=CQV_{y(m)}\left[1+\frac{Q_{3y}\varepsilon_{3y}-Q_{1y}\varepsilon_{1y}}{Q_{3y}-Q_{1y}}-\frac{6Q_{2y}\varepsilon_{2y}-4\bar{Y}\eta_y}{6Q_{2y}-4\bar{Y}}+\frac{36Q_{2y}^2\varepsilon_{2y}^2+16\bar{Y}^2\eta_y^2-48Q_{2y}\bar{Y}\varepsilon_{2y}\eta_y}{\left(6Q_{2y}-4\bar{Y}\right)^2}\right.$$

$$\left.-\frac{6Q_{2y}Q_{3y}\varepsilon_{2y}\varepsilon_{3y}-6Q_{1y}Q_{2y}\varepsilon_{1y}\varepsilon_{2y}-4Q_{3y}\bar{Y}\varepsilon_{3y}\eta_y+4Q_{1y}\bar{Y}\varepsilon_{1y}\eta_y}{\left(Q_{3y}-Q_{1y}\right)\left(6Q_{2y}-4\bar{Y}\right)}+\dots\right]$$

Taking expected value on both sides of (1.9), we get,

$$E\left(\hat{CQV}_{y_{(m)}}\right)=CQV_{y(m)}\left[1+0-0+\frac{1}{\left(6Q_{2y}-4\bar{Y}\right)^2}\left\{36Q_{2y}^2E\left(\varepsilon_{2y}^2\right)+16\bar{Y}^2E\left(\eta_y^2\right)-48Q_{2y}\bar{Y}E\left(\varepsilon_{2y}\eta_y\right)\right\}\right.$$

$$\left.-\frac{1}{\left(Q_{3y}-Q_{1y}\right)\left(6Q_{2y}-4\bar{Y}\right)}\left\{6Q_{2y}Q_{3y}E\left(\varepsilon_{2y}\varepsilon_{3y}\right)-6Q_{1y}Q_{2y}E\left(\varepsilon_{1y}\varepsilon_{2y}\right)-4Q_{3y}\bar{Y}E\left(\varepsilon_{3y}\eta_y\right)+4Q_{1y}\bar{Y}E\left(\varepsilon_{1y}\eta_y\right)\right\}+\dots\right]$$

$$=CQV_{y(m)}\left[1+\frac{1}{\left(6Q_{2y}-4\bar{Y}\right)^2}\left\{36\left(\frac{1}{4}\right)\left(\frac{1-f}{n}\right)\frac{1}{\left\{f_y\left(Q_{2y}\right)\right\}^2}+16\bar{Y}^2\left(\frac{1-f}{n}\right)C_y^2+48Q_{2y}\bar{Y}\left(\frac{1-f}{n}\right)\frac{S_{YQ_{2y}}}{\bar{Y}Q_{2y}}\right\}\right.$$

$$-\frac{1}{\left(Q_{3y}-Q_{1y}\right)\left(6Q_{2y}-4\bar{Y}\right)}\left\{6Q_{2y}Q_{3y}\left(\frac{1-f}{8n}\right)\frac{1}{f_y\left(Q_{2y}\right)f_y\left(Q_{3y}\right)Q_{2y}Q_{3y}}\right.$$

$$\left.\left.-\left(\frac{1-f}{8n}\right)\frac{6Q_{1y}Q_{2y}}{f_y\left(Q_{1y}\right)f_y\left(Q_{2y}\right)Q_{1y}Q_{2y}}+4Q_{3y}\bar{Y}\left(\frac{1-f}{n}\right)\frac{S_{YQ_{3y}}}{\bar{Y}Q_{3y}}-4Q_{1y}\bar{Y}\left(\frac{1-f}{n}\right)\frac{S_{YQ_{1y}}}{\bar{Y}Q_{1y}}\right\}\right] \tag{1.10}$$

$$=CQV_{y(m)}\left[1+\frac{1}{\left(6Q_{2y}-4\bar{Y}\right)^2}\left(\frac{1-f}{n}\right)\left\{\frac{9}{\left\{f_y\left(Q_{2y}\right)\right\}^2}+16S_y^2+48S_{YQ_{2y}}\right\}\right.$$

$$\left.-\frac{1}{\left(6Q_{2y}-4\bar{Y}\right)\left(Q_{3y}-Q_{1y}\right)}\left(\frac{1-f}{n}\right)\left\{\frac{6}{8f_y\left(Q_{2y}\right)f_y\left(Q_{3y}\right)}-\frac{6}{8f_y\left(Q_{1y}\right)f_y\left(Q_{2y}\right)}+4S_{YQ_{3y}}-4S_{YQ_{1y}}\right\}\right]$$

$$=CQV_{y(m)}+\left(\frac{1-f}{n}\right)CQV_{y(m)}\left[\frac{1}{\left(6Q_{2y}-4\bar{Y}\right)^2}\left\{\frac{9}{\left\{f_y\left(Q_{2y}\right)\right\}^2}+16S_y^2+48S_{YQ_{2y}}\right\}\right.$$

$$\left.-\frac{1}{\left(6Q_{2y}-4\bar{Y}\right)\left(Q_{3y}-Q_{1y}\right)}\left\{\frac{3}{4f_y\left(Q_{2y}\right)f_y\left(Q_{3y}\right)}-\frac{3}{4f_y\left(Q_{1y}\right)f_y\left(Q_{2y}\right)}+4S_{YQ_{3y}}-4S_{YQ_{1y}}\right\}\right]$$

By the definition of bias and using (1.10), we have

$$B\left[C\hat{Q}V_{y(m)}\right]=E\left[C\hat{Q}V_{y(m)}\right]-CQV_{y(m)}$$

$$=\left(\frac{1-f}{n}\right)CQV_{y_{(m)}}\left[\frac{1}{\left(6Q_{2y}-4\bar{Y}\right)^2}\left\{\frac{9}{\left\{f_y\left(Q_{2y}\right)\right\}^2}+16S_y^2+48S_{YQ_{2y}}\right\}\right.$$

$$\left.-\frac{1}{\left(Q_{3y}-Q_{1y}\right)\left(6Q_{2y}-4\bar{Y}\right)}\left\{\frac{3}{4f_y\left(Q_{2y}\right)f_y\left(Q_{3y}\right)}-\frac{3}{4f_y\left(Q_{1y}\right)f_y\left(Q_{2y}\right)}+4S_{YQ_{3y}}-4S_{YQ_{1y}}\right\}\right]$$

which proves the theorem.

It is worth declaring that as the sample size $n\rightarrow\infty$, then $B[C\hat{Q}V_{y_{(m)}}]\rightarrow 0$, thus $C\hat{Q}V_{y_{(m)}}$ is a consistent estimator of the parameter $CQV_{y_{(m)}}$.

Theorem 1.2: The mean squared error, to the first order of approximation, of the estimator $C\hat{Q}V_{y_{(m)}}$ is given by

$$MSE\left[C\hat{Q}V_{y(m)}\right]=\left(\frac{1-f}{n}\right)\left(CQV_{y(m)}\right)^2\Psi_y^{*2} \tag{1.11}$$

where

$$\Psi_y^{*2}=\left[\frac{1}{16\left(Q_{3y}-Q_{1y}\right)^2}\left\{\frac{3}{\left\{f_y\left(Q_{3y}\right)\right\}^2}+\frac{3}{\left\{f_y\left(Q_{1y}\right)\right\}^2}-\frac{2}{f_y\left(Q_{1y}\right)f_y\left(Q_{3y}\right)}\right\}\right.$$

$$+\frac{1}{\left(6Q_{2y}-4\bar{Y}\right)^2}\left\{\frac{9}{\left\{f_y\left(Q_{2y}\right)\right\}^2}+16S_y^2+48S_{YQ_{2y}}\right\} \tag{1.12}$$

$$\left.-\frac{2}{\left(6Q_{2y}-4\bar{Y}\right)\left(Q_{3y}-Q_{1y}\right)}\left\{\frac{3}{4f_y\left(Q_{2y}\right)f_y\left(Q_{3y}\right)}-\frac{3}{4f_y\left(Q_{1y}\right)f_y\left(Q_{2y}\right)}+4S_{YQ_{3y}}-4S_{YQ_{1y}}\right\}\right]$$

Proof: We have

$$MSE\left[C\hat{Q}V_{y(m)}\right]=E\left[C\hat{Q}V_{y(m)}-CQV_{y(m)}\right]^2$$

$$=\left(CQV_{y(m)}\right)^2E\left[\frac{Q_{3y}\varepsilon_{3y}-Q_{1y}\varepsilon_{1y}}{Q_{3y}-Q_{1y}}-\frac{6Q_{2y}\varepsilon_{2y}-4\bar{Y}\eta_y}{6Q_{2y}-4\bar{Y}}\right]^2$$

$$=\left(CQV_{y(m)}\right)^2E\left[\left(\frac{Q_{3y}\varepsilon_{3y}-Q_{1y}\varepsilon_{1y}}{Q_{3y}-Q_{1y}}\right)^2+\left(\frac{6Q_{2y}\varepsilon_{2y}-4\bar{Y}\eta_y}{6Q_{2y}-4\bar{Y}}\right)^2-2\left(\frac{Q_{3y}\varepsilon_{3y}-Q_{1y}\varepsilon_{1y}}{Q_{3y}-Q_{1y}}\right)\left(\frac{6Q_{2y}\varepsilon_{2y}-4\bar{Y}\eta_y}{6Q_{2y}-4\bar{Y}}\right)\right]$$

$$=\left(CQV_{y(m)}\right)^2E\left[\frac{1}{\left(Q_{3y}-Q_{1y}\right)^2}\left(Q_{3y}\varepsilon_{3y}-Q_{1y}\varepsilon_{1y}\right)^2+\frac{1}{\left(6Q_{2y}-4\bar{Y}\right)^2}\left(6Q_{2y}\varepsilon_{2y}-4\bar{Y}\eta_y\right)^2\right.$$

$$\left.-\frac{2}{\left(Q_{3y}-Q_{1y}\right)\left(6Q_{2y}-4\bar{Y}\right)}\left(Q_{3y}\varepsilon_{3y}-Q_{1y}\varepsilon_{1y}\right)\left(6Q_{2y}\varepsilon_{2y}-4\bar{Y}\eta_y\right)\right]$$

$$=\left(CQV_{y(m)}\right)^2E\left[\frac{1}{\left(Q_{3y}-Q_{1y}\right)^2}\left(Q_{3y}^2\varepsilon_{3y}^2+Q_{1y}^2\varepsilon_{1y}^2-2Q_{1y}Q_{3y}\varepsilon_{1y}\varepsilon_{3y}\right)\right.$$

$$+\frac{1}{\left(6Q_{2y}-4\bar{Y}\right)^2}\left(36Q_{2y}^2\varepsilon_{2y}^2+16\bar{Y}^2\eta_y^2-48Q_{2y}\bar{Y}\varepsilon_{2y}\eta_y\right)$$

$$-\frac{2}{(Q_{3y}-Q_{1y})(6Q_{2y}-4\bar{Y})}\left(6Q_{2y}Q_{3y}\varepsilon_{2y}\varepsilon_{3y}-6Q_{1y}Q_{2y}\varepsilon_{1y}\varepsilon_{2y}-4Q_{3y}\bar{Y}\varepsilon_{3y}\eta_y+4Q_{1y}\bar{Y}\varepsilon_{1y}\eta_y\right)\Bigg]$$

$$=(CQV_{y(m)})^2\Bigg[\frac{1}{(Q_{3y}-Q_{1y})^2}\left\{Q_{3y}^2E(\varepsilon_{3y}^2)+Q_{1y}^2E(\varepsilon_{1y}^2)-2Q_{1y}Q_{3y}E(\varepsilon_{1y}\varepsilon_{3y})\right\}$$

$$+\frac{1}{(6Q_{2y}-4\bar{Y})^2}\left\{36Q_{2y}^2E(\varepsilon_{2y}^2)+16\bar{Y}^2E(\eta_y^2)-48Q_{2y}\bar{Y}E(\varepsilon_{2y}\eta_y)\right\}$$

$$-\frac{2}{(Q_{3y}-Q_{1y})(6Q_{2y}-4\bar{Y})}\left\{6Q_{2y}Q_{3y}E(\varepsilon_{2y}\varepsilon_{3y})-6Q_{1y}Q_{2y}E(\varepsilon_{1y}\varepsilon_{2y})-4Q_{3y}\bar{Y}E(\varepsilon_{3y}\eta_y)+4Q_{1y}\bar{Y}E(\varepsilon_{1y}\eta_y)\right\}\Bigg]$$

$$=(CQV_{y(m)})^2\Bigg[\frac{1}{(Q_{3y}-Q_{1y})^2}\Bigg\{Q_{3y}^2\left(\frac{1-f}{n}\right)\left(\frac{3}{16}\right)\frac{1}{Q_{3y}^2\{f_y(Q_{3y})\}^2}$$

$$+Q_{1y}^2\left(\frac{1-f}{n}\right)\left(\frac{3}{16}\right)\frac{1}{Q_{1y}^2\{f_y(Q_{1y})\}^2}-2Q_{1y}Q_{3y}\left(\frac{1-f}{16n}\right)\frac{1}{f_y(Q_{1y})f_y(Q_{3y})Q_{1y}Q_{3y}}\Bigg\}$$

$$+\frac{1}{(6Q_{2y}-4\bar{Y})^2}\left\{36Q_{2y}^2\left(\frac{1-f}{n}\right)\left(\frac{1}{4}\right)\frac{1}{Q_{2y}^2\{f_y(Q_{2y})\}^2}+16\bar{Y}^2\left(\frac{1-f}{n}\right)C_y^2+48Q_{2y}\bar{Y}\left(\frac{1-f}{n}\right)\frac{S_{YQ_{2y}}}{\bar{Y}Q_{2y}}\right\}$$

$$-\frac{2}{(Q_{3y}-Q_{1y})(6Q_{2y}-4\bar{Y})}\Bigg\{6Q_{2y}Q_{3y}\left(\frac{1-f}{8n}\right)\frac{1}{f_y(Q_{2y})f_y(Q_{3y})Q_{2y}Q_{3y}}$$

$$-6Q_{1y}Q_{2y}\left(\frac{1-f}{8n}\right)\frac{1}{f_y(Q_{1y})f_y(Q_{2y})Q_{1y}Q_{2y}}+4Q_{3y}\bar{Y}\left(\frac{1-f}{n}\right)\frac{S_{YQ_{3y}}}{\bar{Y}Q_{3y}}-4Q_{1y}\bar{Y}\left(\frac{1-f}{n}\right)\frac{S_{YQ_{1y}}}{\bar{Y}Q_{1y}}\Bigg\}\Bigg]$$

$$=\left(\frac{1-f}{n}\right)(CQV_{y(m)})^2\Bigg[\frac{1}{(Q_{3y}-Q_{1y})^2}\left\{\frac{3}{16\{f_y(Q_{3y})\}^2}+\frac{3}{16\{f_y(Q_{1y})\}^2}-\frac{2}{16f_y(Q_{1y})f_y(Q_{3y})}\right\}$$

$$+\frac{1}{(6Q_{2y}-4\bar{Y})^2}\left\{\frac{9}{\{f_y(Q_{2y})\}^2}+16S_y^2+48S_{YQ_{2y}}\right\}$$

$$-\frac{2}{(Q_{3y}-Q_{1y})(6Q_{2y}-4\bar{Y})}\left\{\frac{6}{8f_y(Q_{2y})f_y(Q_{3y})}-\frac{6}{8f_y(Q_{1y})f_y(Q_{2y})}+4S_{YQ_{3y}}-4S_{YQ_{1y}}\right\}\Bigg]$$

$$=\left(\frac{1-f}{n}\right)(CQV_{y(m)})^2\Bigg[\frac{1}{16(Q_{3y}-Q_{1y})^2}\left\{\frac{3}{\{f_y(Q_{3y})\}^2}+\frac{3}{\{f_y(Q_{1y})\}^2}-\frac{2}{f_y(Q_{1y})f_y(Q_{3y})}\right\}$$

$$+\frac{1}{(6Q_{2y}-4\bar{Y})^2}\left\{\frac{9}{\{f_y(Q_{2y})\}^2}+16S_y^2+48S_{YQ_{2y}}\right\}$$

$$-\frac{2}{(Q_{3y}-Q_{1y})(6Q_{2y}-4\bar{Y})}\left\{\frac{3}{4f_y(Q_{2y})f_y(Q_{3y})}-\frac{3}{4f_y(Q_{1y})f_y(Q_{2y})}+4S_{YQ_{3y}}-4S_{YQ_{1y}}\right\}\Bigg]$$

which proves the theorem.

In the subsequent section, we propose a ratio type estimator of the new measure of coefficient of dispersion $CQV_{y_{(m)}}$ assuming that the new measure of the coefficient of dispersion $CQV_{x_{(m)}}$ of an auxiliary variable is known.

2. Ratio Estimator of the New Measure of Coefficient of Dispersion

Assume the value of the new coefficient of dispersion $CQV_{x_{(m)}}$ of an auxiliary variable is known and is given by

$$CQV_{x(m)}=\frac{Q_{3x}-Q_{1x}}{6Q_{2x}-4\bar{X}} \tag{2.1}$$

From (2.1), a consistent estimator of $CQV_{x_{(m)}}$ can be defined as

$$C\hat{Q}V_{x(m)} = \frac{\hat{Q}_{3x} - \hat{Q}_{1x}}{6\hat{Q}_{2x} - 4\bar{x}} \tag{2.2}$$

Following Cochran (1940), a ratio type estimator of $CQV_{y_{(m)}}$ is defined as,

$$C\hat{Q}V_{rat(m)} = C\hat{Q}V_{y(m)}\left(\frac{CQV_{x(m)}}{C\hat{Q}V_{x(m)}}\right) \tag{2.3}$$

Now we have the subsequent theorem,

Theorem 2.1. The bias, to the first order of approximation, in the proposed ratio type estimator $C\hat{Q}V_{rat_{(m)}}$ is given by

$$\begin{aligned}
B\left[C\hat{Q}V_{rat(m)}\right] = \left(\frac{1-f}{n}\right)CQV_{y(m)}\Bigg[& \frac{1}{\left(6Q_{2y}-4\bar{Y}\right)^2}\left\{\frac{9}{\left\{f_y\left(Q_{2y}\right)\right\}^2}+16S_y^2+48S_{YQ_{2y}}\right\} \\
& +\frac{1}{\left(Q_{3x}-Q_{1x}\right)^2}\left\{\frac{3}{16\left\{f_x\left(Q_{3x}\right)\right\}^2}+\frac{3}{16\left\{f_x\left(Q_{1x}\right)\right\}^2}-\frac{2}{16f_x\left(Q_{1x}\right)f_x\left(Q_{3x}\right)}\right\} \\
& -\frac{1}{\left(Q_{3y}-Q_{1y}\right)\left(6Q_{2y}-4\bar{Y}\right)}\left\{\frac{6}{8f_y\left(Q_{2y}\right)f_y\left(Q_{3y}\right)}-\frac{6}{8f_y\left(Q_{1y}\right)f_y\left(Q_{2y}\right)}+4S_{YQ_{3y}}-4S_{YQ_{1y}}\right\} \\
& +\frac{1}{\left(Q_{3y}-Q_{1y}\right)\left(6Q_{2x}-4\bar{X}\right)}\left\{\frac{6}{f_x\left(Q_{2x}\right)f_y\left(Q_{3y}\right)}\left(P_{Q_{2x}Q_{3y}}-\frac{3}{8}\right)-\frac{6}{f_x\left(Q_{2x}\right)f_y\left(Q_{1y}\right)}\left(P_{Q_{2x}Q_{1y}}-\frac{1}{8}\right)+4S_{XQ_{3y}}-4S_{XQ_{1y}}\right\} \\
& -\frac{1}{\left(6Q_{2y}-4\bar{Y}\right)\left(6Q_{2x}-4\bar{X}\right)}\left\{\frac{36}{f_x\left(Q_{2x}\right)f_y\left(Q_{2y}\right)}\left(P_{Q_{2x}Q_{2y}}-\frac{1}{4}\right)+24S_{YQ_{2x}}+24S_{XQ_{2y}}+16S_{xy}\right\} \\
& -\frac{1}{\left(6Q_{2x}-4\bar{X}\right)\left(Q_{3x}-Q_{1x}\right)}\left\{\frac{6}{8f_x\left(Q_{2x}\right)f_x\left(Q_{3x}\right)}+4S_{XQ_{3x}}-\frac{6}{8f_x\left(Q_{1x}\right)f_x\left(Q_{2x}\right)}-4S_{XQ_{1x}}\right\} \\
& -\frac{1}{\left(Q_{3y}-Q_{1y}\right)\left(Q_{3x}-Q_{1x}\right)}\left\{\frac{1}{f_x\left(Q_{3x}\right)f_y\left(Q_{3y}\right)}\left(P_{Q_{3x}Q_{3y}}-\frac{9}{16}\right)-\frac{1}{f_x\left(Q_{3x}\right)f_y\left(Q_{1y}\right)}\left(P_{Q_{3x}Q_{1y}}-\frac{3}{16}\right)\right. \\
& \left.-\frac{1}{f_x\left(Q_{1x}\right)f_y\left(Q_{3y}\right)}\left(P_{Q_{1x}Q_{3y}}-\frac{3}{16}\right)+\frac{1}{f_x\left(Q_{1x}\right)f_y\left(Q_{1y}\right)}\left(P_{Q_{1x}Q_{1y}}-\frac{1}{16}\right)\right\} \\
& +\frac{1}{\left(6Q_{2y}-4\bar{Y}\right)\left(Q_{3x}-Q_{1x}\right)}\left\{\frac{6}{f_x\left(Q_{3x}\right)f_y\left(Q_{2y}\right)}\left(P_{Q_{3x}Q_{2y}}-\frac{3}{8}\right)+4S_{YQ_{3x}}\right. \\
& \left.-\frac{6}{f_x\left(Q_{1x}\right)f_y\left(Q_{2y}\right)}\left(P_{Q_{1x}Q_{2y}}-\frac{1}{8}\right)-4S_{YQ_{1x}}\right\}\Bigg]
\end{aligned} \tag{2.4}$$

Proof: The estimator $C\hat{Q}V_{rat_{(m)}}$ can be written as,

$$\begin{aligned}
C\hat{Q}V_{rat(m)} &= C\hat{Q}V_{y(m)}\left(\frac{CQV_{x(m)}}{C\hat{Q}V_{x(m)}}\right) \\
&= \frac{\left(\hat{Q}_{3y}-\hat{Q}_{1y}\right)}{\left(6\hat{Q}_{2y}-4\bar{y}\right)}\left[\frac{\dfrac{Q_{3x}-Q_{1x}}{6Q_{2x}-4\bar{X}}}{\dfrac{\hat{Q}_{3x}-\hat{Q}_{1x}}{6\hat{Q}_{2x}-4\bar{x}}}\right] \\
&= \frac{Q_{3y}\left(1+\varepsilon_{3y}\right)-Q_{1y}\left(1+\varepsilon_{1y}\right)}{6Q_{2y}\left(1+\varepsilon_{2y}\right)-4\bar{Y}\left(1+\eta_y\right)}\left[\frac{\dfrac{Q_{3x}-Q_{1x}}{6Q_{2x}-4\bar{X}}}{\dfrac{Q_{3x}\left(1+\varepsilon_{3x}\right)-Q_{1x}\left(1+\varepsilon_{1x}\right)}{6Q_{2x}\left(1+\varepsilon_{2x}\right)-4\bar{X}\left(1+\eta_x\right)}}\right]
\end{aligned}$$

$$=\frac{\left(Q_{3y}+Q_{3y}\varepsilon_{3y}-Q_{1y}-Q_{1y}\varepsilon_{1y}\right)}{\left(6Q_{2y}+6Q_{2y}\varepsilon_{2y}-4\bar{Y}-4\bar{Y}\eta_y\right)}\left[\frac{\left(\dfrac{Q_{3x}-Q_{1x}}{6Q_{2x}-4\bar{X}}\right)}{\dfrac{Q_{3x}+Q_{3x}\varepsilon_{3x}-Q_{1x}-Q_{1x}\varepsilon_{1x}}{6Q_{2x}+6Q_{2x}\varepsilon_{2x}-4\bar{X}-4\bar{X}\eta_x}}\right]$$

$$=\frac{\left(Q_{3y}-Q_{1y}\right)+\left(Q_{3y}\varepsilon_{3y}-Q_{1y}\varepsilon_{1y}\right)}{\left(6Q_{2y}-4\bar{Y}\right)+\left(6Q_{2y}\varepsilon_{2y}-4\bar{Y}\eta_y\right)}\left[\frac{\left(\dfrac{Q_{3x}-Q_{1x}}{6Q_{2x}-4\bar{X}}\right)}{\dfrac{Q_{3x}-Q_{1x}+Q_{3x}\varepsilon_{3x}-Q_{1x}\varepsilon_{1x}}{6Q_{2x}-4\bar{X}+6Q_{2x}\varepsilon_{2x}-4\bar{X}\eta_x}}\right]$$

$$=\frac{\left(Q_{3y}-Q_{1y}\right)\left\{1+\dfrac{Q_{3y}\varepsilon_{3y}-Q_{1y}\varepsilon_{1y}}{Q_{3y}-Q_{1y}}\right\}}{\left(6Q_{2y}-4\bar{Y}\right)\left\{1+\dfrac{6Q_{2y}\varepsilon_{2y}-4\bar{Y}\eta_y}{6Q_{2y}-4\bar{Y}}\right\}}\left[\frac{\left(\dfrac{Q_{3x}-Q_{1x}}{6Q_{2x}-4\bar{X}}\right)\left\{6Q_{2x}-4\bar{X}+6Q_{2x}\varepsilon_{2x}-4\bar{X}\eta_x\right\}}{\left(Q_{3x}-Q_{1x}\right)+Q_{3x}\varepsilon_{3x}-Q_{1x}\varepsilon_{1x}}\right]$$

$$=\left(CQV_{y(m)}\right)\left[1+\frac{Q_{3y}\varepsilon_{3y}-Q_{1y}\varepsilon_{1y}}{Q_{3y}-Q_{1y}}\right]\left[1+\frac{6Q_{2y}\varepsilon_{2y}-4\bar{Y}\eta_y}{6Q_{2y}-4\bar{Y}}\right]^{-1}\left[\frac{\left(\dfrac{Q_{3x}-Q_{1x}}{6Q_{2x}-4\bar{X}}\right)\left(6Q_{2x}-4\bar{X}\right)\left\{1+\dfrac{6Q_{2x}\varepsilon_{2x}-4\bar{X}\eta_x}{6Q_{2x}-4\bar{X}}\right\}}{\left(Q_{3x}-Q_{1x}\right)\left\{1+\dfrac{Q_{3x}\varepsilon_{3x}-Q_{1x}\varepsilon_{1x}}{Q_{3x}-Q_{1x}}\right\}}\right]$$

$$=\left(CQV_{y(m)}\right)\left[1+\frac{Q_{3y}\varepsilon_{3y}-Q_{1y}\varepsilon_{1y}}{Q_{3y}-Q_{1y}}\right]\left[1+\frac{6Q_{2y}\varepsilon_{2y}-4\bar{Y}\eta_y}{6Q_{2y}-4\bar{Y}}\right]^{-1}\left[1+\frac{6Q_{2x}\varepsilon_{2x}-4\bar{X}\eta_x}{6Q_{2x}-4\bar{X}}\right]\left[1+\frac{Q_{3x}\varepsilon_{3x}-Q_{1x}\varepsilon_{1x}}{Q_{3x}-Q_{1x}}\right]^{-1} \tag{2.5}$$

Now assuming $\left|\dfrac{6Q_{2y}\varepsilon_{2y}-4\bar{Y}\eta_y}{6Q_{2y}-4\bar{Y}}\right|<1$ and $\left|\dfrac{Q_{3x}\varepsilon_{3x}-Q_{1x}\varepsilon_{1x}}{Q_{3x}-Q_{1x}}\right|<1$, applying binomial expansion, the expression (2.5), to the first order of approximation, can be written as:

$$\begin{aligned}C\hat{Q}V_{rat(m)}=\left(CQV_{y(m)}\right)&\left[1+\frac{Q_{3y}\varepsilon_{3y}-Q_{1y}\varepsilon_{1y}}{Q_{3y}-Q_{1y}}\right]\left[1-\frac{6Q_{2y}\varepsilon_{2y}-4\bar{Y}\eta_y}{6Q_{2y}-4\bar{Y}}\right.\\&\left.+\left(\frac{6Q_{2y}\varepsilon_{2y}-4\bar{Y}\eta_y}{6Q_{2y}-4\bar{Y}}\right)^2+O_p\left(n^{-1}\right)\right]\left[1+\frac{6Q_{2x}\varepsilon_{2x}-4\bar{X}\eta_x}{6Q_{2x}-4\bar{X}}\right]\left[1-\frac{Q_{3x}\varepsilon_{3x}-Q_{1x}\varepsilon_{1x}}{Q_{3x}-Q_{1x}}\right.\\&\left.+\left(\frac{Q_{3x}\varepsilon_{3x}-Q_{1x}\varepsilon_{1x}}{Q_{3x}-Q_{1x}}\right)^2+O_p\left(n^{-1}\right)\right]\end{aligned}$$

$$\begin{aligned}=\left(CQV_{y(m)}\right)&\left[1+\frac{Q_{3y}\varepsilon_{3y}-Q_{1y}\varepsilon_{1y}}{Q_{3y}-Q_{1y}}-\frac{6Q_{2y}\varepsilon_{2y}-4\bar{Y}\eta_y}{6Q_{2y}-4\bar{Y}}+\left(\frac{6Q_{2y}\varepsilon_{2y}-4\bar{Y}\eta_y}{6Q_{2y}-4\bar{Y}}\right)^2\right.\\&\left.-\left(\frac{Q_{3y}\varepsilon_{3y}-Q_{1y}\varepsilon_{1y}}{Q_{3y}-Q_{1y}}\right)\left(\frac{6Q_{2y}\varepsilon_{2y}-4\bar{Y}\eta_y}{6Q_{2y}-4\bar{Y}}\right)+O_p\left(n^{-1}\right)\right]\left[1+\left(\frac{6Q_{2x}\varepsilon_{2x}-4\bar{X}\eta_x}{6Q_{2x}-4\bar{X}}\right)+\left(\frac{Q_{3x}\varepsilon_{3x}-Q_{1x}\varepsilon_{1x}}{Q_{3x}-Q_{1x}}\right)^2\right.\\&\left.-\left(\frac{6Q_{2x}\varepsilon_{2x}-4\bar{X}\eta_x}{6Q_{2x}-4\bar{X}}\right)\left(\frac{Q_{3x}\varepsilon_{3x}-Q_{1x}\varepsilon_{1x}}{Q_{3x}-Q_{1x}}\right)-\left(\frac{Q_{3x}\varepsilon_{3x}-Q_{1x}\varepsilon_{1x}}{Q_{3x}-Q_{1x}}\right)+O_p\left(n^{-1}\right)\right]\end{aligned}$$

$$\begin{aligned}=\left(CQV_{y(m)}\right)&\left[1+\frac{Q_{3y}\varepsilon_{3y}-Q_{1y}\varepsilon_{1y}}{Q_{3y}-Q_{1y}}-\frac{6Q_{2y}\varepsilon_{2y}-4\bar{Y}\eta_y}{6Q_{2y}-4\bar{Y}}+\left(\frac{6Q_{2y}\varepsilon_{2y}-4\bar{Y}\eta_y}{6Q_{2y}-4\bar{Y}}\right)^2\right.\\&-\left(\frac{Q_{3y}\varepsilon_{3y}-Q_{1y}\varepsilon_{1y}}{Q_{3y}-Q_{1y}}\right)\left(\frac{6Q_{2y}\varepsilon_{2y}-4\bar{Y}\eta_y}{6Q_{2y}-4\bar{Y}}\right)+\left(\frac{6Q_{2x}\varepsilon_{2x}-4\bar{X}\eta_x}{6Q_{2x}-4\bar{X}}\right)-\left(\frac{Q_{3x}\varepsilon_{3x}-Q_{1x}\varepsilon_{1x}}{Q_{3x}-Q_{1x}}\right)\\&-\left(\frac{6Q_{2y}\varepsilon_{2y}-4\bar{Y}\eta_y}{6Q_{2y}-4\bar{Y}}\right)\left(\frac{6Q_{2x}\varepsilon_{2x}-4\bar{X}\eta_x}{6Q_{2x}-4\bar{X}}\right)+\left(\frac{Q_{3x}\varepsilon_{3x}-Q_{1x}\varepsilon_{1x}}{Q_{3x}-Q_{1x}}\right)^2\end{aligned}$$

$$
+\left(\frac{Q_{3y}\varepsilon_{3y}-Q_{1y}\varepsilon_{1y}}{Q_{3y}-Q_{1y}}\right)\left(\frac{6Q_{2x}\varepsilon_{2x}-4\bar{X}\eta_x}{6Q_{2x}-4\bar{X}}\right)-\left(\frac{6Q_{2x}\varepsilon_{2x}-4\bar{X}\eta_x}{6Q_{2x}-4\bar{X}}\right)\left(\frac{Q_{3x}\varepsilon_{3x}-Q_{1x}\varepsilon_{1x}}{Q_{3x}-Q_{1x}}\right)
$$

$$
-\left(\frac{Q_{3y}\varepsilon_{3y}-Q_{1y}\varepsilon_{1y}}{Q_{3y}-Q_{1y}}\right)\left(\frac{Q_{3x}\varepsilon_{3x}-Q_{1x}\varepsilon_{1x}}{Q_{3x}-Q_{1x}}\right)+\left(\frac{6Q_{2y}\varepsilon_{2y}-4\bar{Y}\eta_y}{6Q_{2y}-4\bar{Y}}\right)\left(\frac{Q_{3x}\varepsilon_{3x}-Q_{1x}\varepsilon_{1x}}{Q_{3x}-Q_{1x}}\right)+O_p\left(n^{-1}\right)\Bigg]
$$

$$
\begin{aligned}
&=\left(CQV_{y(m)}\right)\Bigg[1+\left(\frac{Q_{3y}\varepsilon_{3y}-Q_{1y}\varepsilon_{1y}}{Q_{3y}-Q_{1y}}\right)-\left(\frac{6Q_{2y}\varepsilon_{2y}-4\bar{Y}\eta_y}{6Q_{2y}-4\bar{Y}}\right)+\left(\frac{6Q_{2x}\varepsilon_{2x}-4\bar{X}\eta_x}{6Q_{2x}-4\bar{X}}\right)\\
&\quad-\left(\frac{Q_{3x}\varepsilon_{3x}-Q_{1x}\varepsilon_{1x}}{Q_{3x}-Q_{1x}}\right)+\frac{1}{\left(6Q_{2y}-4\bar{Y}\right)^2}\left\{36Q_{2y}^2\varepsilon_{2y}^2+16\bar{Y}^2\eta_y^2-48Q_{2y}\bar{Y}\varepsilon_{2y}\eta_y\right\}\\
&\quad+\frac{1}{\left(Q_{3x}-Q_{1x}\right)^2}\left\{Q_{3x}^2\varepsilon_{3x}^2+Q_{1x}^2\varepsilon_{1x}^2-2Q_{1x}Q_{3x}\varepsilon_{1x}\varepsilon_{3x}\right\}\\
&\quad-\frac{1}{\left(Q_{3y}-Q_{1y}\right)\left(6Q_{2y}-4\bar{Y}\right)}\left\{6Q_{2y}Q_{3y}\varepsilon_{2y}\varepsilon_{3y}-6Q_{1y}Q_{2y}\varepsilon_{1y}\varepsilon_{2y}-4Q_{3y}\bar{Y}\varepsilon_{3y}\eta_y+4Q_{1y}\bar{Y}\varepsilon_{1y}\eta_y\right\}\\
&\quad+\frac{1}{\left(Q_{3y}-Q_{1y}\right)\left(6Q_{2x}-4\bar{X}\right)}\left\{6Q_{2x}Q_{3y}\varepsilon_{3y}\varepsilon_{2x}-6Q_{1y}Q_{2x}\varepsilon_{1y}\varepsilon_{2x}-4Q_{3y}\bar{X}\varepsilon_{3y}\eta_x+4Q_{1y}\bar{X}\varepsilon_{1y}\eta_x\right\}\\
&\quad-\frac{1}{\left(6Q_{2y}-4\bar{Y}\right)\left(6Q_{2x}-4\bar{X}\right)}\left\{36Q_{2y}Q_{2x}\varepsilon_{2y}\varepsilon_{2x}-24\bar{Y}Q_{2x}\varepsilon_{2x}\eta_y-24\bar{X}Q_{2y}\varepsilon_{2y}\eta_x+16\bar{X}\bar{Y}\eta_y\eta_x\right\}\\
&\quad-\frac{1}{\left(6Q_{2x}-4\bar{X}\right)\left(Q_{3x}-Q_{1x}\right)}\left\{6Q_{2x}Q_{3x}\varepsilon_{2x}\varepsilon_{3x}-4Q_{3x}\bar{X}\varepsilon_{3x}\eta_x-6Q_{1x}Q_{2x}\varepsilon_{1x}\varepsilon_{2x}+4Q_{1x}\bar{X}\varepsilon_{1x}\eta_x\right\}\\
&\quad-\frac{1}{\left(Q_{3y}-Q_{1y}\right)\left(Q_{3x}-Q_{1x}\right)}\left\{Q_{3y}Q_{3x}\varepsilon_{3y}\varepsilon_{3x}-Q_{1y}Q_{3x}\varepsilon_{1y}\varepsilon_{3x}-Q_{3y}Q_{1x}\varepsilon_{3y}\varepsilon_{1x}+Q_{1y}Q_{1x}\varepsilon_{1y}\varepsilon_{1x}\right\}\\
&\quad+\frac{1}{\left(6Q_{2y}-4\bar{Y}\right)\left(Q_{3x}-Q_{1x}\right)}\left\{6Q_{2y}Q_{3x}\varepsilon_{2y}\varepsilon_{3x}-4Q_{3x}\bar{Y}\varepsilon_{3x}\eta_y-6Q_{2y}Q_{1x}\varepsilon_{2y}\varepsilon_{1x}+4Q_{1x}\bar{Y}\varepsilon_{1x}\eta_y\right\}\Bigg]
\end{aligned}
\tag{2.6}
$$

Taking expected values on both sides of (2.6), we get,

$$
\begin{aligned}
E\left[CQV_{rat(m)}\right]&=\left(CQV_{y(m)}\right)\Bigg[1+0-0+0-0+\frac{1}{\left(6Q_{2y}-4\bar{Y}\right)^2}\Bigg\{36Q_{2y}^2\left(\frac{1-f}{n}\right)\left(\frac{1}{4}\right)\frac{1}{Q_{2y}^2\left\{f_y\left(Q_{2y}\right)\right\}^2}\\
&\quad+16\bar{Y}^2\left(\frac{1-f}{n}\right)C_y^2+48Q_{2y}\bar{Y}\left(\frac{1-f}{n}\right)\frac{S_{YQ_{2y}}}{\bar{Y}Q_{2y}}\Bigg\}\\
&\quad+\frac{1}{\left(Q_{3x}-Q_{1x}\right)^2}\Bigg\{Q_{3x}^2\left(\frac{1-f}{n}\right)\left(\frac{3}{16}\right)\frac{1}{Q_{3x}^2\left\{f_x\left(Q_{3x}\right)\right\}^2}+Q_{1x}^2\left(\frac{1-f}{n}\right)\left(\frac{3}{16}\right)\frac{1}{Q_{1x}^2\left\{f_x\left(Q_{1x}\right)\right\}^2}\\
&\quad-2Q_{1x}Q_{3x}\left(\frac{1-f}{16n}\right)\frac{1}{f_x\left(Q_{1x}\right)f_x\left(Q_{3x}\right)Q_{1x}Q_{3x}}\Bigg\}\\
&\quad-\frac{1}{\left(Q_{3y}-Q_{1y}\right)\left(6Q_{2y}-4\bar{Y}\right)}\Bigg\{6Q_{2y}Q_{3y}\left(\frac{1-f}{8n}\right)\frac{1}{f_y\left(Q_{2y}\right)f_y\left(Q_{3y}\right)Q_{2y}Q_{3y}}\\
&\quad-6Q_{1y}Q_{2y}\left(\frac{1-f}{8n}\right)\frac{1}{f_y\left(Q_{1y}\right)f_y\left(Q_{2y}\right)Q_{1y}Q_{2y}}\\
&\quad+4Q_{3y}\bar{Y}\left(\frac{1-f}{n}\right)\frac{S_{YQ_{3y}}}{Q_{3y}\bar{Y}}-4Q_{1y}\bar{Y}\left(\frac{1-f}{n}\right)\frac{S_{YQ_{1y}}}{Q_{1y}\bar{Y}}\Bigg\}\\
&\quad+\frac{1}{\left(Q_{3y}-Q_{1y}\right)\left(6Q_{2x}-4\bar{X}\right)}\Bigg\{6Q_{2x}Q_{3y}\left(\frac{1-f}{n}\right)\frac{\left\{f_x\left(Q_{2x}\right)\right\}^{-1}\left\{f_y\left(Q_{3y}\right)\right\}^{-1}}{Q_{2x}Q_{3y}}\left(P_{Q_{2x}Q_{3y}}-\frac{3}{8}\right)
\end{aligned}
$$

$$-6Q_{1y}Q_{2x}\left(\frac{1-f}{n}\right)\frac{\{f_y(Q_{1y})\}^{-1}\{f_x(Q_{2x})\}^{-1}}{Q_{1y}Q_{2x}}\left(P_{Q_{2x}Q_{1y}}-\frac{1}{8}\right)$$

$$\left.+4Q_{3y}\bar{X}\left(\frac{1-f}{n}\right)\frac{S_{XQ_{3y}}}{\bar{X}Q_{3y}}-4Q_{1y}\bar{X}\left(\frac{1-f}{n}\right)\frac{S_{XQ_{1y}}}{\bar{X}Q_{1y}}\right\}$$

$$-\frac{1}{(6Q_{2y}-4\bar{Y})(6Q_{2x}-4\bar{X})}\left\{36Q_{2y}Q_{2x}\left(\frac{1-f}{n}\right)\frac{\{f_x(Q_{2x})\}^{-1}\{f_y(Q_{2y})\}^{-1}}{Q_{2y}Q_{2x}}\left(P_{Q_{2x}Q_{2y}}-\frac{1}{4}\right)\right.$$

$$\left.+24\bar{Y}Q_{2x}\left(\frac{1-f}{n}\right)\frac{S_{YQ_{2x}}}{\bar{Y}Q_{2x}}+24Q_{2y}\bar{X}\left(\frac{1-f}{n}\right)\frac{S_{XQ_{2y}}}{\bar{X}Q_{2y}}+16\bar{X}\bar{Y}\rho_{xy}C_xC_y\right\}$$

$$-\frac{1}{(6Q_{2x}-4\bar{X})(Q_{3x}-Q_{1x})}\left\{6Q_{2x}Q_{3x}\left(\frac{1-f}{8n}\right)\frac{1}{f_x(Q_{2x})f_x(Q_{3x})Q_{2x}Q_{3x}}+4Q_{3x}\bar{X}\left(\frac{1-f}{n}\right)\frac{S_{XQ_{3x}}}{\bar{X}Q_{3x}}\right.$$

$$\left.-6Q_{1x}Q_{2x}\left(\frac{1-f}{8n}\right)\frac{1}{f_x(Q_{1x})f_x(Q_{2x})Q_{1x}Q_{2x}}-4\bar{X}Q_{1x}\left(\frac{1-f}{n}\right)\frac{S_{XQ_{1x}}}{\bar{X}Q_{1x}}\right\}$$

$$-\frac{1}{(Q_{3y}-Q_{1y})(Q_{3x}-Q_{1x})}\left\{Q_{3y}Q_{3x}\left(\frac{1-f}{n}\right)\frac{\{f_x(Q_{3x})\}^{-1}\{f_y(Q_{3y})\}^{-1}}{Q_{3y}Q_{3x}}\left(P_{Q_{3x}Q_{3y}}-\frac{9}{16}\right)\right.$$

$$-Q_{1y}Q_{3x}\left(\frac{1-f}{n}\right)\frac{\{f_x(Q_{3x})\}^{-1}\{f_y(Q_{1y})\}^{-1}}{Q_{1y}Q_{3x}}\left(P_{Q_{3x}Q_{1y}}-\frac{3}{16}\right)$$

$$-Q_{3y}Q_{1x}\left(\frac{1-f}{n}\right)\frac{\{f_x(Q_{1x})\}^{-1}\{f_y(Q_{3y})\}^{-1}}{Q_{3y}Q_{1x}}\left(P_{Q_{1x}Q_{3y}}-\frac{3}{16}\right)$$

$$\left.+Q_{1x}Q_{1y}\left(\frac{1-f}{n}\right)\frac{\{f_x(Q_{1x})\}^{-1}\{f_y(Q_{1y})\}^{-1}}{Q_{1x}Q_{1y}}\left(P_{Q_{1x}Q_{1y}}-\frac{1}{16}\right)\right\}$$

$$+\frac{1}{(6Q_{2y}-4\bar{Y})(Q_{3x}-Q_{1x})}\left\{6Q_{3x}Q_{2y}\left(\frac{1-f}{n}\right)\frac{\{f_x(Q_{3x})\}^{-1}\{f_y(Q_{2y})\}^{-1}}{Q_{3x}Q_{2y}}\left(P_{Q_{3x}Q_{2y}}-\frac{3}{8}\right)\right.$$

$$-6Q_{1x}Q_{2y}\left(\frac{1-f}{n}\right)\frac{\{f_x(Q_{1x})\}^{-1}\{f_y(Q_{2y})\}^{-1}}{Q_{1x}Q_{2y}}\left(P_{Q_{1x}Q_{2y}}-\frac{1}{8}\right)$$

$$\left.\left.+4Q_{3x}\bar{Y}\left(\frac{1-f}{n}\right)\frac{S_{YQ_{3x}}}{\bar{Y}Q_{3x}}-4Q_{1x}\bar{Y}\left(\frac{1-f}{n}\right)\frac{S_{YQ_{1x}}}{\bar{Y}Q_{1x}}\right\}\right] \quad (2.7)$$

From (2.7), the bias, to the first order of approximation, in the proposed ratio estimator $C\hat{Q}V_{rat_{(m)}}$ is given by,

$$B\left[C\hat{Q}V_{rat(m)}\right]=E\left(C\hat{Q}V_{rat(m)}\right)-CQV_{y(m)}$$

$$=\left(\frac{1-f}{n}\right)CQV_{y(m)}\left[\frac{1}{(6Q_{2y}-4\bar{Y})^2}\left\{\frac{9}{\{f_y(Q_{2y})\}^2}+16S_{\bar{y}}^2+48S_{YQ_{2y}}\right\}\right.$$

$$+\frac{1}{(Q_{3x}-Q_{1x})^2}\left\{\frac{3}{16\{f_x(Q_{3x})\}^2}+\frac{3}{16\{f_x(Q_{1x})\}^2}-\frac{2}{16f_x(Q_{1x})f_x(Q_{3x})}\right\}$$

$$-\frac{1}{(Q_{3y}-Q_{1y})(6Q_{2y}-4\bar{Y})}\left\{\frac{6}{8f_y(Q_{2y})f_y(Q_{3y})}-\frac{6}{8f_y(Q_{1y})f_y(Q_{2y})}+4S_{YQ_{3y}}-4S_{YQ_{1y}}\right\}$$

$$+\frac{1}{(Q_{3y}-Q_{1y})(6Q_{2x}-4\bar{X})}\left\{\frac{6}{f_x(Q_{2x})f_y(Q_{3y})}\left(P_{Q_{2x}Q_{3y}}-\frac{3}{8}\right)-\frac{6}{f_x(Q_{2x})f_y(Q_{1y})}\left(P_{Q_{2x}Q_{1y}}-\frac{1}{8}\right)+4S_{XQ_{3y}}-4S_{XQ_{1y}}\right\}$$

$$-\frac{1}{(6Q_{2y}-4\bar{Y})(6Q_{2x}-4\bar{X})}\left\{\frac{36}{f_x(Q_{2x})f_y(Q_{2y})}\left(P_{Q_{2x}Q_{2y}}-\frac{1}{4}\right)+24S_{YQ_{2x}}+24S_{XQ_{2y}}+16S_{xy}\right\}$$

$$-\frac{1}{(6Q_{2x}-4\bar{X})(Q_{3x}-Q_{1x})}\left\{\frac{6}{8f_x(Q_{2x})f_x(Q_{3x})}+4S_{XQ_{3x}}-\frac{6}{8f_x(Q_{1x})f_x(Q_{2x})}-4S_{XQ_{1x}}\right\}$$

$$-\frac{1}{(Q_{3y}-Q_{1y})(Q_{3x}-Q_{1x})}\left\{\frac{1}{f_x(Q_{3x})f_y(Q_{3y})}\left(P_{Q_{3y}Q_{3x}}-\frac{9}{16}\right)-\frac{1}{f_x(Q_{3x})f_y(Q_{1y})}\left(P_{Q_{3x}Q_{1y}}-\frac{3}{16}\right)\right.$$
$$\left.-\frac{1}{f_x(Q_{1x})f_y(Q_{3y})}\left(P_{Q_{1x}Q_{3y}}-\frac{3}{16}\right)+\frac{1}{f_x(Q_{1x})f_y(Q_{1y})}\left(P_{Q_{1x}Q_{1y}}-\frac{1}{16}\right)\right\}$$

$$+\frac{1}{(6Q_{2y}-4\bar{Y})(Q_{3x}-Q_{1x})}\left\{\frac{6}{f_x(Q_{3x})f_y(Q_{2y})}\left(P_{Q_{3x}Q_{2y}}-\frac{3}{8}\right)+4S_{YQ_{3x}}\right.$$
$$\left.\left.-\frac{6}{f_x(Q_{1x})f_y(Q_{2y})}\left(P_{Q_{1x}Q_{2y}}-\frac{1}{8}\right)-4S_{YQ_{1x}}\right\}\right]$$

which proves the theorem.

It may be worth noticing that the $B\left[C\hat{Q}V_{rat(m)}\right]\to 0$ as the sample size $n\to\infty$, hence the proposed ratio type estimator $C\hat{Q}V_{rat_{(m)}}$ is a consistent estimator of $CQV_{y(m)}$.

Further note that all approximate results in this chapter are obtained based on the rules of asymptotic expansions in probability by following Wu (1982) given by:

$$E(\bar{x}-\bar{X})^r(\bar{y}-\bar{Y})^s=\begin{cases}O\left(n^{-\frac{1}{2}(r+s)}\right) & (r+s\ \text{even})\\ O\left(n^{-\frac{1}{2}(r+s+1)}\right) & (r+s\ \text{odd})\end{cases}$$

and

$$(\bar{x}-\bar{X})^r(\bar{y}-\bar{Y})^s=O_p\left(n^{-\frac{1}{2}(r+s)}\right)$$

where r and s are non-negative integers, $\bar{x}$ and $\bar{y}$ are the sample means and $\bar{X}$ and $\bar{Y}$ are the population means of the auxiliary variable and the study variable respectively. One could also refer to David and Srivastava (1974) for such asymptotic approximations.

Theorem 2.2. The mean squared error, to the first order of approximation, of the proposed ratio estimator $C\hat{Q}V_{rat_{(m)}}$ is given by

$$MSE\left[C\hat{Q}V_{rat(m)}\right]=\left(\frac{1-f}{n}\right)\left(CQV_{y(m)}\right)^2\left[\Psi_y^{*2}+\Psi_x^{*2}-2\Psi_{xy}^*\right] \tag{2.8}$$

where

$$\Psi_y^{*2}=\left[\frac{1}{(Q_{3y}-Q_{1y})^2}\left\{\frac{3}{16\{f_y(Q_{3y})\}^2}+\frac{3}{16\{f_y(Q_{1y})\}^2}-\frac{2}{16f_y(Q_{1y})f_y(Q_{3y})}\right\}\right.$$
$$+\frac{1}{(6Q_{2y}-4\bar{Y})^2}\left\{\frac{9}{\{f_y(Q_{2y})\}^2}+16S_y^2+48S_{YQ_{2y}}\right\} \tag{2.9}$$
$$\left.-\frac{2}{(Q_{3y}-Q_{1y})(6Q_{2y}-4\bar{Y})}\left\{\frac{3}{4f_y(Q_{2y})f_y(Q_{3y})}-\frac{3}{4f_y(Q_{1y})f_y(Q_{2y})}+4S_{YQ_{3y}}-4S_{YQ_{1y}}\right\}\right]$$

$$\Psi_x^{*2} = \left[\frac{1}{(Q_{3x}-Q_{1x})^2} \left\{ \frac{3}{16\{f_x(Q_{3x})\}^2} + \frac{3}{16\{f_x(Q_{1x})\}^2} - \frac{2}{16 f_x(Q_{1x}) f_x(Q_{3x})} \right\} + \frac{1}{(6Q_{2x}-4\bar{X})^2} \left\{ \frac{9}{\{f_x(Q_{2x})\}^2} + 16S_x^2 + 48S_{XQ_{2x}} \right\} - \frac{2}{(Q_{3x}-Q_{1x})(6Q_{2x}-4\bar{X})} \left\{ \frac{3}{4f_x(Q_{2x})f_x(Q_{3x})} - \frac{3}{4f_x(Q_{1x})f_x(Q_{2x})} + 4S_{XQ_{3x}} - 4S_{XQ_{1x}} \right\} \right] \tag{2.10}$$

and

$$\begin{aligned}\Psi_{xy}^{*} = \Bigg[& \frac{1}{(Q_{3y}-Q_{1y})(Q_{3x}-Q_{1x})} \left\{ \frac{1}{f_x(Q_{3x})f_y(Q_{3y})}\left(P_{Q_{3x}Q_{3y}} - \frac{9}{16}\right) - \frac{1}{f_x(Q_{3x})f_y(Q_{1y})}\left(P_{Q_{3x}Q_{1y}} - \frac{3}{16}\right) \right. \\ & \left. - \frac{1}{f_x(Q_{1x})f_y(Q_{3y})}\left(P_{Q_{1x}Q_{3y}} - \frac{3}{16}\right) + \frac{1}{f_x(Q_{1x})f_y(Q_{1y})}\left(P_{Q_{1x}Q_{1y}} - \frac{1}{16}\right) \right\} \\ & - \frac{1}{(Q_{3x}-Q_{1x})(6Q_{2y}-4\bar{Y})} \left\{ \frac{6}{f_x(Q_{3x})f_y(Q_{2y})}\left(P_{Q_{3x}Q_{2y}} - \frac{3}{8}\right) \right. \\ & \left. - \frac{6}{f_x(Q_{1x})f_y(Q_{2y})}\left(P_{Q_{1x}Q_{2y}} - \frac{1}{8}\right) + 4S_{YQ_{3x}} - 4S_{YQ_{1x}} \right\} \\ & - \frac{1}{(Q_{3y}-Q_{1y})(6Q_{2x}-4\bar{X})} \left\{ \frac{6}{f_x(Q_{2x})f_y(Q_{3y})}\left(P_{Q_{2x}Q_{3y}} - \frac{3}{8}\right) \right. \\ & \left. - \frac{6}{f_x(Q_{2x})f_y(Q_{1y})}\left(P_{Q_{2x}Q_{1y}} - \frac{1}{8}\right) + 4S_{XQ_{3y}} - 4S_{XQ_{1y}} \right\} \\ & + \frac{1}{(6Q_{2y}-4\bar{Y})(6Q_{2x}-4\bar{X})} \left\{ \frac{36}{f_x(Q_{2x})f_y(Q_{2y})}\left(P_{Q_{2x}Q_{2y}} - \frac{1}{4}\right) + 24S_{YQ_{2x}} + 24S_{XQ_{2y}} + 16S_{xy} \right\} \Bigg]\end{aligned} \tag{2.11}$$

Proof. The mean squared error to the first order of approximation of the ratio estimator $C\hat{Q}V_{rat_{(m)}}$ is given by,

$$MSE\left[C\hat{Q}V_{rat(m)}\right] = E\left[C\hat{Q}V_{rat(m)} - CQV_{y(m)}\right]^2$$

$$\approx E\left[CQV_{y(m)} \left\{ \left(\frac{Q_{3y}\varepsilon_{3y} - Q_{1y}\varepsilon_{1y}}{Q_{3y}-Q_{1y}}\right) - \left(\frac{6Q_{2y}\varepsilon_{2y} - 4\bar{Y}\eta_y}{6Q_{2y}-4\bar{Y}}\right) + \left(\frac{6Q_{2x}\varepsilon_{2x} - 4\bar{X}\eta_x}{6Q_{2x}-4\bar{X}}\right) - \left(\frac{Q_{3x}\varepsilon_{3x} - Q_{1x}\varepsilon_{1x}}{Q_{3x}-Q_{1x}}\right) \right\} \right]^2$$

$$= (CQV_{y(m)})^2 E\left[\left\{ \left(\frac{Q_{3y}\varepsilon_{3y} - Q_{1y}\varepsilon_{1y}}{Q_{3y}-Q_{1y}}\right) - \left(\frac{6Q_{2y}\varepsilon_{2y} - 4\bar{Y}\eta_y}{6Q_{2y}-4\bar{Y}}\right) \right\} - \left\{ \left(\frac{Q_{3x}\varepsilon_{3x} - Q_{1x}\varepsilon_{1x}}{Q_{3x}-Q_{1x}}\right) - \left(\frac{6Q_{2x}\varepsilon_{2x} - 4\bar{X}\eta_x}{6Q_{2x}-4\bar{X}}\right) \right\} \right]^2$$

$$\begin{aligned}= (CQV_{y(m)})^2 E\Bigg[& \left\{ \left(\frac{Q_{3y}\varepsilon_{3y} - Q_{1y}\varepsilon_{1y}}{Q_{3y}-Q_{1y}}\right) - \left(\frac{6Q_{2y}\varepsilon_{2y} - 4\bar{Y}\eta_y}{6Q_{2y}-4\bar{Y}}\right) \right\}^2 \\ & + \left\{ \left(\frac{Q_{3x}\varepsilon_{3x} - Q_{1x}\varepsilon_{1x}}{Q_{3x}-Q_{1x}}\right) - \left(\frac{6Q_{2x}\varepsilon_{2x} - 4\bar{X}\eta_x}{6Q_{2x}-4\bar{X}}\right) \right\}^2 \\ & - 2\left\{ \left(\frac{Q_{3y}\varepsilon_{3y} - Q_{1y}\varepsilon_{1y}}{Q_{3y}-Q_{1y}}\right) - \left(\frac{6Q_{2y}\varepsilon_{2y} - 4\bar{Y}\eta_y}{6Q_{2y}-4\bar{Y}}\right) \right\}\left\{ \left(\frac{Q_{3x}\varepsilon_{3x} - Q_{1x}\varepsilon_{1x}}{Q_{3x}-Q_{1x}}\right) - \left(\frac{6Q_{2x}\varepsilon_{2x} - 4\bar{X}\eta_x}{6Q_{2x}-4\bar{X}}\right) \right\} \Bigg]\end{aligned}$$

$$
\begin{aligned}
&= \left(CQV_{y(m)}\right)^2 E\left[\frac{1}{\left(Q_{3y}-Q_{1y}\right)^2}\left(Q_{3y}\varepsilon_{3y}-Q_{1y}\varepsilon_{1y}\right)^2+\frac{1}{\left(6Q_{2y}-4\bar{Y}\right)^2}\left(6Q_{2y}\varepsilon_{2y}-4\bar{Y}\eta_y\right)^2\right.\\
&\quad -\frac{2}{\left(Q_{3y}-Q_{1y}\right)\left(6Q_{2y}-4\bar{Y}\right)}\left(Q_{3y}\varepsilon_{3y}-Q_{1y}\varepsilon_{1y}\right)\left(6Q_{2y}\varepsilon_{2y}-4\bar{Y}\eta_y\right)\\
&\quad +\frac{1}{\left(Q_{3x}-Q_{1x}\right)^2}\left(Q_{3x}\varepsilon_{3x}-Q_{1x}\varepsilon_{1x}\right)^2+\frac{1}{\left(6Q_{2x}-4\bar{X}\right)^2}\left(6Q_{2x}\varepsilon_{2x}-4\bar{X}\eta_x\right)^2\\
&\quad -\frac{2}{\left(Q_{3x}-Q_{1x}\right)\left(6Q_{2x}-4\bar{X}\right)}\left(Q_{3x}\varepsilon_{3x}-Q_{1x}\varepsilon_{1x}\right)\left(6Q_{2x}\varepsilon_{2x}-4\bar{X}\eta_x\right)\\
&\quad -2\left\{\frac{\left(Q_{3y}\varepsilon_{3y}-Q_{1y}\varepsilon_{1y}\right)\left(Q_{3x}\varepsilon_{3x}-Q_{1x}\varepsilon_{1x}\right)}{\left(Q_{3y}-Q_{1y}\right)\left(Q_{3x}-Q_{1x}\right)}-\frac{\left(6Q_{2y}\varepsilon_{2y}-4\bar{Y}\eta_y\right)\left(Q_{3x}\varepsilon_{3x}-Q_{1x}\varepsilon_{1x}\right)}{\left(6Q_{2y}-4\bar{Y}\right)\left(Q_{3x}-Q_{1x}\right)}\right.\\
&\quad \left.\left.-\frac{\left(Q_{3y}\varepsilon_{3y}-Q_{1y}\varepsilon_{1y}\right)\left(6Q_{2x}\varepsilon_{2x}-4\bar{X}\eta_x\right)}{\left(6Q_{2x}-4\bar{X}\right)\left(Q_{3y}-Q_{1y}\right)}+\frac{\left(6Q_{2y}\varepsilon_{2y}-4\bar{Y}\eta_y\right)\left(6Q_{2x}\varepsilon_{2x}-4\bar{X}\eta_x\right)}{\left(6Q_{2y}-4\bar{Y}\right)\left(6Q_{2x}-4\bar{X}\right)}\right\}\right]
\end{aligned}
$$

$$
\begin{aligned}
&= \left(CQV_{y(m)}\right)^2 E\left[\frac{1}{\left(Q_{3y}-Q_{1y}\right)^2}\left\{Q_{3y}^2\varepsilon_{3y}^2+Q_{1y}^2\varepsilon_{1y}^2-2Q_{1y}Q_{3y}\varepsilon_{1y}\varepsilon_{3y}\right\}\right.\\
&\quad +\frac{1}{\left(6Q_{2y}-4\bar{Y}\right)^2}\left\{36Q_{2y}^2\varepsilon_{2y}^2+16\bar{Y}^2\eta_y^2-48Q_{2y}\bar{Y}\varepsilon_{2y}\eta_y\right\}\\
&\quad -\frac{2}{\left(Q_{3y}-Q_{1y}\right)\left(6Q_{2y}-4\bar{Y}\right)}\left\{6Q_{2y}Q_{3y}\varepsilon_{2y}\varepsilon_{3y}-6Q_{1y}Q_{2y}\varepsilon_{1y}\varepsilon_{2y}-4Q_{3y}\bar{Y}\varepsilon_{3y}\eta_y+4Q_{1y}\bar{Y}\varepsilon_{1y}\eta_y\right\}\\
&\quad +\frac{1}{\left(Q_{3x}-Q_{1x}\right)^2}\left\{Q_{3x}^2\varepsilon_{3x}^2+Q_{1x}^2\varepsilon_{1x}^2-2Q_{1x}Q_{3x}\varepsilon_{1x}\varepsilon_{3x}\right\}\\
&\quad +\frac{1}{\left(6Q_{2x}-4\bar{X}\right)^2}\left\{36Q_{2x}^2\varepsilon_{2x}^2+16\bar{X}^2\eta_x^2-48Q_{2x}\bar{X}\varepsilon_{2x}\eta_x\right\}\\
&\quad -\frac{2}{\left(Q_{3x}-Q_{1x}\right)\left(6Q_{2x}-4\bar{X}\right)}\left\{6Q_{2x}Q_{3x}\varepsilon_{2x}\varepsilon_{3x}-6Q_{1x}Q_{2x}\varepsilon_{1x}\varepsilon_{2x}-4Q_{3x}\bar{X}\varepsilon_{3x}\eta_x+4Q_{1x}\bar{X}\varepsilon_{1x}\eta_x\right\}\\
&\quad -2\left\{\frac{1}{\left(Q_{3y}-Q_{1y}\right)\left(Q_{3x}-Q_{1x}\right)}\left(Q_{3y}Q_{3x}\varepsilon_{3y}\varepsilon_{3x}-Q_{1y}Q_{3x}\varepsilon_{1y}\varepsilon_{3x}-Q_{3y}Q_{1x}\varepsilon_{3y}\varepsilon_{1x}+Q_{1y}Q_{1x}\varepsilon_{1y}\varepsilon_{1x}\right)\right.\\
&\quad -\frac{1}{\left(Q_{3x}-Q_{1x}\right)\left(6Q_{2y}-4\bar{Y}\right)}\left(6Q_{2y}Q_{3x}\varepsilon_{2y}\varepsilon_{3x}-4Q_{3x}\bar{Y}\varepsilon_{3x}\eta_y-6Q_{2y}Q_{1x}\varepsilon_{2y}\varepsilon_{1x}+4Q_{1x}\bar{Y}\varepsilon_{1x}\eta_y\right)\\
&\quad -\frac{1}{\left(Q_{3y}-Q_{1y}\right)\left(6Q_{2x}-4\bar{X}\right)}\left(6Q_{3y}Q_{2x}\varepsilon_{3y}\varepsilon_{2x}-6Q_{1y}Q_{2x}\varepsilon_{1y}\varepsilon_{2x}-4Q_{3y}\bar{X}\varepsilon_{3y}\eta_x+4Q_{1y}\bar{X}\varepsilon_{1y}\eta_x\right)\\
&\quad \left.\left.+\frac{1}{\left(6Q_{2y}-4\bar{Y}\right)\left(6Q_{2x}-4\bar{X}\right)}\left(36Q_{2y}Q_{2x}\varepsilon_{2y}\varepsilon_{2x}-24Q_{2x}\bar{Y}\varepsilon_{2x}\eta_y-24Q_{2y}\bar{X}\varepsilon_{2y}\eta_x+16\bar{Y}\bar{X}\eta_y\eta_x\right)\right\}\right]
\end{aligned}
$$

$$
\begin{aligned}
&= \left(CQV_{y(m)}\right)^2\left[\frac{1}{\left(Q_{3y}-Q_{1y}\right)^2}\left\{Q_{3y}^2\left(\frac{1-f}{16n}\right)\frac{3}{Q_{3y}^2\left\{f_y\left(Q_{3y}\right)\right\}^2}+Q_{1y}^2\left(\frac{1-f}{16n}\right)\frac{3}{Q_{1y}^2\left\{f_y\left(Q_{1y}\right)\right\}^2}\right.\right.\\
&\quad \left.-2Q_{1y}Q_{3y}\left(\frac{1-f}{16n}\right)\frac{1}{f_y\left(Q_{1y}\right)f_y\left(Q_{3y}\right)Q_{1y}Q_{3y}}\right\}\\
&\quad +\frac{1}{\left(6Q_{2y}-4\bar{Y}\right)^2}\left\{36Q_{2y}^2\left(\frac{1-f}{4n}\right)\frac{1}{Q_{2y}^2\left\{f_y\left(Q_{2y}\right)\right\}^2}+16\bar{Y}^2\left(\frac{1-f}{n}\right)C_y^2+48Q_{2y}\bar{Y}\left(\frac{1-f}{n}\right)\frac{S_{YQ_{2y}}}{Q_{2y}\bar{Y}}\right\}
\end{aligned}
$$

$$-\frac{2}{(Q_{3y}-Q_{1y})(6Q_{2y}-4\bar{Y})}\left\{6Q_{2y}Q_{3y}\left(\frac{1-f}{8n}\right)\frac{1}{f_y(Q_{2y})f_y(Q_{3y})Q_{2y}Q_{3y}}+4Q_{3y}\bar{Y}\left(\frac{1-f}{n}\right)\frac{S_{YQ_{3y}}}{Q_{3y}\bar{Y}}\right.$$

$$\left.-6Q_{1y}Q_{2y}\left(\frac{1-f}{8n}\right)\frac{1}{f_y(Q_{1y})f_y(Q_{2y})Q_{1y}Q_{2y}}-4Q_{1y}\bar{Y}\left(\frac{1-f}{n}\right)\frac{S_{YQ_{1y}}}{Q_{1y}\bar{Y}}\right\}$$

$$+\frac{1}{(Q_{3x}-Q_{1x})^2}\left\{Q_{3x}^2\left(\frac{1-f}{n}\right)\frac{3}{16}\frac{1}{Q_{3x}^2\{f_x(Q_{3x})\}^2}+Q_{1x}^2\left(\frac{1-f}{n}\right)\frac{3}{16}\frac{1}{Q_{1x}^2\{f_x(Q_{1x})\}^2}\right.$$

$$\left.-2Q_{1x}Q_{3x}\left(\frac{1-f}{16n}\right)\frac{1}{f_x(Q_{1x})f_x(Q_{3x})Q_{1x}Q_{3x}}\right\}$$

$$+\frac{1}{(6Q_{2x}-4\bar{X})^2}\left\{36Q_{2x}^2\left(\frac{1-f}{4n}\right)\frac{1}{Q_{2x}^2\{f_x(Q_{2x})\}^2}+16\bar{X}^2\left(\frac{1-f}{n}\right)C_x^2+48Q_{2x}\bar{X}\left(\frac{1-f}{n}\right)\frac{S_{XQ_{2x}}}{Q_{2x}\bar{X}}\right\}$$

$$-\frac{2}{(Q_{3x}-Q_{1x})(6Q_{2x}-4\bar{X})}\left\{6Q_{2x}Q_{3x}\left(\frac{1-f}{8n}\right)\frac{1}{f_x(Q_{2x})f_x(Q_{3x})Q_{2x}Q_{3x}}\right.$$

$$\left.-6Q_{1x}Q_{2x}\left(\frac{1-f}{8n}\right)\frac{1}{f_x(Q_{1x})f_x(Q_{2x})Q_{1x}Q_{2x}}+4Q_{3x}\bar{X}\left(\frac{1-f}{n}\right)\frac{S_{XQ_{3x}}}{Q_{3x}\bar{X}}-4Q_{1x}\bar{X}\left(\frac{1-f}{n}\right)\frac{S_{XQ_{1x}}}{Q_{1x}\bar{X}}\right\}$$

$$-2\left\{\frac{1}{(Q_{3y}-Q_{1y})(Q_{3x}-Q_{1x})}\left(Q_{3y}Q_{3x}\left(\frac{1-f}{n}\right)\frac{\{f_x(Q_{3x})\}^{-1}\{f_y(Q_{3y})\}^{-1}}{Q_{3x}Q_{3y}}\left(P_{Q_{3x}Q_{3y}}-\frac{9}{16}\right)\right.\right.$$

$$-Q_{1y}Q_{3x}\left(\frac{1-f}{n}\right)\frac{\{f_x(Q_{3x})\}^{-1}\{f_y(Q_{1y})\}^{-1}}{Q_{3x}Q_{1y}}\left(P_{Q_{3x}Q_{1y}}-\frac{3}{16}\right)$$

$$-Q_{3y}Q_{1x}\left(\frac{1-f}{n}\right)\frac{\{f_x(Q_{1x})\}^{-1}\{f_y(Q_{3y})\}^{-1}}{Q_{3y}Q_{1x}}\left(P_{Q_{1x}Q_{3y}}-\frac{3}{16}\right)$$

$$\left.\left.+Q_{1y}Q_{1x}\left(\frac{1-f}{n}\right)\frac{\{f_x(Q_{1x})\}^{-1}\{f_y(Q_{1y})\}^{-1}}{Q_{1y}Q_{1x}}\left(P_{Q_{1x}Q_{1y}}-\frac{1}{16}\right)\right)\right.$$

$$-\frac{1}{(Q_{3x}-Q_{1x})(6Q_{2y}-4\bar{Y})}\left(6Q_{2y}Q_{3x}\left(\frac{1-f}{n}\right)\frac{\{f_x(Q_{3x})\}^{-1}\{f_y(Q_{2y})\}^{-1}}{Q_{2y}Q_{3x}}\left(P_{Q_{3x}Q_{2y}}-\frac{3}{8}\right)\right.$$

$$+4Q_{3x}\bar{Y}\left(\frac{1-f}{n}\right)\frac{S_{YQ_{3x}}}{Q_{3x}\bar{Y}}-6Q_{2y}Q_{1x}\left(\frac{1-f}{n}\right)\frac{\{f_x(Q_{1x})\}^{-1}\{f_y(Q_{2y})\}^{-1}}{Q_{2y}Q_{1x}}\left(P_{Q_{1x}Q_{2y}}-\frac{1}{8}\right)$$

$$\left.-4Q_{1x}\bar{Y}\left(\frac{1-f}{n}\right)\frac{S_{YQ_{1x}}}{Q_{1x}\bar{Y}}\right)$$

$$-\frac{1}{(Q_{3y}-Q_{1y})(6Q_{2x}-4\bar{X})}\left(6Q_{3y}Q_{2x}\left(\frac{1-f}{n}\right)\frac{\{f_x(Q_{2x})\}^{-1}\{f_y(Q_{3y})\}^{-1}}{Q_{3y}Q_{2x}}\left(P_{Q_{2x}Q_{3y}}-\frac{3}{8}\right)\right.$$

$$-6Q_{1y}Q_{2x}\left(\frac{1-f}{n}\right)\frac{\{f_x(Q_{2x})\}^{-1}\{f_y(Q_{1y})\}^{-1}}{Q_{1y}Q_{2x}}\left(P_{Q_{2x}Q_{1y}}-\frac{1}{8}\right)$$

$$\left.+4Q_{3y}\bar{X}\left(\frac{1-f}{n}\right)\frac{S_{XQ_{3y}}}{Q_{3y}\bar{X}}-4Q_{1y}\bar{X}\left(\frac{1-f}{n}\right)\frac{S_{XQ_{1y}}}{Q_{1y}\bar{X}}\right)$$

$$+\frac{1}{(6Q_{2y}-4\bar{Y})(6Q_{2x}-4\bar{X})}\left(36Q_{2y}Q_{2x}\left(\frac{1-f}{n}\right)\frac{\{f_x(Q_{2x})\}^{-1}\{f_y(Q_{2y})\}^{-1}}{Q_{2y}Q_{2x}}\left(P_{Q_{2x}Q_{2y}}-\frac{1}{4}\right)\right.$$

$$+24Q_{2x}\bar{Y}\left(\frac{1-f}{n}\right)\frac{S_{YQ_{2x}}}{Q_{2x}\bar{Y}}+24Q_{2y}\bar{X}\left(\frac{1-f}{n}\right)\frac{S_{XQ_{2y}}}{Q_{2y}\bar{X}}+16\bar{Y}\bar{X}\left(\frac{1-f}{n}\right)\rho_{xy}C_xC_y\bigg)\bigg\}\bigg]$$

$$\begin{aligned}
&=\left(CQV_{y(m)}\right)^2\left(\frac{1-f}{n}\right)\Bigg[\frac{1}{\left(Q_{3y}-Q_{1y}\right)^2}\left\{\frac{3}{16\{f_y(Q_{3y})\}^2}+\frac{3}{16\{f_y(Q_{1y})\}^2}-\frac{2}{16f_y(Q_{1y})f_y(Q_{3y})}\right\}\\
&\quad+\frac{1}{\left(6Q_{2y}-4\bar{Y}\right)^2}\left\{\frac{9}{\{f_y(Q_{2y})\}^2}+16S_y^2+48S_{YQ_{2y}}\right\}\\
&\quad-\frac{2}{\left(Q_{3y}-Q_{1y}\right)\left(6Q_{2y}-4\bar{Y}\right)}\left\{\frac{6}{8f_y(Q_{2y})f_y(Q_{3y})}-\frac{6}{8f_y(Q_{1y})f_y(Q_{2y})}+4S_{YQ_{3y}}-4S_{YQ_{1y}}\right\}\\
&\quad+\frac{1}{\left(Q_{3x}-Q_{1x}\right)^2}\left\{\frac{3}{16\{f_x(Q_{3x})\}^2}+\frac{3}{16\{f_x(Q_{1x})\}^2}-\frac{2}{16f_x(Q_{1x})f_x(Q_{3x})}\right\}\\
&\quad+\frac{1}{\left(6Q_{2x}-4\bar{X}\right)^2}\left\{\frac{9}{\{f_x(Q_{2x})\}^2}+16S_x^2+48S_{XQ_{2x}}\right\}\\
&\quad-\frac{2}{\left(Q_{3x}-Q_{1x}\right)\left(6Q_{2x}-4\bar{X}\right)}\left\{\frac{6}{8f_x(Q_{2x})f_x(Q_{3x})}-\frac{6}{8f_x(Q_{1x})f_x(Q_{2x})}+4S_{XQ_{3x}}-4S_{XQ_{1x}}\right\}\\
&\quad-2\Bigg\{\frac{1}{\left(Q_{3y}-Q_{1y}\right)\left(Q_{3x}-Q_{1x}\right)}\Bigg(\frac{1}{f_x(Q_{3x})f_y(Q_{3y})}\left(P_{Q_{3x}Q_{3y}}-\frac{9}{16}\right)-\frac{1}{f_x(Q_{3x})f_y(Q_{1y})}\left(P_{Q_{3x}Q_{1y}}-\frac{3}{16}\right)\\
&\qquad-\frac{1}{f_x(Q_{1x})f_y(Q_{3y})}\left(P_{Q_{1x}Q_{3y}}-\frac{3}{16}\right)+\frac{1}{f_x(Q_{1x})f_y(Q_{1y})}\left(P_{Q_{1x}Q_{1y}}-\frac{1}{16}\right)\Bigg)\\
&\quad-\frac{1}{\left(Q_{3x}-Q_{1x}\right)\left(6Q_{2y}-4\bar{Y}\right)}\Bigg(\frac{6}{f_x(Q_{3x})f_y(Q_{2y})}\left(P_{Q_{3x}Q_{2y}}-\frac{3}{8}\right)\\
&\qquad-\frac{6}{f_x(Q_{1x})f_y(Q_{2y})}\left(P_{Q_{1x}Q_{2y}}-\frac{1}{8}\right)+4S_{YQ_{3x}}-4S_{YQ_{1x}}\Bigg)\\
&\quad-\frac{1}{\left(Q_{3y}-Q_{1y}\right)\left(6Q_{2x}-4\bar{X}\right)}\Bigg(\frac{6}{f_x(Q_{2x})f_y(Q_{3y})}\left(P_{Q_{2x}Q_{3y}}-\frac{3}{8}\right)\\
&\qquad-\frac{6}{f_x(Q_{2x})f_y(Q_{1y})}\left(P_{Q_{2x}Q_{1y}}-\frac{1}{8}\right)+4S_{XQ_{3y}}-4S_{XQ_{1y}}\Bigg)\\
&\quad+\frac{1}{\left(6Q_{2y}-4\bar{Y}\right)\left(6Q_{2x}-4\bar{X}\right)}\Bigg(\frac{36}{f_x(Q_{2x})f_y(Q_{2y})}\left(P_{Q_{2x}Q_{2y}}-\frac{1}{4}\right)+24S_{YQ_{2x}}+24S_{XQ_{2y}}+16S_{xy}\Bigg)\Bigg\}\Bigg]
\end{aligned}$$

which proves the theorem.

It would be worth pointing out here that these approximate results are obtained by following Kuk and Mak (1989) and Sedory and Singh (2014).

In the next section, we consider a regression type estimator of the new measure of coefficient of dispersion relative to the empirical mode, hence deriving an approximate bias and an approximate variance expression.

3. Regression Estimator of the New Measure of Coefficient of Dispersion

Following Hansen et al. (1953), we define a new regression type estimator of the newly proposed coefficient of dispersion relative to mode value as:

$$C\hat{Q}V_{reg(m)}=C\hat{Q}V_{y(m)}+\beta\left[CQV_{x(m)}-C\hat{Q}V_{x(m)}\right] \tag{3.1}$$

when β is an aptly selected constant such that the variance of the estimator $C\hat{Q}V_{reg_{(m)}}$ is minimum. Now we have the following theorem.

Theorem 3.1. The bias to the first order of approximation in the regression type estimator $C\hat{Q}V_{reg_{(m)}}$ is given by

$$B\left[C\hat{Q}V_{reg(m)}\right]=\left(\frac{1-f}{n}\right)\left[CQV_{y(m)}\left\{\frac{1}{\left(6Q_{2y}-4\bar{Y}\right)^2}\left(\frac{9}{\left\{f_y\left(Q_{2y}\right)\right\}^2}+16S_y^2+48S_{YQ_{2y}}\right)\right.\right.$$
$$\left.-\frac{1}{\left(Q_{3y}-Q_{1y}\right)\left(6Q_{2y}-4\bar{Y}\right)}\left(\frac{3}{4f_y\left(Q_{2y}\right)f_y\left(Q_{3y}\right)}-\frac{3}{4f_y\left(Q_{1y}\right)f_y\left(Q_{2y}\right)}+4S_{YQ_{3y}}-4S_{YQ_{1y}}\right)\right\}$$
$$-\beta\left(CQV_{x(m)}\right)\left\{\frac{1}{\left(6Q_{2x}-4\bar{X}\right)^2}\left(\frac{9}{\left\{f_x\left(Q_{2x}\right)\right\}^2}+16S_x^2+48S_{XQ_{2x}}\right)\right.$$
$$\left.\left.-\frac{1}{\left(Q_{3x}-Q_{1x}\right)\left(6Q_{2x}-4\bar{X}\right)}\left(\frac{3}{4f_x\left(Q_{2x}\right)f_x\left(Q_{3x}\right)}-\frac{3}{4f_x\left(Q_{1x}\right)f_x\left(Q_{2x}\right)}+4S_{XQ_{3x}}-4S_{XQ_{1x}}\right)\right\}\right] \tag{3.2}$$

Proof. The estimator $C\hat{Q}V_{reg_{(m)}}$ can be written as

$$C\hat{Q}V_{reg(m)}=C\hat{Q}V_{y(m)}+\beta\left[CQV_{x(m)}-C\hat{Q}V_{x(m)}\right]$$
$$=\frac{\hat{Q}_{3y}-\hat{Q}_{1y}}{6\hat{Q}_{2y}-4\bar{y}}+\beta\left[CQV_{x(m)}-\frac{\hat{Q}_{3x}-\hat{Q}_{1x}}{6\hat{Q}_{2x}-4\bar{x}}\right]$$
$$=\frac{Q_{3y}\left(1+\varepsilon_{3y}\right)-Q_{1y}\left(1+\varepsilon_{1y}\right)}{6Q_{2y}\left(1+\varepsilon_{2y}\right)-4\bar{Y}\left(1+\eta_y\right)}+\beta\left[CQV_{x(m)}-\frac{Q_{3x}\left(1+\varepsilon_{3x}\right)-Q_{1x}\left(1+\varepsilon_{1x}\right)}{6Q_{2x}\left(1+\varepsilon_{2x}\right)-4\bar{X}\left(1+\eta_x\right)}\right]$$
$$=\frac{Q_{3y}+Q_{3y}\varepsilon_{3y}-Q_{1y}-Q_{1y}\varepsilon_{1y}}{6Q_{2y}+6Q_{2y}\varepsilon_{2y}-4\bar{Y}-4\bar{Y}\eta_y}+\beta\left[CQV_{x(m)}-\frac{Q_{3x}+Q_{3x}\varepsilon_{3x}-Q_{1x}-Q_{1x}\varepsilon_{1x}}{6Q_{2x}+6Q_{2x}\varepsilon_{2x}-4\bar{X}-4\bar{X}\eta_x}\right]$$
$$=\frac{\left(Q_{3y}-Q_{1y}\right)+Q_{3y}\varepsilon_{3y}-Q_{1y}\varepsilon_{1y}}{\left(6Q_{2y}-4\bar{Y}\right)+6Q_{2y}\varepsilon_{2y}-4\bar{Y}\eta_y}+\beta\left[CQV_{x(m)}-\frac{\left(Q_{3x}-Q_{1x}\right)+Q_{3x}\varepsilon_{3x}-Q_{1x}\varepsilon_{1x}}{\left(6Q_{2x}-4\bar{X}\right)+6Q_{2x}\varepsilon_{2x}-4\bar{X}\eta_x}\right]$$
$$=\frac{\left(Q_{3y}-Q_{1y}\right)\left[1+\frac{Q_{3y}\varepsilon_{3y}-Q_{1y}\varepsilon_{1y}}{Q_{3y}-Q_{1y}}\right]}{\left(6Q_{2y}-4\bar{Y}\right)\left[1+\frac{6Q_{2y}\varepsilon_{2y}-4\bar{Y}\eta_y}{6Q_{2y}-4\bar{Y}}\right]}+\beta\left[CQV_{x(m)}-\frac{\left(Q_{3x}-Q_{1x}\right)\left[1+\frac{Q_{3x}\varepsilon_{3x}-Q_{1x}\varepsilon_{1x}}{Q_{3x}-Q_{1x}}\right]}{\left(6Q_{2x}-4\bar{X}\right)\left[1+\frac{6Q_{2x}\varepsilon_{2x}-4\bar{X}\eta_x}{6Q_{2x}-4\bar{X}}\right]}\right] \tag{3.3}$$
$$=\frac{\left(Q_{3y}-Q_{1y}\right)}{\left(6Q_{2y}-4\bar{Y}\right)}\left[1+\frac{Q_{3y}\varepsilon_{3y}-Q_{1y}\varepsilon_{1y}}{Q_{3y}-Q_{1y}}\right]\left[1+\frac{6Q_{2y}\varepsilon_{2y}-4\bar{Y}\eta_y}{6Q_{2y}-4\bar{Y}}\right]^{-1}$$
$$+\beta\left[CQV_{x(m)}-\frac{\left(Q_{3x}-Q_{1x}\right)}{\left(6Q_{2x}-4\bar{X}\right)}\left\{1+\frac{Q_{3x}\varepsilon_{3x}-Q_{1x}\varepsilon_{1x}}{Q_{3x}-Q_{1x}}\right\}\left\{1+\frac{6Q_{2x}\varepsilon_{2x}-4\bar{X}\eta_x}{6Q_{2x}-4\bar{X}}\right\}^{-1}\right]$$

Assuming $\left|\frac{6Q_{2y}\varepsilon_{2y}-4\bar{Y}\eta_y}{6Q_{2y}-4\bar{Y}}\right|<1$ and $\left|\frac{6Q_{2x}\varepsilon_{2x}-4\bar{X}\eta_x}{6Q_{2x}-4\bar{X}}\right|<1$, and by the binomial expansion, the expression (3.3) can be approximated as:

$$\hat{CQV}_{reg(m)} = CQV_{y(m)}\left\{1+\frac{Q_{3y}\varepsilon_{3y}-Q_{1y}\varepsilon_{1y}}{Q_{3y}-Q_{1y}}\right\}\left\{1-\frac{6Q_{2y}\varepsilon_{2y}-4\bar{Y}\eta_y}{6Q_{2y}-4\bar{Y}}+\left(\frac{6Q_{2y}\varepsilon_{2y}-4\bar{Y}\eta_y}{6Q_{2y}-4\bar{Y}}\right)^2+\ldots\right\}$$
$$+\beta\left[CQV_{x(m)}-CQV_{x(m)}\left\{1+\frac{Q_{3x}\varepsilon_{3x}-Q_{1x}\varepsilon_{1x}}{Q_{3x}-Q_{1x}}\right\}\left\{1-\frac{6Q_{2x}\varepsilon_{2x}-4\bar{X}\eta_x}{6Q_{2x}-4\bar{X}}+\left(\frac{6Q_{2x}\varepsilon_{2x}-4\bar{X}\eta_x}{6Q_{2x}-4\bar{X}}\right)^2+\ldots\right\}\right]$$
$$=CQV_{y(m)}\left\{1+\frac{Q_{3y}\varepsilon_{3y}-Q_{1y}\varepsilon_{1y}}{Q_{3y}-Q_{1y}}-\frac{6Q_{2y}\varepsilon_{2y}-4\bar{Y}\eta_y}{6Q_{2y}-4\bar{Y}}+\left(\frac{6Q_{2y}\varepsilon_{2y}-4\bar{Y}\eta_y}{6Q_{2y}-4\bar{Y}}\right)^2-\left(\frac{Q_{3y}\varepsilon_{3y}-Q_{1y}\varepsilon_{1y}}{Q_{3y}-Q_{1y}}\right)\left(\frac{6Q_{2y}\varepsilon_{2y}-4\bar{Y}\eta_y}{6Q_{2y}-4\bar{Y}}\right)\right\} \quad (3.4)$$
$$-\beta\left(CQV_{x(m)}\right)\left[\frac{Q_{3x}\varepsilon_{3x}-Q_{1x}\varepsilon_{1x}}{Q_{3x}-Q_{1x}}-\frac{6Q_{2x}\varepsilon_{2x}-4\bar{X}\eta_x}{6Q_{2x}-4\bar{X}}+\left(\frac{6Q_{2x}\varepsilon_{2x}-4\bar{X}\eta_x}{6Q_{2x}-4\bar{X}}\right)^2-\left(\frac{Q_{3x}\varepsilon_{3x}-Q_{1x}\varepsilon_{1x}}{Q_{3x}-Q_{1x}}\right)\left(\frac{6Q_{2x}\varepsilon_{2x}-4\bar{X}\eta_x}{6Q_{2x}-4\bar{X}}\right)+\ldots\right]$$

Taking expected values on both sides of (3.4), we get,

$$E\left[\hat{CQV}_{reg(m)}\right]=CQV_{y(m)}\left\{1+0-0+E\left(\frac{6Q_{2y}\varepsilon_{2y}-4\bar{Y}\eta_y}{6Q_{2y}-4\bar{Y}}\right)^2-E\left(\frac{Q_{3y}\varepsilon_{3y}-Q_{1y}\varepsilon_{1y}}{Q_{3y}-Q_{1y}}\right)\left(\frac{6Q_{2y}\varepsilon_{2y}-4\bar{Y}\eta_y}{6Q_{2y}-4\bar{Y}}\right)\right\}$$
$$-\beta\left(CQV_{x(m)}\right)\left[0-0+E\left(\frac{6Q_{2x}\varepsilon_{2x}-4\bar{X}\eta_x}{6Q_{2x}-4\bar{X}}\right)^2-E\left(\frac{Q_{3x}\varepsilon_{3x}-Q_{1x}\varepsilon_{1x}}{Q_{3x}-Q_{1x}}\right)\left(\frac{6Q_{2x}\varepsilon_{2x}-4\bar{X}\eta_x}{6Q_{2x}-4\bar{X}}\right)\right]$$
$$=CQV_{y(m)}+CQV_{y(m)}\left[\frac{1}{\left(6Q_{2y}-4\bar{Y}\right)^2}E\left\{36Q_{2y}^2\varepsilon_{2y}^2+16\bar{Y}^2\eta_y^2-48Q_{2y}\bar{Y}\varepsilon_{2y}\eta_y\right\}\right.$$
$$\left.-\frac{1}{\left(Q_{3y}-Q_{1y}\right)\left(6Q_{2y}-4\bar{Y}\right)}E\left\{6Q_{2y}Q_{3y}\varepsilon_{2y}\varepsilon_{3y}-6Q_{1y}Q_{2y}\varepsilon_{1y}\varepsilon_{2y}-4Q_{3y}\bar{Y}\varepsilon_{3y}\eta_y+4Q_{1y}\bar{Y}\varepsilon_{1y}\eta_y\right\}\right]$$
$$-\beta\left(CQV_{x(m)}\right)\left[\frac{1}{\left(6Q_{2x}-4\bar{X}\right)^2}E\left\{36Q_{2x}^2\varepsilon_{2x}^2+16\bar{X}^2\eta_x^2-48Q_{2x}\bar{X}\varepsilon_{2x}\eta_x\right\}\right.$$
$$\left.-\frac{1}{\left(Q_{3x}-Q_{1x}\right)\left(6Q_{2x}-4\bar{X}\right)}E\left\{6Q_{2x}Q_{3x}\varepsilon_{2x}\varepsilon_{3x}-6Q_{1x}Q_{2x}\varepsilon_{1x}\varepsilon_{2x}-4Q_{3x}\bar{X}\varepsilon_{3x}\eta_x+4Q_{1x}\bar{X}\varepsilon_{1x}\eta_x\right\}\right]$$
$$=CQV_{y(m)}+\left(\frac{1-f}{n}\right)CQV_{y(m)}\left[\frac{1}{\left(6Q_{2y}-4\bar{Y}\right)^2}\left\{\frac{36Q_{2y}^2}{4Q_{2y}^2\left\{f_y\left(Q_{2y}\right)\right\}^2}+16\bar{Y}^2C_y^2+\frac{48Q_{2y}\bar{Y}}{Q_{2y}\bar{Y}}S_{YQ2y}\right\}\right. \quad (3.5)$$
$$\left.-\frac{1}{\left(Q_{3y}-Q_{1y}\right)\left(6Q_{2y}-4\bar{Y}\right)}\left\{\frac{6Q_{2y}Q_{3y}}{8Q_{2y}Q_{3y}f_y\left(Q_{2y}\right)f_y\left(Q_{3y}\right)}-\frac{6Q_{1y}Q_{2y}}{8Q_{1y}Q_{2y}f_y\left(Q_{1y}\right)f_y\left(Q_{2y}\right)}+\frac{4Q_{3y}\bar{Y}}{Q_{3y}\bar{Y}}S_{YQ_{3y}}-\frac{4Q_{1y}\bar{Y}}{Q_{1y}\bar{Y}}S_{YQ_{1y}}\right\}\right]$$
$$-\left(\frac{1-f}{n}\right)\beta\left(CQV_{x(m)}\right)\left[\frac{1}{\left(6Q_{2x}-4\bar{X}\right)^2}\left\{\frac{36Q_{2x}^2}{4Q_{2x}^2\left\{f_x\left(Q_{2x}\right)\right\}^2}+16\bar{X}^2C_x^2+\frac{48Q_{2x}\bar{X}}{Q_{2x}\bar{X}}S_{XQ2x}\right\}\right.$$
$$\left.-\frac{1}{\left(Q_{3x}-Q_{1x}\right)\left(6Q_{2x}-4\bar{X}\right)}\left\{\frac{6Q_{2x}Q_{3x}}{8Q_{2x}Q_{3x}f_x\left(Q_{2x}\right)f_x\left(Q_{3x}\right)}-\frac{6Q_{1x}Q_{2x}}{8Q_{1x}Q_{2x}f_x\left(Q_{1x}\right)f_x\left(Q_{2x}\right)}+\frac{4Q_{3x}\bar{X}}{Q_{3x}\bar{X}}S_{XQ_{3x}}-\frac{4Q_{1x}\bar{X}}{Q_{1x}\bar{X}}S_{XQ_{1x}}\right\}\right]$$
$$=CQV_{y(m)}+\left(\frac{1-f}{n}\right)CQV_{y(m)}\left[\frac{1}{\left(6Q_{2y}-4\bar{Y}\right)^2}\left\{\frac{9}{\left\{f_y\left(Q_{2y}\right)\right\}^2}+16S_y^2+48S_{YQ_{2y}}\right\}\right.$$
$$\left.-\frac{1}{\left(Q_{3y}-Q_{1y}\right)\left(6Q_{2y}-4\bar{Y}\right)}\left\{\frac{3}{4f_y\left(Q_{2y}\right)f_y\left(Q_{3y}\right)}-\frac{3}{4f_y\left(Q_{1y}\right)f_y\left(Q_{2y}\right)}+4S_{YQ_{3y}}-4S_{YQ_{1y}}\right\}\right]$$
$$-\left(\frac{1-f}{n}\right)\beta\left(CQV_{x(m)}\right)\left[\frac{1}{\left(6Q_{2x}-4\bar{X}\right)^2}\left\{\frac{9}{\left\{f_x\left(Q_{2x}\right)\right\}^2}+16S_x^2+48S_{XQ2x}\right\}\right.$$

$$-\frac{1}{(Q_{3x}-Q_{1x})(6Q_{2x}-4\bar{X})}\left\{\frac{3}{4f_x(Q_{2x})f_x(Q_{3x})}-\frac{3}{4f_x(Q_{1x})f_x(Q_{2x})}+4S_{XQ3x}-4S_{XQ1x}\right\}\Bigg]$$

Using (3.5), thus the bias, to the first order of approximation, in the proposed regression type estimator $C\hat{Q}V_{reg_{(m)}}$ is given by,

$$B\left[C\hat{Q}V_{reg(m)}\right]=E\left[C\hat{Q}V_{reg(m)}\right]-CQV_{y(m)}$$

$$=\left(\frac{1-f}{n}\right)\Bigg[CQV_{y(m)}\left\{\frac{1}{(6Q_{2y}-4\bar{Y})^2}\left(\frac{9}{\{f_y(Q_{2y})\}^2}+16S_y^2+48S_{YQ2y}\right)\right.$$

$$\left.-\frac{1}{(Q_{3y}-Q_{1y})(6Q_{2y}-4\bar{Y})}\left(\frac{3}{4f_y(Q_{2y})f_y(Q_{3y})}-\frac{3}{4f_y(Q_{1y})f_y(Q_{2y})}+4S_{YQ3y}-4S_{YQ1y}\right)\right\}$$

$$-\beta(CQV_{x(m)})\Bigg[\frac{1}{(6Q_{2x}-4\bar{X})^2}\left(\frac{9}{\{f_x(Q_{2x})\}^2}+16S_x^2+48S_{XQ2x}\right)$$

$$-\frac{1}{(Q_{3x}-Q_{1x})(6Q_{3x}-4\bar{X})}\left(\frac{3}{4f_x(Q_{2x})f_x(Q_{3x})}-\frac{3}{4f_x(Q_{1x})f_x(Q_{2x})}+4S_{XQ3x}-4S_{XQ1x}\right)\Bigg\}\Bigg]$$

which proves the theorem.

As the sample size n tends to infinity, then $B[C\hat{Q}V_{reg_{(m)}}]\to 0$, thus $C\hat{Q}V_{reg_{(m)}}$ is a consistent estimator of the new coefficient of dispersion $CQV_{Y(m)}$.

Theorem 3.2. The minimum mean squared error of the regression type estimator $C\hat{Q}V_{reg_{(m)}}$ is given by

$$Min.MSE\left[C\hat{Q}V_{reg(m)}\right]=\left(\frac{1-f}{n}\right)(CQV_{y(m)})^2\Psi_y^{*2}\left[1-\frac{\Psi_{xy}^{*2}}{\Psi_y^{*2}\Psi_x^{*2}}\right] \tag{3.6}$$

where Ψ_y^{*2}, Ψ_x^{*2} and Ψ^*_{xy} are the same as defined in (2.9), (2.10) and (2.11) respectively.

Proof: By the definition of mean squared error, we have

$$MSE\left[C\hat{Q}V_{reg(m)}\right]=E\left[C\hat{Q}V_{reg(m)}-CQV_{y(m)}\right]^2$$

$$\approx E\Bigg[CQV_{y(m)}\left\{\frac{Q_{3y}\varepsilon_{3y}-Q_{1y}\varepsilon_{1y}}{Q_{3y}-Q_{1y}}-\frac{6Q_{2y}\varepsilon_{2y}-4\bar{Y}\eta_y}{6Q_{2y}-4\bar{Y}}\right\}$$

$$-\beta(CQV_{x(m)})\left\{\frac{Q_{3x}\varepsilon_{3x}-Q_{1x}\varepsilon_{1x}}{Q_{3x}-Q_{1x}}-\frac{6Q_{2x}\varepsilon_{2x}-4\bar{X}\eta_x}{6Q_{2x}-4\bar{X}}\right\}\Bigg]^2$$

$$\approx(CQV_{y(m)})^2E\left[\left(\frac{Q_{3y}\varepsilon_{3y}-Q_{1y}\varepsilon_{1y}}{Q_{3y}-Q_{1y}}\right)-\left(\frac{6Q_{2y}\varepsilon_{2y}-4\bar{Y}\eta_y}{6Q_{2y}-4\bar{Y}}\right)\right]^2 \tag{3.7}$$

$$+\beta^2(CQV_{x(m)})^2E\left[\left(\frac{Q_{3x}\varepsilon_{3x}-Q_{1x}\varepsilon_{1x}}{Q_{3x}-Q_{1x}}\right)-\left(\frac{6Q_{2x}\varepsilon_{2x}-4\bar{X}\eta_x}{6Q_{2x}-4\bar{X}}\right)\right]^2$$

$$-2\beta(CQV_{y(m)})(CQV_{x(m)})E\left[\left\{\left(\frac{Q_{3y}\varepsilon_{3y}-Q_{1y}\varepsilon_{1y}}{Q_{3y}-Q_{1y}}\right)-\left(\frac{6Q_{2y}\varepsilon_{2y}-4\bar{Y}\eta_y}{6Q_{2y}-4\bar{Y}}\right)\right\}\left\{\left(\frac{Q_{3x}\varepsilon_{3x}-Q_{1x}\varepsilon_{1x}}{Q_{3x}-Q_{1x}}\right)-\left(\frac{6Q_{2x}\varepsilon_{2x}-4\bar{X}\eta_x}{6Q_{2x}-4\bar{X}}\right)\right\}\right]$$

$$=\left(\frac{1-f}{n}\right)\left[(CQV_{y(m)})^2\Psi_y^{*2}+\beta^2(CQV_{x(m)})^2\Psi_x^{*2}-2\beta(CQV_{y(m)})(CQV_{x(m)})\Psi_{xy}^*\right]$$

On minimizing (3.7) with respect to β, that is on setting.

$$\frac{\partial MSE\left(C\hat{Q}V_{reg(m)}\right)}{\partial\beta}=0$$

we get

$$\beta = \frac{\Psi^*_{xy}(CQV_{y(m)})}{\Psi^{*2}_x(CQV_{x(m)})} \tag{3.8}$$

On substituting (3.8) into (3.7), the minimum mean squared error of the proposed regression type estimator is given by

$$Min.MSE\left[C\hat{Q}V_{reg(m)}\right]=\left(\frac{1-f}{n}\right)\left[(CQV_{y(m)})^2\Psi^{*2}_y+\beta^2(CQV_{x(m)})^2\Psi^{*2}_x-2\beta(CQV_{y(m)})(CQV_{x(m)})\Psi^*_{xy}\right]$$

$$=\left(\frac{1-f}{n}\right)\left[(CQV_{y(m)})^2\Psi^{*2}_y+\left(\frac{CQV_{y(m)}\Psi^*_{xy}}{\Psi^{*2}_xCQV_{x(m)}}\right)^2(CQV_{x(m)})^2\Psi^{*2}_x-2\left(\frac{CQV_{y(m)}\Psi^*_{xy}}{\Psi^{*2}_xCQV_{x(m)}}\right)(CQV_{y(m)})(CQV_{x(m)})\Psi^*_{xy}\right]$$

$$=\left(\frac{1-f}{n}\right)\left[(CQV_{y(m)})^2\Psi^{*2}_y-\left(\frac{CQV_{y(m)}\Psi^*_{xy}}{\Psi^{*2}_xCQV_{x(m)}}\right)^2(CQV_{x(m)})^2\Psi^{*2}_x\right]$$

$$=\left(\frac{1-f}{n}\right)(CQV_{y(m)})^2\left[\Psi^{*2}_y-\frac{\Psi^{*2}_{xy}}{\Psi^{*2}_x}\right]$$

$$=\left(\frac{1-f}{n}\right)(CQV_{y(m)})^2\Psi^{*2}_y\left[1-\frac{\Psi^{*2}_{xy}}{\Psi^{*2}_y\Psi^{*2}_x}\right]$$

which proves the theorem.

In the succeeding section, we consider a simulation study to compare the ratio and the regression type estimators with the naïve estimator of the proposed new measure of coefficient of dispersion relative to the empirical mode.

4. Empirical Evidences: Simulation Study

To investigate the percent relative bias and percent relative efficiency values of the proposed regression type estimator $C\hat{Q}V_{reg_{(m)}}$ with regards to the naïve estimator $C\hat{Q}V_{y_{(m)}}$ and the ratio type estimator $C\hat{Q}V_{rat_{(m)}}$ in case of estimating the proposed new coefficient of dispersion $CQV_{y_{(m)}}$, we simulated different populations and situations by making use of different study variables having diverse values of the correlation coefficient with the auxiliary variables. Following Singh and Horn (1998), we used the following transformations to produce our variable pair

$$y_i=\sqrt{(1-\rho^2)}y^*_i+\rho\frac{\sigma_{y^*}}{\sigma_{x^*}}x^*_i \tag{4.1}$$

and

$$x_i=x^*_i \tag{4.2}$$

where y^*_i and x^*_i are the independent Gamma variables generated in the *R* language. Assume if a_y, and b_y are the shape and scale parameters, respectively, for the study variable y^*_i, then $\mu_{y*} = E(y^*_i) = a_yb_y$, $\sigma^2_{y*} = V(y^*_i) = a_yb^2_y$; if a_x, and b_x are the shape and scale parameters, respectively, for the auxiliary variable x^*_i, then $\mu_{x*} = E(x^*_i) = a_xb_x$, $\sigma^2_{x*} = V(x^*_i) = a_xb^2_x$. In (4.1) and (4.2), then

$$E(y_i)=\sqrt{(1-\rho^2)}E(y^*_i)+\rho\frac{\sigma_{y^*}}{\sigma_{x^*}}E(x^*_i) \tag{4.3}$$

and

$$E(x_i) = E(x^*_i) = a_xb_x \tag{4.4}$$

Now

$$V(y_i)=V\left[\sqrt{(1-\rho^2)}y^*_i+\rho\frac{\sigma_{y^*}}{\sigma_{x^*}}x^*_i\right]=\sigma^2_{y^*} \tag{4.5}$$

and

$$V(x_i) = V[x^*_i] = \sigma^2_{x*} \tag{4.6}$$

Now

$$Cov(x_i, y_i) = Cov\left[x_i^*, \sqrt{(1-\rho^2)}y_i^* + \rho\frac{\sigma_{y^*}}{\sigma_{x^*}}x_i^*\right] = \rho\sigma_{y^*}\sigma_{x^*} \tag{4.7}$$

By making use of (4.1) and (4.2), there will be a correlation coefficient $\rho = \frac{Cov(x_i, y_i)}{\sigma_{x^*}\sigma_{y^*}}$ between the study variable y and the auxiliary variable x for any set of independently generated gamma variables $x_i^* \sim G(a_x, b_x)$ and $y_i^* \sim G(a_y, b_y)$.

We generate a population with parameters, such as $N, \rho, \bar{Y}, \bar{X}, S_y$ and S_x, we then compute the percent relative biases and percent relative efficiencies of the proposed regression type estimator and ratio type estimator with respect to the naïve estimator of the new measure of coefficient of dispersion relative to the empirical mode. In each population we also computed the value of the empirical mode, and only those populations showing positive values of the empirical mode for both the study and auxiliary variables were considered. Note that y_i is a linear combination of two independent gamma variables, that is, $y_i^* \sim G(a_y, b_y)$ and $x_i^* \sim G(a_x, b_x)$. So we fit the gamma distribution using fit.gamma<-fitdist(yp, distr="gamma", method ="mle") in R-language, on each one of the generated population values, y_i, $i = 1, 2,\ldots, N$, to find the shape parameter a_y^{fit} and the scale parameter b_y^{fit}, that is, we consider the newly generated population $y_i \sim G(a_y^{fit}, b_y^{fit})$. Also for the generated correlated auxiliary variable, x_i, $i = 1,2,\ldots, N$, we fit gamma distribution to find the shape parameter a_x^{fit} and the scale parameter b_x^{fit}. These best-fit parameters are used to compute the values of $f_y(Q_{1y}), f_y(Q_{2y}), f_y(Q_{3y}), f_x(Q_{1x}), f_x(Q_{2x})$ and $f_x(Q_{3x})$ for a given population.

The empirical population mode of the study variable was computed as

$$Mo_y = 3Q_{2y} - 2\bar{Y} \tag{4.8}$$

and that of the auxiliary variable was computed as

$$Mo_x = 3Q_{2x} - 2\bar{X} \tag{4.9}$$

The percent relative efficiencies of the ratio type estimator $C\hat{Q}V_{rat(m)}$ and the regression type estimator $C\hat{Q}V_{reg(m)}$ with regards to the naïve estimator $C\hat{Q}V_{y(m)}$, respectively, are defined as follows:

$$\text{RE}(1) = \frac{\text{MSE}(C\hat{Q}V_{y(m)})}{\text{MSE}(C\hat{Q}V_{rat(m)})} \times 100\% \tag{4.10}$$

and

$$\text{RE}(2) = \frac{\text{MSE}(C\hat{Q}V_{y(m)})}{\text{MSE}(C\hat{Q}V_{reg(m)})} \times 100\% \tag{4.11}$$

The approximate percent relative biases in the naïve estimator, ratio type estimator and the regression type estimator, respectively, are computed as:

$$\text{RB}(0) = \frac{B(C\hat{Q}V_{y(m)})}{CQV_{y(m)}} \times 100\% \tag{4.12}$$

$$\text{RB}(1) = \frac{B(C\hat{Q}V_{rat(m)})}{CQV_{y(m)}} \times 100\% \tag{4.13}$$

and

$$\text{RB}(2) = \frac{B(C\hat{Q}V_{reg(m)})}{CQV_{y(m)}} \times 100\% \tag{4.14}$$

For different choices of the shape parameters a_x, and a_y and the value of correlation coefficient ρ, the values of the percent relative efficiencies $RE(1)$ and $RE(2)$ are computed by using R codes given in the APPENDIX C. The values of the scale parameters were set as $bx = 1$ and $by = 1$. The value of the sample size $n = 500$, which is 5% of the population size, was noted to be sufficient from each population of size $N = 10{,}001$ such that in all the three estimators under study the absolute values of the percent relative biases were less than 10%. Note that the percent relative efficiencies $RE(1)$ and $RE(2)$ of the ratio and regression type estimators with respect to the naïve estimator are free from the values of sample size. We also varied the values of the correlation coefficient ρ between 0.50 and 0.99 with a step of 0.01 such that the value of the percent relative efficiency of the ratio estimator RE(1) is greater than 100%, while observing the corresponding relative efficiency of the regression estimator RE(2). The simulated empirical evidences are reported in Table B1 in the APPENDIX-B. In Table B1, the value of correlation coefficient ρ are the values which are used as an input to generate a population of two correlated variables and ρ_{xy} is the amount of correlation observed between the two variables at the time of simulation. Thus for each value of the correlation coefficient there is one population for different values of shape parameters a_x and a_y. In the simulation process the values of the shape parameters a_x and a_y were varied between 2.0 and 2.75 with a step of 0.25 and between 3.00 and 3.75 with a step of 0.25 respectively, and thus have created different populations each of size $N = 10{,}001$ for investigation. The change in the values of the parameters Mo_y, Mo_x and $CQV_{y(m)}$, with the value of correlation coefficient for each pair of a_x and a_y is shown in three dimensional Figures 1, 2 and 3 for the different populations, where Figure 1 represents the scatter plot of the mode of y variable versus the shape parameters for the study and the auxiliary variables.

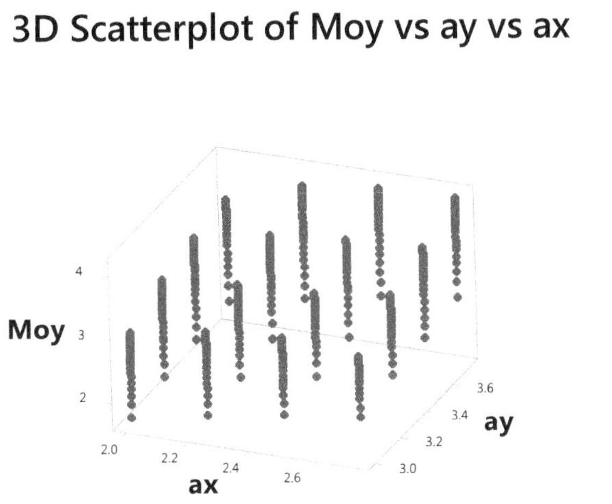

Figure 1: The values of Mo_y for the different values of the a_y and a_x.

The three dimensional figure, Figure 2 displays the scatter plot of Mo_x versus a_x and a_y.

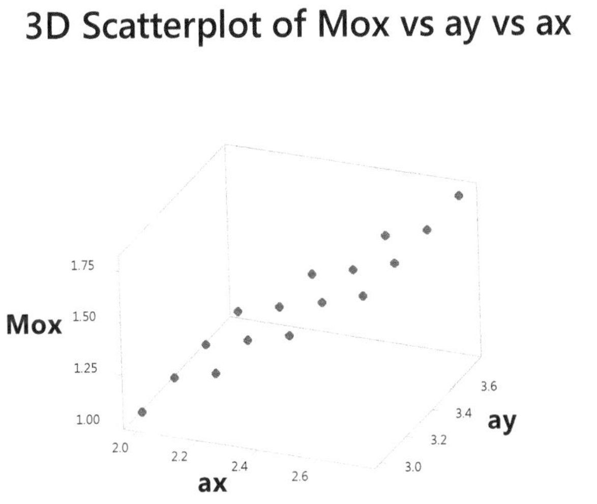

Figure 2: The values of Mo_x for the different values of the a_y and a_x.

Figure 3 is a three dimensiional scatter plot plotting $CQV_{y(m)}$ versus a_y and a_x.

Note that although the values of Mo_y and $CQV_{y(m)}$ are free from the values of auxiliary variables, but the change in their values in the simulation study is due to the different random numbers generated for each combination of a_y, a_x and ρ. Table 1 helps visualize Figures 1 and 3 with more clarity. However, the behavior of Mo_x for different situations as shown in Figure 2 is quite different. In other words, its value remains constant for each combination of shape and scale parameters. For example, for $a_y = 3.0$ and $a_x = 2.0$ there are 34 values of the correlation coefficient between 0.66 and 0.99 with a step of 0.01 where the ratio and regression estimators are more efficient than the naïve estimator. These changes also make changes in the value of $CQV_y(m)$ between 0.370 to 0.620; and

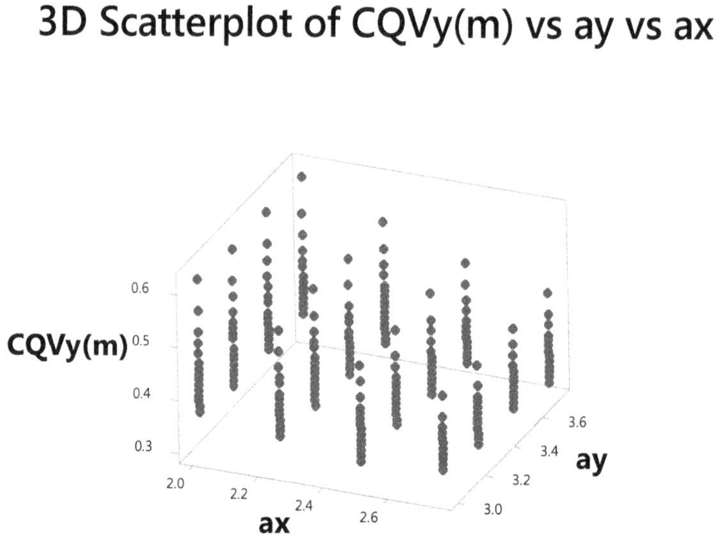

Figure 3: The value of $CQV_{y(m)}$ for the different values of the a_y and a_x.

Table 1: Range of $CQV_{y(m)}$, Mo_y, ρ_{xy}, and Mo_x for different values of a_y and a_x.

			ρ_{xy}		$CQV_{y(m)}$		Mo_y		Mo_x
a_y	a_x	f	**Min**	**Max**	**Min**	**Max**	**Min**	**Max**	**Fixed**
3.00	2.00	34	0.66	0.99	0.37	0.62	1.71	3.05	1.05
	2.25	49	0.51	0.99	0.35	0.55	1.93	3.25	1.29
	2.50	50	0.50	0.99	0.33	0.51	2.13	3.35	1.53
	2.75	16	0.84	0.99	0.34	0.48	2.28	3.23	1.78
3.25	2.00	34	0.66	0.99	0.35	0.61	1.83	3.35	1.05
	2.25	44	0.56	0.99	0.34	0.56	2.00	3.47	1.29
	2.50	50	0.50	0.99	0.33	0.51	2.21	3.52	1.53
	2.75	50	0.50	0.99	0.32	0.47	2.42	3.68	1.78
3.50	2.00	31	0.69	0.99	0.35	0.61	1.88	3.50	1.05
	2.25	40	0.60	0.99	0.33	0.55	2.09	3.72	1.29
	2.50	50	0.50	0.99	0.32	0.51	2.29	3.84	1.53
	2.75	50	0.50	0.99	0.32	0.47	2.48	3.90	1.78
3.75	2.00	21	0.79	0.99	0.35	0.61	1.96	3.57	1.05
	2.25	33	0.67	0.99	0.32	0.55	2.20	3.98	1.29
	2.50	43	0.57	0.99	0.31	0.50	2.42	4.13	1.53
	2.75	50	0.50	0.99	0.30	0.47	2.58	4.15	1.78

Mo_y between 1.71 to 3.05 due to the different set of random numbers being used every time, however the value of Mo_x is always equal to 1.05.

The results reported in Table 2 indicate that if the values of shape parameter $a_x = 2.0$ and scale parameter $b_x = 1$ for the auxiliary variable, and shape parameter $a_y = 3.0$ and scale parameter $b_y = 1$ for the study variable, follow Gamma distributions, then the proposed ratio and the regression estimators are more competent than the naïve estimator if the value of the correlation coefficient ρ_{xy} is between 0.66 and 0.99.

The average percent relative efficiencies of the ratio and the regression estimators are 137.3% and 179.3% with respective standard deviations of 42.8% and 77.0%. The median relative efficiencies of the ratio and the regression type estimators are 124.6% and 150.9% respectively. The maximum value of the percent relative efficiency of the ratio estimator is 313.6% and that of the regression type estimator is 487.2%. In the same way the rest of the results in Table 2 can be interpreted.

Table 2: Summary of percent relative efficiencies RE(1) and RE(2) values.

			RE(1)					RE(2)				
a_y	a_x	f	Mean	SD	Min	Med	Max	Mean	SD	Min	Med	Max
3.00	2.00	34	137.3	42.8	100.8	124.6	313.6	179.3	77.0	123.6	150.9	487.2
	2.25	49	151.0	59.0	100.3	131.3	405.1	161.4	74.5	112.3	134.4	514.6
	2.50	50	157.8	68.0	106.8	133.6	470.6	160.0	72.2	112.3	133.6	510.2
	2.75	16	207.1	91.6	139.7	171.3	484.2	218.0	88.5	149.5	185.2	484.3
3.25	2.00	34	135.2	34.6	100.6	126.9	272.4	179.7	74.0	123.7	152.9	473.8
	2.25	44	149.7	55.5	100.7	132.7	391.7	168.9	80.7	115.6	140.5	549.9
	2.50	50	151.7	60.8	100.6	131.5	421.9	156.7	70.5	109.8	131.8	500.3
	2.75	50	152.1	65.2	107.2	129.1	468.8	154.0	67.5	109.9	130.2	490.6
3.50	2.00	31	131.0	35.8	100.7	122.0	280.2	189.4	84.6	126.7	158.8	514.0
	2.25	40	139.9	40.6	101.0	129.2	310.6	168.1	71.9	117.2	141.9	474.5
	2.50	50	150.9	57.2	103.1	131.5	407.4	158.1	73.7	111.1	132.1	522.2
	2.75	50	152.3	63.8	104.7	130.0	446.2	154.6	69.1	110.2	130.0	495.0
	2.25	33	135.6	37.0	100.4	123.8	272.5	177.2	73.7	121.7	148.7	454.1
	2.50	43	142.2	41.9	100.0	129.0	301.6	164.3	66.5	116.1	139.0	444.3
	2.75	50	149.8	57.1	100.6	129.9	399.3	154.9	68.2	109.3	130.1	486.1

For $a_y = 3$ and $a_x = 2$, from Table 3, the mean values of the percent relative biases in the naïve, ratio and regression type estimators, respectively, are 2.827%, 0.640% and 0.638%, with respective standard deviations of 0.402%, 0.591% and 0.593%. The maximum values of the percent relative biases for the naïve, ratio and regression estimators are negligible, and hence are of not much interest.

It is interesting to note that for a sample of size 500 units out of a population of 10,001 units, which is a 5% sample size, is sufficient to produce an estimate of the proposed coefficient of dispersion based on naïve, ratio and regression type estimators. Note that in Table 1 for $a_y = 3.0$ and $a_x = 2.0$ there are 34 values of the correlation coefficient considered ρ_{xy} between 0.66 and 0.99 with a step of 0.01 when the proposed ratio estimator is more efficient than the naïve estimator and the absolute percent relative biases are less than 10% in the three estimators considered. It is an indication that the performance of the relative biases and ratio estimator are not only dependent on the value the correlation coefficient ρ_{xy} and other parameters such as the shape and scale parameters which are also playing an important role. In the same way the results from Table 1, Table 2 and Table 3 can be interpreted for the different populations created under different combinations of the shape parameters and the correlation coefficient values. Table 4 shows the fitted values of the Gamma parameters for different populations each of 10,001 units. For example, for $a_y = 3.0$, $b_y = 1$, $a_x = 2.0$ and $b_x = 1$, there are 34 fitting showing $a_x^{fit} = 2.01$, $b_x^{fit} = 1.00$, a_y^{fit} lies between 2.89 and 4.91, and b_y^{fit} between 1.02 and 1.27. The variations in the fitted values for the y values are reflected due to the linear combinations of two independent Gamma variables. In a similar way the rest of the fitted Gamma parameters can be found in Table 4.

Figures 4, 5, 6 and 7 are used to visualize the influence of the values of the correlation coefficient on the values of the percent relative efficiencies RE(1) and RE(2) for different combinations of shape parameters used for the generation of the study variable and the auxiliary variable.

Figure 4 is used to visualize the influence of the relative efficiencies of the ratio and regression type estimators versus the shape and scale parameters for $a_y = 3.0$.

Figure 5 is used to visualize the influence of the relative efficiencies of the ratio and regression type estimators versus the shape and scale parameters for $a_y = 3.25$.

Figure 6 is used to visualize the influence of the relative efficiencies of the ratio and regression type estimators versus the shape and scale parameters for $a_y = 3.50$.

Figure 7 is used to visualize the influence of the relative efficiencies of the ratio and regression type estimators versus the shape and scale parameters for $a_y = 3.75$.

The graphical representation shows that there is not a lot of difference in the RE(1) and RE(2) values, thus eight panels per graph are used instead of using four panels. It may be worth pointing out that in an eight panel graphical presentation, it is easy to differentiate between RE(1) and RE(2) values, especially in black and white printing.

Table 3: Summary of RB(0), RB(1) and RB(2) for different situations.

Estimator	a_y	a_x	f	Mean	SD	Min	Med	Max
Naïve	3.00	2.00	34	2.827	0.402	2.160	2.910	3.360
		2.25	49	4.032	0.827	2.120	4.410	4.840
		2.50	50	5.071	1.127	2.060	5.540	6.070
		2.75	16	6.093	2.357	2.270	6.240	9.570
	3.25	2.00	34	2.691	0.397	2.000	2.805	3.130
		2.25	44	3.110	0.527	1.940	3.390	3.590
		2.50	50	4.371	1.009	1.840	4.810	5.290
		2.75	50	5.991	1.621	1.820	6.620	7.410
	3.50	2.00	31	2.339	0.359	1.820	2.360	2.850
		2.25	40	2.661	0.520	1.690	2.885	3.180
		2.50	50	4.015	1.027	1.700	4.480	5.020
		2.75	50	4.625	1.176	1.660	5.045	5.800
	3.75	2.00	21	1.799	0.150	1.600	1.790	2.020
		2.25	33	2.104	0.303	1.590	2.150	2.490
		2.50	43	2.294	0.411	1.490	2.480	2.720
		2.75	50	3.707	0.912	1.490	4.045	4.600
Ratio	3.00	2.00	34	0.640	0.591	−0.340	0.695	1.530
		2.25	49	1.933	1.033	−0.170	2.230	3.020
		2.50	50	2.905	1.252	0.000	3.470	4.040
		2.75	16	3.272	1.967	0.250	3.300	6.310
	3.25	2.00	34	0.534	0.550	−0.410	0.590	1.330
		2.25	44	1.196	0.712	−0.230	1.365	2.080
		2.50	50	2.464	1.144	−0.070	3.070	3.490
		2.75	50	3.910	1.681	0.050	4.800	5.490
	3.50	2.00	31	0.299	0.488	−0.440	0.300	1.060
		2.25	40	0.902	0.655	−0.240	1.035	1.750
		2.50	50	2.161	1.127	−0.130	2.530	3.270
		2.75	50	2.817	1.251	−0.010	3.375	4.020
	3.75	2.00	21	−0.102	0.2330	−0.420	−0.120	0.280
		2.25	33	0.474	0.440	−0.270	0.500	1.110
		2.50	43	0.827	0.541	−0.160	0.940	1.500
		2.75	50	2.107	1.008	−0.060	2.625	3.020
Regression	3.00	2.00	34	0.638	0.593	−0.350	0.695	1.530
		2.25	49	1.920	1.029	−0.180	2.220	3.000
		2.50	50	2.898	1.246	−0.010	3.470	4.030
		2.75	16	3.318	1.989	0.260	3.350	6.380
	3.25	2.00	34	0.533	0.552	−0.410	0.590	1.330
		2.25	44	1.180	0.708	−0.240	1.350	2.050
		2.50	50	2.448	1.137	−0.090	3.065	3.470
		2.75	50	3.916	1.676	0.040	4.790	5.480
	3.50	2.00	31	0.296	0.490	−0.440	0.300	1.060
		2.25	40	0.883	0.653	−0.260	1.020	1.720
		2.50	50	2.143	1.123	−0.150	2.520	3.240
		2.75	50	2.810	1.244	−0.020	3.355	4.010
	3.75	2.00	21	−0.110	0.233	−0.420	−0.130	0.270
		2.25	33	0.453	0.439	−0.290	0.480	1.090
		2.50	43	0.795	0.536	−0.190	0.910	1.450
		2.75	50	2.089	1.000	−0.080	2.620	2.980

Table 4: Fitted values of the Gamma parameters for different situations.

a_y	a_x	Fit	Freq	Min	Max	a_y	a_x	Fit	Freq	Min	Max
3.00	2.00	a_x^{fit}	34	2.01	2.01	3.50	2.00	a_x^{fit}	31	2.01	2.01
	2.25		49	2.24	2.24		2.25		40	2.24	2.24
	2.50		50	2.51	2.51		2.50		50	2.51	2.51
	2.75		16	2.75	2.75		2.75		50	2.75	2.75
	2.00	b_x^{fit}	34	1.00	1.00		2.00	b_x^{fit}	31	1.00	1.00
	2.25		49	1.00	1.00		2.25		40	1.00	1.00
	2.50		50	1.01	1.01		2.50		50	1.01	1.01
	2.75		16	1.01	1.01		2.75		50	1.01	1.01
	2.00	a_y^{fit}	34	2.89	4.91		2.00	a_y^{fit}	31	3.00	5.44
	2.25		49	3.17	5.24		2.25		40	3.24	5.68
	2.50		50	3.45	5.46		2.50		50	3.56	6.03
	2.75		16	3.75	5.60		2.75		50	3.82	6.20
			34	2.89	4.91						
	2.00	b_y^{fit}	34	1.02	1.27		2.00	b_y^{fit}	31	0.96	1.25
	2.25		49	1.06	1.33		2.25		40	1.00	1.27
	2.50		50	1.11	1.35		2.50		50	1.05	1.32
	2.75		16	1.15	1.39		2.75		50	1.08	1.33
3.25	2.00	a_x^{fit}	34	2.01	2.01	3.75	2.00	a_x^{fit}	21	2.01	2.01
	2.25		44	2.24	2.24		2.25		33	2.24	2.24
	2.50		50	2.51	2.51		2.50		43	2.51	2.51
	2.75		50	2.75	2.75		2.75		50	2.75	2.75
	2.00	b_x^{fit}	34	1.00	1.00		2.00	b_x^{fit}	21	1.00	1.00
	2.25		44	1.00	1.00		2.25		33	1.00	1.00
	2.50		50	1.01	1.01		2.50		43	1.01	1.01
	2.75		50	1.01	1.01		2.75		50	1.01	1.01
	2.00	a_y^{fit}	34	2.96	5.22		2.00	a_y^{fit}	21	3.04	5.36
	2.25		44	3.21	5.41		2.25		33	3.29	5.86
	2.50		50	3.50	5.70		2.50		43	3.56	6.05
	2.75		50	3.77	5.99		2.75		50	3.86	6.42
	2.00	b_y^{fit}	34	0.99	1.27		2.00	b_y^{fit}	21	0.94	1.20
	2.25		44	1.03	1.29		2.25		33	0.97	1.24
	2.50		50	1.07	1.33		2.50		43	1.01	1.25
	2.75		50	1.11	1.36		2.75		50	1.05	1.30

Figures 8–11 are used to display the percent relative bias values for the three estimators under different situations. A careful investigation of these graphs indicates that the value of the percent relative bias is less than 10% in all the three estimators.

Figure 8 is used to display the relative biases of the naïve, ratio and regression type estimators versus the shape and scale parameters when $a_y = 3.0$.

Figure 9 is used to display the relative biases of the naïve, ratio and regression type estimators versus the shape and scale parameters when $a_y = 3.25$.

Figure 10 is used to display the relative biases of the naïve, ratio and regression type estimators versus the shape and scale parameters when $a_y = 3.50$.

Figure 11 is used to display the relative biases of the naïve, ratio and regression type estimators versus the shape and scale parameters when $a_y = 3.75$.

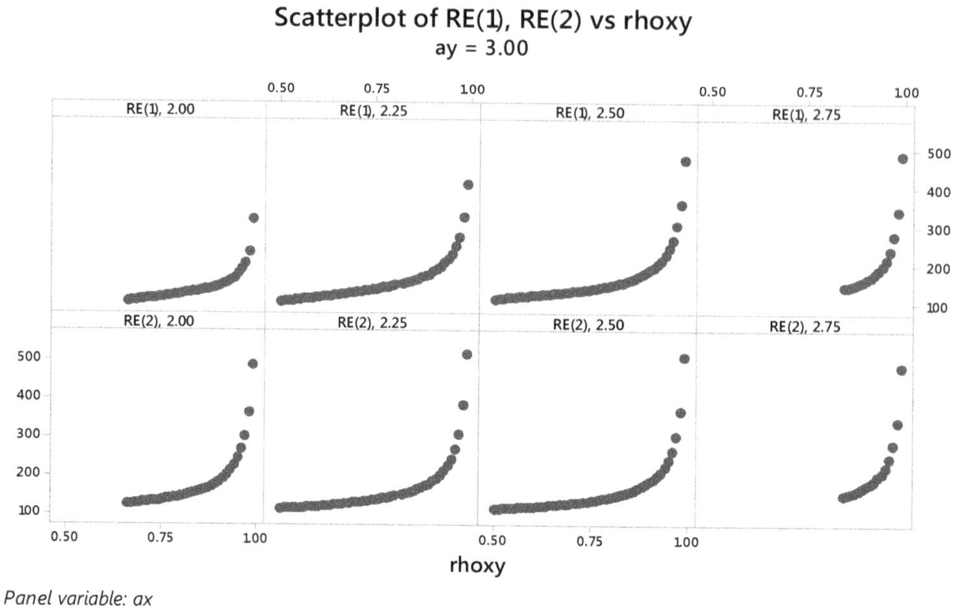

Figure 4: RE(1), RE(2) versus ρ_{xy} for $a_y = 3.0$ and $a_x = 2.0, 2.25, 2.5$ and 2.75.

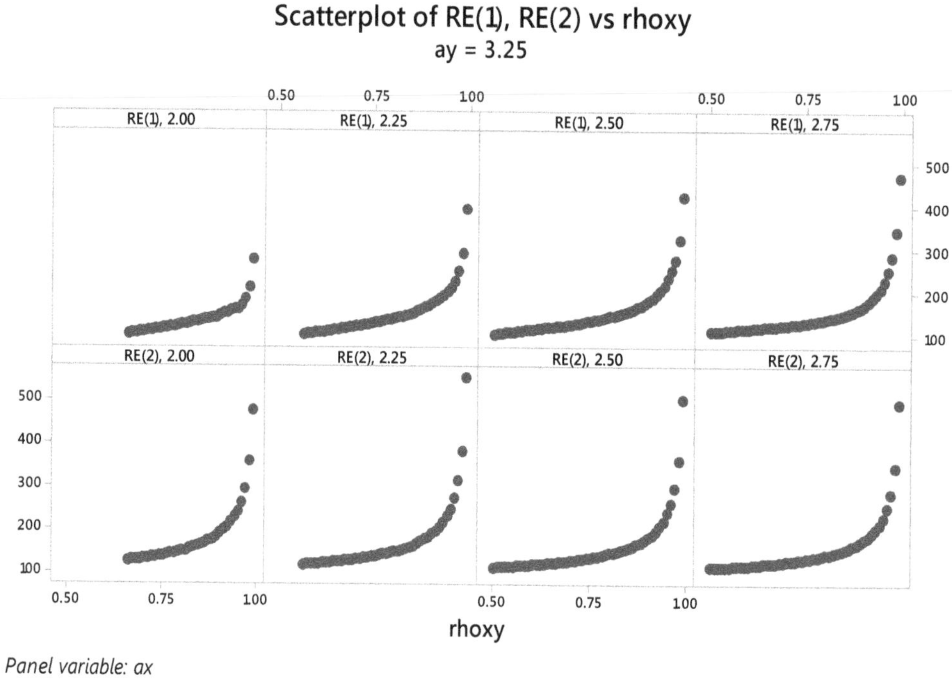

Figure 5: RE(1), RE(2) versus ρ_{xy} for $a_y = 3.25$ and $a_x = 2.0, 2.25, 2.5$ and 2.75.

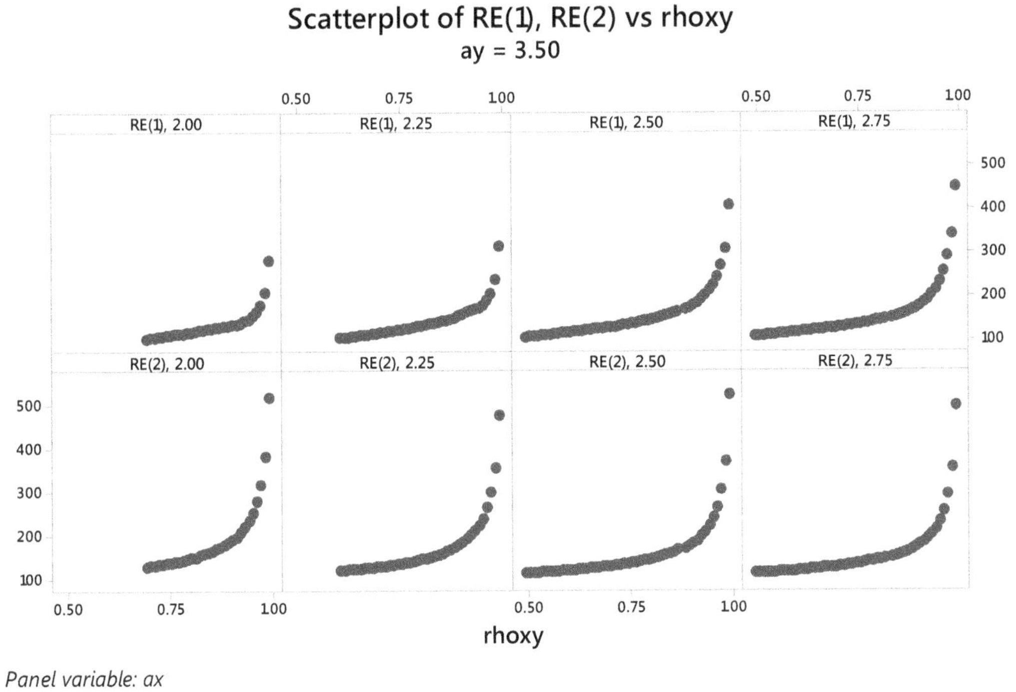

Figure 6: RE(1), RE(2) versus ρ_{xy} for $a_y = 3.50$ and $a_x = 2.0, 2.25, 2.5$ and 2.75.

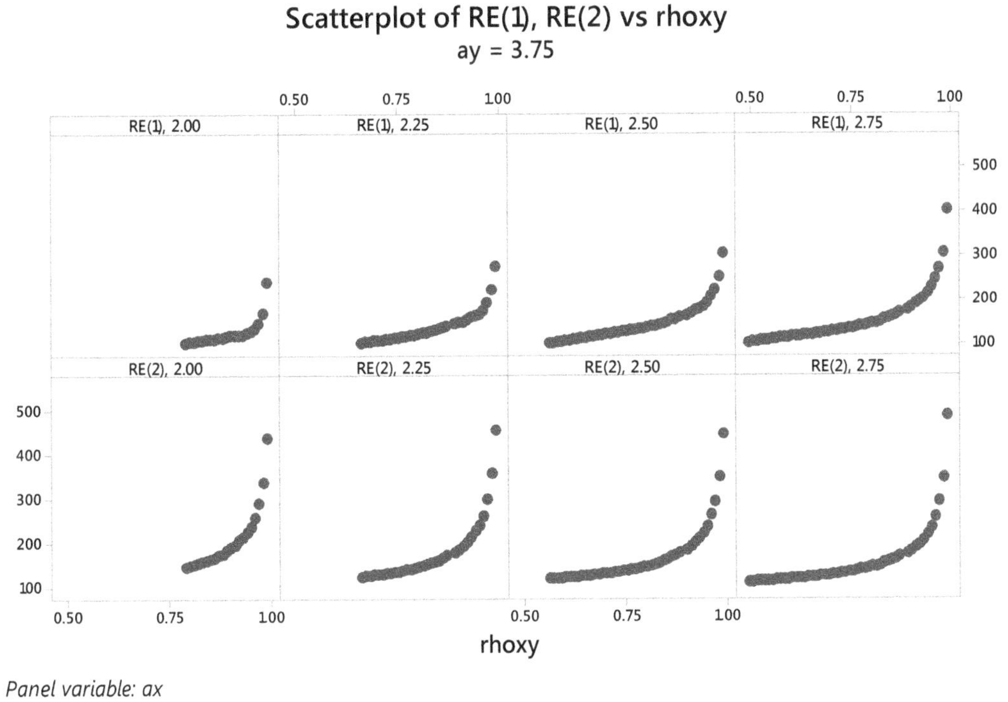

Figure 7: RE(1), RE(2) versus ρ_{xy} for $a_y = 3.75$ and $a_x = 2.0, 2.25, 2.5$ and 2.75.

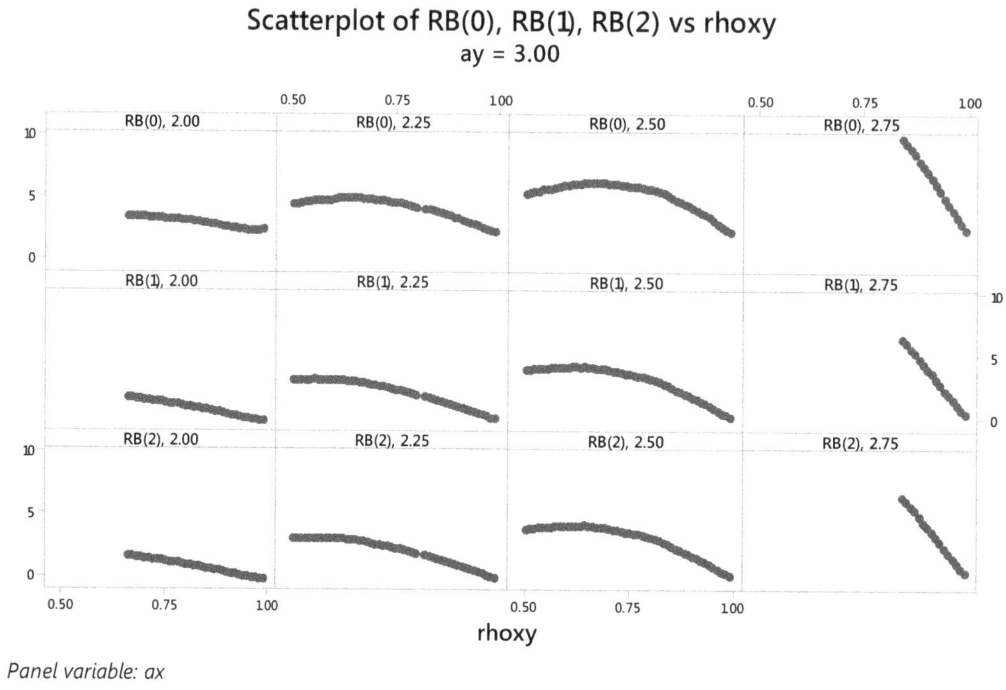

Figure 8: RB(0), RB(1) and RB(2) versus ρ_{xy} for $a_y = 3.0$ and $a_x = 2.0$, 2.25, 2.50 and 2.75.

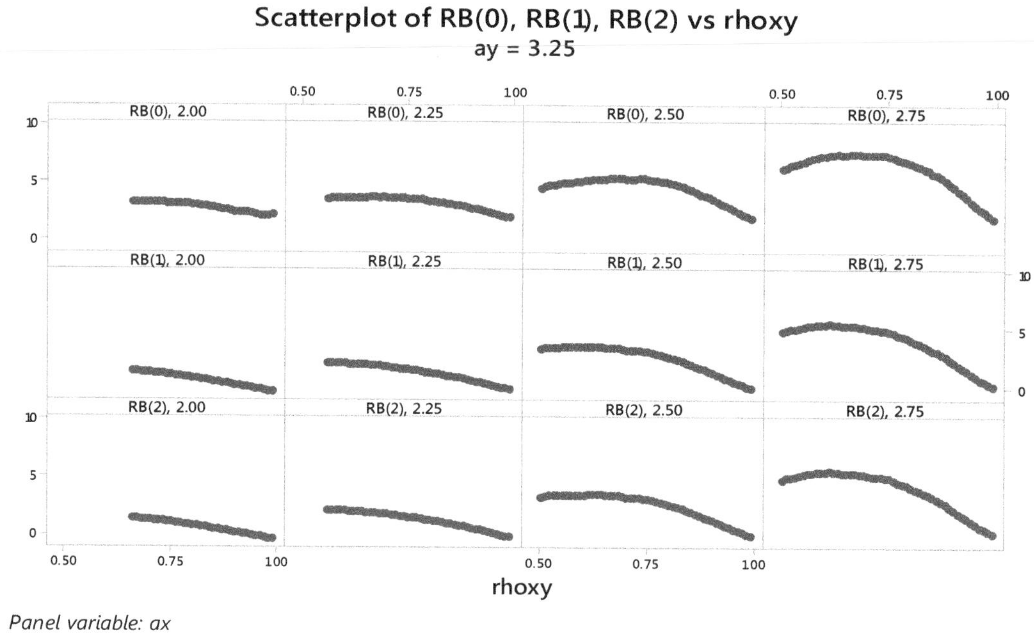

Figure 9: RB(0), RB(1) and RB(2) versus ρ_{xy} for $a_y = 3.25$ and $a_x = 2.0$, 2.25, 2.50 and 2.75.

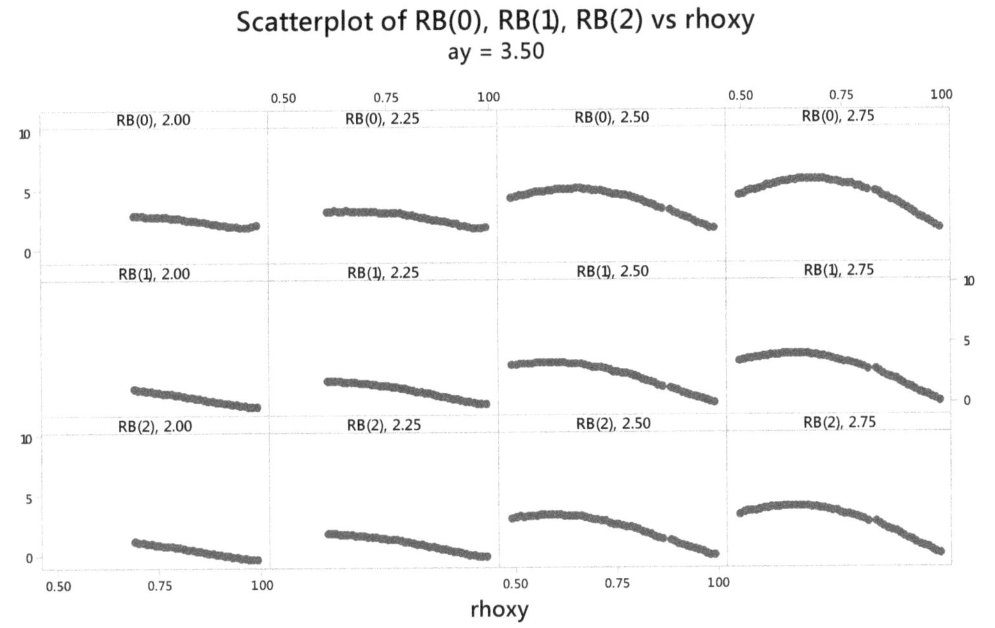

Panel variable: ax

Figure 10: RB(0), RB(1) and RB(2) versus ρ_{xy} for $a_y = 3.50$ and $a_x = 2.0, 2.25, 2.50$ and 2.75.

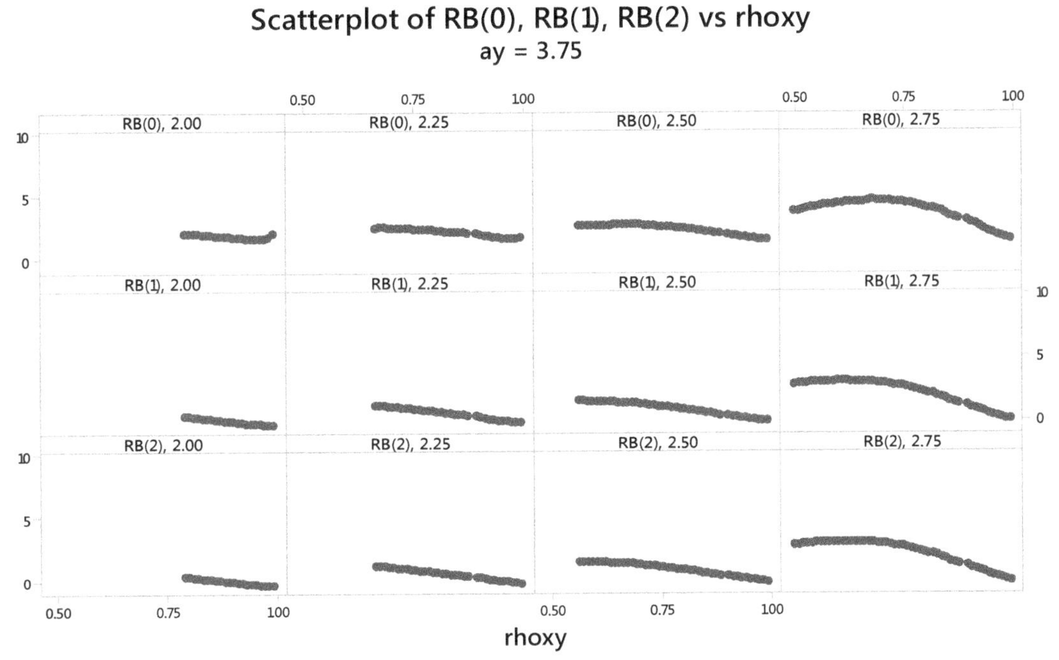

Panel variable: ax

Figure 11: RB(0), RB(1) and RB(2) versus ρ_{xy} for $a_y = 3.75$ and $a_x = 2.0, 2.25, 2.50$ and 2.75.

5. Applications: Real data set

In order to explore the performance of the proposed regression type estimator of the new measure of coefficient of dispersion relative to the empirical mode, we focus on the real data set available in Nash, Sellers, Talbot, Cawthorn, and Ford (1994) about the Abalone population. This data set entails 4176 different instances and 8 attributes. The data is first cleaned in order to fit a gamma distribution. In this study, we consider the logarithmic value of variable Age in years as a study variable; and Diameter, a continuous variable, perpendicular to length, measured in millimeters (mm) as an auxiliary variable. A few descriptive parameters of these eight attributes are given in Table 5.

Table 5: Descriptive Parameters of Diameter, Length, Height, Shucked weight, Whole weight, Viscera weight and log(Age).

Variable	N	Mean	StDev	Min	Q1	Med	Q3	Max	Skew	Kurt
Diameter	4176	0.4079	0.0992	0.055	0.35	0.425	0.48	0.65	–0.6092	–0.0455
Length	4176	0.5240	0.1201	0.075	0.45	0.545	0.615	0.815	–0.6399	0.0646
Height	4176	0.1395	0.0418	0.00	0.115	0.14	0.165	1.13	3.1288	76.0255
Shucked	4176	0.3594	0.2220	0.001	0.186	0.336	0.502	1.488	0.7191	0.5951
Whole	4176	0.8287	0.4904	0.002	0.4415	0.7995	1.1535	2.8255	0.5310	–0.0236
Viscera	4176	0.1806	0.1096	0.0005	0.0932	0.171	0.253	0.76	0.5919	0.0840
Log(Age)	4176	0.9753	0.1388	0.3010	0.903	0.9542	1.041	1.4624	–0.2341	1.3644

The average value of Log(Age) variable is 0.9753 with a standard deviation of 0.1388, minimum value is 0.3010, maximum value is 1.4624, the median value is 0.9542, first quartile is 0.903, third quartile is 1.041, the coefficient of skewness is –0.2341 and that of kurtosis is 1.3644. In the same way the corresponding parameter values for the other variables Length, Diameter, Height, Whole weight, Shucked weight and Viscera weight can be obtained from Table 5. A scatter plot showing the relationship between the study variable Log(Age) and the auxiliary variable Diameter is shown in Figure 12.

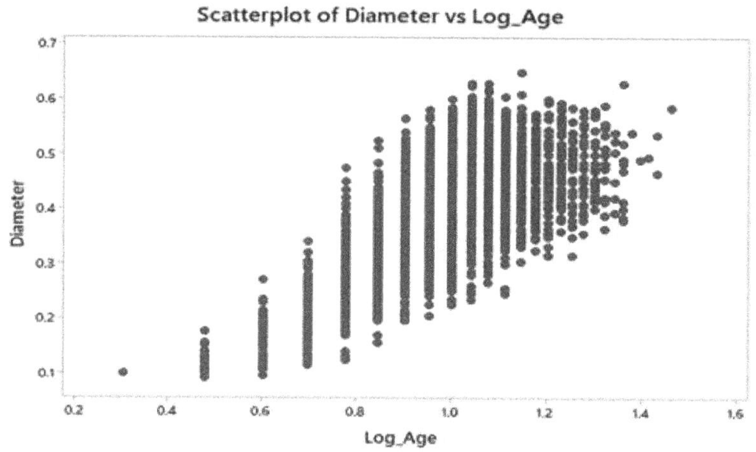

Figure 12: Scatter-Plot of the study variable Log(Age) and the auxiliary variable Diameter.

Based on our simulation results in Section 4, it seems that the regression type estimator is continuously more efficient than the naïve estimator for any value of the correlation coefficient. Figure 13 fits the Gamma distribution on the variables, and provides the values of the shape and the scale parameters. Table 6 is provided with such shape and scale parameters for the convenience of the readers.

Table 6: Shape and Scale parameters for the variables.

Parameter	Log(Age)	Diameter
Shape (α)	47.26	13.79
Scale (β)	0.0206	0.0295

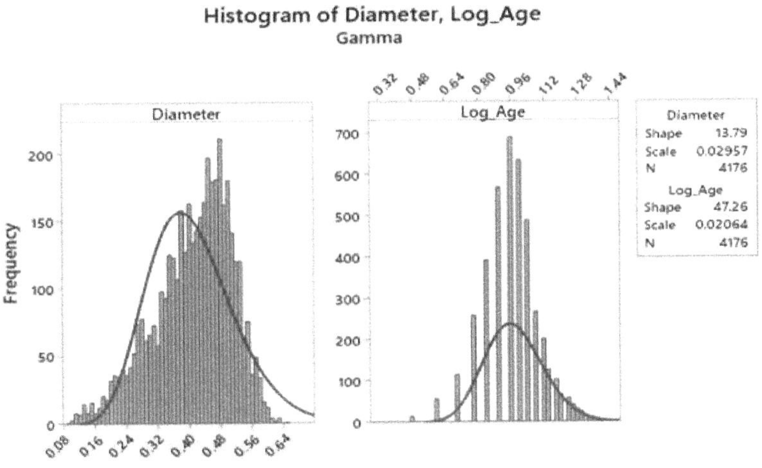

Figure 13: Fitting Gamma distributions for shape and scale parameters.

Thus, for the study variable the value of the shape parameter is 47.26 and that of scale parameter is 0.0206; for the auxiliary variable Diameter the value of the shape parameter is 13.79 and that of the scale parameter is 0.0295. Figure 13 shows the histogram between variables Diameter and logarithmic value of Age.

In APPENDIX-D we provide R-code used in obtaining the results from the real data set with Log(Age) as study variable, and treating Diameter as an auxiliary variable. The R-code explains how to read a data set from an EXCEL file, assign the variables Log(Age) ->Y and Diameter->X. Also change the values of the shape and scale parameters $a_y < -47.26$, $b_y < -0.0206$; $a_x < -13.79$, and $b_x < -0.0295$. It may be worth pointing out that the value of percent relative efficiency of the regression type estimator with respect to the naïve estimator is not based on the value of the sample size. Established from our simulation study experience for hypothetical data sets, we set the sample size $n = 418$, which is equal to 10% of the population size to investigate the value of percent relative biases in the naïve and regression type estimators. Further note that large sample sizes are recommended to estimate such parameters in practice to achieve the goal of reducing absolute percent relative bias below 10%. The results attained by executing the R-code for different variables are reported in Table 7.

Table 7: Results of real data on Log(Age) with $a_y = 47.26$, $b_y = 0.0206$ and $CQV_y = 0.08$.

Variable	a_x	b_x	ρ_{xy}	RE(2)	RB(naive)	RB(ratio)	RB(reg)
Diameter	13.7	0.0297	0.67	450.63	–0.04	0.52	–0.81

The absolute percent relative bias in the proposed regression type estimator is smaller than the naïve estimator, thus a regression type adjustment in the proposed estimator is not only producing efficient estimators but also reduces the value of the percent relative bias. For 10% sample size of the population, the absolute value of the percent relative bias is less than a value of 10% which is quite acceptable by following Cochran (1963) in the field of survey sampling. The value of percent RE(2) is 450.63%, for $\rho_{xy} = 0.67$ between the study variable Log(Age) and the auxiliary variable Diameter, which confirms that the proposed regression type estimator could also be useful in certain situations where the value of correlation coefficient is not very high. Thus other parameters such as shape parameters, scale parameters among others also play a role in the performance of the ratio and regression type estimators. More detail of results and original version of the simulation results can be had from Eappen (2019).

6. Conclusion

We introduce a new coefficient of dispersion that is based on three parameters of the study variable and which provides a relative standing with respect to the empirical mode of the entire data set unlike the traditional measure of the coefficient of dispersion. Then we determine that an appropriate choice of auxiliary information can be used to construct ratio and regression type estimators for the new measure of coefficient of dispersion that could lead to efficient estimates.

Acknowledgements

Thanks are due to R Core Team (2019). R: A language and environment for statistical computing. R Foundation for Statistical Computing, Vienna, Austria. URL https://www.R-project.org/ for using this package in the present investigation.

References

Ambati, S. K., Sedory, S. A. and Singh, S. 2018. Estimation of Coefficient of Dispersion Using Auxiliary Information, Communication in Statistics-Simulation and Computation, 47(7): 1996–2009.

Bonett, D. G. 2006. Confidence interval for a coefficient of quartile variation. Computational Statistics & Data Analysis, 50(11): 2953–2957.

Bonett, D. and Seier, E. 2006. Confidence interval for a coefficient of dispersion in nonnormal distributions. Biometrical Journal, 48(1): 144–148.

Cochran, W. G. 1940. The estimation of the yields of the cereal experiments by sampling for the ratio of grain to total produce. The Journal of Agricultural Science, 30(2): 262–275.

Cochran, W. G. 1963. Sampling Technique. 2nd Ed., John Wiley and Sons Inc., New York.

David, I. P. and Sukhatme, B. V. 1974. On the bias and mean square error of the ratio estimator. J. Am. Statist. Assoc., 69: 464–466.

Doodson, III, T. 1917. Relation of the mode, median and mean in frequency Curves. Biometrika, 11(4): 425–429.

Eappen, Christin Variathu. 2019. Estimation of coefficient of dispersion using auxiliary Information. Unpublished MS Thesis Submitted to the Department of Mathematics, Texas A&M University-Kingsville, Kingsville, TX.

Eappen, Christin, Variathu, Sedory, Stephen, A. and Singh, Sarjinder. 2020. Ratio and regression type estimators of a new measure of coefficient of dispersion. Communications in Statistics: Theory and Methods (Available Online).

Hansen, M. H., Hurwitz, W. N. and Madow, W. G. 1953. Sample Survey Methods and Theory. Vol. 1, Wiley, New York.

Kuk, A. Y. C. and Mak, T. 1989. Median estimation in the presence of auxiliary information. Journal of the Royal Statistical Society, 51(2): 261–269.

Nash, J. W., Sellers, L. T., Talbot, R. S., Cawthorn, J. A. and Ford, B. W. 1994. The Population Biology of Abalone (_Haliotis_ species) in Tasmania. I. Blacklip Abalone (_H. rubra) from the North Coast and Islands of Bass Strait. Sea Fisheries Division, 48(1034–3288).

Sedory, S. A. and Singh, S. 2014. Estimation of mode using auxiliary information. Communications in Statistics: Simulation and Computation, 43: 2390–2402.

Singh, S. and Horn, S. 1998. An Alternative Estimator For Multi-Character Surveys Metrika, 48(2): 99–107.

Wu, Chien-Fu. 1982. Estimation of Variance of the Ratio Estimator. Biometrika, 69(1): 183–189.

APPENDIX A: NOTATIONS AND EXPECTED VALUES

The purpose of this chapter is to develop some standard notations and expected values for investigating the properties of the proposed estimators of a new measure of coefficient of dispersion of a study variable in the presence of some auxiliary information.

Let Y_i and X_i, $i = 1,2,\ldots, N$ be the values of the study variable Y and the auxiliary variable X, respectively, in a population. Also let y_i and x_i, $i = 1,2,\ldots, n$ represent the values of the units selected in a sample s_n of size n taken by using the scheme of simple random sampling without replacement (SRSWOR).

Let,

$\bar{y} = \frac{1}{n}\sum_{i=1}^{n} y_i$ be the sample mean of the study variable y,

$\bar{x} = \frac{1}{n}\sum_{i=1}^{n} x_i$ be the sample mean of the auxiliary variable x,

$\bar{Y} = \frac{1}{N}\sum_{i=1}^{N} Y_i$ be the population mean of the study variable y,

$\bar{X} = \frac{1}{N}\sum_{i=1}^{N} X_i$ be the population mean of the auxiliary variable x,

$S_y^2 = \frac{1}{N-1}\sum_{i=1}^{n} (Y_i - \bar{Y})^2$ be the population variance of the study variable y,

$S_x^2 = \frac{1}{N-1}\sum_{i=1}^{N} \left(X_i - \bar{X}\right)^2$ be the population variance of the auxiliary variable x,

$S_{xy} = \frac{1}{N-1}\sum_{i=1}^{N} \left(Y_i - \bar{Y}\right)\left(X_i - \bar{X}\right)$ be the population covariance between the study the auxiliary variable, and let,

$C_y = \frac{S_y}{\bar{Y}}$ be the coefficient of variation for the study variable,

$C_x = \frac{S_x}{\bar{X}}$ be the coefficient of variation of the auxiliary variable,

and

$\rho_{xy} = \frac{S_{xy}}{S_x S_y}$ be the population correlation coefficient between the study variable and the auxiliary variable.

Let $Y_{(1)} \le Y_{(2)} \le \ldots \le Y_{(n)}$ be the ordered Y values in the sample s_n. Let i_{1y}, i_{2y} and i_{3y} be the integers such that $Y_{(i_{1y})} \le Q_{1y} < Y_{(i_{1y}+1)}$; $Y_{(i_{2y})} \le Q_{2y} < Y_{(i_{2y}+1)}$ and $Y_{(i_{3y})} \le Q_{3y} < Y_{(i_{3y}+1)}$. Let $p_{1y} = i_{1y}/n$, $p_{2y} = i_{2y}/n$ and $p_{3y} = i_{3y}/n$ be the proportion of the Y values in the sample that are less than or equal to Q_{1y}, Q_{2y} and Q_{3y}, respectively. Also let $X_{(1)} \le X_{(2)} \le \ldots \le X_{(n)}$ be the ordered X values in the sample s_n. Let i_{1x}, i_{2x} and i_{3x} be the integers such that $X_{(i_{1x})} \le Q_{1x} < X_{(i_{1x}+1)}$; $X_{(i_{2x})} \le Q_{2x} < X_{(i_{2x}+1)}$ and $X_{(i_{3x})} \le Q_{3x} < X_{(i_{3x}+1)}$. Let, $p_{1x} = i_{1x}/n$, $p_{2x} = i_{2x}/n$ and $p_{3x} = i_{3x}/n$ be the proportions of the X values in the sample that are less than or equal to Q_{1x}, Q_{2x} and Q_{3x}, respectively. Thus Q_{1y}, Q_{2y} and Q_{3y} are approximately the sample p_{1y}, p_{2y} and p_{3y} th quantiles $\hat{Q}_y(p_{1y})$, $\hat{Q}_y(p_{2y})$ and $\hat{Q}_y(p_{3y})$, respectively.

Note that p_{1y}, p_{2y} and p_{3y} are unobservable because the values of Q_{1y}, Q_{2y} and Q_{3y} are unknown and thus have to be estimated from the sample. If $\hat{p}_{1y}$, $\hat{p}_{2y}$ and $\hat{p}_{3y}$ are any estimators of p_{1y}, p_{2y} and p_{3y}, respectively, then naturally $\hat{Q}_y(\hat{p}_{1y})$, $\hat{Q}_y(\hat{p}_{2y})$ and $\hat{Q}_y(\hat{p}_{3y})$ are the estimators of the parameters $Q_y(p_{1y})$, $Q_y(p_{2y})$ and $Q_y(p_{3y})$, respectively. The sample first-quartile $\hat{Q}_{1y}$, the sample second-quartile $\hat{Q}_{2y}$ and the sample third-quartile $\hat{Q}_{3y}$ can be viewed as special cases of the estimators $\hat{Q}_y(\hat{p}_{1y})$, $\hat{Q}_y(\hat{p}_{2y})$ and $\hat{Q}_y(\hat{p}_{3y})$ with $\hat{p}_{1y} = 0.25$, $\hat{p}_{2y} = 0.5$ and $\hat{p}_{3y} = 0.75$, respectively. In other words, $\hat{Q}_{1y} = \hat{Q}_y(0.25)$, $\hat{Q}_{2y} = \hat{Q}_y(0.5)$, and $\hat{Q}_{3y} = \hat{Q}_y(0.75)$.

Again let Q_{1x}, Q_{2x} and Q_{3x} are approximately the sample p_{1x}, p_{2x} and p_{3x} th quantiles $\hat{Q}_x(p_{1x})$, $\hat{Q}_x(p_{2x})$ and $\hat{Q}_x(p_{3x})$, respectively. Note that p_{1x}, p_{2x} and p_{3x} are observable because the values of Q_{1x}, Q_{2x} and Q_{3x} are known. But instead of using their true values, let $\hat{p}_{1x}$, $\hat{p}_{2x}$ and $\hat{p}_{3x}$ be estimators of p_{1x}, p_{2x} and p_{3x}, respectively, then naturally $\hat{Q}_x(\hat{p}_{1x})$, $\hat{Q}_x(\hat{p}_{2x})$ and $\hat{Q}_x(\hat{p}_{3x})$ are the estimators of the parameters $Q_x(p_{1x})$, $Q_x(p_{2x})$ and $Q_x(p_{3x})$, respectively.

The sample first-quartile $\hat{Q}_{1x}$, the sample second-quartile $\hat{Q}_{2x}$ and the sample third-quartile $\hat{Q}_{3x}$ can be viewed as special cases of the estimators $\hat{Q}_x(\hat{p}_{1x})$, $\hat{Q}_x(\hat{p}_{2x})$ and $\hat{Q}_x(\hat{p}_{3x})$ with $\hat{p}_{1x} = 0.25$, $\hat{p}_{2x} = 0.5$ and $\hat{p}_{3x} = 0.75$, respectively. In other words, $\hat{Q}_{1x} = \hat{Q}_x(0.25)$, $\hat{Q}_{2x} = \hat{Q}_x(0.5)$, and $\hat{Q}_{3x} = \hat{Q}_x(0.75)$.

Table A.1: Proportions of data values below and between different quantiles.

	$X \le Q_{1x}$	$X \le Q_{2x}$	$X \le Q_{3x}$	$Y \le Q_{1y}$	$Y \le Q_{2y}$	$Y \le Q_{3y}$
$X \le Q_{1x}$	0.25	0.25	0.25	$P_{Q_{1x}Q_{1y}}$	$P_{Q_{1x}Q_{2y}}$	$P_{Q_{1x}Q_{3y}}$
$X \le Q_{2x}$	0.25	0.50	0.50	$P_{Q_{2x}Q_{1y}}$	$P_{Q_{2x}Q_{2y}}$	$P_{Q_{2x}Q_{3y}}$
$X \le Q_{3x}$	0.25	0.50	0.75	$P_{Q_{3x}Q_{1y}}$	$P_{Q_{3x}Q_{2y}}$	$P_{Q_{3x}Q_{3y}}$
$Y \le Q_{1y}$	$P_{Q_{1x}Q_{1y}}$	$P_{Q_{2x}Q_{1y}}$	$P_{Q_{3x}Q_{1y}}$	0.25	0.25	0.25
$Y \le Q_{2y}$	$P_{Q_{1x}Q_{2y}}$	$P_{Q_{2x}Q_{2y}}$	$P_{Q_{3x}Q_{2y}}$	0.25	0.50	0.50
$Y \le Q_{3y}$	$P_{Q_{1x}Q_{3y}}$	$P_{Q_{2x}Q_{3y}}$	$P_{Q_{3x}Q_{3y}}$	0.25	0.50	0.75

where

$P_{Q_{1x}Q_{1y}} = P(X \le Q_{1x} \cap Y \le Q_{1y})$ is the population proportion of data values such that the auxiliary variable X has a value less than its first-quartile value Q_{1x} and the study variable Y is less than its first-quartile value Q_{1y};

$P_{Q_{1x}Q_{2y}} = P(X \le Q_{1x} \cap Y \le Q_{2y})$ is the population proportion of data values such that the auxiliary variable X has a value less than its first-quartile value Q_{1x} and the study variable Y is less than its second-quartile value Q_{2y};

$P_{Q_{1x}Q_{3y}} = P(X \le Q_{1x} \cap Y \le Q_{3y})$ is the population proportion of data values such that the auxiliary variable X has a value less than its first-quartile value Q_{1x} and the study variable Y is less than its third-quartile value Q_{3y};

$P_{Q_{2x}Q_{1y}} = P(X \le Q_{2x} \cap Y \le Q_{1y})$ is the population proportion of data values such that the auxiliary variable X has a value less than its second-quartile value Q_{2x} and the study variable Y is less than its first-quartile value Q_{1y};

$P_{Q_{2x}Q_{2y}} = P(X \le Q_{2x} \cap Y \le Q_{2y})$ is the population proportion of data values such that the auxiliary variable X has a value less than its second-quartile value Q_{2x} and the study variable Y is less than its second-quartile value Q_{2y};

$P_{Q_{2x}Q_{3y}} = P(X \le Q_{2x} \cap Y \le Q_{3y})$ is the population proportion of data values such that the auxiliary variable X has a value less than its second-quartile value Q_{2x} and the study variable Y is less than its third-quartile value Q_{3y};

$P_{Q_{3x}Q_{1y}} = P(X \le Q_{3x} \cap Y \le Q_{1y})$ is the population proportion of data values such that the auxiliary variable X has a value less than its third-quartile value Q_{3x} and the study variable Y is less than its first-quartile value Q_{1y};

$P_{Q_{3x}Q_{2y}} = P(X \le Q_{3x} \cap Y \le Q_{2y})$ is the population proportion of data values such that the auxiliary variable X has a value less than its third-quartile value Q_{3x} and the study variable Y is less than its second-quartile value Q_{2y};

and

$P_{Q_{3x}Q_{3y}} = P(X \le Q_{3x} \cap Y \le Q_{3y})$ is the population proportion of data values such that the auxiliary variable X has a value less than its third-quartile value Q_{3x} and the study variable Y is less than its third-quartile value Q_{3y};
Let us define,

$$I_{y_i}^{Q_2} = \begin{cases} 1, & \text{if } Y_i \le Q_{2y} \\ 0, & \text{otherwise} \end{cases}; I_{x_i}^{Q_2} = \begin{cases} 1, & \text{if } X_i \le Q_{2x} \\ 0, & \text{otherwise} \end{cases}; I_{y_i}^{Q_1} = \begin{cases} 1, & \text{if } Y_i \le Q_{1y} \\ 0, & \text{otherwise} \end{cases}$$

$$I_{x_i}^{Q_1} = \begin{cases} 1, & \text{if } X_i \le Q_{1x} \\ 0, & \text{otherwise} \end{cases}; I_{y_i}^{Q_3} = \begin{cases} 1, & \text{if } Y_i \le Q_{3y} \\ 0, & \text{otherwise} \end{cases} \text{ and } I_{x_i}^{Q_3} = \begin{cases} 1, & \text{if } X_i \le Q_{3x} \\ 0, & \text{otherwise} \end{cases}$$

Also let us define

$$S_{YQ_{2y}} = \frac{1}{N-1}\sum_{i=1}^{N}\left(Y_i - \bar{Y}\right)\left(I_{y_i}^{Q_2} - 0.5\right),\ S_{XQ_{2x}} = \frac{1}{N-1}\sum_{i=1}^{N}\left(X_i - \bar{X}\right)\left(I_{x_i}^{Q_2} - 0.5\right),$$

$$S_{YQ_{2x}} = \frac{1}{N-1}\sum_{i=1}^{N}\left(Y_i - \bar{Y}\right)\left(I_{x_i}^{Q_2} - 0.5\right), S_{XQ_{2y}} = \frac{1}{N-1}\sum_{i=1}^{N}\left(X_i - \bar{X}\right)\left(I_{y_i}^{Q_2} - 0.5\right),$$

$$S_{YQ_{1y}} = \frac{1}{N-1}\sum_{i=1}^{N}\left(Y_i - \bar{Y}\right)\left(I_{y_i}^{Q_1} - 0.25\right), S_{XQ_{1x}} = \frac{1}{N-1}\sum_{i=1}^{N}\left(X_i - \bar{X}\right)\left(I_{x_i}^{Q_1} - 0.25\right),$$

$$S_{YQ_{1x}} = \frac{1}{N-1}\sum_{i=1}^{N}\left(Y_i - \bar{Y}\right)\left(I_{x_i}^{Q_1} - 0.25\right), S_{XQ_{1y}} = \frac{1}{N-1}\sum_{i=1}^{N}\left(X_i - \bar{X}\right)\left(I_{y_i}^{Q_1} - 0.25\right),$$

$$S_{YQ_{3y}} = \frac{1}{N-1}\sum_{i=1}^{N}\left(Y_i - \bar{Y}\right)\left(I_{y_i}^{Q_3} - 0.75\right), S_{XQ_{3x}} = \frac{1}{N-1}\sum_{i=1}^{N}\left(X_i - \bar{X}\right)\left(I_{x_i}^{Q_3} - 0.75\right),$$

$$S_{YQ_{3x}} = \frac{1}{N-1}\sum_{i=1}^{N}\left(Y_i - \bar{Y}\right)\left(I_{x_i}^{Q_3} - 0.75\right), \text{and } S_{XQ_{3y}} = \frac{1}{N-1}\sum_{i=1}^{N}\left(X_i - \bar{X}\right)\left(I_{y_i}^{Q_3} - 0.75\right).$$

Now defining:

$$\varepsilon_{1x} = \frac{\hat{Q}_{1x}}{Q_{1x}} - 1,\ \varepsilon_{2x} = \frac{\hat{Q}_{2x}}{Q_{2x}} - 1,\ \varepsilon_{3x} = \frac{\hat{Q}_{3x}}{Q_{3x}} - 1,\ \eta_x = \frac{\bar{x}}{\bar{X}} - 1$$

$$\varepsilon_{1y} = \frac{\hat{Q}_{1y}}{Q_{1y}} - 1,\ \varepsilon_{2y} = \frac{\hat{Q}_{2y}}{Q_{2y}} - 1,\ \varepsilon_{3y} = \frac{\hat{Q}_{3y}}{Q_{3y}} - 1 \text{ and } \eta_y = \frac{\bar{y}}{\bar{Y}} - 1$$

such that

$$E(\varepsilon_{1x}) \approx E(\varepsilon_{2x}) \approx E(\varepsilon_{3x}) \approx E(\varepsilon_{1y}) \approx E(\varepsilon_{2y}) \approx E(\varepsilon_{3y}) \approx 0$$

and

$$E(\varepsilon_{1x}^2) \approx E\left(\frac{\hat{Q}_{1x}}{Q_{1x}} - 1\right)^2 = \frac{V(\hat{Q}_{1x})}{Q_{1x}^2} \approx \left(\frac{1-f}{n}\right)\left(\frac{3}{16}\right)\left(\frac{1}{Q_{1x}^2\{f_x(Q_{1x})\}^2}\right)$$

$$E(\varepsilon_{2x}^2) = E\left(\frac{\hat{Q}_{2x}}{Q_{2x}} - 1\right)^2 = \frac{V(\hat{Q}_{2x})}{Q_{2x}^2} \approx \left(\frac{1-f}{n}\right)\left(\frac{1}{4}\right)\left(\frac{1}{Q_{2x}^2\{f_x(Q_{2x})\}^2}\right)$$

$$E(\varepsilon_{3x}^2) \approx E\left(\frac{\hat{Q}_{3x}}{Q_{3x}} - 1\right)^2 \approx \frac{V(\hat{Q}_{3x})}{Q_{3x}^2} \approx \left(\frac{1-f}{n}\right)\left(\frac{3}{16}\right)\left(\frac{1}{Q_{3x}^2\{f_x(Q_{3x})\}^2}\right)$$

$$E(\varepsilon_{1y}^2) \approx E\left(\frac{\hat{Q}_{1y}}{Q_{1y}} - 1\right)^2 \approx \frac{V(\hat{Q}_{1y})}{Q_{1y}^2} \approx \left(\frac{1-f}{n}\right)\left(\frac{3}{16}\right)\left(\frac{1}{Q_{1y}^2\{f_y(Q_{1y})\}^2}\right)$$

$$E(\varepsilon_{2y}^2) \approx E\left(\frac{\hat{Q}_{2y}}{Q_{2y}} - 1\right)^2 \approx \frac{V(\hat{Q}_{2y})}{Q_{2y}^2} \approx \left(\frac{1-f}{n}\right)\left(\frac{1}{4}\right)\left(\frac{1}{Q_{2y}^2\{f_y(Q_{2y})\}^2}\right)$$

$$E(\varepsilon_{3y}^2) \approx E\left(\frac{\hat{Q}_{3y}}{Q_{3y}} - 1\right)^2 \approx \frac{V(\hat{Q}_{3y})}{Q_{3y}^2} \approx \left(\frac{1-f}{n}\right)\left(\frac{3}{16}\right)\left(\frac{1}{Q_{3y}^2\{f_y(Q_{3y})\}^2}\right)$$

$$E(\eta_y^2) = E\left(\frac{\bar{y}}{\bar{Y}} - 1\right)^2 = \frac{V(\bar{y})}{\bar{Y}^2} = \left(\frac{1-f}{n}\right)C_y^2$$

$$E(\eta_x^2) = E\left(\frac{\bar{x}}{\bar{X}} - 1\right)^2 = \frac{V(\bar{x})}{\bar{X}^2} = \left(\frac{1-f}{n}\right)C_x^2$$

$$E(\eta_x \eta_y) = E\left[\left(\frac{\bar{x}}{\bar{X}} - 1\right)\left(\frac{\bar{y}}{\bar{Y}} - 1\right)\right] = \frac{Cov(\bar{y}, \bar{x})}{\bar{Y}\,\bar{X}} = \left(\frac{1-f}{n}\right)\rho_{xy} C_x C_y$$

$$E(\varepsilon_{1x}\eta_y) = E\left[\left(\frac{\hat{Q}_{1x}}{Q_{1x}} - 1\right)\left(\frac{\bar{y}}{\bar{Y}} - 1\right)\right] \approx -\left(\frac{1-f}{n}\right)\frac{S_{YQ_{1x}}}{\bar{Y}Q_{1x}}$$

$$E(\varepsilon_{2x}\eta_y) = E\left[\left(\frac{\hat{Q}_{2x}}{Q_{2x}} - 1\right)\left(\frac{\bar{y}}{\bar{Y}} - 1\right)\right] \approx -\left(\frac{1-f}{n}\right)\frac{S_{YQ_{2x}}}{\bar{Y}Q_{2x}}$$

$$E\left(\varepsilon_{3x}\eta_y\right) = E\left[\left(\frac{\hat{Q}_{3x}}{Q_{3x}} - 1\right)\left(\frac{\bar{y}}{\bar{Y}} - 1\right)\right] \approx -\left(\frac{1-f}{n}\right)\frac{S_{YQ_{3x}}}{\bar{Y}Q_{3x}}$$

$$E\left(\varepsilon_{1y}\eta_y\right) = E\left[\left(\frac{\hat{Q}_{1y}}{Q_{1y}} - 1\right)\left(\frac{\bar{y}}{\bar{Y}} - 1\right)\right] \approx -\left(\frac{1-f}{n}\right)\frac{S_{YQ_{1y}}}{\bar{Y}Q_{1y}}$$

$$E\left(\varepsilon_{2y}\eta_y\right) = E\left[\left(\frac{\hat{Q}_{2y}}{Q_{2y}} - 1\right)\left(\frac{\bar{y}}{\bar{Y}} - 1\right)\right] \approx -\left(\frac{1-f}{n}\right)\frac{S_{YQ_{2y}}}{\bar{Y}Q_{2y}}$$

$$E\left(\varepsilon_{3y}\eta_y\right) = E\left[\left(\frac{\hat{Q}_{3y}}{Q_{3y}} - 1\right)\left(\frac{\bar{y}}{\bar{Y}} - 1\right)\right] \approx -\left(\frac{1-f}{n}\right)\frac{S_{YQ_{3y}}}{\bar{Y}Q_{3y}}$$

$$E\left(\varepsilon_{1x}\eta_x\right) = E\left[\left(\frac{\hat{Q}_{1x}}{Q_{1x}} - 1\right)\left(\frac{\bar{x}}{\bar{X}} - 1\right)\right] \approx -\left(\frac{1-f}{n}\right)\frac{S_{XQ_{1x}}}{\bar{X}Q_{1x}}$$

$$E\left(\varepsilon_{2x}\eta_x\right) = E\left[\left(\frac{\hat{Q}_{2x}}{Q_{2x}} - 1\right)\left(\frac{\bar{x}}{\bar{X}} - 1\right)\right] \approx -\left(\frac{1-f}{n}\right)\frac{S_{XQ_{2x}}}{\bar{X}Q_{2x}}$$

$$E\left(\varepsilon_{3x}\eta_x\right) = E\left[\left(\frac{\hat{Q}_{3x}}{Q_{3x}} - 1\right)\left(\frac{\bar{x}}{\bar{X}} - 1\right)\right] \approx -\left(\frac{1-f}{n}\right)\frac{S_{XQ_{3x}}}{\bar{X}Q_{3x}}$$

$$E\left(\varepsilon_{1y}\eta_x\right) = E\left[\left(\frac{\hat{Q}_{1y}}{Q_{1y}} - 1\right)\left(\frac{\bar{x}}{\bar{X}} - 1\right)\right] \approx -\left(\frac{1-f}{n}\right)\frac{S_{XQ_{1y}}}{\bar{X}Q_{1y}}$$

$$E\left(\varepsilon_{2y}\eta_x\right) = E\left[\left(\frac{\hat{Q}_{2y}}{Q_{2y}} - 1\right)\left(\frac{\bar{x}}{\bar{X}} - 1\right)\right] \approx -\left(\frac{1-f}{n}\right)\frac{S_{XQ_{2y}}}{\bar{X}Q_{2y}}$$

$$E\left(\varepsilon_{3y}\eta_x\right) = E\left[\left(\frac{\hat{Q}_{3y}}{Q_{3y}} - 1\right)\left(\frac{\bar{x}}{\bar{X}} - 1\right)\right] \approx -\left(\frac{1-f}{n}\right)\frac{S_{XQ_{3y}}}{\bar{X}Q_{3y}}$$

$$E\left(\varepsilon_{1x}\varepsilon_{2x}\right) = E\left[\left(\frac{\hat{Q}_{1x}}{Q_{1x}} - 1\right)\left(\frac{\hat{Q}_{2x}}{Q_{2x}} - 1\right)\right] \approx \left(\frac{1-f}{8n}\right)\frac{1}{\left\{f_x\left(Q_{1x}\right)f_x\left(Q_{2x}\right)Q_{1x}Q_{2x}\right\}}$$

$$E\left(\varepsilon_{1x}\varepsilon_{3x}\right) = E\left[\left(\frac{\hat{Q}_{1x}}{Q_{1x}} - 1\right)\left(\frac{\hat{Q}_{3x}}{Q_{3x}} - 1\right)\right] \approx \left(\frac{1-f}{16n}\right)\frac{1}{\left\{f_x\left(Q_{1x}\right)f_x\left(Q_{3x}\right)Q_{1x}Q_{3x}\right\}}$$

$$E\left(\varepsilon_{2x}\varepsilon_{3x}\right) = E\left[\left(\frac{\hat{Q}_{2x}}{Q_{2x}} - 1\right)\left(\frac{\hat{Q}_{3x}}{Q_{3x}} - 1\right)\right] \approx \left(\frac{1-f}{8n}\right)\frac{1}{\left\{f_x\left(Q_{2x}\right)f_x\left(Q_{3x}\right)Q_{2x}Q_{3x}\right\}}$$

$$E\left(\varepsilon_{1y}\varepsilon_{2y}\right) = E\left[\left(\frac{\hat{Q}_{1y}}{Q_{1y}} - 1\right)\left(\frac{\hat{Q}_{2y}}{Q_{2y}} - 1\right)\right] \approx \left(\frac{1-f}{8n}\right)\frac{1}{\left\{f_y\left(Q_{1y}\right)f_y\left(Q_{2y}\right)Q_{1y}Q_{2y}\right\}}$$

$$E\left(\varepsilon_{1y}\varepsilon_{3y}\right)=E\left[\left(\frac{\hat{Q}_{1y}}{Q_{1y}}-1\right)\left(\frac{\hat{Q}_{3y}}{Q_{3y}}-1\right)\right]\approx\left(\frac{1-f}{16n}\right)\frac{1}{\left\{f_y\left(Q_{1y}\right)f_y\left(Q_{3y}\right)Q_{1y}Q_{3y}\right\}}$$

$$E\left(\varepsilon_{2y}\varepsilon_{3y}\right)=E\left[\left(\frac{\hat{Q}_{2y}}{Q_{2y}}-1\right)\left(\frac{\hat{Q}_{3y}}{Q_{3y}}-1\right)\right]\approx\left(\frac{1-f}{8n}\right)\frac{1}{\left\{f_y\left(Q_{2y}\right)f_y\left(Q_{3y}\right)Q_{2y}Q_{3y}\right\}}$$

$$E\left(\varepsilon_{1x}\varepsilon_{1y}\right)=E\left[\left(\frac{\hat{Q}_{1x}}{Q_{1x}}-1\right)\left(\frac{\hat{Q}_{1y}}{Q_{1y}}-1\right)\right]\approx\left(\frac{1-f}{n}\right)\frac{\left\{f_x\left(Q_{1x}\right)\right\}^{-1}\left\{f_y\left(Q_{1y}\right)\right\}^{-1}}{Q_{1x}Q_{1y}}\left(P_{Q_{1x}Q_{1y}}-\frac{1}{16}\right)$$

$$E\left(\varepsilon_{2x}\varepsilon_{1y}\right)=E\left[\left(\frac{\hat{Q}_{2x}}{Q_{2x}}-1\right)\left(\frac{\hat{Q}_{1y}}{Q_{1y}}-1\right)\right]\approx\left(\frac{1-f}{n}\right)\frac{\left\{f_x\left(Q_{2x}\right)\right\}^{-1}\left\{f_y\left(Q_{1y}\right)\right\}^{-1}}{Q_{2x}Q_{1y}}\left(P_{Q_{2x}Q_{1y}}-\frac{1}{8}\right)$$

$$E\left(\varepsilon_{3x}\varepsilon_{1y}\right)=E\left[\left(\frac{\hat{Q}_{3x}}{Q_{3x}}-1\right)\left(\frac{\hat{Q}_{1y}}{Q_{1y}}-1\right)\right]\approx\left(\frac{1-f}{n}\right)\frac{\left\{f_x\left(Q_{3x}\right)\right\}^{-1}\left\{f_y\left(Q_{1y}\right)\right\}^{-1}}{Q_{3x}Q_{1y}}\left(P_{Q_{3x}Q_{1y}}-\frac{3}{16}\right)$$

$$E\left(\varepsilon_{1x}\varepsilon_{2y}\right)=E\left[\left(\frac{\hat{Q}_{1x}}{Q_{1x}}-1\right)\left(\frac{\hat{Q}_{2y}}{Q_{2y}}-1\right)\right]\approx\left(\frac{1-f}{n}\right)\frac{\left\{f_x\left(Q_{1x}\right)\right\}^{-1}\left\{f_y\left(Q_{2y}\right)\right\}^{-1}}{Q_{1x}Q_{2y}}\left(P_{Q_{1x}Q_{2y}}-\frac{1}{8}\right)$$

$$E\left(\varepsilon_{2x}\varepsilon_{2y}\right)=E\left[\left(\frac{\hat{Q}_{2x}}{Q_{2x}}-1\right)\left(\frac{\hat{Q}_{2y}}{Q_{2y}}-1\right)\right]\approx\left(\frac{1-f}{n}\right)\frac{\left\{f_x\left(Q_{2x}\right)\right\}^{-1}\left\{f_y\left(Q_{2y}\right)\right\}^{-1}}{Q_{2x}Q_{2y}}\left(P_{Q_{2x}Q_{2y}}-\frac{1}{4}\right)$$

$$E\left(\varepsilon_{3x}\varepsilon_{2y}\right)=E\left[\left(\frac{\hat{Q}_{3x}}{Q_{3x}}-1\right)\left(\frac{\hat{Q}_{2y}}{Q_{2y}}-1\right)\right]\approx\left(\frac{1-f}{n}\right)\frac{\left\{f_x\left(Q_{3x}\right)\right\}^{-1}\left\{f_y\left(Q_{2y}\right)\right\}^{-1}}{Q_{3x}Q_{2y}}\left(P_{Q_{3x}Q_{2y}}-\frac{3}{8}\right)$$

$$E\left(\varepsilon_{1x}\varepsilon_{3y}\right)=E\left[\left(\frac{\hat{Q}_{1x}}{Q_{1x}}-1\right)\left(\frac{\hat{Q}_{3y}}{Q_{3y}}-1\right)\right]\approx\left(\frac{1-f}{n}\right)\frac{\left\{f_x\left(Q_{1x}\right)\right\}^{-1}\left\{f_y\left(Q_{3y}\right)\right\}^{-1}}{Q_{1x}Q_{3y}}\left(P_{Q_{1x}Q_{3y}}-\frac{3}{16}\right)$$

$$E\left(\varepsilon_{2x}\varepsilon_{3y}\right)=E\left[\left(\frac{\hat{Q}_{2x}}{Q_{2x}}-1\right)\left(\frac{\hat{Q}_{3y}}{Q_{3y}}-1\right)\right]\approx\left(\frac{1-f}{n}\right)\frac{\left\{f_x\left(Q_{2x}\right)\right\}^{-1}\left\{f_y\left(Q_{3y}\right)\right\}^{-1}}{Q_{2x}Q_{3y}}\left(P_{Q_{2x}Q_{3y}}-\frac{3}{8}\right)$$

$$E\left(\varepsilon_{3x}\varepsilon_{3y}\right)=E\left[\left(\frac{\hat{Q}_{3x}}{Q_{3x}}-1\right)\left(\frac{\hat{Q}_{3y}}{Q_{3y}}-1\right)\right]\approx\left(\frac{1-f}{n}\right)\frac{\left\{f_x\left(Q_{3x}\right)\right\}^{-1}\left\{f_y\left(Q_{3y}\right)\right\}^{-1}}{Q_{3x}Q_{3y}}\left(P_{Q_{3x}Q_{3y}}-\frac{9}{16}\right)$$

In the following section, we demonstrate the method of developing the above expected values.

Section A.1 Derivations of Expected Values

The cumulative distribution function F_y for the study variable at the sample first-quartile $\hat{Q}_{1y}$ around the population first-quartile Q_{1y} can be approximated as:

$$\begin{aligned} F_y\left(\hat{Q}_{1y}\right) &= F_y\left[Q_{1y} + \left(\hat{Q}_{1y} - Q_{1y}\right)\right] \\ &= F_y\left(Q_{1y}\right) + \left(\hat{Q}_{1y} - Q_{1y}\right)\frac{\partial F_y}{\partial \hat{Q}_{1y}}\Big|_{\hat{Q}_{1y}=Q_{1y}} + O_p\left(n^{-1}\right) \\ &\approx F_y\left(Q_{1y}\right) + \left(\hat{Q}_{1y} - Q_{1y}\right) f_y\left(Q_{1y}\right) \end{aligned} \tag{A.1}$$

where

$f_y\left(Q_{1y}\right) = \frac{\partial F_y}{\partial \hat{Q}_{1y}}\Big|_{\hat{Q}_{1y}=Q_{1y}}$ indicates the value of the probability density function $f_y(y)$ at the population first-quartile Q_{1y}.

Now the (A.1) can be expressed as:

$$F_y\left(\hat{Q}_{1y}\right) - F_y\left(Q_{1y}\right) \approx \left(\hat{Q}_{1y} - Q_{1y}\right) f_y\left(Q_{1y}\right) \tag{A.2}$$

From (A.2), we have

$$\hat{Q}_{1y} - Q_{1y} = \frac{1}{f_y\left(Q_{1y}\right)}\left[F_y\left(\hat{Q}_{1y}\right) - F_y\left(Q_{1y}\right)\right]$$

Note that the value of $F_y(\hat{Q}_{1y})$ is not known, and can be estimated from the sample with $\hat{F}_y(\hat{Q}_{1y})$. Thus we have

$$\begin{aligned} \hat{Q}_{1y} - Q_{1y} &= \frac{1}{f_y\left(Q_{1y}\right)}\left[\hat{F}_y\left(\hat{Q}_{1y}\right) - \hat{F}_y\left(Q_{1y}\right)\right] + O_p\left(n^{-1}\right) \\ &\approx \{f_y\left(Q_{1y}\right)\}^{-1}\left(\frac{1}{4} - \hat{F}_y\left(Q_{1y}\right)\right) \end{aligned} \tag{A.3}$$

From (A.3), we have

$$\hat{Q}_{1y} \approx Q_{1y} + \left\{f_y\left(Q_{1y}\right)\right\}^{-1}\left(\frac{1}{4} - p_y^{Q_1}\right) \tag{A.4}$$

From (A.4), we have

$$V\left(\hat{Q}_{1y}\right) \approx \left(\frac{1-f}{n}\right)\left\{f_y\left(Q_{1y}\right)\right\}^{-2}\left(\frac{3}{16}\right)$$

Likewise, we have,

$$\hat{Q}_{2y} - Q_{2y} \approx \{f_y\left(Q_{2y}\right)\}^{-1}\left(\frac{1}{2} - \hat{F}_y\left(Q_{2y}\right)\right)$$

Now note that

$$\begin{aligned} \mathrm{cov}\left[\hat{F}_y\left(\hat{Q}_{1y}\right), \hat{F}_y\left(\hat{Q}_{2y}\right)\right] &= \frac{1}{f_y\left(Q_{1y}\right) f_y\left(Q_{2y}\right)} \mathrm{cov}\left[\hat{F}_y\left(Q_{1y}\right), \hat{F}_y\left(Q_{2y}\right)\right] \\ &= \frac{1}{f_y\left(Q_{1y}\right) f_y\left(Q_{2y}\right)}\left[E\left\{\hat{F}_y\left(Q_{1y}\right)\hat{F}_y\left(Q_{2y}\right)\right\} - E\left(\hat{F}_y\left(Q_{1y}\right)\right)E\left(\hat{F}_y\left(Q_{2y}\right)\right)\right] \\ &\approx \left(\frac{1-f}{n}\right)\left(\frac{1}{4} - \frac{1}{4} * \frac{1}{2}\right)\frac{1}{f_y\left(Q_{1y}\right) f_y\left(Q_{2y}\right)} \\ &= \left(\frac{1-f}{8n}\right)\frac{1}{f_y\left(Q_{1y}\right) f_y\left(Q_{2y}\right)} \end{aligned}$$

Hence, we have

$$E\left(\varepsilon_{1y}\varepsilon_{2y}\right) \approx \left(\frac{1-f}{8n}\right)\frac{1}{\left\{f_y\left(Q_{1y}\right) f_y\left(Q_{2y}\right) Q_{1y} Q_{2y}\right\}}$$

In the same way, the other expected values can also be derived.

Using the main outcome of Kuk and Mak (1989), such that, if F_y be the cumulative distribution function of y, then the sample quartile for the study variable can be approximated as:

$$\hat{Q}_{2y} = Q_{2y} + \left(0.5 - p_y^{Q_2}\right) f_y^{-1}(Q_{2y}) + ...$$

where $P_y^{Q_2}$ be the proportion of $I_{y_i}^{Q_2}$ values taking a value of 1 if $y_i \le Q_{2y}$ and 0 otherwise.
The covariance expressions amongst the sample means and the sample second quartiles are given by

$$Cov\left(\bar{y}, \hat{Q}_{2y}\right) \approx -\left(\frac{1-f}{n}\right) S_{YQ_{2y}} \{f_y(Q_{2y})\}^{-1}$$

$$Cov\left(\bar{x}, \hat{Q}_{2x}\right) \approx -\left(\frac{1-f}{n}\right) S_{XQ_{2x}} \{f_x(Q_{2x})\}^{-1}$$

$$Cov\left(\bar{y}, \hat{Q}_{2x}\right) \approx -\left(\frac{1-f}{n}\right) S_{YQ_{2x}} \{f_x(Q_{2x})\}^{-1}$$

and

$$Cov\left(\bar{x}, \hat{Q}_{2y}\right) \approx -\left(\frac{1-f}{n}\right) S_{XQ_{2y}} \{f_y(Q_{2y})\}^{-1}$$

It is easy to verify:

$$(P_{11} - 0.25) = \frac{1}{N-1} \sum_{i=1}^{N} \left(I_{y_i}^{Q_2} - 0.5\right)\left(I_{x_i}^{Q_2} - 0.5\right)$$

thus the Sedory and Singh (2014) method of finding the covariance amid two medians matches with the result of Kuk and Mak (1989).
Likewise, the covariance expressions between the sample means and the sample first quartiles are given by

$$Cov\left(\bar{y}, \hat{Q}_{1y}\right) \approx -\left(\frac{1-f}{n}\right) S_{YQ_{1y}} \{f_y(Q_{1y})\}^{-1}$$

$$Cov\left(\bar{x}, \hat{Q}_{1x}\right) \approx -\left(\frac{1-f}{n}\right) S_{XQ_{1x}} \{f_x(Q_{1x})\}^{-1}$$

$$Cov\left(\bar{y}, \hat{Q}_{1x}\right) \approx -\left(\frac{1-f}{n}\right) S_{YQ_{1x}} \{f_x(Q_{1x})\}^{-1}$$

and

$$Cov\left(\bar{x}, \hat{Q}_{1y}\right) \approx -\left(\frac{1-f}{n}\right) S_{XQ_{1y}} \{f_y(Q_{1y})\}^{-1}$$

On the same lines, the covariance expressions among the sample means and the sample third quartiles are given by

$$Cov\left(\bar{y}, \hat{Q}_{3y}\right) \approx -\left(\frac{1-f}{n}\right) S_{YQ_{3y}} \{f_y(Q_{3y})\}^{-1}$$

$$Cov\left(\bar{x}, \hat{Q}_{3x}\right) \approx -\left(\frac{1-f}{n}\right) S_{XQ_{3x}} \{f_x(Q_{3x})\}^{-1}$$

$$Cov\left(\bar{y}, \hat{Q}_{3x}\right) \approx -\left(\frac{1-f}{n}\right) S_{YQ_{3x}} \{f_x(Q_{3x})\}^{-1}$$

and

$$Cov\left(\bar{x}, \hat{Q}_{3y}\right) \approx -\left(\frac{1-f}{n}\right) S_{XQ_{3y}} \{f_y(Q_{3y})\}^{-1}$$

APPENDIX B: Detail of the results based on artificial data sets

Table B1: Detail of the results based on artificial data sets.

Mo_y	Mo_x	$CQV_{y(m)}$	a_x	a_y	a_x^{fit}	b_x^{fit}	a_y^{fit}	b_y^{fit}	ρ	ρ_{xy}	RE(1)	RE(2)	RB(0)	RB(1)	RB(2)
3.05	1.05	0.37	2	3	2.01	1	4.91	1.27	0.66	0.66	100.89	123.59	3.36	1.53	1.53
3.05	1.05	0.37	2	3	2.01	1	4.91	1.27	0.67	0.67	101.87	124.58	3.34	1.49	1.48
3.04	1.05	0.37	2	3	2.01	1	4.9	1.27	0.68	0.68	103.46	125.89	3.33	1.44	1.44
3.04	1.05	0.37	2	3	2.01	1	4.9	1.27	0.69	0.69	104.6	127.01	3.31	1.39	1.39
3.04	1.05	0.37	2	3	2.01	1	4.89	1.27	0.7	0.7	105.78	128.29	3.28	1.33	1.33
Kindly execute the program to view the complete results															
3.23	1.78	0.38	2.75	3.75	2.75	1.01	4.97	1.18	0.95	0.95	225.94	232.9	2	0.34	0.32
3.13	1.78	0.39	2.75	3.75	2.75	1.01	4.78	1.16	0.96	0.96	243.45	256.38	1.83	0.21	0.19
2.99	1.78	0.41	2.75	3.75	2.75	1.01	4.55	1.14	0.97	0.97	267.15	289.9	1.69	0.11	0.09
2.82	1.78	0.43	2.75	3.75	2.75	1.01	4.26	1.1	0.98	0.98	302.35	344.49	1.57	0.02	0
2.58	1.78	0.47	2.75	3.75	2.75	1.01	3.86	1.05	0.99	0.99	399.35	486.15	1.49	-0.06	-0.08

APPENDIX C: R-Codes used in the simulation study

```
library(fitdistrplus)
library(MASS)
library(survival)
library(npsurv)
library(lsei)
xp<-c()
yp<-c()
ys<-c()
xs<-c()
cqvys<-c()
cqvxs<-c()
cqvrat<-c()
xstar<-c()
ystar<-c()
ypsort<-c()
ypq1ind<-c()
xpsort<-c()
xpq1ind<-c()
ypq1ind1<-c()
xpq1ind1<-c()
ypq1ind2<-c()
xpq1ind2<-c()
ypq1ind3<-c()
xpq1ind3<-c()
ypq1ind4<-c()
xpq1ind4<-c()
ypq1ind5<-c()
xpq1ind5<-c()
ypq1ind6<-c()
xpq1ind6<-c()
ypq1ind7<-c()
xpq1ind7<-c()
ypq1ind8<-c()
```

```
xpq1ind8<-c()
ypq1ind9<-c()
xpq1ind9<-c()
IQ1YI<-c()
IQ2YI<-c()
IQ3YI<-c()
IQ1XI<-c()
IQ2XI<-c()
IQ3XI<-c()
np<-10001
for(ax in seq(2.0,2.75,0.25)){
 for(ay in seq(3.0,3.75,0.25)){
 #ax<-3.2
 bx<-1.0
 #ay<-.6
 by<-1.0
 set.seed(222)
 xstar<-rgamma(np,ax,bx)
 #cat(xstar,'\n')
 #print(c(xstar))
 set.seed(333)
 ystar<-rgamma(np,ay,by)
 meanx<-ax*bx
 meany<-ay*by
 sx<-sqrt(ax*bx*bx)
 sy<-sqrt(ay*by*by)
 #rhoxy<-0.9
 for(rhoxy in seq(0.50,0.99,0.01)){
 yp<-ystar*sqrt(1-rhoxy^2)+rhoxy*sy*xstar/sx
 xp<-xstar
 corrxy<-cor(xp,yp)
 meanyp<-mean(yp)
 meanxp<-mean(xp)
 varyp<-var(yp)
 varxp<-var(xp)
 covxyp<-cov(xp,yp)
 quartyp<-quantile(yp)
 quartxp<-quantile(xp)
 #print(quarty)
 q1yp<-quartyp[2]
 #print(quarty1)
 q2yp<-quartyp[3]
 q3yp<-quartyp[4]
 #print(c(q1yp,q2yp,q3yp))
 cqvyp<-(q3yp-q1yp)/(6*q2yp-4*meanyp)
 q1xp<-quartxp[2]
 #print(quartx1)
 q2xp<-quartxp[3]
 q3xp<-quartxp[4]
```

```
#print(c(q1xp,q2xp,q3xp))
cqvxp<-(q3xp-q1xp)/(6*q2xp-4*meanxp)
fit.gamma<-fitdist(yp, distr="gamma", method ="mle")
ypfit<-summary(fit.gamma)
ypfit1<-data.frame(ypfit1=matrix(ypfit),row.names=names(ypfit))
ypfit2<-ypfit1$ypfit1[1]
ypfit3 <- as.numeric(unlist(ypfit2))
ayfit <- ypfit3[1]
byfit <- ypfit3[2]
#print(c(ayfit,byfit))
fit.gamma<-fitdist(xp, distr="gamma", method ="mle")
xpfit<-summary(fit.gamma)
xpfit1<-data.frame(xpfit1=matrix(xpfit),row.names=names(xpfit))
xpfit2<-xpfit1$xpfit1[1]
xpfit3 <- as.numeric(unlist(xpfit2))
axfit <- xpfit3[1]
bxfit <- xpfit3[2]
EMODEYP<-(3*q2yp-2*meanyp)
EMODEXP<-(3*q2xp-2*meanxp)
fyq2yp<-(q2yp^(ayfit-1)*exp(-q2yp/byfit))/(gamma(ayfit)*byfit^ayfit)
#cat("Fyq2yp=",fyq2yp,'\n')
fyq3yp<-(q3yp^(ayfit-1)*exp(-q3yp/byfit))/(gamma(ayfit)*byfit^ayfit)
#cat("Fyq3yp=",fyq3yp,'\n')
fyq1yp<-(q1yp^(ayfit-1)*exp(-q1yp/byfit))/(gamma(ayfit)*byfit^ayfit)
#cat("Fyq1yp=",fyq1yp,'\n')
fxq2xp<-(q2xp^(axfit-1)*exp(-q2xp/bxfit))/(gamma(axfit)*bxfit^axfit)
#cat("Fxq2xp=",fxq2xp,'\n')
fxq3xp<-(q3xp^(axfit-1)*exp(-q3xp/bxfit))/(gamma(axfit)*bxfit^axfit)
#cat("Fxq3xp=",fxq3xp,'\n')
fxq1xp<-(q1xp^(axfit-1)*exp(-q1xp/bxfit))/(gamma(axfit)*bxfit^axfit)
#cat("Fxq1xp=",fxq1xp,'\n')
ns<-500
####biasratio#####
ypsort<-yp
xpsort<-xp
for(i in seq(1,np,1)){
if(ypsort[i]<q1yp) ypq1ind1[i]<-1 else ypq1ind1[i]<-0
if(xpsort[i]<q1xp) xpq1ind1[i]<-1 else xpq1ind1[i]<-0
if(ypsort[i]<q2yp) ypq1ind2[i]<-1 else ypq1ind2[i]<-0
if(xpsort[i]<q1xp) xpq1ind2[i]<-1 else xpq1ind2[i]<-0
if(ypsort[i]<q3yp) ypq1ind3[i]<-1 else ypq1ind3[i]<-0
if(xpsort[i]<q1xp) xpq1ind3[i]<-1 else xpq1ind3[i]<-0
if(ypsort[i]<q1yp) ypq1ind4[i]<-1 else ypq1ind4[i]<-0
if(xpsort[i]<q2xp) xpq1ind4[i]<-1 else xpq1ind4[i]<-0
if(ypsort[i]<q2yp) ypq1ind5[i]<-1 else ypq1ind5[i]<-0
if(xpsort[i]<q2xp) xpq1ind5[i]<-1 else xpq1ind5[i]<-0
if(ypsort[i]<q3yp) ypq1ind6[i]<-1 else ypq1ind6[i]<-0
if(xpsort[i]<q2xp) xpq1ind6[i]<-1 else xpq1ind6[i]<-0
if(ypsort[i]<q1yp) ypq1ind7[i]<-1 else ypq1ind7[i]<-0
```

```
if(xpsort[i]<q3xp) xpq1ind7[i]<-1 else xpq1ind7[i]<-0
if(ypsort[i]<q2yp) ypq1ind8[i]<-1 else ypq1ind8[i]<-0
if(xpsort[i]<q3xp) xpq1ind8[i]<-1 else xpq1ind8[i]<-0
if(ypsort[i]<q3yp) ypq1ind9[i]<-1 else ypq1ind9[i]<-0
if(xpsort[i]<q3xp) xpq1ind9[i]<-1 else xpq1ind9[i]<-0
#cat("yp indicator=",ypq1ind[i],"xp indicator=",xpq1ind[i],"Multiplied",xyind,'\n')
}#i
pq1xq1y<-mean((ypq1ind1*xpq1ind1))
pq1xq2y<-mean((ypq1ind2*xpq1ind2))
pq1xq3y<-mean((ypq1ind3*xpq1ind3))
pq2xq1y<-mean((ypq1ind4*xpq1ind4))
pq2xq2y<-mean((ypq1ind5*xpq1ind5))
pq2xq3y<-mean((ypq1ind6*xpq1ind6))
pq3xq1y<-mean((ypq1ind7*xpq1ind7))
pq3xq2y<-mean((ypq1ind8*xpq1ind8))
pq3xq3y<-mean((ypq1ind9*xpq1ind9))
#print(c(pq1xq1y,pq1xq2y,pq1xq3y,pq2xq1y,pq2xq2y,pq2xq3y,pq3xq1y,pq3xq2y,pq3xq3y))
for(i in seq(1,np,1)){
if(ypsort[i]<q1yp) IQ1YI[i]<-1 else IQ1YI[i]<-0
if(ypsort[i]<q2yp) IQ2YI[i]<-1 else IQ2YI[i]<-0
if(ypsort[i]<q3yp) IQ3YI[i]<-1 else IQ3YI[i]<-0
if(xpsort[i]<q1xp) IQ1XI[i]<-1 else IQ1XI[i]<-0
if(xpsort[i]<q2xp) IQ2XI[i]<-1 else IQ2XI[i]<-0
if(xpsort[i]<q3xp) IQ3XI[i]<-1 else IQ3XI[i]<-0
}
SYQ1Y<-sum((yp-mean(yp))*(IQ1YI-0.25))/(np-1)
SYQ2Y<-sum((yp-mean(yp))*(IQ2YI-0.5))/(np-1)
SYQ3Y<-sum((yp-mean(yp))*(IQ3YI-0.75))/(np-1)
SYQ1X<-sum((yp-mean(yp))*(IQ1XI-0.25))/(np-1)
SYQ2X<-sum((yp-mean(yp))*(IQ2XI-0.5))/(np-1)
SYQ3X<-sum((yp-mean(yp))*(IQ3XI-0.75))/(np-1)
SXQ1X<-sum((yp-mean(yp))*(IQ1YI-0.25))/(np-1)
SXQ2X<-sum((yp-mean(yp))*(IQ2YI-0.5))/(np-1)
SXQ3X<-sum((yp-mean(yp))*(IQ3YI-0.75))/(np-1)
SXQ1Y<-sum((yp-mean(yp))*(IQ1XI-0.25))/(np-1)
SXQ2Y<-sum((yp-mean(yp))*(IQ2XI-0.5))/(np-1)
SXQ3Y<-sum((yp-mean(yp))*(IQ3XI-0.75))/(np-1)
#cat(SYQ1Y,SYQ2Y,SYQ3Y,SYQ1X,SYQ2X,SYQ3X,SXQ1X,SXQ2X,SXQ3X,SXQ1Y,SXQ2Y,SXQ3Y,'\n')
# NAIVE ESTIMATOR;
f<-ns/np
fact<-((1-f)/ns)
tb1<- (9/fyq2yp^2+16*varyp+48*SYQ2Y)
tb2<-3/(4*fyq2yp*fyq3yp)-3/(4*fyq1yp*fyq2yp)+4*SYQ3Y-4*SYQ1Y
rbcqvyapprox<-fact*( tb1/(6*q2yp-4*meanyp)^2 -tb2/((q3yp-q1yp)*(6*q2yp-4*meanyp)))*100
####MSE APPROXIMATE - NAIVE ESTIMATOR
term1<-3/fyq3yp^2+3/fyq1yp^2-2/(fyq1yp*fyq3yp)
term2<-9/(fyq2yp^2)+16*varyp+48*SYQ2Y
term3<-3/(4*fyq2yp*fyq3yp)-3/(4*fyq1yp*fyq2yp)+4*SYQ3Y-4*SYQ1Y
```

```
mseapprox1<-((1-f)/ns)*cqvyp^2*( term1/(16*(q3yp-q1yp)^2) + term2/(6*q2yp-4*meanyp)^2 -2*term3/((q3yp-q1yp)*(6*q2yp-4*meanyp)) )
#cat(rbcqvyapprox,mseapprox1, '\n')

#### MSE OF THE RATIO TYPE ESTIMATOR
ratt1<-( 3/(16*fyq3yp^2) + 3/(16*fyq1yp^2) - 2/(16*fyq1yp*fyq3yp) )/((q3yp-q1yp)^2)
ratt2<-( 9/(fyq2yp^2)+16*varyp+48*SYQ2Y )/((6*q2yp-4*meanyp)^2)
ratt3<-2*(6/(8*fyq2yp*fyq3yp)-6/(8*fyq1yp*fyq2yp) +4*SYQ3Y -4*SYQ1Y )/((q3yp-q1yp)*(6*q2yp-4*meanyp))
SHIY2<-ratt1+ratt2-ratt3
ratt4<-( 3/(16*fxq3xp^2)+3/(16*fxq1xp^2)-2/(16*fxq1xp*fxq3xp))/((q3xp-q1xp)^2)
ratt5<-(9/fxq2xp^2+16*varxp+48*SXQ2X)/((6*q2xp-4*meanxp)^2)
ratt6<-2*( 6/(8*fxq2xp*fxq3xp)-6/(8*fxq1xp*fxq2xp)+4*SXQ3X-4*SXQ1X)/((q3xp-q1xp)*(6*q2xp-4*meanxp))
SHIX2<-ratt4+ratt5-ratt6
ratt71<-( (pq3xq3y-9/16)/(fxq3xp*fyq3yp)-(pq3xq1y-3/16)/(fxq3xp*fyq1yp)-(pq1xq3y-3/16)/(fxq1xp*fyq3yp) +
(pq1xq1y-1/16)/(fxq1xp*fyq1yp) )/((q3yp-q1yp)*(q3xp-q1xp))
ratt72<-(6*(pq3xq2y-3/8)/(fxq3xp*fyq2yp)-6*(pq1xq2y-1/8)/(fxq1xp*fyq2yp)+4*SYQ3X-4*SYQ1X)/((6*q2yp-4*meanyp)*(q3xp-q1xp))
ratt73<-(6*(pq2xq3y-3/8)/(fxq2xp*fyq3yp) -6*(pq2xq1y-1/8)/(fxq2xp*fyq1yp)+4*SXQ3Y-4*SXQ1Y)/((q3yp-q1yp)*(6*q2xp-4*meanxp))
ratt74<-(36*(pq2xq2y-1/4)/(fxq2xp*fyq2yp)+24*SYQ2X+24*SXQ2Y+16*covxyp)/((6*q2yp-4*meanyp)*(6*q2xp-4*meanxp))
SHIXY<-(ratt71-ratt72-ratt73+ratt74)
mseratio<-((1-f)/ns)*cqvyp^2*(SHIY2+SHIX2-2*SHIXY)

#Relative bias in the ratio estimator
bt1<-( 9/(fyq2yp^2) + 16*varyp+48*SYQ2Y )/((6*q2yp-4*meanyp)^2)
bt2<-(3/(16*(fxq3xp^2)) + 3/(16*(fxq1xp^2)) -2/(16*fxq1xp*fxq3xp) )/((q3xp-q1xp)^2)
bt3<-(6/(8*fyq2yp*fyq3yp)-6/(8*fyq1yp*fyq2yp)+4*SYQ3Y-4*SYQ1Y)/( (q3yp-q1yp)*(6*q2yp-4*meanyp) )
bt4<-( 6*(pq2xq3y-3/8)/(fxq2xp*fyq3yp) - 6*(pq2xq1y-1/8)/(fxq2xp*fyq1yp) + 4*SXQ3Y - 4*SXQ1Y )/ ( (q3yp-q1yp)*(6*q2xp-4*meanxp) )
bt5<-(36*(pq2xq2y-1/4)/(fxq2xp*fyq2yp) + 24*SYQ2X + 24*SXQ2Y + 16*covxyp) / ( (6*q2yp-4*meanyp)*(6*q2xp-4*meanxp) )
bt6<-( 6/(8*fxq2xp*fxq3xp)+4*SXQ3X-6/(8*fxq1xp*fxq2xp)-4*SXQ1X )/ ( (6*q2xp-4*meanxp)*(q3xp-q1xp) )
bt7<- ( (pq3xq3y-9/16)/(fxq3xp*fyq3yp) - (pq3xq1y-3/16)/(fxq3xp*fyq1yp) -(pq1xq3y-3/16)/(fxq1xp*fyq3yp) +
(pq1xq1y-1/16)/(fxq1xp*fyq1yp) )/((q3yp-q1yp)*(q3xp-q1xp))
bt8<-(6*(pq3xq2y-3/8)/(fxq3xp*fyq2yp) + 4*SYQ3X-6*(pq1xq2y-1/8)/(fxq1xp*fyq2yp)-4*SYQ1X)/((6*q2yp-4*meanyp)*(q3xp-q1xp))
rbratio<-fact*(bt1+bt2-bt3+bt4-bt5-bt6-bt7+bt8)*100
#cat("RB(ratio)=",rbratio, '\n')
#cat("Ratio MSE=",mseratio, "RB(ratio)=",rbratio, '\n')
re2<-mseapprox1*100/mseratio
BETA<-SHIXY*cqvyp/(SHIX2*cqvxp)
tbreg1y<- (9/fyq2yp^2+16*varyp+48*SYQ2Y)
tbreg2y<-3/(4*fyq2yp*fyq3yp)-3/(4*fyq1yp*fyq2yp)+4*SYQ3Y-4*SYQ1Y
tbreg1x<- (9/fxq2xp^2+16*varxp+48*SXQ2X)
tbreg2x<-3/(4*fxq2xp*fxq3xp)-3/(4*fxq1xp*fxq2xp)+4*SXQ3X-4*SXQ1X
rbreg<-fact*( cqvyp* ( tbreg1y/(6*q2yp-4*meanyp)^2 -tbreg2y/((q3yp-q1yp)*(6*q2yp-4*meanyp))) -BETA*cqvxp*( tbreg1x/
(6*q2xp-4*meanxp)^2 -tbreg2x/((q3xp-q1xp)* (6*q2xp-4*meanxp))) )*100/cqvyp
msereg<-((1-f)/ns)*cqvyp^2*SHIY2*(1-SHIXY^2/(SHIY2*SHIX2))
```

```
re3<-mseapprox1*100/msereg
EMODEYP<-round(EMODEYP,2)
EMODEXP<-round(EMODEXP,2)
cqvyp<-round(cqvyp,2)
axfit<-round(axfit,2)
bxfit<-round(bxfit,2)
ayfit<-round(ayfit,2)
byfit<-round(byfit,2)
corrxy<-round(corrxy,2)
re2<-round(re2,2)
re3<-round(re3,2)
rbcqvyapprox<-round(rbcqvyapprox,2)
rbratio<-round(rbratio,2)
rbreg<-round(rbreg,2)
if (EMODEYP>0 & EMODEXP>0 & re2>100 & abs(rbratio) < 10 & abs(rbreg)<10 & abs(rbcqvyapprox)<10)
cat(EMODEYP,EMODEXP,cqvyp, ax, ay, axfit, bxfit, ayfit, byfit, rhoxy, corrxy, re2, re3, rbcqvyapprox, rbratio, rbreg,'\n')
if (EMODEYP>0 & EMODEXP>0 & re2>100 & abs(rbratio) < 10 & abs(rbreg)<10 & abs(rbcqvyapprox)<10)data1<-
c(EMODEYP,EMODEXP,cqvyp, ax, ay, axfit, bxfit, ayfit, byfit, rhoxy, corrxy, re2, re3, rbcqvyapprox, rbratio, rbreg)
if (EMODEYP>0 & EMODEXP>0 & re2>100 & abs(rbratio) < 10 & abs(rbreg)<10 & abs(rbcqvyapprox)<10) cat(data1, file =
"C:/Users/Sarjinder/Desktop/christin1out.csv", sep=",", fill = TRUE, append = TRUE)
}#rhoxy
}#ay
}#ax
```

APPENDIX D: R-Codes using Real Data

```
library("readxl")
D<-read_excel("C:\\Users\\kittyve\\Desktop\\CHRISTIN_THESIS\\shellfish.xls")
#attach(D)
ypin<-D$Log_Age
xpin<-D$Diameter
yp<-ypin
xp<-xpin
np<-length(yp)
ys<-c()
xs<-c()
cqvys<-c()
cqvxs<-c()
cqvrat<-c()
xstar<-c()
ystar<-c()
ypsort<-c()
ypq1ind<-c()
xpsort<-c()
xpq1ind<-c()
ypq1ind1<-c()
xpq1ind1<-c()
ypq1ind2<-c()
xpq1ind2<-c()
ypq1ind3<-c()
xpq1ind3<-c()
ypq1ind4<-c()
```

```
xpq1ind4<-c()
ypq1ind5<-c()
xpq1ind5<-c()
ypq1ind6<-c()
xpq1ind6<-c()
ypq1ind7<-c()
xpq1ind7<-c()
ypq1ind8<-c()
xpq1ind8<-c()
ypq1ind9<-c()
xpq1ind9<-c()
IQ1YI<-c()
IQ2YI<-c()
IQ3YI<-c()
IQ1XI<-c()
IQ2XI<-c()
IQ3XI<-c()
corrxy<-cor(xp,yp)
meanyp<-mean(yp)
meanxp<-mean(xp)
varyp<-var(yp)
varxp<-var(xp)
covxyp<-cov(xp,yp)
quartyp<-quantile(yp)
quartxp<-quantile(xp)
#print(quarty)
q1yp<-quartyp[2]
#print(quarty1)
q2yp<-quartyp[3]
q3yp<-quartyp[4]
#print(c(q1yp,q2yp,q3yp))
cqvyp<-(q3yp-q1yp)/(6*q2yp-4*meanyp)
q1xp<-quartxp[2]
#print(quartx1)
q2xp<-quartxp[3]
q3xp<-quartxp[4]
#print(c(q1xp,q2xp,q3xp))
cqvxp<-(q3xp-q1xp)/(6*q2xp-4*meanxp)
EMODEYP<-(3*q2yp-2*meanyp)
EMODEXP<-(3*q2xp-2*meanxp)
ay<-47.26
by<-0.02064
ax<-13.7
bx<-0.02978
fyq2yp<-(q2yp^(ay-1)*exp(-q2yp/by))/(gamma(ay)*by^ay)
#cat("Fyq2yp=",fyq2yp,'\n')
fyq3yp<-(q3yp^(ay-1)*exp(-q3yp/by))/(gamma(ay)*by^ay)
#cat("Fyq3yp=",fyq3yp,'\n')
fyq1yp<-(q1yp^(ay-1)*exp(-q1yp/by))/(gamma(ay)*by^ay)
```

```
#cat("Fyq1yp=",fyq1yp,'\n')
fxq2xp<-(q2xp^(ax-1)*exp(-q2xp/bx))/(gamma(ax)*bx^ax)
#cat("Fxq2xp=",fxq2xp,'\n')
fxq3xp<-(q3xp^(ax-1)*exp(-q3xp/bx))/(gamma(ax)*bx^ax)
#cat("Fxq3xp=",fxq3xp,'\n')
fxq1xp<-(q1xp^(ax-1)*exp(-q1xp/bx))/(gamma(ax)*bx^ax)
#cat("Fxq1xp=",fxq1xp,'\n')
ns<-round(0.10*np)
####biasratio#####
ypsort<-yp
xpsort<-xp
for(i in seq(1,np,1)){
if(ypsort[i]<q1yp) ypq1ind1[i]<-1 else ypq1ind1[i]<-0
if(xpsort[i]<q1xp) xpq1ind1[i]<-1 else xpq1ind1[i]<-0
if(ypsort[i]<q2yp) ypq1ind2[i]<-1 else ypq1ind2[i]<-0
if(xpsort[i]<q1xp) xpq1ind2[i]<-1 else xpq1ind2[i]<-0
if(ypsort[i]<q3yp) ypq1ind3[i]<-1 else ypq1ind3[i]<-0
if(xpsort[i]<q1xp) xpq1ind3[i]<-1 else xpq1ind3[i]<-0
if(ypsort[i]<q1yp) ypq1ind4[i]<-1 else ypq1ind4[i]<-0
if(xpsort[i]<q2xp) xpq1ind4[i]<-1 else xpq1ind4[i]<-0
if(ypsort[i]<q2yp) ypq1ind5[i]<-1 else ypq1ind5[i]<-0
if(xpsort[i]<q2xp) xpq1ind5[i]<-1 else xpq1ind5[i]<-0
if(ypsort[i]<q3yp) ypq1ind6[i]<-1 else ypq1ind6[i]<-0
if(xpsort[i]<q2xp) xpq1ind6[i]<-1 else xpq1ind6[i]<-0
if(ypsort[i]<q1yp) ypq1ind7[i]<-1 else ypq1ind7[i]<-0
if(xpsort[i]<q3xp) xpq1ind7[i]<-1 else xpq1ind7[i]<-0
if(ypsort[i]<q2yp) ypq1ind8[i]<-1 else ypq1ind8[i]<-0
if(xpsort[i]<q3xp) xpq1ind8[i]<-1 else xpq1ind8[i]<-0
if(ypsort[i]<q3yp) ypq1ind9[i]<-1 else ypq1ind9[i]<-0
if(xpsort[i]<q3xp) xpq1ind9[i]<-1 else xpq1ind9[i]<-0
#cat("yp indicator=",ypq1ind[i],"xp indicator=",xpq1ind[i],"Multiplied",xyind,'\n')
}#i
pq1xq1y<-mean((ypq1ind1*xpq1ind1))
pq1xq2y<-mean((ypq1ind2*xpq1ind2))
pq1xq3y<-mean((ypq1ind3*xpq1ind3))
pq2xq1y<-mean((ypq1ind4*xpq1ind4))
pq2xq2y<-mean((ypq1ind5*xpq1ind5))
pq2xq3y<-mean((ypq1ind6*xpq1ind6))
pq3xq1y<-mean((ypq1ind7*xpq1ind7))
pq3xq2y<-mean((ypq1ind8*xpq1ind8))
pq3xq3y<-mean((ypq1ind9*xpq1ind9))
#print(c(pq1xq1y,pq1xq2y,pq1xq3y,pq2xq1y,pq2xq2y,pq2xq3y,pq3xq1y,pq3xq2y,pq3xq3y))
for(i in seq(1,np,1)){
if(ypsort[i]<q1yp) IQ1YI[i]<-1 else IQ1YI[i]<-0
if(ypsort[i]<q2yp) IQ2YI[i]<-1 else IQ2YI[i]<-0
if(ypsort[i]<q3yp) IQ3YI[i]<-1 else IQ3YI[i]<-0
if(xpsort[i]<q1xp) IQ1XI[i]<-1 else IQ1XI[i]<-0
if(xpsort[i]<q2xp) IQ2XI[i]<-1 else IQ2XI[i]<-0
```

```
if(xpsort[i]<q3xp) IQ3XI[i]<-1 else IQ3XI[i]<-0
}
SYQ1Y<-sum((yp-mean(yp))*(IQ1YI-0.25))/(np-1)
SYQ2Y<-sum((yp-mean(yp))*(IQ2YI-0.5))/(np-1)
SYQ3Y<-sum((yp-mean(yp))*(IQ3YI-0.75))/(np-1)
SYQ1X<-sum((yp-mean(yp))*(IQ1XI-0.25))/(np-1)
SYQ2X<-sum((yp-mean(yp))*(IQ2XI-0.5))/(np-1)
SYQ3X<-sum((yp-mean(yp))*(IQ3XI-0.75))/(np-1)
SXQ1X<-sum((yp-mean(yp))*(IQ1YI-0.25))/(np-1)
SXQ2X<-sum((yp-mean(yp))*(IQ2YI-0.5))/(np-1)
SXQ3X<-sum((yp-mean(yp))*(IQ3YI-0.75))/(np-1)
SXQ1Y<-sum((yp-mean(yp))*(IQ1XI-0.25))/(np-1)
SXQ2Y<-sum((yp-mean(yp))*(IQ2XI-0.5))/(np-1)
SXQ3Y<-sum((yp-mean(yp))*(IQ3XI-0.75))/(np-1)
#cat(SYQ1Y,SYQ2Y,SYQ3Y,SYQ1X,SYQ2X,SYQ3X,SXQ1X,SXQ2X,SXQ3X,SXQ1Y,SXQ2Y,SXQ3Y,'\n')
# NAIVE ESTIMATOR;
f<-ns/np
fact<-((1-f)/ns)
tb1<- (9/fyq2yp^2+16*varyp+48*SYQ2Y)
tb2<-3/(4*fyq2yp*fyq3yp)-3/(4*fyq1yp*fyq2yp)+4*SYQ3Y-4*SYQ1Y
rbcqvyapprox<-fact*( tb1/(6*q2yp-4*meanyp)^2 -tb2/((q3yp-q1yp)*(6*q2yp-4*meanyp)))*100
####MSE APPROXIMATE - NAIVE ESTIMATOR
term1<-3/fyq3yp^2+3/fyq1yp^2-2/(fyq1yp*fyq3yp)
term2<-9/(fyq2yp^2)+16*varyp+48*SYQ2Y
term3<-3/(4*fyq2yp*fyq3yp)-3/(4*fyq1yp*fyq2yp)+4*SYQ3Y-4*SYQ1Y
mseapprox1<-((1-f)/ns)*cqvyp^2*( term1/(16*(q3yp-q1yp)^2) + term2/(6*q2yp-4*meanyp)^2 -2*term3/((q3yp-q1yp)*(6*q2yp-
4*meanyp)) )
#cat(rbcqvyapprox,mseapprox1, '\n')
#### MSE OF THE RATIO TYPE ESTIMATOR
ratt1<-( 3/(16*fyq3yp^2) + 3/(16*fyq1yp^2) - 2/(16*fyq1yp*fyq3yp) )/((q3yp-q1yp)^2)
ratt2<-( 9/(fyq2yp^2)+16*varyp+48*SYQ2Y )/((6*q2yp-4*meanyp)^2)
ratt3<-2*(6/(8*fyq2yp*fyq3yp)-6/(8*fyq1yp*fyq2yp) +4*SYQ3Y -4*SYQ1Y )/((q3yp-q1yp)*(6*q2yp-4*meanyp))
SHIY2<-ratt1+ratt2-ratt3
ratt4<-(3/(16*fxq3xp^2)+3/(16*fxq1xp^2)-2/(16*fxq1xp*fxq3xp))/((q3xp-q1xp)^2)
ratt5<-(9/fxq2xp^2+16*varxp+48*SXQ2X)/((6*q2xp-4*meanxp)^2)
ratt6<-2*(6/(8*fxq2xp*fxq3xp)-6/(8*fxq1xp*fxq2xp)+4*SXQ3X-4*SXQ1X)/((q3xp-q1xp)*(6*q2xp-4*meanxp))
SHIX2<-ratt4+ratt5-ratt6
ratt71<-( (pq3xq3y-9/16)/(fxq3xp*fyq3yp)-(pq3xq1y-3/16)/(fxq3xp*fyq1yp)-(pq1xq3y-3/16)/(fxq1xp*fyq3yp) +
(pq1xq1y-1/16)/(fxq1xp*fyq1yp) )/((q3yp-q1yp)*(q3xp-q1xp))
ratt72<-(6*(pq2xq3y-3/8)/(fxq2xp*fyq3yp)-6*(pq2xq1y-1/8)/(fxq2xp*fyq1yp)+4*SXQ3Y-4*SXQ1Y)/((6*q2xp-
4*meanxp)*(q3yp-q1yp))
ratt73<-(6*(pq3xq2y-3/8)/(fxq3xp*fyq2yp) -6*(pq1xq2y-1/8)/(fxq1xp*fyq2yp)+4*SYQ3X-4*SYQ1X)/((q3xp-q1xp)*(6*q2yp-
4*meanyp))
ratt74<-(36*(pq2xq2y-1/4)/(fxq2xp*fyq2yp) + 24*SYQ2X + 24*SXQ2Y+16*covxyp)/((6*q2yp-4*meanyp)*(6*q2xp-
4*meanxp))
SHIXY<-(ratt71-ratt72-ratt73+ratt74)
testing<-((1-f)/ns)*cqvyp^2*SHIY2
ratio<-mseapprox1/testing
mseratio<-((1-f)/ns)*cqvyp^2*(SHIY2+SHIX2-2*SHIXY)
```

```
#Relative bias in the ratio estimator
bt1<-( 9/(fyq2yp^2) + 16*varyp+48*SYQ2Y )/((6*q2yp-4*meanyp)^2)
bt2<-(3/(16*(fxq3xp^2)) + 3/(16*(fxq1xp^2)) -2/(16*fxq1xp*fxq3xp) )/((q3xp-q1xp)^2)
bt3<-(6/(8*fyq2yp*fyq3yp)-6/(8*fyq1yp*fyq2yp)+4*SYQ3Y-4*SYQ1Y)/( (q3yp-q1yp)*(6*q2yp-4*meanyp) )
bt4<-( 6*(pq2xq3y-3/8)/(fxq2xp*fyq3yp) - 6*(pq2xq1y-1/8)/(fxq2xp*fyq1yp) + 4*SXQ3Y - 4*SXQ1Y )/ ( (q3yp-
q1yp)*(6*q2xp-4*meanxp) )
bt5<-(36*(pq2xq2y-1/4)/(fxq2xp*fyq2yp) + 24*SYQ2X + 24*SXQ2Y + 16*covxyp) / ( (6*q2yp-4*meanyp)*(6*q2xp-
4*meanxp) )
bt6<-( 6/(8*fxq2xp*fxq3xp)+4*SXQ3X-6/(8*fxq1xp*fxq2xp)-4*SXQ1X )/ ( (6*q2xp-4*meanxp)*(q3xp-q1xp) )
bt7<- ( (pq3xq3y-9/16)/(fxq3xp*fyq3yp) - (pq3xq1y-3/16)/(fxq3xp*fyq1yp) -(pq1xq3y-3/16)/(fxq1xp*fyq3yp) +
(pq1xq1y-1/16)/(fxq1xp*fyq1yp) )/((q3yp-q1yp)*(q3xp-q1xp))
bt8<-(6*(pq3xq2y-3/8)/(fxq3xp*fyq2yp) + 4*SYQ3X-6*(pq1xq2y-1/8)/(fxq1xp*fyq2yp)-4*SYQ1X)/((6*q2yp-
4*meanyp)*(q3xp-q1xp))
rbratio<-fact*(bt1+bt2-bt3+bt4-bt5-bt6-bt7+bt8)*100
#cat("RB(ratio)=",rbratio, '\n')
#cat("Ratio MSE=",mseratio, "RB(ratio)=",rbratio, '\n')
re2<-mseapprox1*100/mseratio
BETA<-SHIXY*cqvyp/(SHIX2*cqvxp)
tbreg1y<- (9/fyq2yp^2+16*varyp+48*SYQ2Y)
tbreg2y<-3/(4*fyq2yp*fyq3yp)-3/(4*fyq1yp*fyq2yp)+4*SYQ3Y-4*SYQ1Y
tbreg1x<- (9/fxq2xp^2+16*varxp+48*SXQ2X)
tbreg2x<-3/(4*fxq2xp*fxq3xp)-3/(4*fxq1xp*fxq2xp)+4*SXQ3X-4*SXQ1X
rbreg<-fact*( cqvyp* ( tbreg1y/(6*q2yp-4*meanyp)^2 -tbreg2y/((q3yp-q1yp)*(6*q2yp-4*meanyp))) -BETA*cqvxp* ( tbreg1x/
(6*q2xp-4*meanxp)^2 -tbreg2x/((q3xp-q1xp)*(6*q2xp-4*meanxp))))*100/cqvyp
rho<-SHIXY^2/(SHIY2*SHIX2)
print(c('rho=',rho))
msereg<-((1-f)/ns)*cqvyp^2*SHIY2*(1-SHIXY^2/(SHIY2*SHIX2))
re3<-mseapprox1*100/msereg
EMODEYP<-round(EMODEYP,2)
EMODEXP<-round(EMODEXP,2)
cqvyp<-round(cqvyp,2)
corrxy<-round(corrxy,2)
re2<-round(re2,4)
re3<-round(re3,4)
rbcqvyapprox<-round(rbcqvyapprox,2)
rbratio<-round(rbratio,2)
rbreg<-round(rbreg,2)
mseapprox1<-round(mseapprox1,5)
msereg<-round(msereg,5)
if (abs(rbratio) < 10 & abs(rbreg)<10 & abs(rbcqvyapprox)<10)
 cat(EMODEYP,EMODEXP,cqvyp, ax, ay, corrxy, ns, re2, re3, rbcqvyapprox, rbratio, rbreg,mseapprox1,msereg,'\n')
```

CHAPTER 12

Class of Exponential Ratio Type Estimator for Population Mean in Adaptive Cluster Sampling

Akingbade Toluwalase Janet[1,*] and *Balogun Oluwafemi Samson*[2]

1. Introduction

One of the areas of survey for statisticians is to determine efficient estimators for estimating population characteristics to avoid waste of resources. The conventional design, otherwise known as traditional fixed sampling procedure like simple random sampling has been in use for so many years and the designs have not been given a desirable result in some studies where the variable of interest is hidden or rare in the population. Researchers usually encounter difficulties in areas such as environmental pollution studies, surveys of rare animal and plant species, studies of contagious diseases, drug use epidemiology, marketing surveys, and assessments of mineral and fossil fuel reserves, because members of the populations are rare, hidden, or hard to reach. Adaptive Cluster sampling (ACS) was first proposed by Thompson (1990) as a data driven sampling method that can provide higher sampling yields to provide better estimates of the population characteristics. The procedure of sampling in ACS thus goes like this : an initial sample of n units is selected from the population of N units using simple random sampling with or without replacement and, whenever the variable of interest (y) for a unit in the sample satisfies a pre-specified condition (C), i.e., $y_i>C$, neighbouring units are added to the sample until no more units satisfy the criterion. Even though the initial sample is selected without replacement, some units in the sample may be selected more than once because the initial sample may contain more than one unit in a given network of units satisfying the criterion. The set of all units meeting the criterion in the neighbourhood of one another is called a **network** and the units that were adaptively sampled that did not meet the pre specified condition are called **edge units**. Combination of networks and its associated edge units are called a **cluster**.

Thompson (1990) developed an unbiased sample mean estimator in ACS using variable of interest only by modifying estimators by Hansen Hansen-Hurwitz (1943) and Horvitz-Thompson (1952). Ever since then ACS has received considerable attention in practical surveys because of its flexibility in application and ability to provide better estimates as well as a more meaningful sample. It is well known that higher precision of estimates could be achieved by using the auxiliary information in formulating estimators such as ratio, regression, difference or product for estimating the population means. Ratio estimator is used for estimation where there exists positive correlation between the study and auxiliary variable while product estimator is suitable when negative correlation exists between the variables. Ratio estimator is known to be as efficient as classical linear regression estimator when the regression line passes through the origin. This is practically impossible in real life situations for the regression line to pass through the origin; however this limitation has motivated many authors like Sissodia and Dwivedi(1981), Upadhyaya and Singh (1999) Singh and Tailor (2003), Singh et al. (2017) and Akingbade and Okafor (2019) to use known population parameters such as coefficient of variation, coefficient of skewness, coefficient of kurtosis of an auxiliary variable, to obtain better estimates in modifying a classical ratio estimator by Cochran (1940) under simple random sampling scheme. With these developments in simple random sampling, Dryver and Chao (2007) introduced an auxiliary variable to develop a classical ratio estimator in adaptive cluster sampling by modifying Cochran's (1940) classical ratio estimator in simple random sampling. They obtained the Mean Square Error (MSE) up to the first order of approximation and their estimator was an improvement on Thompson's (1990) sample mean estimator. For further improvement on the classical ratio estimator in ACS, Chutiman (2013), and Yadav et al. (2016) made use of some known parameters of the auxiliary variable to develop ratio type estimators in ACS while Noor-ul-Amin and Hanif (2013) extended the exponential ratio estimator in simple random sampling by Bahl and Tuteja (1991) to exponential ratio estimator in adaptive cluster sampling.

[1] Department of Mathematical Science, Faculty of Natural Sciences, Kogi State University Anyigba, Nigeria.
[2] School of Computing, University of Eastern Finland, FI-70211, Kuopio, Finland.
Email: samson.balogun@uef.fi
* Corresponding author: afuapetoluse@yahoo.com

These authors obtained the MSEs of their estimators and compared them with some other estimators. The existing ratio estimators in adaptive cluster sampling assumes that the population mean of the auxiliary variable is known and some with additional information of other known parameters of the auxiliary variable. In this work, a class of exponential ratio type estimator under the same situations was proposed and comparisons were made with some of the existing estimators using their mean squared error.

2. Some Existing Estimators in Adaptive Cluster Sampling

Suppose a finite population of size N is labeled as $1,2,3,\ldots,N$. Let the auxiliary variable (x) be correlated with the variable of interest (y), such that $Y = \{Y_1, Y_2 \ldots Y_N\}$ and $X = \{X_1, X_2, \ldots X_N\}$. Let an initial sample of size n units be selected with a simple random sample without replacement from a finite population size N. $\bar{x}$ and $\bar{y}$ are sample means of the auxiliary variable and variable of interest respectively. It is assumed that the population mean $\bar{X}$ of the auxiliary variable X is known. Let A_i denote the network that includes unit i and m_i be the number of units in that network. Let w_{yi} and w_{xi} denotes the average y values and average x values in the network respectively, where $w_{yi} = \frac{1}{m_i}\sum_{j\in A_i} y_j$ and $w_{xi} = \frac{1}{m_i}\sum_{j\in A_i} x_j$.

Thompson (1990) developed an unbiased estimator in adaptive cluster sampling based on Hansen-Hurwitz's (1943) estimator for estimating the population mean using only variable of interest. The estimator was given as

$$t_T = \bar{w}_y = \frac{1}{n}\sum_{i=1}^{n}(w_{yi}) \tag{1}$$

An unbiased estimator of $V(t_{hh})$ is

$$V(t_T) = \frac{N-n}{N(N-1)}\sum_{i=1}^{N}\left(w_{yi} - t_T\right)^2 \tag{2}$$

The suggested classical ratio estimator in adaptive cluster sampling by Dryver and Chao (2007) is given as

$$t_{DC} = \bar{w}_y\left(\frac{\bar{X}}{\bar{w}_x}\right) \tag{3}$$

The mean square error of the estimator is given as

$$MSE(t_{DC}) = f_1\bar{Y}^2\left[C_{wy}^2 + C_{wx}^2 - 2\rho_{wx.wy}C_{wy}C_{wx}\right] \tag{4}$$

Where C_{wx} is the population coefficient of variation of wx, C_{wy} is the population coefficient of variation of wy and $\rho_{wx.wy}$ is the population correlation coefficient of wx and wy. $f_1 = \frac{1-f}{n}$, $f = \frac{n}{N}$ (Sampling fraction)

Chutiman (2013) utilized the population mean of auxiliary variable ($\bar{X}$) with addition of population coefficient variation and kurtosis of wx to formulate ratio type estimators under adaptive cluster sampling. The estimators are given as:

$$t_{C1} = \bar{w}_y\left(\frac{\bar{X} + C_{wx}}{\bar{w}_x + C_{wx}}\right) \tag{5}$$

$$t_{C2} = \bar{w}_y\left(\frac{\beta_{2(wx)}\bar{X} + C_{wx}}{\beta_{2(wx)}\bar{w}_x + C_{wx}}\right) \tag{6}$$

$$t_{C3} = \bar{w}_y\left(\frac{C_{wx}\bar{X} + \beta_{2(wx)}}{C_{wx}\bar{w}_x + \beta_{2(wx)}}\right) \tag{7}$$

where C_{wx} and $\beta_{2(wx)}$ are the population coefficient of variation and population coefficient of kurtosis of w_x respectively and the other symbols have their usual meanings. They obtained the mean square errors of the above estimators using Taylor series method up to first order of approximation and are respectively given as

$$MSE(t_{C1}) = f_1\bar{Y}^2\left[C_{wy}^2 + \theta_{C1}^2C_{wx}^2 - 2\theta_{C1}\rho_{wx.wy}C_{wy}C_{wx}\right] \tag{8}$$

$$MSE(t_{C2}) = f_1\bar{Y}^2\left[C_{wy}^2 + \theta_{C2}^2C_{wx}^2 - 2\theta_{C2}\rho_{wx.wy}C_{wy}C_{wx}\right] \tag{9}$$

$$MSE(t_{C3}) = f_1\bar{Y}^2\left[C_{wy}^2 + \theta_{C3}^2C_{wx}^2 - 2\theta_{C3}\rho_{wx.wy}C_{wy}C_{wx}\right] \tag{10}$$

where $\theta_{C1} = \frac{\overline{X}}{\overline{X} + C_{wx}}$, $\theta_{C2} = \frac{\beta_{2(wx)}\overline{X}}{\beta_{2(wx)}\overline{X} + C_{wx}}$ and $\theta_{C3} = \frac{C_{wx}\overline{X}}{C_{wx}\overline{X} + \beta_{2(wx)}}$ and the other terms have their usual meanings.

Yadav et al.'s (2016) ratio estimators under adaptive cluster sampling were obtained by modifying the ratio estimators by Kadilar and Cingi (2003) and Yan and Tian's (2010) estimators using coefficients of skewness ($\beta_{1(wx)}$) and kurtosis ($\beta_{2(wx)}$). The estimators are given as

$$t_{YMMC1} = \overline{w}_y\left(\frac{\overline{X}^2}{\overline{w}_x^2}\right) \tag{11}$$

$$t_{YMMC2} = \overline{w}_y\left(\frac{\beta_{2(wx)}\overline{X} + \beta_{1(wx)}}{\beta_{2(wx)}\overline{w}_x + \beta_{1(wx)}}\right) \tag{12}$$

$$t_{YMMC3} = \overline{w}_y\left(\frac{\beta_{1(wx)}\overline{X} + \beta_{2(wx)}}{\beta_{1(wx)}\overline{w}_x + \beta_{2(wx)}}\right) \tag{13}$$

The MSE of the estimators up to the first order approximation were given as

$$MSE(t_{YMMC1}) = f_1\overline{Y}^2\left(C_{wy}^2 + 4C_{wx}^2 - 4\rho_{wx.wy}C_{wy}C_{wx}\right) \tag{14}$$

$$MSE(t_{YMMC2}) = f_1\overline{Y}^2\left[C_{wy}^2 + \theta_{YMMC2}^2C_{wx}^2 - 2\theta_{YMMC2}\rho_{wx.wy}C_{wy}C_{wx}\right] \tag{15}$$

$$MSE(t_{YMMC3}) = f_1\overline{Y}^2\left[C_{wy}^2 + \theta_{YMMC3}^2C_{wx}^2 - 2\theta_{YMMC3}\rho_{wx.wy}C_{wy}C_{wx}\right] \tag{16}$$

Where, $\theta_{YMMC2} = \frac{\beta_{2(wx)}\overline{X}}{\beta_{2(wx)}\overline{X} + \beta_{1(wx)}}$ and $\theta_{YMMC3} = \frac{\beta_{1(wx)}\overline{X}}{\beta_{1(wx)}\overline{X} + \beta_{2(wx)}}$.

The exponential ratio estimator in ACS by Noor-ul-Amin and Hanif (2013) is given as

$$t_{NH} = \overline{w}_y \exp\left(\frac{\overline{X} - \overline{w}_x}{\overline{X} + \overline{w}_x}\right) \tag{17}$$

The MSE of the estimator up to the first order approximation is given as

$$\text{MSE}(t_{NH}) = f_1\overline{Y}^2\left[C_{wy}^2 + \frac{1}{4}C_{wx}^2 - \rho_{wx.wy}C_{wy}C_{wx}\right] \tag{18}$$

3. The Proposed Class of Exponential Ratio-type Estimator in ACS

In this section, a class of estimators is developed by combining the concept of Housila et al.'s (2017) ratio estimator and Noor-ul-Amin and Hanif's (2013) exponential ratio estimator in ACS. The exponential ratio-type estimator for estimating the population means ($\overline{Y}$) in adaptive cluster sampling is proposed as

$$t_p = \overline{w}_y\left[\frac{a\overline{X} + b\overline{w}_x}{c\overline{w}_x + d\overline{X}}\right]^{\delta} \exp\phi\left(\frac{\overline{X} - \overline{w}_x}{\overline{X} + \overline{w}_x}\right) \tag{19}$$

where a, b, c and d are suitably chosen constants, such that $a = c$ and $b = d$. They may take real values ?? where population parameters such as C_{wx} is the population coefficient of variation of wx, C_{wy} is the population coefficient of variation of wy, $\rho_{wx.wy}$ is the population correlation coefficient of wx and wy, $\beta_{2(wx)}$ is the population coefficient of kurtosis of wx, $\beta_{1(wx)}$ is the population coefficient of skewness of wx, population mean of the auxiliary variable ($\overline{X}$), population variance of wx (S_{wx}), and ϕ can either be 0 or 1 and δ is an arbitrary constant or may be chosen the statistic.

Derivation of the Bias and Mean Square Error of t_p

To obtain the Bias and MSE of t_p up to the first order of approximation, let us define the relative sampling errors of the auxiliary variable and study variable in ACS as $\overline{e}_{wx} = \frac{\overline{w}_x - \overline{X}}{\overline{X}}$ and $\overline{e}_{wy} = \frac{\overline{w}_y - \overline{Y}}{\overline{Y}}$ respectively then $\overline{w}_x = \overline{X}(1 + \overline{e}_{wx})$, $\overline{w}_y = \overline{Y}(1 + \overline{e}_{wy})$ such that $\text{E}(\overline{e}_{wy}) = \text{E}(\overline{e}_{wx}) = 0$, $\text{E}(\overline{e}_{wx}^2) = f_1C_{wx}^2$, $\text{E}(\overline{e}_{wy}^2) = f_1C_{wy}^2$, $\text{E}(\overline{e}_{wx}\overline{e}_{wy}) = f_1\rho_{wx.wy}C_{wx}C_{wy}$ $f_1 = \frac{1-f}{n}$, $f = \frac{n}{N}$ (Sampling fraction)

Expressing the proposed estimator t_p in terms of $\overline{e}$'s and simplifying, we have

$$t_p = \bar{Y}(1+\bar{e}_{wy})\left[\frac{(a+b)\left[1+\left(\dfrac{b}{a+b}\right)\bar{e}_{wx}\right]}{(c+d)\left[1+\left(\dfrac{c}{c+d}\right)\bar{e}_{wx}\right]}\right]^{\delta} \exp\phi\left[\frac{-\bar{e}_{wx}}{2}\left(1+\frac{\bar{e}_{wx}}{2}\right)^{-1}\right] \tag{20}$$

$$t_p = \bar{Y}(1+\bar{e}_{wy})\left[\frac{(a+b)[1+\lambda_1\bar{e}_{wx}]}{(c+d)[1+\lambda_2\bar{e}_{wx}]}\right]^{\delta} \exp\phi\left[\frac{-\bar{e}_{wx}}{2}\left(1+\frac{\bar{e}_{wx}}{2}\right)^{-1}\right] \tag{21}$$

where $\left(\dfrac{a+b}{c+d}\right)^{\delta}=1$, since $a=c$ and $b=d$.

Expanding (21) to the first order of approximation using binomial series expansion, stopping at order 2, we have

$$t_p = \bar{Y}\begin{pmatrix} 1+\delta\lambda_1\bar{e}_{wx}+\dfrac{\delta(\delta-1)\lambda_1^2\bar{e}_{wx}^2}{2}-\delta\lambda_2\bar{e}_{wx}-\delta^2\lambda_1\lambda_2\bar{e}_{wx}^2+\dfrac{\delta(\delta+1)\lambda_2^2\bar{e}_{wx}^2}{2}-\dfrac{\phi\bar{e}_{wx}}{2} \\ -\dfrac{\phi\delta\lambda_1\bar{e}_{wx}^2}{2}+\dfrac{\phi\delta\lambda_2\bar{e}_{wx}^2}{2}+\dfrac{\phi^2\bar{e}_{wx}^2}{8}+\dfrac{\phi\bar{e}_{wx}^2}{4}+\bar{e}_{wx}+\delta\lambda_1\bar{e}_{wx}\bar{e}_{wy}-\delta\lambda_2\bar{e}_{wx}\bar{e}_{wy}-\dfrac{\phi\bar{e}_{wx}\bar{e}_{wy}}{2} \end{pmatrix} \tag{22}$$

where $\lambda_1=\dfrac{b}{a+b}$ and $\lambda_2=\dfrac{c}{c+d}$

$$t_p-\bar{Y} = \bar{Y}\begin{pmatrix} \delta\lambda_1\bar{e}_{wx}+\dfrac{\delta(\delta-1)\lambda_1^2\bar{e}_{wx}^2}{2}-\delta\lambda_2\bar{e}_{wx}-\delta^2\lambda_1\lambda_2\bar{e}_{wx}^2+\dfrac{\delta(\delta+1)\lambda_2^2\bar{e}_{wx}^2}{2}-\dfrac{\phi\bar{e}_{wx}}{2} \\ -\dfrac{\phi\delta\lambda_1\bar{e}_{wx}^2}{2}+\dfrac{\phi\delta\lambda_2\bar{e}_{wx}^2}{2}+\dfrac{\phi^2\bar{e}_{wx}^2}{8}+\dfrac{\phi\bar{e}_{wx}^2}{4}+\bar{e}_{wx}+\delta\lambda_1\bar{e}_{wx}\bar{e}_{wy}-\delta\lambda_2\bar{e}_{wx}\bar{e}_{wy}-\dfrac{\phi\bar{e}_{wx}\bar{e}_{wy}}{2} \end{pmatrix} \tag{23}$$

After simplification, the bias is

$$\mathrm{B}(t_p)=\mathrm{E}(t_p-\bar{Y})$$

$$= f_1\bar{Y}\begin{pmatrix} \dfrac{\delta(\delta-1)\lambda_1^2C_{wx}^2}{2}-\delta^2\lambda_1\lambda_2C_{wx}^2+\dfrac{\delta(\delta+1)\lambda_2^2C_{wx}^2}{2}-\dfrac{\phi\delta\lambda_1C_{wx}^2}{2}+\dfrac{\phi\delta\lambda_2C_{wx}^2}{2}+\dfrac{\phi^2C_{wx}^2}{8} \\ +\dfrac{\phi C_{wx}^2}{4}+\delta\lambda_1\rho_{wx.wy}C_{wx}C_{wy}-\delta\lambda_2\rho_{wx.wy}C_{wx}C_{wy}-\dfrac{\phi\rho_{wx.wy}C_{wx}C_{wy}}{2} \end{pmatrix} \tag{24}$$

The mean square error of the estimator is

$$MSE(t_p)=E(t_p-\bar{Y})^2$$

Taking the expectation of the square of Equation (23) and ignoring orders higher than 2 and after simplification we have

$$\mathrm{MSE}(t_p)=f_1\bar{Y}^2\left(C_{wy}^2+\frac{1}{4}(\phi-2\lambda\delta)^2C_{wx}^2-(\phi-2\lambda\delta)\rho_{wx.wy}C_{wx}C_{wy}\right) \tag{25}$$

where $\lambda=\lambda_1-\lambda_2$

Further simplification of (25), we have

$$\mathrm{MSE}(t_p)=f_1\bar{Y}^2\left(C_{wy}^2+\frac{1}{4}V^2C_{wx}^2-V\rho_{wx.wy}C_{wx}C_{wy}\right) \tag{26}$$

where $V=\phi-2\lambda\delta$

To obtain the approximate optimum value of δ we differentiate (26) with respect to δ, equate to zero and solve the equation, we have

$$\delta_o=\frac{1}{\lambda}\left(\frac{1}{2}-\frac{\rho_{wx.wy}C_{wy}}{C_{wx}}\right)$$

Substituting the optimum value of δ (δ_o) in (26), we obtain the minimum mean square error of t_p as

$$\mathrm{MSE}_{opt}(t_p)=f_1\bar{Y}^2C_{wy}^2(1-\rho_{wx.wy}^2) \tag{27}$$

This is equivalent to the approximate variance of the classical linear regression estimator in adaptive cluster sampling by Chutiman (2013) whose estimator was given as $t_{lr} = \overline{w}_y + b_w(\overline{X} - \overline{w}_x)$

This implies that, the proposed class of exponential ratio estimator in ACS at t optimum is equally efficient as the classical linear regression estimator in ACS,

where b_w is the sample regression coefficient of wy on wx in the network. $b_w = \frac{s_{wx.wy}}{s_{wx}^2}$, $s_{wx.wy} = (n-1)^{-1}\sum_{i=1}^{n}(w_{xi} - \overline{w}_x)(w_{yi} - \overline{w}_y)$ and $s_{wx}^2 = (n-1)^{-1}\sum_{i=1}^{n}(w_{xi} - \overline{w}_x)$.

4. Members of the Proposed Class of Exponential Ratio Type Estimator in ACS

From the suggested class of exponential ratio type estimator in (19), we can obtain some members of the proposed class of ratio type estimator in ACS at different values of *a, b, c, d,* ϕ, δ.

Setting $\delta = 1$, $\phi = 0$, $a - a_j$, $b = b_j$, $c = c_j$, $d = d_j$ in (19), we have the jth member of the ratio type estimator for estimating the population mean of the variable of interest and is given as follows:

$$t_{pj} = \overline{w}_y \left[\frac{a_j\overline{X} + b_j\overline{w}_x}{c_j\overline{w}_x + d_j\overline{X}}\right]^{\delta}, \quad j = 1,2, \ldots, 5 \tag{28}$$

This *j*th member of the ratio type estimator can be seen in Table 1. Substituting these parameters in (24) and (25) we have the bias and MSE as

$$B(t_{pj}) = f_1\overline{Y}\left((\lambda_{2j}^2 - \lambda_{1j}\lambda_{2j})C_{wx}^2 + (\lambda_{1j} - \lambda_{2j})\rho_{wx.wy}C_{wx}C_{wy}\right) \tag{29}$$

Table 1: Members of the proposed class of exponential ratio type estimators.

j	Members of the proposed class of exponential ratio type estimator t_{pj}	a_j	b_j	c_j	d_j	ϕ	δ
1	$t_{p1} = \overline{w}_y\left[\frac{C_{wx}^2\overline{X}}{C_{wx}^2\overline{w}_x}\right]$	C_{wx}^2	0	C_{wx}^2	0	0	1
2	$t_{p2} = \overline{w}_y\left[\frac{C_{wx}^2\overline{X} - \rho_{wx.wy}\overline{w}_x}{C_{wx}^2\overline{w}_x - \rho_{wx.wy}\overline{X}}\right]$	C_{wx}^2	$-\rho_{wx.wy}$	C_{wx}^2	$-\rho_{wx.wy}$	0	1
3	$t_{p3} = \overline{w}_y\left[\frac{\beta_2(wx)\overline{X} + \beta_1(wx)\overline{w}_x}{\beta_2(wx)\overline{w}_x + \beta_1(wx)\overline{X}}\right]$	$\beta_2(wx)$	$\beta_1(wx)$	$\beta_2(wx)$	$\beta_1(wx)$	0	1
4	$t_{p4} = \overline{w}_y\left[\frac{C_{wx}^2\overline{X} + \rho_{wx.wy}\overline{w}_x}{C_{wx}^2\overline{w}_x + \rho_{wx.wy}\overline{X}}\right]$	C_{wx}^2	$\rho_{wx.wy}$	C_{wx}^2	C_{wx}^2	0	1
5	$t_{p4} = \overline{w}_y\left[\frac{C_{wx}^2\overline{X} + \rho_{wx.wy}\overline{w}_x}{C_{wx}^2\overline{w}_x + \rho_{wx.wy}\overline{X}}\right]$	S_{wx}	$\rho_{wx.wy}$	S_{wx}	$\rho_{wx.wy}$	0	1
6	$t_{p6} = \overline{w}_y\left[\frac{C_{wx}^2\overline{X}}{C_{wx}^2\overline{w}_x}\right]\exp\left(\frac{\overline{X} - \overline{w}_x}{\overline{X} + \overline{w}_x}\right)$	C_{wx}^2	0	C_{wx}^2	0	1	1
7	$t_{p7} = \overline{w}_y\left[\frac{C_{wx}^2\overline{X} + \rho_{wx.wy}\overline{w}_x}{C_{wx}^2\overline{w}_x + \rho_{wx.wy}\overline{X}}\right]\exp\left(\frac{\overline{X} - \overline{w}_x}{\overline{X} + \overline{w}_x}\right)$	C_{wx}^2	$-\rho_{wx.wy}$	C_{wx}^2	$-\rho_{wx.wy}$	1	1
8	$t_{p8} = \overline{w}_y\left[\frac{\beta_2(wx)\overline{X} + \beta_1(wx)\overline{w}_x}{\beta_2(wx)\overline{w}_x + \beta_1(wx)\overline{X}}\right]\exp\left(\frac{\overline{X} - \overline{w}_x}{\overline{X} + \overline{w}_x}\right)$	$\beta_2(wx)$	$\beta_1(wx)$	$\beta_2(wx)$	$\beta_1(wx)$	1	1
9	$t_{p9} = \overline{w}_y\left[\frac{C_{wx}^2\overline{X} + \rho_{wx.wy}\overline{w}_x}{C_{wx}^2\overline{w}_x + \rho_{wx.wy}\overline{X}}\right]\exp\left(\frac{\overline{X} - \overline{w}_x}{\overline{X} + \overline{w}_x}\right)$	C_{wx}^2	$\rho_{wx.wy}$	C_{wx}^2	C_{wx}^2	1	1
10	$t_{p10} = \overline{w}_y\left[\frac{S_{wx}\overline{X} + \rho_{wx.wy}\overline{w}_x}{S_{wx}\overline{w}_x + \rho_{wx.wy}\overline{X}}\right]\exp\left(\frac{\overline{X} - \overline{w}_x}{\overline{X} + \overline{w}_x}\right)$	S_{wx}	$\rho_{wx.wy}$	S_{wx}	$\rho_{wx.wy}$	1	1

$$\text{MSE}(t_{pj}) = f_1\bar{Y}^2\left(C_{wy}^2 + \lambda_j^2 C_{wx}^2 - 2\lambda_j \rho_{wx.wy} C_{wx} C_{wy}\right) \tag{30}$$

where $\lambda_j = \lambda_{1j} - \lambda_{2j}$ and $\lambda_{1j} = \dfrac{b_j}{a_j + b_j}$, $\lambda_{2j} = \dfrac{c_j}{c_j + d_j}$

a_j, b_j, c_j, d_j are suitable chosen constants or known parameters in the jth estimator

Also setting $\delta = 1$, $\phi = 0$, $a - a_j$, $b = b_j$, $c = c_j$, $d = d_j$ in (19), we have the jth member of exponential ratio type estimator t_{pj} in ACS as;

$$t_{pj} = \bar{w}_y \left[\frac{a_j \bar{X} + b_j \bar{w}_x}{c_j \bar{w}_x + d_j \bar{X}}\right]^{\delta} \exp\phi\left(\frac{\bar{X} - \bar{w}_x}{\bar{X} + \bar{w}_x}\right), \quad j = 6, 7, \ldots, 10 \tag{31}$$

This *j*th member of the exponential ratio type estimator can be seen in Table 1.
Substituting these parameters in (24) and (25) we have the bias and MSE are,

$$B(t_{pj}) = f_1\bar{Y}\left((\lambda_{2j}^2 - \lambda_{1j}\lambda_{2j})C_{wx}^2 - \frac{\lambda_j C_{wx}^2}{2} + \frac{3C_{wx}^2}{8} + \lambda_j \rho_{wx.wy} C_{wx} C_{wy} - \frac{\rho_{wx.wy} C_{wx} C_{wy}}{2}\right) \tag{32}$$

$$\text{MSE}(t_{pj}) = f_1\bar{Y}^2\left(C_{wy}^2 + \frac{1}{4}(1 - 4\lambda_j + 4\lambda_j^2)C_{wx}^2 - (1 - 2\lambda_j)\rho_{wx.wy} C_{wx} C_{wy}\right) \tag{33}$$

5. Theoretical Comparison

In this section, the theoretical performances of the proposed estimator with other existing estimators are compared through their mean square errors.

From Equations (2) and (27), the proposed exponential ratio type estimator in ACS at optimum value will perform better than the sample mean estimator in ACS by Thompson (1990)

$MSE_{opt}(t_p) \leq V(t_N)$ iff

$-\rho_{wx.wy}^2 C_{wy}^2 \leq 0$

This condition will always be satisfied. Therefore, the proposed estimator at optimum value will always be better than the usual sample mean estimator in ACS by Thompson (1990).

Using Equations (4) and (27), the proposed exponential ratio type estimator in ACS at optimum value will perform better than the ratio estimator in ACS by Dryver and Chao (2007).

$MSE_{opt}(t_p) \leq MSE(t_{DC})$ iff

$(C_{wx} - C_{wy}\rho_{wx.wy})^2 \geq 0$

The proposed exponential ratio type estimator in ACS at optimum value will perform better than the ratio estimators in ACS by Chutiman (2013) if the following conditions are satisfied
From Equations (8) and (27), $MSE_{opt}(t_p) \leq MSE(t_{C1})$ iff

$(\theta_{C1}C_{wx} - C_{wy}\rho_{wx.wy})^2 \geq 0$

Using Equations (9) and (27), $MSE_{opt}(t_p) \leq MSE(t_{C2})$ iff

$(\theta_{C2}C_{wx} - C_{wy}\rho_{wx.wy})^2 \geq 0$

Also, from Equations (10) and (27), $MSE_{opt}(t_p) \leq MSE(t_{C3})$ iff

$(\theta_{C3}C_{wx} - C_{wy}\rho_{wx.wy})^2 \geq 0$

The proposed exponential ratio type estimator in ACS at optimum value will perform better than the ratio estimators in ACS by Yadav et al. (2016) if the following conditions are satisfied.
From Equations (14) and (27), $MSE_{opt}(t_p) \leq MSE(t_{YMMC1})$ iff

$-C_{wy}^2\rho_{wx.wy}^2 \leq 4(C_{wx}^2 - \rho_{wx.wy}C_{wx}C_{wy})$

Using Equations (15) and (27), $MSE_{opt}(t_p) \leq MSE(t_{YMMC2})$ iff

$(\theta_{YMMC2}C_{wx} - C_{wy}\rho_{wx.wy})^2 \geq 0$

Also, from Equations (16) and (27), the $MSE_{opt}(t_p) \leq MSE(t_{YMMC3})$ iff

$(\theta_{YMMC3}C_{wx} - C_{wy}\rho_{wx.wy})^2 \geq 0$

From Equations (18) and (27), the proposed exponential ratio type estimator in ACS at optimum value will perform better than the exponential ratio estimator in ACS by Noor-ul-Amin and Hanif (2013), if $MSE_{opt}(t_p) \leq MSE(t_{NH})$, if and only if

$$-C_{wy}\rho_{wx.wy} \leq \left(\frac{C^2_{wx}}{4} + \rho_{wx.wy}C_{wx}C_{wy}\right)$$

The smaller the mean square error, the more efficient the estimator becomes.

The Percent Relative Efficiencies (PREs) of the existing ratio estimators in ACS mentioned in (1), (3), (5), (6), (7), (11), (12), (13), (17) and the jth members of the proposed estimators t_{pj}, j = 1,2,3,...,10 given in Table 1 and $t_{p(opt)}$ with respect to the usual sample mean estimator in adaptive cluster sampling, is of the form;

PRE $= \frac{MSE(t_T)}{MSE(.)} * 100$ where (.) is any of the existing members of the proposed estimator.

The higher the percent relative efficiencies, the more efficient the estimator becomes.

6. Empirical Comparison

In this section, the mean square errors (MSEs), and percent relative efficiencies (PREs) of the existing and the proposed estimators in ACS with respect to the usual sample mean t_T were computed. The simulation x–values and y–values from Chutiman and Kumphon (2008) were used and the statistics of the population data sets are shown below. The results are given in Table 2.

Table 2: The MSE and PRE with respect to t_T of the existing and proposed estimators.

S/N	Existing Estimators	MSE	PRE
1	t_T (Thompson (1900))	0.6028	100
2	t_{DC} (Dryver and Chao (2007))	0.1326	454.3341
3	t_{C1} (Chutiman (2013))	0.4291	140.4801
4	t_{C2} (Chutiman (2013))	0.1102	547.0054
5	t_{C3} (Chutiman (2013))	0.5751	104.8166
6	t_{YMMC1} (Yadav et al. (2016))	1.4117	42.70029
7	t_{YMMC2} (Yadav et al. (2016))	0.0939	641.9595
8	t_{YMMC3} (Yadav et al. (2016))	0.5427	111.0743
9	t_{NH} (Noor-ul-Amin and Hanif (2013))	0.1491	404.2924
10	t_{p1}	**0.1326**	454.6003
12	t_{p2}	**0.2215**	272.1445
13	t_{p3}	**0.0903**	667.5526
14	t_{p4}	**0.0932**	646.7811
15	t_{p5}	**0.2352**	256.2925
16	t_{p6}	**0.5535**	108.907
17	t_{p7}	**0.3918**	153.854
18	t_{p8}	**0.3715**	162.2611
19	t_{p9}	**0.3918**	153.854
20	t_{p10}	**0.0925**	651.6757
21	**Proposed estimator at optimum value** $t_{p_{opt}}$	**0.0859**	**701.7462**

$N = 400$, $n = 20$, $\bar{Y} = 1.2225$ $\bar{X} = 0.5550$ $S_{wy} = 3.562$ $S_{wx} = 1.948$ $S_{wx.wy} = 6.428$ $\rho_{wx.wy} = 0.926$

$C_{wy} = 2.914$ $C_{wx} = 3.510$ $\beta_1(wx) = 7.953$ $\beta_2(wx) = 91.369$

6. Results Discussion from Table 2

The existing estimator t_{YMMC1} suggested by Yadav et al. (2016) has the worst performance because it has the highest value of MSE of 1.4117 and lowest value of PRE of 42.70029. The usual sample mean estimator, t_T in adaptive cluster sampling by Thompson (1990) gave the MSE and PRE of 0.6028 and 100 respectively. This estimator has the second highest value of MSE and second lowest PRE out of the estimators considered in this study. This estimator utilizes only the variable of interest but it still performs better than the existing estimator t_{YMMC1} by Yadav et al. (2016) which uses an auxiliary variable but with no significant improvement. The classical ratio estimator t_{DC} in adaptive cluster sampling by Dryver and Chao (2007), and other existing estimators t_{C1}, t_{C2} and t_{C3}, by Chutiman (2013); t_{YMMC2} and t_{YMMC2} by Yadav et al. (2016); and t_{NH} by Noor-ul-Amin and Hanif (2013) are all more efficient than the usual sample mean estimator t_T by Thompson (1990) and estimator t_{YMMC1} by Yadav et.al (2016) because they have a smaller MSE and a higher PRE.

All the members of the proposed class of exponential ratio type estimators t_{p1}, t_{p2}, t_{p3}, t_{p4}, t_{p5}, t_{p6}, t_{p7}, t_{p8}, t_{p9} and t_{p10} in ACS have significant improvements on the usual sample mean estimator in ACS because they have smaller MSEs and higher PREs. The proposed estimators t_{p3}, t_{p4} and t_{p10} have close a performance to the existing estimator, t_{YMMC2} by Yadav et al. (2016) because they have mutual values of MSEs and PREs. The proposed class of exponential ratio type estimator t_p, at δ optimum value is efficient as a regression estimator as indicated in (27) and it is also the most efficient estimator compared to the existing estimators considered in this study because it has the least MSE of 0.0859 and the highest PRE of 701.7462. This implies that the performance of this estimator at optimum value is seven times higher than the usual sample mean estimator t_T in ACS by Thompson (1990).

7. Conclusion

A class of exponential ratio type estimator in adaptive cluster sampling for finite population means using a single auxiliary variable with some known population parameters is developed and members of the proposed class of estimators are identified. The theoretical conditions under which the proposed estimators perform better than the existing estimators are obtained.

In conclusion, the proposed class of exponential ratio type estimator t_p, at δ optimum value performs better than the existing estimators by Thompson (1990), Chutiman (2013), Dryver and Chao (2007), Noor-ul-Amin and Hanif (2013) and Yadav et.al (2016). The proposed class of exponential ratio-type estimator in adaptive cluster sampling is recommended for practical application when the object of study is considered to be rare, clustered and hidden in the population.

References

Akingbade, T. J. and Okafor, F. C. 2019. A generalized class of difference-type ratio estimators for estimating the population mean using some known population parameters of the auxiliary variable. Pakistan Journal of Statistics, 35(3): 197–215.

Bahl, S. and Tuteja, R. K. 1991. Ratio and product type exponential estimator. Journal of Information and Optimization Sciences, 12(1): 159–163.

Chutiman, N. and Chutiman, B. 2008. Ratio estimator using two auxiliary variables for adaptive cluster sampling. Journal of Thai Statistical Association, 6(2): 241–256.

Chutiman, N. 2013. Adaptive Cluster Sampling using auxiliary variable. Journal of Mathematics and Statistics, 9(3): 249–255.

Cochran, W. G. 1940. The estimation of the yields of the cereal experiments by sampling for the ratio of grain to total produce. Journal of Agricultural Sciences, 59: 1225–1226.

Dryver, A. L. and Chao, C. T. 2007. Ratio estimators in adaptive cluster sampling. Environmetrics, 18(6): 607–620.

Hansen, M. M. and Hurwitz, W. N. 1943. On the theory of sampling from finite population. Annals of Mathematical Statistics, 1: 333–362.

Horvitz, D. G. and Thompson, D. J. 1952. A generalization of sampling without replacement from a finite universe. Journal of American Statistical Association, 47: 663–685.

Singh, H. P., Pal, S. K. and Solanki, R. S. 2017. A new class of estimators of finite population mean in sample surveys. Communications in Statistics-Theory and Methods, 46(6): 2630–2637.

Kadilar, C. and Cingi, H. 2003. A study on the chain ratio-type estimator. Hacettepe Journal of Mathematics and Statistics, 32(1): 105–108.

Noor-ul-Amin, M. and Hanif, M. 2013. Exponential Estimators in Survey Sampling. Unpublished Thesis. Submitted to National College of Business Administration and Economics, Lahore.

Sisodia, B. V. S. and Dwivedi, V. K. 1981. A modified ratio estimator using coefficient of variation of auxiliary variable. Journal of the Indian Society of Agricultural Statistics, 33(1): 13–18.

Singh, H. P. and Tailor, R. 2003. Use of known correlation coefficient in estimating the finite population mean, Statistics in Transition, 6: 555–560.

Thompson, S. K. 1990. Adaptive cluster sampling, Journal of the American Statistical Association, 85: 1050–1059.

Upadhyaya, L. N. and Singh, H. P. 1999. Use of transformed auxiliary variable in estimating the finite population mean. Biometrical Journal, 41(5): 627–636.

Yadav, K. S., Misra, S., Mishra, S. S. and Chutiman, N. 2016. Improved Ratio Estimators of Population Mean in Adaptive Cluster Sampling. Journal of Statistics & Probability Letters, 3(1): 1–6.

Yan, Z. and Tian, B. 2010. Ratio method to the mean estimation using coefficient of skewness of auxiliary variable. International Conference on Information Computing and Application 2010, Part II. Communications in Computer and Information Science, 106: 103–110.

Chapter 13

An Inventory Model for Substitutable Deteriorating Products under Fuzzy and Cloud Fuzzy Demand Rate

*Nita H Shah** and *Milan B Patel*

1. Introduction

Management of inventory has been a subject of interest for many researchers and decision makers since long as it directly affects the profit of an organisation. Many super markets and big retail stores face the situation of stock-out of some most consumed items specially on holidays due to a heavy rush of customers. It is commonly seen that in today's fast and busy life many customers choose the substitute product in place of the original one and avoid going to other stores. The study carried out by Anupindi et al. (1998) reveals that 82.88% of customers would switch over to buy the substitute product if the original product they wish to buy is not available instead of visiting different stores. This study is an attempt to model this phenomenon and offers a joint ordering inventory policy for deteriorating substitute products under crisp, fuzzy and cloud fuzzy environment.

Since Harris (1913) developed the first economic order quantity (EOQ) model, many researchers developed mathematical models by considering various parameters with different cases. Whitin (1957) developed the first inventory model which includes the effect of deterioration. A brief review on perishable inventory having stochastic and deterministic demand has been done by Nahmias (1982). A comprehensive survey on continuously deteriorating inventory has been done by Raafat (1991), Shah and Shah (2000) and, Goyal and Giri (2001). Shah (2006) has formulated an inventory model for deteriorating items for a finite time horizon with the condition of permissible delay in payments. Min et al. (2010) developed a lot-sizing model for deteriorating items when demand is dependent on current stock. Tayal et al. (2015) derived an inventory model for seasonal products having Weibull rate of deterioration. Jaggi et al. (2017) worked on an inventory model for deteriorating items by allowing imperfect quality items and one level of trade credit. Khanna et al. (2020) studied the effect of preservation technology in controlling the life span of deteriorating items having time-varying holding cost and stock dependent demand.

In general practices, there are very few cases when the demand remains constant with time. In most of the cases, it depends on many parameters. The stock of inventory displayed in large quantity has a significant effect on the demand and sale of the product. Balakrishnan et al. (2004) developed an optimal policy by considering the cases when inventories stimulate demand. Levin et al. (1972) suggested that inventory displayed in large piles will tempt the customer to buy the product. Mandal and Maiti (1999) developed a deterministic inventory model having stock-dependent demand. An inventory model with stock and time varying demand has been presented by Prasad and Mukherjee (2016). Mishra et al. (2017) investigated the inventory problem by considering price and stock dependent demand for a controllable deterioration rate. Bardhan et al. (2019) derived an optimal replenishment policy for the items with non-instantaneous deterioration rate with stock dependent demand.

McGillivray and Silver (1978) made the first attempt to formulate an inventory model for substitutable products by assuming that they all have the same unit variable cost.

An EOQ model was presented by Drenzner et al. (1995) by comparing substitutable and non-substitutable products. Maity and Maiti (2009) investigated inventory policies for deteriorating complementary and substitute items with stock-dependent demand. Salameh et al. (2014) developed an inventory model for two substitutable items with deterministic demand rate. Krommyda et al. (2015) worked out an inventory control problem to find the optimal order quantity of substitute products having stock dependent demand.

Department of Mathematics, Gujarat University, Ahmedabad, Gujarat 380009, India.
Email: milanmath314@gmail.com
* Corresponding author: nitahshah@gmail.com

In most of the inventory problems, crisp parameters are considered which may not represent the reality always since it can be calculated only up to some extent. Therefore, in the area of decision making where the intuition plays a vital role, fuzzy set theory introduced by Zadeh (1965) provides a more suitable approach to deal with the uncertainties.

Bellman and Zadeh (1970) first worked on decision making in a fuzzy environment. Since then many researchers devoted their work in the field of fuzzy decision making. A vast literature on fuzzy inventory modelling can be found from the works of Ouyang and Yao (2002), Dutta et al. (2007), Soni (2011), Shah and Soni (2012), Dash and Sahoo (2015) and their references.

After analysing the above-mentioned literature, it can be seen that most of the studies ignore the fuzziness associated with the role of the human factor in decision making. These literatures considered the degree of fuzziness as a crisp value instead of taking into account the fact that human decisions are subject to change with time. Bera et al. (2009) developed an inventory model under a finite time period with the consideration of learning effects. Glock et al. (2012) worked on an EOQ model having fuzzy demand under learning considerations. Owing to the fact that after a long-time fuzzy numbers converge to a singleton crisp set, a new concept of cloudy fuzzy numbers has been introduced by De and Mahata (2017). In their work, the defuzzification method used was developed by Yager (1981). Karmakar et al. (2018) apply the concept of cloudy fuzzy number to solve the EOQ model with cloudy fuzzy demand rate. Again, De and Mahata (2019) have developed an EOQ model for imperfect quality items by applying the concept of cloudy fuzzy numbers. Maity et al. (2019) studied backorder EOQ models for cloud type intuitionistic dense fuzzy demand rate. Giri et al. (2020) invented a joint ordering inventory policy for deteriorating substitute products by considering price and stock dependent demand. The present work is an attempt to extend the joint ordering inventory policy developed by Giri et al. (2020) under a fuzzy and cloud fuzzy environment. This chapter contributes to the area of fuzzy inventory modelling by assuming that after a long-time fuzzy parameters converge to the crisp one. Yager's ranking index method is used for defuzzification. Critical inventory parameters have been detected through sensitivity analysis under different environments. A comparative analysis among the results obtained in crisp, fuzzy and cloud fuzzy environments is presented graphically as well as with the help of numerical experiments.

2. Notations and Assumptions

The proposed model is formulated under following notations and assumptions.

2.1 Notations

Decision Variables

T	total cycle time (in years)
t_1	time at which one product is exhausted and the substitute product is available (in years)
s_i	sales price of ith product per unit (in \$)

Parameters

	$i = 1,2$
A	ordering cost for both products together per replenishment (in \$)
h_i	holding cost for i^{th} product per unit (in \$)
l	lost sale penalty cost for product P_1 per unit (in \$)
c_i	purchase cost of ith product per unit (in \$)
θ	deterioration rate for each product; $0 \leq \theta < 1$
r_i	scale demand for i^{th} product per year
$\bar{r}_i$	fuzzy scale demand for i^{th} product per year
$\tilde{r}_i$	cloud fuzzy scale demand for i^{th} product per year
H_1	amount of holding inventory of product P_1 during time interval $[0, T]$
H_2	amount of holding inventory of product P_2 during time interval $[0, t_1]$
H_3	amount of holding inventory of product P_2 during time interval $[t_1, T]$
$I_i(t)$	inventory level of ith product at time t
ϕ	a portion of substitution from product P_1 to product P_2
ψ	a portion of P_1 customers who leave the system
a_1	rate of linear increase in demand due to its own price; $0 \leq a_1 \leq 1$
a_2	rate of linear increase in demand due to price of substitute product; $0 \leq a_2 \leq 1$

b_1 rate of linear increase in demand due to its own stock; $0 \leq b_1 \leq 1$

b_2 rate of linear increase in demand due to stock of substitute product; $0 \leq b_2 \leq 1$

2.2 Assumptions

i) The proposed problem deals with substitute products only.
ii) Rate of deterioration of both products are considered to be equal and constant. Further, no replenishment or repair occurs during the planning horizon.
iii) Product P_1 is exhausted before product P_2 at time t_1 and a known portion of excess demand of product P_1 is fulfilled by product P_2.
iv) Shortages are allowed and backlogged partially.
v) Lead time is negligible.
vi) Planning horizon is considered to be infinite.

3. Preliminary Concepts

3.1 Triangular Fuzzy Number (TFN)

A triangular fuzzy number (TFN) defined on the set of real numbers $\mathbb{R}$ can be expressed as $\overline{R} = (r - \Delta_1, r, r + \Delta_2)$ and defined by its membership function as,

$$f(\overline{R}) = \begin{cases} \dfrac{x-(r-\Delta_1)}{r-(r-\Delta_1)}, (r-\Delta_1) \leq x \leq r \\ \dfrac{x-(r+\Delta_2)}{r-(r+\Delta_2)}, r \leq x \leq (r+\Delta_2) \\ 0, \quad otherwise \end{cases} \tag{1}$$

3.2 α-cut of TFN

α-cut of $\overline{R} = (r - \Delta_1, r, r + \Delta_2)$ is a crisp set $\alpha_{\overline{R}} = [L_\alpha, R_\alpha]$ where, $L_\alpha = r - \Delta_1(1-\alpha)$ is known as left α-cut and $R_\alpha = r + \Delta_2(1-\alpha)$ is known as right α-cut. $(0 \leq \alpha \leq 1)$

3.3 Cloud Triangular Fuzzy Number (CTFN)

A triangular fuzzy number is known as cloud triangular fuzzy number (CTFN) if the set converges to a crisp number as time tends to infinity.

$$\tilde{R} = \left(r\left(1 - \frac{\beta}{1+t}\right), r, r\left(1 + \frac{\gamma}{1+t}\right) \right) \tag{2}$$

where, $\beta, \gamma \in (0,1)$ and $t > 0$. From Equation (2) it can be seen that as $t \to \infty$, $\tilde{R} \to \{r\}$
Membership function of CTFN can be defined as follow:

$$g(\tilde{R},t) = \begin{cases} \dfrac{x - r\left(1-\dfrac{\beta}{1+t}\right)}{\dfrac{\beta r}{1+t}}, r\left(1-\dfrac{\beta}{1+t}\right) \leq x \leq r \\ \dfrac{r\left(1+\dfrac{\gamma}{1+t}\right) - x}{\dfrac{\gamma r}{1+t}}, r \leq x \leq r\left(1+\dfrac{\gamma}{1+t}\right) \\ 0 \quad , otherwise \end{cases} \tag{3}$$

3.4 Left and Right α-cut of CTFN

As per the definition of Left and Right α–cut of TFN (Section 3.2), they can be expressed as,

$$L_{\alpha,t} = r\left(1 - \frac{\beta}{1+t}\right) + \frac{\alpha\beta}{1+t} r \quad \& \quad R_{\alpha,t} = r\left(1 + \frac{\gamma}{1+t}\right) - \frac{\alpha\gamma}{1+t} r \tag{4}$$

respectively.

3.5 Yager's Ranking Index Method

According to Yager's Ranking Index method (1981), defuzzification for a TFN can be given by,

$$YRI(\bar{R}) = \frac{1}{2}\int_0^1 (L_\alpha + R_\alpha)\, d\alpha \tag{5}$$

where, L_α and R_α are left and right α–cut of TFN respectively. By substituting the value of L_α and R_α, Equation (5) reduces to

$$YRI(\bar{R}) = \frac{1}{4}(r - \Delta_1 + 2r + r + \Delta_2) \tag{6}$$

3.6 Yager's Ranking Index Method for CTFN

It is an extension of Yager's Ranking Index method for TFN given by De and Mahata (2017). By this method, defuzzification of CTFN can be given by,

$$YRI(\tilde{R}) = \frac{1}{2T}\int_{\alpha=0}^{\alpha=1}\int_{t=0}^{t=T} (L_{\alpha,t} + R_{\alpha,t})\, d\alpha\, dt \tag{7}$$

Substituting the value of left and right α–cut of CTFN from Equation (4), Equation (7) reduces to,

$$YRI(\tilde{R}) = r\left(1 - \frac{(\beta-\gamma)}{4}\frac{\log(1+T)}{T}\right) \tag{8}$$

4. Formulation of Crisp Model

Consider an organisation which sells two substitute products P_1 and P_2. At the beginning of each replenishment cycle, a joint order of both products is placed which depletes with time due to demand and the effect of deterioration.

When we go to any shopping mall, we get attracted to the product having more stock as compared to the one having lesser stock. Further, in the case of the substitute product, customer always compares the prices of substitute products. Therefore, along with the scale demand, demand rate of either products depends upon piece and stock of both products. During the recent COVID-19 pandemic, this type of demand pattern has been observed with many products such as bakery products, dairy products, and grocery items among others.

Demand functions of products P_1 and P_2 are given by,

$$R_1(s_1, s_2, I_1, I_2) = r_1 + b_1 I_1(t) - b_2 I_2(t) - a_1 s_1 + a_2 s_2 \tag{9}$$

$$R_2(s_1, s_2, I_1, I_2) = r_2 + b_1 I_2(t) - b_2 I_1(t) - a_1 s_2 + a_2 s_1 \tag{10}$$

respectively.

During time interval $[0, t_1]$,the change in inventory level of products P_1 and P_2 can be modelled by the following linear differential equations:

$$\frac{dI_1}{dt} = -R_1 - \theta I_1(t), \quad 0 \le t \le t_1 \tag{11}$$

$$\frac{dI_2}{dt} = -R_2 - \theta I_2(t), \quad 0 \le t \le t_1 \tag{12}$$

Assuming $(1-\psi)r_1$ to be demand of product P_1 and $(r_2 + \phi r_1)$ for product P_2 during the stockout period $[t_1, T]$, the change in inventory level of either product can be explained by the following linear differential equations:

$$\frac{dI_1}{dt} = (1-\psi)r_1, \quad I_1(t_1) = 0, \quad t_1 \le t \le T \tag{13}$$

$$\frac{dI_2}{dt} = -(r_2 + \phi r_1) - \theta I_2(t), \quad I_2(T) = 0, \quad t_1 \le t \le T \tag{14}$$

By solving Equations (13) and (14), the inventory level of products P_1 and P_2 at any time t during $[t_1, T]$ can be obtained from the following equations:

$$I_1(t) = (1-\psi) r_1 (t - t_1) \tag{15}$$

$$I_2(t) = \left(\frac{r_2 + \phi r_1}{\theta}\right)(e^{\theta(T-t)} - 1) \tag{16}$$

respectively.

By putting $t = T$ in Equation (15), the amount of shortage of the product P_1 is given by,

$$L = (1-\psi) r_1 (T - t_1) \tag{17}$$

When product P_1 is exhausted, the inventory level of product P_2 can be obtained by putting $t = t_1$ in Equation (16).

$$I_2(t_1) = \left(\frac{r_2 + \phi r_1}{\theta}\right)(e^{\theta(T-t_1)} - 1) \tag{18}$$

Now by solving Equations (11) and (12) with the help of boundary conditions $I_1(t_1) = 0$ and $I_2(t_1)$ as given in Equation (18), we get the inventory level of products P_1 and P_2 at any time t as per the following equations:

$$I_1(t) = B_1 e^{-(\theta + b_1 + b_2)(t - t_1)} + B_2 e^{-(\theta + b_1 - b_2)(t - t_1)} - B_3 \tag{19}$$

$$I_2(t) = -B_1 e^{-\theta(t - t_1)} + B_2 e^{-\theta(t - t_1)} - B_4 \tag{20}$$

Respectively, where,

$$B_1 = \frac{(r_1 - r_2) - (a_1 + a_2)(s_1 - s_2) - (\theta + b_1 + b_2) I_2(t_1)}{2(\theta + b_1 + b_2)}$$

$$B_2 = \frac{(r_1 + r_2) - (a_1 - a_2)(s_1 + s_2) + (\theta + b_1 - b_2) I_2(t_1)}{2(\theta + b_1 - b_2)}$$

$$B_3 = \frac{(\theta + b_1)(r_1 - a_1 s_1 + a_2 s_2) + b_2 (r_2 + a_2 s_1 - a_1 s_2)}{(\theta + b_1 + b_2)(\theta + b_1 - b_2)}$$

$$B_4 = \frac{(\theta + b_1)(r_2 + a_2 s_1 - a_1 s_2) + b_2 (r_1 - a_1 s_1 + a_2 s_2)}{(\theta + b_1 + b_2)(\theta + b_1 - b_2)}$$

Following are the various costs associated with the total profit of the inventory:

i) Total Sales Revenue;

$$TSR = s_1 \left[\int_0^{t_1} R_1 dt + \int_{t_1}^{T} (1-\psi) r_1 dt\right] + s_2 \left[\int_0^{t_1} R_2 dt + \int_{t_1}^{T} (r_2 + \phi r_1) dt\right]$$

$$= s_1 \left[(r_1 - a_1 s_1 + a_2 s_2) t_1 + b_1 H_1 - b_2 H_2 + (1-\psi) r_1 (T - t_1)\right]$$
$$+ s_2 \left[(r_2 + a_2 s_1 - a_1 s_2) t_1 + (r_2 + \phi r_1)(T - t_1)\right]$$

ii) Holding Cost;

$$HC = h_1 H_1 + h_2 (H_2 + H_3)$$

$$= h_1 \left(\int_0^{t_1} I_1(t) dt\right) + h_2 \left(\int_0^{t_1} I_2(t) dt + \int_{t_1}^{T} I_2(t) dt\right)$$

$$= B_1(h_1-h_2)\left(\frac{e^{(\theta+b_1+b_2)t_1}-1}{\theta+b_1+b_2}\right)+B_2(h_1+h_2)\left(\frac{e^{(\theta+b_1-b_2)t_1}-1}{\theta+b_1-b_2}\right)$$
$$-t_1(h_1B_3+h_2B_4)+\frac{h_2(r_2+\phi r_1)}{\theta^2}\left(e^{\theta(T-t_1)}-\theta(T-t_1)-1\right)$$

iii) Lost Sales Cost for product P_1 during $[0,t_1]$;

$$LSC = l\int_{t_1}^{T} I_1(t)dt = \frac{(1-\psi)r_1(T-t_1)^2 l}{2}$$

iv) Purchase Cost;

$$PC = c_1\left(I_1(0)+L\right)+c_2\left(I_2(0)\right)$$
$$= B_1(c_1-c_2)(e^{(\theta+b1+b2)t_1})+B_2(c_1+c_2)(e^{(\theta+b_1-b_2)t_1})$$
$$+c_1\left((1-\psi)r_1(T-t_1)-B_3\right)-c_2B_4$$

v) Ordering Cost; $OC = A$

Total Crisp Profit per replenishment cycle is given by,

$$\Pr(s_1,s_2,t_1,T)=\frac{1}{T}\left(TSR-HC-LSC-PC-OC\right) \tag{21}$$

4.1 Solution Algorithm

Classical optimization technique is used to solve the objective function in order to get the values of decision variables and through them the maximum joint profit. Following are the steps for the solution of the crisp model.

Step 1: Assign the values of inventory parameters except decision variables in their respective units.

Step 2: Differentiate the objective function (Equation (21)) partially with respect to the decision variables and equate them to zero.

$$\frac{\partial \Pr}{\partial s_1}=0,\ \frac{\partial \Pr}{\partial s_2}=0,\ \frac{\partial \Pr}{\partial t_1}=0\ \&\ \frac{\partial \Pr}{\partial T}=0 \tag{22}$$

Derive the values of decision variables (i.e., s_1, s_2, t_1, T) from Equation (22) and put in the objective function (Equation (21)) to get the maximum value of profit.

Step 3: Verify the concavity of the objective function by the method of Hessian Matrix.

$$\text{Hessian matrix can be defined as, } H=\begin{pmatrix}\frac{\partial^2 \Pr}{\partial s_1^2} & \frac{\partial^2 \Pr}{\partial s_1\partial s_2} & \frac{\partial^2 \Pr}{\partial s_1\partial t_1} & \frac{\partial^2 \Pr}{\partial s_1\partial T}\\ \frac{\partial^2 \Pr}{\partial s_2\partial s_1} & \frac{\partial^2 \Pr}{\partial s_2^2} & \frac{\partial^2 \Pr}{\partial s_2\partial t_1} & \frac{\partial^2 \Pr}{\partial s_2\partial T}\\ \frac{\partial^2 \Pr}{\partial t_1\partial s_1} & \frac{\partial^2 \Pr}{\partial t_1\partial s_2} & \frac{\partial^2 \Pr}{\partial t_1^2} & \frac{\partial^2 \Pr}{\partial t_1\partial T}\\ \frac{\partial^2 \Pr}{\partial T\partial s_1} & \frac{\partial^2 \Pr}{\partial T\partial s_2} & \frac{\partial^2 \Pr}{\partial T\partial t_1} & \frac{\partial^2 \Pr}{\partial T^2}\end{pmatrix}$$

For the objective function to be concave, Hessian matrix must be negative definite.
i.e., $D_1<0,\ D_2>0,\ D_3<0,\ D_4>0$, where, D_1, D_2, D_3, D_4 are principal minors of the Hessian matrix H defined as follows:

$$D_1=\frac{\partial^2 \Pr}{\partial s_1^2},\quad D_2=\begin{vmatrix}\frac{\partial^2 \Pr}{\partial s_1^2} & \frac{\partial^2 \Pr}{\partial s_1\partial s_2}\\ \frac{\partial^2 \Pr}{\partial s_2\partial s_1} & \frac{\partial^2 \Pr}{\partial s_2^2}\end{vmatrix},\quad D_3=\begin{vmatrix}\frac{\partial^2 \Pr}{\partial s_1^2} & \frac{\partial^2 \Pr}{\partial s_1\partial s_2} & \frac{\partial^2 \Pr}{\partial s_1\partial t_1}\\ \frac{\partial^2 \Pr}{\partial s_2\partial s_1} & \frac{\partial^2 \Pr}{\partial s_2^2} & \frac{\partial^2 \Pr}{\partial s_2\partial t_1}\\ \frac{\partial^2 \Pr}{\partial t_1\partial s_1} & \frac{\partial^2 \Pr}{\partial t_1\partial s_2} & \frac{\partial^2 \Pr}{\partial t_1^2}\end{vmatrix}\ \text{and } D_4=\det(H)$$

5. Formulation of Fuzzy and Cloud Fuzzy Model

While formulating the crisp mathematical model, we consider scale demand as a fixed quantity. But this phenomenon rarely takes place in practice. Hence to formulate the fuzzy model we consider scale demand of both products to be a triangular fuzzy number $\bar{r}_1 = (r_1 - \Delta_1, r_1, r_1 + \Delta_2)$ and $\bar{r}_2 = (r_2 - \Delta_3, r_2, r_2 + \Delta_4)$ for products P_1 and P_2 respectively. To formulate cloud fuzzy model, scale demand for both products are considered to be cloud triangular fuzzy numbers.

$\tilde{r}_1 = \left(r_1\left(1-\frac{\beta}{1+T}\right), r_1, r_1\left(1+\frac{\gamma}{1+T}\right)\right)$ and $\tilde{r}_2 = \left(r_2\left(1-\frac{\beta}{1+T}\right), r_2, r_2\left(1+\frac{\gamma}{1+T}\right)\right)$ for products P_1 and P_2 respectively.

Therefore, demand functions for products P_1 and P_2 in the fuzzy model reduces to;

$$\overline{R}_1(s_1, s_2, I_1, I_2) = \bar{r}_1 + b_1 I_1(t) - b_2 I_2(t) - a_1 s_1 + a_2 s_2 \tag{23}$$

$$\overline{R}_2(s_1, s_2, I_1, I_2) = \bar{r}_2 + b_1 I_2(t) - b_2 I_1(t) - a_1 s_2 + a_2 s_1 \tag{24}$$

respectively.

For the cloud fuzzy model demand functions for products P_1 and P_2 are;

$$\widetilde{R}_1(s_1, s_2, I_1, I_2) = \tilde{r}_1 + b_1 I_1(t) - b_2 I_2(t) - a_1 s_1 + a_2 s_2 \tag{25}$$

$$\widetilde{R}_2(s_1, s_2, I_1, I_2) = \tilde{r}_2 + b_1 I_2(t) - b_2 I_1(t) - a_1 s_2 + a_2 s_1 \tag{26}$$

Proceeding similarly as in crisp mathematical modelling (Section 4), the fuzzy and cloud fuzzy models can be derived. The defuzzified value of the fuzzy objective function can be expressed as follow:

$$\overline{\Pr}(s_1, s_2, t_1, T) = \frac{1}{4T}\left(\Pr_1 + 2\Pr_2 + \Pr_3\right) \tag{27}$$

where, $\Pr_1$ can be obtained by replacing $\bar{r}_1$ and $\bar{r}_2$ with $(r_1 - \Delta_1)$ and $(r_2 - \Delta_3)$ respectively, $\Pr_2$ can be obtained by replacing $\bar{r}_i$ with r_i and $\Pr_3$ can be obtained by replacing $\bar{r}_1$ and $\bar{r}_2$ with $(r_1 - \Delta_2)$ and $(r_2 - \Delta_4)$ in fuzzy demand functions (Equations 23 and 24). Similarly, defuzzified value of cloud fuzzy objective function can be written as follow:

$$\widetilde{\Pr}(s_1, s_2, t_1, T) = \frac{1}{4T}\int_0^T (\Pr_1 + 2\Pr_2 + \Pr_3)\,dt \tag{28}$$

where, $\Pr_1$ can be obtained by replacing $\tilde{r}_i$ with $r_i\left(1-\frac{\beta}{1+T}\right)$, $\Pr_2$ can be obtained by replacing $\tilde{r}_i$ with r_i and $\Pr_3$ can be obtained by replacing $\tilde{r}_i$ with $r_i\left(1+\frac{\gamma}{1+T}\right)$ in cloud fuzzy demand functions (Equations 25 and 26). Membership function for fuzzy objective function can be defined as follow:

$$\xi_1(\Pr) = \begin{cases} \dfrac{\Pr - \Pr_1}{\Pr_2 - \Pr_1}, & \Pr_1 \le \Pr \le \Pr_2 \\ \dfrac{\Pr - \Pr_3}{\Pr_2 - \Pr_3}, & \Pr_2 \le \Pr \le \Pr_3 \\ 0, & \Pr < \Pr_1 \,\&\, \Pr > \Pr_3 \end{cases} \tag{29}$$

Membership function for cloud fuzzy objective function can be defined as follow:

$$\xi_2(\Pr, t) = \begin{cases} \dfrac{\Pr - \Pr_1}{\Pr_2 - \Pr_1}, & \Pr_1 \le \Pr \le \Pr_2 \\ \dfrac{\Pr - \Pr_3}{\Pr_2 - \Pr_3}, & \Pr_2 \le \Pr \le \Pr_3 \\ 0, & \Pr < \Pr_1 \,\&\, \Pr > \Pr_3 \end{cases} \tag{30}$$

Defuzzified value of fuzzy and cloud fuzzy objective function can be calculated using the methods explained in sections 3.5 and section 3.6 respectively.

6. Numerical Experiment with Proof of Concavity

For the crisp model let

$\theta = 0.05,\ r_1 = 50,\ r_2 = 40,\ a_1 = 0.3,\ a_2 = 0.2,\ b_1 = 0.27,\ b_2 = 0.35,\ A = 100,$
$s_1 = 5,\ s_2 = 5.5,\ h_1 = 3,\ h_2 = 1.6,\ \phi = 0.3,\ \psi = 0.33,\ l = 25$. For the fuzzy model consider $\Delta_1 = 20, \Delta_2 = 40, \Delta_3 = 30, \Delta_4 = 20$ and keep other parameters as in the crisp model. For cloud fuzzy model consider $\beta = 0.10,\ \gamma = 0.70$.

By using the algorithm described in Section 4.1, optimal values of decision variables are calculated under crisp, fuzzy and cloud fuzzy environments and shown in Table 1. Concavity of the objective function is verified using the method of Hessian matrix as explained in Step 3 (Section 4.1). Hessian matrix for the objective function is worked out and the values of D_1, D_2, D_3, D_4 are calculated. We obtained $D_1 = -1.66 < 0,\ D_2 = 1.54 > 0,\ D_3 = -2.17\times10^7 < 0,\ D_4 = 1.30\times10^{11} > 0$ which ensures concavity of the objective function. Concavity of the objective functions of different environments is also shown graphically in Figure 1. Table 2 is presented to show the values of decision variables under cloud fuzzy environment over several cycle times. Further, for better understanding of the difference among the results in crisp, fuzzy and cloud fuzzy environments, Figure 2 shows the value of total profit over different cycle time under different environments.

Table 1: Optimal solutions under different environments.

Environment	Sales Price of Product 1 (S_1) (in \$)	Sales Price of Product 2 (S_2) (in \$)	Critical Time (t_1) (in year)	Cycle Time (T) (in year)	Total Joint Profit (PR) (in \$)
Crisp	320.83	323.87	0.30	0.43	13625.95
Fuzzy	334.37	327.98	0.28	0.41	14415.97
Cloud Fuzzy	358.08	360.72	0.39	0.56	16518.79

Table 2: Values of decision variables under cloud fuzzy environment over several cycle time.

Cycle Time (T) (in year)	Sales Price of Product 1 (S_1) (in \$)	Sales Price of Product 2 (S_2) (in \$)	Critical Time (t_1) (in year)	Total Joint Profit (PR) (in \$)
0.53	358.73	361.17	0.371	16516.31
0.54	358.55	361.05	0.378	16517.52
0.55	358.37	360.92	0.385	16518.31
0.56	358.19	360.80	0.392	**16518.72**
0.57	358.01	360.68	0.399	16518.76
0.58	357.83	360.55	0.406	16518.46
0.59	357.66	360.43	0.413	16517.83

N.B.: Bold font in Table 2 represents the optimal solution under cloud fuzzy environment.

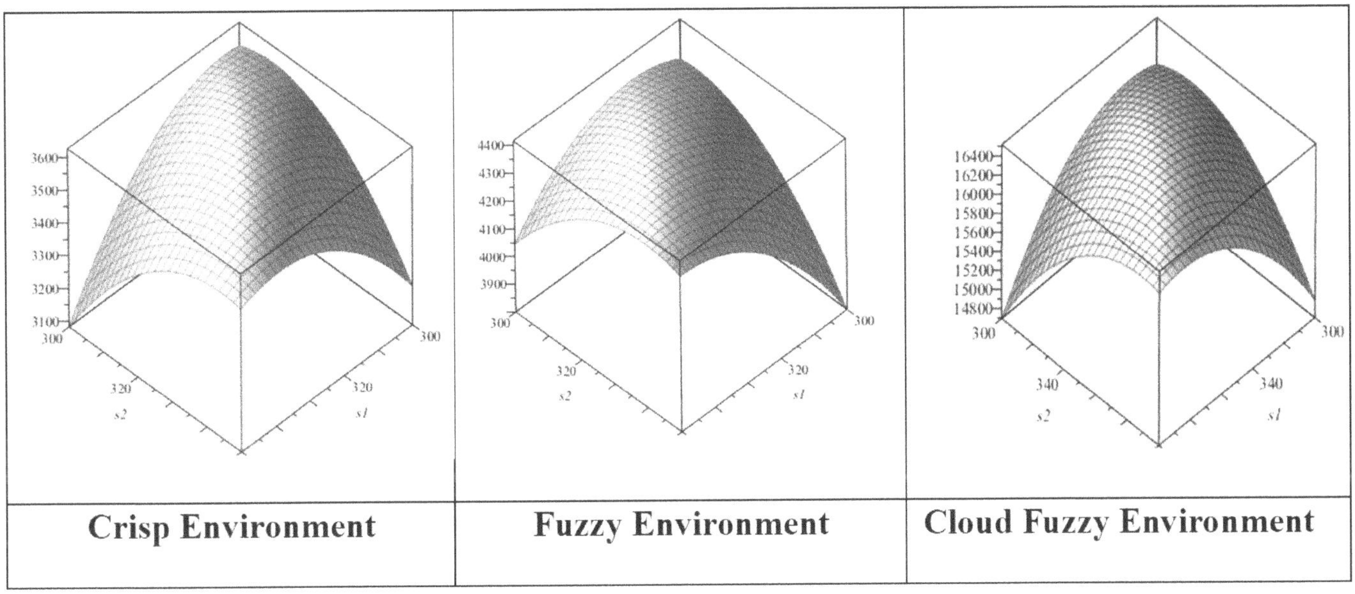

Figure 1: Concavity of objective functions under different environments.

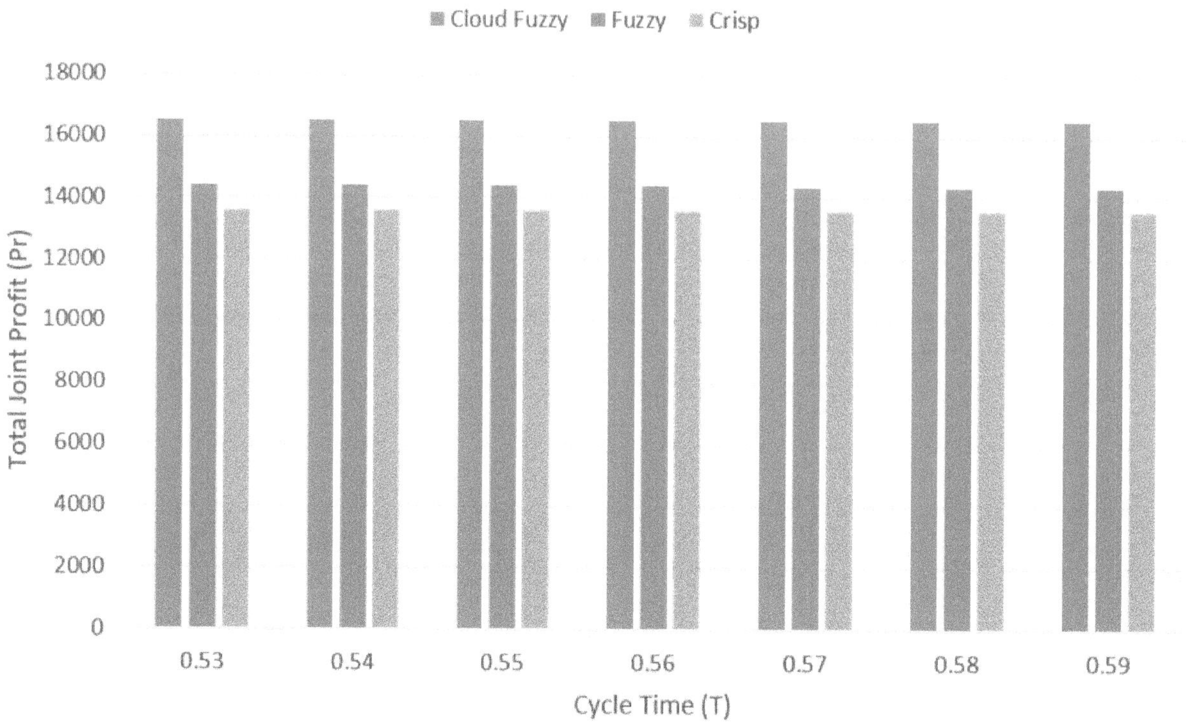

Figure 2: Value of Total Profit under different environments over several cycle time.

7. Sensitivity Analysis and Managerial Insights

To identify the most critical parameters of the model, sensitivity analysis on total join profit is performed by changing one inventory parameter from –20% to 20% and keeping other parameters as it is. This analysis is carried out under crisp, fuzzy and cloud fuzzy environments and presented graphically in Figure 3, Figure 4 and Figure 5 respectively.

It is observed from the sensitivity analysis under crisp environment that ordering cost and purchase cost of both the products shows the maximum sensitivity against change in cycle time. Therefore, the manager should carefully choose ordering cost and purchase cost so that the profit will not be affected. The rate of deterioration has lesser sensitivity as compare to ordering cost and purchase cost of both products and parameters such as penalty cost, and holding cost of both products have negligible sensitivity as

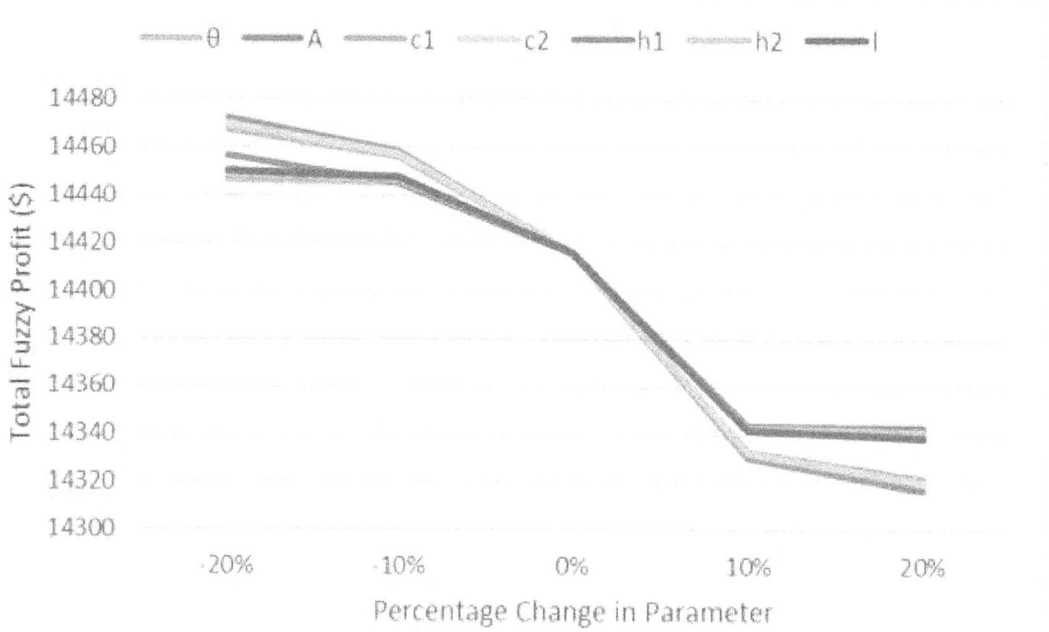

Figure 3: Sensitivity analysis under crisp environment.

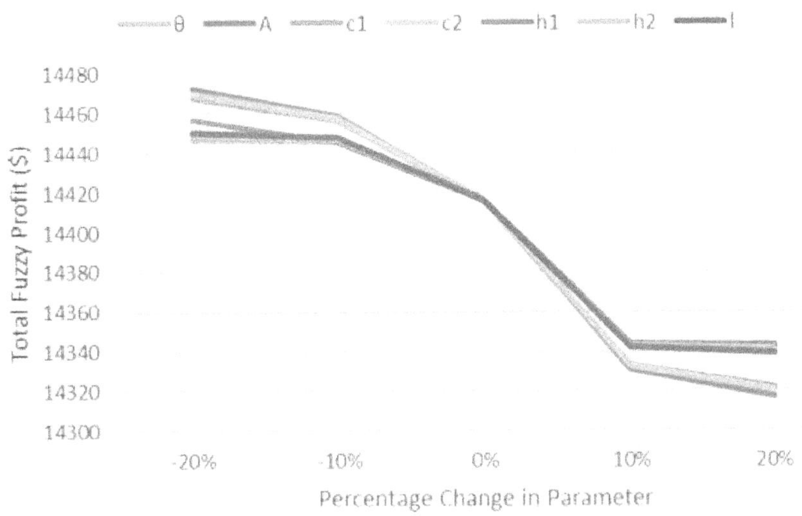

Figure 4: Sensitivity analysis under fuzzy environment.

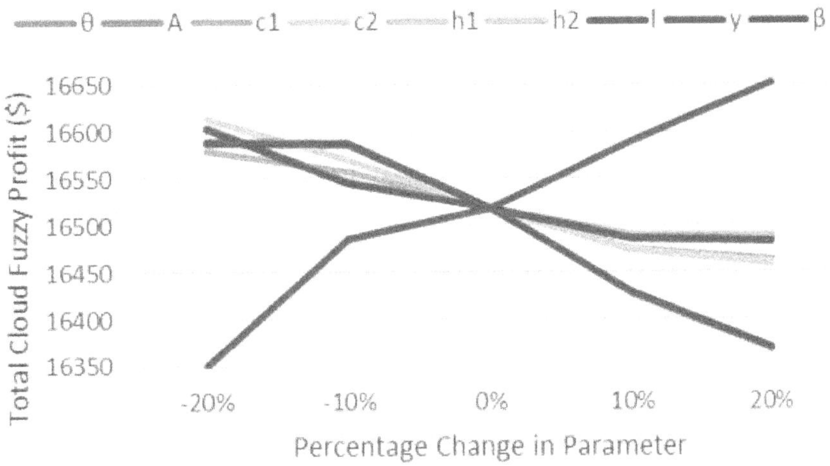

Figure 5: Sensitivity analysis under cloud fuzzy environment.

compare to other inventory parameters. It can be concluded that if possible, the retailer should go for non-deteriorating items instead of deteriorating ones. Further, the same behaviours in the change of inventory parameters are observed under fuzzy environment, i.e., there is no effect of environment on the sensitivity of parameters. Under cloud fuzzy environment the parameter γ is the most sensitive parameter. Other parameters behave similarly as in crisp and fuzzy environments. The present study advised the manager to place a joint order of both the products which can reduce cost components such as transportation cost. Decision makers (DM) are advised to choose cloud fuzzy environments as they gain more and more experience at the end of each cycle. DM should choose the solution where there is less fuzziness.

8. Conclusion

This study extended the inventory model for deteriorating substitute products under fuzzy and cloud fuzzy environment. The analysis from the proposed model suggested that the cloud fuzzy is the more realistic approach for such a problem as compared to crisp and fuzzy environments. The cloud fuzzy demand rate was a realistic assumption made for this model which helps decision makers to make precise predictions. Extension of Yager's ranking index method was used for defuzzification under cloud fuzzy environment. Sensitivity analysis and numerical experiment were carried out and results were presented graphically for better visualisation of the difference among the different environments. This model can be extended by considering uncertainty in more than one parameter such as holding cost, ordering cost, and deterioration rate among others. It would be an interesting extension to study the effect of uncertainty of various parameters on the total profit of the inventory.

Acknowledgement

The authors thank DST-FIST File # MSI-097 for technical support to the Department of Mathematics. The authors thank DST-FIST File # MSI-097 for technical support to Department of Mathematics.

References

Anupindi, R., Dada, M. and Gupta, S. 1998. Estimation of consumer demand with stock out based substitution: an application to vending machine products. Mark. Sci., 17: 406–423.

Balakrishnan, A., Pangburn, M. S. and Stavrulaki, E. 2004. "Stack them high, let'em fly": Lot-sizing policies when inventories stimulate demand. Manage. Sci., 50: 630–644.

Bardhan, S., Pal, H. and Giri, B. C. 2019. Optimal replenishment policy and preservation technology investment for a non-instantaneous deteriorating item with stock-dependent demand. Oper. Res., 19: 347–368.

Bellman, R. E. and Zadeh, L. A. 1970. Decision-making in a fuzzy environment. Manage. Sci., 17: B141–B164.

Bera, U. K., Mahapatra, N. K. and Maiti, M. 2009. An imperfect fuzzy production-inventory model over a finite time horizon under the effect of learning. Int. J. Mathe. Oper. Res., 1: 351–371.

Dash, J. K. and Sahoo, A. 2015. Optimal solution for a single period inventory model with fuzzy cost and demand as a fuzzy random variable. J. Inte. and Fuzzy Sys., 28: 1195–1203.

De, S. K. and Mahata, G. C. 2017. Decision of a fuzzy inventory with fuzzy backorder model under cloudy fuzzy demand rate. Int. J. Appl. And Comp. Mathe., 3: 2593–2609.

De, S. K. and Mahata, G. C. 2019. A cloudy fuzzy economic order quantity model for imperfect-quality items with allowable proportionate discounts. J. Indus. Eng. Int., 15: 571–583.

Drezner, Z., Gurnani, H. and Pasternack, B. A. 1995. An EOQ model with substitutions between products. J. Oper. Res. Soc., 46: 887–891.

Dutta, P., Chakraborty, D. and Roy, A. R. 2007. Continuous review inventory model in mixed fuzzy and stochastic environment. Appl. Mathe. and Comp., 188: 970–980.

Giri, R. N., Mondal, S. K. and Maiti, M. 2020. Joint ordering inventory policy for deteriorating substitute products with prices and stocks dependent demand. Int. J. Indus. Eng., 27: 58–71.

Glock, C. H., Schwindl, K. and Jaber, M. Y. 2012. An EOQ model with fuzzy demand and learning in fuzziness. Int. J. Serv. and Oper. Manag., 12: 90–100.

Goyal, S. K. and Giri, B. C. 2001. Recent trends in modeling of deteriorating inventory. Eur. J. Oper. Res., 134: 1–16.

Harris, F. W. 1913. How many parts to make at once. Factory. The Magazine of Management, 10: 135–136.

Jaggi, C. K., Cárdenas-Barrón, L. E., Tiwari, S. and Shafi, A. 2017. Two-warehouse inventory model for deteriorating items with imperfect quality under the conditions of permissible delay in payments. Sci. Iran., 24: 390–412.

Karmakar, S., De, S. K. and Goswami, A. 2018. A study of an EOQ model under cloudy fuzzy demand rate. Commu. in Comp. and Infor. Sci., 834: 149–163.

Khanna, A., Pritam, P. and Jaggi, C. K. 2020. Optimizing preservation strategies for deteriorating items with time-varying holding cost and stock-dependent demand. Yugos. J. Ope. Res., 30: 237–250.

Krommyda, I. P., Skouri, K. and Konstantaras, I. 2015. Optimal ordering quantities for substitutable products with stock-dependent demand. Appl. Mathe. Mode., 39: 147–164.

Levin, R. I., McLaughlin, C. P., Lamone, R. P. and Kottas, J. F. 1972. Production operations management: contemporary policy for managing operating system. McGraw-Hill, New York.

Maity, K. and Maiti, M. 2009. Optimal inventory policies for deteriorating complementary and substitute items. Int. J. Sys. Sci., 40: 267–276.

Maity, S., De, S. K. and Mondal, S. P. 2019. A study of an EOQ model under lock fuzzy environment. Mathe., 7: 75–98.

Mandal, M. and Maiti, M. 1999. Inventory of damageable items with variable replenishment rate, stock-dependent demand and some units in hand. Appl. Mathe. Mode., 23: 799–807.

Mcgillivray, R. and Silver, E. 1978. Some concepts for inventory control under substitutable demand. Info. Sys. and Oper. Res., 16: 47–63.

Min, J., Zhou, Y. W. and Zhao, J. 2010. An inventory model for deteriorating items under stock-dependent demand and two-level trade credit. Appl. Mathe. Mode., 34: 3273–3285.

Mishra, U., Cárdenas-Barrón, L. E., Tiwari, S., Shaikh, A. A. and Treviño-Garza, G. 2017. An inventory model under price and stock dependent demand for controllable deterioration rate with shortages and preservation technology investment. Ann. Oper. Res., 254: 165–190.

Nahmias, S. 1982. Perishable inventory theory: A review. Oper. Res., 30: 680–708.

Ouyang, L. Y. and Yao, J. S. 2002. A minimax distribution free procedure for mixed inventory model involving variable lead time with fuzzy demand. Comp. and Oper. Res., 29: 471–487.

Prasad, K. and Mukherjee, B. 2016. Optimal inventory model under stock and time dependent demand for time varying deterioration rate with shortages. Ann. Oper. Res., 243: 323–334.

Raafat, F. 1991. Survey of literature on continuously deteriorating inventory models. J. Oper. Res. Soci., 42: 27–37.

Salameh, M. K., Yassine, A. A., Maddah, B. and Ghaddar, L. 2014. Joint replenishment model with substitution. Appl. Mathe. Model., 38: 3662–3671.

Shah, Nita, H. 2006. Inventory model for deteriorating items and time value of money for a finite time horizon under the permissible delay in payments. Int. J. Sys. Sci., 37: 9–15.

Shah, N. H. and Soni, H. N. 2012. Continuous review inventory model with fuzzy stochastic demand and variable lead time. Int. J. Appl. Indus. Eng., 1: 7–24.

Shah, N. H. and Shah, Y. K. 2000. Literature survey on inventory models for deteriorating items. Eco. Ann., 44: 221–237.

Soni, H. 2011. (Q, R) inventory model with service level constraint and variable lead time in fuzzy-stochastic environment. Int. J. Indus. Engi. Compu., 2: 901–912.

Tayal, S., Singh, S. and Sharma, R. 2015. An inventory model for deteriorating items with seasonal products and an option of an alternative market. Uncer. Sup. Cha. Manag., 3: 69–86.

Whitin, T. M. 1957. Theory of inventory Management, Princeton University Press, Princeton, NJ, 62–77.

Yager, R. R. 1981. A procedure for ordering fuzzy subsets of the unit interval. Info. Sci., 24: 143–161.

Zadeh, L. A. 1965. Fuzzy sets. Info. and Cont., 8: 338–353.

Chapter 14

Co-ordinated Selling Price and Replenishment Policies for Duopoly Retailers under Quadratic Demand and Deteriorating Nature of Items

Nita H Shah[1,*] and *Monika K Naik*[2]

1. Introduction

As such, deterioration is a process of becoming gradually worse, to a state of damage, decay, spoilage, and loss of utility resulting in a declination of the usefulness of the original one. Deterioration reduces the quality and physical quantity of inventory. The product's selling price is one of the key factors in uplifting a customer's demand for a product which is directly influenced by the customer's satisfaction level. Recent studies include selling price-based rate of market demand and much weightage on deterioration is drawn to highlight the shorter life cycles of goods. Therefore, appropriate inventory control of deterioration of items is considered to be a crucial matter to elaborate.

Moreover, the retailer's influence has been raised intensely by development of huge supermarkets and chain-stores. In the retailer's sell-list, a product diversity has been observed including the branded items. The comparable products with diverse brands sold by various other retailers are compatible items for customers. For example, the two big online retailers, Grofers and bigbasket, selling perishable items such as vegetables and fruits. The main factor in dominating the share market is the selling price of these two competing retailers. As such selling price strategies are one of the main components influencing inventory management. So, a pricing and inventory policy is needed for consideration of these retailers. This article provides us a better approach to reduce the wastage occurred by the spoilage of deteriorating items by analysing and considering the problem of pricing and inventory decisions in a duopoly setting.

As such the key subject to this work is related to pricing-inventory decisions, and among them mostly the research is about a duopoly environment. Cohen (1977) initially introduced the strategy of joint pricing and inventory decisions for deteriorating items with an assumption of exponential decay and in cases of no shortages and complete backordering by analysing the optimal pricing and ordering policy. Thereafter, a generalized study of Cohen (1977) was considered by Rajan et al. (1992) where the concept of dynamic pricing was encountered. Then, Rajan et al.'s (1992) model was extended by Abad (1996) including partial backlogging by estimating the optimal policy for joint pricing.

There are few papers in which the study of a joint pricing-inventory problem in a duopoly environment is done. Zhu and Thonemann (2009) considered joint pricing and inventory control problem for a retailer who sells two non-deteriorating products with cross-price effects on demand and concluded that the retailer can significantly improve its profit by managing the two products jointly as opposed to independently, especially when the cross-price demand elasticity is high. An analysis of the joint dynamic pricing and inventory control problem of non-deteriorating items in a duopoly setting was done by Adida and Perakis (2010) and showed the existence of a Nash equilibrium in continuous time.

Two competing retailers with different cost structures selling non-deteriorating products, and addressing the problem of dynamic quantity competition was considered by Transchel and Minner (2011). An analysis is done by Chen et al. (2014) on the joint pricing and inventory control problem for perishable products with a fixed lifetime under an assumption that the inventories can be intentionally disposed of before they reach their maximum shelf life. A joint pricing, delivery study was proposed by Chen and Shi (2017) where in an inventory problem the customers can strategize their purchasing times.

[1] Department of Mathematics, Gujarat University, Ahmedabad-380009, Gujarat, India.
[2] Vadodara-390010, Gujarat, India.
Email: monikaknaik@gmail.com
* Corresponding author: nitahshah@gmail.com

Many studies are carried out based on demand rate of inventory models. The time-based demand rate was initially introduced by Silver et al. (1969). After that various studies were carried out in the direction of variety of demand structures by Silver (1979), Chung et al. (1993), Bose et al. (1995), Shah et al. (2009), who have described market demand rate as time based of linear, exponential or quadratic nature.

Datta and Pal (2001) proposed an inventory model with stock dependent and price-based demand rate. Further, joint pricing inventory model for deteriorating items with quantity discounts and time-dependent partial backlogging by the application of Weibull distribution was derived by Papachristos and Skouri (2003). Singh (2006) studied an inventory model for deteriorating items with price-based demand. Chang et al. (2010) proposed an optimal replenishment policy for non-instantaneous deteriorating items with stock dependent demand. Under a progressive payment scheme, for price-stock based demand for obtaining an optimal ordering and pricing policy a model was proposed by Shah et al. (2011).

Hou et al. (2011) derived an inventory model for perishable products under partial backlogging, inflation having stock-based selling rate. A two warehouses inventory model for non-instantaneous perishable products for stock dependent demand was proposed by Singh and Malik (2011).

On considering a joint pricing policy, partial backlogging and time-price based market demand rate for a non-instantaneous deteriorating item was proposed by Maihami and Kamalabadi (2012). Panda et al. (2013) explained an inventory model for deteriorating products by utilizing ramp-type demand. An inventory model by Qin et al. (2014) estimates an optimal solution by formulating an algorithm for calculating selling price, where deterioration is considered as quality and physical quantity dependent.

An inventory model by Rabbani et al. (2016) was proposed dealing with immediate deterioration for quality and delayed deterioration for physical quantity for estimating optimal dynamic pricing and replenishment policies. An inventory model under price and stock dependent demand for controllable deterioration rate with shortages and preservation technology investment was represented by Mishra et. al (2017). An integrated inventory model for deteriorating items with price-dependent demand under two-level trade credit policy was derived by Rameswari and Uthayakumar (2018). An inventory control problem by considering Joint pricing of duopoly retailers with deteriorating items and linear demand was considered by Mahmoodi (2019). An economic order quantity model under two-level partial trade credit for time varying deteriorating items was described by Mahata et al. (2020).

The uniqueness of this article is that it consists of time-price difference and stock dependent quadratic demand rate for deteriorating items for duopoly retailers. The computation of selling price and replenishment cycle lengths simultaneously, for both the retailers are done to maximize the joint total profit of the inventory system. Finally, the model is validated through hypothetical data.

The article contains five sections. In Section 2, notations and assumptions are explained. Section 3 deals with the development of a mathematical model. Numerical examples and sensitivity analysis are given in Section 4. Section 5 consists of the conclusion.

2. Notations and Assumptions

2.1 Notations

Inventory Parameters	
A_i	The fixed replenishment cost of product i (in \$)
B_i	The unit purchasing cost of product i (in \$)
a_i	The scale demand of product i *where*, $a_i > 0$, $i = 1,2$
b_i	The price sensitivity co-efficient of retailer i with respect to price *where*, $0 \leq b_i < 1$, $i = 1,2$
c	The price sensitivity co-efficient of retailer i with respect to the difference between the prices of two products/ retailers *where*, $0 \leq c < 1$
θ_i	The deterioration co-efficient of retailer i *where*, $0 \leq \theta_i \leq 1$, $i = 1,2$
h_i	The holding cost co-efficient of retailer i, where $i = 1,2$
d_i	The stock availability co-efficient of product i, where $i = 1,2$
e_i	Linear rate of change of demand of product i, where $i = 1,2$
f_i	Quadratic rate of change of demand of product i, where $i = 1,2$
Decision variables	
p_i	The selling price of product i per unit (in \$), where $i = 1,2$
T_i	The length of product cycle i (in years), where $i = 1,2$

Functions	
$D_i(t,p_i,p_j,I_i)$	The demand rate of retailer/product i; where $i = 1,2$ and $j = 3 - i$
$I_i(t)$	The on-hand inventory of product i at time t, where, $0 \le t \le T_i$, $i = 1, 2$
$TP_i(t,p_i,I_i)$	The total profit of retailer/product i per unit time (in \$), where $i = 1, 2$

2.2 Assumptions

1. Shortages and discounts are not allowed.
2. The competing retailers are selling two similar products which are subject to deterioration and are of different brands.
3. Deterioration is taken instantaneously.
4. The deterioration co-efficient is a constant denoted by θ_0, where $0 \le \theta_0 \le 1$
5. Each retailer faces a market demand rate as,

$$D_i(t, p_i, p_j) = a_i - b_i p_i + c(p_j - p_i) + d_i I_i(t) + e_i t - f_i t^2;\ where\ i = 1,2\ and\ j = 3 - i$$

The co-efficient a_i represents the total potential market size of retailer i, in case if both the prices and stock levels of both retailers are zero and when there is no rate of change of demand with respect to the time period. The co-efficients b_i and c, where $b_i \ge 0$, $c \ge 0$ representing the effect of prices on the retailer's demand. The demand of each retailer is a decreasing function of its price and increasing function of the competitor's price. If a retailer reduces its own price by a unit, her demand decreases by $b_i + c$ and demand for her competitor increases by c. The co-efficient d_i is the stock availability co-efficient, e_i and f_i are Linear rate of change of demand of product, Quadratic rate of change of demand respectively of product i, where $i = 1,2$. This type of demand function holds true for Seasonal food items, such as milk products, vegetables, and fruits.

3. Mathematical formulation of the Model

Initially, retailer i orders the quantity Q_i which depletes completely over the cycle due to realized demand and product deterioration, then another cycle begins. Each retailer is the decision maker for the price and replenishment cycle.

Let $\theta u = \theta o$, $0 \le \theta o \le 1$ be the constant deterioration rate. The inventory level $I_i(t)$ for $i = 1,2$ at any time instant, t, declines due to the combined effect of the rate of demand and the deterioration governed by the following differential equations for each time period :

$$\frac{dI_i(t)}{dt} = -D_i - \theta_i I_i(t)\ ; \qquad 0 \le t \le T_i\ , \quad I_i(T_i) = 0 \tag{1}$$

Thus, $I_i(t) = -(t + T_i)(-b_i p_i - cp_i + cp_j + a_i)$; where $i = 1,2$ and $j = 3 - i$

The procurement quantity for the retailer in the the beginning of each cycle, Q_i, will be

$$Q_i = I_i(0) = -T_i(-b_i p_i - cp_i + cp_j + a_i);\ where\ i = 1,2\ and\ j = 3 - i \tag{2}$$

The total profit of each retailer (TP_i) per unit time where $i = 1,2$ comprises of the following cost components.

(a) **Sales revenue:** With the unit sale price p_i, the sales revenue per unit time is $SR_i = \dfrac{p_i \int_0^{T_i} D_i\, dt}{T_i}$, where $i = 1,2$

(b) **Holding cost:** The holding cost per unit time is, $HC_i = \dfrac{h_i \int_0^{T_i} I_i\, dt}{T_i}$, where $i = 1,2$

(c) **Ordering cost:** With A_i, as ordering cost per order, the ordering cost per unit time is $OC_i = \dfrac{A_i}{T_i}$, where $i = 1,2$

(d) **Purchasing cost:** The purchasing cost per unit time is, $PC_i = \dfrac{Q_i B_i}{T_i}$, where $i = 1,2$

Therefore, the total profit per unit time is calculated by,

$$Total\ Profit\ (TP_i) = \left\{ \frac{SR_i - (HC_i + OC_i + PC_i)}{T_i} \right\},\ \text{where } i = 1,2 \tag{3}$$

Considering Equation (3), $TP = TP_1 + TP_2$ is the joint total profit of the inventory system.

Now, by the classical optimization technique to maximize the total profit stated in Equation (3), we apply the necessary and sufficient conditions as under :

$$\frac{\partial TP}{\partial p_1}=0, \frac{\partial TP}{\partial p_2}=0, \frac{\partial TP}{\partial T_1}=0, \frac{\partial TP}{\partial T_2}=0 \tag{4}$$

To check the concavity of the total profit function of the obtained solution, we adopt the below stated algorithm,

Step 1: Assigning the inventory parameters some specific hypothetical values.

Step 2: Obtaining the solutions by solving simultaneous equations stated in Equation (4), utilising the mathematical software Maple XVIII.

Step 3: For concavity with respect to optimal replenishment cycle length, check the condition

$$\Delta=\left(\frac{\partial^2 TP}{\partial T_1^2}\right)\left(\frac{\partial^2 TP}{\partial T_2^2}\right)-\left(\frac{\partial^2 TP}{\partial T_1 \partial T_2}\right)>0 \ and \left(\frac{\partial^2 TP}{\partial T_1^2}\right)<0 .$$

Computing all the Eigen values of the below stated hessian matrix H for the optimal selling price point obtained from Equation (4),

$$H=\begin{bmatrix} \frac{\partial^2 TP}{\partial p_1^2} & \frac{\partial^2 TP}{\partial p_1 p_2} \\ \frac{\partial^2 TP}{\partial p_2 p_1} & \frac{\partial^2 TP}{\partial p_2^2} \end{bmatrix}$$

If all the eigenvalues are negative, the total profit is maximum at the optimal selling price.

Then, we can conclude that total profit function is concave in nature with respect to both the replenishment cycle length as well as the selling price.

4. Numerical Example and Sensitivity Analysis

4.1 Numerical Analysis

Considering the specified values in the appropriate units:

$a1=8000, a2=5000, b1=0.2, b2=01, c=0.8, \theta 1=10\%, \theta 2=10\%, h1=\$20/unit/year,$
$h2=\$20/unit/year, A_1=\$40/unit/year, A_2=\$40/unit/year, B_1=\$10/unit/year,$
$B_2=\$20/unit/year, d_1=0.7, d_2=0.9, e_1=0.1, e_2=0.1, f_1=0.2, f2=0.2$

Solution: Optimum replenishment cycle length for first retailer $T_1=0.94752\ years$

Optimum replenishment cycle length for second retailer $T_2=0.73305\ years$

Optimum selling price per unit for first product $p_1=\$\ 30.3140$

Optimum selling price per unit for second product $p_2=\$\ 30.6506$

Total Profit per unit time for first retailer $TP_1=\$\ 308311.5$

Total Profit per unit time for second retiler $TP_2=\$\ 210928.5$

Total Profit of inventory system per unit time $TP=\$\ 519240.1$

If all the eigenvalues are negative, the total profit is maximum at the optimal selling price. Here,

$$H=\begin{bmatrix} -0.00965 & 0.00994 \\ 0.00994 & -0.01368 \end{bmatrix}$$

$$\lambda_1=-0.021817<0\ ;\ \lambda_2=-0.001522<0$$

are two eigenvalues of the Hessian matrix, and both the eigenvalues are negative. So, it is a negative-definite matrix and hence, total profit function is concave in nature with respect to selling price.

Also, $\Delta=\left(\frac{\partial^2 TP}{\partial T_1^2}\right)\left(\frac{\partial^2 TP}{\partial T_2^2}\right)-\left(\frac{\partial^2 TP}{\partial T_1 \partial T_2}\right)=2.104407 \text{ X } 10^7>0$ and $\left(\frac{\partial^2 TP}{\partial T_1^2}\right)=-5.089115\ X 10^5<0$

Therefore, the total profit function is concave in nature with respect to replenishment cycle lengths as shown in Figure (1).

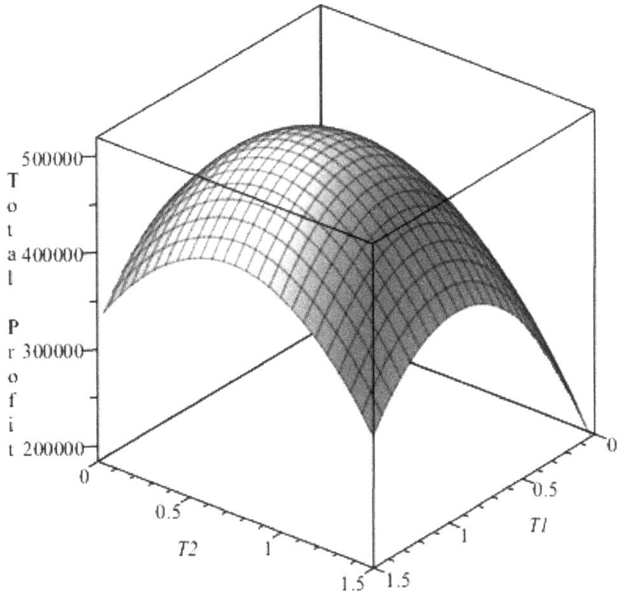

Figure 1: Concavity of Total Profit function

4.2 Sensitivity Analysis on the Optimal Inventory Policy

In this part, the sensitivity analysis of the decision variables with respect to various inventory parameters is carried out. Table 1 demonstrates the values of decision variables on varying the various inventory parameters in the range of –20% to 20%. From Table 1 the following observations are extracted:

(a) Sensitivity analysis of the total potential market size of first retailer (a_1):

It is observed that the increase in basic market demand a_1 of the first retailer, by lowering the selling price of deteriorating items its replenishment cycle length increases. There are no remarkable variations in selling price and replenishment cycle length of the second retailer. The total profit of the first retailer increases but the total profit of the second retailer decreases and hence, the total profit of the inventory system rises.

(b) Sensitivity analysis of the total potential market size of the second retailer (a_2)

The variation in scale demand a_2 of the second retailer, by reducing the selling price of deteriorating items the replenishment cycle length increases. There are no remarkable variations in the selling price and replenishment cycle length of first retailer. The total profit of the second retailer increases but the total profit of first retailer decreases by the total potential market size of second retailer and hence, the total profit of the inventory system rises.

(c) Sensitivity analysis of price sensitivity co-efficient of first retailer with respect to price (b_1)

An increase in price sensitivity co-efficient of first retailer with respect to price b_1, by increasing the selling price of deteriorating items, it's the replenishment cycle length decreases. There are no remarkable variations in the selling price and replenishment cycle length of the second retailer. The total profit of the first retailer decreases but the total profit of second retailer increases by the change in the price sensitivity co-efficient of the first retailer and reduces the total profit per unit time.

(d) Sensitivity analysis of the price sensitivity co-efficient of second retailer with respect to its price (b_2)

The variation in demand b_2, lowers the selling price of deteriorating items of the second retailer, and its replenishment cycle length increases. There are no remarkable variations in the selling price and the replenishment cycle length of the first retailer. The total profit of the second retailer increases but the total profit of the first retailer decreases by the changes in the price sensitivity co-efficient of second retailer and hence, the total profit of the inventory system decreases.

(e) Sensitivity analysis of the price sensitivity co-efficient (c)

With respect to increase in price sensitivity co-efficient c, by up-lifting the selling price of deteriorating items of both retailers, their replenishment cycle time decreases simultaneously. The total profit of the first retailer increases but the total profit of the second retailer decreases by the change in the price sensitivity co-efficient with respect to the difference between the prices of two products/retailers and hence, the total profit of the inventory system increases.

Table 1: Sensitivity Analysis of Decision variables with respect to various inventory parameters.

Inventory Parameter	Decision Variable	–20%	–10%	0	10%	20%
a_1	T_1	0.9463	0.9469	0.9475	0.9479	0.9483
	T_2	0.7330	0.7330	0.7330	0.7330	0.7330
	p_1	30.3942	30.3496	30.3140	30.2851	30.2610
	p_2	30.6506	30.6506	30.6506	30.6506	30.6506
	TP_1	246595.7	277453.7	308311.5	339169.2	370026.8
	TP_2	210931.2	210929.7	210928.5	210927.5	210926.7
	TP	457527.0	488383.5	519240.1	550096.8	580953.6
a_2	T_1	0.9475	0.9475	0.9475	0.9475	0.9475
	T_2	0.7311	0.7321	0.7330	0.7337	0.7343
	p_1	30.3140	30.3140	30.3140	30.3140	30.3140
	p_2	30.8201	30.7256	30.6506	30.5897	30.5392
	TP_1	308316.8	308313.8	308311.5	308309.7	308308.1
	TP_2	168703.1	189816.0	210928.5	232040.8	253152.9
	TP	477019.9	498129.9	519240.1	540350.5	561461.0
b_1	T_1	0.9477	0.9476	0.9475	0.9474	0.9473
	T_2	0.7330	0.7330	0.7330	0.73305	0.7330
	p_1	30.3014	30.3077	30.3140	30.3203	30.3266
	p_2	30.6506	30.6506	30.6506	30.6506	30.6506
	TP_1	308358.3	308334.9	308311.5	308288.1	308264.8
	TP_2	210928.1	210928.3	210928.5	210928.7	210928.9
	TP	519286.4	519263.3	519240.1	519216.9	519193.7
b_2	T_1	0.9475	0.9475	0.9475	0.9475	0.9475
	T_2	0.7332	0.7331	0.7330	0.7329	0.7328
	p_1	30.3140	30.3140	30.3140	30.3140	30.3140
	p_2	30.6360	30.6433	30.6506	30.6580	30.6653
	TP_1	308311.1	308311.3	308311.5	308311.8	308312.0
	TP_2	210954.4	210941.4	210928.5	210915.6	210902.6
	TP	519265.5	519252.8	519240.1	519227.4	519214.7
c	T_1	0.9483	0.9479	0.9475	0.9471	0.9467
	T_2	0.7344	0.7337	0.7330	0.7323	0.7316
	p_1	30.2640	30.2890	30.3140	30.3391	30.3643
	p_2	30.5345	30.5924	30.6506	30.7092	30.7681
	TP_1	308306.4	308308.8	308311.5	308314.5	308317.8
	TP_2	210929.0	210928.8	210928.5	210928.2	210927.9
	TP	519235.5	519237.7	519240.1	519242.8	519245.7
θ_1	T_1	0.9475	0.9475	0.9475	0.9475	0.9475
	T_2	0.7330	0.7330	0.7330	0.7330	0.7330
	p_1	30.3140	30.3140	30.3140	30.3140	30.3140
	p_2	30.6506	30.6506	30.6506	30.6506	30.6506
	TP_1	308311.5	308311.5	308311.5	308311.5	308311.5
	TP_2	210928.5	210928.5	210928.5	210928.5	210928.5
	TP	519240.1	519240.1	519240.1	519240.1	519240.1

Table 1 contd. ...

...Table 1 contd.

Inventory Parameter	Decision Variable	–20%	–10%	0	10%	20%
θ_2	T_1	0.9475	0.9475	0.9475	0.9475	0.9475
	T_2	0.7330	0.7330	0.7330	0.7330	0.7330
	p_1	30.3140	30.3140	30.3140	30.3140	30.3140
	p_2	30.6506	30.6506	30.6506	30.6506	30.6506
	TP_1	308311.5	308311.5	308311.5	308311.5	308311.5
	TP_2	210928.5	210928.5	210928.5	210928.5	210928.5
	TP	519240.1	519240.1	519240.1	519240.1	519240.1
h_1	T_1	0.94825	0.9478	0.9475	0.9471	0.9468
	T_2	0.7330	0.7330	0.7330	0.7330	0.7330
	p_1	24.2151	27.2622	30.3140	33.3706	36.4319
	p_2	30.6513	30.6510	30.6506	30.6503	30.6500
	TP_1	262828.3	285578.6	308311.5	331027.1	353725.2
	TP_2	210722.5	210825.4	210928.5	211031.8	211135.2
	TP	473550.8	496404.1	519240.1	542058.9	564860.4
h_2	T_1	0.9475	0.9475	0.9475	0.9475	0.9475
	T_2	0.7338	0.7334	0.7330	0.7326	0.7322
	p_1	30.3142	30.3141	30.3140	30.3139	30.3138
	p_2	24.4679	27.5557	30.6506	33.7527	36.8620
	TP_1	308120.7	308216.0	308311.5	308407.3	308503.2
	TP_2	188926.9	199933.8	210928.5	221910.8	232880.8
	TP	497047.7	508149.9	519240.1	530318.1	541384.1
A_1	T_1	0.9475	0.9475	0.9475	0.9475	0.9475
	T_2	0.7330	0.7330	0.7330	0.7330	0.7330
	p_1	30.3129	30.3135	30.3140	30.3146	30.3152
	p_2	30.6506	30.6506	30.6506	30.6506	30.6506
	TP_1	308320.0	308315.8	308311.5	308307.3	308303.1
	TP_2	210928.5	210928.5	210928.5	210928.5	210928.5
	TP	519248.5	519244.3	519240.1	519235.9	519231.7
A_2	T_1	0.9475	0.9475	0.9475	0.9475	0.9475
	T_2	0.7330	0.7330	0.7330	0.7330	0.7330
	p_1	30.3140	30.3140	30.3140	30.3140	30.3140
	p_2	30.6476	30.6491	30.6506	30.6522	30.6537
	TP_1	308311.4	308311.5	308311.5	308311.6	308311.6
	TP_2	210939.4	210934.0	210928.5	210923.0	210917.6
	TP	519250.9	519245.5	519240.1	519234.7	519229.3
b_1	T_1	0.9477	0.9476	0.9475	0.9474	0.9472
	T_2	0.7330	0.7330	0.7330	0.7330	0.7330
	p_1	30.2978	30.3059	30.3140	30.3222	30.3303
	p_2	30.6506	30.6506	30.6506	30.6506	30.6506
	TP_1	292323.1	300317.3	308311.5	316305.7	324299.9
	TP_2	210928.0	210928.2	210928.5	210928.8	210929.1
	TP	503251.1	511245.6	519240.1	527234.6	535229.0

Table 1 contd. ...

...Table 1 contd.

Inventory Parameter	Decision Variable	–20%	–10%	0	10%	20%
b_2	T_1	0.9475	0.9475	0.9475	0.9475	0.9475
	T_2	0.7337	0.7334	0.7330	0.7326	0.7323
	p_1	30.3140	30.3140	30.3140	30.3140	30.3140
	p_2	30.5886	30.6196	30.6506	30.6818	30.7131
	TP_1	308309.6	308310.6	308311.5	308312.5	308313.5
	TP_2	190941.7	200935.1	210928.5	220921.8	230915.1
	TP	499251.4	509245.8	519240.1	529234.4	539228.6
d_1	T_1	1.1847	1.0529	0.9475	0.8612	0.7893
	T_2	0.7330	0.7330	0.7330	0.7330	0.7330
	p_1	30.2956	30.3048	30.3140	30.3234	30.3329
	p_2	30.6506	30.6506	30.6506	30.6506	30.6506
	TP_1	365422.1	333694.5	308311.5	287542.8	270234.7
	TP_2	210927.9	210928.2	210928.5	210928.8	210929.1
	TP	576350.0	544622.8	519240.1	498471.6	481163.9
d_2	T_1	0.9475	0.9475	0.9475	0.9475	0.9475
	T_2	0.9172	0.8149	0.7330	0.6660	0.3311
	p_1	30.3140	30.3140	30.3140	30.3140	30.3140
	p_2	30.5830	30.6166	30.6506	30.6850	31.0911
	TP_1	308309.4	308310.5	308311.5	308312.6	3.0832
	TP_2	238701.4	223272.7	210928.5	200827.6	150288.1
	TP	547010.9	531583.2	519240.1	509140.3	458613.2
e_1	T_1	0.9475	0.9475	0.9475	0.9475	0.9475
	T_2	0.7330	0.7330	0.7330	0.7330	0.7330
	p_1	30.3140	30.3140	30.3140	30.3140	30.3140
	p_2	30.6506	30.6506	30.6506	30.6506	30.6506
	TP_1	308311.3	308311.4	308311.5	308311.7	308311.8
	TP_2	210928.5	210928.5	210928.5	210928.5	210928.5
	TP	519239.8	519240.0	519240.1	519240.2	519240.4
e_2	T_1	0.9475	0.9475	0.9475	0.9475	0.9475
	T_2	0.7330	0.7330	0.7330	0.7330	0.7330
	p_1	30.3140	30.3140	30.3140	30.3140	30.3140
	p_2	30.6506	30.6506	30.6506	30.6506	30.6506
	TP_1	308311.5	308311.5	308311.5	308311.5	308311.5
	TP_2	210928.3	210928.4	210928.5	210928.6	210928.7
	TP	519239.9	519240.0	519240.1	519240.2	519240.3
f_1	T_1	0.9475	0.9475	0.9475	0.9475	0.9475
	T_2	0.7330	0.7330	0.7330	0.7330	0.7330
	p_1	30.3140	30.3141	30.3140	30.3140	30.3140
	p_2	30.6506	30.6506	30.6506	30.6506	30.6506
	TP_1	308311.9	308311.7	308311.5	308311.4	308311.2
	TP_2	210928.5	210928.5	210928.5	210928.5	210928.5
	TP	519240.4	519240.3	519240.1	519239.9	519239.7
f_2	T_1	0.9475	0.9475	0.9475	0.9475	0.9475
	T_2	0.7330	0.7330	0.7330	0.7330	0.7330
	p_1	30.3140	30.3140	30.3140	30.3140	30.3140
	p_2	30.6507	30.6507	30.6506	30.6506	30.6506
	TP_1	308311.5	308311.5	308311.5	308311.5	308311.5
	TP_2	210928.7	210928.6	210928.5	210928.4	210928.3
	TP	519240.3	519240.2	519240.1	519240.0	519239.9

(f) Sensitivity analysis of the holding cost co-efficient of first retailer (h_1)

The variation in the holding cost co-efficient of the first retailer h_1, by up-lifting the selling price of deteriorating items, its replenishment cycle length decreases. There are no remarkable variations in replenishment cycle length of second retailer but its selling price decreases. The total profit of both the retailers increases by the variation in the holding cost co-efficient of first retailer and hence, the total profit of the inventory system rises.

(g) Sensitivity analysis of the holding cost co-efficient of second retailer (h_2)

With respect to increase in the holding cost co-efficient of second retailer h_2, by up-lifting the selling price of deteriorating items its replenishment cycle length decreases. There are no remarkable variations in replenishment cycle length of second retailer but its selling price decreases. The total profit of both the retailers increases by the variation in the holding cost co-efficient of second retailer and hence, the total profit of the inventory system up-lifts.

(h) Sensitivity analysis of the fixed ordering cost co-efficient of first product (A_1)

The variation in the ordering cost co-efficient of first product, A_1, by up-lifting the selling price of deteriorating items of first retailer there are no remarkable variations in replenishment cycle length of both retailers and the selling price of the second retailer. The total profit of first retailer decreases by the variation in the ordering cost co-efficient and hence, the total profit of the inventory system decreases.

(i) Sensitivity analysis of the fixed ordering cost co-efficient of the second product (A_2)

With respect to increase in the ordering cost co-efficient of second product, A_2, by up-lifting the selling price of deteriorating items of second retailer there are no remarkable variations in the replenishment cycle length of both retailers but its selling price decreases. The total profit of first retailer increases by the variation in the ordering cost co-efficient and hence, the total profit of the second retailer as well as the total profit of the inventory system decreases.

(j) Sensitivity analysis of the purchasing cost co-efficient of first product (B_1)

The variation in purchasing cost co-efficient of first product B_1, uplifts the selling price of deteriorating items of first retailer, and its replenishment cycle length decreases. There are no remarkable variations in selling price and replenishment cycle length of the second retailer. The total profit of the both retailers increases and hence, the total profit of the inventory system rises.

(k) Sensitivity analysis of the purchasing cost co-efficient of second product (B_2)

With respect to increase in the purchasing cost co-efficient of second product, B_2, by up-lifting the selling price of deteriorating items and reduction of the replenishment cycle length of second retailer there are no remarkable variations in replenishment cycle length and selling price of the first retailer but its selling price decreases. The total profit of the both retailers increases and hence, the total profit of the inventory system rises.

(l) Sensitivity analysis of the Linear rate of change of demand of the first product (e_1)

With respect to increase in the linear rate of change of demand of the first product, e_1, the total profit of the first retailer increases and hence, the total profit of the inventory system rises. There are no remarkable variations in the rest of the parameters.

(m) Sensitivity analysis of the Linear rate of change of demand of the second product (e_2)

With respect to increase in the linear rate of change of demand of the second product, e_2, the total profit of the second retailer increases and hence, the total profit of the inventory system rises. There are no remarkable variations in the rest of the parameters.

(n) Sensitivity analysis of the Quadratic rate of change of demand of first product (f_1)

With respect to increase in the Quadratic rate of change of demand of first product, f_1, the total profit of the first retailers increases and hence, the total profit of the inventory system decreases. There are no remarkable variations in rest other parameters.

(o) Sensitivity analysis of the Quadratic rate of change of demand of the second product (f_2)

With respect to increase in the Quadratic rate of change of demand of the second product, f_2, the total profit of the second retailer increases and hence, the total profit of the inventory system decreases. There are no remarkable variations in the rest of the parameters.

5. Conclusion

This article develops an inventory model dealing with the selling of deteriorating products by duopoly retailers and the market demand rate is expressed as a time-price-stock dependent quadratic in nature. The simultaneous computation of selling price and replenishment cycle lengths, for both the retailers is done to maximize the total profit of the inventory system. Finally, in order to validate the derived models, numerical examples along with sensitivity analysis is undertaken, which extracts some fruitful managerial insights like the following parameters; 1. The total market potential size for both the retailers, price sensitivity co-efficient with respect to the difference between the prices of the two products/retailers, the Linear rate of change of demand for both the retailers is to be raised in order to uplift the total profit of the inventory system. 2. This study provides several interesting insights in pricing and inventory decisions of competing agents. For instance, the results show that while the retailer with bigger market size, longer cycle length and with lower price makes more profit benefits compared to the retailer with smaller market size, smaller replenishment cycle length and higher selling price.

This work can be extended in the following dimensions. One immediate possible extension could be allowable shortages, cash discounts, and quantity discounts. Additionally, preservation technology can be inserted to reduce the deterioration effect of the product in future research.

Acknowledgements

The authors are very thankful to the reviewers for their deep and thorough review. The authors thank DST-FIST File # MSI-097 for technical support to the department.

References

Abad, P. 1996. Optimal pricing and lot-sizing under conditions of perishability and partial backordering. Management Science, 42(8): 1093–1104.

Adida, E. and Perakis, G. 2010. Dynamic pricing and inventory control: uncertainty and competition. Operations Research, 58(2): 289–302.

Bose, S., Goswami, A. and Chaudhari, K. 1995. An EOQ model for deteriorating items worth linear time-dependent demand rate and shortages under inflation and time discounting. Journal of the Operational Research Society, 46(6): 771–782.

Chang, C. T., Teng, J. T. and Goyal, S. K. 2010. Optimal replenishment policies for non-instantaneous deteriorating items with stock-dependent demand. International Journal of Production Economics, 123(1): 62–68.

Chen, X., Pang, Z. and Pan, L. 2014. Coordinating inventory control and pricing strategies for perishable products. Operations Research, 62(2): 284–300.

Chen, Y., and Shi, C. 2017. Joint pricing and inventory management with strategic customers. Operations Research, 67(6): http://dx.doi.org/10.2139/ssrn.2770242.

Chung, K. J. and Ting, P. S. 1993. A heuristic for replenishment of deteriorating items with a linear trend in demand. Journal of the Operational Research Society, 44(12).

Cohen, M. A. 1997. Joint pricing and ordering policy for exponentially decaying inventory with known demand. Naval Research Logistics, 24(2): 257–268.

Datta, T. K. and Pal, A. K. 2001. An inventory system stock-dependent, price sensitive demand rate. Production Planning and Control, 12: 13–20.

Goyal, S. K. and Giri, B. C. 2001. Recent trends in modeling of deteriorating inventory. European Journal of Operational Research, 134: 1–16.

Hou, K. L., Huang, Y. F. and Lin, L. C. 2011. An inventory model for deteriorating items with stock-dependent selling rate and partial backlogging under inflation. African Journal of Business Management, 5(10): 3834–3843.

Mahata, P., Mahata, G. C. and De, S. K. 2020. An economic order quantity model under two-level partial trade credit for time varying deteriorating items, International Journal of Systems Science: Operations & Logistics, 7(1): 1–17, doi: 10.1080/23302674.2018.1473526.

Mahmoodi, A. 2019. Joint pricing and inventory control of duopoly retailers with deteriorating items and linear demand. Computers & Industrial Engineering, doi: https://doi.org/10.1016/j.cie.2019.04.017.

Maihami, R. and Kamalabadi, I. N. 2012. Joint pricing and inventory control for non-instantaneous deteriorating items with partial backlogging and time and price dependent demand. International Journal of Production Economics, 136(1): 116–122.

Mishra, U., Barrón, L. E. C., Tiwari, S., Shaikh, A. A. and Garza, G. T. 2017. An inventory model under price and stock dependent demand for controllable deterioration rate with shortages and preservation technology investment. Annals of Operations Research, Springer, 254(1): 165–190.

Mukhopadhyay, S., Mukherjee, R. and Chaudhuri, K. 2004. Joint pricing and ordering policy for a deteriorating inventory. Computers & Industrial Engineering, 47(4): 339–349.

Panda, S., Saha, S. and Basu, M. 2013. Optimal pricing and lot-sizing for perishable inventory with price and time dependent ramp-type demand. International Journal of Systems Science, 44(1): 127–138.

Papachristos, S. and Skouri, K. 2003. An inventory model with deteriorating items, quantity discount, pricing and time-dependent partial backlogging. International Journal of Production Economics, 83(3): 247–256.

Qin, Y., Wang, J. and Wei, C. 2014. Joint pricing and inventory control for fresh produce and foods with quality and physical quantity deteriorating simultaneously. International Journal of Production Economics, 152: 42–48.

Rabbani, M., Zia, N. P. and Rafiei, H. 2016. Joint optimal dynamic pricing and replenishment policies for items with simultaneous quality and physical quantity deterioration. Applied Mathematics and Computation, 287: 149–160.

Rajan, A., Rakesh and Steinberg, R. 1992. Dynamic pricing and ordering decisions by a monopolist. Management Science, 38(2): 240–262.

Rameswari, M. and Uthayakumar, R. 2018. An integrated inventory model for deteriorating items with price-dependent demand under two-level trade credit policy. International Journal of Systems Science: Operations & Logistics, 5(3): 253–267, doi: 10.1080/23302674.2017.1292432.

Shah, N. H. and Pandey, P. 2009. Deteriorating inventory model when demand depends on advertisement and stock display. International Journal of Operations Research, 6(2): 33–44.

Shah, N. H., Patel, A. R. and Louc, K. R. 2011. Optimal ordering and pricing policy for price sensitive stock-dependent demand under progressive payment scheme. International Journal of Industrial Engineering Computations, 2: 523–532.
Silver, E. A. and Meal, H. C. 1969. A simple modification of the EOQ for the case of a varying demand rate. Production of Inventory Management, 10: 52–65.
Silver, E. A. 1979. A simple inventory replenishment decision rule of a linear trend in demand. Journal of the Operational Research Society, 30: 71–75.
Singh, S. R. 2006. Production inventory model for perishable product with price dependent demand. Ref. des ERA, 1(1): 1–10.
Singh, S. R. and Malik, A. K. 2011. Two warehouses inventory model for non-instantaneous deteriorating items with stock dependent demand. International Transactions in Applied Sciences, 3(4): 911–920.
Transchel, S. and Minner, S. 2011. Economic lot-sizing and dynamic quantity competition. International Journal of Production Economics, 133(1): 416–422.
Zhu, K. and Thonemann, U. 2009. Coordination of pricing and inventory control across products. Naval Research Logistics, 56(2): 175–190.

CHAPTER 15

Quadratic Programming Approach for the Optimal Multi-objective Transportation Problem

Masar Al-Rabeeah,[1,*] *Ali Al-Hasani*[1] and *M G M Khan*[2]

1. Introduction

The transportation problem is a major part of operations research because of its importance in economics and industry, Dantzig and Thapa (2006). The main goal of the transportation problem is distributing demand from different sources to different destinations with minimum cost. The single-objective transportation problem is the classical transportation problem which contains only one objective function, set of constraints and bounds. There are many well-known methods that solve the transportation problem such as Northwest Corner method, Minimum cost method, Vogel's approximation method, Row Minimum Method, and Column Minimum Method Zhao et al. (1999).

It is not sufficient nowadays to tackle the real-world decision-making problems in the market with high level competition by using the single-objective transportation problem. To handle most of these problems, we are dealing with Multi-objective Transportation Problems (MOTP) that involve many objective functions to be optimized simultaneously, Ignizio (1978); Opricovic and Tzeng (2004); Ulungu and Teghem (1994); Ustun (2012). In multi-objective optimization finding an optimal solution that can optimize all objective functions at the same time is not possible so researchers have a tendency to find another solution that is called a non-dominated solution Al-Hasani et al. (2019); Al-Rabeeah et al. (2019a,b); Aneja and Nair (1979). The non-dominated solutions are those where one cannot improve one objective function in value without decreasing the value of the other objective. The non-dominated solutions are not a single point but instead are a set of points, hence, the decision maker can select from the non-dominated solutions set according to personal preference. However, in large problems, the non-dominated solution set will be very large which causes a difficulty for the decision maker to select the best. Therefore, another side of the study to find a solution compromise becomes of interest to the researchers Bit et al. (1993); El-Wahed and Lee (2006); Ulungu and Teghem (1994).

Many researchers have developed a procedure to find a solution compromise for a Multi-objective Linear Problem (MOLP) where the decision variables are real. If the decision variables are integer restricted, the problem becomes more challenging, which is termed as Multi-objective Integer Problem (MOIP) Ulungu and Teghem (1994). The MOLP transportation model is a special case of MOIP when it is without any integer restrictions on the variables.

Here, we review some programming method to solve MOTP like quadratic, with goal and fuzzy programming. In Maleki and Khodaparasti (2008) a method had been conducted to solve MOTP based on fuzzy goal programming using a non-linear membership function (a special quadratic membership function as hyperbolic). More discussion about the relationship between goal programming and fuzzy programming is displayed in the paper. In addition to considering a membership function in Rani et al. (2016) a non-membership function has also been of interest, where the authors formulated the problem as parabolic MOTP under a fuzzy environment. Their method had been tested using a transportation problem and another problem from an industrial system.

Fuzzy parameters values have been considered in Gupta and Kumar (2012) to solve linear MOTP. These parameters include cost of transporting the product, the availability source and demand required. Fuzzy programming again used in Maity and Kumar Roy (2016), when the authors find the compromise solution for MOTP in a non-linear case with multi-choice demand and non-linear cost. Many researchers have developed a technique using fuzzy goal programming approach for MOTP Bit et al. (1993); El-Wahed and Lee (2006); Hu et al. (2007); Li and Lai (2000).

[1] Department of Mathematics, Faculty of Sciences, Basrah University, Basrah, Iraq.
[2] School of Computing, Information and Mathematical Sciences, The University of the South Pacific, Suva, Fiji.
Emails: ali.alhasani1309@gmail.com, khan_mg@usp.ac.fj
* Corresponding author: masar.alrabeeah23@gmail.com

On the other side, goal programming (GP) has been widely used to solve MOTP. The original goal programming concept was developed in Charnes et al. (1955, 1961). The essential element in GP model is the attainable function which is expressed as a mathematical formulation of the undesirable deviation from the goals. Each type of attainable function leads to a variant of the GP model Nunkaew and Phruksaphanrat (2009); Tamiz (2012). In Chang (2007) a new approach for the GP, called multi-choice goal programming, where the decision maker needs to show the importance of each goal to avoid underestimation. Further improvement has been discussed in Chang (2008).

Our goal in this paper is to find a compromise solution for MOTP by formulating the problem as a quadratic program which seeks a minimization of the sum of square of the relative increase in the cost due to deviation from the individual optimum solution. The proposed programming quadratic approach has been compared with the one of the goal programming approach that is called weighted goal programming, Deb (1999). We have implemented our approach using Cplex Callable library in Lubuntu operating system. To demonstrate the efficiency of the proposed approach different problems are implemented in addition to the transportation problem like an assignment problem and the travailing salesman problem. In all these experiments our proposed approach outperformed the current approach significantly.

This chapter is organized as follow: Section 2 contains mathematical formulations which are important and relevant for the rest of this paper. Section 3 presents a review of the weighted goal programming approach. Section 4 gives details of the proposed approach and numerical examples have been illustrated too. Computational experiments are presented in Section 5. Finally, the paper is concluded in Section 6.

2. Multi-objective Transportation Problem Formulation

Consider a multi-objective transportation problem with p linear objective functions as given in Equation (1).

Let $a_i(i = 1,2,\cdots, m)$ be the number of available units at the ith origin, $b_j(\mathrm{j} = 1,2,\cdots, n)$ be the required number of units at the jth destination and x_{ij} be the number of units to be transported from the ith origin to the j destination, then the MOTP maybe formulated as follows:

$$min\left[Z_1 = \sum_{i=1}^{m}\sum_{j=1}^{n} c_{ij1}x_{ij}, Z_2 = \sum_{i=1}^{m}\sum_{j=1}^{n} c_{ij2}x_{ij}, \cdots, Z_p = \sum_{i=1}^{m}\sum_{j=1}^{n} c_{ijp}x_{ij} \right] \tag{1}$$

Subject to

$$\sum_{i=1}^{m} x_{ij} = b_j; j = 1,2,\cdots n,$$

$$\sum_{j=1}^{n} x_{ij} = a_i; j = 1,2,\cdots m,$$

and $x_{ij} \geq 0$, $x_{ij} \in \mathbb{Z}\ \forall\ i, j$.

A feasible point $z^c \in Y$ is defined as a compromise solution when it becomes closest to the individual optimal solution.

Note that, usually, in multi-objective optimization, it is not possible to use the optimal solution for any individual optimal solution of the kth *objective where* $k = 1,2,\cdots p$, from all p objectives. In this situation, there is a need to find a solution which is near optimal for all p objectives in MOTP (1). Such a solution may be called a compromise solution; therefore, the aim of this paper is to find one that is closest to its intended optimal solution.

3. Review of Weighted Goal Programming Approach

Weighted goal programming approach (WGP) Deb (1999) is an efficient method for determining an optimal compromise solution based on the rank priority of each objective function provided by the decision maker. In the main objective of this method all deviation variables are scalarized with weight. These weights show the relative importance of the goals. The weighted goal programming model can be presented as follows:

$$min\sum_{i=1}^{p} \omega_k \sum_{i=1}^{m}\sum_{j=1}^{n} c_{ijk}x_{ij}(d_k^+ + d_k^-), \tag{2}$$

Subject to

$$\sum_{i=1}^{m}\sum_{j=1}^{n} c_{ijk}x_{ij} - d_k^+ + d_k^- = \sum_{i=1}^{m}\sum_{j=1}^{n} c_{ijk}x_{ij}^*$$

$$\sum_{i=1}^{m} x_{ij} = b_j; j = 1,2,\cdots n,$$

$$\sum_{j=1}^{n} x_{ij} = a_i; j = 1,2,\cdots m,$$

$$d_k^+, d_k^- \geq 0, \qquad x_{ij} \geq 0 \; x_{ij} \in \mathbb{Z} \forall i, j$$

Where $\sum_{i=1}^{m} \sum_{j=1}^{n} c_{ijk} x_{ij}^*$ is a known individual optimal value with respect to the kth objective function, and d_k^+, d_k^- represent the over-achievement and under-achievement of the kth goal where $k = 1,2,\cdots p$ and ω represent a nonnegative real weight assigned to goal which shows the importance of each deviation.

4. The Proposed Quadratic Programming Approach

Let x_{ijk}^* denote the individual optimum solutions for the kth objectives in (1) and $Z_k^* = \sum_{i=1}^{m} \sum_{j=1}^{n} c_{ijk} x_{ij}^*$ be the optimum value of the objectives at x_{ijk}^*.

Since x_{ijk}^* is the individual optimum solution of kth objectives, there is no other solution except x_{ijk}^* that can provide the value of Z_k less than Z_k^*. Therefore, if the problem is to find an optimal solution to MOTP (1) that can minimize all the kth objectives, it will be known as the compromise solution.

Let x_{ij}^* be the optimum compromise solution, that is, x_{ij}^* is the solution of MOTP (1) with the value of kth objectives:

$$Z_k = \sum_{i=1}^{m} \sum_{j=1}^{n} c_{ijk} x_{ij}^* \tag{3}$$

Obviously, $Z_k \geq Z_k^*$ and $Z_k - Z_k^*$ is the cost due to the compromise. Thus, the relative increase in cost for optimizing k^{th} objectives $k = 1,2,\cdots p$ due to the non use of an individual optimum solution is obtained by:

$$\frac{Z_k - Z_k^*}{Z_k^*} \tag{4}$$

Substituting (3) in (4), we get

$$\frac{\sum_{i=1}^{m} \sum_{j=1}^{n} c_{ijk} x_{ij}^* - Z_k^*}{Z_k^*} \tag{5}$$

Then, by minimizing the sum of the relative increase in total cost of all k^{th} objectives $k = 1,2,\cdots p$ for using x_{ij}^* instead of individual optimum solution x_{ijk}^* we can determine a new optimum compromise solution x_{ij}^*.

However, as the quantity in (5) may turn out to be negative for a Z_k^*, we propose a chi-square type objective function , which turns out to be a quadratic multi-objective programming approach to determine the compromise solution by minimizing the sum of square of the relative increase in the cost due to deviation from the individual optimum solution. Thus, a reasonable criterion to determine the optimum compromise solution to the quadratic multi-objective transportation problem (QMOTP) is given by:

$$\min \sum_{k=1}^{p} \frac{\left(\sum_{i=1}^{m} \sum_{j=1}^{n} c_{ijk} x_{ij}^* - Z_k^*\right)^2}{Z_k^*} \tag{6}$$

Subject to

$$\sum_{i=1}^{m} x_{ij} = b_j; j = 1,2,\cdots n,$$

$$\sum_{j=1}^{n} x_{ij} = a_i; j = 1,2,\cdots m,$$

and $x_{ij} \geq 0$, $x_{ij} \in \mathbb{Z} \; \forall \; i, j$

Using (6), one can find a compromise solution for any multi-objective transportation problem with any number of objective functions.

4.1 Illustration using Simple Numerical Example

In this section, we present an algorithm to solve a multi-objective problem using the proposed quadratic programming approach.

Suppose that we have two objectives z_1 and z_2 of a (3×4) transportation problem, all their coefficients are illustrated in Table 1. Then, an algorithm to solve the problem is outlined in the following steps:

Step 1: Optimize z_1 and get z_1^*, so for our example above $z_1^* = 12057$.

Step 2: Optimize z_2 and get z_2^*, so for our example above $z_2^* = 13450$.

Step 3: For the individual optimal z_1^* and z_2^* obtained in step 2, the quadratic programming problem formulated in Equation (6), that is:

$$\min \sum_{k=1}^{2} \frac{\left(\sum_{i=1}^{3}\sum_{j=1}^{4} c_{ijk} x_{ij}^{*} - Z_k^{*}\right)^2}{Z_k^{*}}$$

and get the compromise value. For this example, the compromise value is: $z^c = 249223$ which represents the closest point to the optimal point for both first and second objective functions.

5. Computational Experiments

In this section, we will illustrate the efficiency of the proposed algorithm. We implemented it using Cplex Callable library that is the C programming language linked with CPLEX 12.5 as a solver for quadratic programming problems. All our experiments are done on a Dell Inc. OptiPlex 9020 with Processor Intel (R) Core (TM) i7-4770 CPU@ 3.40 GH and RAM 8.00 GB. Lubuntu operating system 16.04.01 is used for the implementation.

For all experiments below, randomly generated instances have been used as we have developed a simple code that can generate any instance like: Transportation Problem (TP), Assignment Problem (AP) and Traveling Salesman Problem (TSP), all these instances include multiple objectives. We have tested our algorithm using these problems with different size (n) and different number of objectives (p).

To investigate the efficiency of the proposed algorithm, we compare it with the weighted goal programming approach Deb (1999). The proposed algorithm (QMOTP) gives significant enhancement on the compromise solution over the weighted goal programming (WGP) approach as presented in Tables 2, 3, 4. In all these tables, n refers to the number of variables (problem size) and p assigned to the number of objectives sized. Our algorithm obviously gives a better compromise solution than the current algorithm.

Table 1: Transportation numerical example with two objective functions as below.

	Destination				**Supply**
Origin	11	13	17	14	250
	16	18	14	10	300
	21	24	13	10	400
Demand	200	225	275	250	
Origin	19	14	12	14	250
	10	13	14	17	300
	15	21	16	22	400
Demand	200	225	275	250	

Table 2: Comparison in terms of the compromise solution, when p = 2.

N	**Problem**	**QMOTP**	**WGP**
5 × 5	TP	16698	18821
5 × 5	TP	20132	29005
10 × 10	TP	72418	79517
10 × 10	TP	94993	106432
20 × 20	TP	385181	404059
20 × 20	**TP**	**213454**	**298543**
10 × 10	AP	2754	2950
10 × 10	AP	6599	7200
20 × 20	AP	15647	16327
20 × 20	AP	21872	22321
30 × 30	AP	33975	35029
30 × 30	AP	45657	47666
8 × 8	TSP	294263	303680
8 × 8	TSP	367898	398764
12 × 12	TSP	863069	880335
12 × 12	TSP	932001	960254

Table 3: Comparison in terms of the compromise solution, when $p = 3$.

N	Problem	QMOTP	WGP
5 × 5	TP	28708	31718
5 × 5	TP	56398	73245
10 × 10	TP	119281	129610
10 × 10	TP	234987	267455
20 × 20	TP	574464	601128
20 × 20	**TP**	**876645**	**920333**
10 × 10	AP	4962	5311
10 × 10	AP	7300	8209
20 × 20	AP	24485	25493
20 × 20	AP	19235	20333
30 × 30	AP	45323	46713
30 × 30	AP	37665	39636
8	TSP	462580	475839
8	TSP	328919	348777

Table 4: Comparison in terms of the compromise solution, when $p = 4,5,6$.

N	*P*	Problem	QMOTP	WGP
10 × 10	4	TP	91233	99424
10 × 10	4	TP	190162	203244
10 × 10	5	TP	248431	263655
10 × 10	5	TP	328989	341233
10 × 10	6	TP	261069	278755
10 × 10	**6**	**TP**	**442123**	**478934**
10 × 10	4	AP	8185	8675
10 × 10	4	AP	9234	9982
20 × 20	4	AP	32975	34251
20 × 20	4	AP	55987	61983
10 × 10	5	AP	10923	11529
10 × 10	5	AP	12365	13425
15 × 15	5	AP	9144	9613
15 × 15	5	AP	9800	1029
10 × 10	6	AP	11233	11844
10 × 10	6	AP	14567	15111

For performance illustration, we discussed some transportation instances (in bold) from Tables 2, 3 and 4. One can see in Table 2, in the two objectives transportation problem size 20 × 20, the proposed algorithm gives 213454 as a compromise solution while the weighted goal programming gives 298543 as a compromise solution. Also, in Table 3, in the three objectives transportation problem size 20 × 20, the proposed algorithm gives 876645 as a compromise solution while the weighted goal programming gives 920333 as a compromise solution. In addition, in Table 4, in six objectives transportation problem size 10 × 10, the proposed algorithm gives 442123 as a compromise solution while the weighted goal programming gives 478934 as a compromise solution.

6. Conclusion

In this research, we have proposed a sensible quadratic programming technique for dealing with multi-objective transportation problem (MOTP). The proposed quadratic programming multi-objective transportation (QMOTP) approach determines an optimum solution that can minimize the increase in deviations of the optimum solution from the individual optimum solution. The algorithm for the proposed technique written in C programming is implemented with CPLEX as a solver. The algorithm is tested using different multi-objectives problems such as Transportation, Assignment and Traveling Salesman with different size and different number of objectives. The efficiency of the proposed algorithm (QMOTP) is also compared with the weighted goal programming (WGP) approach, which revealed that QMOTP gives significant enhancement over the WGP approach and can be used as an efficient tool for dealing with an MOTP.

References

Al-Hasani, A., Al-Rabeeah, M., Kumar, S. and Eberhard, A. 2019. Finding all non-dominated points for a bi-objective generalized assignment problem. In Journal of Physics: Conference Series, volume 1218, page 012004. IOP Publishing.

Al-Rabeeah, M., Elias, M., Al-Hasani, A., Kumar, S. and Eberhard, E. 2019a. Computational enhancement in the application of the branch and bound method for linear integer programs and related models. International Journal of Mathematical, Engineering and Management Sciences, 4(5): 1140–1153.

Al-Rabeeah, M., Kumar, S., Al-Hasani, A., Munapo, E. and Eberhard, A. 2019b. Bi-objective integer programming analysis based on the characteristic equation. International Journal of System Assurance Engineering and Management, 10(5): 937–944.

Aneja, Y. P. and Nair, K. P. 1979. Bicriteria transportation problem. Management Science, 25(1): 73–78.

Bit, A., Biswal, M. and Alam, S. 1993. Fuzzy Programming Approach to Multiobjective Solid 2.

Chang, C.-T. 2007. Multi-choice goal programming. Omega, 35(4): 389–396.

Chang, C.-T. 2008. Revised multi-choice goal programming. Applied Mathematical Modelling, 32(12): 2587–2595.

Charnes, A., Cooper, W. W. and Ferguson, R. O. 1955. Optimal estimation of executive compensation by linear programming. Management Science, 1(2): 138–151.

Charnes, A. C. et al. 1961. Management models and industrial applications of linear programming. Technical report.

Dantzig, G. B. and Thapa, M. N. 2006. Linear programming 2: theory and extensions. Springer Science & Business Media.

Deb, K. 1999. Solving goal programming problems using multi-objective genetic algorithms. In Proceedings of the 1999 Congress on Evolutionary Computation-CEC99 (Cat. No. 99TH8406), 1: 77–84. IEEE.

El-Wahed, W. F. A. and Lee, S. M. 2006. Interactive fuzzy goal programming for multi-objective transportation problems. Omega, 34(2): 158–166.

Gupta, A. and Kumar, A. 2012. A new method for solving linear multiobjective transportation problems with fuzzy parameters. Applied Mathematical Modelling, 36(4): 1421–1430.

Hu, C.-F., Teng, C.-J. and Li, S.-Y. 2007. A fuzzy goal programming approach to multi-objective optimization problem with priorities. European Journal of Operational Research, 176(3): 1319–1333.

Ignizio, J. P. 1978. A review of goal programming: A tool for multiobjective analysis. Journal of the Operational Research Society, pp. 1109–1119.

Li, L. and Lai, K. K. 2000. A fuzzy approach to the multiobjective transportation problem. Computers & Operations Research, 27(1): 43–57.

Maity, G. and Kumar Roy, S. 2016. Solving a multi-objective transportation problem with nonlinear cost and multi-choice demand. International Journal of Management Science and Engineering Management, 11(1): 62–70.

Maleki, H. and Khodaparasti, S. 2008. Non-linear multi-objective transportation problem: A fuzzy goal programming approach. New Aspects of Urban Planning and Transportation.

Nunkaew, W. and Phruksaphanrat, B. 2009. A multiobjective programming for transportation problem with the consideration of both depot to customer and customer to customer relationships. In Proceedings of the International Multiconference of Engineers and Computer Scientists, 2: 439–445.

Opricovic, S. and Tzeng, G.-H. 2004. Compromise solution by mcdm methods: A comparative analysis of vikor and topsis. European Journal of Operational Research, 156(2): 445–455.

Paksoy, T. and Chang, C.-T. 2010. Revised multi-choice goal programming for multi-period, multi-stage inventory-controlled supply chain model with popup stores in guerrilla marketing. Applied Mathematical Modelling, 34(11): 3586–3598.

Rani, D., Gulati, T. and Garg, H. 2016. Multi-objective non-linear programming problem in intuitionistic fuzzy environment: optimistic and pessimistic viewpoint. Expert Systems with Applications, 64: 228–238.

Romero, C. 2004. A general structure of achievement function for a goal programming model. European Journal of Operational Research, 153(3): 675–686.

Tamiz, M. 2012. Multi-objective programming and goal programming: theories and applications, volume 432. Springer Science & Business Media.

Thie, P. R. and Keough, G. E. 2011. An introduction to linear programming and game theory. John Wiley & Sons.

Ulungu, E. L. and Teghem, J. 1994. Multi-objective combinatorial optimization problems: A survey. Journal of Multi-Criteria Decision Analysis, 3(2): 83–104.

Ustun, O. 2012. Multi-choice goal programming formulation based on the conic scalarizing function. Applied Mathematical Modelling, 36(3): 974–988.

Zhao, Y. Q., Braun, W. J. and Li, W. 1999. Northwest corner and banded matrix approximations to a markov chain. Naval Research Logistics (NRL), 46(2): 187–197.

CHAPTER 16

Analyzing Multi-Objective Fixed-Charge Solid Transportation Problem under Rough and Fuzzy-Rough Environments

Sudipta Midya **and** ***Sankar Kumar Roy****

1. Introduction

In the ancient *transportation problem* (TP), the distribution process is performed from distinct sources to destinations in a way that the objective function is minimized while it fulfils the supply and demand limits. The *solid transportation problem* (STP) is an enhanced version of the classical TP. In the conventional TP, supply constraints and demand constraints are considered whereas the STP deals with supply constraints, demand constraints and extra constraints, namely conveyance constraints (i.e., mode of transportation).

Most of the practical situations arise where an additional cost function (i.e., fixed-charge) is calculated in addition to the shipping cost in each path to move the product from the supply centre to the demand centre. For these cases, STP turns into *fixed-charge STP* (FSTP). The fixed penalty may be contemplated as booking fees of freight trains, toll costs for highway transportation, aircraft landing fees of an air station to name a few. For the industrial shipping problem, it is more acceptable to treat more than one goal function relative to one goal function. To take these goals, multiple goal functions are fitted contemporaries in the FSTP. For specification, the goal functions can be taken as the total cost of transportation including fixed-charge, delivery time, and damage rate of product, which are minimized. To manipulate this kind of problem in practical circumstances, we consider *multi-objective FSTP* (MFSTP) in the suggested study.

Generally, in a realistic situation, the parameters in a STP/FSTP are vague and inexact in nature due to various aspects such as insufficient information about input variables, fluctuation of economic market, weather condition and other inconsiderable components. To manoeuvre uncertainty in real-world distribution problems, researchers have studied STP/FSTP in numerous inexact environments, like stochastic, interval, and fuzzy, amongst others.

However, a rough set is a key mathematical aid to analyse vague and inexact information about variables connected to practical FSTP. In this view, a MFSTP model is designed with parameters which are rough variables. Apart from these circumstances some practical cases appear in industrial sectors for distribution where an uncertain framework is not suitable to deal with the state of events. From this outlook, a fuzzy-rough MFSTP model is suggested where the parameters are treated as fuzzy-rough variables. Moreover, two individual circumstances are presented for treating rough and fuzzy-rough variables as the parameters of the designed MFSTP's. Thus, this study retains a realistic significance in this consideration.

From publications view, numerous research articles have been published on STP/FSTP. As for our knowledge until now, this is the first attempt to develop rough MFSTP and fuzzy-rough MFSTP models and to solve the models by intuitionistic fuzzy TOPSIS. The major outlines of the chapter are listed as below.

- Rough MFSTP and fuzzy-rough MFSTP models are designed.
- Using expected value operator rough MFSTP and fuzzy-rough MFSTP models are transformed to their similar crisp forms.
- Crisp MFSTP's are solved by a new methodology intuitionistic fuzzy TOPSIS to derive Pareto-optimal solutions.
- Two application examples are provided in separate circumstances to analyse how rough and fuzzy-rough variables tackle vagueness and inexactness in industrial sectors for distribution.

The remainder of the chapter is structured as follows. A literature survey is discussed in Section 2. Section 3 presents background behind this study. The primary concept of rough set, fuzzy numbers and fuzzy-rough variables is depicted in Section 4. Section 5

Department of Applied Mathematics with Oceanology and Computer Programming, Vidyasagar University, Midnapore-721102, West Bengal, India.
Email: sudiptamidya932@gmail.com
* Corresponding author: sankroy2006@gmail.com

is represented by introducing notations and assumptions. Rough MFSTP model and fuzzy-rough MFSTP model and their identical deterministic forms are designed in Section 6. The solution approach for multiple objective FSTP is presented in Section 7. In Section 8, two practical examples on MFSTP are stated. Results and discussions are displayed in Section 9. Finally, conclusion with forward-looking studies are given in Section 10.

2. Related Literature Survey

STP was first initiated by Haley 1962 whereas FCTP was originated from Hirsch and Dantzig 1968. Thereafter, various evolvements and moderations on FCTP and FSTP have been made. In the last decade, numerous research articles have been published on multiple objective TP/FCTP by researchers. Specifically, Midya and Roy 2014 solved a single-sink stochastic FCTP with multiple objective grounds. Maity et al. 2019b formulated a multimodal TP for application of artificial intelligence. Atteya 2016 stated rough programming to a multiple objective problem while Xu and Zhao 2008 designed a fuzzy rough multiple objective policy-making problem and solved an inventory model. Roy et al. 2017a, b investigated a two-stage grey TP by considering a multiple objective function, and proposed a conic scalarization methodology to solve a TP under multiple goals and multiple choices frameworks. Ebrahimnejad and Verdegay 2018 proposed a novel technique to solve intuitionistic fuzzy TP. Maity et al. 2019a solved a dual hesitant fuzzy TP by a new method. Roy et al. 2018a presented a novel method to solve a TP with multiple goal functions under intuitionistic fuzzy circumstances.

Uncertainty is the primary factor of practical distribution systems. To tackle inexactness in industrial distribution systems, researchers have studied STP and FCTP/FSTP in several imprecise environments. Some of the articles are described as: Verdegay 1999 solved a fuzzy STP by an evolutionary procedure. Midya and Roy 2017 formulated FCTP under interval and rough interval frameworks. Gupta et al. 2016 studied a fuzzy STP with a fixed-cost and multiple items ground. Roy et al. 2018b designed a FCTP with multiple goal functions under rough and random rough frameworks. Zhang et al. 2016 proposed a methodology to solve a FSTP with uncertain variables. Midya and Roy 2020 presented a multiple objective FCTP with rough variables. Tao and Xu 2012 proposed a multiple objective model with rough variables and solved a STP. Roy and Midya 2019 formulated an intuitionistic fuzzy MFSTP with product blending. Zavardehi et al. 2013 solved a fuzzy FSTP by metaheuristics. Recently, Roy et al. 2019 studied a FSTP with multiple goals and multiple items grounds under twofold uncertain frameworks and Midya et al. 2020 developed an intuitionistic fuzzy FSTP model in a green supply chain.

Generally, in practical FSTPs, the objective functions are of the conflicting type. So, the notion of Pareto-optimal solution appears to handle this state of events. Intuitionistic fuzzy TOPSIS is a novel methodology to solve multiple objective FSTPs that provides a Pareto-optimal solution. Basically, TOPSIS transfers a multiple objective optimization problem into its bi-objective form. Abo-Sinna et al. 2008 used TOPSIS to solve nonlinear programming problems with multiple goal functions. After that, Damghani et al. 2013 solved project selection problems with multiple periods using TOPSIS.

3. Motivation for this Study

Traditional FCTPs/STPs require exact input variables, whereas a real-life FSTP is usually modelled under inexact framework, hence, input information may not always be obtainable for application of practical distribution systems. In this view, two distinct models of MFSTP under rough and fuzzy rough frameworks are formulated.

Rough set Pawlak 1982 and fuzzy set Zadeh 1965 are two familiar mathematical equipments for tackling different kinds of inaccurate, inexact, and vague data. However, these theories include individual incidental outlines. Rough set is mainly varied to contrast with exactness whereas fuzzy set is defined by the membership function. Rough set is described with separate approaches for vagueness and uncertainty. Usually, rough sets are the consequences of approximating exact sets using similar classes.

Aside from fuzziness, roughness is a different kind of uncertainty. The problem appears when *decision maker* (DM) does not have exact input variables, and the coefficients in MFSTP are taken approximately. Based on this fact, the feasible domain of MFSTP is mutable. To fit this status, rough interval Rebelledo 2006 is incorporated in MFSTP when the feasible domain of MFSTP is more pliable. To realize the fact, an example is discussed as below.

Assuming D indicates "the demand of sugar" of a metropolis, which fluctuates regularly between 7 to 12 quintals. Usually, it changes in a day within 8 to 10 quintals. Other values are taken outside of the interval [8,10] in special events such as carnivals, festive seasons and others. This implies that every day demand of sugar can be represented as: D = ([8, 10], [7, 12]) a rough interval. From this view, the parameters of MFSTPs are considered as rough variables.

In many intricate real-world situations, which arise in a delivery system we cannot tackle the situation by an uncertain framework. Under this circumstance, the notion of twofold uncertain framework Dubois and Prade 1987 is initiated. Thus, fuzzy-rough variables are treated as the parameters of MFSTP in this chapter. A separate example is provided to present this phenomenon.

Woollen blanket is a perfect seasonal product and it is required mainly in the winter season, i.e., demand of blanket occurs for a period of time. For prediction of the demand of blankets for a winter season, one may employ a fuzzy number (i.e., m-α, m, m+β) to approximate it. Here, α and β are the left and right spreads of the fuzzy number. According to the previous knowledge or the experiences of experts or affluence of information, the mid value m of the fuzzy number is not exact and it fluctuates year to year. So, *the DM is unable to achieve a preferable decision to foretell the demand of woollen blankets in the winter season of a year. Hence, it is suitable to depict a rough interval to represent the mid value of m. Therefore, the demand of woollen blankets can be expressed as a fuzzy-rough variable to develop a more acceptable strategy.*

Thus, from a practical point of view we investigate a rough MFSTP and fuzzy-rough MFSTP in our suggested study. In this view, MFSTP models are analyzed under rough and fuzzy-rough frameworks separately.

4. Preliminaries

This section presents the primary definitions of rough set, rough intervals and fuzzy numbers. Several definitions and theorems associated to a rough variable and fuzzy-rough variable are described.

Definition 4.1 [Liu 2002]: Assume Λ be a nonempty set, A be a σ-algebra of subsets of Λ, Δ be an object in A and π be a nonnegative, real and additive set function. Then, $(\Lambda, \Delta, A, \pi)$ is called a rough space.

Rough sets and their approximations: Consider a set of elements U which is called the "universe". The deficiency of knowledge about the objects of U is represented by an indiscernibility relation, $R \subseteq U \times U$. Because of simplicity, an equivalence relation R is considered. Assuming X is a subset of U. Our intention is to identify the set X to relate to R. The primary concepts of a rough set are described as below.

Definition 4.2 Set X is said to be *precise* (i.e., *crisp*) to relate to R, if the border area of X is null. Set X is said to be *imprecise* (i.e., *rough*) to relate to R, if the frontier area of X is nonempty.

$R(x)$ is indicated as an equivalence class of R which is denoted by any of its elements x. Our insufficiency knowledge about universe in a sure sense is expressed by an indiscernibility relation. Equivalence classes of an indiscernibility relation, are said to be granules of knowledge, which are generated by R and they represent a basic part of knowledge.

Definition 4.3 [Xu and Tao 2012]: *Lower approximation* of X to relate to R is symbolized by $\underline{R}(X)$ and is characterized as below:

$$\underline{R}(X) := \bigcup_{x\in U}\{R(x) : R(x) \subseteq X\}.$$

Upper approximation of X to relate to R is indicated by $\overline{R}(X)$ and is depicted as stated below:

$$\overline{R}(X) := \bigcup_{x\in U}\{R(x) : R(x) \cap X \neq \phi\}.$$

Boundary region of X to relate to R is symbolized by $BN_R(X)$ and is characterized as follows:

$$BN_R(X) := \overline{R}(X) \cap \underline{R}(X).$$

The pictorial presentation of rough set is displayed in Figure 1.

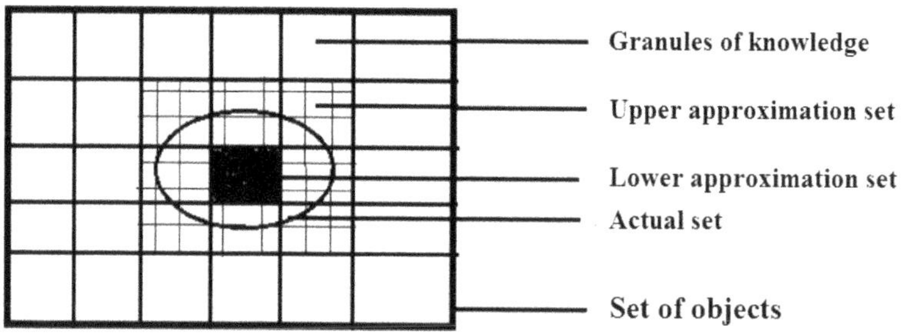

Figure 1: Pictorial presentation of rough set.

Arithmetic of rough intervals: Arithmetic of rough intervals (RIs) are identical with arithmetic of ordinary intervals. Arithmetic of RIs [Xu and Tao 2012] are described in the following way.

Assume RI = ([b2, b3],[b1, b4]), b1 $\leq$ b2 $<$ b3 $\leq$ b4, and RI' = ([d2, d3], [d1, d4]), d1 $\leq$ d2 $<$ d3 $\leq$ d4, be two rough intervals, $o \in \{+, -, \times \div\}$ indicates a binary operator on the set of ordinary intervals. So, the arithmetic of rough intervals is symbolized by $RI\ o\ RI'$ and presented as:

Addition: $RI + RI'$ = ([b2+ d2, b3+ d3], [b1+ d1, b4+ d4]).

Subtraction and multiplication on rough intervals are defined identically as addition.

Division: $RI \div RI'$ = ([b2 $\div$ d3, b3 $\div$ d2], [b1 $\div$ d4, b4 $\div$ d1]) if $0 \notin$ [d1, d4].

Expected-value on rough interval: Expected-value on rough interval is determined in the similar way as of expected value of a random variable in probability theory.

Definition 4.4 [Liu 2002]: Assuming ξ be a rough variable on the rough space (Λ, Δ, *A*, π). Then expected value of ξ is described as:

$$E(\xi) = \int_0^\infty Tr\{\xi \geq r\}dr - \int_{-\infty}^0 Tr\{\xi \leq r\}dr,$$

provided that out of the two integrals one exists, where *E* is an operator of expected-value and *Tr* indicates trust measure.

Theorem 4.1 [Liu 2002]: Assuming $\xi = ([a, b], [c, d])$ $(c \leq a < b \leq d)$ indicates a rough interval. Then expected value of ξ is symbolized by $E(\xi)$ and depicted as:

$$E(\xi) = \frac{1}{4}(a+b+c+d).$$

Theorem 4.2 [Xu and Tao 2012]: Assuming $\xi = ([x2, x3], [x1, x4])$ and $\zeta = ([y2, y3], [y1, y4])$ indicates the rough variables with definite expected values. Then for arbitrary *a* and *b* (real numbers), it can be written as: $E[a\xi + b\zeta] = aE[\xi] + bE[\zeta]$.

Definition 4.5 (Fuzzy Number) [Zimmerman 1978]: Assuming $\tilde{A}$ be a fuzzy set defined on *R*; (real number set) is said to be a fuzzy number, if its membership function $\mu_{\tilde{A}} : R \to [0,1]$ has the following criteria:

1) $\mu_{\tilde{A}}$ is convex, i.e., $\mu_{\tilde{A}}\{(1-\lambda)x_1 + \lambda x_2\} \geq \min\{\mu_{\tilde{A}}(x_1), \mu_{\tilde{A}}(x_2)\}\ \forall x_1, x_2 \in R,\ 0 \leq \lambda \leq 1.$
2) $\mu_{\tilde{A}}$ is normal, i.e., there exists an $x \in R$ such that $\mu_{\tilde{A}}(x) = 1$.
3) $\mu_{\tilde{A}}$ is piecewise continuous.

Definition 4.6 (Triangular Fuzzy Number) [Zimmerman 1978]: A triangular fuzzy number is symbolized as $\tilde{A} = (a_1, a_2, a_3)$. This triplet is indicated as a membership function which follows the criteria stated as below.

- membership function increases between a_1 to a_2.
- membership function decreases between a_2 to a_3, and
- $a_1 \leq a_2 \leq a_3$.

The membership function of $\tilde{A}$ is symbolized as $\mu_{\tilde{A}}(x)$ (as displayed in Figure 2) and presented by:

$$\mu_{\tilde{A}}(x) = \begin{cases} 0, & x < a_1, \\ (x-a_1)/(a_2-a_1), & a_1 \leq x \leq a_2, \\ (a_3-x)/(a_3-a_2), & a_2 \leq x \leq a_3, \\ 0, & x > a_3. \end{cases}$$

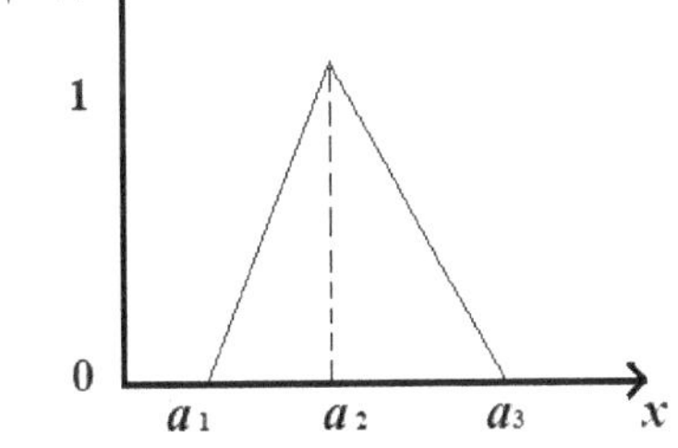

Figure 2: Graph of a triangular fuzzy number.

Definition 4.7 (Fuzzy-rough variable) [Xu and Zhou 2010]: A fuzzy-rough variable is a mapping ζ from a rough space (Λ, Δ, *A*, π) to a cluster of fuzzy variables such that for an arbitrary Borel set *B* of *R*, the function $\zeta^*(B)(\lambda) = Pos\{\zeta(\lambda) \in B\}$ is a measurable function of λ, where *Pos* indicates possibility.

Example 4.1: Assuming ζ = (ρ-1, ρ, ρ+1) with ρ = ([*a*, *b*], [*c*, *d*]) $(0 < c \leq a < b \leq d)$, where the triplex indicates a triangular fuzzy number and ρ is a rough variable; then ζ is a fuzzy-rough variable.

For usability, a fuzzy-rough variable ζ = (ρ-1, ρ, ρ+1) with ρ = ([*a*, *b*], [*c*, *d*]), represented by $\rho \dashv ([a, b], [c, d])$.

Definition 4.8 [Xu and Zhou 2010]: Choosing ζ be a rough variable on the rough space (Λ, Δ, *A*, π). Then expected value of ζ is described as:

$$E(\zeta)=\int_{0}^{\infty}Tr\{\lambda\in\Lambda:E[\zeta(\lambda)]\geq r\}dr-\int_{-\infty}^{0}Tr\{\lambda\in\Lambda:E[\zeta(\lambda)]\leq r\}dr,$$

provided that out of the two integrals one exists, where *E* and *Tr* indicate expected-value operator and trust measure respectively.

Theorem 4.2 [Xu and Zhou 2010]: Considering ζ = (ρ-α, ρ, ρ+β) with ρ = ([*a*, *b*], [*c*, *d*]) be a fuzzy-rough variable where α, β > 0 and $0 < c \leq a < b \leq d$. Then the expected value of ζ is symbolized by $E(\zeta)$ and is depicted as:

$$E(\zeta)=\frac{1}{4}[a+b+c+d+\alpha+\beta].$$

Theorem 4.3 [Xu and Zhou 2010]: Letting ζ and ς be the fuzzy-rough variables with definite expected values. Then, for arbitrary *u* and *v* (real numbers), it is written as: $E[u\zeta+v\varsigma]=uE[\zeta]+vE[\varsigma]$.

5. Notations and Assumptions

The notations and assumptions used to formulate the models are listed below.

Notations

m	:	number of supply centres,
n	:	number of demand centres,
p	:	number of carriers (i.e., different shipping modes),
x_{ijk}	:	unit quantity of the commodity to be moved from the ith supply centre to the jth demand centre by the kth carrier,
$\eta(x_{ijk})$	:	binary variable receives the value "1" if supply centre *i* is used, and else "0",
$\bar{c}_{ijk}$, $\bar{\tilde{c}}_{ijk}$	:	rough and fuzzy-rough cost of transportation respectively for unit quantity of the goods moved from the ith supply centre to the jth demand centre by the kth carrier,
$\bar{f}_{ijk}$, $\bar{\tilde{f}}_{ijk}$	:	rough and fuzzy-rough fixed cost respectively connected to the ith supply centre to the jth demand centre by the kth carrier,
$\bar{t}_{ijk}$, $\bar{\tilde{t}}_{ijk}$	:	rough and fuzzy-rough shipping time of the commodity respectively from the ith supply centre to the jth demand centre by the kth carrier,
$\bar{d}_{ijk}$	:	rough damaged rate of products for unit quantity of goods from the ith supply centre to the jth demand centre by the kth carrier,
$\bar{\tilde{g}}_{ijk}$	:	fuzzy-rough packing cost for unit packing of the commodity from the ith supply centre to the jth demand centre by the kth carrier,
$\bar{a}_i$,$\bar{\tilde{a}}_i$	:	rough and fuzzy-rough availability of the commodity respectively at the ith supply centre,
$\bar{b}_j$,$\bar{\tilde{b}}_j$	:	rough and fuzzy-rough requirement of the goods respectively at the jth demand centre,
$\bar{e}_k$,$\bar{\tilde{e}}_k$	:	rough and fuzzy-rough capacity of the kth carrier respectively,
$\bar{Z}_K$,$\bar{\tilde{Z}}_K$	:	objective functions in rough and fuzzy-rough types (K = 1, 2, 3),
Z_K	:	objective functions in crisp type (K = 1, 2, 3), where $Z_K = E[\bar{Z}_K]$ or $E[\bar{\tilde{Z}}_K]$ and E indicates the operator of expected value.

Assumptions

1. $\bar{a}_i$, and $\bar{b}_i$, $\bar{\tilde{b}}_i > 0\ \forall\ i, j$.
2. Suspect that no products deteriorate during the movement of the product for Model 1.
3. Rough variable and triangular fuzzy-rough variable are positive in all of its parts.

6. Mathematical model

We spilt this section mainly into two subsections. Firstly, we describe the MFSTP model with parameters taken as rough intervals and the rest of the MFSTP model is represented with parameters treated as fuzzy-rough variables.

a. Rough MFSTP model

In our suggested rough MFSTP model, three objective functions are depicted in which the first goal presents the total cost of transportation (direct cost and fixed cost), the second goal considers the delivery time and the third goal refers to the deterioration rate of commodities; all goal functions are minimization type. Assuming there are m plants (i = 1, 2,..., m) from which the goods are moved to the n demand centres (j = 1, 2, ..., n) by p (k = 1, 2,..., p) different kinds of carriers. The problem is to determine the unspecified quantity x_{ijk} to be moved from the ith supply centre to the jth demand centre by the kth carrier in a way that the objective functions have taken minimum values. The designed MFSTP can be presented as below:

Model 1

$$\min \bar{Z}_1 = \sum_{i=1}^{m}\sum_{j=1}^{n}\sum_{k=1}^{p}[\bar{c}_{ijk}x_{ijk} + \bar{f}_{ijk}\eta(x_{ijk})],$$

$$\min \bar{Z}_2 = \sum_{i=1}^{m}\sum_{j=1}^{n}\sum_{k=1}^{p}[\bar{t}_{ijk}\eta(x_{ijk})],$$

$$\min \bar{Z}_3 = \sum_{i=1}^{m}\sum_{j=1}^{n}\sum_{k=1}^{p}\bar{d}_{ijk}x_{ijk},$$

$$\text{subject to} \sum_{j=1}^{n}\sum_{k=1}^{p}x_{ijk} \le \bar{a}_i \ (i=1, 2, ..., m),$$

$$\sum_{i=1}^{m}\sum_{k=1}^{p}x_{ijk} \ge \bar{b}_j \ (j=1, 2, ..., n),$$

$$\sum_{i=1}^{m}\sum_{j=1}^{n}x_{ijk} \le \bar{e}_k \ (k=1, 2, ..., p),$$

$$x_{ijk} \ge 0 \ \forall i, j, k,$$

$$\eta(x_{ijk}) = 0 \text{ if } x_{ijk} = 0,$$

$$\eta(x_{ijk}) = 1 \text{ if } x_{ijk} = 1.$$

Feasibility criteria of Model 1 are:

$$\sum_{i=1}^{m}\bar{a}_i \ge \sum_{j=1}^{n}\bar{b}_j, \text{ and } \sum_{k=1}^{p}\bar{e}_k \ge \sum_{j=1}^{n}\bar{b}_j.$$

Equivalent deterministic form of Model 1:

The designed rough MFSTP is a theoretical model. So, it is required to transform Model 1 into its similar crisp form. Employing E (expected-value operator) to transform Model 1 into its crisp version (i.e., Model 2) by using Theorems 4.1 and 4.2. The crisp version of rough MFSTP can be formulated in the following manner.

Model 2

$$\min \ E[\bar{Z}_1] = \sum_{i=1}^{m}\sum_{j=1}^{n}\sum_{k=1}^{p}\left[E[\bar{c}_{ijk}]x_{ijk} + E[\bar{f}_{ijk}]\eta(x_{ijk})\right], \tag{6.1}$$

$$\min \ E[\bar{Z}_2] = \sum_{i=1}^{m}\sum_{j=1}^{n}\sum_{k=1}^{p}E[\bar{t}_{ijk}]\eta(x_{ijk}), \tag{6.2}$$

$$\min \ E[\bar{Z}_3] = \sum_{i=1}^{m}\sum_{j=1}^{n}\sum_{k=1}^{p}E[\bar{d}_{ijk}]x_{ijk}, \tag{6.3}$$

$$\text{subject to} \sum_{j=1}^{n}\sum_{k=1}^{p}x_{ijk} \le E[\bar{a}_i] \ (i=1, 2, ..., m), \tag{6.4}$$

$$\sum_{i=1}^{m}\sum_{k=1}^{p}x_{ijk} \ge E[\bar{b}_j] \ (j=1, 2, ..., n), \tag{6.5}$$

$$\sum_{i=1}^{m}\sum_{j=1}^{n} x_{ijk} \le E[\bar{e}_k]\ (k=1, 2, ..., p), \tag{6.6}$$

$$x_{ijk} \ge 0\ \forall i, j, k,$$
$$\eta(x_{ijk}) = 0 \text{ if } x_{ijk} = 0, \tag{6.7}$$
$$\eta(x_{ijk}) = 1 \text{ if } x_{ijk} = 1.$$

Herein, feasibility criteria of Model 2 are $\sum_{i=1}^{m} E[\bar{a}_i] \ge \sum_{j=1}^{n} E[\bar{b}_j]$, and $\sum_{k=1}^{p} E[\bar{e}_k] \ge \sum_{j=1}^{n} E[\bar{b}_j]$.

Definition 6.1: A feasible solution $x^o = (x^o_{ijk} : i=1, 2, ..., m, j=1, 2, ..., n, k=1, 2, ..., p)$ of Model 2 is called a Pareto-optimal solution if another feasible solution does not exist $x = (x_{ijk} : i=1, 2, ..., m, j=1, 2, ..., n, k=1, 2, ..., p)$ in such a way

$$E[\bar{Z}_K(x)] \le E[\bar{Z}_K(x^\circ)] \text{ for } K = 1, 2, 3, \text{ and}$$

$$E[\bar{Z}_K(x)] < E[\bar{Z}_K(x^\circ)] \text{ for at least one } K.$$

Definition 6.2: If Model 2 has a Pareto-optimal solution x^o, then it is referred to as a Pareto-optimal solution of the rough expected-value.

b. Fuzzy-rough MFSTP model

In our designed MFSTP, we assign three goal functions in which the first goal indicates the total cost of transportation, the second goal represents the delivery time of shipping and the third goal considers the packing cost. Considering that there are m plants ($i = 1, 2, \ldots, m$) from which the commodity is moved to the n demand centres ($j = 1, 2, \ldots, n$) by p ($k=1,2, ..., p$) distinct types of carriers. We determine the unspecific quantity x_{ijk} to ship from the ith supply centre to the jth demand centre by the kth carrier in a way that the objective functions take a minimum value. The suggested MFSTP can be constructed as below:

Model 3

$$\min \tilde{\bar{Z}}_1 = \sum_{i=1}^{m}\sum_{j=1}^{n}\sum_{k=1}^{p}[\tilde{\bar{c}}_{ijk} x_{ijk} + \tilde{\bar{f}}_{ijk}\eta(x_{ijk})],$$

$$\min \tilde{\bar{Z}}_2 = \sum_{i=1}^{m}\sum_{j=1}^{n}\sum_{k=1}^{p}[\tilde{\bar{t}}_{ijk}\eta(x_{ijk})],$$

$$\min \tilde{\bar{Z}}_3 = \sum_{i=1}^{m}\sum_{j=1}^{n}\sum_{k=1}^{p}\tilde{\bar{d}}_{ijk} x_{ijk},$$

$$\text{subject to } \sum_{j=1}^{n}\sum_{k=1}^{p} x_{ijk} \le \tilde{\bar{a}}_i\ (i=1, 2, ..., m),$$

$$\sum_{i=1}^{m}\sum_{k=1}^{p} x_{ijk} \ge \tilde{\bar{b}}_j\ (j=1, 2, ..., n),$$

$$\sum_{i=1}^{m}\sum_{j=1}^{n} x_{ijk} \le \tilde{\bar{e}}_k\ (k=1, 2, ..., p),$$

the constraints (6.7).

Feasibility conditions of Model 3 are $\sum_{i=1}^{m} \tilde{\bar{a}}_i \ge \sum_{j=1}^{n} \tilde{\bar{b}}_j$ and $\sum_{k=1}^{p} \tilde{\bar{e}}_k \ge \sum_{j=1}^{n} \tilde{\bar{b}}_j$.

Similar crisp form of Model 3:

The formulated MFSTP is a theoretical model because of the presence of fuzzy-rough variables. So, we employ E (expected-value operator) to convert Model 3 into a similar expected value model (i.e., Model 4) by using Theorems 4.2 and 4.3. The deterministic form of fuzzy-rough MFSTP can be designed as:

Model 4

$$\min E[\tilde{\bar{Z}}_1] = \sum_{i=1}^{m}\sum_{j=1}^{n}\sum_{k=1}^{p}\Big[E[\tilde{\bar{c}}_{ijk}]x_{ijk} + E[\tilde{\bar{f}}_{ijk}]\eta(x_{ijk})\Big], \tag{6.8}$$

$$\min E[\tilde{\bar{Z}}_2] = \sum_{i=1}^{m}\sum_{j=1}^{n}\sum_{k=1}^{p} E[\tilde{\bar{t}}_{ijk}]\eta(x_{ijk}), \quad (6.9)$$

$$\min E[\tilde{\bar{Z}}_3] = \sum_{i=1}^{m}\sum_{j=1}^{n}\sum_{k=1}^{p} E[\tilde{\bar{d}}_{ijk}]x_{ijk}, \quad (6.10)$$

$$\text{subject to } \sum_{j=1}^{n}\sum_{k=1}^{p} x_{ijk} \le E[\tilde{\bar{a}}_i] \ (i=1, 2, ..., m), \quad (6.11)$$

$$\sum_{i=1}^{m}\sum_{k=1}^{p} x_{ijk} \ge E[\tilde{\bar{b}}_j] \ (j=1, 2, ..., n), \quad (6.12)$$

$$\sum_{i=1}^{m}\sum_{j=1}^{n} x_{ijk} \le E[\tilde{\bar{e}}_k] \ (k=1, 2, ..., p), \quad (6.13)$$

the constraints (6.7).

Feasibility criteria of Model 4 are $\sum_{i=1}^{m} E[\tilde{\bar{a}}_i] \ge \sum_{j=1}^{n} E[\tilde{\bar{b}}_j]$ and $\sum_{k=1}^{p} E[\tilde{\bar{e}}_k] \ge \sum_{j=1}^{n} E[\tilde{\bar{b}}_j]$.

Definition 6.3: A feasible solution $x^* = (x^*_{ijk} : i=1, 2, ..., m, j=1, 2, ..., n, k=1, 2, ..., p)$ of Model 4 is called a Pareto-optimal solution if another feasible solution $x = (x_{ijk} : i=1, 2, ..., m, j=1, 2, ..., n, k=1, 2, ..., p)$ does not exist in such a way that $E[\tilde{\bar{Z}}_K(x)] \le E[\tilde{\bar{Z}}_K(x^*)]$ for $K = 1, 2, 3$, and $E[\tilde{\bar{Z}}_K(x)] < E[\tilde{\bar{Z}}_K(x^*)]$ for at least one K.

Definition 6.4: If Model 4 has a Pareto-optimal solution x^*, then it is stated as a Pareto-optimal solution of fuzzy-rough expected-value.

7. Solution Methodology

Herein, two crisp versions of MFSTP (i.e., Model 2 and Model 4) are solved by intuitionistic fuzzy TOPSIS (a new methodology).

Intuitionistic fuzzy TOPSIS: Hwang and Yoon 1981 originated TOPSIS for multiple characteristics decision making problems to extract Pareto-optimal solutions. Primary concept of TOPSIS is: the alternatives are selected in a way in which the smallest path from the *Positive Ideal Solution* (PIS) and the greatest path from the *Negative Ideal Solution* (NIS) are chosen. Intuitionistic fuzzy TOPSIS is a hybrid methodology where the notion of intuitionistic fuzzy programming is included with the TOPSIS. It was initiated by Roy and Midya 2019 to extract Pareto optimal solutions from multiple objective STPs. The important feature of intuitionistic fuzzy TOPSIS is that it produces a preferable Pareto-optimal solution and fulfills the satisfactory and dissatisfactory degrees of the objective functions. From an application point of view, it is simple to apply (due to a lesser calculation burden) for solving multiple objective TPs/STPs and similar kinds of problems. Intuitionistic fuzzy TOPSIS is presented as below:

Step 1: Calculate separately the minimum and maximum values for the objective functions of crisp MFSTP model (i.e., Model 4).

Step 2: Assign PIS (Z^+) and NIS (Z^-) for Model 4, which are expressed as below:

$$Z^+ = (Z_1^+, Z_2^+, Z_3^+),$$
$$Z^- = (Z_1^-, Z_2^-, Z_3^-),$$

we describe Z_K^+ and Z_K^- for each K ($K = 1; 2; 3$) are in the following manner:

$Z_K^+ = \min Z_K$ subject to constraints (6.11) and (6.13); (6.7),

$Z_K^- = \max Z_K$ subject to constraints (6.11) and (6.13); (6.7).

It is noticed that the objective functions scale in MFSTP with $Z_K^+ < Z_K^-$ are approximated by $Z_K^- - Z_K^+$ ($K = 1; 2; 3$).

Step 3: Determine the distance functions separately from PIS (i.e., $d_u^{PIS}(x)$) and from NIS (i.e., $d_u^{NIS}(x)$) by using PIS and NIS which are depicted as:

$$d_q^{PIS}(x) = \left[\sum_{K=1}^{3}\left(W_K \frac{Z_K(x) - Z_K^+}{Z_K^- - Z_K^+}\right)^q\right]^{\frac{1}{q}}, \quad (7.14)$$

$$d_q^{NIS}(x) = \left[\sum_{K=1}^{3} \left(W_K \frac{Z_K^- - Z_K(x)}{Z_K^- - Z_K^+} \right)^q \right]^{\frac{1}{q}}, \tag{7.15}$$

$$\sum_{K=1}^{3} W_K = 1,\ W_K \geq 0\ \forall\ K.$$

Here, W_K ($K = 1, 2, 3$) in (7.14) and (7.15) indicate the weights corresponding to the objective functions. Taking the weights as priorities for the three objective functions which are $W_1 = 0.5$, $W_2 = 0.3$, $W_3 = 0.2$ respectively, for Model 4. The role of indices $q(= 1, 2, \cdots, \infty)$ is to lead the solution in TOPSIS. Usually, $q = 1$, $q = 2$, and $q = \infty$ are broadly used to tackle optimization problems with multiple goals. Distinct values of q represent different distances, like, $q = 1$ depicts the Manhattan distance, $q = 2$ represents the Euclidean distance, and $q = \infty$ describes the Tchebycheff distance. Other distance functions have no physical meaning in application view so, they are useless.

Step 4: Put $q = 2$ into the Equations (7.14) and (7.15) model 4 is converted into a bi-objective problem to produce Pareto-optimal solutions which are stated below.

minimize $d_2^{PIS}(x)$,

maximize $d_2^{NIS}(x)$,

subject to constraints (6.11) – (6.13) and (6.7).

Step 5: Determine the values of minimize $d_2^{PIS}(x)$ and to maximize $d_2^{NIS}(x)$ separately with the constraints (6.11) and (6.13), and (6.7). The payoff table of ideal solutions (displayed in Table 1) is composed as below:
For suitability, we use the symbols as: $(d_2^{PIS})^* = d_2^{PIS}(x^{PIS})$, $(d_2^{NIS})^* = d_2^{NIS}(x^{NIS})$, $(d_2^{PIS})^\circ = d_2^{PIS}(x^{NIS})$, $(d_2^{NIS})^\circ = d_2^{NIS}(x^{PIS})$.

Table 1: Payoff table of ideal solutions.

	$d_2^{PIS}(x)$	$d_2^{NIS}(x)$
x^{PIS}	$(d_2^{PIS})^*$	$(d_2^{NIS})^\circ$
x^{NIS}	$(d_2^{PIS})^\circ$	$(d_2^{PIS})^*$

Step 6: Using the preference idea, the membership functions $\mu_1(x)$ and $\mu_2(x)$, and non-membership functions $v_1(x)$ and $v_2(x)$ are constructed for two goal functions respectively from **Step 5**. They are presented as:

$$\mu_1(x) = \begin{cases} 1, & if\ d_2^{PIS}(x) \leq \left(d_2^{PIS}\right)^*, \\ \left(\left(d_2^{PIS}\right)^\circ - d_2^{PIS}(x)\right) \Big/ \left(\left(d_2^{PIS}\right)^\circ - \left(d_2^{PIS}\right)^*\right) & if\ \left(d_2^{PIS}\right)^* \leq d_2^{PIS}(x) \leq \left(d_2^{PIS}\right)^\circ, \\ 0 & if\ \left(d_2^{PIS}\right)^\circ \leq d_2^{PIS}(x), \end{cases}$$

and

$$\mu_2(x) = \begin{cases} 1, & if\ \left(d_2^{NIS}\right)^* \leq d_2^{NIS}(x), \\ \left(d_2^{NIS}(x) - \left(d_2^{NIS}\right)^\circ\right) \Big/ \left(\left(d_2^{NIS}\right)^* - \left(d_2^{NIS}\right)^\circ\right) & if\ \left(d_2^{NIS}\right)^\circ \leq d_2^{NIS}(x) \leq \left(d_2^{NIS}\right)^*, \\ 0 & if.d_2^{NIS} \leq \left(d_2^{NIS}\right)^\circ, \end{cases}$$

$$v_1(x) = \begin{cases} 0, & if.d_2^{PIS}(x) \leq \left(d_2^{PIS}\right)^*, \\ \left(d_2^{PIS}(x) - \left(d_2^{PIS}\right)^*\right) \Big/ \left(\left(d_2^{PIS}\right)^* + \beta_1 - \left(d_2^{PIS}\right)^*\right) & if\ \left(d_2^{PIS}\right)^* \leq d_2^{PIS}(x) \leq \left(d_2^{PIS}\right)^\circ + \beta_1, \\ 1 & if\ \left(d_2^{PIS}\right)^\circ + \beta_1 \leq d_2^{PIS}(x), \end{cases}$$

and

$$v_2(x) = \begin{cases} 0, & if \left(d_2^{NIS}\right)^* \leq d_2^{NIS}(x), \\ \left(\left(d_2^{NIS}\right)^* - d_2^{NIS}(x)\right) \Big/ \left(\left(d_2^{NIS}\right)^* - \left(\left(d_2^{PIS}\right)^* - \beta_2\right)\right) & if \left(d_2^{NIS}\right)^\circ - \beta_2 \leq d_2^{NIS}(x) \leq \left(d_2^{PIS}\right)^*, \\ 1 & if\, d_2^{PIS}(x) \leq \left(d_2^{PIS}\right)^\circ - \beta_2, \end{cases}$$

where β_1 and β_2 indicate tolerances of lower and upper bounds respectively related to the goal functions.

Step 7: The intuitionistic fuzzy linear form of the bi-objective model in **Step 4** can be rewritten (by the help of max-min operator) as:

Model 5

$$\text{maximize } (\lambda_1 - \eta_1)$$
$$\text{subject to } \mu_1(x) \geq \lambda_1, \mu_2(x) \geq \lambda_1,$$
$$v_1(x) \leq \eta_1,\ v_2(x) \leq \eta_1,$$

constraints (6.11)–(6.13) and (6.7),

$$\lambda_1 - \eta_1 \leq 1,$$
$$\lambda_1, \eta_1 \leq 0$$

Where, λ_1 = min $\{\mu_1(x), \mu_2(x)\}$ is the belonging grade and η_1 = max$\{v_1(x), v_2(x)\}$ is the non-belonging grade for the norms of the shortest distance from the PIS and the farthest distance from the NIS of Model 4.

Step 8: The Pareto-optimal solution of the designed MFSTP (i.e., Model 4) is extracted finally from **Step** 7.

Theorem 7.1 [Roy and Midya 2019]: If Model 5 has an optimal solution, then it is a Pareto-optimal solution of Model 4. It is noted that the Pareto-optimal solution can be obtained from Model 2 by using **Steps 1 to 8** with a similar procedure.

8. Application Examples

This section describes two numerical examples on industrial transportation systems. In the first example we present a practical transportation system wherein a rough variable is treated to tackle inexactness of the parameters. Moreover, another example is depicted to represent a real-world distribution problem where a fuzzy-rough variable is considered to deal with the impreciseness of the parameters.

a. Application example with real-life data for Model 1

Khaitan Sugar Co. Ltd is a reliable sugar production company in India which has two factories situated at Nadia (West Bengal), and Samastipur (Bihar) in India from where sugars are supplied to three distinct demand centres situated at Midnapore, Tata, and Siliguri in India through two kinds of carriers by road (large size track) and rail way (goods train). The DM requires to minimize: the total cost for shipping the product, time of transportation of commodity (from plants to demand centres) and deterioration rate of product. The shipping cost is taken in rupees per bag (1 bag = 50 kg), fixed cost in rupees for an unlock path, deterioration rate of commodity in kg per hour. The supply and demand are taken in tonnes. The DM also desires to determine the quantity of sugars in tonnes to move from the ith plant to the jth demand centre so as to fulfil the total obligation.

Table 2: Table displays the cost of transportation in rough nature ($\bar{c}_{ijk}$).

	$j=1$	$j=2$	$j=3$	$j=1$	$j=2$	$j=3$
$i=1$	([4, 10], [3,12])	([6,13], [5,15])	([10, 16], [8,18])	([3, 6], [2,7])	([6, 9], [5,11])	([8, 12], [7,15])
$i=2$	([8, 11], [7,14])	([12,16],[10,18])	([12, 18], [9,20])	([5, 8], [4,9])	([7,10],[6,12])	([10, 12], [9,17])
	$k=1$			$k=2$		

Table 3: Table presents the rough fixed-charge ($\bar{f}_{ijk}$).

	$j=1$	$j=2$	$j=3$	$j=1$	$j=2$	$j=3$
$i=1$	([40, 42], [38,44])	([42,43], [41,44])	([44, 48], [43,49])	([30, 34], [28,36])	([35, 36], [34,38])	([30, 32], [29,33])
$i=2$	([41, 43], [40,44])	([42,45], [40,46])	([43, 45], [44,48])	([31, 32], [29,33])	([32,35], [31,37])	([38, 39], [36,40])
	$k=1$			$k=2$		

Table 4: Table contains the time of transportation in rough nature $(\bar{t}_{ijk})$.

	$j = 1$	$j = 2$	$j = 3$	$j = 1$	$j = 2$	$j = 3$
$i = 1$	([10, 12], [9,15])	([12,13], [11,18])	([12, 15], [9,19])	([7, 9], [6,11])	([10, 14], [9,16])	([8, 13], [7,15])
$i = 2$	([16, 20], [14,22])	([14,18], [12,20])	([9, 11], [8,14])	([14, 18], [12,20])	([9,15], [8,16])	([7, 10], [6,12])
	$k = 1$			$k = 2$		

Table 5: Table displays the deterioration rate of products in rough nature $(\bar{d}_{ijk})$.

	$j = 1$	$j = 2$	$j = 3$	$j = 1$	$j = 2$	$j = 3$
$i = 1$	([10, 12], [9,14])	([12,16], [10,18])	([11, 13], [10,15])	([12, 14], [10,16])	([14, 18], [12,21])	([15, 20], [14,22])
$i = 2$	([12, 14], [11,15])	([14,16], [13,17])	([9, 11], [8,14])	([16, 20], [14,22])	([16,18], [14,24])	([11, 13], [10,18])
	$k = 1$			$k = 2$		

Table 6: Table presents the rough supply, demand and capacity of carriers.

Rough supply at plants $(\bar{a}_i)$		Rough demand at destinations $(\bar{b}_j)$			Rough capacity of conveyances $(\bar{e}_k)$	
$\bar{a}_1$	$\bar{a}_2$	$\bar{b}_1$	$\bar{b}_2$	$\bar{b}_3$	$\bar{e}_1$	$\bar{e}_2$
([18, 20],[16, 22])	([14, 16],[12, 20])	([8,10], [6, 11])	([10,12], [9, 15])	([7,9], [6, 12])	([14, 15],[12, 16])	([16, 17],[14, 18])

b. Numerical example based on practical problem for Model 3

Kashmir Woollen Ltd. is a reputed woollen items production company in India has two factories situated in Kashmir (Jammu & Kashmir) and Ahmedabad (Gujarat) in India from where woollen blankets are supplied to three distinct demand centres situated in Kolkata, Jamshedpur, Delhi in India through two kinds of carriers by road (medium and large size of commodity carriers). The DM requires to minimize: the total shipping cost, delivery time of products (from plants to demand centres) and the packing cost for blankets. The shipping cost is taken in rupees per item, fixed cost in rupees for an unlock path, packing cost for blanket in rupees and time in hours. The DM desires to determine the number of blankets to move from the ith plant to the jth demand centre so as to fulfil the total obligation.

Table 7: Table displays the cost of transportation in fuzzy-rough nature $(\bar{\bar{c}}_{ijk})$.

	$j = 1$	$j = 2$	$j = 3$	$j = 1$	$j = 2$	$j = 3$
$i = 1$	$(\rho_{11}$-1, ρ_{11}, ρ_{11}+1), $\rho_{11}\dashv$ ([5, 6], [4,9])	$(\rho_{12}$-1, ρ_{12}, ρ_{12}+1), $\rho_{12}\dashv$([4,8], [3,10])	$(\rho_{13}$-1, ρ_{13}, ρ_{13}+1), $\rho_{13}\dashv$([5,9], [4,10])	$(\alpha_{11}$-2, α_{11}, α_{11}+2), $\alpha_{11}\dashv$([6, 8], [5,11])	$(\alpha_{12}$-2, α_{12}, α_{12}+2), $\alpha_{12}\dashv$([7, 11], [6,12])	$(\alpha_{13}$-2, α_{13}, α_{13}+2), $\alpha_{13}\dashv$([5, 9], [4,10])
$i = 2$	$(\rho_{21}$-1, ρ_{21}, ρ_{21}+1), $\rho_{21}\dashv$([4, 8], [3,9])	$(\rho_{22}$-1, ρ_{22}, ρ_{22}+1), $\rho_{22}\dashv$([7,9], [6,12])	$(\rho_{23}$-1, ρ_{23}, ρ_{23}+1), $\rho_{23}\dashv$([5, 7], [4,11])	$(\alpha_{21}$-2, α_{21}, α_{21}+2), $\alpha_{21}\dashv$([7, 10], [5,13])	$(\alpha_{22}$-2, α_{22}, α_{22}+2), $\alpha_{22}\dashv$([6,7], [4,12])	$(\alpha_{23}$-2, α_{23}, α_{23}+2), $\alpha_{23}\dashv$ ([5,10], [4,14])
	$k = 1$			$k = 2$		

Table 8: Table presents the fixed cost in fuzzy-rough nature $(\bar{\bar{f}}_{ijk})$.

	$j = 1$	$j = 2$	$j = 3$	$j = 1$	$j = 2$	$j = 3$
$i = 1$	$(\gamma_{11}$-5, γ_{11}, γ_{11}+5), $\gamma_{11}\dashv$ ([15, 20],[13,25])	$(\gamma_{12}$-5, γ_{12}, γ_{12}+5), $\gamma_{12}\dashv$([20,24], [19,26])	$(\gamma_{13}$-5, γ_{13}, γ_{13}+5), $\gamma_{13}\dashv$([21, 25],[20,28])	$(\chi_{11}$-4, χ_{11}, χ_{11}+4), $\chi_{11}\dashv$([20, 24], [18,26])	$(\chi_{12}$-4, χ_{12}, χ_{12}+2), $\chi_{12}\dashv$([22, 26], [20,28])	$(\chi_{13}$-4, χ_{13}, χ_{13}+2), $\chi_{13}\dashv$([22, 27], [21,29])
$i = 2$	$(\gamma_{21}$-5, γ_{21}, γ_{21}+5), $\gamma_{21}\dashv$([18, 22],[16,24])	$(\gamma_{22}$-5, γ_{22}, γ_{22}+5), $\gamma_{22}\dashv$([21,24], [20,27])	$(\gamma_{23}$-5, γ_{23}, γ_{23}+5), $\gamma_{23}\dashv$([23, 25],[22,28])	$(\chi_{21}$-4, χ_{21}, χ_{21}+4), $\chi_{21}\dashv$([24, 26], [23,27])	$(\chi_{22}$-4, χ_{22}, χ_{22}+4), $\chi_{22}\dashv$([23,27], [22,28])	$(\chi_{23}$-4, χ_{23}, χ_{23}+4), $\chi_{23}\dashv$([25, 27], [24,30])
	$k = 1$			$k = 2$		

Table 9: Table contains the time of transportation in fuzzy-rough nature $(\bar{\bar{t}}_{ijk})$.

	$j = 1$	$j = 2$	$j = 3$	$j = 1$	$j = 2$	$j = 3$
$i = 1$	$(\delta_{11}$-1, δ_{11}, δ_{11}+1), $\delta_{11}\dashv$ ([7, 9],[6,11])	$(\delta_{12}$-1, δ_{12}, δ_{12}+1), $\delta_{12}\dashv$([9,10], [8,12])	$(\delta_{13}$-1, δ_{13}, δ_{13}+1), $\delta_{13}\dashv$([9, 10],[7,11])	$(\phi_{11}$-2, ϕ_{11}, ϕ_{11}+2), $\phi_{11}\dashv$([7, 10], [6,12])	$(\phi_{12}$-2, ϕ_{12}, ϕ_{12}+2), $\phi_{12}\dashv$([10, 11], [9,13])	$(\phi_{13}$-2, ϕ_{13}, ϕ_{13}+2), $\phi_{13}\dashv$([9, 12], [8,16])
$i = 2$	$(\delta_{21}$-1, δ_{21}, δ_{21}+1), $\delta_{21}\dashv$([7, 8],[5,10])	$(\delta_{22}$-1, δ_{22}, δ_{22}+1), $\delta_{22}\dashv$([8,10], [7,13])	$(\delta_{23}$-1, δ_{23}, δ_{23}+1), $\delta_{23}\dashv$([10, 11],[9,14])	$(\phi_{21}$-2, ϕ_{21}, ϕ_{21}+2), $\phi_{21}\dashv$([7, 9], [8,15])	$(\phi_{22}$-2, ϕ_{22}, ϕ_{22}+2), $\phi_{22}\dashv$([11,12], [10,14])	$(\phi_{23}$-2, ϕ_{23}, ϕ_{23}+2), $\phi_{23}\dashv$([8, 11], [7,15])
	$k = 1$			$k = 2$		

Table 10: Table presents the packing cost in fuzzy-rough nature ($\tilde{\bar{g}}_{ijk}$).

	$j = 1$	$j = 2$	$j = 3$	$j = 1$	$j = 2$	$j = 3$
$i = 1$	$(\psi_{11}$-0.5,ψ_{11}, ψ_{11}+0.5), $\psi_{11}\dashv$([3, 4],[2, 5])	$(\psi_{12}$-0.5, ψ_{12}, ψ_{12}+0.5), $\psi_{12}\dashv$([3.5, 4.5],[2.5,5.5])	$(\psi_{13}$-0.5, ψ_{13}, ψ_{13}+0.5), $\psi_{13}\dashv$([2.5, 3.5],[2,4])	$(\theta_{11}$-0.5, θ_{11}, θ_{11}+0.5), $\theta_{11}\dashv$([3.5, 4.5], [3,5.5])	$(\theta_{12}$-0.5, θ_{12}, θ_{12}+0.5), $\theta_{12}\dashv$([3, 5.5], [2.5,6])	$(\theta_{13}$-0.5, θ_{13}, θ_{13}+0.5), $\theta_{13}\dashv$([4, 5], [3,6.5])
$i = 2$	$(\psi_{21}$-0.5,ψ_{21}, ψ_{21}+0.5), $\psi_{21}\dashv$([2, 3],[1.5,4.5])	$(\psi_{22}$-0.5, ψ_{22}, ψ_{22}+0.5), $\psi_{22}\dashv$([3, 6],[2, 6.5])	$(\psi_{23}$-0.5, ψ_{23}, ψ_{23}+0.5), $\psi_{23}\dashv$([3.5, 4],[2, 5])	$(\theta_{21}$-0.5, θ_{21}, θ_{21}+0.5), $\theta_{21}\dashv$([4.5, 5], [3.5,6])	$(\theta_{22}$-0.5, θ_{22}, θ_{22}+0.5), $\theta_{22}\dashv$([5.5,6], [3,6.5])	$(\theta_{23}$-0.5, θ_{23}, θ_{23}+0.5), $\theta_{23}\dashv$([4.5, 5.5], [3,6])
	$k = 1$			$k = 2$		

Table 11: Table represents the fuzzy-rough supply, demand and capacity of carriers.

Rough supply at plants ($\tilde{\bar{a}}_i$)		**Rough demand at demand centre** ($\tilde{\bar{b}}_j$)			**Rough capacity of carriers** ($\tilde{\bar{e}}_k$)	
$\tilde{\bar{a}}_1$	$\tilde{\bar{a}}_2$	$\tilde{\bar{b}}_1$	$\tilde{\bar{b}}_2$	$\tilde{\bar{b}}_3$	$\tilde{\bar{e}}_1$	$\tilde{\bar{e}}_2$
$(\kappa_1$-3, κ_1, κ_1+3), $\kappa_1\dashv$([24,30], [22, 34])	$(\kappa_2$-3, κ_2, κ_2+3), $\kappa_2\dashv$([24,30], [22, 34])	$(\tau_1$-3, τ_1, τ_1+3), $\tau_1\dashv$([14,19], [12, 21])	$(\tau_2$-3, τ_2, τ_2+3), $\tau_2\dashv$([16,17], [14, 24])	$(\tau_3$-3, τ_3, τ_3+3), $\tau_3\dashv$([18,21],[16, 23])	$(\omega_1$-3, ω_1, ω_1+3), $\omega_1\dashv$([26,28], [24, 36])	$(\omega_2$-3, ω_2, ω_2+3), $\omega_2\dashv$([27,31], [25, 37])

9. Results and Discussion

Herein, we discuss the optimal solutions of two similar crisp models of our designed rough MFSTP and fuzzy-rough MFSTP (i.e., Model 2 and Model 4).

a. Results of Model 2 (i.e., crisp form of rough MFSTP)

Using Theorem 4.1, the expected-value of each rough interval of Tables 2 to 6 are calculated. Thereafter, utilising the crisp data in the designed Model 2 and using the methodology depicted in Section 7 and applying LINGO software, we get the Pareto-optimal solutions of Model 2 by intuitionistic fuzzy TOPSIS. The results are exhibited in Table 12.

b. Results of Model 4 (i.e., crisp form of fuzzy-rough MFSTP)

We calculate the expected-value of each fuzzy-rough variable of Tables 7 to 11 by using Theorem 4.3. Thereafter, using the crisp data in the formulated Model 4 and applying the methodology described in Section 7 and using LINGO optimizer, the Pareto-optimal solutions of Model 4 are obtained by intuitionistic fuzzy TOPSIS. The results are listed in Table 12.

Table 12: Pareto-optimal solutions of the designed MFSTP models.

Names of the models	**Optimal value of the objective functions**		
	Z_1	Z_2	Z_3
Rough MFSTP model	452.37	53.75	388.12
Fuzzy-rough MFSTP model	586.91	66.25	247.66

It is noticed from Table 12 that intuitionistic fuzzy TOPSIS gives the Pareto-optimal solutions for two distinct MFSTP models under two separate situations. Optimal values of the objective functions for rough MFSTP model are: total shipping cost is Rs. 452.37, total delivery time is 53.75 hr. and deterioration rate of products is 388.12 kg/hour for moving sugars from the plants to the demand centres through several modes of transportation. The minimized values of the objective functions for the fuzzy-rough MFSTP model are: total cost of transportation is Rs. 586.91, total shipping time is 66.25 hrs, and packing cost is Rs. 247.66 for moving woollen blankets from the factories to the demand centres through various modes of transportation.

10. Conclusion and Future Research Directions

In this chapter, we have basically investigated MFSTP models which arise in different practical situations. To tackle vague and uncertainty information in the industrial sector for transportation, we have adopted rough and fuzzy-rough variables in our proposed studies. Expected-value operator is used to transform rough and fuzzy-rough MFSTP models into similar crisp forms. New methodology namely, intuitionistic fuzzy TOPSIS is successfully used to extract Pareto-optimal solutions from similar deterministic models. In this chapter, a novel idea has appeared for MFSTP which is based on the representation and tackling of vague and inexact information under rough and fuzzy-rough frameworks of practical distribution systems.

For future research studies, one can consider the MFCTP under intuitionistic fuzzy-rough frameworks. This chapter can encompass a new concept to make step fixed-charge STP with multiple goals in another twofold uncertain ground.

References

Abo-Sinna, M. A., Amer, A. H. and Ibrahim, A. S. 2008. Extension of TOPSIS for large scale multi-objective non-linear programming problems with block angular structure. Appl. Math. Model., 32: 292–302.

Aggarwal, S. and Gupta, C. 2016. Solving intuitionistic fuzzy solid transportation problem via new ranking method based on signed distance. Int. J. Uncer. Fuzzi. Knowledge-Based Syst., 24: 483–501.

Atteya, T. E. M. 2016. Rough multiple objective programming. Eur. J. Ope. Res., 248(1): 204–210.

Damghani, K. K., Nezhad, S. S. and Tavana, M. 2013. Solving multi-period project selection problems with fuzzy goal programming based on TOPSIS and a fuzzy preference relation. Inf. Sci., 252: 42–61.

Dubois, D. and Prade, H. 1987. Twofold fuzzy sets and rough sets-some issues in knowledge representation. Fuzzy Sets Syst., 23(1): 3–18.

Ebrahimnejad, A. and Verdegay, J. L. 2018. A new approach for solving fully intuitionistic fuzzy transportation problems. Fuzzy Opt. Deci. Maki., 17(4): 447–474.

Gupta, G., Kaur, J. and Kumar, A. 2016. A note on fully fuzzy fixed charge multi-item solid transportation problem. Appl. Soft Comp., 41: 418–419.

Haley, K. B. 1962. The solid transportation problem. Ope. Res., 10: 448–463.

Hirsch, W. M. and Dantzig, G. B. 1968. The fixed charge problem. Nav. Res. Log. Quart., 15: 413–424.

Hwang, C. L. and Yoon, K. 1981. Multiple attribute decision making: Methods and Applications, Springer, New York.

Jimenez, F. and Verdegay, J. L. 1999. Solving fuzzy solid transportation problems by an evolutionary algorithm based parametic approach. Eur. J. Ope. Res., 117: 485–510.

Liu, B. 2002. Theory and practice of uncertain programming, Springer, Physica-Verlag, Heidelberg.

Maity, G., Mardanya, D., Roy, S. K. and Weber, G. W. 2019a. A new approach for solving dual-hesitant fuzzy transportation problem with restrictions. Sadhana, 44(4): 1–11.

Maity, G., Roy, S. K. and Verdegay, J. L. 2019b. Analyzing multimodal transportation problem and its application to artificial intelligence. Neu. Comp. Appl., 32: 2243–2256.

Midya, S. and Roy, S. K. 2014. Solving single-sink fixed-charge multi-objective multi-index stochastic transportation problem. American J. Math. Manag. Sci., 33(4): 300–314.

Midya, S. and Roy, S. K. 2017. Analysis of interval programming in different environments and its application to fixed-charge transportation problem. Disc. Math. Alg. Appl., 9(3): 1750040 (17 pages).

Midya, S. and Roy, S. K. 2020. Multi-objective fixed-charge transportation problem using rough programming. Int. J. Ope. Res., 37(3): 377–395.

Midya, S., Roy, S. K. and Yu, V. F. 2021. Intuitionistic fuzzy multi-stage multi-objective fixed-charge solid transportation problem in a green supply chain. Int. J. Machine Lear and Cyber, 12(3): 699–717.

Pawlak, Z. 1982. Rough sets. Int. J. Inf. Comp. Sci., 11(5): 341–356.

Rebolledo, M. 2006. Rough intervals-enhancing intervals for qualitative modeling of technical systems. Arti. Intel., 170: 667–685.

Roy, S. K., Maity, G. and Weber, G. W. 2017a. Multi-objective two-stage grey transportation problem using utility function with goals. Cent. Eur. J. Ope. Res., 25: 417–439.

Roy, S. K. Maity, G., Weber, G. W. and Gok, S. Z. A. 2017b. Conic scalarization approach to solve multi-choice multi-objective transportation problem with interval Goal. Ann. Ope. Res., 253(1): 599–620.

Roy, S. K., Ebrahimnejad, A., Verdegay, J. L. and Das, S. 2018a. New approach for solving intuitionistic fuzzy multi-objective transportation problem. Sadhana, 43: 3, Indian Academy of Sciences, doi.org/10.1007/s12046-017-0777-7.

Roy, S. K., Midya, S. and Yu, V. F. 2018b. Multi-objective fixed-charge transportation problem with random rough variables. Int. J. Uncer. Fuzzi. Knowledge-Based Syst., 26(6): 971–996.

Roy, S. K. and Midya, S. 2019. Multi-objective fixed-charge solid transportation problem with product blending under intuitionistic fuzzy environment. Appl. Intel., 49(10): 3524–3538.

Roy, S. K., Midya, S. and Weber, G. W. 2019. Multi-objective multi-item fixed-charge solid transportation problem under twofold uncertainty. Neu. Comp. Appl., 31(12): 8593–8613.

Tao, Z. and Xu, J. 2012. A class of rough multiple objective programming and its application to solid transportation problem. Inf. Sci. 188: 215–235.

Xu, J. and Zhao, L. 2008. A class of fuzzy rough expected value multi-objective decision making model and its application to inventory problems. Comp. Math. Appl., 56: 2107–2119.

Xu, J. and Zhou, X. 2010. Fuzzy-like multiple-objective decision making, Springer, Berlin, Germany.

Xu, J. and Tao, Z. 2012. Rough multiple objective decision making, Taylor and Francis Group, CRC Press, USA.

Zavardehi, S. M. A., Nezhad, S. S., Moghaddam, R. T. and Yazdani, M. 2013. Solving a fuzzy fixed charge solid transportation problem by metaheuristics. Fuzzy Sets Syst., 57: 183–194.

Zhang, B., Peng, J., Li, S. and Chen, L. 2016. Fixed charge solid transportation problem in uncertain environment and its algorithm. Comp. Ind. Eng., 102: 186–197.

Zimmermann, H. J. 1978. Fuzzy programming and linear programming with several objective functions. Fuzzy Sets Syst., 1: 45–55.

CHAPTER 17

Overall Shale Gas Water Management

A Neutrosophic Optimization Approach

Ahmad Yusuf Adhami,[1] *Firoz Ahmad*[1,*] and *Nahida Wani*[2]

1. Introduction

Sources of energy are the most important gifts from nature. Various energy sources are utilized for energy generation to fulfill the domestic demand including that in every sector. Currently, energy-production is done by various conventional resources. However, among all sources of energy, the shale gas energy is the most promising source in producing the energy. The shale gas is one of the major sources of natural gas that are trapped within shale-rocks. Presently, shale gas energy is emerging as the most promising natural gas production resource across the globe. Usually, huge amounts of shale rocks are found in China followed by the United States having them in abundance. According to the Energy Information Administration (EIA) report, about 47% of the natural gas production can be done by the year 2035 from shale gas (Stevens 2012). The very first shale gas generation was done in 1821 in the city of New York called Fredonia. The process consisted of shallow and low-pressure f racking. Horizontal drilling was initiated in 1930 in the US at commercial level.

In the last few decades, lots of researchers and practitioners have taken deep interest in designing wholesome production strategies, sustainable supply chain networks, shale rocks selection area, and considering the socio-economic and environmental aspects at the commercial level. Lutz et al. (2013) developed an empirical study that describes the production processes from the beginning to the end. Yang et al. (2014) presented a study comprising the optimal use of the whole water cycle in boring wells and proposing bi-stage stochastic mixed-integer linear programming in a vague environment. Yang et al. (2015) suggested the entire framework for the water cycle and propounded the mixed-integer linear fractional programming model for the shale gas production process. Li et al. (2014) presented a study on the adsorption process during fracking and a methodology is designed to overcome the wastes during shale gas production. Gao and You (2015) suggested the supply chain network for the flow of produced natural gas from shale rocks. The outcomes reflected the co-ordination between the economic and environmental objectives. Moreover, Sang et al. (2014) also made an attempt to reveal the optimal configuration of the shale gas production process and designed the integrated supply chain network linked with the flow of water. Gao and You (2015a) designed the multiobjective mixed-integer linear programming model for shale gas water resilience over the multi-period time horizons. Zhang et al. (2016) performed an analytical study on shale gas water management in an uncertain environment. Guerra et al. (2016) addressed the mathematical models describing the consumption of water resources in hydraulic fracturing processes throughout the shale gas production. Some advanced and relevant studies on shale gas water management system can be found in Lan et al. (2019), Ren et al. (2019), Denham et al. (2019), Al-Aboosi and El-Halwagi (2018), Guo et al. (2016), and Chebeir et al. (2017), Bartholomew and Mauter (2016), Lira-Barragán et al. (2016), Chen et al. (2017), and Knee and Masker (2019).

The above discussed studies are restricted to the uncertain model formulations, and do not considers the uncertainty linked with the parameters. Although, Zhang et al. (2016) performed a study on shale gas water management system by depicting the uncertain parameters as fuzzy and stochastic approaches it has two major limitations from the practical points of view. First, the fuzzy quantification of uncertain parameters may not be appropriate estimates due to the existence of only the membership function of the element into the feasible set. Similarly, the historical data may not be available every time for which the stochastic approaches can be applied to obtain the estimated values. Thus, both sorts of vagueness may not be more effective uncertainty quantification techniques in practical scenarios. Therefore, the actual or more realistic representation of uncertainty can be depicted by the intuitionistic fuzzy

[1] Department of Statistics and Operations Research, Aligarh Muslim University, Aligarh.
[2] Department of Mathematics, University of Kashmir, Jammu and Kashmir.
* Corresponding author: firoz.ahmad02@gmail.com

approach that considers the membership and non-membership functions of the element into the feasible set. Also, the historical data is not required while using the intuitionistic fuzzy set theory. Secondly, Zhang et al. (2016) discussed the shale gas waste water management system and did not consider the associated study with fresh water such as freshwater acquirement, demand, supply and optimal usage in each planning horizons. The requirement of fresh water in shale gas production is very high due to the hydraulic fracturing process. So, the fresh water management aspects have much importance along with waste water management. Therefore, this chapter considers both the important aspects related to shale gas water management system. Also, the opportunity for available on-site treatment facility for waste water and capacity expansion of treatment plants are also included that would be quite beneficial for Pennsylvania because the opportunity for underground water disposal facilities are very rare and most often wastewater is delivered to nearby cities in Ohio.

The remaining part of the chapter is summarized as follows: In Section 2, basic definitions regarding fuzzy, intuitionistic fuzzy and neutrosophic sets are presented whereas Section 3 describes the problem descriptions and formulation. Section 4 discusses the different neutrosophic optimization models with linear-type membership functions. In Section 5, a case study is presented to reveal the validity and applicability of the modeling and optimization approaches. At last, conclusions along with the future research scope are described in Section 6.

2. Problem Descriptions and Modeling Under Uncertainty

The propounded shale gas water management system is designed to achieve cost-effective production policies linked with movement of freshwater and wastewater. In addition, a lesser capital investment incurred over the expansion of the treatment plant is depicted in different time horizons. The presented overall water supply chain designed comprises the various inter-connected parts in identifying the most prominent aspects. The modeling and optimization framework are developed under the well-defined dynamic constraints along with the potential objectives. The useful notions and descriptions are presented in Table 1.

Table 1: Nomenclature.

Indices	**Descriptions**
i	Index set for shale sites
j	Index set for disposal sites and treatment plants
m	Index set for available option for the expansion capacity of treatment plant
o	Index set for on-site treatment technology
s	Index set for source of freshwater
t	Index set for time period
Parameters	
LO_o	Recovery factor associated with wastewater using on-site treatment technology o
$FDW_{i,t}$	Demand of freshwater at shale site i in time period t
$FCA_{s,t}$	Supply capacity of freshwater at source s in time period t
RF_o	Blending ratio of freshwater to wastewater after applying on-site treatment technology o
$WWDS_{j,t}$	Wastewater capacity at disposal site j in time period t
$WWTP_{j,t}$	Wastewater capacity at treatment plant j in time period t
$WDW_{j,t}$	Total wastewater capacity at treatment plant and disposal site j in time period t
$EO_{j,m,t}$	Increased capacity of wastewater treatment plant j using expansion option m in time period t
$CAQ_{s,t}$	Unit freshwater acquisition cost at source s in time period t
$CTF_{s,t}$	Unit freshwater transportation cost from source s to shale site i in time period t
$CTW_{i,j,t}$	Unit wastewater transportation cost from shale site i to treatment plant j in time period t
$CTR_{j,t}$	Unit wastewater treatment cost at treatment plant j in time period t
$CD_{j,t}$	Unit wastewater treatment cost at disposal site j in time period t
$RE_{j,t}$	Revenues generated from wastewater reuse from treatment plant j in time period t
$RR_{j,t}$	Reuse rate from wastewater treatment plant j in time period t
$CEX_{j,m,t}$	Capital investment for expanding treatment plant j using expansion option m in time period t
Decision variables	
$fw_{s,i,t}$	Acquisition of freshwater from source s at shale site i in time period t
$wto_{i,o,t}$	Amount of wastewater treated at shale site i using on-site technology o in time period t
$ww_{i,j,t}$	Total amount of wastewater generation at shale site i to the facility j in time period t
$wwd_{i,j,t}$	Total amount of wastewater generation at shale site i to disposal site j in time period t
$wwt_{i,j,t}$	Total amount of wastewater generation at shale site i to treatment plant j in time period t
$y_{j,m,t}$	Binary variable for expanding treatment plant j using expansion option m in time period t

2.1 Multiple Objectives

The first objective function represents the total cost associated with freshwater and represented in Equation (1).

$$Minimize\ O_1 = \sum_{s=1}^{S}\sum_{i=1}^{I}\sum_{t=1}^{T}\left(CAQ_{s,t} + CTF_{s,i,t}\right) fw_{s,i,t} \tag{1}$$

The total economic cost associated with wastewater depicts the second objective function and represented in Equation (2).

$$Minimize\ O_2 = \sum_{i=1}^{I}\sum_{j=1}^{J}\sum_{t=1}^{T}\left(CTR_{j,t} + CD_{j,t} + CTW_{i,j,t} + RR_{j,t} \times RE_{j,t}\right) ww_{i,j,t} \tag{2}$$

The total capital investment is depicted in the third objective and represented in Equation (3).

$$Minimize\ O_3 = \sum_{j=1}^{J}\sum_{m=1}^{M}\sum_{t=1}^{T}\left(CEX_{j,m,t}\right) y_{j,m,t} \tag{3}$$

2.2 Constraints

Constraints (4) ensure the freshwater demand at various shale sites.

$$\sum_{s=1}^{S} fw_{s,i,t} + \sum_{o=1}^{O} LO_o \times wto_{i,o,t} \geq FDW_{i,t} \qquad \forall\, i,t \tag{4}$$

Constraints (5) represents the capacity restrictions of freshwater acquired from various sources which should be satisfied at each shale site over time horizons.

$$\sum_{i=1}^{I} fw_{s,i,t} \leq FCA_{s,t} \qquad \forall\, s,t \tag{5}$$

The wastewater released from each shale sites to the disposal facility must be less than or equal to the maximum capacity of the disposal facility centers over time periods and can be depicted in constraint (6).

$$\sum_{i=1}^{I} wwd_{i,j,t} \leq WWDS_{j,t} \qquad \forall\, j,t \tag{6}$$

The total amount of wastewater produced must satisfy its maximum level and can be depicted in constraint (7).

$$\sum_{s=1}^{S} wwt_{i,j,t} \leq WWTP_{j,t} + \sum_{m=1}^{M} EO_{j,m,t} \times y_{j,m,t} \qquad \forall\, j,t \tag{7}$$

Constraint (8) ensures that the produced wastewater must relax the demand for wastewater treatment.

$$\sum_{s=1}^{S} ww_{i,j,t} \leq WDW_{j,t} \qquad \forall\, j,t \tag{8}$$

Constraint (9) represents the total wastewater capacity at water different facility centers

$$\sum_{m=1}^{M} EO_{j,m,t} \times y_{j,m,t} + WWDS_{j,t} + WWTP_{j,t} \leq WDW_{j,t} \qquad \forall\, i,j,t \tag{9}$$

The reuse specification for the fracking process with blending ratio after applying each freshwater source over time periods can be represented in constraint (10).

$$\sum_{m=1}^{M} RF_o \times LO_o \times wto_{i,o,t} \leq fw_{s,i,t} \qquad \forall\, s,i,t \tag{10}$$

Constraint (11) ensures the total sum of wastewater at disposal site and treatment plant must be equal to the maximum allocated capacity at its respective facility centers over time periods.

$$\sum_{i=1}^{I}\sum_{j=1}^{J}\sum_{t=1}^{T} wwd_{i,j,t} + \sum_{i=1}^{I}\sum_{j=1}^{J}\sum_{t=1}^{T} wwt_{i,j,t} = \sum_{i=1}^{I}\sum_{j=1}^{J}\sum_{t=1}^{T} ww_{i,j,t} \tag{11}$$

Thus the overall shale gas water management model with intuitionistic fuzzy parameters can be depicted in (12).

$$
\begin{aligned}
&\text{Minimize } O_1 = \sum_{s=1}^{S}\sum_{i=1}^{I}\sum_{t=1}^{T}\left(\widetilde{CAQ}_{s,t} + \widetilde{CTF}_{s,i,t}\right) fw_{s,i,t} \\
&\text{Minimize } O_2 = \sum_{i=1}^{I}\sum_{j=1}^{J}\sum_{t=1}^{T}\left(\widetilde{CTR}_{j,t} + \widetilde{CD}_{j,t} + \widetilde{CTW}_{i,j,t} + RR_{j,t} \times RE_{j,t}\right) ww_{i,j,t} \\
&\text{Minimize } O_3 = \sum_{j=1}^{J}\sum_{m=1}^{M}\sum_{t=1}^{T}\left(\widetilde{CEX}_{j,m,t}\right) y_{j,m,t} \\
&\text{subject to} \\
&\sum_{s=1}^{S} fw_{s,i,t} + \sum_{o=1}^{O} LO_o \times wto_{i,o,t} \geq \widetilde{FDW}_{i,t} \\
&\sum_{i=1}^{I} fw_{s,i,t} \leq \widetilde{FCA}_{s,t} \\
&\sum_{i=1}^{I} wwd_{i,j,t} \leq \widetilde{WWDS}_{j,t} \\
&\sum_{s=1}^{S} wwt_{i,j,t} \leq \widetilde{WWTP}_{j,t} + \sum_{m=1}^{M} EO_{j,m,t} \times y_{j,m,t} \\
&\sum_{s=1}^{S} ww_{i,j,t} \leq \widetilde{WDW}_{j,t} \\
&\sum_{m=1}^{M} EO_{j,m,t} \times y_{j,m,t} + \widetilde{WWDS}_{j,t} + \widetilde{WWTP}_{j,t} \leq \widetilde{WDW}_{j,t} \\
&\sum_{m=1}^{M} RF_o \times LO_o \times wto_{i,o,t} \leq fw_{s,i,t} \\
&\sum_{i=1}^{I}\sum_{j=1}^{J}\sum_{t=1}^{T} wwd_{i,j,t} + \sum_{i=1}^{I}\sum_{j=1}^{J}\sum_{t=1}^{T} wwt_{i,j,t} = \sum_{i=1}^{I}\sum_{j=1}^{J}\sum_{t=1}^{T} ww_{i,j,t}
\end{aligned}
\tag{12}
$$

Where (~) over different parameters depicts the triangular intuitionistic fuzzy number.

2.3 Treating Intuitionistic Fuzzy Parameters

In this section, some important definitions regarding fuzzy, intuitionistic fuzzy and neutrosophic sets are discussed.

Definition 1: (Ahmad et al. 2019) (Fuzzy set) Let W denote a universe of discourse; then a fuzzy set X in W can be defined by a function X: → [0,1]
$X = \{w, \mu_X(w) \mid w \in W\}$ such that $\mu_X(w): W \to [0,1]$ with conditions $0 \leq \mu_X(w) \leq 1$
where $\mu_X(w)$ denotes the membership function of the element $w \in W$, in to the set X respectively.

Definition 2: (Ahmad et al. 2019) (Intuitionistic Fuzzy Set (IFS)) Let W denote a universe of discourse; then an IFS X in W is given by the ordered triplets as follows:
$X = \{w, \mu_X(w)\, \nu_X(w) \mid w \in W\}$ with $\mu_X(w): W \to [0,1]$; $\nu_X(w): W \to [0,1]$ such that
$0 \leq \mu_X(w) + \nu_X(w) \leq 1$ where $\mu_X(w)$ and $\nu_X(w)$ denote the membership and non-membership function of the element $w$in to the set X.

Definition 3: (Ahmad et al. 2019) (Triangular intuitionistic fuzzy number) An intuitionistic fuzzy number $\widetilde{X}$ is said to be a triangular intuitionistic fuzzy number if the membership function $\mu_{\widetilde{X}}(w)$ and non-membership function $\nu_{\widetilde{X}}(w)$ are given by

$$
\mu_{\widetilde{X}}(w) = \begin{cases} \dfrac{w - a_1}{b - a_1}, & \text{if } a_1 \leq w \leq b \\ 1, & \text{if } w = b \\ \dfrac{a_2 - w}{a_2 - b}, & \text{if } b \leq w \leq a_2 \end{cases} \quad \text{and } \nu_{\widetilde{X}}(w) = \begin{cases} \dfrac{b - w}{b - a_3}, & \text{if } a_3 \leq w \leq b \\ 0, & \text{if } w = b \\ \dfrac{w - b}{a_4 - b}, & \text{if } b \leq w \leq a_4 \end{cases}
$$

Where $a_3 \leq a_1 \leq b \leq a_2 \leq a_4$ and is denoted by $\widetilde{X} = \{(a_1, b, a_2, \mu_{\widetilde{X}}(w)), (a_3, b, a_4, \nu_{\widetilde{X}}(w))\}$.

Definition 4: (Ahmad et al. 2019) (Expected interval and value for triangular intuitionistic fuzzy number) Suppose that $\widetilde{X} = \{(a_1, b, a_2, \mu_{\widetilde{X}}(w)), (a_3, b, a_4, \nu_{\widetilde{X}}(w))\}$ be a triangular intuitionistic fuzzy number with membership and non-membership functions $\mu_{\tilde{X}}(w)$ and $v_{\tilde{X}}(w)$; then the expected interval of the triangular intuitionistic fuzzy number by using the above definition can be obtained as follows:

$$EI(\tilde{X}) = \left[E_1(\tilde{X}),\ E_2(\tilde{X})\right]$$

$$E_1(\tilde{X}) = \frac{b_1 - a_2}{2} + \int_{b_1}^{b_2} h_{\tilde{X}}(w) - \int_{a_1}^{a_2} f_{\tilde{X}}(w) = \frac{3a + b_1 + (a - b_1)v_{\tilde{X}} - (a - a_1)\mu_{\tilde{X}}}{4} \tag{13}$$

$$E_2(\tilde{X}) = \frac{a_3 - b_4}{2} + \int_{a_3}^{a_4} g_{\tilde{X}}(w) - \int_{b_3}^{b_4} k_{\tilde{X}}(w) = \frac{3a + b_2 + (a_2 - a)\mu_{\tilde{X}} + (a - b_2)v_{\tilde{X}}}{4} \tag{14}$$

where $h_{\tilde{X}}(w) = \dfrac{w - b_1}{b_2 - b_1}$, $f_{\tilde{X}}(w) = \dfrac{w - a_1}{a_2 - a_1}$, $g_{\tilde{X}}(w) = \dfrac{w - a_4}{a_3 - a_4}$ and $k_{\tilde{X}}(w) = \dfrac{w - b_4}{b_3 - b_4}$ respectively.

Grzegrorzewski (2003) derived the concept ranking function using (Equation 13) and (Equation 14) which can be given as follows:

$$EV(\tilde{X}) = \left[\frac{E_1(\tilde{X}) + E_2(\tilde{X})}{2}\right] \tag{15}$$

3. Solution Approach

Mathematical programing problems are widely used in modeling the real-life problems under uncertainty. The multiobjective optimization model is an extensively used form of mathematical programming problems and can be solved using some multiobjective optimization techniques. A recent advance shows that a neutral thought in decision-making processes exists by having a severe effect. Thus the neutrosophic set (NS) is introduced by Smarandache (1999). The NS manages the indeterminacy degrees efficiently. It contains three different members such as; truth, indeterminacy and falsity degrees corresponding to each element into the feasible set. Interested researchers or practitioners can visit the suggested readings on neutrosophic set theory in Smarandache (1999); Ahmad et al. (2018); Ahmad and Adhami (2019); and Ahmad et al. (2019). Many more research works are available on NS theory that describe the advantages and advances in the contemporary research domain. Ahmad et al. (2020) presented a CLSC planning model and solved using modified neutrosophic programming problems. The presented chapter discusses the modeling configuration of shale gas water management system and takes the form of neutrosophic fuzzy optimization models. A different model has been presented to solve the multiple objective programming under the set of constraints. The neutrosophic approach is reliable in quantifying the membership function based on three aspects such as truth, indeterminacy, and falsity degrees, respectively.

Definition 5: (Ahmad et al. 2019) (Neutrosophic Set (NS)) Let W be the universe of discourse with its generic element w, then a neutrosophic set X in W can be defined by a function $X: \rightarrow [0,1]$

$X = \{w, \mu_X(w), \lambda_X(w), v_X(w) \mid w \in W\}$ where $\mu_X(w): W \rightarrow [0^-,1^+]$, $\lambda_X(w): W \rightarrow [0^-,1^+]$ and $v_X(w): W \rightarrow [0^-,1^+]$ with the conditions that $^-0 \le \mu_X(w) + \lambda_X(w) + v_X(w) \le 3^+$

Where $\mu_X(w)$, $\lambda_X(w)$ and $v_X(w)$denotes the truth, indeterminacy and a falsity membership functions of the element $w \in W$, in to the set X respectively.

Definition 6: (Ahmad et al. 2019) (Single valued neutrosophic set) A single valued neutrosophic set X over universe of discourse W is defined as follows:

$X = \{w, \mu_X(w), \lambda_X(w), v_X(w) \mid w \in W\}$ where $\mu_X(w): W \rightarrow [0,1]$, $\lambda_X(w): W \rightarrow [0,1]$ and $v_X(w): W \rightarrow [0,1]$ with the conditions that $0 \le \mu_X(w) + \lambda_X(w) + v_X(w) \le 3$

where $\mu_X(w)$, $\lambda_X(w)$ and $v_X(w)$ denote the truth, indeterminacy and a falsity membership function of the element $w \in W$, in to the set X respectively.

Definition 7: (Ahmad et al. 2019) (Neutrosophic score function) A neutrosophic score function is defined as follows:
$S^N = \{\mu_X(w) + \lambda_X(w) - v_X(w) \mid x \in X\}$ where $\mu_X(w)$, $\lambda_X(w)$ and $v_X(w) \in [0,1]$ such that $0 \le \mu_X(w) + \lambda_X(w) + v_X(w) \le 3$ for each $x \in X$.

Definition 8: (Ahmad et al. 2019) (Pareto-Optimal Solution) A solution x^* is said to be a Pareto-optimal solution to the multiobjective linear programming problem if and only if there does not exist another $x \in X$ such that $Z_o(x^*) \le or \ge Z_o(x)$ (for minimization or maximization case) for $o = 1,2,\cdots,O$ and $Z_o(x^*) \ne Z_o(x)$ for at least one o, $o \in (1,2,\cdots,O)$.

The general form of multiobjective programming problem (MOPP) can be represented as follows (16):

$$\begin{aligned}
&\text{Maximize } Z_o(x) = (Z_1, Z_2, Z_3, \ldots, Z_o) \quad \forall\ o = 1,2,\ldots,O_1 \\
&\text{Minimize } Z_o(x) = (Z_1, Z_2, Z_3, \ldots, Z_o) \quad \forall\ o = O_1+1, O_1+2,\ldots,O \\
&\text{Subject to} \\
&g_i(x) \le b_i, \ \ \forall\ i = 1,2,\ldots,I_1, \\
&g_i(x) \ge b_i, \ \ \forall\ i = I_1+1, I_1+2,\ldots,I_2, \\
&g_i(x) = b_i, \ \ \forall\ i = I_2+1, I_2+2,\ldots,I. \\
&x = (x_1, x_2, \ldots, x_j) \in X, \ x \ge 0.
\end{aligned} \tag{16}$$

where $Z_o(x)$ is the Oth objective function. $g_i(x)$, b_i are the real valued function and numbers respectively. $x = (x_1, x_2, \ldots, x_j)$ represents the set of decision variables.

To determine the various membership functions for MOPPs, we have defined lower and upper bounds L_o and U_o that can be derived as follows (Equation 17):

$$U_o = \max\{Z_o(X^o)\} \text{ and } L_o = \min\{Z_o(X^o)\} \quad \forall o = 1,2,\ldots,O \tag{17}$$

3.1 Characterization of Marginal Evaluations

In multiobjective programming problems, each objective function's marginal evaluation is depicted by its respective membership functions. The linear membership functions are defined for the truth, indeterminacy and a falsity membership function which seems to be more realistic. In general, the most extensive and widely used membership function is a linear one due to its simple structure and more straightforward implications. The linear membership function contemplates over the constant marginal rate of satisfaction, up to some extent and dissatisfaction degrees towards an objective.

The bounds for the oth objective function under the neutrosophic environment can be obtained as follows (Equation 18):

$$\begin{aligned}
&U_o^T = U_o, \qquad L_o^T = L_o \\
&U_o^I = L_o^T + s_o(U_o^T - L_o^T), \ \ L_o^I = L_o^T \\
&U_o^F = U_o^T, \ \ L_o^F = L_o^T + t_o(U_o^T - L_o^T)
\end{aligned} \tag{18}$$

where s_o and t_o are the predetermined real numbers such that s_o, $t_o \in (0,1)$are assigned by the decision maker [See, Ahmad et al. (2019)].

By using the above lower and upper bounds, we have defined the linear membership functions for truth, indeterminacy and a falsity membership function under a neutrosophic environment. Thus the linear-type truth $\mu_o(Z_o(x))$, indeterminacy $\lambda_o(Z_o(x))$ and falsity $v_o(Z_o(x))$ membership functions under neutrosophic environment can be furnished as follows (Equation 19):

$$\begin{aligned}
&\mu_o(Z_o(x)) = \begin{cases} 1, & \text{if } Z_o(x) \le L_o^T \\ 1 - \dfrac{Z_o(x) - L_o^T}{U_o^T - L_o^T}, & \text{if } L_o^T \le Z_o(x) \le U_o^T \\ 0, & \text{if } Z_o^T \ge U_o^T \end{cases} \\
&\lambda_o(Z_o(x)) = \begin{cases} 1, & \text{if } Z_o(x) \le L_o^I \\ 1 - \dfrac{Z_o(x) - L_o^I}{U_o^I - L_o^I}, & \text{if } L_o^I \le Z_o(x) \le U_o^I \\ 0, & \text{if } Z_o^I \ge U_o^I \end{cases} \\
&v_o(Z_o(x)) = \begin{cases} 1, & \text{if } Z_o(x) \ge U_o^F \\ 1 - \dfrac{U_o^F - Z_o(x)}{U_o^F - L_o^F}, & \text{if } L_o^F \le Z_o(x) \le U_o^F \\ 0, & \text{if } Z_o^T \le L_o^T \end{cases}
\end{aligned} \tag{19}$$

where $\mu_o(Z_o(x))$, $\lambda_o(Z_o(x))$ and $v_o(Z_o(x))$ are the linear–type membership functions for truth, indeterminacy and falsity membership functions under a neutrosophic environment. If for any membership $U_o = L_o$, then the value of these memberships will be equal to 1.

3.2 Proposed Neutrosophic Optimization Techniques

The concept of fuzzy decision (D), fuzzy goal (G) and fuzzy constraints (C) was first discussed by Zimmermann (1978) and extensively used in many real life decision making problems under fuzziness. Therefore, the fuzzy decision set can be defined as follows:

$$D = (Z \cap C)$$

Equivalently, the neutrosophic decision set D_N with the set of neutrosophic objectives and constraints can be defined as:

$$D_N = \left(\bigcap_{o=1}^{O} Z_o\right)\left(\bigcap_{i=1}^{I} C_i\right) = \left(x,\ \mu_D(x),\ \lambda_D(x),\ \nu_D(x)\right)$$

where

$$\mu_D(x) = \begin{Bmatrix} \mu_{D_1}(x),\ \mu_{D_2}(x),\ \mu_{D_3}(x),\dots,\mu_{D_O}(x) \\ \mu_{C_1}(x),\ \mu_{C_2}(x),\ \mu_{C_3}(x),\dots,\mu_{C_I}(x) \end{Bmatrix} \quad \forall\ x \in X$$

$$\lambda_D(x) = \begin{Bmatrix} \lambda_{D_1}(x),\ \lambda_{D_2}(x),\ \lambda_{D_3}(x),\dots,\lambda_{D_O}(x) \\ \lambda_{C_1}(x),\ \lambda_{C_2}(x),\ \lambda_{C_3}(x),\dots,\lambda_{C_I}(x) \end{Bmatrix} \quad \forall\ x \in X$$

$$\nu_D(x) = \begin{Bmatrix} \nu_{D_1}(x),\ \nu_{D_2}(x),\ \nu_{D_3}(x),\dots,\nu_{D_O}(x) \\ \nu_{C_1}(x),\ \nu_{C_2}(x),\ \nu_{C_3}(x),\dots,\nu_{C_I}(x) \end{Bmatrix} \quad \forall\ x \in X$$

Where $\mu_D(x)$, $\lambda_D(x)$ and $\nu_D(x)$ are the truth, indeterminacy and a falsity membership function of neutrosophic fuzzy decision set D_N respectively.

By utilizing the concept of Bellman and Zadeh (1970), our intention is to maximize the minimum truth (degree of belongingness) and minimize the maximum of indeterminacy (belongingness up to some extent) and falsity (degree of non-belongingness) degrees at a time. Therefore an overall achievement function can be defined as the differences of truth, indeterminacy and falsity degrees to reach the optimal solution of each objective under a neutrosophic environment. Thus the mathematical expression for the achievement function is defined as follows (Equation):

$$\begin{aligned} & Max\ \min_{o=1,2,\dots,O}\ \mu_o(Z_o(x)) \\ & Min\ \max_{o=1,2,\dots,O}\ \lambda_o(Z_o(x)) \\ & Min\ \max_{o=1,2,\dots,O}\ \nu_o(Z_o(x)) \\ & subject\ to \\ & all\ the\ restrictions\ of\ (Eq.\ 16) \end{aligned} \tag{20}$$

Using the auxiliary variables α, β and γ, the problem (Equation 20) can be transformed into the following problem (Equation 21):

$$\begin{aligned} & Max\ (\alpha - \beta - \gamma) \\ & subject\ to \\ & \mu_o(Z_o(x)) \ge \alpha, \\ & \lambda_o(Z_o(x)) \le \beta, \\ & \nu_o(Z_o(x)) \le \gamma, \\ & \alpha \ge \beta,\ \beta \ge \gamma,\ 0 \le \alpha + \beta + \gamma \le 3, \\ & \alpha, \beta, \gamma \in [0,1] \\ & all\ the\ restrictions\ of\ (Eq.\ 16) \end{aligned} \tag{21}$$

Now, we maximize a minimal neutrosophic satisfactory degree to reach the optimal solution of each objective named as interactive neutrosophic compromise programming approach (INCPA). The mathematical expression for the achievement function is defined as follows (22):

$$\begin{aligned} & Max\ \phi(x) = \omega(\alpha - \beta - \gamma) + (1-\omega)\left[\sum_o \left(\mu_o(Z_o(x)) - \lambda_o(Z_o(x)) - \nu_o(Z_o(x))\right)\right] \\ & subject\ to \\ & \mu_o(Z_o(x)) \ge \alpha, \\ & \lambda_o(Z_o(x)) \le \beta, \\ & \nu_o(Z_o(x)) \le \gamma, \\ & \alpha \ge \beta,\ \beta \ge \gamma,\ 0 \le \alpha + \beta + \gamma \le 3, \\ & \alpha, \beta, \gamma \in [0,1] \\ & all\ the\ restrictions\ of\ (Eq.\ 16) \end{aligned} \tag{22}$$

Where ω is the compensation co-efficient assigned to each individual satisfaction level in the achievement function.

Definition 11: A vector $x^* \in X$ is said to be an optimal solution to the proposed INCPA (Equation 22) or an efficient solution to the crisp MOOP (Equation 16) if and only if there does not exist any $x \in X$ such that, $\mu_o(x) \geq \mu_o(x^*)$, $\lambda_o(x) \leq \lambda_o(x^*)$ and $v_o(x) \leq v_o(x^*)$, $\forall\, o = 1, 2,\ldots.O$.

Theorem 1: A unique optimal solution of proposed INCPA (Equation 22) is also an efficient solution to the crisp MOOP (Equation 16).

Proof: Consider that x^* be a unique optimal solution of proposed INCPA (Eq.) which is not an efficient solution to crisp MOOP (Eq.). It means that there must be an efficient solution, say x^{**}, for the crisp MOOP (Eq.) so that we can have: $\mu_o(x^{**}) \geq \mu_o(x^*)$, $\lambda_o(x^{**}) \leq \lambda_o(x^*)$ and $v_o(x^{**}) \leq v_o(x^*)$, $\forall\, o = 1, 2,\ldots.O$. Also, there exists $k \mid \mu_k(x^{**}) \geq \mu_k(x^*)$, $\lambda_k(x^{**}) \leq \lambda_k(x^*)$and $v_k(x^{**}) \leq v_k(x^*)$, for at least one k. Thus for the overall satisfaction level of each objective functions in x^* and x^{**} solutions, we would have $(\alpha - \beta - \gamma)(x^{**}) \geq (\alpha - \beta - \gamma)(x^*)$, and concerning the related objective values we would have the following inequalities:

$$\begin{aligned}\phi(x^*) &= \omega(\alpha-\beta-\gamma)(x^*)+(1-\omega)\left[\sum_o \left(\mu_o(Z_o(x^*)\right) - \lambda_o(Z_o(x^*)) - v_o(Z_o(x^*))\right)\Big] \\ &= \omega(\alpha-\beta-\gamma)(x^*)+(1-\omega)\left[\sum_o \left(\mu_o(Z_o(x^*)\right) - \lambda_o(Z_o(x^*)) - v_o(Z_o(x^*))\right)\Big] \\ &< \omega(\alpha-\beta-\gamma)(x^{**})+(1-\omega)\left[\sum_o \left(\mu_o(Z_o(x^{**})\right) - \lambda_o(Z_o(x^{**})) - v_o(Z_o(x^{**}))\right)\Big] \\ &= \phi(x^{**})\end{aligned}$$

Hence, we have arrived at a contradiction that x^* is not a unique optimal solution of the proposed INCPA (Equation 22). This completes the proof of Theorem 8.

4. Computational Study

The inter-dependent and inter-connected overall shale gas water optimization framework is studied in matching the consistency with the real-life scenario, hypothetical data-set, and a fast review of the published articles; Zhang et al. (2016); Ahmad et al. (2019); Ahmad et al. (2020); Bartholomew and Mauter (2016); Lira-Barragán et al. (2016); Chen et al. (2017); Knee and Masker (2019); Lan et al. (2019); Ren et al. (2019) and Denham et al. (2019). The proposed modeling and optimization approach is designed to facilitate the decision-maker(s) in making the optimal distribution policies and strategies during the shale gas extraction processes. Various inter-connected components have been potentially optimized under the intuitionistic fuzzy uncertainty. Different costs associated with freshwater and waste water are the critical aspects of the proposed model because of the generated water and are a matter of prime concern. In budget allocation, freshwater acquisitions have significant impacts as the unit cost of acquisition is quite high. On-site treatment technology for waste water is also a prominent aspect for the reuse of water within the shale sites. The expansion option for the enhancement of the treatment plant capacity has been adopted. The use of pipelines for the flow of water is not considered due to economic reasons. The movement of water is allowed only through the roadways. All three planning periods are designed with five years time horizons. The dimensions of the study have been depicted in Table 1. All the parameters are depicted as triangular intuitionistic fuzzy numbers and summarized in Tables 2 and 3, respectively. The objective is to find the optimal distribution of overall shale gas water during the natural gas extraction processes. The minimum economic costs associated with the water management system and maximum revenue should be extracted from selling the shale gas at commercial level.

4.1 Results and Discussions

The addressed overall shale gas water management system is coded in AMPL language and the solution results are obtained using the BARON solver available on NEOS server version 5.0. The access is permitted by Dolan (2001); Drud (1994). The propounded model is solved at different compensation co-efficient parameters assigned to each objective function with respect to their marginal membership degree. The outcomes can be analyzed based on five different categories: (i) the optimal allocation of freshwater from different sources to shale sites, (ii) the promising on-site treatment technologies after hydraulic fracking processes, (iii) wholesome distribution of waste water for different purposes, (iv) various expansion strategies to expand the wastewater treatment plant capacity, and (v) the optimal compromise objectives under the different compensation co-efficient. The optimal allocation of freshwater quantity is presented in Table 4. About 89% of freshwater is acquired in the planning period 1, whereas in time period 2 and 3, the required amount of freshwater is calculated as 73% and 78%, respectively. This is because of the huge amount of freshwater required to initiate the procedures and the initial shale wells are bored with the efficiency factor of shale sites. The feasible quantity of waste

Table 2: Different cost parameters ($/bbl).

Transportation cost		**Time period**		
Source	**Destination**	$t = 1$	$t = 2$	$t = 3$
Shale site 1	Disposal site	(45,47,49; 43,47,51)	(52,54,56; 50,54,58)	(53,55,57; 51,55,59)
Shale site 1	Treatment plant 1	(65,67,69; 63,67,71)	(22,24,26; 20,24,28)	(94,96,97; 92, 96,98)
Shale site 1	Treatment plant 2	(53,55,57; 51,55,59)	(35,37,39; 33,37,42)	(82,84,86; 80, 84,88)
Shale site 2	Disposal site	(94,96,97; 92, 96,98)	(69,71,73; 67, 69,75)	(82,84,86; 80, 84,88)
Shale site 2	Treatment plant 1	(82,84,86; 80, 84,88)	(73,75,77; 71,75,79)	(84,86,87; 82, 86,88)
Shale site 2	Treatment plant 2	(74,76,77; 72, 76,78)	(62,64,66; 60,64,68)	(69,71,73; 67, 69,75)
Shale site 3	Disposal site	(84,86,87; 82, 86,88)	(53,55,57; 51,55,59)	(73,75,77; 71,75,79)
Shale site 3	Treatment plant 1	(69,71,73; 67, 69,75)	(82,84,86; 80, 84,88)	(53,55,57; 51,55,59)
Shale site 3	Treatment plant 2	(73,75,77; 71,75,79)	(74,76,77; 72, 76,78)	(82,84,86; 80, 84,88)
Shale site 4	Disposal site	(52,54,56; 50,54,58)	(45,47,49; 43,47,51)	(53,55,57; 51,55,59)
Shale site 4	Treatment plant 1	(22,24,26; 20,24,28)	(65,67,69; 63,67,71)	(74,76,77; 72, 76,78)
Shale site 4	Treatment plant 2	(35,37,39; 33,37,42)	(53,55,57; 51,55,59)	(94,96,97; 92, 96,98)
Shale site 5	Disposal site	(82,84,86; 80, 84,88)	(94,96,97; 92, 96,98)	(74,76,77; 72, 76,78)
Shale site 5	Treatment plant 1	(74,76,77; 72, 76,78)	(84,86,87; 82, 86,88)	(84,86,87; 82, 86,88)
Shale site 5	Treatment plant 2	(53,55,57; 51,55,59)	(73,75,77; 71,75,79)	(69,71,73; 67, 69,75)
Operational cost				
Disposal site		(8.2,8.4,8.6; 8.0, 8.4,8.8)	(9.4,9.6,9.7; 9.2, 9.6,9.8)	(7.4,7.6,7.7; 7.2, 7.6,7.8)
Treatment plant 1		(7.4,7.6,7.7; 7.2, 7.6,7.8)	(8.4,8.6,8.7; 8.2, 8.6,8.8)	(8.4,8.6,8.7; 8.2, 8.6,8.8)
Treatment plant 2		(5.3,5.5,5.7; 5.1,5.5,5.9)	(7.3,7.5,7.7; 7.1,7.5,7.9)	(6.9,7.1,7.3; 6.7, 6.9,7.5)
Capital investment	**Expansion option**			
Treatment plant 1	1	(9.4,9.6,9.7; 9.2, 9.6,9.8)	(6.9,7.1,7.3; 6.7, 6.9,7.5)	(8.2,8.4,8.6; 8.0, 8.4,8.8)
Treatment plant 1	2	(8.2,8.4,8.6; 8.0, 8.4,8.8)	(7.3,7.5,7.7; 7.1,7.5,7.9)	(8.4,8.6,8.7; 8.2, 8.6,8.8)
Treatment plant 1	3	(7.4,7.6,7.7; 7.2, 7.6,7.8)	(6.2,6.4,6.6; 6.0,6.4,6.8)	(6.9,7.1,7.3; 6.7, 6.9,7.5)
Treatment plant 2	1	(8.4,8.6,8.7; 8.2, 8.6,8.8)	(5.3,5.5,5.7; 5.1,5.5,5.9)	(7.3,7.5,7.7; 7.1,7.5,7.9)
Treatment plant 2	2	(6.9,7.1,7.3; 6.7, 6.9,7.5)	(8.2,8.4,8.6; 8.0, 8.4,8.8)	(5.3,5.5,5.7; 5.1,5.5,5.9)
Treatment plant 2	3	(7.3,7.5,7.7; 7.1,7.5,7.9)	(7.4,7.6,7.7; 7.2, 7.6,7.8)	(8.2,8.4,8.6; 8.0, 8.4,8.8)
Increased treatment capacity				
Treatment plant 1	1	850	350	830
Treatment plant 1	2	630	655	720
Treatment plant 1	3	450	925	610
Treatment plant 2	1	900	425	780
Treatment plant 2	2	870	550	810
Treatment plant 2	3	560	680	930

water which are distributed to different facility centers are depicted in Table 6. Almost 47% of generated wastewater is treated with on-site treatment technology which ensures the effective usage of the on-site treatment option and reduces the cost and time of transporting the waste water away from the shale sites. Thus the strategies for allocating the on-site treated waste water for the repeated fracking processes optimizes the time and other related restrictions in all the planning periods. The opportunity to enhance the waste water treatment plant capacity is depicted in Table 5. Almost all the three expansion options are utilized with significant reduction in the capital investment of expanding the treatment plant capacity. The total quantities of waste water are distributed through proper channels for underground injection disposal and the treatment facility is summarized in Table 6. Massive amounts of waste water are supplied to treatment plants and very less amounts are transported to underground disposal facilities which shows the less harmful impact on the underground and surface water resources. The optimal objectives are obtained by tuning the compensation

Table 3: Capacity restrictions on freshwater and wastewater (bbl/day).

Freshwater acquisition capacity	**Time period**		
	$t = 1$	$t = 2$	I = 3
Source	(500,600,700; 400,600,800)	(750,850,850; 700, 850,900)	(425,470,480; 410,470,490)
Freshwater demand at shale site			
Shale site 1	(1600,1700,1800; 1500,1700,1900)	(1450,1500,1550; 1400,1500,1600)	(1780,1810,1830; 1750,1810,1850)
Shale site 2	(2650,2750,2850; 2600,2750, 2900)	(2860,2870,2880; 2850,2870,2900)	(2800,2820,2840; 2780,2820,2860)
Shale site 3	(3520,3530,3540; 3500,3530,3550)	(3900,3920,3940; 3880,3920,3980)	(3620,3640,3660; 3600,3640,3660)
Shale site 4	(4910,4920,4930; 4900,4920,4930)	(4560,4570,4580; 4550,4570,4590)	(4600,4700,4800; 4500,4700,4900)
Shale site 5	(6650,6750,6850; 6600,6750, 6900)	(6800,6820,6840; 6780,6820,6860)	(6450,6500,6550; 6400,6500,6600)
Treatment capacity of wastewater			
Disposal site	(600,700,800; 500,700,900)	(450,500,550; 400,500,600)	(780,810,830; 750,810,850)
Treatment plant 1	(650,750,850; 600,750, 900)	(860,870,880; 850,870,900)	(800,820,840; 780,820,860)
Treatment plant 2	(520,530,540; 500,530,550)	(900,920,940; 880,920,980)	(620,640,660; 600,640,660)
Overall wastewater storage capacity			
Disposal site	(910,920,930; 900,920,930)	(560,570,580; 550,570,590)	(600,700,800; 500,700,900)
Treatment plant 1	(650,750,850; 600,750, 900)	(800,820,840; 780,820,860)	(450,500,550; 400,500,600)
Treatment plant 2	(520,530,540; 500,530,550)	(620,640,660; 600,640,660)	(860,870,880; 850,870,900)
Revenues generation			
Treatment plant 1	1.32	1.95	1.54
Treatment plant 2	1.65	1.02	1.65
Reuse rate			
Treatment plant 1	0.87	0.92	0.88
Treatment plant 2	0.78	0.65	0.69
	On-site treatment technology		
	$o = 1$	$o = 2$	$o = 3$
Recovery factor	0.32	0.56	0.78
Blending ratio	0.37	0.44	0.53

co-efficient and summarized in Table 7. All the values of objective values are obtained at different compensation co-efficient values between 0 and 1. At $\omega = 0.1$, the total amount of acquisition and transportation costs associated with freshwater are obtained USD \$542651. Similarly, the overall costs linked with the wastewater management system are depicted as USD \$48256.32. Furthermore, the managing policies for the expansion capacity of treatment plants using different expansion options are designed optimally and the total capital investment is obtained as USD \$8523.25 that reflects the ample opportunity to utilize the expansion options of treatment plants. Thus, a wholesome optimizing framework for overall shale gas water management system ensures the sustainable production of shale gas energy over various planning horizons.

The presented overall shale gas water management system inevitably exhibits the realistic scenarios of decision-making processes. Various uncertain parameters are dealt with intuitionistic fuzzy set theory and depicted as triangular intuitionistic fuzzy numbers. All these parameters comprise the membership and non-membership functions which represent the degree of vagueness and hesitation simultaneously. The intuitionistic fuzzy parameters have the tendency violate the risks due to hesitation to a lesser degree. The entire configuration of various restrictions imposed over different parameters also reflects the real situation of Pennsylvania. In Pennsylvania, most often the wastewaters are transported to a nearby city in Ohio due to the scarcity of the underground disposal facilities. A similar scenario has been depicted by the solution results and trivial amounts of wastewater are shipped to different underground disposal facilities. The decision maker(s) can unanimously adopt the proposed overall shale gas water modeling framework to determine the optimal distribution of fresh and wastewater in different planning horizons at the commercial level.

Table 4: Optimal amount of wastewater allocation.

	Total amount of wastewater $ww_{i,j,t}$	Amount of wastewater at disposal site $wwd_{i,j,t}$	Amount of wastewater at treatment plant $wwt_{i,j,t}$	Amount of wastewater for on-site treatment $wto_{i,o,t}$
1 1 1	6.75	6.75	3.80	150
1 1 2	17.25	17.25	95.80	150
1 1 3	13.25	13.25	215.06	150
1 2 1	645	0	645	200
1 2 2	842.5	0	842.5	200
1 2 3	0	0	0	200
1 3 1	0	0	0	1551.71
1 3 2	0	0	0	2401.65
1 3 3	0	0	0	2701.62
2 1 1	6.75	6.75	0	127.352
2 1 2	17.25	17.25	0	127.352
2 1 3	13.25	13.25	0	127.352
2 2 1	0	0	0	200
2 2 2	0	0	0	6250
2 2 3	0	0	0	200
2 3 1	0	0	0	525.313
2 3 2	0	0	0	4554.66
2 3 3	0	0	0	527.01
3 1 1	6.75	6.75	0	150
3 1 2	17.25	17.25	0	150
3 1 3	13.25	13.25	0	150
3 2 1	0	0	0	200
3 2 2	0	0	0	200
3 2 3	0	0	0	2865.32
3 3 1	137.71	0	137.71	1551.32
3 3 2	675	0	675	2060.51
3 3 3	850	0	850	2208.04
4 1 1	6.75	6.75	0	147.779
4 1 2	17.25	17.25	0	2023.26
4 1 3	13.25	13.25	0	147.779
4 2 1	645	0	645	200
4 2 2	842.5	0	842.5	272.266
4 2 3	937.5	0	937.5	730.788
4 3 1	137.71	0	137.71	751.275
4 3 2	675	0	675	300
4 3 3	850	0	850	300
5 1 1	6.75	6.75	95.60	150
5 1 2	17.25	17.25	85.890	150
5 1 3	13.25	13.25	95.630	150
5 2 1	645	65.30	645	342.801
5 2 2	842.5	52.60	842.5	861.73
5 2 3	937.5	65.620	937.5	861.73
5 3 1	137.71	983.60	137.71	300
5 3 2	675	25.60	675	360.539
5 3 3	850	94.680	850	360.539

Table 5: Strategy for expansion of treatment plant.

Increased treatment plant capacity (*EO*)	Expansion option (*m*)	Time period		
		$t = 1$	$t = 2$	$t = 3$
Treatment plant 1	1	600	-	600
Treatment plant 1	2	750	-	750
Treatment plant 1	3	850	-	850
Treatment plant 2	1	550	550	-
Treatment plant 2	2	650	-	-
Treatment plant 2	3	800	-	-

Table 6: Optimal amount of acquisition of freshwater.

Sources to destinations over time period	Amount of freshwater $fw_{s,i,t}$
1 1 1	700
1 1 2	1125
1 1 3	1275
1 2 1	186.765
1 2 2	1125
1 2 3	187.613
1 3 1	700
1 3 2	654.419
1 3 3	528.153
1 4 1	300.48
1 4 2	131.542
1 4 3	131.542
1 5 1	74.1
1 5 2	212.553
1 5 3	212.553

Table 7: Optimal objective values.

Objective values	Compensation co-efficient				
	$\omega = 0.1$	$\omega = 0.3$	$\omega = 0.5$	$\omega = 0.7$	$\omega = 0.9$
Minimum O_1	542651	542783	542968	543128	543295
Minimum O_2	48256.32	48356.82	48523.95	48852.36	48921.41
Minimum O_3	8523.25	8692.56	8702.96	8763.24	8791.65

5. Conclusions

This chapter has studied the multi-objective shale gas water management problem under intuitionistic fuzzy parameters in order to optimize integrated decisions on facility location and distribution problems under the uncertainty. The proposed shale gas water management model determines the distribution of freshwater from different sources to satisfy the demand at various shale sites for hydraulic fracking purposes during shale gas extraction processes. Also, we have successfully captured the economic objectives in the proposed model which also signifies the wholesome management of overall water in the different planning horizons. Minimization of total economic cost depicts the financial and budget concerns under a set of dynamic constraints which consequently results in a powerful modeling approach for overall shale gas water management problem. The deterministic version of the proposed model is obtained with the aid of expected interval and expected values of the intuitionistic fuzzy parameters. An interactive neutrosophic compromise optimization technique has been suggested to solve the proposed shale gas water management model by maximizing the minimal satisfactory degrees of the decision-makers and also the score function used to select the best compromise solution set.

A real case study in Pennsylvania, USA, is used to show the applicability and validity of the proposed shale gas water planning model. The opportunity to select several solutions for decision makers by maximizing the satisfaction degree is also a benchmarking contribution to this research work. The selection of an efficient solution can be done by tuning the compensation co-efficient. Moreover, we have designed the model according to the input information of the parameters which result in the model under intuitionistic fuzzy uncertainty. Hence the proposed model with socio-economic objectives is quite worthy of importance to reveal the actual real life scenario and could assist the decision makers to design fruitful policies and strategies to reduce the impact of wastewater on environmental points of view.

In the future, the proposed shale gas water management problem may be extended with other environmental objectives such as road closures or traffic congestions, as well as mode of shipping that may affect the efficiency of overall water management. It is also possible to extend the modeling concept of shale gas water management problems with the stochastic and uncertain programming domain. Various nature-inspired algorithms can also be considered to solve the proposed shale gas water management model as a future research scope.

References

Adhami, A. Y. and Ahmad, F. 2020. Interactive pythagorean hesitant fuzzy computational algorithm for multiobjective transportation problems under uncertainty. International Journal of Management Science and Engineering Management, 15(4): 288–297.

Ahmad, F. and Adhami, A. Y. 2019a. Neutrosophic programming approach to multiobjective nonlinear transportation problem with fuzzy parameters. International Journal of Management Science and Engineering Management, 14(3): 218–229.

Ahmad, F. and Adhami, A. Y. 2019b. Total cost measures with probabilistic cost function under varying supply and demand in transportation problem. Opsearch, 56(2): 583–602.

Ahmad, F., Adhami, A. Y. and Smarandache, F. 2018. Single valued neutrosophic hesitant fuzzy computational algorithm for multiobjective nonlinear optimization problem. Neutrosophic Sets and Systems, 22: 76–86.

Ahmad, F., Adhami, A. Y. and Smarandache, F. 2019. Neutrosophic optimization model and computational algorithm for optimal shale gas water management under uncertainty. Symmetry, 11(4): 544.

Ahmad, F., Adhami, A. Y. and Smarandache, F. 2020. Modified neutrosophic fuzzy optimization model for optimal closed-loop supply chain management under uncertainty. Optimization Theory Based on Neutrosophic and Plithogenic Sets (pp. 343–403). Academic Press.

Ahmad, F., Hussain, A. and Adhami, A. Y. 2020. Modeling texture of shale gas water management under risk factor.

Al-Aboosi, F. and El-Halwagi, M. 2018. An integrated approach to water-energy nexus in shale-gas production. Processes, 6: 52.

Bartholomew, T. V. and Mauter, M. S. 2016. Multiobjective optimization model for minimizing cost and environmental impact in shale gas water and wastewater management. ACS Sustain. Chem. Eng., 4: 3728–3735.

Chebeir, J., Geraili, A. and Romagnoli, J. 2017. Development of shale gas supply chain network under market uncertainties. Energies, 10: 246.

Chen, Y., He, L., Guan, Y., Lu, H. and Li, J. 2017. Life cycle assessment of greenhouse gas emissions and water-energy optimization for shale gas supply chain planning based on multi-level approach: Case study in Barnett, Marcellus, Fayetteville, and Haynesville shales. Energy Convers. Manag, 134: 382–398.

Denham, A., Willis, M., Zavez, A. and Hill, E. 2019. Unconventional natural gas development and hospitalizations: Evidence from Pennsylvania, United States, 2003–2014. Public Health, 168: 17–25.

Drouven, M. G. and Grossmann, I. E. 2017. Mixed-integer programming models for line pressure optimization in shale gas gathering systems. J. Pet. Sci. Eng., 157: 1021–1032.

Gao, J. and You, F. 2015. Optimal design and operations of supply chain networks for water management in shale gas production: MILFP model and algorithms for the water-energy nexus. AIChE J., 61: 1184–1208.

Gao, J. and You, F. 2015a. Shale gas supply chain design and operations toward better economic and life cycle environmental performance: MINLP model and global optimization algorithm. ACS Sustain. Chem. Eng., 3: 1282–1291.

Guerra, O. J., Calderón, A. J., Papageorgiou, L. G., Siirola, J. J. and Reklaitis, G. V. 2016. An optimization framework for the integration of water management and shale gas supply chain design. Comput. Chem. Eng., 92: 230–255.

Guo, M., Lu, X., Nielsen, C. P., McElroy, M. B., Shi, W., Chen, Y. and Xu, Y. 2016. Prospects for shale gas production in China: Implications for water demand. Renew. Sustain. Energy Rev., 66: 742–750.

Knee, K. L. and Masker, A. E. 2019. Association between unconventional oil and gas (UOG) development and water quality in small streams overlying the Marcellus Shale. Freshw. Sci., 38: 113–130.

Lan, Y., Yang, Z., Wang, P., Yan, Y., Zhang, L. et al. 2019. A review of microscopic seepage mechanism for shale gas extracted by supercritical CO2 flooding. Fuel, 238: 412–424.

Li, L. G. and Peng, D. H. 2014. Interval-valued hesitant fuzzy Hamacher synergetic weighted aggregation operators and their application to shale gas areas selection. Math. Probl. Eng., 181050.

Lira-Barragán, L. F., Ponce-Ortega, J. M., Serna-González, M. and El-Halwagi, M. M. 2016. Optimal reuse of flowback wastewater in hydraulic fracturing including seasonal and environmental constraints. AIChE J., 62: 1634–1645.

Lutz, B. D., Lewis, A. N. and Doyle, M. W. 2013. Generation, transport, and disposal of wastewater associated with Marcellus Shale gas development. Water Resour. Res., 49: 647–656.

Ren, K., Tang, X., Jin, Y., Wang, J., Feng, C. et al. 2019. Bi-objective optimization of water management in shale gas exploration with uncertainty: A case study from Sichuan, China. Resour. Conserv. Recycl., 143: 226–235.

Sang, Y., Chen, H., Yang, S., Guo, X., Zhou, C. et al. 2014. A new mathematical model considering adsorption and desorption process for productivity prediction of volume fractured horizontal wells in shale gas reservoirs. J. Nat. Gas Sci. Eng., 19: 228–236.

Smarandache, F. (Ed.). 2003. A unifying field in logics: Neutrosophic logic. neutrosophy, neutrosophic set, neutrosophic probability: Neutrosophic logic: neutrosophy, neutrosophic set, neutrosophic probability. Infinite Study.

Stevens, P. 2012. The Shale Gas Revolution: Developments and Changes; Chatham House: London, UK.

Yang, L., Grossmann, I. E. and Manno, J. 2014. Optimization models for shale gas water management. AIChE J., 60: 3490–3501.

Yang, L., Grossmann, I. E., Mauter, M. S. and Dilmore, R. M. 2015. Investment optimization model for freshwater acquisition and wastewater handling in shale gas production. AIChE J., 61: 1770–1782.

Zhang, X., Sun, A. Y. and Duncan, I. J. 2016. Shale gas wastewater management under uncertainty. J. Environ. Manag., 165: 188–198.

CHAPTER 18

Memory Effect on an EOQ Model with Price Dependant Demand and Deterioration

Mostafijur Rahaman,[1,*] *Sankar Prasad Mondal*[2] and *Shariful Alam*[1]

1. Introduction

The much celebrated conversation between two renowned scholars Leibnitz and L. Hospital regarding the existence of ½ order derivative as a logical continuation (expansion) of integer calculus towards the fractional counterpart is considered to be the first initiation to formulate the sense of fractional calculus in literature. This generalizing approach was very obvious and inclusive as well. However, the literature of fractional calculus grew at a very slow rate because of its ambiguity in philosophical relevance on the real ground. Researchers were confused about the physical and geometrical interpretation of the fractional derivatives and integrations. The classical integer order calculus has a clear geometrical and physical meaning. On comparison, the fractional calculus does not have any straight forward physical meaning. Many researchers considered the idea to be an abstract one. This was the main reason of hesitations and dilemmas concerning the development of the theory and application of fractional calculus. However, some practical investigation proved the superiority of fractional calculus over the integer order calculus to describe the memory motivated dynamical model. Then, memory effect is regarded as one of the physical significances of fractional calculus. The Riemamn-Louville and Caputo's definitions of the fractional derivative and integral are considered to be the two most popular approaches to describe memory involved in the problems with given initial conditions. However, the initial state given for the initial valued fractional differential equation with Riemann-Liouville fractional derivative is of fractional order whereas that of for the Caputo fractional derivative is of integer order. This issue proves the greatness of the Caputo fractional derivative over the Riemann-Liouville fractional derivative to describe real world problems because the fractional order initial conditions are abstract in realty. So, in the last two or three decades, the situation has changed and the notion of fractional calculus has been prominently utilized for the demonstrating problems in an extended domain of theoretical science as well as the application counterpart (Diethelm et al. 2012, Mainardi et al. 2007, Agila et al. 2016, Podlubny 1999). Using the definitions of fractional differentiations and integrations, the fractional differential equation has established itself as a better alternative of the integer order differential equation to describe many physical situations (Machado and Mata 2015, Mainardi et al. 2007, Arikoglu and Ozkol 2009, Duan et al. 2013, Hajipour et al. 2019, Jajarmi and Baleanu 2018).

The economic order/production quantity modeling is a very popular approach in an inventory management system. Here, optimal lot size and optimal time cycle is a matter of concern with respect to the cost minimization or profit maximization objective. After the introduction of the classical EOQ model by Harris (1913) under the assumption of constant demand and no shortage, the study of inventory modeling has been made mature through the inclusion of more realistic assumptions like price, stock, deterioration, and preservation. Generally, the dynamical systems describing the supply-chain model are given in terms of differential equations of integer order. But, the human associations in the production or retailing procedures make the system too complex to trace the actual dynamical behavior of the system. In a real-world marketing phenomena, the customer's present demand about some items must be motivated by the experiences earned with the previous deals (in terms of the quality, durability of the same or similar kind of products as well as the attitude of the dealing agency). Good customer-service in the past creates enthusiasm among customers to buy products in the present time. On the contrary, negative rating and review decrease the demand rate. Again, the producers or the retailers have their own assessment on the demand pattern from their dealing experiences which may contribute to making appropriate decisions regarding the optimal lot-size of the inventory. These obvious phenomena were neglected by the mathematical modeling described by the integer order differential equation. The memory dependent situations can be tackled aptly through the replacement of integer order

[1] Department of Mathematics, Indian Institute of Engineering Science and Technology, Shibpur, Howrah-711103, India.
[2] Department of Applied Science, Maulana Abul Kalam Azad University of Technology, West Bengal, Haringhata, Nadia-741249, West Bengal, India.
* Corresponding author: imostafijurrahaman@gmail.com

differential equations by the fractional differential equation in the sense of Caputo differentiation and Riemann-Liouville integration. An initiative study by Das and Roy (2014) accounted for the role of fractional calculus on the generalized economic order quantity model. Few more studies by them (Das and Roy 2015, 2017) also addressed the sense of fractional calculus on the study of inventory control problems. Pakhira et al. (2018a, 2018b, 2018c, 2019a, 2019b, 2019c, 2019d, 2019e, 2019f, 2020) contributed several worthy experimentations on the memory effect of the EOQ model with demand to linear and quadratic functions of time. The inventory with deteriorating products in the fractional framework has been discussed rarely in the literature. In this context, Rahaman et al. (2020a) developed an EPQ model with deterioration assuming uniform demand and production rate under the memory sensitive phenomena. In another study, Rahaman et al. (2020b) formulated an EPQ model with deterioration, stock dependent production rate and stock and price dependent demand and introduced the numerical optimization technique using the ABC algorithm. Very recently, the production phase of an EPQ model with deterioration of items has been analyzed in a fuzzy uncertain environment by Rahaman et al. (2020c).

This current chapter aims to develop an EOQ model under a memory affected scenario. The demand is considered to be selling price dependant. Generally, in a developing country like India, the customers are looking for cheap products compromising with their quality. So, demand rate in this situation is inversely proportional to the selling price. This obvious fact is incorporated here in the modeling procedure. Besides that, a very small count of deterioration rate is allowed here. And the overall situation is considered to be memory motivated.

The rest of this chapter is organized with the following sections: Section 2 describes the fractional calculus (Riemann-Liouville and Caputo's approaches) in brief. Section 3 stands for describing the notations and assumptions concerning the model respectively. The mathematical formulation as well as analysis of the proposed model is represented by Section 4. The classical integer order model and others are depicted as particular cases in Section 5. The numerical analysis of the fractional modeling has been illustrated in Section 6. Major managerial insights of the study are scripted in Section 7. Finally, the conclusion over the present paper is made in Section 8.

2. Basics of Fractional Calculus

In this section, some fundamental definitions and theorems of fractional calculus are described briefly which are very useful for the present documentation. The memory sensitive approach of inventory modelling and its optimization in the fractional domain requires few preliminary components of fractional calculus inspired by Caputo and Riemann-Liouville approaches.

In the below mentioned definitions of fractional differentiations and integrations, the origin is chosen to be at the point $x = a$. Here, basically, the left sided derivatives and integrals are defined.

Definition 2.1: (Podlubny 1999) Riemann- Liouville integral of $f(x)$ of order $m - \alpha$ is denoted by ${}_{a}^{RL}I_{x}^{m-\alpha}\{f(x)\}$ and is defined as

$${}_{a}^{RL}I_{x}^{m-\alpha}\{f(x)\} = \frac{1}{\Gamma(m-\alpha)}\int_{a}^{x}(x-v)^{m-\alpha-1} f(v)dv, \text{ where } m-1<\alpha\leq m \text{ and } m \text{ is an integer.}$$

In the next definition, the Riemann-Liouville derivative of order α $(m - 1 < \alpha \leq m)$ is given by the -th order derivative of ${}_{a}^{RL}I_{t}^{m-\alpha}\{f(x)\}$.

Definition 2.2: (Podlubny 1999) Riemann-Liouville derivative of $f(x)$ of order α is denoted by ${}_{a}^{RL}D_{x}^{\alpha}\{f(x)\}$ and is defined as

$${}_{a}^{RL}D_{x}^{\alpha}x(t) = \frac{1}{\Gamma(m-\alpha)}\frac{d^m}{dx^m}\int_{a}^{x}(x-v)^{m-\alpha-1} f(v)dv \text{ where } m-1<\alpha\leq m \text{ and } m \text{ is an integer.}$$

The Riemann-Liouville derivative is considered to be an inverse of the Riemann-Liouville integral. However, this definition of the derivative creates difficulties for portraying many real phenomena. Thus, the next definition stands for a better alternative of the Riemann-Liouville derivative. Taking $f(x)$ to be m-th differential and then interchanging the integration and differentiation operators in definition 2.2, the definition of Caputo derivative is obtained.

Definition 2.3: (Podlubny 1999) Caputo derivative of $f(x)$ of order α is denoted by ${}_{a}^{C}D_{x}^{\alpha}\{f(x)\}$ and is defined as

$${}_{a}^{C}D_{x}^{\alpha}f(x) = \frac{1}{\Gamma(m-\alpha)}\int_{a}^{x}(x-v)^{m-\alpha-1}\left(\frac{d^m f(v)}{dv^m}\right)dv, \text{ where } m-1<\alpha\leq m, m \text{ is an integer and } \frac{d^m f(v)}{dv^m} \text{ is the } m\text{-th derivative of } f(x).$$

Theorem 2.1: (Podlubny 1999) Laplace Transformation of the Riemann-Liouville fractional derivative of ${}_{a}^{RL}D_{x}^{\alpha}f(x)$ of order α is given by

$$L\{{}_{a}^{RL}D_{x}^{\alpha}f(x);s\} = s^{\alpha}F(s) - \sum_{k=1}^{m-1}s^{m-1-k}\,{}_{a}^{RL}D_{x}^{\alpha}F(a) - s^{m-1}\,{}_{a}^{RL}I_{x}^{m-\alpha}F(a),$$

where $F(s)$ is the Laplace transform of $f(x)$ and $m - 1 < \alpha \leq m$.

Theorem 2.2: (Podlubny 1999) Laplace Transformation of the Caputo fractional derivative of ${}_{a}^{C}D_{x}^{\alpha}f(x)$ of order α is given by

$$L\{{}_{a}^{C}D_{x}^{\alpha}f(x);s\} = s^{\alpha}F(s) - \sum_{k=0}^{m-1}s^{\alpha-1-k}f^{k}(a), \text{ where } F(s) \text{ is the Laplace transform of } f(x) \text{ and } m-1<\alpha\leq m.$$

Remarks 2.1: The Laplace transform of the Riemann-Liouville fractional derivatives claim for the fractional initial state which seems to be a little abstract. However, for the case of the Caputo fractional derivative, the initial state is of integer order having a proper physical meaning. This advantage makes the later definition to be smarter to deal with a problem like inventory control. Here we consider the Caputo fractional derivative and the Riemann-Liouville integral of order $\alpha(0 < \alpha \leq 1)$ to describe the model. Throughout the remainder of this chapter, $\frac{d^\alpha f(x)}{dx^\alpha}$ stands for the Caputo fractional derivative of $f(x)$ of the order $\alpha(0 < \alpha \leq 1)$ and $D^{-\beta}f(x)$ stands for the Riemann-Liouville fractional integral of $f(x)$ of order $\alpha(0 < \alpha \leq 1)$.

The Mittag-Leffler Function

The Mittag-Leffelr function is considered to be the fractional generalization of the exponential function of the integer order calculus. The one-parameter Mittag-Leffler function is denoted by $E_\alpha(x)$ and is given by the infinite Taylor's series expansion $E_\alpha(x) = \sum_{k=0}^{\infty} \frac{x^k}{\Gamma(\alpha k + 1)}$, where $\alpha > 0$. A more extensive form of the two parameter Mittag-Leffler function is given by $E_{\alpha,\beta}(x) = \sum_{k=0}^{\infty} \frac{x^k}{\Gamma(\alpha k + \beta)}$, where $\alpha, \beta > 0$. By definitions of the one and two parameters Mittag-Leffler function we have,

$$E_1(x) = e^x = E_{1,1}(x) \text{ and } E_\alpha(x) = E_{\alpha,1}(x).$$

3. Notations and Assumptions

The following basic notations are used to formulate and describe the mathematical model in this chapter:

Notations	*Units*	*Descriptions*
h	\$/unit	Holding cost per unit time
oc	\$/unit	Ordering cost per unit time
p	\$/unit	Selling Price
D	Units	Demand rate per cycle
T	Months	Total time cycle
Q	Units	Lot size
$TAP_{\alpha,\beta}$	\$/months	Total generalized average profit
α	Constant	The order of integration
β	Constant	The order of differentiation
Decision variables		
T	Months	Total time cycle
Q	Units	Lot size
Objective function		
$TAP_{\alpha,\beta}$	\$/months	Total generalized average profit

The proposed EOQ model is developed under the following assumptions:

i. Demand of the production depends on price. When selling price is low then the demand increases, i.e., $D(p) = \frac{a}{p}$, where is a positive constant and p is the price of the product.
ii. The constant deterioration $\varepsilon \ll 1$ is present throughout the whole production cycle.
iii. Shortage is not allowed.
iv. The retailing procedure is memory motivated.

4. Detailed Discussion of Mathematical Modelling

Initially, an inventory system starts with the lot size Q. The inventory level decays as the time grows fulfilling the customer demand and reaches zero level completing the cycle length T.

Then the fractional differential equation of fractional order $\alpha(0 < \alpha \leq 1)$ under Caputo approach to describe the proposed EOQ model is

$$\begin{cases} \frac{d^{\alpha} I(t)}{dt^{\alpha}} + \varepsilon I(t) = -\frac{a}{p}, 0 \leq t \leq T \\ \text{With the ending conditions } I(0) = Q \text{ and } I(T) = 0 \end{cases} \tag{1}$$

Now taking the Laplace transformation of (1) we have,

$$s^{\alpha}\, \tilde{I}(s) - s^{\alpha-1} I(0) + \varepsilon\, \tilde{I}(s) = -\frac{a}{ps}$$

$$\Rightarrow \left(s^{\alpha} + \varepsilon\right) \tilde{I}(s) = s^{\alpha-1} Q - \frac{a}{ps} \tag{2}$$

$$\Rightarrow \tilde{I}(s) = \frac{s^{\alpha-1} Q}{\left(s^{\alpha} + \varepsilon\right)} - \frac{a}{ps\left(s^{\alpha} + \varepsilon\right)}$$

Therefore, using the inverse Laplace transformation of (2) we have

$$I(t) = L^{-1}\{\tilde{I}(s)\} = L^{-1}\{\frac{s^{\alpha} Q}{s\left(s^{\alpha} + \varepsilon\right)}\} - \frac{a}{\varepsilon p} L^{-1}\{\frac{\varepsilon}{s\left(s^{\alpha} + \varepsilon\right)}\}$$

$$= QE_{\alpha}(-\varepsilon t^{\alpha}) - \frac{a}{\varepsilon p}\{1 - E_{\alpha}\left(-\varepsilon t^{\alpha}\right)\} \tag{3}$$

$$= (Q + \frac{a}{\varepsilon p}) E_{\alpha}(-\varepsilon t^{\alpha}) - \frac{a}{\varepsilon p}$$

Again, $I(T) = 0$ gives

$$Q = \frac{a}{\varepsilon p}\{E_{\alpha}(\varepsilon T^{\alpha}) - 1\} \tag{4}$$

Then (3) takes the form

$$I(t) = [\frac{a}{\varepsilon p}\{E_{\alpha}(\varepsilon T^{\alpha}) - 1\} + \frac{a}{\varepsilon p}] E_{\alpha}(-\varepsilon t^{\alpha}) - \frac{a}{\varepsilon p}$$

$$= \frac{a}{\varepsilon p}[E_{\alpha}\{\varepsilon\left(T^{\alpha} - t^{\alpha}\right)\} - 1] \tag{5}$$

The corresponding holding cost,

$$HC_{\alpha,\beta}(T) = hD^{-\beta} I(T) = \frac{h}{\Gamma(\beta)} \int_0^T (T-x)^{\beta-1} I(x) dx$$

$$= \frac{h}{\Gamma(\beta)} \frac{a}{\varepsilon p} \int_0^T (T-x)^{\beta-1} [E_{\alpha}\{\varepsilon\left(T^{\alpha} - x^{\alpha}\right)\} - 1] dx$$

$$\approx \frac{h}{\Gamma(\beta)} \frac{a}{p} \int_0^T (T-x)^{\beta-1} \left\{ \frac{\left(T^{\alpha} - x^{\alpha}\right)}{\Gamma(\alpha+1)} + \frac{\varepsilon\left(T^{\alpha} - x^{\alpha}\right)^2}{\Gamma(2\alpha+1)} \right\} dx$$

$$= \frac{h}{\Gamma(\beta)} \frac{a}{p} \left[\left\{ \frac{T^{\alpha}}{\Gamma(\alpha+1)} + \frac{\varepsilon T^{2\alpha}}{\Gamma(2\alpha+1)} \right\} \int_0^T (T-x)^{\beta-1} dx - \left\{ \frac{1}{\Gamma(\alpha+1)} + \frac{2\varepsilon T^{\alpha}}{\Gamma(2\alpha+1)} \right\} \int_0^T (T-x)^{\beta-1} x^{\alpha} dx + \frac{\varepsilon}{\Gamma(2\alpha+1)} \int_0^T (T-x)^{\beta-1} x^{2\alpha} dx \right]$$

$$= \frac{h}{\Gamma(\beta)} \frac{a}{p} \left[\left\{ \frac{T^{\alpha}}{\Gamma(\alpha+1)} + \frac{\varepsilon T^{2\alpha}}{\Gamma(2\alpha+1)} \right\} \frac{T^{\beta}}{\beta} - \left\{ \frac{1}{\Gamma(\alpha+1)} + \frac{2\varepsilon T^{\alpha}}{\Gamma(2\alpha+1)} \right\} T^{\alpha+\beta} \mathrm{B}(\alpha+1, \beta) + \frac{\varepsilon}{\Gamma(2\alpha+1)} T^{2\alpha+\beta} \mathrm{B}(2\alpha+1, \beta) \right]$$

$$=\frac{h}{\Gamma(\beta)}\frac{a}{p}\left[\left\{\frac{1}{\beta}-B(\alpha+1,\beta)\right\}\frac{T^{\alpha+\beta}}{\Gamma(\alpha+1)}+\left\{\frac{1}{\beta}-2B(\alpha+1,\beta)+B(2\alpha+1,\beta)\right\}\frac{\varepsilon T^{2\alpha+\beta}}{\Gamma(2\alpha+1)}\right]$$

Sales revenue,

$$SR_{\alpha,\beta}=pD^{-\beta}D=\frac{p}{\Gamma(\beta)}\int_0^T (T-x)^{\beta-1}\frac{a}{p}dx$$

$$=\frac{a}{\Gamma(\beta)}\int_0^T (T-x)^{\beta-1}dx=\frac{aT^{\beta}}{\Gamma(\beta+1)}$$

Total profit of the system during the entire circle is given by

$$TP_{\alpha,\beta}(T)=SR_{\alpha,\beta}-c_0-HC_{\alpha,\beta}$$

Average profit of the system during the entire cycle is given by

$$TAP_{\alpha,\beta}(T)=\frac{TP_{\alpha,\beta}(T)}{T}$$

$$=\frac{aT^{\beta-1}}{\Gamma(\beta+1)}-\frac{c_0}{T}-\frac{h}{\Gamma(\beta)}\frac{a}{p}\left[\left\{\frac{1}{\beta}-B(\alpha+1,\beta)\right\}\frac{T^{\alpha+\beta-1}}{\Gamma(\alpha+1)}+\left\{\frac{1}{\beta}-2B(\alpha+1,\beta)+B(2\alpha+1,\beta)\right\}\frac{\varepsilon T^{2\alpha+\beta-1}}{\Gamma(2\alpha+1)}\right]$$

So, the optimization problem can be written as

$$\begin{cases} Maximize\,TAP_{\alpha,\beta}(T) \\ TAP_{\alpha,\beta}(T)=\frac{aT^{\beta-1}}{\Gamma(\beta+1)}-\frac{c_0}{T}-\frac{h}{\Gamma(\beta)}\frac{a}{p}\left[\left\{\frac{1}{\beta}-B(\alpha+1,\beta)\right\}\frac{T^{\alpha+\beta-1}}{\Gamma(\alpha+1)}+\left\{\frac{1}{\beta}-2B(\alpha+1,\beta)+B(2\alpha+1,\beta)\right\}\frac{\varepsilon T^{2\alpha+\beta-1}}{\Gamma(2\alpha+1)}\right] \\ Q=\frac{a}{\varepsilon p}\{E_{\alpha}(\varepsilon T^{\alpha})-1\} \\ \text{Subject to} T>0 \\ 0<\alpha,\beta<1 \end{cases}$$

5. Particular Cases

This present section points to two special phenomena of the generalized study presented in the previous section.

5.1 When both the Memory Indexes Tend to 1

Then the optimization model will be

$$\begin{cases} Maximize\,TAP_{1,1}(T) \\ TAP_{1,1}(T)=a-\frac{c_0}{T}-\frac{ha}{p}\left[\frac{T}{2}+\frac{\varepsilon T^2}{3}\right] \\ Q=\frac{a}{\varepsilon p}(e^{\varepsilon T}-1) \\ \text{Subject to} T>0 \end{cases}$$

The above is the optimization model corresponding to the memory less situation of the classical lot-sizing approach for a model with small rate of deterioration.

5.2 When both the Memory Indexes tend to 1 and Deterioration Tends to zero

Then the optimization model will be

$$\begin{cases} Maximize TAP_{1,1}(T) \\ TAP_{1,1}(T) = a - \dfrac{c_0}{T} - \dfrac{haT}{2p} \\ Q = \dfrac{aT}{p} \\ \text{Subject to} T > 0 \end{cases}$$

The above is the optimization model corresponding to the memory less situation of the classical lot-sizing approach for a model with no deterioration. This establishes the remarks that, fractional modelling is a more generalized approach. Also, the integer value of the memory index represents the memory less cases.

6. Numerical and Graphical Illustration

6.1 Numerical Simulation

We use LINGO software for numerical optimization of the problem given in Section 4. Here, we consider the following numerical values of the related input parameters:
$a = 1000$, $p = 25$, $c_0 = 500$, $h = 1.5$ and $\varepsilon = 0.05$.

Then, the optimal values of the objective functions (profit function to be minimized), the optimal lot size and total cycle time are given by Table 2, 3, and 4 for a different memory grid. Table 1 presents different values of gamma and beta functions for different integral and differential memory indexes.

Table 1: Values of Gamma and Beta function.

Differential memory index α	**Integral memory index β**	$\Gamma(\beta)$	$\Gamma(\beta+1)$	$\Gamma(\alpha+1)$	$\Gamma(2\alpha+1)$	$\Gamma(\alpha+1,\beta)$	$\Gamma(2\alpha+1,\beta)$
1	1	1	1	1	2	0.5	0.33333
0.9	0.9	1.06863	0.96177	0.96177	1.67649	0.61305	0.42956
0.8	0.8	1.64230	0.93138	0.93138	1.42962	0.75848	0.55830
0.7	0.7	1.29806	0.90864	0.90864	1.24217	0.94952	0.73370
0.6	0.6	1.48919	0.89352	0.89352	1.10180	1.20767	0.97871
0.5	0.5	1.77245	0.88623	0.88623	1	1.57080	1.33333
0.4	0.4	2.21816	0.88726	0.88726	0.93138	2.11308	1.87507
0.3	0.3	2.99157	0.89747	0.89747	0.89352	3.00481	2.77928
0.2	0.2	4.59084	0.91817	0.91817	0.88726	4.75075	4.55872
0.1	0.1	9.51351	0.95135	0.95135	0.91817	9.85732	9.73291
1	0.9	1.06863	0.96177	1	2	0.58480	0.40331
1	0.8	1.64230	0.93138	1	2	0.69444	0.49603
1	0.7	1.29806	0.90864	1	2	0.84034	0.62247
1	0.6	1.48919	0.89352	1	2	1.04167	0.80128
1	0.5	1.77245	0.88623	1	2	1.33333	1.06667
1	0.4	2.21816	0.88726	1	2	1.78571	1.48810
1	0.3	2.99157	0.89747	1	2	2.56410	2.22965
1	0.2	4.59084	0.91817	1	2	4.16667	3.78788
1	0.1	9.51351	0.95135	1	2	9.09091	8.65801
0.9	1	1	1	0.96177	1.67649	0.52631	0.42956
0.8	1	1	1	0.93138	1.42962	0.55556	0.45970
0.7	1	1	1	0.90864	1.24217	0.58824	0.49467
0.6	1	1	1	0.89352	1.10180	0.625	0.53577
0.5	1	1	1	0.88623	1	0.66667	0.58480
0.4	1	1	1	0.88726	0.93138	0.71429	0.64434
0.3	1	1	1	0.89747	0.89352	0.76923	0.71828
0.2	1	1	1	0.91817	0.88726	0.83333	0.81267
0.1	1	1	1	0.95135	0.91817	0.90909	0.93760

In the theory of fractional calculus to demonstrate the memory, it is convenient to consider the lower values of the memory index to represent stronger memory and vice-versa. Accordingly, for the values of the memory index to be 1, the system becomes memory less.

Table 2: Variation of total average profit, time cycle and lot size with respect to both the memory indexes.

Differential memory index (α)	Integral memory index (β)	T	Q	$TAP_{\alpha,\beta}$
1	1	3.844	168.52	747.22
0.9	0.9	3.173	127.10	691.80
0.8	0.8	2.490	95.13	659.14
0.7	0.7	1.696	67.10	613.46
0.6	0.6	1.180	51.66	601.87
0.5	0.5	0.785	41.56	619.93
0.4	0.4	0.473	34.60	695.20
0.3	0.3	0.229	29.57	929.98
0.2	0.2	0.063	25.78	1989.74
0.1	0.1	0.017	22.84	32506.87

The Table 2 represents the variation of total average profit, time cycle and lot size with respect to both the memory indexes (the differential as well as the integral memory index). From Table 2, we observe the following facts:

i. When the memory is going to be stronger, the total average profit decreases initially. However, after crossing the memory index 0.5 and goes through the bottom of the table the total average profit gradually increases. For the least values of the memory index, the values of the total average profit grows faster.
ii. The memory strength maintains a straight forward pattern for the lot size throughout the table. As the memory strength increases, the lot size decreases gradually.
iii. The total cycle time also gradually decreases with the increasing nature of the memory.

Table 3: Variation of total average profit, time cycle and lot size with respect to the differential memory index.

Differential memory index (α)	T	Q	$TAP_{\alpha,\beta}$
1	3.844	168.52	747.22
0.9	4.312	171.52	764.58
0.8	5.043	175.32	786.64
0.7	6.139	177.23	811.96
0.6	7.894	176.31	840.39
0.5	10.960	171.34	871.40
0.4	17.018	160.81	903.88
0.3	31.460	143.14	935.91
0.2	78.754	117.25	964.74
0.1	385.576	83.43	987.02

The Table 3 represents the variation of total average profit, time cycle and lot size with respect to the differential memory index. From Table 3, we observe the following facts:

i. The memory strength maintains a straight forward pattern for the total average profit throughout the table. As the memory strength increases, the total average profit also increases accordingly.
ii. When the memory is going to be stronger, the lot size increases initially. However, after crossing the memory index 0.5 and goes through the bottom of the table the lot size gradually decreases. Secondly, the memory strength maintains a straight forward pattern for the lot size throughout the table. As the memory strength increases, the lot size decreases gradually.
iii. The total cycle time also gradually increases with the increasing nature of the memory.

Table 4: Variation of total average profit, time cycle and lot size with respect to the integral memory index.

Integral memory index (β)	T	Q	$TAP_{\alpha,\beta}$
1	3.844	168.52	747.22
0.9	2.932	125.90	681.64
0.8	2.315	97.96	650.56
0.7	1.585	65.92	604.59
0.6	1.127	46.34	595.73
0.5	0.761	31.03	616.54
0.4	0.464	18.77	694.75
0.3	0.226	9.10	933.47
0.2	0.062	2.49	2002.35
0.1	0.002	0.07	32659.46

The Table 4 represents the variation of total average profit, time cycle and lot size with respect to the integral memory index. From Table 4, we observe the following facts:

i. The variation pattern for the Total average profit, the lot size and the total cycle time is almost the same as that of Table 2.
ii. The rates of increase of the values of total average profit and rate of decrease of the values of total cycle time and lot size for the integral memory index only are greater than the cases corresponding to the simultaneous presence of the integral and memory index.

6.2 Graphical Visualization

In this subsection, the results from the Table 2, 3, 4 are plotted to make a graphical illustration of various insights of the proposed optimization technique. The variation of the total average profit with respect to the differential memory index is given by Figure 1.

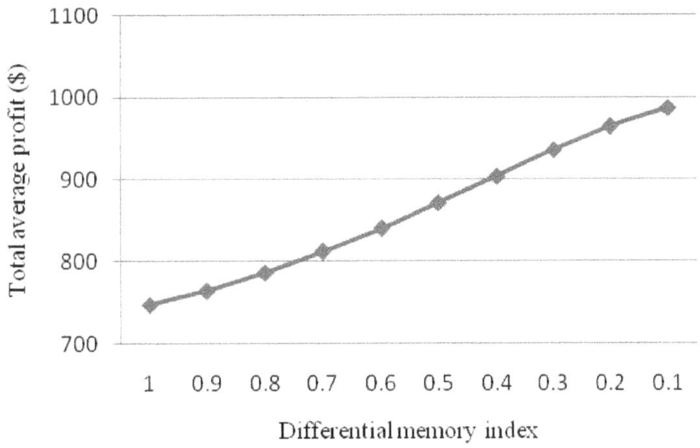

Figure 1: Total average versus Differential memory index.

The Figure 1 describes a strictly increasing curve of total average profit with respect to the improvement of memory. The Figure 2 represents the relationships between the integral memory index and total average profit.

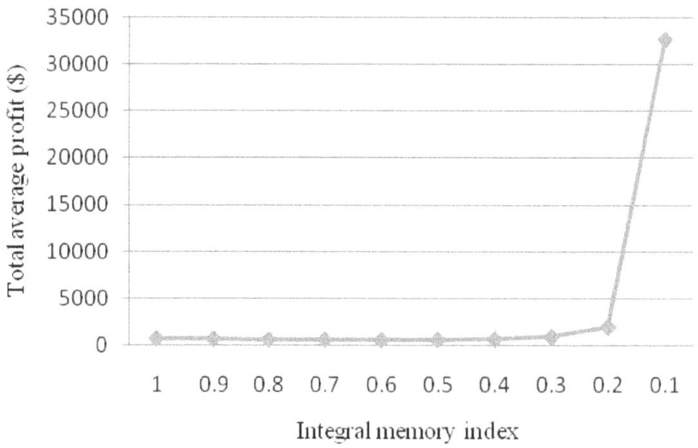

Figure 2: Total average profit versus integral memory index.

In the Figure 2, there is a decreasing initial pattern of the total average profit with respect to the integral memory index. However, the curve ultimately increases exponentially according as the integral memory.

The Figure 3 plots the graph of Total average profit with respect to both the integral and differential memory index simultaneously.

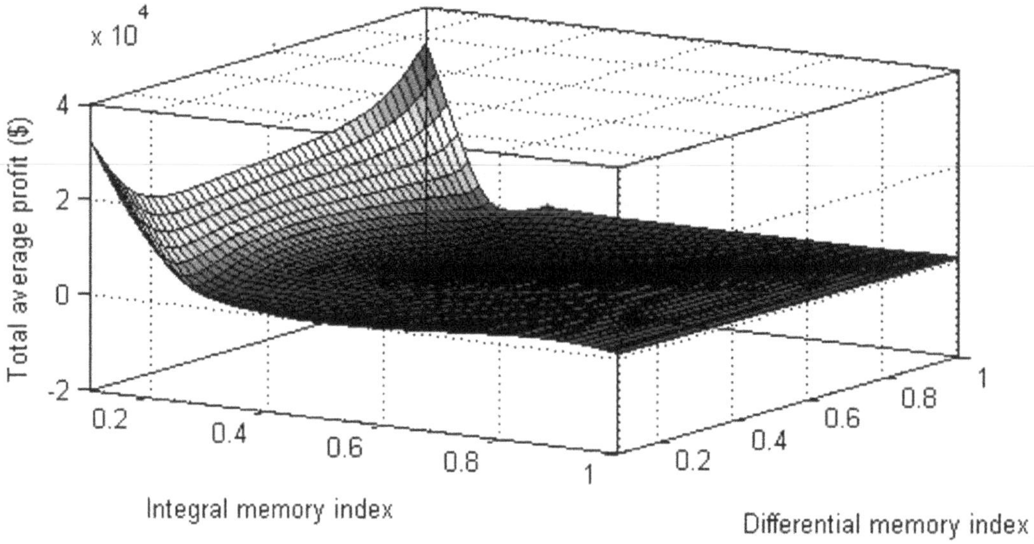

Figure 3: Total average profit versus memory index.

7. Managerial Insight

This chapter makes a hypothetical data based experiment on the memory effect on an EOQ model with deterioration. Though the data was not collected from real grounds of the retailing sector, several observations from this current theoretical investigation points to some business oriented insight for the decision maker. In this context, it is noted that bigger values of memory index stand for weak sense of memory and vice versa. So, from the numerical simulation, it is observed that stronger system memory always creates good phenomena with an attitude for profit maximization in the proposed EOQ model. However, much stronger sense of memory can lead the system towards a non-feasible scenario making the values of the time cycle and optimal lot size arbitraryly small or large. Thus, the system is very much memory sensitive. Though memory through past experience advocates a better chance of profit, a smart manager must be careful about the feasibility of the decision on real ground.

8. Conclusion

This present chapter is focused on the development of an inventory control problem of deteriorating items considering the price dependent demand. Here, the study is concerned about the memory of the retailing system. One clear observation is that stronger memory leads the system towards more gain or profits. Moreover, presence of the differential memory index only gives a stable relationship with the variation of total average profit. However, when both the memory indexes are considered, then the integral memory index dominates the differential memory index in this model. In this study, we assume rate of deterioration to be very small which makes the analysis little easier by approximating the Mittag-Leffler function with its few terms. For an arbitrary amount of deterioration, tackling the optimization function having Mittag-Leffler function in integration symbols will be a challenge for future research. Also, feasible data from the real world markets make these types of studies more fruitful validating the mathematical derivations.

References

Agila, A. Baleanu, D., Eid, R. and Iranfoglu, B. 2016. Applications of the extended fractional Euler-Lagrange equations model to freely oscillating dynamical systems. Romanian J. Phys., 61: 350–359.

Arikoglu, A. and Ozkol, I. 2009. Solution of fractional integro-differential equations by using fractional differential transform method. Chaos Solitons Fractals, 40: 521–529.

Das, A. K. and Roy, T. K. 2014. Role of fractional calculus to the generalized inventory model. Journal of Global Research in Computer Science, 5(2): 11–23.

Das, A. K. and Roy, T. K. 2015. Fractional order EOQ model with linear trend of time-dependent demand. I.J. Intelligent Systems and Applications, 03: 44–53.

Das, A. K. Roy and T. K. 2017. Fractional order generalized EPQ model. International Journal of Computational and Applied Mathematics, 12(2): 525–536.

Diethelm, K., Baleanu, D. and Scalas, E. 2012. Fractional calculus: Models and numerical methods. World Scientific. DOI: 10.1142/10044.

Duan, J. S., Chaolu, T., Rach, R. and Lu, L. 2013. The Adomian decomposition method with convergence acceleration techniques for nonlinear fractional differential equations. Computers & Mathematics with Applications, 66: 728–736.

Hajipour, M., Jajarmi, A., Baleanu, D. and Guang Sun, H. 2019. On an accurate discretization of a variable-order fractional reaction-diffusion equation. Communications in Nonlinear Science and Numerical Simulation, 69: 119–133.

Harris, F. W. 1913. How many parts to make at once. Factory, The magazine of Management, 10(2): 135–136.

Jajarmi, A. and Baleanu, D. 2018. A new fractional analysis on the interaction of HIV with CD4+ T-cells, Chaos. Solitons & Fractals, 113: 221–229.

Machado, J. A. and Mata, M. E. 2015. Pseudo phase plane and fractional calculus modeling of western global economic downturn. Commun. Nonlinear Sci. Numer. Simul., 22: 396–406.

Mainardi, F., Pagnini, G. and Gorenflo, R. 2007. Some aspects of fractional diffusion equations of single and distributed order. Appl. Math. Comput., 187: 295–305.

Pakhira, R., Ghosh, U. and Sarkar, S. 2018a. Application of Memory Effects in an inventory model with linear demand and no shortage. International Journal of Research in Advent Technology, 6(8): 1853–1871.

Pakhira, R., Ghosh, U. and Sarkar, S. 2018b. Study of Memory Effects in an inventory model with linear demand and salvage value. International Journal of Applied Engineering Research, 13(20): 14741–14751.

Pakhira, R., Ghosh, U. and Sarkar, S. 2018c. Study of Memory Effects in an inventory model. Applied Mathematical Sciences, 12(17): 797–824.

Pakhira, R., Ghosh, U. and Sarkar, S. 2019a. Study of Memory Effects in an inventory model with quadratic type demand rate and salvage value. Applied Mathematical Sciences, 13(5): 209–223.

Pakhira, R., Ghosh, U. and Sarkar, S. 2019b. Study of Memory Effects in an economic order quantity model with quadratic type demand rate. CMST, 25(2): 71–80.

Pakhira, R., Ghosh, U. and Sarkar, S. 2019c. Study of Memory Effects in a fuzzy EOQ Model with no shortage. International Journal of Intelligent Systems and Applications, 11: 58–68.

Pakhira, R., Ghosh, U. and Sarkar, S. 2019d. Study of Memory Effects in an inventory model with linear demand and shortage. International Journal of Mathematical Sciences and Computing, 2: 54–70.

Pakhira, R., Ghosh, U. and Sarkar, S. 2019e. Study of Memory Effects in an inventory model with price dependent demand. Journal of Applied Economic Sciences, XIV, 2(64): 360–367

Pakhira, R., Ghosh, U. and Sarkar, S. 2019f. Application of Memory Effects in an inventory model with price dependent demand rate during shortage. International Journal of Education and Management Engineering, 3: 51–64.

Pakhira, R., Ghosh, U. and Sarkar, S. 2020. Study of memory effect between two memory dependent inventory models. Journal of Fractional Calculus and Applications, 11(1): 203–218.

Podlubny, I. 1999. Fractional Differential Equations, Academic Press, San Diego, CA.

Rahaman, M., Mondal, S. P. Shaikh, A. A. Ahmadian, A. Norazak, S. et al. 2020. Arbitrary-order economic production quantity model with and without deterioration: generalized point of view. Adv Differ Equ, 16(2020). https://doi.org/10.1186/s13662-019-2465-x.

Rahaman, M., Mondal, S. P., Shaikh, A. A., Pramanik, P. Roy, S. et al. 2020. Artificial bee colony optimization-inspired synergetic study of fractional-order economic production quantity model. Soft Comput. https://doi.org/10.1007/s00500-020-04867-y.

Rahaman, M., Mondal, S. P., Alam, S., Khan, N. A. and Biswas, A. 2020. Interpretation of exact solution for fuzzy fractional non-homogeneous differential equation under the Riemann-Liouville sense and its application on the inventory management control problem. Granular Computing, https://doi.org/10.1007/s41066-020-00241-3.

CHAPTER 19

Optimality Conditions of an Unconstrained Imprecise Optimization Problem via Interval Order Relation

*Md Sadikur Rahman** and *Asoke Kumar Bhunia*

1. Introduction

Because of the occurrence of randomness and inexactness, most of the parameters of real-life problems, especially, of the optimization problems are not precise. So, the optimization problem with imprecise parameters is an interesting research problem. Representing the impreciseness of a parameter involved in an optimization problem by a random variable or fuzzy set/number or interval, several researchers studied the imprecise optimization problems. The optimization problem with random variable/fuzzy set valued/interval valued parameters is called a stochastic/fuzzy/interval optimization problem. In stochastic optimization, Bridge and Louveaux [3], Prekopa [15], and Kall [10] introduced some useful techniques to solve stochastic optimization problems. In fuzzy optimization, Pathak and Pirzada [14], Heidari et al. [8], Umamaneswari and Ganesan [20], Effati et al. [4] established several fuzzy optimization techniques.

On the other hand, interval analysis is an important tool for the tackling the uncertainty which appears in optimization problems. In his book, Moore [12] first introduced the concept of interval analysis for handling uncertainty. Again, Moore at el. [13] introduced this concept for the computational optimization purpose only. Also, the notions of the calculus of interval-valued functions are already reported by several researchers. Among them the works of Markov [11], Stefanini and Bede [19] and Ramezanadeh et al. [18] are worth mentioning. Besides these, several researchers, with the help of the calculus of interval valued objective functions, developed the optimality conditions for interval optimization problems. Ishibuchi and Tanaka [9] established the optimality conditions for interval-valued functions from the view point of the optimality of multi-objective functions. Then the optimality criteria of interval-valued non-linear programming problem are studied by Wu [20]. After that Bhurjee and Panda [2] developed an efficient methodology to solve an interval optimization problem. Recently, Ghosh [5–6], Aguirre et al. [1], Han and Liu [7], Rahman et al. [16] contributed their works to solve interval optimization problems.

In this chapter, the optimality conditions of unconstrained imprecise optimization problem in interval parametric form are discussed. For this purpose, using the concept of parametric form, the definition of interval ranking/order relation is introduced. Then based on this order relation, the definitions of maximizer, minimizer and optimizer search function are defined. Then the optimality conditions (necessary and sufficient) are derived for the said unconstrained imprecise optimization problem. Then a set of numerical examples is solved by using these conditions. This work is organised in the following manner: The concepts of parametric representation, interval mathematics, interval-valued functions with some of its analytical properties are presented in Section 2 whereas in Section 3, the definition of interval order relation is proposed with numerical examples. Then, in Section 4, standard form of unconstrained interval maximization as well as minimization problems are presented in both interval and parametric forms along with the definitions of maximizer, minimizer, optimizer search function. Also, both the necessary and sufficient conditions for optimality are established in the same section. The numerical illustrations are presented in Section 5.

2. Basic Concepts

In this section, we have discussed parametric representation of interval, preliminary definitions and arithmetic operations of intervals, interval-valued functions in parametric form with some analytical properties. First of all, we have clarified that here an interval means a closed and bounded interval which is defined mathematically as $C = [c_L, c_U] = \{x \in \mathbb{R} : c_L \le x \le c_U\}$.

Department of mathematics, The University of Burdwan, Burdwan, India-713104.
Email: akbhunia@math.buruniv.ac.in
* Corresponding author: mdsadikur.95@gmail.com

2.1 Parametric Representation and Interval Arithmetic

Let $C = [c_L, c_U] \in K_c$, the set of all closed and bounded intervals. Then the parametric forms of C be defined as follows:

(i) $C = \{c(p): c(p) = c_L + p(c_U - c_L), p \in [0,1]\}$ (Increasing form (IF)),

(ii) $C = \{c(p): c(p) = c_U - p(c_U - c_L), p \in [0,1]\}$ (Decreasing form (DF)).

Definition 1. Let $\{c(p): p \in [0,1]\}$ and $\{d(p): p \in [0,1]\}$ be the IFs (DFs) of $C = [c_L, c_U]$ and $D = [d_L, d_U]$ respectively, and λ be any real number. Then addition, subtraction, multiplication, division, scalar multiplication and parametric difference of two intervals in the parametric form are defined as follows:

$$1. C + D = \{c(p_1) + d(p_2): p_1, p_2 \in [0,1]\},$$
$$2. C - D = \{c(p_1) - d(p_2): p_1, p_2 \in [0,1]\}$$
$$3. CD = \{c(p_1)d(p_2): p_1, p_2 \in [0,1]\},$$
$$4. C/D = \{c(p_1)/d(p_2): d(p_2) \neq 0, p_1, p_2 \in [0,1]\},$$
$$5. \lambda.C = \{\lambda c(p_1): p_1 \in [0,1]\},$$
$$6. C(-)_p D = \{c(p) - d(p): p \in [0,1]\}.$$

Remark 1. Let $C = \{c(p): p \in [0,1]\}$ and $D = \{d(p): p \in [0,1]\}$ be two intervals in parametric form. Then $C = D \Leftrightarrow c(p) = d(p), \forall p \in [0,1]$.

Proposition 1. Let $C = [c_L, c_U] = \{c(p): p \in [0,1]\}$ and $D = [d_L, d_U] = \{d(p): p \in [0,1]\}$ be two intervals. Then $C = D \Leftrightarrow c(p) = d(p), for\ p = 0,1$.

Proof.

Since $C = D \Leftrightarrow c_L = d_L$ and $c_U = d_U$.

Again, from parametric form of interval, it is observed that

$c_L = c(0), c_U = c(1)\ and$ similar for d_L, d_U. Thus, $C = D \Leftrightarrow c(0) = d(0)$ and $c(1) = d(1)$.

Hence, the proof is completed.

2.2 Parametric form of Interval-valued Function

Let $K_c = \{[c_L, c_U]: c_L, c_U \in \mathbb{R}\}$. Now we have defined a metric on K_c in such a way, so that it becomes a complete metric space. The function, $D_P: K_c \times K_c \to \mathbb{R}$ defined by

$d_P(C,D) = \max\left\{\left|\min_p (c(p) - d(p))\right|, \left|\max_p (c(p) - d(p))\right|\right\}$ is clearly a metric on the set of closed and bounded intervals K_c, where $C = \{c(p): p \in [0,1]\}$, $D = \{d(p): p \in [0,1]\} \in K_c$.

Let $H: T \subseteq \mathbb{R}^n \to K_c$ be an interval-valued function in lower and upper bounds form, $H(x) = [h_L(x), h_U(x)]$. Then a parametric form of $H(x)$ is $\{h(p,x) = h_L(x) + p(h_U(x) - h_L(x)): p \in [0,1]\}$.

Example 1.

(i) Consider an interval-valued function $H: \mathbb{R} \to K_c$ defined by $H(x) = [xe^x, 2xe^x + 1]$. Then the parametric form of $H(x)$ is $\{h(p,x) = xe^x + p(xe^x + 1): p \in [0,1]\}$.

(ii) Let $H: \mathbb{R}^n \to K_c$ defined by $H(x) = [a_L, a_U]h(x)$, where $h: \mathbb{R}^n \to \mathbb{R}$. Then the parametric form of $F(x)$ is $\{h(p,x) = (a_L + p(a_U - a_L))h(x): p \in [0,1]\}$.

Definition 2. Let $H: T \subseteq \mathbb{R}^n \to K_c$ be interval-valued function of n variables and x_0 be limit point of T. Then $L \in K_c$ is said to be the limit of H at x_0, if for every $\varepsilon > 0, \exists \delta > 0$ such that $D_P(H(x), L) < \varepsilon$, whenever $0 < d_u(x, x_0) < \delta$, where d_u is the usual metric on $\mathbb{R}^n$. It is denoted by $\lim_{x \to x_0} H(x) = L$.

Definition 3. Let $H: T \subseteq \mathbb{R}^n \to K_c$ be an interval-valued function of several variables and $x_0 \in T$. Then H is said to be continuous at x_0, if for every $\varepsilon > 0, \exists \delta > 0$ such that $D_P(H(x), H(x_0)) < \varepsilon$, whenever $d_u(x, x_0) < \delta$, where d_u is the usual metric on $\mathbb{R}^n$. It is denoted by $\lim_{x \to x_0} H(x) = H(x_0)$.

Proposition 2.
The interval-valued function $H: T \subseteq \mathbb{R}^n \to K_c$ is continuous at $x_0 \in T$, then $h(p,x)$ is continuous at $x_0, \forall\, p \in [0,1]$, where $H(x) = \{h(p,x) = h_L(x) + p(h_U(x) - h_L(x)) : p \in [0,1]\}$.

Proof. Let H is continuous at $x_0 \in T$.
Then from the definition of continuity of an interval-valued function at a point, we get for every $\varepsilon > 0, \exists \delta > 0$ such that

$$D_P(H(x), H(x_0)) < \varepsilon, \text{whenever } d_u(x, x_0) < \delta,$$

$$\text{i.e., } \max\left\{\left|\min_p (h(p,x) - h(p,x_0))\right|, \left|\max_p (h(p,x) - h(p,x_0))\right|\right\} < \varepsilon, \text{ whenever } d_u(x, x_0) < \delta,$$

$$\text{i.e., } \left|\min_p (h(p,x) - h(p,x_0))\right| < \varepsilon \text{ and } \left|\max_p (h(p,x) - h(p,x_0))\right| < \varepsilon, \text{ whenever } d_u(x, x_0) < \delta,$$

$$\text{i.e., } \left|(h(p,x) - h(p,x_0))\right| < \varepsilon \text{ whenever } |x - x_0| < \delta \text{ and } \forall p \in [0,1].$$

Thus $h(p, x)$ is continuous at $x_0, \forall\, p \in [0,1]$.

Definition 4. The interval-valued function $H: T \subseteq \mathbb{R}^n \to K_c$ is said to be p-differentiable at $x_0 \in T$, if $h(p,x)$ is differentiable at x_0 in the usual sense. where $h(p,x) = h_L(x) + p(h_U(x) - h_L(x))$, $p \in [0,1]$. Furthermore, $\nabla_p H(x^*) = \{\nabla h(x,p) : p \in [0,1]\}$, where $\nabla_p H(x^*)$ denotes the total p-derivative of H at $x = x^*$.

Definition 5. The interval-valued function $H: T \subseteq \mathbb{R}^n \to K_c$ is said to be p-differentiable up to second order at $x_0 \in T$, if $h(p,x)$ is differentiable up to the second order at x_0 in the usual sense. Where $h(p,x)$, $p \in [0,1]$ is the parametric representation of H. Furthermore, $\nabla_p^2 H(x^*) = \{\nabla^2 h(x,p) : p \in [0,1]\}$, where $\nabla_p^2 H(x^*)$ denotes the total Hessian metrix of H at $x = x^*$.

3. Interval Order Relations

In this section, a new interval order relation is proposed in parametric form and the definition is justified by some numerical examples.

Definition 6. Let $C = [c_L, c_U] = \{c(p) : p \in [0,1]\}$ and $D = [d_L, d_U] = \{d(p) : p \in [0,1]\}$ be two intervals in parametric form and k $(>1) \in \mathbb{N}$ be given. Then the order relation between C and D is defined as follows:

$$C \leq_k D \Leftrightarrow \begin{cases} \psi_s(r_0, r_1, \ldots, r_{k-1}) \leq \eta_s(r_0, r_1, \ldots, r_{k-1}), \text{ if } \psi_s(r_0, r_1, \ldots, r_{k-1}) \neq \eta_s(r_0, r_1, \ldots, r_{k-1}) \\ \psi_s(r_1, \ldots, r_{k-1}) \leq \eta_s(r_1, \ldots, r_{k-1}), \text{ if } \psi_s(r_0, r_1, \ldots, r_{k-1}) = \eta_s(r_0, r_1, \ldots, r_{k-1}) \,\&\, \psi_s(r_1, \ldots, r_{k-1}) \neq \eta_s(r_1, \ldots, r_{k-1}) \\ \psi_s(r_2, \ldots, r_{k-1}) \leq \eta_s(r_2, \ldots, r_{k-1}), \text{ if } \psi_s(r_1, \ldots, r_{k-1}) = \eta_s(r_1, \ldots, r_{k-1}) \text{ and } \psi_s(r_2, \ldots, r_{k-1}) \neq \eta_s(r_2, \ldots, r_{k-1}) \\ \vdots \\ \psi_s(r_{k-1}) \leq \eta_s(r_{k-1}), \text{ if } \psi_s(r_{k-2}, r_{k-1}) = \eta_s(r_{k-2}, r_{k-1}) \end{cases}$$

where

$$\psi_s(r_j, r_{j+1}, \ldots, r_{k-1}) = \frac{\sum_{i=j}^{k-1} c(r_i)}{k-j} \;\&\; \eta_s(r_j, r_{j+1}, \ldots, r_{k-1}) = \frac{\sum_{i=j}^{k-1} d(r_i)}{k-j}, \; j = 0, 1, \ldots, k-1, r_0 = 0, r_{k-1} = 1$$

$$\text{and } r_i = \frac{1}{k-1} + r_{i-1}, i = 1, 2, \ldots, k-2.$$

Remark 2.
The proposed interval order relation satisfies the properties of reflexivity, anti-symmetry and transitivity, i.e., $(K_c, \leq_k)$ is a partial ordered set. Also, any two elements of K_c are comparable by $\leq_k$, i.e., $(K_c, \leq_k)$ is a totally ordered set.

Corollary 1. Taking $k = 3$, the **Definition 4** can be rewritten as follows:

$$C \leq_3 D \Leftrightarrow \begin{cases} \frac{c(0)+c(0.5)+c(1)}{3} \leq \frac{d(0)+d(0.5)+d(1)}{3}, \text{ if } \frac{c(0)+c(0.5)+c(1)}{3} \neq \frac{d(0)+d(0.5)+d(1)}{3} \\ \frac{c(0.5)+c(1)}{2} \leq \frac{d(0.5)+d(1)}{2}, \; if \; \frac{c(0)+c(0.5)+c(1)}{3} = \frac{c(0)+d(0.5)+d(1)}{3} \text{ and } \frac{c(0.5)+c(1)}{2} \neq \frac{d(0.5)+d(1)}{2} \\ c(1) \leq d(1), \; if \; \frac{c(0.5)+c(1)}{2} = \frac{d(0.5)+d(1)}{2} \end{cases}$$

Example 2.
Let us consider the following pair of interval numbers and compare those by using Corollary 1.

(i) $C = [2,8]$ and $D = [4,6]$ (ii) $C = [-2,3]$ and $D = [-1,2]$

(i) For first pair, the parametric representation of $C = [2,8]$ and $D = [4,6]$ are $c(p) = 2 + 6p$ and $d(p) = 4 + 2p$, $p \in [0,1]$.

Since $\frac{c(0)+c(0.5)+c(1)}{3} = \frac{14}{3} < \frac{15}{3} = \frac{c(0)+c(0.5)+c(1)}{3}$, Thus $C <_3 D$

For the second pair, the parametric representation of $C = [-2,3]$ and $D = [-1,2]$ are $c(p) = -2 + 5p$ and $d(p) = -1 + 3p$, $p \in [0,1]$.
Since

$$\frac{c(0)+c(0.5)+c(1)}{3} = \frac{d(0)+d(0.5)+d(1)}{3} = \frac{1}{2} \text{ and } \frac{c(0.5)+c(1)}{2} = \frac{7}{4} > \frac{5}{4} = \frac{c(0.5)+c(1)}{2}.$$

Hence $D <_3 C$

4. Optimality Conditions of Unconstrained Interval Optimization Problem

In this section, standard forms of unconstrained interval maximization as well as minimization problems with the definitions of maximizer, minimizer and optimization search function are discussed. Then optimality conditions of both minimization and maximization problems are derived.

4.1 Unconstrained Interval Optimization Problem

The standard form of unconstrained interval maximization problem can be written as

$$\begin{aligned} &\textit{Maximize } H(x) \\ &\text{subject to } \quad x \in S \subseteq \mathbb{R}^n \end{aligned} \tag{1}$$

where $H(x)$ may be represented by either

$$H(x) = [h_L(x), h_U(x)] \text{ or } H(x) = \sum [a_{iL}, a_{iU}] g_i(x) \text{ and } h_L, h_U, g_i : S \to \mathbb{R}, i = 1, \ldots, n.$$

The corresponding parametric form of (1) can be written as follows:

$$\begin{aligned} &\textit{Maximize } h(p,x) \\ &\text{subject to } x \in S \subseteq \mathbb{R}^n, \; p \in [0,1] \end{aligned} \tag{2}$$

where

$$\text{either, } h(p,x) = h_L(x) + p(h_U(x) - h_L(x)) \text{ or } h(p,x) = \sum_{i=1}^{n} a_i(p) g_i(x) \text{ and } a_i(p) = a_{iL} + p(a_{iU} - a_{iL})$$

The standard form of unconstrained interval minimization problem can be written as

$$\begin{aligned} &\textit{Minimize } H(x) \\ &\text{subject to } \quad x \in S \subseteq \mathbb{R}^n \end{aligned} \tag{3}$$

where

$$\text{either, } H(x) = [h_L(x), h_U(x)] \text{ or } H(x) = \sum_{i=1}^{n} [a_{iL}, a_{iU}] g_i(x) \text{ and } h_L, h_U, g_i : S \to \mathbb{R}, i = 1, \ldots, n.$$

The corresponding parametric form of (3) can be written as follows:

$$\begin{aligned} &Minimize\ h(p,x) \\ &\text{subject to } x \in S \subseteq \mathbb{R}^n,\ p \in [0,1] \end{aligned} \tag{4}$$

where

$$h(p,x) = h_L(x) + p(h_U(x) - h_L(x)) \text{ or } h(p,x) = \sum_{i=1}^{n} a_i(p) g_i(x) \text{ and } a_i(p) = a_{iL} + p(a_{iU} - a_{iL})$$

Proposition 3. The interval optimization problems (1) and (3) are equivalent to the optimization problem in parametric form (2) and (4) respectively.

Proof. The proof follows from **Remark 1.**

4.2 Maximizer, Minimizer and Optimizer Search function (OSF)

Definition 7.
The point $x = x^* \in S$ is said to be a maximizer of the maximization problem (1) if $H(x^*) \geq_k H(x)\ \forall x \in B(x^*, s) \cap S$, where $B(x^*, s)$ is the open ball with centre at $x = x^*$ and radius $s(>0)$.

Definition 8.
The point $x = x^* \in S$ is said to be a the minimizer of the minimization problem (2) if $H(x^*) \leq_k H(x)\ \forall x \in B(x^*, s) \cap S$, where $B(x^*, s)$ is the open ball with centre at $x = x^*$ and radius $s(>0)$.

Definition 9. A real-valued function $f_{s_i} : S \to \mathbb{R}$ is said to be an optimizer search function (OSF) of an interval optimization problem if the optimizers of f_{s_i} be the same as that of the optimization problem.

Now, for a given natural number k (>1) and $r_i \in [0,1]$ with $r_0 = 0, r_i = r_{i-1} + \frac{1}{k-1}, i = 1, ..., k-1$, We have defined a real-valued function $h_j(x)$ as follows:

$$h_j(x) = \frac{\sum_{i=j}^{k-1} h(r_i, x)}{k-1},\ j = 0, 1, ..., k-1.$$

where $h(r_i, x)$ is given in (2) or (4)

In the next proposition, we have established the conditions under which $h_j(x)$ will be an OSF of the interval optimization problem (1) or (3).

Proposition 4. The function $h_j(x)$ will be an OSF of the interval optimization problem (1) or (3) if $h_j(x) \neq$ constant, $j = 0, 1, ..., k-1$.

Proof. The proof follows from the definition of interval order relation.

4.3 Optimality Conditions

Theorem 1. If the point $x = x^* \in S \subseteq \mathbb{R}^n$ is the maximizer (minimizer) of the interval maximization (minimization) problem 1(or 3)then

$$\begin{cases} \nabla h_0(x^*) = 0_n, \text{ if } h_0(x) \neq \text{constant} \\ \nabla h_1(x^*) = 0_n, \text{ if } \nabla h_0(x) = \text{constant and } \nabla h_1(x) \neq \text{constant} \\ \nabla h_2(x^*) = 0_n, \text{ if } \nabla h_1(x) = \text{constant and } \nabla h_2(x) \neq \text{constant} \\ \vdots \\ \nabla h_{k-1}(x^*) = 0_n, \text{ if } \nabla h_{k-2}(x) = \text{constant} \end{cases}$$

where $0_n = (0, 0, ..., 0) \in \mathbb{R}^n$

Proof.
As the proofs of both maximization and minimization case can be derived in a similar way, we have proved this theorem for the maximization case only.

If $x = x^* \in S$ be the maximiser of the problem (1), then by the definition of maximiser we can say that $H(x^*) \geq_k H(x)$, $\forall x \in B(x^*, s) \cap S$.

i.e., from the definition of interval order relation, we can say that $\forall x \in B(x^*, s) \cap S$, $H(x^*) \geq_k H(x)$ implies

$$\begin{cases} h_0(x^*) \geq h_0(x), \text{ if } h_0(x^*) \neq h_0(x) \\ h_1(x^*) \geq h_1(p_1, \ldots, p_{k-1}, x), \text{ if } h_0(x^*) = h_0(x) \,\&\, h_1(x^*) \neq h_1(x) \\ h_2(x^*) \geq h_2(x), \text{ if } h_1(x^*) = h_1(x) \,\&\, h_2(x^*) \neq h_2(x) \\ \vdots \\ h_{k-1}(x^*) \geq h_{k-1}(x), \text{ if } h_{k-2}(x^*) = h_{k-2}(x) \end{cases}$$

From the necessary conditions of optimality, the above condition implies that

$$\begin{cases} \nabla h_0(x^*) = 0_n, \text{ if } h_0(x) \neq \text{constant} \\ \nabla h_1(x^*) = 0_n, \text{ if } \nabla h_0(x) = \text{constant and } \nabla h_1(x) \neq \text{constant} \\ \nabla h_2(x^*) = 0_n, \text{ if } \nabla h_1(x) = \text{constant and } \nabla h_2(x) \neq \text{constant} \\ \vdots \\ \nabla h_{k-1}(x^*) = 0_n, \text{ if } \nabla h_{k-2}(x) = \text{constant} \end{cases}$$

This proves the theorem.

Theorem 2. Let $x = x^* \in S \subseteq \mathbb{R}^n$ be the point satisfying the conditions of **Theorem 1.** Then $x = x^*$ will be the maximizer of (1) if

$$\begin{cases} \nabla^2 h_0(x^*) \text{ is negative definite, if } h_0(x) \neq \text{constant} \\ \nabla^2 h_1(x^*) \text{ is negative definite, if } \nabla h_0(x) = \text{constant and } \nabla h_1(x) \neq \text{constant} \\ \nabla^2 h_2(x^*) \text{ is negative definite, if } \nabla h_1(x) = \text{constant and } \nabla h_2(x) \neq \text{constant} \\ \vdots \\ \nabla^2 h_{k-1}(x^*) \text{ is negative definite, if } \nabla h_{k-2}(x) = \text{constant} \end{cases}$$

Proof. when $h_0(x) \neq$ constant,
then from $\Delta h_0(x^*) = 0_n$ and the negative definiteness of $\nabla^2 h_0(x^*)$, we can easily say that $h_0(x^*) \geq h_0(x)$, $\forall x \in B(x^*, s) \cap S$ (according to the sufficient optimality conditions of crisp optimization problem).
and when $h_0(x) =$ constant and $h_0(x) \neq$ constant,
then, from $\nabla h_1(x^*) = 0_n$ and the negative definiteness of $\nabla^2 h_1(x^*)$, we can say that $h_1(x^*) \geq h_0(x)$, $\forall x \in B(x^*, s) \cap S$
and so on.
Finally, when $h_{k-2}(x) =$ constant, we can obtain
$h_{k-1}(x^*) \geq h_{k-1}(x)$, $\forall x \in B(x^*, s) \cap S$
Combining all of the above we can obtain $\forall x \in B(x^*, s) \cap S$,

$$\begin{cases} h_0(x^*) \geq h_0(x), \text{ if } h_0(x^*) \neq h_0(x) \\ h_1(x^*) \geq h_1(x), \text{ if } h_0(x^*) = h_0(x) \text{ and } h_1(x^*) \neq h_1(x) \\ h_2(x^*) \geq h_2(x), \text{ if } h_1(x^*) = h_1(x) \text{ and } h_2(x^*) \neq h_2(x) \\ \vdots \\ h_{k-1}(x^*) \geq h_{k-1}(x), \text{ if } h_{k-2}(x^*) = h_{k-2}(x) \end{cases}$$

which implies, $H(x^*) \geq H(x)$, $\forall x \in B(x^*, s) \cap S$,
Hence, $x = x^*$ is the maximizer of (1).
This completes the proof.

Theorem 3. Let $x = x^* \in S \subseteq \mathbb{R}^n$ be the point with satisfying the conditions of **Theorem 1.** Then $x = x^*$ be the minimizer of (3) if

$$\begin{cases} \nabla^2 h_0\left(x^*\right) \text{is positive definite, if } h_0\left(x\right) \neq \text{constant} \\ \nabla^2 h_1\left(x^*\right) \text{ is positive definite, if } \nabla h_0\left(x\right) = \text{constant and } \nabla h_1\left(x\right) \neq \text{constant} \\ \nabla^2 h_2\left(x^*\right) \text{ is positive definite, if } \nabla h_1\left(x\right) = \text{constant and } \nabla h_2\left(x\right) \neq \text{constant} \\ \vdots \\ \nabla^2 h_{k-1}\left(x^*\right) \text{ is positive definite, if } \nabla h_{k-2}\left(x\right) = \text{constant} \end{cases}$$

Proof. The proof is similar as for **Theorem 2.**

4.3 Computational Procedure

Step-1: Set a natural number $k(>1)$.

Step 2: Calculate $h(p_j, p_{j+1},\ldots, p_{k-1}, x)$ for $j = 0$ to $k - 1$.

Step 3: Check $h(p_j, p_{j+1},\ldots, p_{k-1}, x) \neq$ constant or not, for $j = 0,1,\ldots k - 1$.

Step 4: If $h(p_j, p_{j+1},\ldots, p_{k-1}, x) \neq$ constant, go to step 5. Otherwise go to step 3.

Step 5: Maximize(Minimize) $h(p_j, p_{j+1},\ldots, p_{k-1}, x)$

5. Numerical Examples

In order to check the justification of all the theoretical results, two numerical examples are solved.

Example 3:
Let us consider the following function for optimization.

$$\text{Optimize } H\left(x_1,x_2\right) = \left[-2\left(x_1^2+x_2^2\right)+3, 2\left(x_1^2+x_2^2\right)+4\right].$$
$$\text{subject to } x_1, x_2 \in \mathbb{R}^n$$

Solution: The corresponding parametric form of the objective of (5) can be written as:

$$h\left(p,x_1,x_2\right) = -2\left(x_1^2+x_2^2\right)+3+p\left\{4\left(x_1^2+x_2^2\right)+1\right\}, p\in\left[0,1\right]$$

Here, we take $k=3$ to solve this problem.
Now,

$$h_0\left(x_1,x_2\right) = \frac{h\left(0,x_1,x_2\right)+h\left(0.5,x_1,x_2\right)+h\left(1,x_1,x_2\right)}{3} = \frac{7}{2} = \text{constant}$$

$$\text{and } h_1\left(x_1,x_2\right) = \frac{h\left(0.5,x_1,x_2\right)+h\left(1,x_1,x_2\right)}{2} = 2\left(x_1^2+x_2^2\right)+\frac{15}{2} \neq \text{constant}$$

So, using **Theorem 1** and **Theorem 3** it can be concluded that
$h_1\left(x_1,x_2\right) = 2\left(x_1^2+x_2^2\right)+\frac{15}{2}$ be the OFS for this problem.
The optimizers of this problem are obtained from $\nabla h_1(x_1, x_2) = (0,0)$
which implies $(x_1, x_2) = (0,0)$.
Clearly, $\nabla^2 h_0(0,0)$ is positive definite.
Hence H has minimum at $(x_1, x_2) = (0,0)$ and the minimum value is $H(0,0) = [3,4]$.

Example 4: Let us take the following interval optimization problem

$$\text{Minimize } H\left(x_1,x_2\right) = \left[1,4\right]x_1^2 - \left[2,3\right]x_1x_2 + \left[1,3\right]x_2^2 + \left[0,1\right]$$
$$\text{subject to } \left(x_1,x_2\right)\in\mathbb{R}^2$$

Solution: The corresponding parametric form of the objective of (6) can be written as

$$h\left(p,x_1,x_2\right) = \left(1+3p\right)x_1^2 - \left(2+p\right)x_1x_2 + \left(1+2p\right)x_2^2 + p, p\in\left[0,1\right]$$

Here $h_0(x_1,x_2)=\dfrac{h(0,x_1,x_2)+h(0.5,x_1,x_2)+h(1,x_1,x_2)}{3}=\dfrac{(15x_1^2-15x_1x_2+12x_2^2+3)}{6}\neq 0,$

So, using **Theorem 1** and **Theorem 3** we can say that

$h_0(x_1,x_2)=\dfrac{(15x_1^2-15x_1x_2+12x_2^2+3)}{6}$ be the OFS for this problem.

The optimizers of this problem are obtained from $\nabla h_0(x_1, x_2) = (0,0)$

which implies $(x_1, x_2) = (0,0)$.

Clearly, $\nabla^2 h_0(0,0)$ is positive definite.

Hence H has minimum at $(x_1, x_2) = (0,0)$ and the minimum value is $H(0,0) = [0,1]$.

6. Conclusion

In this work, the theories of optimality conditions of an unconstrained interval optimization problem are derived by an alternative approach. Here, all the theoretical results are derived in the parametric form by using the proposed interval order relation, the definitions of maximizer, minimizer and optimization search function. All the results throughout the works are validated with numerical examples.

In future, the Karush-Kuhn-Tucker conditions of a constrained interval optimization problem can be derived by using these concepts. The same concepts of this chapter may be extended for the optimization problem in Type-2 interval environment [17]. Also, the concepts of this work can be applied to solve several real-life optimization problems such as modelling problems,and reliability problems.

References

[1] Aguirre-Cipe, I., López, R., Mallea-Zepeda, E. and Vásquez, L. 2019. A study of interval optimization problems. Optimization Letters, 1–19.
[2] Bhurjee, A. K. and Panda, G. 2012. Efficient solution of interval optimization problem. Mathematical Methods of Operations Research, 76(3): 273–288.
[3] Birge, J. and Louveaux, R. F. 1997. Introduction to Stochastic Programming, Physica-Verlag, NY.
[4] Effati, S., Mansoori, A. and Eshaghnezhad, M. 2020. Linear quadratic optimal control problem with fuzzy variables via neural network. Journal of Experimental & Theoretical Artificial Intelligence, 1–14.
[5] Ghosh, D. 2016. A Newton method for capturing efficient solutions of interval optimization problems. Opsearch, 53(3): 648–665.
[6] Ghosh, D. 2017. Newton method to obtain efficient solutions of the optimization problems with interval-valued objective functions. Journal of Applied Mathematics and Computing, 53(1-2): 709–731.
[7] Han, X. and Liu, J. 2020. Introduction to Uncertain Optimization Design. In Numerical Simulation-based Design (pp. 215-220). Springer, Singapore.
[8] Heidari, M., Zadeh, M. R., Fard, O. S. and Borzabadi, A. H. 2016. On unconstrained fuzzy-valued optimization problems. International Journal of Fuzzy Systems, 18(2): 270–283.
[9] Ishibuchi, H. and Tanaka, H. 1990. Multi-objective programming in optimization of the interval objective function. Eur. J. Oper. Res., 48(2): 219–22.
[10] Kall, P. 1976. Stochastic Linear Programming, Springer-Verlag, NY.
[11] Markov, S. 1979. Calculus for interval functions of a real variable. Computing, 22(4): 325–337.
[12] Moore, R. E. 1966. Interval analysis (Vol. 4). Englewood Cliffs, NJ: Prentice-Hall.
[13] Moore, R. E., Kearfott, R. B. and Cloud, M. J. 2009. Introduction to interval analysis (Vol. 110). Siam.
[14] Pathak, V. D. and Pirzada, U. M. 2011. Necessary and sufficient optimality conditions for nonlinear fuzzy optimization problem.
[15] Prekopa, A. 1995. Stochastic Programming, Kluwer Academic Publishers, Boston.
[16] Rahman, M. S., Shaikh, A. A. and Bhunia, A. K. 2020. Necessary and Sufficient Optimality Conditions for non-linear Unconstrained and Constrained optimization problem with Interval valued objective function. Computers & Industrial Engineering, 106634.
[17] Rahman, M. S., Shaikh, A. A. and Bhunia, A. K. 2020. On Type-2 interval with interval mathematics and order relations: its applications in inventory control. International Journal of Systems Science: Operations & Logistics, 1–13.
[18] Ramezanadeh, M., Heidari, M., Fard, O. S. and Borzabadi, A. H. 2015. On the interval differential equation: novel solution methodology. Advances in Difference Equations, 2015(1): 338.
[19] Stefanini, L. and Bede, B. 2009. Generalized Hukuhara differentiability of interval-valued functions and interval differential equations. Nonlinear Analysis: Theory, Methods & Applications, 71(3-4): 1311–1328.
[20] Wu, H. C. 2008. On interval-valued nonlinear programming problems. J. Math Anal. Appl. 338(1): 299–316.

Chapter 20

Power Comparison of Different Goodness of Fit Tests for Beta Generalized Weibull Distribution

Kanchan Jain,* Neetu Singla **and** ***Suresh K Sharma***

1. Introduction

Goodness of fit tests are used to verify whether a sample can be taken as belonging to a population following a specified distribution, that is, whether the empirical distribution fits the theoretical distribution closely or not. Empirical Distribution Function (EDF) tests or goodness of fit tests (GOFTs) measuring the distance between empirical and theoretical distribution functions were explored by Dufour et al., 1998. According to Seier 2002 and Arshad et al., 2003, most commonly used tests are Kolmogorov-Smirnov (KS), Cramer-von Mises (CVM) and Anderson-Darling (AD) for testing goodness of fit.

Many resesarchers have explored the goodness of fit tests. Chi-square test for continuous distributions and exponential family was studied by Nikulin, 1973a, 1973b and Rao and Robson, 1974 respectively. Lemeshko et al., 2009, 2010a compared power of goodness of fit tests for testing simple versus complex hypotheses. Lemeshko et al., 2010b explored GOFTs for Inverse Gaussian family and Lemeshko and Lemeshko, 2011 worked on these tests for composite hypotheses using double exponential distribution. For Weighted Gamma distribution, Singla et al., 2016 explored GOFTs and compared their powers. Betsch and Ebner, 2019 studied Steinian characterization of the family of Gamma distributions and associated GOFTs. Few other references in this area are D'Agostino and Stephens, 1986; Ampadu, 2008; Bagdonavicius and Nikulin, 2012; Henze et al., 2012; Voinov et al., 2013; Villasen and Estrada, 2015; Plubin and Siripanich, 2017.

Beta Generalized Weibull (BGW) distribution with five parameters was introduced by Singla et al., 2012 using the cumulative distribution function of Generalized Weibull (GW) distribution. Importance of BGW also lies in the fact that skewness and tail weight get controlled with addition of new parameters. Mudholkar and Srivastava, 1993; Mudholkar and Hutson 1996 discussed applications of GW family (with bathtub shaped failure rate) using data on bus motor failure and head and neck cancer clinical trials. Many distributions used in Reliability Engineering are special cases of the BGW distribution. These distributions are Beta Generalized Exponential (BGE) (Barreto-Souza et al., 2009), Beta Exponential (BE) (Nadarajah and Kotz, 2006), Generalized or Exponentiated Weibull (GW or EW) (Nassar and Eissa, 2003; Mudholkar et al., 1995; Mudholkar and Srivastava,, 1993), Generalized Rayleigh (GR) (Kundu and Rakab, 2005), Beta Weibull (BW) (Cordeiro et al., 2011; Lee et al., 2007; Famoye et al., 2005), Generalized Exponential (GE) (Gupta and Kundu, 1999), Exponential, Weibull and Rayleigh distributions.

For different combinations of parameters, the failure (hazard) rate of BGW distribution is increasing or decreasing or bathtub shaped or upside down bathtub shaped. Researchers have been attracted by distributions possessing different shapes of failure rates. Failure (hazard) rate is often non-monotonic (bathtub or upside-down bathtub shaped (Lai and Xie, 2006) for mechanical and electronic systems. In firmware (useful for operation of hard drives in big computers) in software reliability, one often comes across bathtub shaped failure rate. This shape of failure rate is also experienced in space crafts, aircraft control, weapons and safety critical control systems in chemical and nuclear plants (Zhang et al., 2005). Shape of the failure rate is upside down bathtub for data on bus motor failures (Mudholkar et al., 1995) and exploration of ageing properties in reliability (Gupta and Gupta, 1996; Jiang et al., 2003).

BGW distribution has great applications in reliability and biological sciences. Due to its flexibility, it models a number of real data sets. Singla et al., 2016 observed that even though failure rates of few other variants of Weibull or Beta G distributions also have different shapes, BGW distribution provided a better fit than all its sub models. This was corroborated through simulations and two real life data sets.

Department of Statistics, Panjab University, Chandigarh- 160014 (India).
Emails: neetu.singla12@gmail.com; ssharma643@yahoo.co.in
* Corresponding author: jaink14@gmail.com

In this work, GOFTs namely Kolmogorov-Smirnov (KS), Cramer-von Mises (CVM) and Anderson-Darling (AD) have been employed for Beta Generalized Weibull (BGW) distribution versus Generalized Weibull (GW) (Mudholkar and Srivastava, 1993), Beta Weibull (BW) (Famoye et al., 2005) and Beta Generalized Exponential (BGE) (Barreto-Souza et al., 2009) distributions.

For x, α, β, λ, a and b > 0, BGW (x, α, β, λ, a and b) distribution has the cumulative distribution function (cdf) given by

$$F(x)=\frac{1}{B(a,b)}\int_0^{(1-e^{-(\lambda x)^\beta})^\alpha} w^{a-1}(1-w)^{b-1}dw. \tag{1}$$

The probability density and hazard (failure) rate functions are

$$f(x)=\frac{\alpha\beta\lambda^\beta x^{\beta-1}}{B(a,b)}\left(1-e^{-(\lambda x)^\beta}\right)^{\alpha a-1}\left\{1-\left(1-e^{-(\lambda x)^\beta}\right)^\alpha\right\}^{b-1}e^{-(\lambda x)^\beta}, x>0, \tag{2}$$

and

$$h(x)=\frac{\alpha\beta\lambda^\beta x^{\beta-1}}{B_{\left\{1-\left(1-e^{-(\lambda x)^\beta}\right)^\alpha\right\}}(a,b)}\left(1-e^{-(\lambda x)^\beta}\right)^{\alpha a-1}\left\{1-\left(1-e^{-(\lambda x)^\beta}\right)^\alpha\right\}^{b-1}e^{-(\lambda x)^\beta} \tag{3}$$

where B(a,b) is the Beta function and B_y(a,b) is incomplete Beta function.

Plots of probability density and hazard rate functions of BGW distribution for few combinations of parametric values are displayed in Figures 1 and 2.

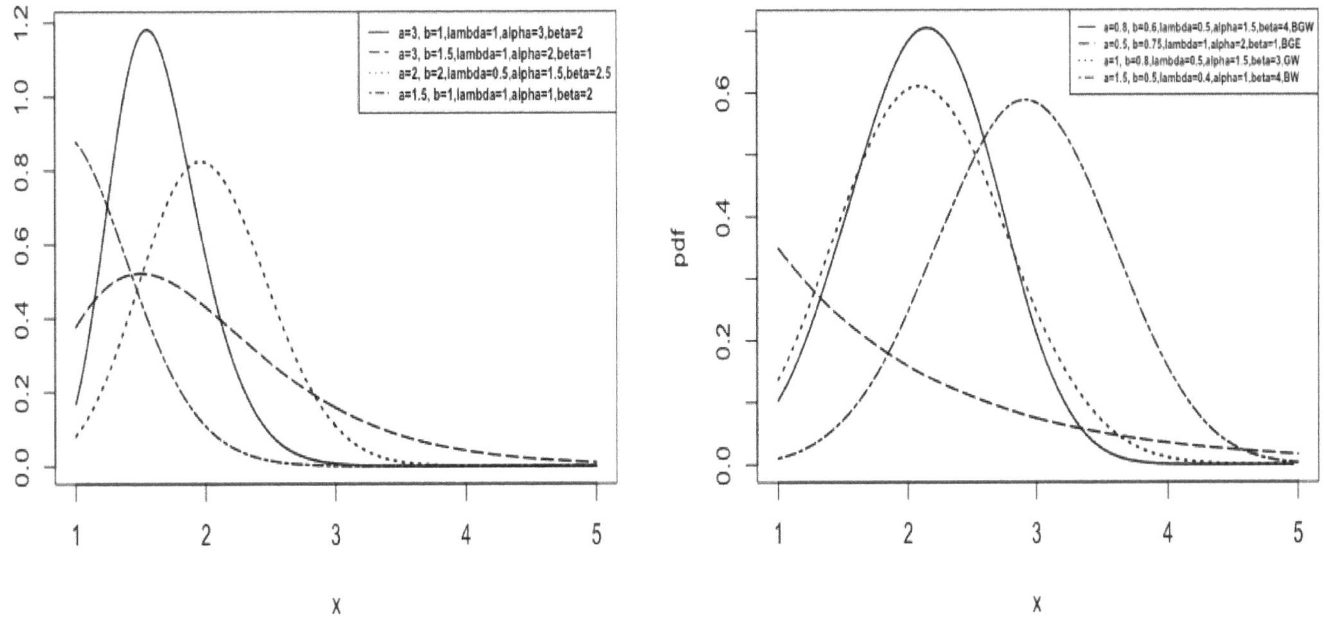

Figure 1: Plots of BGW Probability Density Function.

It is observed that

- BGW α, 1, λ, a and b) is a Beta Generalized Exponential distribution,
- BGW (1, β, λ, a and b) is a Beta Weibull (BW) distribution,
- BGW (α, β, λ, 1 and 1) is a Generalized Weibull (GW) distribution.

Other submodels of a BGW distribution are depicted in Figure 3.

For finding the maximum likelihood estimators (MLEs) of the parameters, the non-linear equations have been given in Singla et al. (2012). But since it was difficult to solve these equations analytically, numerical approximation technique such as quasi-Newton method was employed to find the ML estimates.

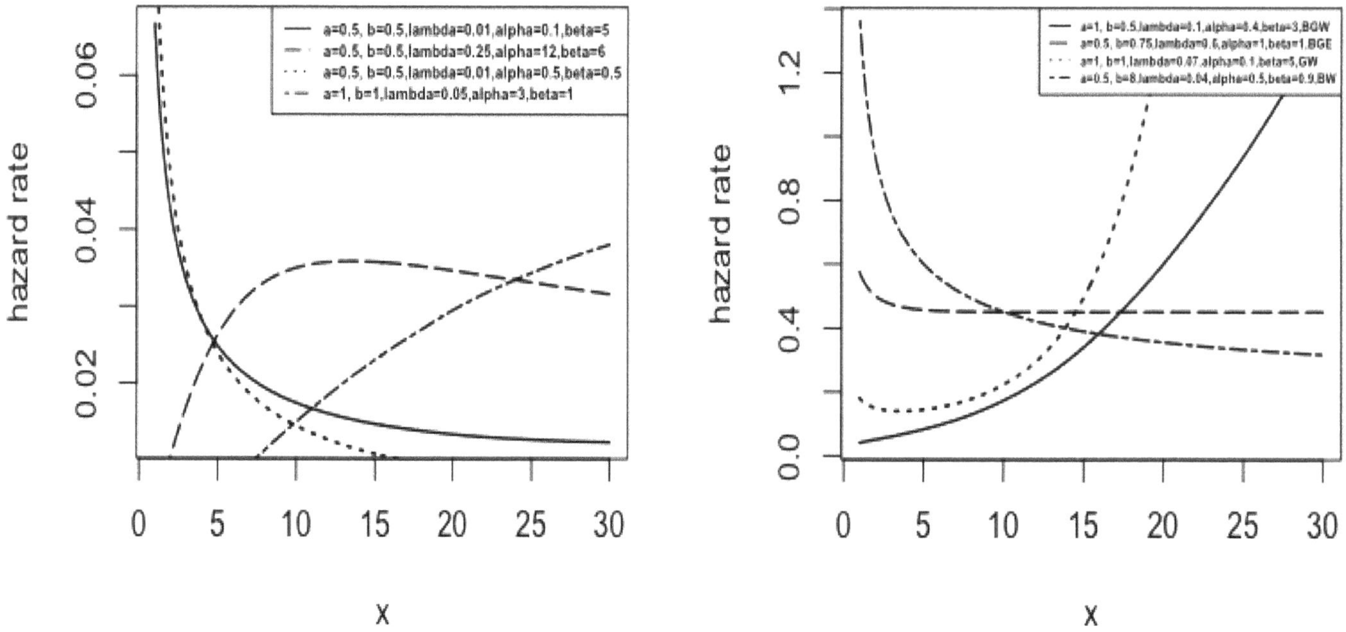

Figure 2: Plots of BGW Hazard Rates.

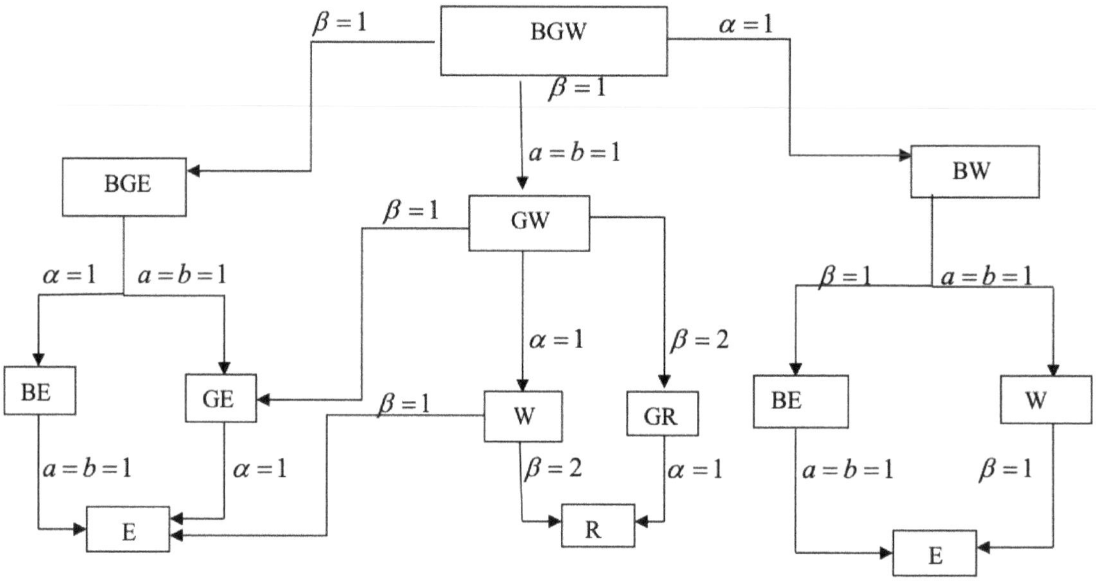

Figure 3: Sub models of BGW.

The goal of this study is comparative analysis of powers of the most popular GOFTs based on some pairs of sufficiently near competing hypotheses. On the basis of powers, the optimal sample sizes have been suggested for differentiating among various distributions.

The chapter is organized as follows. Various goodness of fit tests have been discussed in Section 2. Section 3 consists of testing for BGW distribution versus GW, BW and BGE distributions. Simulations and power studies are included in Section 4. Conclusions are reported in Section 5.

2. Tests for Goodness of Fit

For checking whether empirical distribution fits well to the theoretical distribution or not, GOFTs have been used.

Let $x_1, x_2, \ldots, x_n$ be ordered observations corresponding to a random sample of size n. The empirical distribution function $F_n(x)$ is a step function taking values between 0 and 1 with steps of height 1/n corresponding to each ordered sample observation and let $F(x, \theta)$ denote the theoretical distribution.

These tests can be used for testing simple hypothesis

$H_0 : F(x) = F(x, \theta)$ where θ is the known value of the parameter

and composite hypothesis

$H'_0 : F(x) \in F(x, \theta)$ where $\theta \in \Theta$.

The parameter is estimated using the sample on which GOFTs are applied. These tests are described in the sequel.

Kolmogorov Smirnov (KS) Test:

Test statistic is defined as

$$D_n = sup_{|n|<\infty} \{|F_n(x) - F(x, \theta)|\}, \qquad \theta \in \Theta$$

which is based upon the maximum vertical distance between $F_n(x)$, the empirical distribution function and $F(x, \theta)$, the theoretical distribution function.

Large value of KS statistic leads to rejection of null hypothesis (Gibbons and Chakraborti (2011)).

Stephens (1986) defined class of quadratic EDF statistics as

$$n \int_{-\infty}^{\infty} (F_n(x) - F(x))^2 \, w(x) \, dF(x) \tag{4}$$

where w(x) is a weight function.

Cramer-von Mises and Anderson-Darling statistics belong to this class.

Cramer-von Mises (CVM) Test:

If w(x)=1, (4) is same as n times the CVM test statistic. Value of CVM test statistic can be obtained through the form (Anderson and Darling 1954) given below

$$CVM = \frac{1}{12n} + \sum_{i=1}^{n}\left(F(x_i,\theta) - \frac{2i-1}{2n} \right)^2$$

If value of test statistic exceeds the critical point, null hypothesis is rejected. As CVM test uses greater amount of sample data, hence it is more powerful than KS test (Conover 1999).

Anderson-Darling (AD) Test

This test grants more weightage to the tails of the distribution (Farrel and Stewart 2006) and is a modified form of the CVM test.
Letting

$w(x) = F(x)(1 - F(x))]^{-1}$ in (4), AD test statistic (Anderson and Darling (1954)) is written as

$$n\int_{-\infty}^{\infty} \frac{(F_n(x) - F(x))^2}{F(x)(1 - F(x))} dF(x).$$

An alternative form given by Lewis (1961) is

$$AD = -n - 2\sum_{i=1}^{n}\left\{ \frac{2i-1}{2n} \ln(F(x_i,\theta)) + (1 - \frac{2i-1}{2n}) \ln(1 - F(x_i,\theta)) \right\}.$$

If calculated AD test statistic is more than the value of the critical point, then it leads to the rejection of the null hypothesis.

Critical points (CP) have been computed on the basis of random samples generated from the distribution under null hypothesis, calculation of values of test statistics and then their arrangement in increasing order. For level of significance as α, critical point is obtained as $(\alpha)^{th}$ order test statistic.

CPs have been calculated taking n as 100, 200, 500, 800, 1000, 2000 with α = .05, .10, .15 and .20 using 1000 replications and are displayed in Table 1.

In the next section, simulations have been carried out for finding the powers of different GOFTs. Beta Generalized Exponential (BGE), Beta Weibull (BW) and Generalized Weibull (GW) distributions are considered as competing distributions. The capacity of these tests to differentiate among the models has been assessed by comparing the powers. An optimal sample size has been decided for differentiating BGW distribution from BGE, BW and GW distributions.

Table 1: Critical Points for Goodness of Fit Tests.

n	*KS* *α*				*CVM* *α*				*AD* *α*			
	.20	.15	.10	.05	.20	.15	.10	.05	.20	.15	.10	.05
100	.107	.114	.122	.136	.244	.286	.361	.475	1.388	1.603	1.909	2.412
200	.076	.081	.086	.096	.237	.282	.339	.453	1.39	1.59	1.92	2.49
500	.048	.051	.055	.061	.232	.275	.334	.444	1.405	1.609	1.932	2.50
800	.038	.040	.043	.048	.241	.286	.347	.469	1.410	1.617	1.904	2.399
1000	.034	.036	.039	.043	.245	.286	.347	.449	1.423	1.638	1.945	2.514
2000	.024	.026	.027	.030	.241	.287	.349	.476	1.395	1.588	1.914	2.438

3. Simulation Study

To study the comparison of powers of tests,a data set of size 200 is generated from BGW ($\alpha = 0.8, \beta = 1.2, \lambda = 2, a = 0.9, b = 0.8$) distribution.
For the generated observations, empirical and theoretical cumulative distribution functions are plotted in Figure 4. As plots in Figure 4 are quite close, it can be concluded that BGW distribution fits the data set well.

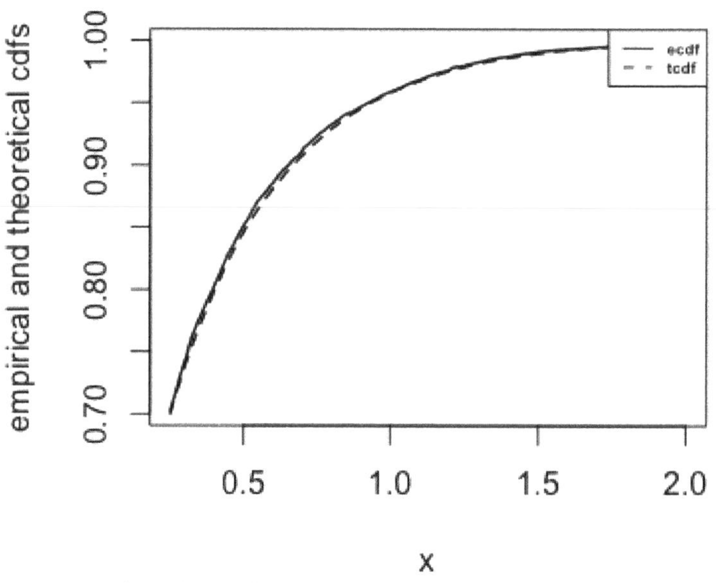

Figure 4: Plots for empirical and theoretical cdf of BGW.

For the generated data set, parameters of the BGW, BGE, BW and GW distributions are estimated using quasi-newton method in R package. The values of corresponding Akaike Information Criterion (AIC) (Akaike (1974)) are calculated for all distributions. The maximum likelihood estimates (MLEs) and the values of AIC for different distributions are given in Table 2.
On the basis of the values in Table 2, it can be concluded that

- BGW fits best since AIC is least in case of BGW distribution;
- BGW and BGE are closely related since BGE has the second least value of AIC. This means that a large sample is required to distinguish between BGW and BGE distributions;
- BGW and BW distributions are not close as AIC is highest in case of BW distribution. Hence, even a small sample will be able to differentiate BGW from BW distribution.

In the following discussion, we test the simple and composite hypotheses for verifying whether BGW distribution (with assumed and estimated parameters), BGE, BW and GW distributions fit the generated observations closely.

Table 2: Maximum Likelihood Estimates of Parameters and AIC for Generated Data Set.

Distribution	$\hat{\alpha}$	$\hat{\beta}$	$\hat{\lambda}$	$\hat{a}$	$\hat{b}$	AIC
BGW	0.7925	1.2614	2.0057	0.8930	0.8122	96.7171
BGE	0.2463	1	4.0568	3.3915	0.5272	98.4352
BW	1	1.088	2.0119	0.8262	0.8262	100.8097
GW	1.1406	0.9064	2.4346	1	1	99.1229

Testing Simple Hypothesis

The aim is to test simple hypothesis
H_0 : Fitting of BGW ($\alpha = 0.8, \beta = 1.2, \lambda = 2, a = 0.9, b = 0.8$) to the generated data set is good
versus
H_1 : Fitting of BGW distribution to the generated data set is not good.

Using MATLAB software, values of three GOFT statistics have been obtained for implementing the testing procedure. Table 3 consists of test statistic values and critical points at 0.05 level of significance for n=200 (extracted from Table 1).

Table 3: Test Statistics Values and CPs for Simple Hypothesis H_0 versus H_0.

GOFT	KS	CVM	AD
Critical Point at $\alpha = 0.05$	0.096	0.453	2.49
Test Statistic Value	0.0379	0.0519	0.4683

On the basis of Table 3, it is inferred that BGW with assumed values of parameters fits the data well since the computed value of test statistic is smaller than the critical point at 0.05 level of significance in all cases.

Testing of Composite Hypotheses

To test the composite null hypothesis
H_{01} : Fitting of BGW ($\hat{\alpha}$ **= 0.7925,** $\hat{\beta}$ **= 1.2614,** $\hat{\lambda}$ **= 2.0057,** $\hat{a}$ **= 0.8930,** $\hat{b}$ **= 0.8122**) to the generated data set is good.
versus
H_{11} : Fitting of BGW distribution with estimated parameters is not good.
Values of KS, CVM and AD test statistics have been computed and compiled in Table 4.

Table 4: Test Statistics Values and CPs for Simple Hypothesis H_{01} versus H_{11}.

GOFT	KS	CVM	AD
Critical Point at $\alpha = 0.05$	0.096	0.453	2.49
Test Statistic Value	0.0484	0.1002	0.7009

Thus the BGW with estimated values of the parameters also fits well in each case, as the calculated values of test statistic is less than the critical point at 0.05 level of significance.

We also consider the testing of following composite hypotheses:

- H_{02} : Fitting of BGE ($\hat{\alpha}$ **= 0.2463,** $\hat{\beta}$ **= 1,** $\hat{\lambda}$ **= 4.0568,** $\hat{a}$ **= 3.3915,** $\hat{b}$ **= 0.5272**) to the generated data set is good

versus
H_{12} : Fit of BGE distribution with estimated parameters is not good.

- H_{03} : Fitof BW ($\hat{\alpha}$ **= 1,** $\hat{\beta}$ **= 1.0888,** $\hat{\lambda}$ **= 2.0119,** $\hat{a}$ **= 0.8674,** $\hat{b}$ **= 0.8262**) is good

versus
H_{13} : BW distribution with estimated parameters does not fit nicely.

- H_{04} : Fitting of GW ($\hat{\alpha}$ **= 1.1406,** $\hat{\beta}$ **= 0.9064,** $\hat{\lambda}$ **= 2.4346,** $\hat{a}$ **= 1,** $\hat{b}$ **= 1**) to the generated data set is good

versus
H_{14} : Fit of GW distribution with estimated parameters is not good.
Table 5 shows calculated values of KS, CVM and AD test statistics and critical points for testing H_{02}, H_{03} and H_{04}.

From Table 5, it is inferred that GOFT statistics for all the considered distributions are less than critical points at 0.05 level of significance. Hence all the distributions fit well to the data set. Therefore, power analysis is carried out for differentiating among these distributions.

Table 5: Test Statistics Values and CPs for H_{02} versus H_{12}, H_{03} versus H_{13}, H_{04} versus H_{14}.

GOFT	KS	CVM	AD
Critical Point at $\alpha = 0.05$	0.096	0.453	2.49
	H_{02} versus H_{12}		
Test Statistic Value	0.0484	0.1002	0.7009
	H_{03} versus H_{13}		
Test Statistic Value	0.0567	0.1852	1.5045
	H_{04} versus H_{14}		
Test Statistic Value	0.0547	0.1307	0.9128

In the next section, GOFTs for BGW distribution against BGE, BW and GW distributions are explored. Power has been found for KS, CVM and AD tests. Comparison of these values helps in differentiating among BGW, BGE, BW and GW distributions.

4. Powers of Goodness of Fit Tests for BGW Distribution

To find the power, random samples have been generated under alternative hypothesis with sample sizes taken as 100, 500, 800, 1000 and 2000. Values of test statistics have been computed using estimates of parameters. These values are then compared with critical points for taking a decision regarding rejection of the null hypothesis. This process is replicated 1000 times and power is calculated as proportion of rejections.

Although we have computed powers for AD, CVM and KS tests, the power of AD test is more as compared to CVM and KS test in general (Lemeshko et al., 2009, 2010a) but this ordering is not rigid. In order to maintain uniformity for sake of comparison, type II error has been computed for 0.10 level of significance for AD as power is more as compared to other two tests.

We now consider the testing of composite hypotheses

H_{05} : BGW ($\hat{\alpha} = \mathbf{0.7925}, \hat{\beta} = \mathbf{1.2614}, \hat{\lambda} = \mathbf{2.0057}, \hat{a} = \mathbf{0.8930}, \hat{b} = \mathbf{0.8122}$) distribution fits the data set well

versus

(i) H_{15} : BG ($\hat{\alpha} = \mathbf{0.2463}, \hat{\beta} = \mathbf{1}, \hat{\lambda} = \mathbf{4.0568}, \hat{a} = \mathbf{3.3915}, \hat{b} = \mathbf{0.5272}$) distribution fits the data set well,

(ii) H'_{15} : BW ($\hat{\alpha} = \mathbf{1}, \hat{\beta} = \mathbf{1.0888}, \hat{\lambda} = \mathbf{2.0119}, \hat{a} = \mathbf{0.8674}, \hat{b} = \mathbf{0.8262}$) distribution is good,

(iii) H''_{15} : Fitting of GW ($\hat{\alpha} = \mathbf{1.1406}, \hat{\beta} = \mathbf{0.9064}, \hat{\lambda} = \mathbf{2.4346}, \hat{a} = \mathbf{1}, \hat{b} = \mathbf{1}$) is good.

Tables 6–8 display powers of three goodness of fit tests.

Table 6: Powers for testing H_{05} versus H_{15}.

α	n					
	100	**500**	**800**	**1000**	**2000**	**3000**
	Power of KS Test					
0.20	.041	.137	.239	.301	.447	
0.15	.037	.082	.142	.167	.348	
0.10	.011	.072	.074	.126	.258	
0.05	.006	.015	.037	.076	.126	
	Power of CVM Test					
0.20	.052	.205	.293	.314	.57	
0.15	.062	.116	.207	.25	.446	
0.10	.023	.098	.104	.183	.361	
0.05	.004	.032	.049	.1	.201	
	Power of AD Test					
0.20	.093	.291	.456	.52	.83	
0.15	.092	.199	.335	.446	.753	
0.10	**.044**	**.158**	**.237**	**.339**	**.66**	**.889**
0.05	.011	.07	.147	.199	.486	

Table 7: Powers for testing H_{05} versus H'_{15}.

α	n				
	100	**500**	**800**	**1000**	**2000**
	Power of KS Test				
0.20	.208	.83	.962	.997	1
0.15	.182	.768	.95	.994	1
0.10	.099	.711	.92	.988	1
0.05	.061	.572	.87	.967	.999
	Power of CVM Test				
0.20	.328	.872	.981	.999	1
0.15	.274	.839	.976	1	1
0.10	.163	.786	.962	1	1
0.05	.106	.675	.929	.992	.999
	Power of AD Test				
0.20	.419	.949	.994	1	1
0.15	.365	.936	.997	1	1
0.10	**.28**	**.896**	**.993**	**1**	**1**
0.05	.182	.818	.975	.998	1

Table 8: Powers for testing H_{05} versus H''_{15}.

α	n				
	100	**500**	**800**	**1000**	**2000**
	Power of KS Test				
0.20	.081	.171	.278	.346	.711
0.15	.035	.162	.206	.255	.61
0.10	.021	.10	.134	.174	.456
0.05	.014	.034	.055	.073	.256
	Power of CVM Test				
0.20	.096	.253	.352	.429	.791
0.15	.059	.214	.269	.349	.687
0.10	.031	.124	.162	.235	.538
0.05	.014	.047	.091	.119	.331
	Power of AD Test				
0.20	.158	.461	.729	.82	.989
0.15	.101	.408	.598	.75	.968
0.10	**.053**	**.273**	**.52**	**.617**	**.957**
0.05	.038	.139	.333	.422	.889

Values in Tables 6–8 lead to the conclusion that

- power of AD test is highest in all the cases followed by CVM and power of KS test is the least.

For differentiating among different distributions, it is observed from Tables 6–8 that for having low probability of Type II error at 0.10 level of significance

- the sample size should be at least 3000 for differentiating between BGW and BGE distributions;
- to differentiate between BW and BGW distributions, the sample size should be at least 800;
- a minimum sample size 2000 is needed to distinguish GW from BGW.

Hence the required sample size for differentiating among different distributions is

- highest for BGW versus BGE distribution;
- smallest for BGW versus BW distribution.

5. Conclusions

In this chapter, tables of critical points have been provided for testing hypotheses concerning BGW family. It is concluded that among three considered goodness of fit tests, AD has the maximum power and KS has the lowest power. The sample size n is chosen in such a way that it is able to differentiate among closely related distributions.

Acknowledgements

All the authors are grateful to the worthy editors for suggestions that led to an improvement in the contents. Second author expresses her thanks to University Grants Commission, Government of India, for financial assistance provided to complete this piece of work.

References

Akaike, H. 1974. A new look at the statistical model identification. IEEE Transactions on Automatic Control, 19(6): 716–723.

Anderson, T. W. and Darling, D. A. 1954. A test of goodness of fit. Journal of The American Statistical Association, 49: 765–769.

Arshad, M., Rasool, M. T. and Ahmad, M. I. 2003. Anderson darling and modified anderson darling tests for generalized pareto distribution. Pakistan Journal of Applied Sciences, 3(2): 85–88.

Barreto-Souza, W., Santos, A. H. S. and Cordeiro, G. M. 2009. The beta generalized exponential distribution. Journal of Statistical Computation and Simulation, 80: 159–172.

Betsch, S. and Ebner, B. 2019. A New Characterization of the Gamma Distribution and Associated Goodness of Fit Tests, 82(7): 779–806.

Conover, W. J. 1999. Practical Nonparametric Statistics. Third Edition, John Wiley & Sons, Inc. New York.

Cordeiro, G. M., Simas, A. B. and Stosic, B. 2011. Closed form expressions for moments of the beta Weibull distribution. Annals of the Brazilian Academy of Sciences, 83(2): 357–373.

Dufour, J. M., Farhat, A., Gardiol, L. and Khalaf, L. 1998. Simulation-based finite sample normality tests in linear regressions. Econometrics Journal, 1: 154–173.

Famoye, F., Lee, C. and Olumolade, O. 2005. The beta—Weibull distribution. Journal of Statistical Theory and Applications, 4(4.2.2): 121–136.

Farrel, P. J. and Stewart, K. R. 2006. Comprehensive study of tests for normality and symmetry: extending the spiegelhalter test. Journal of Statistical Computation and Simulation, 76(9): 803–816.

Gibbons, J. D. and Chakraborti, S. 2011. Nonparametric statistical inference. Chapman and Hall/ CRC, Florida.

Gupta, R. D. and Kundu, D. 1999. Generalized Exponential Distributions, Australian and New Zealand Journal of Statistics, 41: 173–188.

Gupta, G. L and Gupta, R. C. 1996. Aging characteristics of the Weibull mixtures. Probability in the Engineering and Information Science, 10: 591–600.

Henze, N., Meintanis, S. G. and Ebner, B. 2012. Goodness-of-fit tests for the gamma distribution based on the empirical Laplace transform. Communications in Statistics - Theory and Methods, 41(9): 1543–1556.

Jiang, R., Ji, P. and Xiao, X. 2003. Aging property of unimodal failure rate models. Reliability Engineering System Safety, 79: 113–116.

Kundu, D. and Rakab, M. Z. 2005. Generalized Rayleigh distribution: different methods of estimation, Computational Statistics and Data Analysis, 49: 187–200.

Lai, C. D. and Xie, M. 2006. Stochastic Ageing and Dependence for Reliability. Germany: Springer.

Lee, C., Famoye, F. and Olumolade, O. 2007. Beta-Weibull distribution: some properties and applications to censored data. Journal of Modern Applied Statistical Methods, 6: 173–186.

Lemeshko, B. Y., Lemeshko, S. B. and Postovalov, S. N. 2009. Comparative Analysis of the Power of Goodness-of-Fit Tests for Near Competing Hypotheses. I. The verification of simple hypotheses. Journal of Applied and Industrial Mathematics, 3(4): 462–475.

Lemeshko, B. Y., Lemeshko, S. B. and Postovalov, S. N. 2010a. Comparative Analysis of the Power of Goodness-of-Fit Tests for Near Competing Hypotheses. II. Verification of complex hypotheses. Journal of Applied and Industrial Mathematics, 4(1): 79–93.

Lemeshko, B. Y., Lemeshko, S. B., Akushkina, K. A., Nikulin, M. S. and Saaidia, N. 2010b. Inverse Gaussian Model and its Applications in Reliability and Survival Analysis, Mathematical and Statistical Models and Methods in Reliability: Applications to Medicine, Finance and Quality Control, Statistics for Industry and Technology. V.V. Rykov et al. (eds.): 433–453.

Lemeshko, B. Y. and Lemeshko, S. B. 2011. Models of Statistic Distributions of Nonparametric Goodness-of-Fit Tests in Composite Hypotheses Testing for Double Exponential Law Cases. Communications in Statistics—Theory and Methods, 40: 2879–2892.

Lewis, P. A. W. 1961. Distribution of the anderson-darling statistic. The Annals of Mathematical Statistics, 32(4): 1118–1124.

Mudholkar, G. S. and Srivastava, D. K. 1993. Exponentiated Weibull family for analyzing bathtub failure data. IEEE Transactions on Reliability, 42: 299–302.

Mudholkar, G. S., Srivastava, D. K. and Freimer, M. 1995. The exponentiated Weibull family. A reanalysis of the bus-motor-failure data. Technometrics, 37: 436–45.

Nassar M. M. and Eissa F. H. 2003. On the exponentiated Weibull distribution. Commun. Stat-Theor. M., 32: 1317–1336.

Nikulin, M. S. 1973a. On a chi-squared test for continuous distributions. Theory of Probability and its Applications, 18(3): 638–639.

Nikulin, M. S. 1973b. Chi-squared test for continuous distributions with shift and scale parameters. Theory of Probability and its Applications, 18(3): 559–568.

Rao, K. C. and Robson, D. S. 1974. A Chi-squared statistic for goodness-of-fit tests within the exponential family. Communications in Statistics Theory and Methods, 3: 1139–1153.

Seier, E. 2002. Comparison of Tests for Univariate Normality. InterStat Statistical Journal, 1: 1–17.

Singla, N., Jain, K. and Sharma, S. K. 2012. The Beta Generalized Weibull distribution: Properties and applications, Reliability Engineering and System Safety, 102: 5–15.

Singla, N., Jain, K. and Sharma, S. K. 2016. Goodness of Fit Tests and Power Comparisons for Weighted Gamma Distribution. Revstat-Statistical Journal, 14(1): 29–48.

Stephens, M. A. 1986. Tests based on EDF statistics. pp. 97–193. *In*: D'Agostino, R. B. and Stephens, M. A. (Eds.). Goodness-of-Fit Techniques. Marcel Dekker, New York.

Voinov, V., Nikulin, M. and Balakrishnan, N. 2013. Chi Squared Goodness of Fit Tests with Applications, Academic Press, MA, USA.

Zhang, T., Xie, M., Tang, L. C. and Ng, S. H. 2005. Reliability and modeling of systems integrated with firmware and hardware. International Journal of Reliability, Quality and Safety Engineering, 12(3): 227–239.

CHAPTER 21

On the Transmuted Modified Lindley Distribution

Theory and Applications to Lifetime Data

Lishamol Tomy,[1] *Christophe Chesneau*[2,*] and *Jiju Gillariose*[3]

1. Introduction

In many applied sciences, modeling and analyzing lifetime data using lifetime distributions have received the attention of several researchers. Undoubtedly, the one-parameter Lindley distribution (Lindley, 1958, 1965) is one of the most attractive distributions in Statistics. There are many extensions of the Lindley distribution to provide flexibility for modeling data. In particular, see, Zakerzadeh and Dolati, 2009; Nadarajah et al., 2011; Shanker and Mishra, 2013a; Ghitany et al., 2013; Singh et al., 2014; Shanker and Mishra, 2013b; Sharma et al., 2015; Sharma et al., 2016. The interested reader can find a comprehensive review on the Lindley distribution in Tomy 2018. More recently, a new modified Lindley (ML) distribution has been proposed by Chesneau et al., 2020a as a simple one-parameter alternative to the exponential and Lindley distributions. It is defined with the following survival function:

$$\bar{G}(x) = \left[1 + \frac{\theta x}{1+\theta} e^{-\theta x}\right] e^{-\theta x}, \quad x > 0, \tag{1}$$

with $\theta > 0$. An important property of the ML distribution is that its probability density function (pdf) can be expressed as a linear combination of exponential and gamma pdfs. In practical contexts, the ML distribution is a strong one-parameter competitor to the Lindley and exponential distributions. In addition to this, Chesneau et al., 2020b; Chesneau et al., 2020c further studied two extensions for the ML distribution, such as the inverse ML and wrapped ML distributions, respectively, and presented their statistical properties.

In the last decades, several researchers have added new parameters to expanding classical distributions in order to improve the modeling of survival data. In this regard, numerous meaningful new families of distributions have been built towards the generalization of well-established classical lifetime distributions. The transmuted family was first proposed by Shaw and Buckley, 2007 based on the transmutation method and the theory was further clearly mentioned in Shaw and Buckley, 2009. The transmuted generated (T-G) family of distributions is characterized by the cumulative density function (cdf) given by

$$F(x) = G(x)[1 + \beta - \beta G(x)], \quad x \in \mathbb{R}, \tag{2}$$

where $\beta \in [-1,1]$, β introduces the skewness and varies the corresponding tail weights, and $G(x)$ denotes the cdf of a parent continuous distribution. For more details about the quadratic rank transmutation map, see Shaw and Buckley, 2009. Subsequently, the T-G transformation was applied to several well-known distributions from its inception. For example, Aryal and Tsokos, 2009; Aryal and Tsokos, 2011 derived the two transmuted transformed distributions such as, transmuted extreme value and transmuted Weibull distributions. Aryal, 2013 suggested the transmuted log-logistic distribution and its various properties. Merovci, 2013a introduced the transmuted Lindley distribution and applied it to bladder cancer data. Merovci, 2013b proposed the transmuted exponentiated exponential distribution. The transmuted Lindley-geometric distribution has been discussed by Merovci and Elbatal, 2013. Ashour and Eltehiwy, 2013a deduced the transmuted Lomax distribution and Ashour and Eltehiwy, 2013b created the transmuted exponentiated Lomax distribution. Eltehiwy and Ashour, 2013 developed transmuted exponentiated modified Weibull distribution. The transmuted exponentiated gamma distribution has been discussed by Hussian, 2014. Elbatal et al., 2014 suggested various estimation methods

[1] Department of Statistics, Deva Matha College, Kuravilangad, Kerala- 686633, India.
Email: lishatomy@gmail.com
[2] Université de Caen, LMNO, Campus II, Science 3, 14032, Caen, France.
[3] Department of Statistics, St. Thomas College, Pala, Kerala-686574, India.
Email: jijugillariose@yahoo.com
* Corresponding author: christophe.chesneau@unicaen.fr

for the transmuted exponentiated Fréchet distribution. The exponentiated transmuted Weibull distribution has been studied by Abd El Hady, 2014. Merovci and Puka, 2014 introduced the transmuted Pareto distribution and Elbatal and Aryal, 2016 studied the transmuted additive Weibull distribution. The transmuted Rayleigh distribution has been proposed and analyzed by Merovci, 2016. Bourguignon et al., 2017 pioneered the transmuted Birnbaum-Sauders distribution and the transmuted Burr type X discussed Khan et al., 2019. Recently, the cubic transmuted Pareto distribution has been proposed by Rahman et al., 2020a. Over the last decade, more than hundred T-G distributions have been derived from existing well-known distributions. For a more detailed discussion, see Rahman et al., 2020b.

The above mentioned short review about T-G distributions and references therein show that the theory on the T-G family is enriching and rapidly growing. Motivated by previous works, in this chapter, we propose and study a new extension of the ML distribution based on the T-G family. The second motivation is related to the hazard rate function (hrf) of the new model; it can be increasing, upside-down bathtub (unimodal) and increasing-decreasing-constant-decreasing. In addition, another motivation is related to the extensive applicability of the new distribution. The proposed distribution is a flexible distribution, thus it can be applied in many practical data. Merovci (2013a) highlighted several advantages of the transmuted Lindley distribution over the transmuted exponential distribution, namely; its resemblance to analytic expressions for pdf and hrf, flexibility. In order to establish the relevance of the study, we compare the performance of the proposed distribution with the transmuted Lindley (TL) and transmuted exponential (TE) distributions based on some data sets of interest. In view of the obtained results, we hope that the new proposed distribution can illuminate the literature of distribution theory.

The rest of the chapter is organized as follows. In Section 2, we discuss the transmuted ML distribution. In Section 3, some theoretical results are discussed. In Section 4, we investigate the maximum likelihood estimation of the parameters. Applications of this new distribution are discussed in Section 5. Two data sets are considered and shown that for all data sets, the proposed distribution is the appropriate model. Conclusions are presented in Section 6.

2. Transmuted ML Distribution

An unexplored way to flexibilize the ML distribution is to use the structure of the transmuted family. Thus, by substituting (1) into (2), we introduce the transmuted ML (TML) distribution defined by the following cdf:

$$F(x)=\left\{1-\left[1+\frac{\theta x}{1+\theta}e^{-\theta x}\right]e^{-\theta x}\right\}\left\{1+\beta\left[1+\frac{\theta x}{1+\theta}e^{-\theta x}\right]e^{-\theta x}\right\},\quad x>0. \tag{3}$$

Just to illustrate the importance of the parameter with possible values in, the following identified interpretations of the TML distribution are valuable:

- When $\beta = 0$, $F(x)$ becomes the cdf of the ML distribution,
- when $\beta = -1$, we get the cdf of $\max(X, Y)$, where X and Y are two independent random variables having the ML distribution as common distribution,
- when $\beta = 1$, we get the cdf of $\min(X, Y)$, where X and Y are two independent random variables having the ML distribution as common distribution.

All the intermediary values of β in $(-1, 0)$ and $(0, 1)$ are, of course possible, making the TML distribution of high level of pliancy. Figure 1 illustrates some of the possible shapes of the cdf of the TML distribution for selected values of the parameters θ and β, respectively.

By differentiating $F(x)$ with respect to x, the corresponding pdf is given by

$$f(x)=\frac{\theta}{1+\theta}e^{-2\theta x}[(1+\theta)e^{\theta x}+2\theta x-1]\left\{1-\beta+2\beta\left[1+\frac{\theta x}{1+\theta}e^{-\theta x}\right]e^{-\theta x}\right\},\quad x>0.$$

With mathematical efforts, we can write $f(x)$ as a linear combination of polynomial-exponential functions. More precisely, we have

$$\begin{aligned}f(x)=&\frac{4\beta\theta^3}{(1+\theta)^2}x^2e^{-4\theta x}+\frac{2\beta\theta^3}{(1+\theta)^2}xe^{-3\theta x}+\frac{2\beta\theta^2}{1+\theta}e^{-2\theta x}-\frac{\beta\theta^2}{1+\theta}e^{-\theta x}+\frac{4\beta\theta^2}{1+\theta}xe^{-3\theta x}\\&-\frac{2\beta\theta^2}{1+\theta}xe^{-2\theta x}-\frac{2\beta\theta^2}{(1+\theta)^2}xe^{-4\theta x}+\frac{2\beta\theta^3}{(1+\theta)^2}xe^{-3\theta x}+\frac{2\beta\theta}{1+\theta}e^{-3\theta x}+\frac{3\beta\theta}{1+\theta}e^{-2\theta x}\\&-\frac{\beta\theta}{1+\theta}e^{-\theta x}+\frac{\theta^2}{1+\theta}e^{-\theta x}+\frac{2\theta^2}{1+\theta}xe^{-2\theta x}-\frac{\theta}{1+\theta}e^{-2\theta x}+\frac{\theta}{1+\theta}e^{-\theta x}.\end{aligned} \tag{4}$$

Another point of view: with a work on the coefficient, one can prove that $f(x)$ is a finite general mixture of gamma pdfs. This decomposition is useful to express several probabilistic measures.

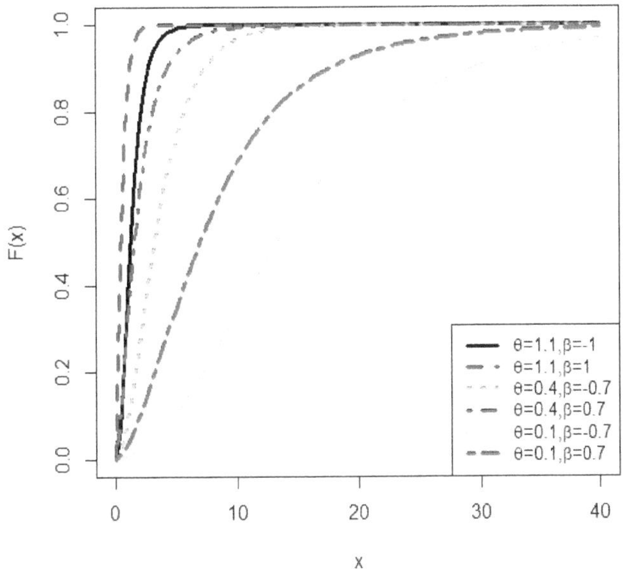

Figure 1: Graph of cdf of the TML distribution for different values of θ and β.

The corresponding hrf is given by

$$\begin{aligned} h(x) &= \frac{f(x)}{1-F(x)} \\ &= \frac{\theta e^{-2\theta x}[(1+\theta)e^{\theta x}+2\theta x-1]\{1-\beta+2\beta[1+\theta x e^{-\theta x}/(1+\theta)]e^{-\theta x}\}/(1+\theta)}{1-\{1-[1+\theta x e^{-\theta x}/(1+\theta)]e^{-\theta x}\}\{1+\beta[1+\theta x e^{-\theta x}/(1+\theta)]e^{-\theta x}\}},\ x>0. \end{aligned} \tag{5}$$

These pdfs and hrfs will be the object of analytical studies in the next section, revealing the pliancy of the TML model.

3. Some Theoretical Results

3.1 Properties of the pdf

We now derive some theoretical results satisfied by the TML distribution. Here, some basic properties of the pdf are discussed. Firstly, we have $f(0) = \theta^2(1+\beta)/(1+\theta)$. Therefore, thanks to the possible values of β, we have $f(0) \in [0, 2\theta^2/(1+\theta)]$, showing a certain

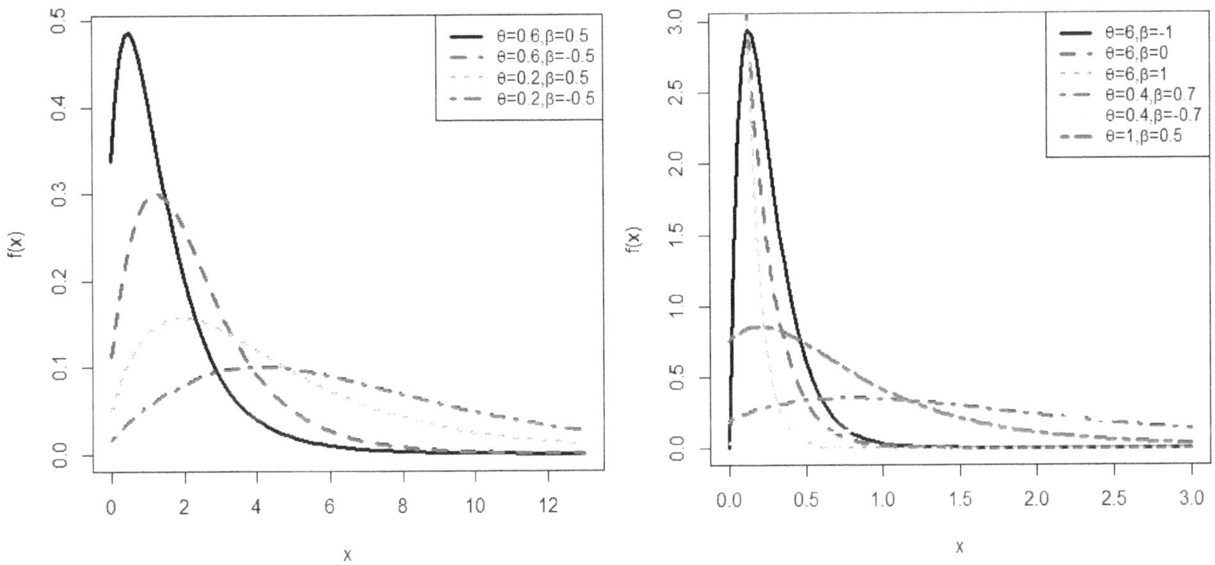

Figure 2: Graphs of pdf of the TML distribution for different values of θ and β.

flexibility in this initial value. Also, by applying $x \to +\infty$, $f(x) \to 0$ with the following equivalence: $f(x) \sim \theta e^{-\theta x}(1 - \beta + 2\beta e^{-\theta x})$. The convergence therefore follows an exponential rate, whatever or $\beta \in [-1,1)$ or $\beta = 1$. An extremum for $f(x)$ satisfies $d\log f(x)/dx = 0$, with

$$\frac{d}{dx}\log[f(x)] = -2\theta + \theta\frac{(\theta+1)e^{\theta x}+2}{(1+\theta)e^{\theta x}+2\theta x-1} - \frac{2\beta\theta}{\theta+1}e^{-2\theta x}\frac{(\theta+1)e^{\theta x}+2\theta x-1}{1-\beta+2\beta[1+\theta x e^{-\theta x}/(1+\theta)]e^{-\theta x}}. \tag{6}$$

We thus observe that the parameter β influences the last terms, which can play a crucial role in the determination of the extrema. To have a more comprehensible approach, a graphical analysis showing diverse curves for $f(x)$ is presented in Figure 2. The figure shows

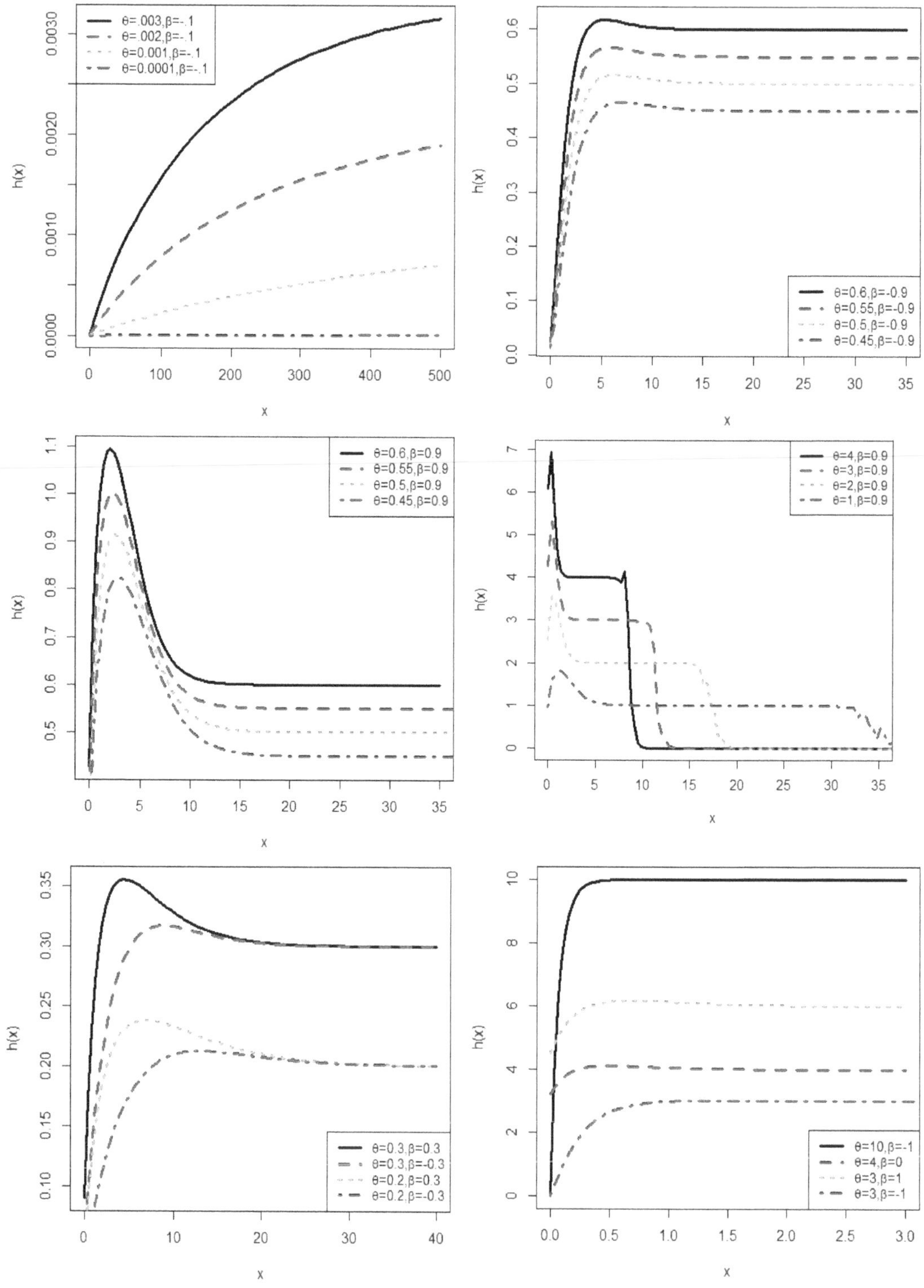

Figure 3: Graphs of hrf of the TML distribution for different values of θ and β.

that the TML distribution is decreasing and unimodal with small and large values of skewness and kurtosis for different parameter values. Moreover, the shape of pdf approaches to flatten out for the negative values of β.

3.2 Properties of the hrf

Now, we focus on the analytical properties of the hrf. First, we have $h(0) = f(0) = \theta^2(1 + \beta)/(1 + \theta)$, showing the influence of β in this important initial value. Also, when $x \to +\infty$, we have $h(x) \to \theta$ if $\beta \in [-1,1)$, and $h(x) \to +\infty$ if $\beta = 1$, with the following equivalence:

$$h(x) \sim \frac{\theta(1-\beta+2\beta e^{-\theta x})}{1-\beta+2\theta\beta x e^{-2\theta x}/(1+\theta)}.$$

In particular, for $\beta = 1$, the following equivalence holds: $h(x) \sim (1 + \theta)e^{\theta x}/x$. Here again, we see the complex role of β in the asymptotic properties. An extremum for $h(x)$ satisfies $d\log h(x)/dx = 0$, with $d\log h(x)/dx = d\log f(x)/dx + h(x)$, $d\log f(x)/dx$ being expressed in (6) and $h(x)$ in (5). So, it is clear that β impacts the numbers and values of the extrema. We illustrate the flexibility of $h(x)$ through Figure 3. The figures exhibit increasing, upside-down bathtub (unimodal) and increasing-decreasing-constant-decreasing shapes for the selected values of the model parameters. These atypical shapes make the hrf useful for non-monotonic empirical hazard behaviors which are found in numerous real life situations.

3.3 Moments

Let X be a random variable following the TML distribution. Then, the uth raw moments of X are defined by

$$\begin{aligned} m'_u &= \mathbb{E}(X^u) = \int_0^{+\infty} x^u f(x)dx \\ &= \frac{\theta}{1+\theta}\int_0^{+\infty} x^u e^{-2\theta x}[(1+\theta)e^{\theta x}+2\theta x-1]\left\{1-\beta+2\beta\left[1+\frac{\theta x}{1+\theta}e^{-\theta x}\right]e^{-\theta x}\right\}dx. \end{aligned} \tag{7}$$

By using the linear combination given as (4), this integral can be expressed via a linear combinations of gamma functions, but the expression remains massive and not really tractable. The integral (7) can be easily computed numerically in software such as **MAPLE**, **MATLAB**, **MATHEMATICA**, **Ox** and **R**. From it, one can, however, express various measures depending on these moments, such as the skewness coefficients and kurtosis coefficients.
The variance of X is given by

$$\mathbb{V}(X) = m_2 = m'_2 - (m'_1)^2.$$

Other ordinary moments (m_3 and m_4) can be calculated from well-known relationships. The skewness and kurtosis coefficients of X are respectively given by

$$\sqrt{\beta_1} = \frac{m_3}{m_2^{\frac{3}{2}}}$$

and

$$\beta_2 = \frac{m_4}{m_2^2}.$$

Table 1 indicates numerical values of the four first moments, variance, skewness and kurtosis for selected values for θ and β. From the table, the values of variance increase as $\beta(\beta > -1)$ increases for all values of θ. In addition, the values of the coefficient of skewness increase as θ increases, except for $\theta = 1$. The positive values of skewness mean that the TML is skewed to the right. Moreover, the values of coefficient of kurtosis are increasing as the values of β get larger, except for $\beta = 1$. Thus, the distribution tail gets heavier for large values of β.

4. Estimation

Let $x_1,\ldots, x_n$ be observed values from the TML distribution. We now discuss the estimation of β and θ through the maximum likelihood (ML) technique. First, the log-likelihood function is specified by

$$\begin{aligned} \ell(\beta,\theta) &= \sum_{i=1}^n \log f(x_i) = n\log\theta - n\log(1+\theta) - 2\theta\sum_{i=1}^n x_i + \sum_{i=1}^n \log[(1+\theta)e^{\theta x_i}+2\theta x_i - 1] \\ &+ \sum_{i=1}^n \log\left\{1+\beta+2\beta\left[1+\frac{\theta x_i}{1+\theta}e^{-\theta x_i}\right]e^{-\theta x_i}\right\}. \end{aligned}$$

Table 1: Numerical values of the TML distribution for some selected values of parameter.

$\theta = 2$	β	−1	−0.8	−0.6	−0.4	−0.2	0	0.2	0.4	0.6	0.8	1
	m_1'	0.7946	0.7693	0.6934	0.6428	0.5922	0.5417	0.4911	0.4405	0.3899	0.3394	0.2888
	m_2'	0.9323	0.8933	0.7761	0.6979	0.6198	0.5417	0.4635	0.3854	0.3073	0.2291	0.1510
	m_3'	1.4805	1.4121	1.2071	1.0703	0.9336	0.7969	0.6601	0.5234	0.3867	0.2500	0.1132
	m_4'	3.0140	2.8688	2.4334	2.1431	1.8528	1.5625	1.2722	0.9819	0.6916	0.4013	0.1110
	$\mathbb{V}(X)$	0.301	0.3015	0.2953	0.2847	0.2690	0.2483	0.2224	0.1913	0.1552	0.114	0.0676
	$\sqrt{\beta_1}$	1.5825	1.5773	1.6171	1.6827	1.7760	1.8959	2.0419	2.211	2.3849	2.4656	1.7408
	β_2	7.1132	7.0920	7.2356	7.5143	7.963	8.6192	9.5394	10.7982	12.4284	13.8997	7.6369
$\theta = 0.2$	m_1'	8.5489	8.0475	7.546	7.0446	6.5431	6.0417	5.5402	5.0388	4.5373	4.0359	3.5344
	m_2'	101.3466	93.1607	84.9747	76.7886	68.6026	60.4166	52.2306	44.0446	35.8586	27.6726	19.4866
	m_3'	1588.22	1444.013	1299.806	1155.599	1011.392	867.1851	722.978	578.771	434.5639	290.3569	146.1498
	m_4'	31724.61	28692.13	25659.66	22627.19	19594.72	16562.25	13529.77	10497.3	7464.829	4432.357	1399.885
	$\mathbb{V}(X)$	28.263	28.3991	28.0324	27.1628	25.7903	23.9149	21.5366	18.6554	15.2713	11.3843	6.9944
	$\sqrt{\beta_1}$	1.5879	1.5675	1.5869	1.6385	1.718	1.823	1.9508	2.0954	2.2333	2.2593	1.5046
	β_2	7.3004	7.2247	7.2929	7.5093	7.8895	8.4627	9.272	10.3666	11.7304	12.7582	6.6653

Then, the ML estimates of β and θ are given by $\hat{\beta}$ and $\hat{\theta}$ such that

$$(\hat{\beta}, \hat{\theta}) = argmax_{(\beta,\theta) \in [-1,1] \times (0,+\infty)} \ell(\beta, \theta).$$

They can be determined by solving via the score equations defined by $\partial \ell(\beta, \theta)/ \partial \beta = 0$ and $\partial \ell(\beta, \theta)/ \partial \theta = 0$, where

$$\frac{\partial}{\partial \beta} \ell(\beta,\theta) = \sum_{i=1}^{n} \frac{2[1+\theta x_i e^{-\theta x_i}/(1+\theta)]e^{-\theta x_i} - 1}{1-\beta+2\beta[1+\theta x_i e^{-\theta x_i}/(1+\theta)]e^{-\theta x_i}}$$

and

$$\frac{\partial}{\partial \theta} \ell(\beta,\theta) = \frac{n}{\theta} - \frac{n}{1+\theta} - 2\sum_{i=1}^{n} x_i + \sum_{i=1}^{n} \frac{e^{\theta x_i} x_i(\theta+1)]+1+2x_i}{(1+\theta)e^{\theta x_i} + 2\theta x_i - 1}$$
$$- \frac{2\beta}{(\theta+1)^2} \sum_{i=1}^{n} x_i e^{-2\theta x_i} \frac{e^{\theta x_i}(\theta+1)^2 + 2\theta(\theta+1)x_i - 1}{1-\beta+2\beta[1+\theta x_i e^{-\theta x_i}/(1+\theta)]e^{-\theta x_i}}.$$

As usual, numerical methods can be used to determine the numerical values of $\hat{\beta}$ and $\hat{\theta}$. Also, the observed information matrix can be expressed in view of determining confidence intervals or diverse kinds of likelihood tests.

5. Application

In what follows, we shall illustrate the potentiality of the new model by means of real data applications. Here, we examine the versatility of the TML distribution in comparison with the TL distribution having the following cdf:

$$F(x) = \left\{1 - \left[1 + \frac{\theta x}{1+\theta}\right] e^{-\theta x}\right\}\left\{1 + \beta\left[1 + \frac{\theta x}{1+\theta}\right] e^{-\theta x}\right\}, \quad x > 0$$

and the TE distribution having the following cdf:

$$F(x) = [1 - e^{-\theta x}][1 + \beta e^{-\theta x}], \quad x > 0.$$

For checking the goodness of fit, we derive the unknown parameters by the maximum likelihood method and then standard error (SE), –log likelihood (–logL), the values of the AIC (Akaike Information Criterion) and BIC (Bayesian Information Criterion), the values of the Kolmogorov-Smirnov (K-S) statistic, the corresponding p-values and the values of the Anderson-Darling (A^*) and Cramér von Mises (W^*) are compared. They are evaluated using the **R** software via the commands *fitdist(), ks.test(), ad.test() and cvm.test()*.

5.1 Data Set 1 (Bladder Cancer Data)

The real data set presents the remission times (in months) of a random sample of 128 bladder cancer patients, which has been given by Lee and Wang, 2003. Several authors studied this data set. According to Merovci, 2013a, the TL distribution works quite well for this data in comparison to the Lindley distribution. The data are given below:

0.08, 2.09, 3.48, 4.87, 6.94, 8.66, 13.11, 23.63, 0.20, 2.23, 3.52, 4.98, 6.97, 9.02,13.29, 0.40, 2.26, 3.57, 5.06, 7.09, 9.22, 13.80, 25.74, 0.50, 2.46, 3.64, 5.09, 7.26, 9.47, 14.24, 25.82, 0.51, 2.54, 3.70, 5.17, 7.28, 9.74, 14.76, 26.31, 0.81, 2.62, 3.82, 5.32, 7.32, 10.06, 14.77, 32.15, 2.64, 3.88, 5.32, 7.39, 10.34, 14.83, 34.26, 0.90, 2.69, 4.18, 5.34, 7.59, 10.66, 15.96, 36.66, 1.05, 2.69, 4.23, 5.41, 7.62, 10.75, 16.62, 43.01, 1.19, 2.75, 4.26, 5.41, 7.63, 17.12, 46.12, 1.26, 2.83, 4.33, 5.49, 7.66, 11.25, 17.14, 79.05, 1.35, 2.87, 5.62, 7.87, 11.64, 17.36, 1.40, 3.02, 4.34, 5.71, 7.93, 11.79, 18.10, 1.46, 4.40, 5.85, 8.26, 11.98, 19.13, 1.76, 3.25, 4.50, 6.25, 8.37, 12.02, 2.02, 3.31, 4.51, 6.54, 8.53, 12.03, 20.28, 2.02, 3.36, 6.76, 12.07, 21.73, 2.07, 3.36, 6.93, 8.65, 12.63, 22.69.

Table 2 summarizes the results of descriptive study for the fitted TML, TL and TE models for the current data set. From the study, the smallest –logL, AIC, BIC, K-S statistic, A^*, W^* and the highest -values are obtained for the TML distribution. In Figure 4a, we present the estimated pdfs against fitted pdfs. The figure sketches the fitted density for the TML model closer to the empirical histogram than the fits of the TL and TE models. The empirical distribution function against fitted distribution function is also given in Figure 4b. An inspection of these plots reveals that, the TML model is superior to the TL and TE models in terms of model fit. This certifies the performance of the TML model and the eminence of the extra transmuted parameter β.

Table 2: Estimated values, –logL, AIC, BIC, K-S statistics, p-value, A^* and W^* for data set 1.

Model	Estimates(SE)	–logL	AIC	BIC	K-S	p-value	A^*	W^*
TML	$\hat{\theta} = 0.1105(0.0128)$	410.6611	825.3222	831.0263	0.0667	0.6196	0.9590	0.1358
	$\hat{\beta} = 0.5299(0.2066)$							
TL	$\hat{\theta} = 0.1567(0.0148)$	412.3712	828.7424	834.4465	0.0831	0.3389	1.2996	0.2258
	$\hat{\beta} = 0.6367(0.1651)$							
TE	$\hat{\theta} = 0.0623(0.0100)$	411.2578	826.5156	832.2197	0.0987	0.1645	1.4648	0.2288
	$\hat{\beta} = 0.8607(0.1678)$							

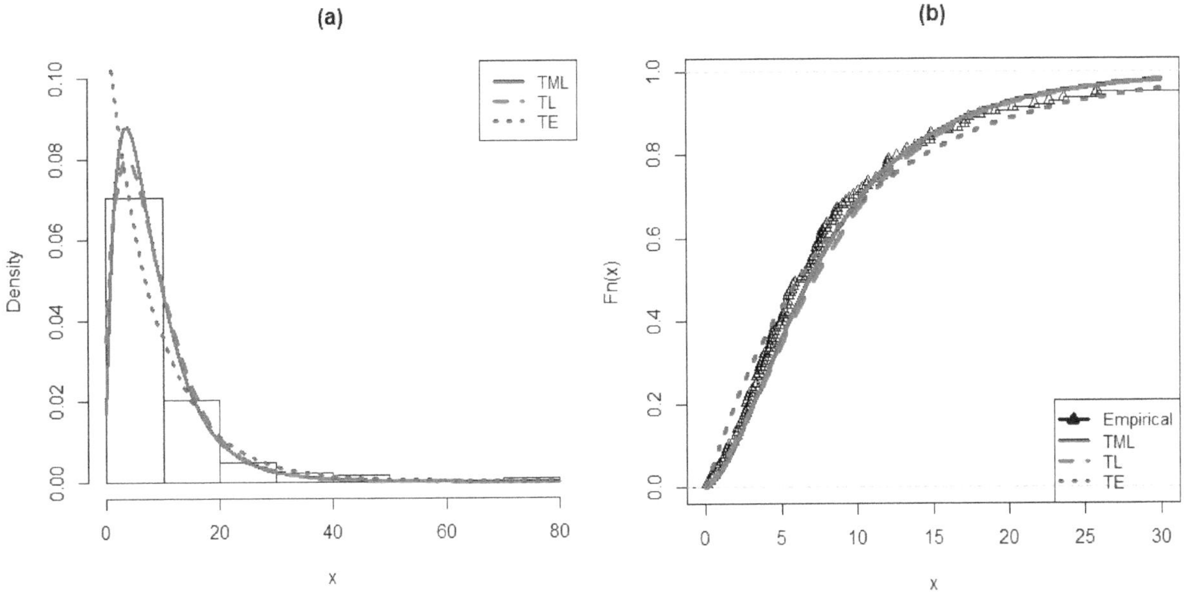

Figure 4: Plots of (a) estimated pdfs and (b) estimated cdfs of the considered distributions for data set 1.

5.2 Data Set 2 (COVID-19 Data)

The data set represents the numbers of new daily deaths due to COVID-19 in Brazil during the period of 1st April 2020 to the 30th of May 2020. The data set has been extracted from the site www.who.int. The data are given below:

23, 42, 40, 58, 60, 73, 54, 67, 114, 133, 141, 115, 68, 99, 105, 204, 204, 188, 217, 206, 115, 113, 166, 165, 407, 357, 346, 189, 338, 474, 449, 435, 428,421, 275, 296, 600, 615, 610, 751, 730, 496, 396, 881, 749, 844, 824, 816, 485, 674, 1179, 888, 1188, 1001, 965, 653, 807, 1039, 1086, 1156.

The performances of fitted models have been compared by computing values of the selection criterion's such as, –logL, AIC, BIC, K-S statistics, *p*-value, A^* and W^*. The summary of the data study is given in Table 3. The smallest values of –logL, AIC, BIC, K-S statistics, A^* and W^* and the highest *p*-value of the TML model give confirmation in favor of the proposed distribution. In this regard, this result can be proved graphically, see Figure 5. From this figure, as expected, the TML model outperforms its competitors.

Table 3: Estimated values, –logL, AIC, BIC, K-S statistics, *p*-value, A^* and W^* for data set 2.

Model	Estimates(SE)	–logL	AIC	BIC	K-S	*p*-value	A^*	W^*
TML	$\hat{\theta}$ = 0.0027(0.0003)	427.1417	858.2833	862.472	0.13856	0.1995	2.2952	0.3132
	$\hat{\beta}$ = 0.1063(0.2403)							
TL	$\hat{\theta}$ = 0.0043(0.0005)	427.3365	858.673	862.8617	0.15549	0.1099	2.7437	0.3596
	$\hat{\beta}$ = 0.1662(0.2532)							
TE	$\hat{\theta}$ = 0.1662(0.2532)	427.3365	858.673	862.8617	0.1702	0.0619	3.1527	0.3258
	$\hat{\beta}$ =0.0042(0.0005)							

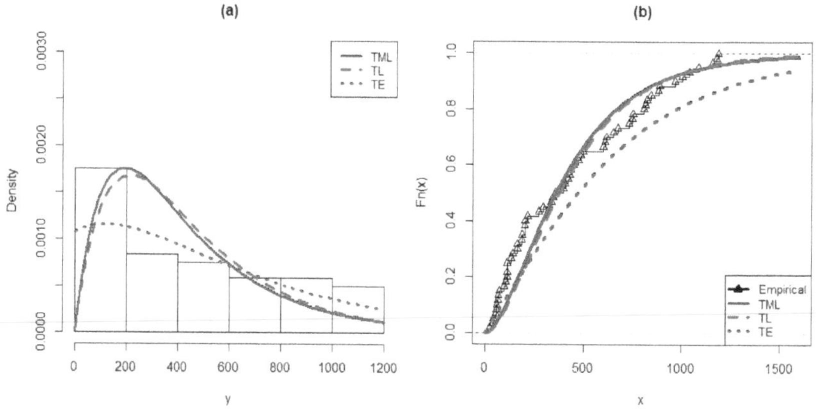

Figure 5: Plots of (a) estimated pdfs and (b) estimated cdfs of the considered distributions for data set 2.

6. Conclusions

Statistical lifetime distributions are widely applied in various fields such as reliability engineering, survival analysis, social sciences and a host of other applications. Consequently, there has been a great interest among statisticians in pioneering new distributions. In this chapter, we introduced a two-parameter distribution, named the TML distribution using the transmutation method, as a simple generalization of the ML distribution. Some interesting properties of the proposed distribution are provided. In addition, the maximum likelihood estimation of the parameters is discussed. The suitable performance of the TML distribution to two real data sets are analyzed and discussed. According to the descriptive study for real data sets, we can conclude that the proposed TML distribution achieves more success in modeling than well-known TL and TE models. In this regard, this study will bring new musings in addition to the existing works.

References

Abd El Hady, N. E. 2014. Exponentiated transmuted weibull distribution a generalization of the weibull distribution. International Scholarly and Scientific Research & Innovation, 8: 903–911.

Aryal, G. R. and Tsokos, C. P. 2009. On the transmuted extreme value distribution with application. Nonlinear Analysis: Theory, Methods & Applications, 71: 1401–1407.

Aryal, G. R. and Tsokos, C. P. 2011. Transmuted weibull distribution: a generalization of the weibull probability distribution. European Journal of Pure and Applied Mathematics, 4: 89–102.

Aryal, G. R. 2013. Transmuted log-logistic distribution. Journal of Statistics Applications & Probability, 2: 11–20.

Ashour, S. K. and Eltehiwy, M. A. 2013a. Transmuted Lomax distribution. American Journal of Applied Mathematics and Statistics, 1: 121–127.

Ashour, S. K. and Eltehiwy, M. A. 2013b. Transmuted exponentiated Lomax distribution. Australian Journal of Basic and Applied Sciences, 7: 658–667.

Bourguignon, M., Leão, J., Leiva, V. and Santos-Neto, M. 2017. The Transmuted Birnbaum-Saunders distribution, REVSTAT - Statistical Journal, 15: 601–628.

Chesneau, C., Tomy, L. and Gillariose, J. 2020a. A new modified Lindley distribution with properties and applications. Preprint.

Chesneau, C., Tomy, L. and Gillariose, J. 2020b. The inverted modified Lindley distribution, Journal of Statistical Theory and Practice, 14: 1–17.

Chesneau, C., Tomy, L. and Jose, M. 2020c. Wrapped modified Lindley distribution, Preprint.

Elbatal, I. and Aryal, G. 2016. On the Transmuted Additive Weibull Distribution. Austrian Journal of Statistics, 42: 117–132.

Elbatal, I., Asha, G. and Raja, V. 2014. Transmuted exponentiated Fréchet distribution: properties and applications. Journal of Statistics and Applied Probability, 3: 379–394.

Eltehiwy, M. and Ashour, S. 2013. Transmuted exponentiated modified Weibull distribution. International Journal of Basic and Applied Sciences, 2: 258–269.
Ghitany, M. E., Al-Mutairi, D. K., Balakrishnan, N. and Al-Enezi, L. J. 2013. Power Lindley distribution and associated inference. Computational Statistics and Data Analysis, 6: 20–33.
Hussian, M. A. 2014. Transmuted exponentiated gamma distribution: A generalization of the exponentiated gamma probability distribution. Applied Mathematical Sciences, 8: 1297–1310.
Khan, M. S., King, R. and Hudson, I. L. 2019. Transmuted Burr Type X distribution with covariates regression modeling to analyze reliability data. American Journal of Mathematical and Management Sciences, 39: 99–121.
Lee, E. T. and Wang, J. W. 2003. Statistical Methods for Survival Data Analysis, New York, Wiley.
Lindley, D. V. 1958. Fiducial distributions and Bayes theorem, Journal of the Royal Statistical Society, A, 20: 102–107.
Lindley, D. V. 1965. Introduction to Probability and Statistics from a Bayesian Viewpoint, Part II : inference, Combridge University Press, New Yourk.
Merovci, F. 2013a. Transmuted Lindley distribution. International Journal of Open Problems in Computer Science and Mathematics, 6: 63–72.
Merovci, F. 2013b. Transmuted exponentiated exponential distribution. Mathematical Sciences and Applications E-Notes, 1: 112–121.
Merovci, F. and Elbatal, I. 2014. Transmuted Lindley-geometric distribution and its applications. Journal of Statistics and Applied Probability, 1: 77–91.
Merovci, F. and Puka, L. 2014. Transmuted Pareto distribution. ProbStat Forum, 7: 1–11.
Merovci, F. 2016. Transmuted Rayleigh distribution. Austrian Journal of Statistics, 42: 21–31.
Nadarajah, S., Bakouch, H. S. and Tahmasbi, R. 2011. A generalized Lindley distribution, Sankhya B - Applied and Interdisciplinary Statistics, 73: 331–359.
Rahman, M. M., Al-Zahrani, B. and Shahbaz, M. Q. 2020a. Cubic transmuted pareto distribution. Annals of Data Science, 7: 91–108.
Rahman, M. M., Al-Zahrani, B., Shahbaz, S. H. and Shahbaz, M. Q. 2020b. Transmuted probability distributions: a review. Pakistan Journal of Statistics and Operation Research, 16: 83–94.
Shanker, R. and Mishra, A. 2013. A two parameter Lindley distribution. Statistics in Transition New Deries, 14: 45–56.
Shanker, R. and Mishra, A. 2013. A quasi Lindley distribution. African Journal of Mathematics and Computer Science Research, 6: 64–71.
Sharma, V., Singh, S., Singh, U. and Agiwal, V. 2015. The inverse Lindley distribution: a stress-strength reliability model with applications to head and neck cancer data. Journal of Industrial and Production Engineering, 32: 162–173.
Sharma, V. K., Singh, S. K., Singh, U. and Merovci, F. 2016. The generalized inverse Lindley distribution: a new inverse statistical model for the study of upside-down bathtub data, Communications in Statistics-Theory and Methods, 45: 5709–5729.
Shaw, W. T. and Buckley, I. R. 2007. The alchemy of probability distributions: beyond Gram-Charlier expansions and a skew-kurtotic-normal distribution from a rank transmutation map. Research report.
Shaw, W. T. and Buckley, I. R. 2009. The alchemy of probability distributions: beyond Gram-Charlier expansions and skew-kurtotic-normal distribution from a rank transmutation map, http://arxiv.org/abs/0901.0434v1:1-8.
Singh, S. K., Singh, U. and Sharma, V. K. 2014. The truncated Lindley distribution-Inference and application. Journal of Statistics Applications and Probability, 3: 219–228.
Singh, S. K., Singh, U. and Sharma, V. K. 2016. Estimation and prediction for Type-I hybrid censored data from generalized Lindley distribution. Journal of Statistics and Management Systems, 19: 367–396.
Tomy, L. 2018. A retrospective study on Lindley distribution. Biometrics and Biostatistics International Journal, 7: 163–169.
Zakerzadeh, H. and Dolati, A. 2009. Generalized Lindley distribution. Journal of Mathematical Extension, 3: 13–25.

Chapter 22

Adjusted Bias and Risk for Estimating Treatment Effect after Selection with an Application in Idiopathic Osteoporosis

Omer Abdalghani,[2] *Mohd Arshad,*[1,2,*] *K R Meena*[3] and *A K Pathak*[4]

1. Introduction

In clinical researches, when comparing the effects of different treatments (therapies or drugs), usually a physician would like to select the most effective treatment among k (≥ 2) active treatments. The classical statistical approach to such a problem are the statistical significance tests (such as the test of homogeneity), where we examine the hypothesis of equality of treatment effects. If this hypothesis is rejected, we have the information that the effects are not equal, but we do not have the information about the best (most effective) treatment. Therefore, statistical tests (whether or not they yield statistically significant results) do not supply the information about the selection of the most effective treatment. To this end, one statistical inference problem concerned with the correct selection objective is the ranking and selection problem which concentrates on selecting the most effective treatment among the k available treatments, using some selection rules. The quality of a treatment is assessed in terms of the characteristic (or parametric function) associated with it. Often, a primary characteristic of interest is the mean effect of a treatment. Moreover, the treatment that corresponds to the largest mean effect will be selected using some selection rule. Further, the problem of interest is the estimation of treatment mean effect after selection. Some relevant selection problems in medicine are represented in finding the optimal dose of treatment or identifying subgroups of patients that respond better to specific therapies than to populations at large.

In clinical trials, most of the work carried out for evaluation of new treatments mainly based on designs that compare a single or number of experimental treatments with a standard therapy or a placebo, then one or two treatments will be selected, based on their observed data, for further investigations. Such a design is called 'select and test' design due to Tall et al., 1998; Stallard and Todd, 2003. Most randomized comparative clinical trials including well-designed trials can produce bias in conventional treatment estimation. For example, in the process of randomization, if the allocation of patients is not completely blinded, so that, experimenters or patients have a preconceived idea about their allocation, then the process would be a form of selection bias (intervention allocation bias). If a physician has prior knowledge of how a new treatment might work, then, their evaluation of the patient's responses could be a source of bias. However, they are often potential sources of bias that might not be so apparent, for example follow-up bias, measurement bias, and exclusion bias.

The bias of estimators may occur when the maximum mean effect of several treatments has to be determined, or the mean effect of the selected treatment has to be estimated. It is so because these estimators may contribute to the decision as to whether to continue a drug development program or to select a specific treatment. Bias is likely to be high if the experimental treatments have similar mean effects. The risk of overestimating mean effect after selection may present in these situations as well. Some theoretical results were constructed for adjusting the selection bias that may arise in these situations as discussed in Shen, 2001; Stallard and Todd, 2005. In some situations, the experimenter may wish to estimate the treatment mean effect after selection. In the literature, the problem of estimating mean effect after selection has been studied by many authors. Most discussions focused on obtaining estimators of the parameters associated with the treatment (population) after selection and deriving various results using different loss functions. For some recent contributions on these problems, the reader may refer to Sackrowitz and Samuel-Cahn, 1986; Misra and Meulen, 2001;

[1] Department of Mathematics, Indian Institute of Technology Indore, Simrol, Indore, India.
[2] Department of Statistics and Operations Research, Aligarh Muslim University, Aligarh, India.
Email: abdalghani.amu@gmail.com
[3] Department of Mathematics, Acharya Narendra Dev College,University of Delhi, New Delhi, India.
Email: kmeena.iitr@gmail.com
[4] Department of Mathematics and Statistics, Central University of Punjab, Bathinda, India.
Email: ashokiitb09@gmail.com
* Corresponding author. arshad@iiti.ac.in, arshad.iitk@gmail.com

Sill and Sampson, 2007; Al-Mosawi and Khan, 2016; Nematollahi and Jozani, 2016; Arshad et al., 2015, 2019; Arshad and Misra, 2016; Meena and Gangopadhyay, 2017; Amini and Nematollahi, 2017; Mohammadi and Towhidi, 2017; Meena et al., 2018; Arshad and Abdalghani, 2019, 2020.

In the situations where the available observed data are considered to be normally distributed, a study by Putter and Rubinstein, 1968; Vallaisamy, 2009 proved the non-existence of unbiased estimators for the mean. Alternative estimators, i.e., bias-corrected estimators have been presented in the literature. Attempts to adjust the bias and risk resulting from point estimation (such as maximum likelihood) were carried-out by Dahiya, 1974; Parsian and Farsipour, 1999; Misra and Van der Muelen, 2003. The estimators derived by these authors are discussed in Section 3. For more references on these problems, see Hseih, 1981; Lu et al., 2013; Kimani et al., 2015; Meena et al., 2015; Mohammadi and Towhidi, 2016; Fuentes et al., 2018; Mazarei and Nematollahi, 2019. Stallard et al., 2008 considered a simple setting of a clinical trial in which they compared two experimental treatments, with response outcomes considered normally distributed with common variance. They have shown that a conditionally unbiased estimator of the treatment differences does not exist. Most of the works for normally distributed variables are theoretic and less attention has been given for real-life applications. In this chapter, we consider an application developed by Johansson et al., 1996, in which, a single-stage trial was conducted for comparing two active treatments, Insulin-like growth factor-I(IGF-I) versus growth hormone (GH) on men with idiopathic osteoporosis. One of the treatments was selected based on their effectiveness and no further investigations were conducted. Herein, we demonstrate the theoretical results of estimation after selection in evaluating a study of clinical relevance to assert a correct selection of the treatment that is considered as the most effective one and a more precise estimate of the mean effect of the selected treatment. The response outcomes of the two treatments are considered normally distributed as these responses are compatible with the normality assumption.

The remainder discussions of this chapter are organized as follows: Estimation after selection and natural selection rules are discussed in Section 2. Bias corrected, and risk reduced estimators suggested in the literature are also presented. In Section 3, a review of the experiment conducted by Johansson et al., 1996 is provided, and the computation of the estimators using idiopathic osteoporosis data is obtained. Formal discussion of this work is presented in Section 4.

2. Formulation of the Problem

Let T_1 and T_2 be two experimental treatments. Suppose that n patients are randomized to receive two experimental treatments. Let $Y_{11}, ..., Y_{1n}$ and $Y_{21}, ..., Y_{2n}$, be responses from patients receiving treatments T_1 and T_2, respectively. Assume that these responses are normally distributed with a common known variance τ^2 and different unknown means θ_1 and θ_2. Let $\overline{Y}_i$ denote the mean response for the treatment T_i, $i = 1,2$, and $\overline{Y}_{max} = \max(\overline{Y}_1, \overline{Y}_2)$.

For the aim of selecting the treatment associated with the largest mean response, a natural selection rule selects the treatment that corresponds to the largest mean responses. In case of ties for the most effective treatment, since both treatments are equally good, one of the treatments will arbitrarily be tagged as the most effective one. Although this argument is mathematically suitable it is not ethically right for a physician in practice. For selecting the most effective treatment, a non-randomized selection rule $\psi = (\psi_1, \psi_2)$ is a map from the sample space χ (= R^2) to $\{0,1\}^2$ such that $\psi_1(\mathbf{y}) + \psi_2(\mathbf{y}) = 1$, $\forall\ \mathbf{y} \in \chi$, where R^2 denotes the 2-dimensional Euclidean space. Note that the estimator $\overline{Y}_i$, $i = 1, 2$ is a complete sufficient statistic for the parameter θ_i. Therefore, the natural selection rule $\psi = (\psi_1, \psi_2)$ can be expressed as

$$\psi_1(y) = \begin{cases} 1, & \text{if } \overline{Y_1} > \overline{Y_2} \\ 0, & \text{if } \overline{Y_1} \leq \overline{Y_2} \end{cases} \tag{1.1}$$

and $\psi_2(y) = 1 - \psi_1(y)$.

We are interested in estimating the mean responses θ_T that corresponds to the treatment selected using ψ, and can be expressed as follows

$$\theta_T = \theta_1 \psi_1(y) + \theta_2 \psi_2(y)$$

$$\begin{cases} \theta_1, & \text{if } \overline{Y_1} > \overline{Y_2} \\ \theta_2, & \text{if } \overline{Y_1} \leq \overline{Y_2} \end{cases}.$$

Several different estimators of θ_T of the selected normal population have been proposed in the literature. Here, we consider estimators of θ_T proposed by Dahiya, 1974; Parsian and Farsipour, 1999; Misra and Van der Meulen, 2003. The authors studied the estimation of θ_T using the criterion of bias and linear-exponential (LINEX) loss function. They have suggested various estimators for θ_T and evaluated the performances of these estimators using their biases and risk functions. We consider the best choices of estimators suggested by the authors for use in practical situations. Misra and Van der Meulen, 2003 considered the LINEX loss function and obtained five improved estimators over some of the estimators proposed by Parsian and Farsipour, 1999. Qomi et al., 2012 also studied the above-discussed problems under the reflected normal loss function.

3. Competing Estimators

3.1 Bias Corrected Estimators

Correction of bias of an estimator can be obtained by subtracting the estimate by its bias. Such estimators are called bias reduced estimators or adjusted bias estimators. These types of estimators have been extensively discussed in the literature. We present some important estimators and their biases discussed in the literature.

It is intuitively clear that the estimator $\bar{Y}_{max}$ is biased and an over estimate the θ_T. A bias-corrected estimator $\delta_{1,\lambda}$ of θ_T is investigated by Putter and Rubinstein (1968) is given by

$$\delta_{1,\lambda}(\boldsymbol{Y}) = \bar{Y}_{max} - \sqrt{2}\,\lambda\sigma\phi\,(\bar{Y}/\sqrt{2}\sigma), \tag{3.1}$$

where $\lambda \geq 0$ is an arbitrary pre-specified value, $\bar{Y} = \bar{Y}_1 - \bar{Y}_2$, $\sigma^2 = V(\bar{Y}) = 2\tau^2/n$ and $\phi(y)$ denote a usual normal density function. The expression on the right-hand side of Equation (3.1) is simply $\bar{Y}_{max}$ minus its estimated bias multiplied by λ.

Bias of $\delta_{1,\lambda}$ can be obtained using the usual bias form $(\delta_{1,\lambda}) - \theta_T$, is given by

$$B(\delta_{1,\lambda}) = \sigma[\phi\,(\alpha) - \lambda/\sqrt{2}\phi\,(\alpha/\sqrt{2})],$$

where $\alpha = \sigma/\theta$ and $\theta = \theta_1 - \theta_2$.

An estimator $\delta_{2,c}$ proposed by Dahiya 1974, is given by

$$\delta_{2,c}(\boldsymbol{Y}) = \bar{Y}_2 + \bar{Y}\Phi(c\bar{Y}/\sigma) - c\sigma\phi(c\bar{Y}/\sigma), \tag{3.2}$$

where $c > 0$ is an arbitrary prespecified value that quantifies the size of bias reduction, $\Phi(y)$ and $\phi(y)$ are the standard normal cumulative distribution function and density function, respectively.

The bias of $\delta_{2,c}$ is given by

$$B(\delta_{2,c}) = \sigma\alpha[\Phi\,(c\alpha/\sqrt{1 + c^2}) - \Phi(\alpha)].$$

An estimator δ_3 suggested by Dahiya 1974, is given by

$$\delta_3\,(Y) = \bar{Y}_2 + \bar{Y}\Phi(\bar{Y}/\sigma). \tag{3.3}$$

Note that δ_3 is the maximum likelihood estimator of $E(T_\theta)$.

The bias of δ_3 is

$$B(\delta_3) = \sigma[(\phi(\alpha/\sqrt{2})/\sqrt{2} - \delta\{\Phi(\alpha) - \Phi(\alpha/\sqrt{2})\}].$$

Another estimator suggested by Dahiya 1974, is given by

$$\delta_{4,\lambda}(Y) = \delta_3(Y) - \lambda[\phi(\bar{Y}/\sqrt{2}\sigma)\,\sigma/\sqrt{2} + \bar{Y}\{\Phi(\bar{Y}/\sqrt{2}\sigma) - \Phi(\bar{Y}/\sigma)] \tag{3.4}$$

where $\lambda \geq 0$ is an arbitrary prespecified value. The estimator $\delta_{4,\lambda}$ is simply $\bar{Y}_{max}$ minus its estimated bias multiplied by λ. The estimator $\delta_{4,\lambda}$ is called a bias reducing estimator, which depends on the constant λ that determines the size of bias reduction.

The bias of $\delta_{4,\lambda}(\boldsymbol{Y})$ is

$$B(\delta_{4,\lambda}) = \sigma[(1 + \lambda)\,\phi(\alpha/\sqrt{2})/\sqrt{2} - 2\lambda\phi(\alpha/\sqrt{3})/\sqrt{3} + \alpha\{(1 + \lambda)\Phi(\alpha/\sqrt{2}) - \Phi(\alpha) - \lambda\Phi(\alpha/\sqrt{3})\}].$$

3.2 Risk Reduced Estimators

In some situations, underestimation (negative bias) may be more severe than over estimation (positive bias) or vice versa. For instance, in medicine, an overestimation of the dose levels is considered less severe than underestimation. To deal with such a phenomenon, Varian 1975 suggested the linear exponential (LINEX) loss function

$$L(\theta, \delta) = e^{a(\delta - \theta)} - a(\delta - \theta) - 1, \quad \boldsymbol{\theta} \in \Omega,\ \delta \in \mathbb{D},$$

where $a \neq 0$ is a parameter that determines the shape of LINEX loss, and δ is an estimator of the unknown parameter of interest θ. The characteristics of LINEX loss are discussed in details by Zellner 1986.

Parsian and Farsipour 1999 proposed an estimator of θ_T under the LINEX loss function, which is given by

$$\delta_5(\sigma) = \bar{Y}_{max} - \frac{a\sigma^2}{2}, \tag{3.5}$$

where $\bar{Y}_{max} = max\,(\bar{Y}_1, \bar{Y}_2)$, and $\sigma = \tau/\sqrt{n}$. The estimator given in (3.5) is a generalized Bayes estimator of θ_T using a uniform prior.

Also, concerning the LINEX loss function, Misra and Van der Muelen, 2003 obtained improved estimators over some of the estimators suggested by Parsian and Farsipour, 1999, given by

$$\delta_6(\boldsymbol{Y}) = \begin{cases} \bar{Y}_2, & \text{if } \bar{Y} \le -\frac{|\alpha|}{2}\sigma^2 \\ \frac{\bar{Y}_1+\bar{Y}_2}{2} - \frac{a\sigma^2}{4}, & \text{if } \bar{Y} > -\frac{|\alpha|}{2}\sigma^2. \end{cases} \tag{3.6}$$

$$\delta_7(\boldsymbol{Y}) = \begin{cases} max[\delta_0(\boldsymbol{Y}), \bar{Y}_2 + \varphi^*(\boldsymbol{Y})], & \text{if } a < 0 \\ min[\delta_0(\boldsymbol{Y}), \bar{Y}_2 + \varphi^{**}(\boldsymbol{Y})], & \text{if } a > 0, \end{cases} \tag{3.7}$$

where

$$\delta_0(\boldsymbol{Y}) = \bar{Y}_2 + \frac{1}{a} ln\left[1 + \left(e^{a\bar{Y}} - 1\right)\left(\frac{\bar{Y}}{\sigma\sqrt{2}}\right)\right],$$

$$\varphi^*(\boldsymbol{Y}) = \frac{\bar{Y}}{2} - \frac{a\sigma^2}{4},$$

and

$$\varphi^{**}(\boldsymbol{Y}) = \begin{cases} \infty, & \text{if } \bar{Y} < -\frac{a\sigma^2}{2} \\ \frac{\bar{Y}}{2} - \frac{a\sigma^2}{2}, & \text{if } \bar{Y} > -\frac{a\sigma^2}{2} \end{cases}$$

For $a < 0$, an improved estimator of θ_T is given by

$$\delta_{8,c}(\boldsymbol{Y}) = \begin{cases} \frac{\bar{Y}_1+\bar{Y}_2}{2} - \frac{a\sigma^2}{4}, & \text{if } \bar{Y} \ge min\left(-\sqrt{2}c\sigma,\ \frac{a\sigma^2}{2}\right) \\ \bar{Y}_2, & \text{if } \bar{Y} < min\left(-\sqrt{2}c\sigma,\ \frac{a\sigma^2}{2}\right). \end{cases} \tag{3.8}$$

For $0 < a < 2\sqrt{2}\frac{c}{\sigma}$, an improved estimator of θ_T is given by

$$\delta_{9,c}(\boldsymbol{Y}) = \begin{cases} \frac{\bar{Y}_1+\bar{Y}_2}{2} - \frac{a\sigma^2}{4}, & \text{if } \bar{Y} \ge -\frac{a\sigma^2}{2} \\ \frac{\bar{Y}_1+\bar{Y}_2}{2}, & \text{if} -\sqrt{2}c\sigma \le \bar{Y} < -\frac{a\sigma^2}{2} \\ \bar{Y}_2, & \text{if } \bar{Y} < -\sqrt{2}c\sigma. \end{cases} \tag{3.9}$$

For $a \ge 2\sqrt{2}\frac{c}{\sigma}$ an improved estimator of θ_T is given by

$$\delta_{10}(\boldsymbol{Y}) = \begin{cases} \frac{\bar{Y}_1+\bar{Y}_2}{2} - \frac{a\sigma^2}{4}, & \text{if } \bar{Y} \ge -\frac{a\sigma^2}{2} \\ \bar{Y}_2, & \text{if } \bar{Y} < -\frac{a\sigma^2}{2}. \end{cases} \tag{3.10}$$

4. Illustration Using Idiopathic Osteoporosis Data

4.1 A review of the Problem Conducted by Johansson et al., 1996

Johansson et al., 1996 conducted a single-stage design for comparison between two experimental treatments, insulin-like growth factor I (IGF-I) versus growth hormone (GH) on patients with idiopathic osteoporosis. Twelve patients were randomized between the two treatments and grouped into two groups. Patients in group 1 received the first placebo for IGF-I and GH injections for some specified period. Later, they received a placebo for GH and IGF-I injections. Another group received the treatments in reverse order. This implies that the actual treatment regimes were complex and not simply GH and IGF-I. Previous studies have proven the necessity of GH secretion for bone mass maintenance. Also, the GH and IGF treatments are positively correlated to bone density. Initially, the main purpose of the consideration of IGF-I was to conciliate GH effects. If GH has a significant remark on constructions of IGF-I, then the serum levels may increase during both treatments.

The biomarker plasma levels were measured repeatedly over a specified period. Among the response outcomes, are the serum concentrations of IGF binding protein-3. The serum concentrations are one of the primary variables that stimulate bone metabolism and significantly increases with GH and IGF-I injections. Note that, the serum is not the primary outcome of the study nor the primary biomarker variable that stimulates bone metabolism. However, the observed difference among IGF-I and GH is the serum concentrations of IGF binding protein-3. Other biochemical markers of bone metabolism and biological relevances are discussed in Johansson et al. 1996 (see result and discussion sections) and hence omitted.

Johansson et al. 1996 tested differences within treatments by using ANOVA,and used paired t-test for the comparison between the two treatments. The authors have pointed out the following results: The serum binding protein level increases and becomes relatively higher after GH injections. However, the levels of serum were above normal during the administration of both treatments. There were fewer side effects reported during GH than during IGF-I treatment. The difference between the two treatments was marginal and might depend on different timing of sampling injections and the number of doses.

Johansson et al. 1996 concluded that though GH has shown significant results on bone metabolism, the study avoids rejecting the effectiveness of IGF-I, and suggested the study of IGF-I in long-term osteoporosis treatment.

4.2 Analysis of serum Concentrations of IGF Binding Protein-3 Observed Data after GH and IGF-I Injections

Johansson et al., 1996 compared the treatment effects of growth hormone GH and insulin-like growth factor I (IGF-I) on biochemical markers of bone metabolism in men with idiopathic osteoporosis. Patients were randomly assigned to receive the two treatments. Subjects ranged from age 32 to 57 years. In this study, outcomes are the serum concentrations of IGF binding protein-3 after injections for each treatment. For illustration, the longitudinal data are reduced to a single number for each treatment, so that the comparison will remain between treatments for the serum concentrations of IGF binding protein-3. We use this data from a sample of individuals to draw some statistical inferences about the two treatments. The data-set used by Johansson et al., 1996 are reported in the book Daniel and Gross 2010 and are presented in Table 1.

Note that, the standard error of the differences between two treatments means is to be minimized, observations may be allocated in proportion to the standard deviation (so that, when the variances are equal, allowing for an equal number of observations from each

Table 1: Serum concentrations of binding protein-3 after GH and IGF-I treatments.

	Treatments Observations						
GH	4507	4072	3036	2484	3540	3480	2055
	4095	2315	1840	2483	2354	3178	3574
	3196	2365	4136	3088	3464	5874	2929
	3903	3367	2938	4142	4465	3967	4213
	4321	4990	3622	6800	6185	4247	4450
	4199	5390	5188	4788	4602	4926	5793
	3161	4942	3222	2699	3514	2963	3228
	5995	3315	2919	3235	4379	5628	6152
	4415	5251	3334	3910	2304	4721	3700
	3228	2440	2698				
IGF-I	3480	3515	4003	3667	4263	4797	2354
	3570	3630	3666	2700	2782	3088	3405
	3309	3444	2357	3831	2905	2888	2797
	3083	3376	3464	4990	4590	2989	4081
	4806	4435	3504	3529	4093	4114	4445
	3622	5130	4784	4093	4852	4943	5390
	3074	2691	2614	3003	3145	3161	4379
	3548	3339	2379	2783	3000	5838	5025
	4137	5777	5659	5628	2698	2621	3072
	2383	3075	2822				

treatment may be possible). In such a situation, it is required to combine past information with the present experimental data. Such randomized clinical trials will produce bias in conventional treatment effect estimation based on the final data. Moreover, we restrict our-selves to a single-stage procedure involving a fixed number of observations from each of the two treatments, where the outcomes from the two treatments are normal with common variance and different means.

Let Π_1 represent the population under GH treatment and Π_2 represents the population under IGF-I treatment. Let $Y_{i1},..., Y_{in}$, $i = 1,2$, be random observations from Π_1 and Π_2. The data are assumed to follow the normal distribution. To check the validity of the normal distribution for the given observations, we apply the Kolmogorov-Smirnov (K-S) goodness of fit test (to measure the distances between the empirical distribution function and the fitted normal model). In Table 2, we present the K-S distances, the estimated parameters, and their corresponding p-values.

From Table 2, we may conclude that, at 0.05 level of significance, the observed data sets from the two populations are typically normally distributed. Additionally, we plot the fitted normal distributions and the empirical distribution functions in Figure 1. This figure provides more evidence to our conclusion that the available observations satisfy the normality assumption.

Table 2: Estimated parameters, p-values, and K-S distances.

Population	Mean	SD	K-S distance	p-value
Π_1	3877.50	1038.143	0.1364	0.5715
Π_2	3706.29	1038.143	0.1212	0.7174

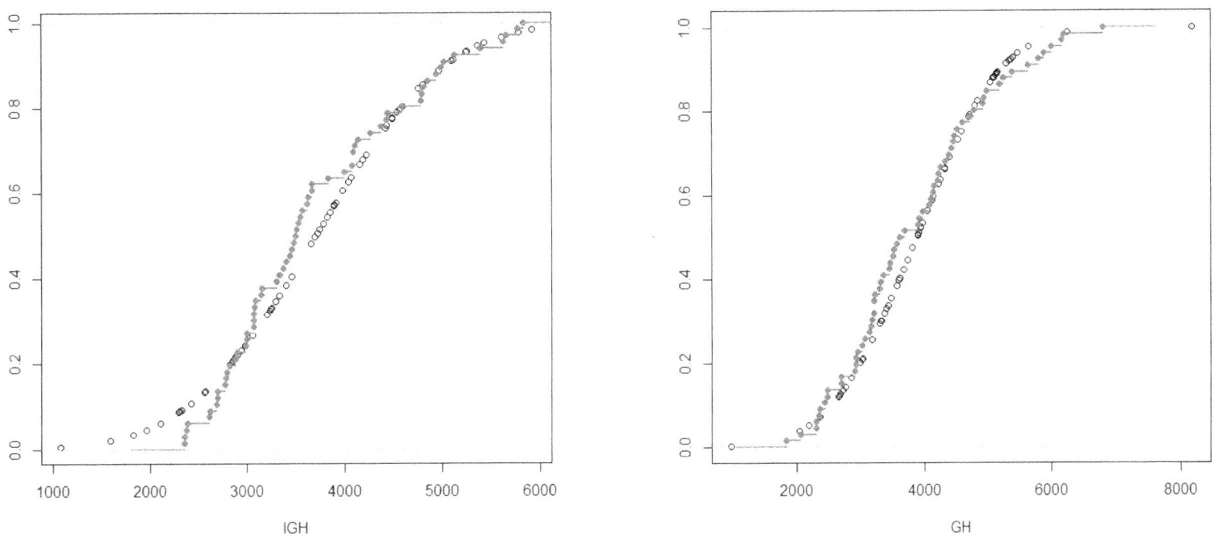

Figure 1: The fitted normal distribution and the empirical distribution function for GH and IGF data.

Note that, the two treatments are assessed for their quality based on their average effects, so that, the population $\Pi_1 \equiv N(\theta_1, \tau^2)$ is considered more effective than $\Pi_2 \equiv N(\theta_2, \tau^2)$ if $\theta_1 > \theta_2$, similarly the population Π_2 is considered more effective than Π_1 if $\theta_2 > \theta_2$. For the aim of selecting the most effective treatment, we employ the natural selection rule ψ, defined in (1.1), according to which the treatment corresponds to the largest mean effects selected. From the data, the observed mean effects in serum concentrations for patients under GH and IGF-I treatments are $\bar{Y}_1 = 3877.485$ and $\bar{Y}_2 = 3706.288$, respectively. $\bar{Y}_{max} = \bar{Y}_1 = 3877.485$, $\bar{Y} = \bar{Y}_2 - \bar{Y}_1 = -171.197$. Therefore, the observed data suggest that the treatment GH would be a correct selection by using the natural selection rule defined in (1.1). Further, the computation of the estimates of θ_T of the selected treatment is constructed in two scenarios. In Scenario 1, the treatment GH is selected as the most effective one by using the natural selection rule defined in (1.1). In Scenario 2, we assume that the treatment IGF-I is selected as the most effective one. The purpose of considering two scenarios is to quantify the extent of the adjusted estimators that neglect the assumed hypothesis and select the treatment that is most likely to give plausible estimates for the true parameter value. The various bias-corrected estimates of θ_T and their biases are presented in Tables 3 and 4. The risk reduced estimates of θ_T are reported in Tables 5, 6 and 7.

We can see from Table 3 that, for all combinations of λ and c, the bias-corrected estimates in Scenario 1 are larger than in Scenario 2. It is also noticed in both the scenarios that the values of estimates decrease as λ and c increase. From Table 4, we noticed that, the bias of $\delta_{1,\lambda}$ and $\delta_{4,\lambda}$ is positive for $\lambda = 1$, and for $\lambda > 1$ is negative and decreases as λ increases. The bias of $\delta_{2,c}$ is non-positive and gets close to zero as c increases. The bias of δ_4 is positive and substantial. Note that the bias of these estimators are equal for Scenario 1 and 2.

From Table 5, 6 and 7 we can see that, for the selected values of a, the estimates δ_5 and δ_7 in Scenario1 are larger than in Scenario 2, while the estimates δ_6 are equal in both scenarios. The estimates $\delta_{8,c}$ are equal in both scenarios for the selected values of a and c,

Table 3: Various bias-corrected estimates of θ_T.

Scenarios	$\delta_{1,\lambda}$			$\delta_{2,\lambda}$			$\delta_{3,\lambda}$			$\delta_{4,\lambda}$
	$\lambda = 1$	$\lambda = 4 - 2\sqrt{2}$	$\lambda = \sqrt{2}$	$c = 1$	$c = 1.5$	$c = \sqrt{3}$	$\lambda = 1$	$\lambda = 1.25$	$\lambda = 1.5$	
1	3796	3782.04	3762.3	3869.6	3785.4	3742.8	3992.2	3985.4	3978.7	4019.3
2	3624.8	3610.8	3591.1	3645.9	3671.9	3681.9	3649.8	3643.03	3636.3	3676.9

Table 4: Bias of various corrected estimates of θ_T.

$B(\delta_{1,\lambda})$			$B(\delta_{2,c})$			$B(\delta_{3,\lambda})$			$B(\delta_4)$
$\lambda = 1$	$\lambda = 4 - 2\sqrt{2}$	$\lambda = \sqrt{2}$	$c = 1$	$c = 1.5$	$c = \sqrt{3}$	$\lambda = 1$	$\lambda = 1.25$	$\lambda = 1.5$	
5.296	−1.6933	−11.577	−13.651	−7.455	−5.864	3.108	−2.886	−8.880	27.084

Table 5: Various risk reduced estimates of θ_T for different values of a.

Scenarios	δ_5				δ_6		δ_7	
	$a = 0.1$	$a = -0.1$	$a = 0.01$	$a = -0.01$	$a = 0.01$	$a = -0.01$	$a = 0.01$	$a = -0.01$
1	3061.01	4693.96	3795.838	3959.13	1097.53	6486.24	4002.21	3936.87
2	2889.82	4522.76	3624.64	3787.94	1097.53	6486.24	3659.81	3594.47

Table 6: Estimates $\delta_{8,c}$ of θ_T for different values of a and c.

Scenarios	$\delta_{8,c}$					
	$c = 1$			$c = 0.1$		
	$a = -0.01$	$a = -0.001$	$a = -0.0001$	$a = -0.01$	$a = -0.001$	$a = -0.0001$
1	6486.239	4061.322	3818.83	6486.239	4061.322	3818.83
2	6486.239	4061.322	3818.83	6486.239	4061.322	3706.288

Table 7: Various risk reduced estimates of θ_T for different values of a and c.

Scenarios	$\delta_{9,c}$			δ_{10}		
	$a = 0.01$	$a = 0.001$	$a = 0.0001$	$a = 0.01$	$a = 0.001$	$a = 0.0001$
1	1097.534	3522.451	3764.943	1097.534	3522.451	3764.943
2	1097.534	3522.451	3706.288	1097.534	3522.451	3706.288

but for $a = -0.0001$ and $c = 0.1$ the estimates in Scenario 1 are larger than in Scenario 2. For all selected values of a, the estimates $\delta_{9,c}$ and δ_{10} in Scenarios 1 and 2 are equal except for $a = 0.0001$, and both the estimates are larger in Scenario 1 than in Scenario 2. Given the performance among theses estimators, the estimator δ_5 seems to be the best choice because of its simple form.

We can deduce from an overall assessment of the estimators from both the scenarios when treatment GH is selected as the most effective one; all the adjusted estimators have the larger values as compared to the scenario where treatment IGF-I is selected. So, the observed data have provided more evidence that the correct selection for both scenarios would be to select treatment GH and hence motivate the use of natural selection rule which is required.

4. Discussion

This chapter focused on the problem of selecting the most effective treatment and estimating the mean of the treatment after selection. The motivation is a single-setting design in a clinical trial. We have considered an application conducted by Johansson et al., 1996 in which two active treatments are compared. The response outcomes of the two treatments are considered to be normally distributed. In the literature, studies have shown that for the selected normal population, no unbiased estimators of the mean exist. Alternatively, various bias-adjusted and risk-reduced estimators were suggested. To this end, we have considered the estimators derived by Dahiya, 1974; Parsian and Farsipour, 1999; Misra and Van der Meulen, 2003 for the normal population case.

Bias is likely to be high, especially in the case of similar treatments effects, so that either treatment would be a correct selection, and overestimation of the effects could be very large. When means are identical, the bias will be substantial and can partially be reduced by the bias corrections methods. The main reason for bias correction after selection is to obtain estimates that do not misrepresent the effects as large or small. In Section 4.2, we can see that the estimators appear to provide a considerable correction to the observed mean effect of the selected treatment. Johansson et al., 1996 tested changes within treatments by using ANOVA and used paired t-tests for the comparison between the two treatments. The above two tests can only tell us whether treatments are equivalent or not, but they do not provide any information about which is the best (most effective) one. To this end, for comparison, we demonstrated the estimation after selection framework alongside the statistical significance approach to supply additional information (magnitude of the effects and the significance of the trial outcome) that may provide sufficient evidence to the correct selection of the most effective treatment. The results obtained in Section 4.2 support the conclusions reported by Johansson et al., 1996, so that, the treatment GH is selected as the most effective one.

Estimation after selection procedure can also be adapted in more general situations involving two/multiple stages of adaptive designs, and in particular when a treatment has to be selected, and a more realistic estimate of the mean effect of the selected treatment is to be determined.

Finally, we state some future problems of interest that related to the problem discussed in this chapter, and in practice, they are much prominent. Suppose that treatment A depends linearly on factor B. Then, the experimenter may be interested in the rate of changes in the effect of the treatment A as B changes. A typical formulation of this problem is the regression model between A and B, in which the experimenter might wish to estimate the regression coefficient associated with the treatment after selection. This problem could be seen as a general case of the problem discussed in this chapter.

Furthermore, in practice, it is almost always critical to consider adjustment for the covariate. Adjusting for a variable Z (say), will decrease (often) the standard error which indicates a more precise treatment effect estimate. For example, in the study described in Section 4, a covariate could be a patient variability in dose, several injections, or any other adverse events.

References

Al-Mosawi, R. R. and Khan, S. 2018. Estimating moments of a selected Pareto population under asymmetric scale invariant loss function. Stat. Papers., 59: 183–198.

Amini, M. and Nematollahi, N. 2017. Estimation of the parameters of a selected multivariate population. Sankhya A., 79: 13–38.

Arshad, M. and Misra, N. 2016. Estimation after selection from exponential population with unequal scale parameters. Stat. Papers., 57: 605–621.

Arshad, M. and Abdalghani, O. 2019. Estimation after selection from uniform populations under an asymmetric loss function. Amer. Journ. Math Manage. Sci., 38: 349–362.

Arshad, M. and Abdalghani, O. 2020. On estimating the location parameter of the selected exponential population under the LINEX loss function. Brazil Journ. Probab. Stat., 34: 167–182.

Arshad, M., Abdalghani, O., Meena, K. R. and Pathak. A. K. 2021. Estimation after selection from bivariate normal population using LINEX loss function. arXiv preprint arXiv: 1911.05422.

Arshad, M., Misra, N. and Vellaisamy, P. 2015. Estimation after selection from gamma populations with unequal known shape parameters. Journ. Stat. Theor. Prac., 9: 395–418.

Daniel, W. W. and Cross, C. L. 2010. Biostatistics: basic concepts and methodology for the health sciences. New York: John Wiley and Sons.

Dahiya, R. C. 1974. Estimation of the mean of the selected population. Journ. Amer. Stat. Assoc., 69: 226–230.

Fuentes, C., Casella, G. and Wells, M. T. 2018. Confidence intervals for the mean of the selected populations. Elect. Journ. Stat., 12: 58–79.

Johansson, A. G., Lindh, E. R., Blum, W. F., Kollerup, G. I., Srensen, O. H. et al. 1996. Effects of growth hormone and insulin-like growth factor I in men with idiopathic osteoporosis. Journ. Clinic. Endocri. Metabol., 81: 44–48.

Kimani, P. K., Todd, S. and Stallard, N. 2015. Estimation after subpopulation selection in adaptive seamless trials. Stat. Medic., 34: 2581–2601.

Lu, X., Sun, A. and Wu, S. S. 2013. On estimating the mean of the selected normal population in two-stage adaptive designs. Journ. Stat. Plan Infer., 143: 1215–1220.

Mazarei, H. and Nematollahi, N. 2019. Admissible and minimax estimation of the parameters of the selected normal population in two-stage adaptive designs under reflected normal loss function. Probab. Math Stat., 39: 361–383.

Meena, K. R., Gangopadhyay, A. K. and Mandal, S. 2015. Estimation of the mean of the selected population. Intern. Journ. Math Compu. Scie., 8: 1480–1485.

Meena, K. R. and Gangopadhyay, A. K. 2017. Estimating volatility of the selected security. American Journ. Math Manage Scie., 36: 177–187.

Meena, K. R., Arshad, M. and Gangopadhyay, A. K. 2018. Estimating the parameter of selected uniform population under the squared log error loss function. Commun. Stati. Theo. Meth., 47: 1679–1692.

Misra, N. and van der Meulen, E.C. 2001. On estimation following selection from non-regular distributions. Commun. Stat. Theo. Meth., 30: 2543–2561.

Misra, N. and van der Meulen, E. C. 2003. On estimating the mean of the selected normal population under the LINEX loss function. Metrika., 58: 173–183.

Mohammadi, Z. and Towhidi, M. 2016. Empirical Bayes estimation for the mean of the selected normal population when there are additional data. Commun. Stat. Theo. Meth., 45: 3675–3691.

Nematollahi, N. and Jozani, M. J. 2016. On risk unbiased estimation after selection. Brazil Journ. Probab. Stat., 30: 91–106.

Parsian, A. and Farsipour, N. S. 1999. Estimation of the mean of the selected population under asymmetric loss function. Metrika., 50: 89–107.

Putter, J. and Rubinstein, D. 1968. On estimating the mean of a selected population. Technical Report. Department of Statistics, Univ. of Wisconsin, June (165).

Qomi, M. N., Nematollahi, N. and Parsian, A. 2012. Estimation after selection under reflected normal loss function. Commu. Stat. Theo. Meth., 41: 1040–1051.

Sackrowitz, H. B. and Samuel-Cahn, E. 1986. Evaluating the chosen population: Bayes and minimax approach. Lecture Notes-Monograph Series, 386–399.

Shen, L. 2001. An improved method of evaluating drug effect in a multiple-dose clinical trial. Stat. Medic., 20: 1913–1929.

Sill, M. W. and Sampson, A. R. 2007. Extension of a two-stage conditionally unbiased estimator of the selected population to the bivariate normal case. Commu. Stat. Theo. Meth., 36: 801–813.

Stallard, N. and Todd, S. 2005. Point estimates and confidence regions for sequential trials involving selection. Journ. Stat. Plann Infer., 135: 402–419.

Stallard, N., Todd, S. and Whitehead, J. 2008. Estimation following selection of the largest of two normal means. Journ. Stat. Plann. Infer., 138: 1629–1638.

Thall, P. F., Simon, R. and Ellenberg, S. S. 1988. Two-stage selection and testing designs for comparative clinical trials. Biometrika, 303–310.

Varian, H. R. 1975. A Bayesian approach to real estate assessment. Studies in Bayesian Econometric and Statistics in honour of Leonard J. Savage, 195–208.

Zellner, A. 1986. Bayesian estimation and prediction using asymmetric loss functions. Jour. Amer. Stat. Assoc., 81: 446–451.

Chapter 23

Validity Judgement of an EOQ Model using Phi-coefficient

Suman Maity, [1,*] ***Sujit Kumar De,*** [2] ***Madhumangal Pal*** [1] **and *Sankar Prasad Mondal*** [3]

1. Introduction

1.1 General Overview

Inventory management protects a company against unexpected fluctuations in consumption and supply and support for moving the production process continuously. Nowadays modern researchers are involved in handling inventory in a fuzzy environment. But none of the researchers give attention to public judgement. Arrow et al., 1951 proposed a method to derive optimal rules of inventory policy for finished goods. After that Arrow et al., 1958 developed some mathematical theory of inventory and production. Kenkel, 1981 studied introductory statistics for management and economics which is specially designed for business, economics and management. Kennedy and Bush 1985 gave an introduction for designing and analysis of experiments for behavioral research. Mehra et al., 1991 have shown that the validity of the EOQ formula will be different under inflationary conditions. Fraud is a critical issue in the finance industry, government and corporate sectors. In recent years, many researchers are working on fraud inventory. Debreceny and Gray 2010 studied on applying data mining techniques to journal entries. They also found the dataset with fraud indicators and compared it with other data mining techniques. Hesse and Jr., 2016 offered seven practical recommendations to fraud investigators for preventing and detecting fraud in this unique environment. Kanapickiene and Grundiene, 2015 developed a logistic regression model of fraud detection in financial statements. They also distinguished financial ratios which indicate the fraud in financial statements. Kim et al., 2016 developed a multi class financial misstatement detection model for detecting misstatements with fraud intention. Manufacturing firms face different risks and the most devastating risk lies in internal fraud. Mu and Carroll, 2016 developed a fraud risk decision model for prioritizing fraud risk cases in manufacturing firms. West and Bhattacharya, 2016 gave a comprehensive classification of existing fraud detection literature using a detection algorithm. Burton 1995 found numerous ways in which Leslie Fay ensured that quarterly sales met pre-established budgets.

Fahrmeir and Tutz, 1994 mainly focused on a multivariate Statistical Model that may be estimated similarly to generalized linear models. Angulo et al., 1998 utilized semiparametric statistical approaches to estimate and predict space-time processes. Pruscha and Gottlein, 2002 analyzed the data of forest inventory focusing on the space and time dependencies of the data. Besta et al., 2012 have shown the utility of some selected statistical tools in the area of inventory management. Ayloo et al., 2015 evaluated the relationship between beck depression inventory and bariatric surgical procedures. Iqbal et al., 2017 studied how to include statistical methods and sensitivity analysis for verifying the statistical significance of multi-criteria inventory models with respect to customer order fill rates. Mutschler, 2018 included higher-order statistics for DSGE models.

De and Pal, 2015 developed an intelligent decision-making model for book producer's conflict demand. Beyond the deterministic approach, researchers are growing interested in fuzzy uncertainty modeling. De and Sana, 2013a, 2013b, 2015, 2016 studied fuzzy backlogging model. Concepts of dense fuzzy set developed by De and Beg and its contemporary applications in inventory modelling are discussed by De and Mahata, 2016, 2019a,b. The heptagonal fuzzy set and its application in modelling explained by Maity et al., 2018a, also for the pioneering works on intuitionistic dense fuzzy set and its application through the utilization of score function was analyzed by Maity et al., 2019b. After this invention many articles on inventory modelling have been developed by eminent researchers [karmakar et al., 2017, 2018; Maity et al., 2019a; Maity et al., 2018b; Garg and Kumar, 2019; Chakraborty et al., 2020; Nobil et al., 2020; Khan et al., 2020; Rahman et al., 2020].

[1] Department of Applied Mathematics with Oceanology and Computer Programming, Vidyasagar University, W. B. India.
Email: mmpalvu@gmail.com
[2] Department of Mathematics, Midnapore College (Autonomous), W. B. India.
Email: skdemamo2008.com@gmail.com
[3] Department of Applied Science, Maulana Abul Kalam Azad University of Technology, W. B. India.
Email: sankar.mondal02@gmail.com
* Corresponding author: maitysuman2012@gmail.com

In this article we have studied an EOQ model where customers' judgement is the main focus of interest. Here we have taken a statistical survey over public agreement and disagreement on retailer's view. Then we have solved the EOQ model and done the sensitivity analysis of this model. After calculating mean, median and mode of order quantity, backorder quantity and average profit from the sensitivity analysis table we have obtained the Phi-coefficient of order quantity, backorder quantity and average profit. The intersection of these three Phi-coefficient curves in the positive region of the coordinate axes indicates the actual range of the validity of our proposed model.

1.2 Specific Study

The main objectives of an inventory practitioner are to develop a model first. Then they optimize the model along with proposed constraints under some methodologies. The traditional inventory managers usually assumed that any kind of model can be fitted into any category of customers. That is the customers from different social setups might have similar responses in favors of a particular inventory. But in practice, it is quite unnatural and hence framing a model is much more challenging which is consistent with the customer's needs/attitude/behavior over the inventory of that particular place. The common people play a vital role because the major parts of the customers from among them and are responsible for the acceptance or rejection of an inventory set up socially in respective places. Though the managers have the supportive documents and survey/audit reports in favor of installing an inventory in a particular place in many cases it is found to be fraudulent. So, the truth must come from public choice. The truth of reality and fate of an inventory go side by side silently over the sphere of insights of the DM of any kind of inventory process. The inventory practitioner however knew the actual facts which have been ongoing whether right or wrong in the inventory process itself. To sustain the real-world competition and to avoid maximum profit in many cases the DMs usually hide the actual facts from their customers. An alternative logic is that though some inventories have poor demand (fewer customers) but they continuously declare that they have huge number of customers to buy their products because of their good will and legal government transactions. Instead, the facts don't agree with their claim in the actual sense. In this way the DM's are usually promoting their policy to motivate customers. There are five cases which may arise from the customers' side

(i) Common people are interested in knowing the maximum profit of an inventory practitioner rather than minimum cost.

(ii) Common people usually believe the verbal message declared from the DM's desk or from his/her representatives during the item's delivery on demand.

(iii) Learned customers usually test or judge the forecasted message with real practice whether it is true or false.

(iv) Based on motivational study the learned customers might agree with the DM's view or otherwise might have a disagreement with the DM's declaration.

(v) If the customers are satisfied through the agreement of positive motivational strategies then we may say the inventory model is socially valid otherwise the model is invalid. In other words we may call, the inventory model has social existence else the model is of virtual inventory or pseudo inventory or dummy inventory without social existence.

However, from the DM's behavior on inventory set up the following views may arise:

a) All transactions are of white paper (money) and government taxations are neat and clean and up to date. This can only be possible if profits are usually high and no other inventory set up is available in the market as the market is less competitive.

b) Some transactions are of white papers and some others are of black papers. Here the govt. generates revenue partly. The inventory policy is a mixture of a dummy and actual transaction. That is a mixture of true and false. In this situation it is troublesome to qualify the models for the betterment of society. So, it is emergent that the model should be judged under the social aspect to know its validation.

c) If the case arises such that it is impossible to earn a considerable amount (very high) of profit so that the inventory can last in crucial competition then all transactions might be performed in black papers (statement manipulation). Such types of inventory are usually dummy inventories. The business under study may be called an underworld business in general. In this case the govt. generates no revenue [the cases of parallel economy, terrorism.] and has no social acceptance and hence the model is invalid and has no social existence by virtue of truth.

Thus, for DM's point of view, as the real world is not less competitive, so it is impossible to run dummy inventory to sustain within the society. Therefore, it is more practical to consider the inventory model having partly genuine and partly dummy (pseudo inventory) inventory which is the main focus of attention of the paper over here.

Suppose the DM frames the policy such that some parts are true and the rest of them are false in practical sense. It means, the ultimate profit will not be declared by the DM in the paper work because of exemption from the income tax department of the concerned government. But general public does know the fact, that the inventory managers are making more profit or are not in genuine transactions.

Inspired by a case study we develop a backorder inventory model. Generally, an inspection team (or a person who is not a customer of that inventory) set a common questionnaire for the inventory practitioner as well as for the common public (generally customers) of society first to know the facts of the concerned case. Since central tendencies are the best popular measure of any kind of statistical information so Mean, Median and Mode and the model optimum is taken here to get the customer's feedback. Against each claim of profit the customer have a response like agreement and disagreement. The claims of the inventory practitioner may be assumed by the response of getting/earning profit value either of the model optimum, or mean, median and mode any one of these which may be true or false. Thus, we have a set of responses of the form (agree, disagree) against the claims which may be (true, false). We may view these responses as dichotomous variables and we use phi-coefficient to know correlation between DM's claims and public judgments. Generally positive correlation indicates the DM's claim over public judgment is true and hence the model is valid and that of negative correlation indicates that the DM's claim over the public judgment is not true and hence the model is invalid. In this paper we compute phi-coefficient for order quantity, backorder quantity and the average profit itself over all possible cases. Considering several validation hypotheses over phi coefficient tests we discuss the actual validity of the model. Then we draw the corresponding curves for each of them. Finally, we made a conclusion followed by the scope of future work.

1.3 Case Study

We visited a cosmetic company 'Elco Cosmetics Pvt. Ltd' situated at Kolkata. After a long discussion with the manager we came to know that they are selling high-quality products and customers are fully satisfied with their products. The products basically are hair shampoos, face washes, body lotions, suns crims, and face creams. They strongly suggested that these products work nicely and customers are totally satisfied with them. This company has different selling counters situated in Kolkata city and abroad. We came back and met with many customers at every counter. They gave us different feedback regarding the quality of the products. The summary of the data so obtained is given below.

Table 1: Data available from the case study.

	Company's forecast			Customer's feedback		
	Order Quantity	Backorder Quantity	Profit	Order Quantity	Backorder Quantity	Expected Profit
Product A	462	137	531	450	145	500
Product B	426	173	561	410	160	523
Product C	446	153	519	426	145	490

Holding cost (c_1) = 5, Shortage cost (c_1) = 1.5, Set up cost (c_1) = 500, Demand (D_1) = 100

2. Preliminaries

Statistics is a way to get information from data. It is a tool for creating a new understanding from a set of numbers. **Population** is the group of all items of interest to a statistics practitioner. **Sample** is a set of data drawn from the population. To get more information about the sample we calculate mean, median, and mode of the sample.

2.1 Mean

The sample mean is the average and is computed as the sum of all the observed outcomes from the sample divided by the total number of events. We use $\bar{x}$ as the symbol for the sample mean. The formula is given below: $\bar{x} = \frac{1}{n}\sum_{i=1}^{n} x_i$, where n is the sample size. This is the most popular and useful measure of central location. It may be called the arithmetic mean to distinguish from the other mean. When each number x_i is to have weight w_i, the weighted mean can be defined as $\bar{x} = \frac{w_1x_1 + w_1x_1 + w_3x_3 + \cdots + w_nx_n}{w_1 + w_2 + w_3 + \cdots + w_n}$. In this paper to calculate mean we use the weighted mean formula.

2.2 Median

The *median* is the middle score. To find the median, first we have to rearrange the set of numerical data in ascending order. Then the median is the middle observation for an odd number of observations, and is the average of the two middle observations for an even number of observations. The formula which we have used in this paper to calculate median is given below. $Median = l + \left[\frac{\frac{n}{2} - cf}{f}\right] \times h$

Where, l = lower bound of median class, n = number of observations, f = frequency of median class. cf = Cumulative frequency of the class before the median class, h = length of median class.

2.3 Mode

The *mode* of a set of data is the number with the highest frequency, the one that occurs the maximum number of times. The formula which we have used in this paper to calculate the mode is given below. $Mode = l + \left(\frac{f_1 - f_0}{2f_1 - f_0 - f_2}\right) \times h$. Where, l = lower bound of modal class, f_0 = frequency of the class before of modal class, f_1 = frequency of modal class, f_2 = frequency of the class after the modal class, h = length of the modal class.

2.4 Phi-(φ) Coefficient

In statistics, the phi coefficient is an important concept. Sometimes, phi coefficient is also called the "**mean square contingency coefficient**". It is generally denoted by r_φ or simply φ. Karl Pearson first discovered Phi coefficient. It measures the amount of association between two binary variables. The significance of the correlation coefficient and measure of phi coefficient are nearly the same. Actually, Pearson correlation coefficient for two binary variables is nothing but the phi coefficient. Also, Pearson product-moment correlation coefficient and the numerical result of phi correlation coefficient both are similar. Phi coefficient is based on chi-square coefficient which determines the statistical relationship between two variables. As, Phi coefficient has a known sampling distribution, it is very easy for computation of significance and standard deviation of the phi coefficient.

Table 2: Phi-coefficient matrix.

Y \ X	True	False
Agree	A_1	A_2
Disagree	B_1	B_2

If A_1, A_2, B_1, and B_2 are the frequencies of the observations then φ is determined by the formula

$$\varphi = \frac{A_1B_2 - A_2B_1}{\sqrt{(A_1 + A_2)(B_1 + B_2)(A_2 + B_2)(A_1 + B_1)}}$$

The phi coefficient is very useful in many fields. Particularly, it is used for educational as well as psychological testing in which we frequently use a dichotomous continuous variable. When the variables having categories like pass or fail, success or failure, yes or no, agree or disagree, are observed, Phi coefficient can be utilized.

2.5 Considerations on Model Validation

We shall draw a logical decision for which we can conclude whether a model is practically valid or not. The following are the possible reasons of model validations

i) The model is valid if φ assumes positive values.
ii) The model is invalid if φ assumes negative values.
iii) The model is valid if φ coefficient curves for several decision variables must intersect at least at one positive point.
iv) The model is invalid if φ coefficient curves for several decision variables do not intersect at all or all of them meet at a negative point.
v) No decision can be made if some of the φ coefficient curves for several decision variables intersect and few of them do not intersect at all.

3. Assumptions and Notations

The following notations are used to develop the proposed model.

C_1 : Holding cost per quantity per unit time ($)

C_2 : Shortage cost per unit quantity per unit time ($)

C_3 : Set up cost per unit time period per cycle ($)

D_1 : Demand per unit time.

t_1 : Inventory run time (days)

t_1 : Shortage time (days)

T : Inventory cycle time(days)

Q_1 : Order quantity at time

Q_2 : Shortage during the time

S : Selling price per unit quantity (\$)

Assumptions

The assumptions for our proposed model are given below

1) Lead time is zero/negligible.
2) Demand rate is uniform and known.
3) Shortage are allowed and fully backlogged.
4) Rate of replenishment is finite.

4. Formulation of Natural Mathematical Model

Let the inventory starts at time $t = 0$ with order quantity Q_1 and demand rate D_1. After time $t = t_1$ the inventory reaches zero level and the shortage starts and it continues up to time $t = t_1 + t_2$. Let Q_2 be the shortage quantity during that time period t_2. Therefore, the mathematical problem associated to the proposed model is shown in Figure 1 and the several costs are given below.

$$\text{Inventory Holding cost} = \frac{1}{2}C_1Q_1t_1 = \frac{1}{2}C_1D_1t_1^2 \tag{1}$$

$$\text{Shortage cost} = \frac{1}{2}C_2Q_2t_3 = \frac{1}{2}C_2D_1t_2^2 \tag{2}$$

$$\text{Set up cost} = C_3 \tag{3}$$

$$\text{Total selling price} = SQ_1 \tag{4}$$

$$\text{and} \begin{cases} Q_1 = D_1t_1 \\ Q_2 = D_1t_2 \\ T = t_1 + t_2 \end{cases} \tag{5}$$

Therefore, the total average inventory cost

$$\begin{aligned} &= \frac{1}{T}\left(\frac{1}{2}C_1D_1t_1^2 + \frac{1}{2}C_2D_1t_2^2 + C_3\right) = \frac{1}{t_1+t_2}\left(\frac{1}{2}C_1D_1t_1^2 + \frac{1}{2}C_2D_1t_2^2 + C_3\right) \\ &= \frac{1}{2}C_1D_1\frac{t_1^2}{t_1+t_2} + \frac{1}{2}C_2D_1\frac{t_2^2}{t_1+t_2} + \frac{C_3}{t_1+t_2} \end{aligned} \tag{6}$$

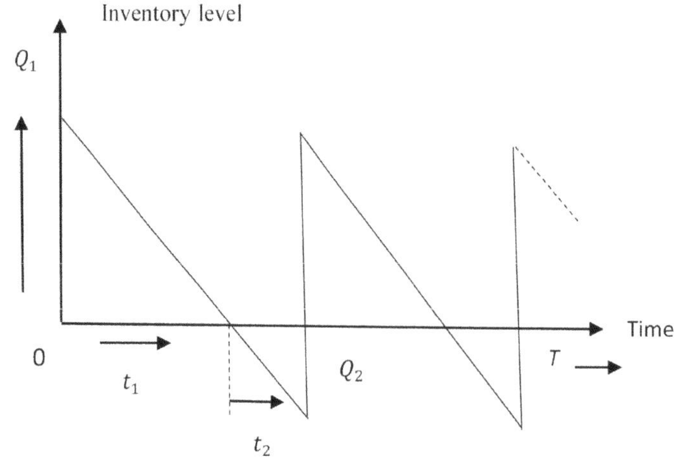

Figure 1: Classical backorder EOQ model.

Therefore, total average profit = (Average Selling price- Average Inventory cost)

$$Z = \frac{SQ_1}{t_1+t_2} - \left(\frac{1}{2}C_1D_1\frac{t_1^2}{t_1+t_2} + \frac{1}{2}C_2D_1\frac{t_2^2}{t_1+t_2} + \frac{C_3}{t_1+t_2}\right) \tag{7}$$

4.1 Numerical Example 1

Let us consider $C_1 = 5, C_2 = 1.5$, $C_3 = 500$, $T = 6$, $S = 20$, $D_1 = 100$ then we get the following result

$$Z_* = 544.87,\ t_1^* = 4.46,\ t_2^* = 1.538,\ Q_1^* = 446.153,\ Q_2^* = 153.846$$

Sensitivity Analysis

Based on the numerical example considered above for the EOQ mode, we now calculate the corresponding outputs for changing inputs parameter one by one. Again the sensitivity analysis is performed by changing of each parameter S, C_1, C_2, C_3, D_1, and T by +50%, +30%, –30% and –50%, considering one at a time and keeping the remaining parameters unchanged.

Table 3: Sensitivity analysis of the backorder EOQ model.

Parameter	% Change	Time (t_1^*)	Time (t_2^*)	Q_1^*	Q_2^*	Maximum Profit (Z^*)	$\frac{Z^* - Z_*}{Z_*}$ 100%
S	+50	5.998	0.001	599.875	0.124	1416.667	160
	+30	5.384	0.615	538.461	61.538	1037.179	90.353
	–30	3.538	2.461	353.846	246.153	144.871	-73.41
	–50	…..	……	……	……	….	…..
D_1	+50	4.461	1.538	669.230	230.769	858.974	57.64
	+30	4.461	1.538	579.999	200.000	733.333	34.58
	–30	4.461	1.538	312.307	107.692	356.410	–34.58
	–50	4.461	1.538	223.076	76.923	230.769	–57.64
C_1	+50	3.222	2.777	322.222	277.777	245.370	–54.96
	+30	3.625	2.375	362.500	237.500	342.708	–37.10
	–30	5.799	0.200	579.999	20.000	868.333	59.36
	–50	…..	…..	…..	…..	…..	…..
C_2	+50	4.620	1.379	462.069	137.931	531.609	–2.43
	+30	4.561	1.438	456.115	143.884	536.570	–1.52
	–30	4.347	1.652	434.710	165.289	554.407	1.75
	–50	4.260	1.739	426.087	173.913	561.594	3.06
C_3	+50	4.461	1.538	446.153	153.846	503.205	–7.64
	+30	4.461	1.538	446.153	153.846	519.871	–4.58
	–30	4.461	1.538	446.153	153.846	569.871	4.58
	–50	4.461	1.538	446.153	153.846	586.538	7.64
T	+50	5.153	3.846	515.384	384.615	228.632	–58.03
	+30	4.876	2.923	487.692	292.307	341.913	–37.24
	–30	4.046	0.153	404.614	15.385	832.783	52.84
	–50	…..	…..	…..	…..	…..	…..

Now from Table 3, we have calculated the frequency distribution of the average profit, average order quantity and backorder quantity which are shown in Table 4, Table 5 and Table 6 respectively. Then we have calculated mean, median and mode of order quantity, backorder quantity and average profit which is given in Table 7.

Table 4: Frequency distribution of average Profit.

Class intervals	Frequency f_i	Cumulative frequency	Mid interval Value (x_i)	$f_i x_i$
140–340	4	4	240	960
340–540	7	11	440	3080
540–740	5	16	640	3200
740–940	3	19	840	2520
940–1140	1	20	1040	1040
1140–1340	0	20	1240	0
1340–1540	1	21	1440	1440
Total	$\Sigma f_i = 21$			$\Sigma f_i x_i = 12240$

Table 5: Frequency distribution of average order quantity.

Class intervals	Frequency f_i	Cumulative frequency	Mid interval Value (x_i)	$f_i x_i$
200–300	1	1	250	250
300–400	4	5	350	1400
400–500	10	15	450	4500
500–600	5	20	550	2750
600–700	1	21	650	650
Total	21			9550

Table 6: Frequency distribution of Backorder quantity.

Class intervals	Frequency f_i	Cumulative frequency	Mid interval Value (x_i)	$f_i x_i$
1–100	5	5	50	250
100–200	10	15	150	1500
200–300	5	20	250	1250
300–400	1	21	350	350
Total	21			3350

Table 7: Central measures of order and backorder quantity.

Parameters	Expected Profit of the model	Q_1^*	Q_2^*
Mean (M_n)	582.85	454.7	159.5
Median (M_d)	525.71	455	155
Mode (M_o)	460	454.54	150

4.2 The Following Graph Shows Fraudulent Cases over 100 Inspections

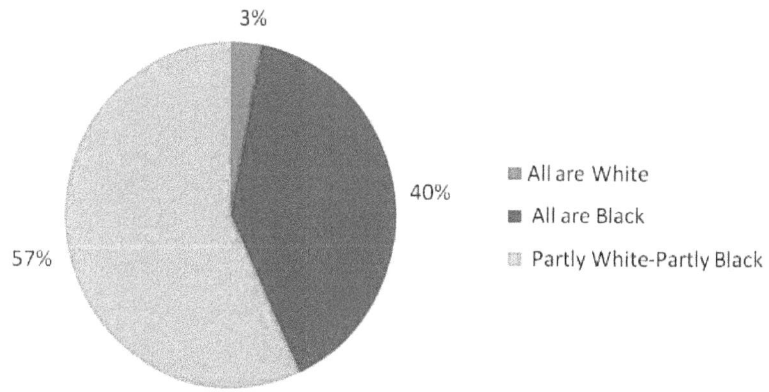

Figure 2: Cases of fraudulent inventory.

5. Claim of Decision Maker under Social Judgment

Here we assume, the decision maker may have the following choices shown in Table 8; i.e., the average profit be {Mean (M_n), Median (M_d), Mode (M_o), Optimum (O_p)}. Each response of the decision maker might be judged by the public domain by responses agree and disagree only. Since there are four options in the response matrix made by the decision maker so, we must have at most 24 decision tests criteria and it can be minimized to 6 criteria by applying the similar terms taken once from the possible permutations. The following are the social judgments on the decision maker's claim.

5.1 Discussion on Table 6

Table 6 shows six criteria, first one: DM claims: model optimum True, agreed; mode value True, Disagreed by public. This is quite valid for profit aspects but casts doubts over claiming on ordering and backordering quantity subsequently under phi coefficients tests.

Table 8: Phi-coefficients over several decision variables.

Criteria	Decision maker's claim				Phi-coeff. Of Q_1^*	Phi-coeff. Of Q_2^*	Phi-coeff. Of Z^*
	Public Judgment Agree		Public Judgment Disagree				
	True	False	True	False			
1	O_p	M_n	M_o	M_d	–0.0044	–0.0008	0.0164
2	O_p	M_n	M_d	M_o	–0.0049	–0.0172	–0.0500
3	O_p	M_o	M_D	M_n	–0.0048	0.0134	0.0679
4	M_n	O_p	M_o	M_d	0.0049	0.0172	0.0500
5	M_n	O_p	M_D	M_o	0.0044	0.0008	–0.0160
6	M_o	O_p	M_n	M_d	0.0048	–0.0134	–0.0679

Secondly, DM claims: model optimum True, agreed; median value True, Disagreed by public. This is invalid for all (profit, ordering and backordering quantity) aspects under phi coefficients tests. Criteria 4 shows, DM claims: median value True, agreed; mode value True, Disagreed by public. This is valid for all (profit, ordering and backordering quantity) aspects under phi coefficients tests. Similar comments can be drawn from other criteria also.

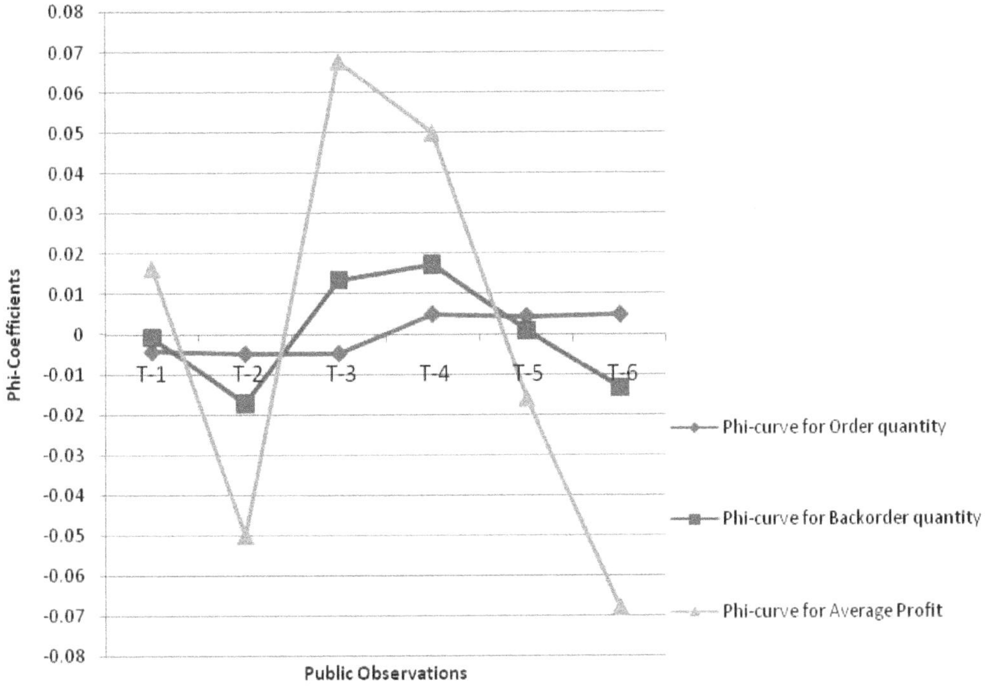

Figure 3: Model validation curve under public judgement.

5.2 Graphical Illustration of the Proposed Model

Discussion on Figure 3

From the above figure we see that the phi-coefficient curves for average profit function, order quantity and backorder quantity intersect at three points. Out of them, two points are negative and one is positive. Negative points are lying from the first and second criteria, second and third criteria respectively. These mean that the DM's 'True' claim on model optimum (actual profit) is rejected by the public judgments and hence the model is invalid. That is the claim of the profit \$ 544.87 against the order quantity 446.153 units and the backorder quantity 153.846 units is false. However, the positive point lies within criteria 4 and criteria 5. This indicates that, the 'true' claim on mean profit (coming from sensitivity analysis) is accepted by the public judgments. Hence the model is valid and the actual profit of the DM is \$ 582.85 against the order quantity 454.7 units and the backorder quantity 159.5 units exclusively.

5.3 Main Contributions

In this article we have checked the validity of an inventory model under social aspect. For this purpose, we have collected public judgement over the declaration of inventory practitioner and used Phi-coefficient to find the actual range of validity of this model. We have shown that the decision maker's declaration is not true always. So, in this way we can check the validity of any other model to find the actual reality behind those models. The basic novelties are stated as follows:

- Concept of Model validation is new, not a single article has been studied yet.
- Post Sensitivity Analysis is performed which is the first-time application as per literature concerned.
- Introduction of Dichotomous variable in inventory management problem is a totally new concept.
- Use of Phi-coefficient in inventory modeling is new.
- Simultaneous work of "Modeling and Validating" is the new concept in inventory control problems.
- Fraud work can be detected by this approach.

6. Conclusion

In this article we develop a classical backorder EOQ model. Constructing a profit function of the proposed model we optimize it with the help of Lingo 16.0 Software. Taking the sensitivity analysis of the proposed model we compute the mean, median and mode using those data sets and get their corresponding average inventory profits. Then we make a public judgment through the responses true or false over any one of the profit values from model optimum, mean, median and mode values of the model itself. Incorporating phi-coefficient curves for the decision variables we have searched their point of intersections for validity testing. In this study we get positive and negative phi-coefficient values. Positive values correspond to model validation and negative values correspond to no validation of the model.

Again, by phi-coefficients we are intending to study the overall intention of the different views through the intersections of different curves. So, no matter how much the value of phi-coefficients gets closer to one another the results may vary if we consider the quartile points of the data set or the upper and lower bounds of the profit function instead. In addition, the model validity can be judged by another non-traditional method which we shall discuss in our future works.

Conflict of interest: The authors declare that there are no conflicts of interest regarding the publication of this article.

Acknowledgements

The authors are grateful to the anonymous reviewers for their valuable comments and suggestions to improve the quality of this Chapter.

References

Angulo, J. M., Gonzales-Manteiga, W., Febrero-Bande, M. and Alonso, F. J. 1998. Semiparametric statistical approaches for space-time process prediction. Environmental and Ecological Statistics, 5: 297–316.

Arrow, K., Karlin, J. and Scarf, H. (ed.). 1958. Studies in the mathematical theory of inventory and production, Stanford, Calif, Stanford University Press.

Arrow, K., Harris, T. and Marschak, J. 1951. Optimal inventory policy, Econometrica, XIX.

Ayloo, S., Thompson, K., Choudhary, N. and Sheriffdeen, R. 2015. Correlation between the beck depression inventory and bariatric surgical procedures. Surgery for Obesity and Related Disease, 3(11): 637–642.

Besta, P., Janovska, K., Vilamova, S., Voznakova, I. and Kutac, J. 2012. The utilization of statistical methods in the area of inventory management. Jesenik, Czech Republic, EU.

Burton, L. 1995. Audit Report Details Fraud at Leslie Fay. The Wall Street Journal. March 28, p. B1.

Chakraborty, A., Maity, S., Jain, S., Mondal, S. P. and Alam, S. 2020. Hexagonal fuzzy number and its distinctive representation, ranking, defuzzification technique and application in production inventory management problem. Granular Computing, DOI: 10.1007/s41066-020-00212-8.

Debreceny, R. S. and Gray, G. L. 2010. Data mining journal entries for fraud detection: An exploratory study. Int. J. of Accounting Information Systems. 11: 157–181.

De, S. K. and Mahata, G. C. 2019a. A cloudy fuzzy economic order quantity model for imperfect-quality items with allowable proportionate discounts. Journal of Industrial Engineering International, DOI: 10.1007/s40092-019-0310-1.

De, S. K. and Mahata, G. C. 2019b. A comprehensive study of an economic order quantity model under fuzzy monsoon demand. Sadhana, DOI:10.1007/s12046-019-1059–3.

De, S. K. and Mahata, G. C. 2016. Decision of a fuzzy inventory with fuzzy backorder model under cloudy fuzzy demand rate. International Journal of Applied and Computational Mathematics, DOI: 10.1007/s40819-016-0258-4.

De, S. K. and Pal, M. 2015. An intelligent decision for a bi-objective inventory problem. Int. Journal of System Science: Operation: Logistics, DOI: 10.1080/23302674.2015.1043363.

De, S. K. and Sana S. S. 2015. An EOQ model with backlogging. Int. Journal of Management Sciences and Engineering Management, DOI:10.1080/17509653.201 4.995736.

De, S. K. and Sana, S. S. 2013b. Backlogging EOQ model for promotional effort and selling price sensitive demand- an intuitionistic fuzzy approach. Annals of Operations Research. Advance Online Publication, DOI: 10.1007/s10479-013-1476-3.

De, S. K. and Sana S. S. 2013a. Fuzzy order quantity inventory model with fuzzy shortage quantity and fuzzy promotional index. Economic Modelling., 31: 351–358.

De, S. K. and Sana S. S. 2016. The (p, q, r, l) model for stochastic demand under intuitionistic fuzzy aggregation with Bonferroni mean. Journal of Intelligent Manufacturing, DOI: 10.1007/s10845-016-1213-2.

Fahrmeir, L. and Tutz, G. 1994. Multivariate Statistical Modeling Based on Generalized Linear Models. Springer-Verlag: New York.

Hesse, M. F. and Jr., J. H. C. 2016. Fraud risk management: A small Business Perspective. Business Horizons, 59: 13–18.

Iqbal, Q. Malzahn, D. and Whitman, L. 2017. Statistical analysis of multi-criteria inventory classification models in the presence of forecast upsides. Production & Manufacturing Research, 5(1): 15–39.

Kanapickiene, R. and Grundiene, Z. 2015. The model of fraud detection in financial statements by means of financial ratios. Procedia-Social and Behavioral Sciences, 213: 321–327.

Karmakar, S., De, S. K. and Goswami, A. 2017. A pollution sensitive dense fuzzy economic production quantity model with cycle time dependent production rate. Journal of Cleaner Production, 154: 139–150.

Karmakar, S., De, S. K. and Goswami, A. 2018. A pollution sensitive remanufacturing model with waste items: Triangular dense fuzzy lock set approach. Journal of Cleaner Production, DOI: 10.1016/j.jclepro.2018.03.161.

Kenkel, J. L. 1981. Introductory Statistics for Management and Economics, Boston: Prindil, Weber & Schmidt.

Kennedy, J. J. and Bush, A. J. 1985. An introduction to the design and Analysis of Experiments of Behavioral Research. Newyork: University Press of America.

Khan, M. A., Shaikh, A. A., Konstantaras, L., Bhunia, A. K. and Cárdenas-Barrón, L. E. 2020. Inventory models for perishable items with advanced payment, linearly time-dependent holding cost and demand dependent on advertisement and selling price. International Journal of Production Economics, 230:107804. DOI: 10.1016/j.ijpe.2020.107804.

Kim, Y. J., Baik, B. and Cho, S. 2016. Detecting financial misstatements with fraud intension using multi-class cost sensitive learning. Expert Systems with Applications, 62: 32–43.

Mehra, S., Agrawal, S. P. and Rajagopalan, M. 1991. Some comments on the validity of EOQ formula under inflationary conditions. Decision Sciences, 22(1): 206–212. DOI: 10.1111/j.1540-5915.1991.tb01272.x.

Mu, E. and Carroll, J. 2016. Development of a fraud risk decision model for prioritizing fraud risk cases in manufacturing firms. Int. J. of Production Economics, 173: 30–42.

Mutschler, W. 2018. Higher-order statistics for DSGE models. Econometrics and Statistics, 6: 44–56.

Maity, S., Chakraborty, A., De, S. K., Mondal, S. P. and Alam, S. 2018a. A comprehensive study of a backlogging EOQ model with nonlinear heptagonal dense fuzzy environment. RAIRO-operations Research, DOI:10.1051/ro/201811.

Maity, S., De, S. K. and Mondal, S. P. 2019a. A Study of an EOQ Model under Lock Fuzzy Environment. Mathematics, DOI: 10.3390/math7010075.

Maity, S., De, S. K. and Pal, M. 2018b. Two Decision Makers' Single Decision over a Back Order EOQ Model With Dense Fuzzy Demand Rate. Finance and Market, 3: 1–11.

Maity, S., De, S. K. and Mondal, S. P. 2019b. A study of a back order EOQ model for cloud type intuitionistic dense fuzzy demand rate. Int. Journal of Fuzzy System, DOI: 10.1007/s40815-019-00756-1.

Nobil, A. M., Sedigh, A. H. A. and Barron, L. E. C. 2020. Reorder point for the EOQ inventory model with imperfect quality items. Ain Shams Engineering Journal, DOI: 10.1016/j.asej.2020.03.004.

Pruscha, H. and Gottlein, A. 2002. Regression analysis of forest inventory data with time and space dependencies. Environmental and Ecological Statistics, 9(1): 43–56.

Rahman, M. S., Shaikh, A. A. and Bhunia, A. K. 2020. Necessary and Sufficient Optimality Conditions for non-linear Unconstrained and Constrained optimization problem with Interval valued objective function. Computers & Industrial Engineering, DOI: 10.1016/j.cie.2020.106634.

West, J. and Bhattacharya, M. 2016. Intelligent Financial Fraud Detection: A Comprehensive Review. Computers & Security, 57: 47–66.

CHAPTER 24

Uncertain Chance-Constrained Multi-Objective Geometric Programming Problem

Sahidul Islam

1. Introduction

Nonlinear investigation assumes an ever-expanding part in hypothetical and applied arithmetic, just as in numerous different regions of science, for example, designing, measurements, software engineering, financial aspects, funds, and medication. The majority issue in our environment is non-linear issue. There are numerous strategies for tackling this nonlinear issue. Geometric programming is one of the best techniques to deal with these nonlinear optimization problems. Geometric programming was presented in 1967 by Duffin, Peterson and Zener. It is exceptionally helpful in the utilizations of the applications of a variety of optimization problems, and falls under the overall class of signomial problems. Geometric programming is a special technique used to solve a class of nonlinear programming problems; predominantly we utilize this problem to solve optimal design problems where we minimize cost and /or weight, maximize volume and/or efficiency . It is a significant strategy to solve special types of nonlinear optimization problems. The global optimum of a convex problem is accomplished more rapidly than the result of a non-linear problem. Since its inception, GP has been firmly connected with applications in engineering analysis and design problems.

Imprecision in the problem data often cannot be avoided when dealing with real world problems. Errors happen in real-world data for a large group of reasons. Nonetheless, in the course of the most recent thirty years, the fuzzy set methodology has ended up being helpful in these circumstances. Many research papers on geometric programming are created under a fluffy climate.

Multi-objective mathematical programming is an incredible improved method created by researchers to take care of different non-linear programming problems subject to linear and non-linear constraints. MOGP has been applied by numerous scientists to several optimizations, structural optimization and engineering problems, for example integrated circuit design, engineering design, project management and inventory management. MOGP is a special kind of non-linear programming problem with multiple objective functions.

2. Literature Review

Geometric programming (GP) is probably the best technique to solve non-linear optimization programming problems subject to linear and additionally non-linear constraints. In 1967, Duffin et al. showed the fundamental hypotheses of geometric programming in their book "Geometric Programming". Beightler and Philips (1979) gave a full account of the then entire current theory of geometric programming (GP) and numerical applications of GP to real-world problems in their book "Applied Geometric Programming".

In many real-life optimization problems, different goals must be considered, which might be identified with the social, economical and technical aspects of real optimization problems. Chankong and Haimes, (2008) presented a multi-objective decision making problem. In 2003, Liu et. al. introduced multi-objective decision making in their book "Multi-Objective Optimization and Control". In 2010, Ojha and Das proposed a method to solve specific types of multi-objective geometric programming (MOGP) problems. In 1992, Bishal and in 1990 Vermahave presented a fuzzy programming technique to solve multi-objective geometric programming problems. In 2008 Islam presented the multi-objective marketing planning inventory model and solved it by geometric programming techniques. In 2010 Islam and Roy have considered multi-objective geometric programming (MOGP) problems and their applications. Das and Roy (2014) demonstrated multi-objective geometric programming and its application in a gravel box problem. Over the last 20 years, a tremendous number of research papers have expanded the theory and practice of multi-objective decision making problems.

Department of Mathematics, University of Kalyani, Kalyani, Nadia, W.B., India.
Email: sahidul.math@gmail.com, sahidulmath18@klyuniv.ac.in

Uncertainty theory is a new branch of mathematics established by Liu (2015). Liu proposed an uncertain stock model and a European option price formula in (2009). Following this, Peng and Yao (2011) considered another uncertain stock model and some option price formulas. Additionally, Liu (2010) and Wang et al. (2012) applied uncertainty theory to uncertain statistics. Risk analysis, reliability theory analysis, and control under uncertainty were introduced by Liu (2010, 2014) and Zhu (2010,1994), separately. Li et al. (2013) used risk as a non-negative uncertain variable and predominantly examined the uncertain risk within the framework of uncertainty theory. Hang et al.(2014)showed that uncertainty theory can fill in as an incredible asset to depict the maximum flow in a network under uncertainty. Ojha and Biswas (2014) introduced the ε-constraint strategy for taking care of multi-objective geometric programming problems. Ojha and Ota (2014) solved multi-objective geometric programming problems with Karush–Kuhn–Tucker conditions utilizing the ϵ-imperative method. Ding (2015) explained the maximum flow problem under uncertainty and formulated the maximum flow and the α-maximum flow in an uncertainty based framework. In 2016 Shiraz et. al. considered geometric programming with linear, normal and zigzag uncertainty and in 2017, Shiraz et. al. presented fuzzy chance-constrained programming under the possibility, credibility and necessity approaches. Yang and Bricker, (1997) introduced the path-following method for signomial geometric programming problems. Worral and Hall (1982) analyzed an inventory model by using geometric programming. Lin and Tsai, 2012b, 2012a; Tsai, 2009; Tsai et al., 2007, 2002; Tsai and Lin, 2006 did some interesting work on geometric programming. Scott and Jefferson, (1995) did research on allocation of resources in project management. Samadi et al., (2013) applied geometric programming approach in a fuzzy inventory model. Peng and Yao, (2011) did research on a new pricing model for uncertainty stocks markets. The book by Miettinen, (1999) on Non-linear Multi-objective Optimization is very useful. Maranas and Floudas, (1997) have done good research on global optimization in generalized geometric programming. Biswal, (1992) did an interesting research on fuzzy programming methods to solve some multi-objective mathematical programming problems.

In this chapter, we have used uncertain variables (UVs) to account for the ambiguity of the parameters characterizing real-world MOGP problems. Precisely, we explain three different chance-constrained MOGPs that can be executed when the coefficients considered are uncertain variables (UVs) with linear, normal and zigzag distributions. All proposed MOGPs under uncertainty can be converted into conventional MOGPs, and calculate optimal solutions by using their corresponding dual problems. We derive the Uncertain Chance Constrained Multi-Objective Geometric Programming (UCCMOGP) problem. Here we develop a solution process to solve this problem by using a conventional MOGP method on the basis of weighted-sum method. Finally, the solution process is illustrated by some examples.

3. Mathematical Preliminaries

3.1 Geometric Programming Problem

3.1.1 Unconstrained Problem

Primal Programming Problem:

$$Min\ \ g(t)=\sum_{j=1}^{T_0} c_j \prod_{i=1}^{m} t_i^{\alpha_{ji}} \tag{3.1}$$

subject to $t_i>0$, (i=1,2,….,m).

Here c_j (>0), α_{ji} (i=1,2,….,m; j=1,2,..,T_0) are real numbers.

Here Degrees of Difficulty GP = No. of terms in primal problem –(1+ No. of variables in primal problem). The problem (3.1) is unconstrained GP problem with DD = T_0-(m+1).

Dual Programming Problem (DPP):

$$Max\ \ v(\lambda)=\prod_{j=1}^{T_0}\left(\frac{c_j}{\lambda_j}\right)^{\lambda_j} \tag{3.2}$$

subject to

$\sum_{j=1}^{T_0}\lambda_j = 1$ (Normality condition)

$\sum_{j=1}^{T_0}\alpha_{ji}\lambda_j = 0$, (i=1,2,….,m) (Orthogonality conditions)

$\lambda_j > 0$, (j=1,2,…….,T_0). (Positivity conditions)

Where $\lambda = (\lambda_1, \lambda_2,\ldots\ldots, \lambda_{T_0})^T$.

Case I: DD = 0, i.e., T_0 = m+1. So, DPP gives a system of equations in dual variables and unique solution exists.

Case II: DD>0, i.e., $T_0 > m+1$. So, the DPP presents a system of equations in dual variables. Here the no. of dual variables is greater than the no. of linear equations. Many solutions exist. We have to use algorithmic methods, to find the optimal value of DPP.

Case III: DD<0 i.e., $T_0 < m+1$. So, the DPP presents a system of equations in dual variables. Here the no. of dual variables is less than the no. of linear equations. No solutions for dual variables exist. However, by using Least Square (LS) and Min-Max (MM) technique we can get approximate solution of this system.
Ones optimal value of dual vector λ^* is known, corresponding value of the primal vector t is found from the relations:

$$c_j \prod_{i=1}^{n} t_i^{\alpha_{ji}} = \lambda_j^* v^*(\lambda^*), \quad (j=1,2,\ldots,T_0) \tag{3.3}$$

Taking logarithm both side in (3), we get T_0 log-linear equations are as follows:

$$\sum_{i=1}^{n} \alpha_{ji} (\log t_i) = \log\left(\frac{\lambda_j^* v^*(\lambda^*)}{c_j} \right), \quad (j=1,2,\ldots,T_0) \tag{3.4}$$

This is the system of T_0 linear simultaneous equations in x_i (=log t_i) for i=1,2,...,m.

3.1.2 Constrained Problem

Primal Programming Problem:

$$Min \ \ g_0(t) = \sum_{j=1}^{T_0} c_{0j} \prod_{i=1}^{n} t_i^{\alpha_{0ji}} \tag{3.5}$$

subject to

$$g_r(t) = \sum_{j=1+T_{r-1}}^{T_r} c_{rj} \prod_{i=1}^{n} t_i^{\alpha_{rji}} \leq 1$$

$$t_i > 0, \ i=1,2,\ldots,n$$

where $c_{rj}(>0)$, α_{rji} ($j=1,2,..,1+T_{r-1},..,T_r; r=0,1,2,..,l$ $i=1,2,\ldots,n$) are real numbers.

It is a constrained GP problem. The no. of terms in each constraint function varies and denoted by T_r for each r=0,1,2,…,l. Let $T=T_0+T_1+T_2+\ldots\ldots\ldots+T_l$, the total no. of terms in the primal problem. Here, Degree of Difficulty =T-(n+1).

Dual Programming Problem (DPP):

$$\text{Maximize } Max \ \ v(\lambda) = \prod_{r=0}^{l} \prod_{j=1}^{T_r} \left(\frac{c_{rj}}{\lambda_{rj}} \right)^{\delta_{rj}} \left(\sum_{s=1+T_{r-1}}^{T_r} \lambda_{rs} \right)^{\delta_{rj}} \tag{3.6}$$

subject to

$\sum_{j=1}^{T_0} \lambda_{0j} = 1$, Normality condition

$\sum_{r=0}^{l} \sum_{j=1}^{T_r} \alpha_{rji} \lambda_r = 0$, (i=1,2,...,n) Orthogonality conditions

$\lambda_{rj} > 0$, ($r=0,1,2,\ldots,l$; $j=1,2,..,T_r$). Positivity conditions

Case I: T = n+1, So DPP presents a system of simultaneous equations in dual variables and unique solution exists.

Case II: T>n+1, So the DPP presents a system of simultaneous equations in dual variables. Here the no. of dual variables is greater than the no. of linear equations. Many solutions for dual variables exist.

Case III: T<n+1, So the DPP presents a system of simultaneous equations in dual variables. Here the no. of dual variables is less than the no. of linear equations. No solutions for dual variables exist. However, by using Least Square (LS) and Min-Max (MM) technique we can get approximate solution of this system.
The solution process of this constrained problem is same as unconstrained problem.

3.2 Some basic Concepts of Uncertainty Theory

Def^n 3.1: Let M : Γ [0,1] be a set function defined on the universal set Γ. M is called uncertain measure if and only if it holds the following axioms.

Axiom 1: $M(\Gamma) = 1$, (Normality)

Axiom 2: $\forall\, \Lambda \subseteq \Gamma$, $M(\Lambda) + M(\Lambda^c) = 1$, (Self-Duality)

Axiom 3: $\forall$ countable sequences of Λ_i (i = 1, 2,….., ∞) countable sequence $M(\cup_{i=1}^{\infty} \Lambda_i) \leq \sum_{i=1}^{\infty} M(\Lambda_i)$ (Countable sub-additivity).

Note: Axioms 1-3 also signify monotonicity (i.e., $M(\Lambda_2) \leq M(\Lambda_2)$ when $\Lambda_1 \leq \Lambda_2$).

Def^n 3.2: A triplet (Γ, Λ, M) is called the uncertainty space if and only if L is σ-algebra on and M is uncertain measure.

Def^n 3.3: The UV λ is non-negative if and only if $M\ \{\lambda < 0\} = 0$ and positive if and only if $M\ \{\lambda \leq 0\} = 0$.

Def^n 3.4: Let $\lambda_1, \lambda_2, \ldots \ldots \ldots, \lambda_n$ be UVs, then $\forall\, \Lambda \subseteq \Gamma$

$$(\lambda_1 + \lambda_2 + \cdots + \lambda_n)\Lambda = \lambda_1(\Lambda) + \lambda_1(\Lambda) + \cdots + \lambda_n(\Lambda) \text{ and } (\lambda_1 \,.\, \lambda_2 \,.\, \cdots \,.\, \lambda_n)\Lambda$$
$$= \lambda_1(\Lambda) \,.\, \lambda_2(\Lambda). \cdots \,.\, \lambda_n(\Lambda),$$

Proposition 3.1: If $\lambda_1, \lambda_2, \ldots \ldots \ldots, \lambda_n$ are UVs and f is a measurable real-valued function, then $f(\lambda_1, \lambda_2, \ldots \ldots \ldots, \lambda_n)$ is UV. Particularly, sum and product of UVs are UVs.

Def^n 3.5: Given UV λ, the function: ϕ_λ:IR $\rightarrow [0, 1]$, defined by $\phi_\lambda(x) := M\{\xi \leq x\}$ for each $x \in$ IR, is called uncertainty distribution (in short: UD) of λ.

Def^n 3.6: A UV λ is called linear if and only if it has linear UD.
Symbolically: $\phi_\lambda(t)$

$$\phi_\lambda(t) = \begin{cases} 0, & t > a \\ \dfrac{t-a}{b-a}, & a \leq t \leq b\ 1 \\ 1, & t > b \end{cases}$$

To indicate that ξ has a linear UD, we shall write ξ: $L(a, b)$.

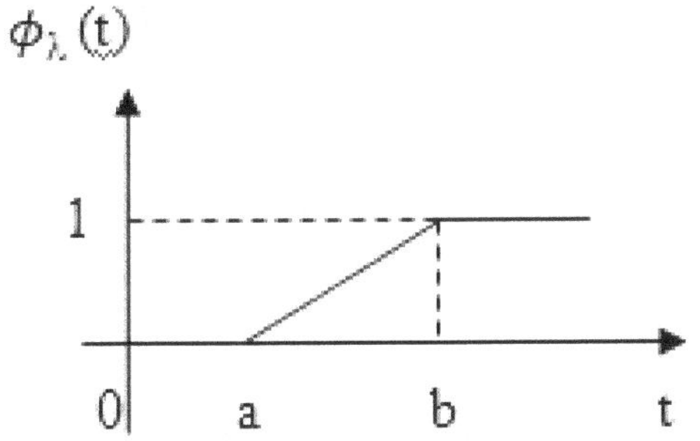

Figure 1: Linear UD.

Def^n 3.7: A UV λ is called normal if and only if it has normal UD.

Symbolically: $\phi_\lambda(t) = \left(1 + \exp\left(\dfrac{\pi(e - \log t)}{\sqrt{3}\sigma}\right)\right)^{-1}, \quad t \geq 0.$

To indicate that λ has normal UD, we write λ: $N(a, b)$.

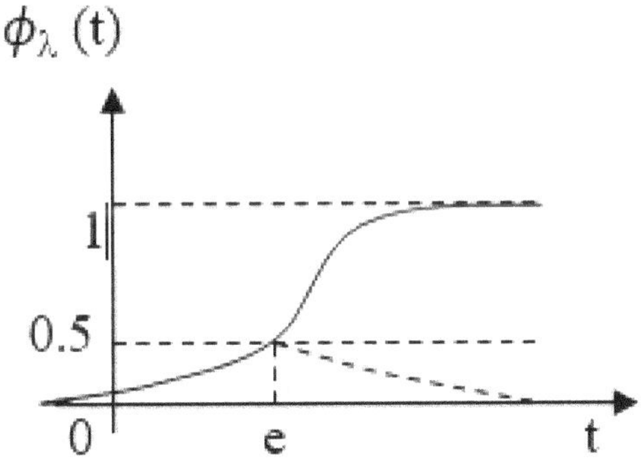

Figure 2: Normal UD.

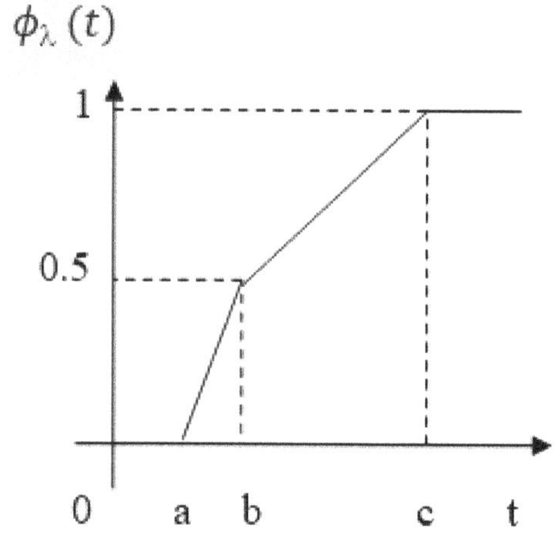

Figure 3: Zigzag UD.

Def^n^ 3.8: A UV λ is called zigzag if and only if it has zigzag UD. $\phi_\lambda(t)$
Symbolically:

$$\phi_\lambda(t) = \begin{cases} 0, & t < a \\ \frac{t-a}{2(b-a)}, & a \le t \le b \\ \frac{t+c-2b}{2(c-b)}, & b < t \le c \\ 1, & t > c \end{cases}$$

This shows λ has zigzag UD, we write λ: *Z(a, b, c)*.

Def^n^ 3.9: Let λ be UV. The expected value of λ is defined by $E(\lambda) = \int_0^\infty M(\lambda \ge r)dr - \int_{-\infty}^0 M(\lambda \le r)dr$, provided at least one of two integrals is finite.
It shows that $E(\lambda) = \int_0^\infty (1 - \phi_\lambda(r))\, dr - \int_{-\infty}^0 \phi_\lambda(r)dr$.

4. Mathematical Analysis

4.1 Multi-Objective Non-linear Programming Problem

A multi-objective geometric programming (MOGP) problem be written as

$$\text{Find } t = (t_1, t_2, \dots \dots \dots \dots \dots, t_m)^T \tag{4.1}$$

so as to

$\text{Min } f_{10}(t) = \sum_{j=1}^{p_{10}} c_{10j} \prod_{k=1}^{m} t_k^{\alpha_{k10j}}$

$\text{Min } f_{20}(t) = \sum_{j=1}^{p_{20}} c_{20j} \prod_{k=1}^{m} t_k^{\alpha_{k20j}}$

…….. …………… ………………… ……………..

$\text{Min } f_{n0}(t) = \sum_{j=1}^{p_{n0}} c_{n0j} \prod_{k=1}^{m} t_k^{\alpha_{kn0j}}$

subject to $f_r(t) = \sum_{j=1}^{p_r} c_{rj} \prod_{k=1}^{m} t_k^{\alpha_{krj}} \le c_r$, r=1,2,………,q,

$tk > 0$, k=1,2,…………,m,

where

c_{i0j} be +ve real numbers ∀ i=1, 2 …n; j=1, 2...,p_r,

α_{ki0j} and α_{krj} be real numbers ∀ k=1, 2 …m; i=1, 2 …n; j=1, 2.... p_r,

p_{j0} = No. of terms in the j_0th objective function.

p_r = No. of terms in the rth constraint.

c_r = Boundary value of the rth constraint.

In the above multi-objective geometric programming problem, there are m minimize objective functions, q inequality constraints and n strictly positive decision variables.

Here, we build up a MOGP problem under uncertainty whose related chance-constrained version admits an equivalent crisp formulation. To start with, we change the traditional MOGP problem in Equation (4.1) into an MOGP problem under uncertainty, where , $\widetilde{c_{i0j}}$, $\widetilde{c_{ir}}$ (i=1, 2 …n; j =1, 2...,p_r) be UVs. The above problem is as follows:

$$\text{Find } t = (t_1, t_1, \dots \dots \dots, t_m)^T \tag{4.2}$$

so as to

$\text{Min } f_{10}(t) = \sum_{j=1}^{P_{10}} \tilde{c}_{10j} \prod_{k=1}^{m} t_k^{\alpha_{k10j}}$

$\text{Min } f_{20}(t) = \sum_{j=1}^{P_{20}} \tilde{c}_{20j} \prod_{k=1}^{m} t_k^{\alpha_{k20j}}$

…….. …………… ………………… ……………..

$\text{Min } f_{n0}(t) = \sum_{j=1}^{P_{n0}} \tilde{c}_{n0j} \prod_{k=1}^{m} t_k^{\alpha_{kn0j}}$

subject to $f_r(t) = \sum_{j=1}^{P_r} \tilde{c}_{rj} \prod_{k=1}^{m} t^{\alpha_{krj}} \le c_r$, r=1,2,………,q,

$tk > 0$, k=1,2,…………,m,

where

$\tilde{c}_{i0j}$ = Uncertain positive real numbers ∀ i=1, 2 …n; j=1, 2..., p_r,

$\tilde{c}_{rj}$ = Uncertain boundary of rth constraint.

In the above multi-objective geometric programming problem, there are m minimize objective functions, q inequality constraints and n strictly positive decision variables.

In light of the model characterized by Equation (4.2) with constraints, we can formulate the following multi-objective GP problem, which is a variation of the UCCMOGP problem:

$$\text{Find } t = (t_1, t_1, \dots \dots \dots, t_m)^T \tag{4.3}$$

so as to

$\text{Min } E(f_{10})(t)) = E[\sum_{j=1}^{P_{10}} \tilde{c}_{10j} \prod_{k=1}^{m} t_k^{\alpha_{k10j}}]$,

$\text{Min } E(f_{20})(t)) = E[\sum_{j=1}^{P_{20}} \tilde{c}_{20j} \prod_{k=1}^{m} t_k^{\alpha_{k20j}}]$,

…….. …………… ………………… ……………..

$\text{Min } E(f_{n0})(t)) = E[\sum_{j=1}^{P_{n0}} \tilde{c}_{n0j} \prod_{k=1}^{m} t_k^{\alpha_{kn0j}}]$,

subject to $m(f_r(t)) = m\,(\sum_{j=1}^{P_r} \tilde{c}_{rj} \prod_{k=1}^{m} t^{\alpha_{krj}} \le c_r) \ge \alpha$, r=1,2,………,q,

$tk > 0$, k=1,2,…………,m, $\alpha \in]0,1[$.

4.2 Proposed UCCMOGP Problem using Uncertainty Linear Distributions

Let coefficients $\tilde{c}_{i0j}$, $\tilde{c}_{rj}$ in Equation (4.3) be positive linear independent UVs. That is, $\tilde{c}_{i0j} : L(c^a{}_{i0j}, c^b{}_{i0j})$, with $0 < c^a{}_{i0j} < c^b{}_{i0j}$ and $\tilde{c}_{rj} : L(c^a{}_{rj}, c^b{}_{rj})$, with $0 < c^a{}_{rj} < c^b{}_{rj}$.

Lemma 4.1 Let λ_i (i=1, ..., n) be linear independent UVs, then we say $\tilde{\lambda}_i$: L(a_i, b_i) with $a_i < b_i$. If U_i be non-negative variables. Then

$$M\left(\sum_{i=1}^{n} \tilde{\lambda}_i U_i \le 1\right) \ge \alpha \Leftrightarrow \sum_{i=1}^{n} ((1-\alpha)a_i + \alpha b_i)U_i \le 1, \text{ for every } \alpha \in]0,1[.$$

Lemma 4.2 $E(\tilde{\lambda}) = [\frac{a+b}{2}]$ for linear UV $\tilde{\lambda} : L(a, b)$.

We obtain the following objective function of the UCCMOGP problem in Equation (4.3) by using Lemma 4.2

$$E\left[\sum_{j=1}^{p_{i0}} \tilde{c}_{i0j} \prod_{k=1}^{m} t_k^{\alpha_{ki0j}}\right] = \sum_{j=1}^{p_{i0}} \left(\tilde{c}_{i0j}\right) \prod_{k=1}^{m} t_k^{\alpha_{ki0j}} = \sum_{j=1}^{p_{i0}} \left(\frac{c^a{}_{i0j} + c^b{}_{i0j}}{2}\right) \prod_{k=1}^{m} t_k^{\alpha_{ki0j}}.$$

$$i = 1, 2, \dots \dots \dots n.$$

The constraints in Equation (4.3) reduced to the following deterministic form by using Lemma 4.1

$$M\left(\sum_{j=1}^{p_r}\tilde{c}_{rj}\prod_{k=1}^{m}t_k^{\alpha_{krj}}\le c_r\right)\ge\alpha\Leftrightarrow\sum_{j=1}^{p_r}((1-\alpha)c^a_{rj}+\alpha c^a_{rj})\prod_{k=1}^{m}t_k^{\alpha_{krj}}\le 1.$$

When the coefficients of the problem are UVs enriched with uncertain linear distributions, then the problem in Equation (4.3) is reduced to:

$$\text{Find } t=(t_1, t_2,\ldots\ldots\ldots, t_m)^T \tag{4.4}$$

so as to

$$\text{Min } E(f_{10}(t))=\sum_{j=1}^{p_{10}}\left(\frac{c^a_{10j}+c^b_{10j}}{2}\right)\prod_{k=1}^{m}t^{\alpha_{k10j}}$$

$$\text{Min } E(f_{20}(t))=\sum_{j=1}^{p_{20}}\left(\frac{c^a_{20j}+c^b_{20j}}{2}\right)\prod_{k=1}^{m}t^{\alpha_{k20j}}$$

....

$$\text{Min } E(f_{n0}(t))\sum_{j=1}^{p_{n0}}\left(\frac{c^a_{n0j}+c^b_{n0j}}{2}\right)\prod_{k=1}^{m}t^{\alpha_{kn0j}}$$

subject to $\sum_{j=1}^{p_r}((1-\alpha)\,c^a_{rj}, \alpha c^a_{rj})\prod_{k=1}^{m}t_k^{\alpha_{krj}}\le 1$, r=1, 2,.........,q,

$t_k > 0$, k=1,2,............,m, $\alpha \in\,]0,1[$.

4.2.1 Optimal Solution of UCCMOGP problem by using weighted-sum method

Let w be set of positive weights. The above multi-objective problem (4.4) can be reduced to a single-objective geometric programming problem by using the weighted sum technique as follows

$$E(f(t))=\sum_{i=1}^{n}\sum_{j=1}^{p_{i0}}\left(\frac{c^a_{i0j}+c^b_{i0j}}{2}\right)\prod_{k=1}^{m}t_k^{\alpha_{ki0j}} \tag{4.5}$$

subject to $\sum_{j=1}^{p_r}((1-\alpha)\,c^a_{rj}, \alpha c^a_{rj})\prod_{k=1}^{m}t_k^{\alpha_{krj}}\le 1$,

$t_k > 0$, $\alpha \in\,]0,1[$, r=1, 2,.........,q, k=1,2,............,m.

Def^n^ 4.1 The feasible solution t^* is said to be a Pareto optimal solution of the problem under uncertainty (4.5), if there exists no feasible solution t such that

$E[f(t, \xi)] \le E[f(t^*, \xi)]$,

and $E[f(t, \xi)] < E[f(t, \xi)]$, for at least one index.

Def^n^ 4.2 The feasible solution t^* is said to be a weak Pareto optimal solution of the problem under uncertainty (4.5), if there exists no solution t such that

$E[f(t, \xi)] < E[f(t, \xi)]$,

Theorem 4.1 The solution of the problem (4.5) under uncertainty based on weighted-sum method is Pareto optimal solution if $w_i > 0$ for all i = 1,2... ... ..., n.

Proof: Let x^* be the feasible solution of the problem (4.5), obtained by minimizes the function $(f(t))\ \sum_{i=1}^{n} w_i\, E(f_{i0}(t)) = \sum_{i=1}^{n} w_i \sum_{j=1}^{p_{i0}}\left(\frac{c^a_{i0j}+c^b_{i0j}}{2}\right)\prod_{k=1}^{m}t_k^{\alpha_{ki0j}}$. It follows $Ef(t^*, \xi\lambda)] \le Ef(t, \lambda)$, which implies

$$\begin{gathered}\sum_{i=1}^{n}w_i\sum_{j=1}^{p_{i0}}\left(\frac{c^a_{i0j}+c^b_{i0j}}{2}\right)\prod_{k=1}^{m}t_k^{*\,\alpha_{ki0j}}\le\sum_{i=1}^{n}w_i\sum_{j=1}^{p_{i0}}\left(\frac{c^a_{i0j}+c^b_{i0j}}{2}\right)\prod_{k=1}^{m}t_k^{*\,\alpha_{ki0j}}\\ \Leftrightarrow\sum_{i=1}^{n}w_i\sum_{j=1}^{p_{i0}}\left(\frac{c^a_{i0j}+c^b_{i0j}}{2}\right)\prod_{k=1}^{m}\left(t_k^{\alpha_{ki0j}}-t_k^{*\,\alpha_{ki0j}}\right)\ge 0.\end{gathered} \tag{4.6}$$

Suppose that the solution t^* of problem (4.5) is not Pareto optimal solution. Then there exists solution t' of problem (4.5) satisfies $Ef_{i0}(t', \lambda) \leq Ef_{i0}(t^*, \lambda)$, which implies

$$Ef_{i0}(t', \lambda) \leq Ef_{i0}(t^*, \lambda) < 0 \;\forall\; i = 1,2.........,n.$$

$$\sum_{i=1}^{n}\sum_{j=1}^{p_{i0}}\left(\frac{c^a_{i0j}+c^b_{i0j}}{2}\right)\prod_{k=1}^{m} t'^{\alpha_{ki0j}}_k - \sum_{j=1}^{p_{i0}}\left(\frac{c^a_{i0j}+c^b_{i0j}}{2}\right)\prod_{k=1}^{m} t^{*\,\alpha_{ki0j}}_k < 0$$

$$\Leftrightarrow \sum_{j=1}^{p_{i0}}\left(\frac{c^a_{i0j}+c^b_{i0j}}{2}\right)\prod_{k=1}^{m}\left(t'_k - t^*_k\right)^{\alpha_{ki0j}} < 0.$$

By summing the above inequalities and taking the assumption of the theorem that the weights w_i are all positive, we obtain

$$\sum_{i=1}^{n} w_i \sum_{j=1}^{p_{i0}}\left(\frac{c^a_{i0j}+c^b_{i0j}}{2}\right)\prod_{k=1}^{m}\left(t'^{\alpha_{ki0j}}_k - t^{*\,\alpha_{ki0j}}_k\right) < 0.$$

The above inequality contradicts the inequality (4.6). Thus, the solution t^* is a Pareto optimal solution for $w_i > 0$.

4.3 Proposed UCCMOGP Problem using Uncertainty Normal Distributions

Let coefficients $\tilde{c}_{i0j}$, $\tilde{c}_{rj}$ in Equation (4.3) be positive independent normal UVs. That is, $\tilde{c}_{i0j} : N(c_{i0j}, \sigma_{i0j})$, and $\tilde{c}_{rj} : N(c_{rj}, \sigma_{rj})$, where c_{i0j}, c_{rj}, σ_{i0j} are σ_{rj} all positive real values.

Lemma 4.3 Let λ_i (i=1, ..., n) be normal independent UVs, then we say $\tilde{\lambda}_i : N(\lambda_i, \sigma_i)$ where λ_i, σ_i are all +ve real numbers. Then

$$M\left(\sum_{i=1}^{n}\tilde{\lambda}_i U_i \leq 1\right) \geq \alpha \Leftrightarrow \sum_{i=1}^{n}\left(\xi_i + \frac{\sigma_i\sqrt{3}}{\pi}log\left(\frac{\alpha}{1-\alpha}\right)\right)U_i \leq 1, \text{ for every } \alpha \in]0,1[.$$

Lemma 4.4 $E(\tilde{\lambda}) = e$ for a normal UV $\tilde{\lambda} : N(e, \sigma)$

We obtain the following objective function of the UCCMOGP problem in Equation (4.3) by using Lemma 4.4

$$E\left[\sum_{j=1}^{p_{i0}}\tilde{c}_{i0j}\prod_{k=1}^{m} t_k^{\alpha_{ki0j}}\right] = \sum_{j=1}^{p_{i0}} E\left(\tilde{c}_{i0j}\right)\prod_{k=1}^{m} t_k^{\alpha_{ki0j}} = \sum_{j=1}^{p_{i0}} c_{i0j}\prod_{k=1}^{m} t_k^{\alpha_{ki0j}}.\ i = 1,2,........n.$$

The constraints in Equation (4.3) reduced to the following deterministic form by using Lemma 4.3

$$M\left(\sum_{j=1}^{p_r}\tilde{c}_{rj}\prod_{k=1}^{m} t_k^{\alpha_{krj}} \leq c_r\right) \geq \alpha \Leftrightarrow \sum_{j=1}^{p_r}\left(c_{rj} + \frac{\sigma_{rj}\sqrt{3}}{\pi}log\left(\frac{\alpha}{1-\alpha}\right)\right)\prod_{k=1}^{m} t_k^{\alpha_{krj}} \leq 1.$$

When the coefficients of the problem are UVs enriched with uncertain normal distributions, then the problem in Equation (4.3) is reduced to:

$$\text{Find } t = (t_1, t_2, \ldots \ldots \ldots \ldots, t_m)^T \tag{4.7}$$

so as to

Min $E(f_{10})(t)) = \sum_{j=1}^{P_{10}} (c_{10j}) \prod_{k=1}^{m} t_k^{\alpha_{k10j}}$

Min $E(f_{20})(t)) = \sum_{j=1}^{P_{20}} (c_{20j}) \prod_{k=1}^{m} x_k^{\alpha_{k20j}}$

....

Min $E(f_{n0})(t)) = \sum_{j=1}^{P_{n0}} (c_{n0j}) \prod_{k=1}^{m} t_k^{\alpha_{kn0j}}$

subject to $\sum_{j=1}^{P_r}\left(c_{rj} + \frac{\sigma_{rj} + \sqrt{3}}{\pi}\, log\,(\frac{\alpha}{1-\alpha})\right) \prod_{k=1}^{m} t_k^{\alpha_{krj}} \leq 1$, r=1, 2,.........,q,

$t_k > 0$, k=1,2,...........,m, $\alpha \in\ ]0,1[$.

4.3.1 Optimal solution of UCCMOGP problem using weighted-sum method

Let $w = (w_i;\ w \in \mathbb{R}^n,\ w_i > 0,\ \sum_{i=1}^{n} w_i = 1)$ be set of positive weights. The above multi-objective problem (4.7) can be reduced to a single-objective geometric programming problem by using the weighted sum technique as follows

$$\text{Min } \sum_{i=1}^{n} w_i \sum_{j=1}^{P_{i0}} (c_{i0j}) \prod_{k=1}^{m} t_k^{\alpha_{ki0j}} \tag{4.8}$$

$$\text{subject to } \sum_{j=1}^{P_r} \left(c_{rj} + \frac{\sigma_{rj} + \sqrt{3}}{\pi} \log\left(\frac{\alpha}{1-\alpha}\right)\right) \prod_{k=1}^{m} t_k^{\alpha_{krj}} \le 1,$$

$t_k > 0$, $\alpha \in]0,1[$, r=1, 2,.........,q, k=1,2,............,m.

4.4 Proposed UCCMOGP Problem using Uncertainty Zigzag Distributions

Let coefficients $\tilde{c}_{i0j}$, $\tilde{c}_{rj}$ in Equation (4.3) are positive independent zigzag UVs. That is, $\tilde{c}_{i0j} : Z(c^a_{i0j}, c^b_{i0j}, c^c_{i0j})$ with $0 < c^a_{i0j} < c^b_{i0j} < c^c_{i0j}$ and $\tilde{c}_{rj} : Z(c^a_{rj}, c^b_{rj}, c^c_{rj})$ with $0 < c^a_{rj} < c^b_{rj} < c^c_{rj}$.

Lemma 4.5 Let λ_i *(i*=1, ..., *n)* be zigzag independent UVs, then we say, $\tilde{\lambda}_i : Z(a_i, b_i, c_i)$ where $a_i < b_i < c_i$ and U_i be non-negative variables. Then

$$M\left(\sum_{i=1}^{n} \tilde{\lambda}_i U_i \le 1\right) \ge \alpha \Leftrightarrow \begin{cases} \sum_{i=1}^{n} ((1-2\alpha)a_i + 2\alpha b_i)U_i \le 1, & \text{if } \alpha \in]0, 0.5[; \\ \sum_{i=1}^{n} ((2\alpha-1)c_i + 2(1-\alpha)b_i)U_i \le 1, & \text{if } \alpha \in]0.5, 1[. \end{cases}$$

Lemma 4.6 $E(\tilde{\xi}) = [\frac{a+2b+c}{4\alpha}]$ for a zigzag *UV* $\tilde{\lambda} : Z(a, b, c)$.

We obtain the following objective function of the UCCMOGP problem in Equation (4.3) by using Lemma 4.6

$$E\left[\sum_{j=1}^{P_{i0}} \tilde{c}_{i0j} \prod_{k=1}^{m} t_k^{\alpha_{ki0j}}\right] = \sum_{j=1}^{P_{i0}} E(\tilde{c}_{i0j}) \prod_{k=1}^{m} t_k^{\alpha_{ki0j}} = \sum_{j=1}^{P_{i0}} \left(\frac{a+2b+c}{4}\right) \prod_{k=1}^{m} t_k^{\alpha_{ki0j}}.\ i = 1,2,\ldots\ldots n.$$

The constraints in Equation (4.3) reduced to the following deterministic form by using Lemma 4.5

$$\forall i = 1,\ldots,n,$$

$$M\left(\sum_{j=1}^{P_r} \tilde{c}_{rj} \prod_{k=1}^{m} t_k^{\alpha_{krj}} \le c_r\right) \ge \alpha \Leftrightarrow$$

$$\begin{cases} \sum_{j=1}^{P_r} ((1-2\alpha)c^a_{rj} + 2\alpha c^b_{rj}) \prod_{k=1}^{m} t_k^{\alpha_{krj}} \le 1, & \text{if } \alpha \in]0, 0.5[; \\ \sum_{j=1}^{P_r} ((2\alpha-1)c^c_{rj} + 2(1-\alpha)c^b_{rj}) \prod_{k=1}^{m} t_k^{\alpha_{krj}} \le 1, & \text{if } \alpha \in]0.5, 1[. \end{cases}$$

When the coefficients of the problem are UVs enriched with uncertain zigzag distributions, then the problem in Equation (4.3) is reduced to:

<u>When $\alpha < 0.5$</u>

$$\text{Find } t = (t_1, t_2, \ldots \ldots \ldots \ldots, t_m)^T \tag{4.9}$$

so as to

$$\text{Min } E(f_{10})(t)) = \sum_{j=1}^{P_{10}} \left(\frac{c^a_{10j} + 2c^b_{10j} + c^c_{10j}}{4}\right) \prod_{k=1}^{m} t_k^{\alpha_{k10j}}$$

$$\text{Min } E(f_{20})(t)) = \sum_{j=1}^{P_{20}} \left(\frac{c^a_{20j} + 2c^b_{20j} + c^c_{20j}}{4}\right) \prod_{k=1}^{m} t_k^{\alpha_{k20j}}$$

....

$$\text{Min } E(f_{n0})(t)) = \sum_{j=1}^{P_{n0}} \left(\frac{c^a_{n0j} + 2c^b_{n0j} + c^c_{n0j}}{4}\right) \prod_{k=1}^{m} t_k^{\alpha_{kn0j}}$$

subject to $\sum_{j=1}^{P_r} ((1-2\alpha)c^a_{rj} + 2\alpha c^b_{rj}) \prod_{k=1}^{m} t_k^{\alpha_{krj}} \le 1$, r=1, 2,.........,q. k=1,2,............,m,

$t_k > 0$, $\alpha \in]0,1[$.

When $\alpha > 0.5$

$$\text{Find t } (t_1, t_2, \ldots \ldots \ldots \ldots, t_m)^T \tag{4.10}$$

so as to

$\text{Min } E(f_{10})(t)) = \sum_{j=1}^{p_{10}} (\frac{c^a{}_{10j} + 2c^b{}_{10j} + c^c{}_{10j}}{4}) \prod_{k=1}^{m} t_k{}^{\alpha_{k10j}}$

$\text{Min } E(f_{20})(t)) = \sum_{j=1}^{p_{20}} (\frac{c^a{}_{20j} + 2c^b{}_{20j} + c^c{}_{20j}}{4}) \prod_{k=1}^{m} t_k{}^{\alpha_{k20j}}$

………………………………………………

$\text{Min } E(f_{n0})(t)) = \sum_{j=1}^{p_{n0}} (\frac{c^a{}_{n0j} + 2c^b{}_{n0j} + c^c{}_{n0j}}{4}) \prod_{k=1}^{m} t_k{}^{\alpha_{kn0j}}$

subject to $\sum_{j=1}^{p_r} ((1-2\alpha)c^c{}_{rj} + 2(1-\alpha)c^b{}_{rj}) \prod_{k=1}^{m} t_k{}^{\alpha_{krj}} \leq 1,$

$t_k > 0,\ \alpha \in]0,1[,\ r=1, 2,\ldots\ldots\ldots,q;\ k=1,2,\ldots\ldots\ldots\ldots,m.$

4.4.1 Optimal Solution of UCCMOGP problem using weighted-sum method

Let $w = (w_i;: w \in \mathbb{R}^n, w_i > 0, \sum_{i=1}^{n} w_i = 1)$ be set of positive weights. The above multi-objective problems ((4.9) & (4.10)) can be reduced to a single-objective geometric programming problem by using the weighted sum technique as follows:

When $\alpha < 0.5$

$$\text{Min } \sum_{i=1}^{n} w_i \sum_{j=1}^{p_{i0}} (\frac{c^a{}_{10j} + 2c^b{}_{10j} + c^c{}_{10j}}{4}) \prod_{k=1}^{m} t_k{}^{\alpha_{ki0j}} \tag{4.11}$$

subject to $\sum_{j=1}^{p_r} ((1-2\alpha)c^a{}_{rj} + 2\alpha c^b{}_{rj}) \prod_{k=1}^{m} t_k{}^{\alpha_{krj}} \leq 1,$

$t_k > 0,\ \alpha \in]0,1[,\ r=1, 2,\ldots\ldots\ldots,q;\ k=1,2,\ldots\ldots\ldots\ldots,m.$

When $\alpha > 0.5$

$$\text{Min } \sum_{i=1}^{n} w_i \sum_{j=1}^{p_{i0}} (\frac{c^a{}_{10j} + 2c^b{}_{10j} + c^c{}_{10j}}{4}) \prod_{k=1}^{m} t_k{}^{\alpha_{ki0j}} \tag{4.12}$$

subject to $\sum_{j=1}^{p_r} ((2\alpha-1)c^c{}_{rj} + (1-2\alpha)c^b{}_{rj}) \prod_{k=1}^{m} t_k{}^{\alpha_{krj}} \leq 1,$

$t_k > 0,\ \alpha \in]0,1[,\ r=1, 2,\ldots\ldots\ldots,q;\ k=1,2,\ldots\ldots\ldots\ldots,n.$

5. Numerical Examples

Example:

$$\text{Min } f_{10}(t) = \frac{\widetilde{c_{101}}}{t_1 t_2 t_3} + \widetilde{c_{102}} t_2 t_3 \tag{5.1}$$

$\text{Min } f_{20}(t) = \frac{\widetilde{c_{201}}}{t_1 t_2 t_3}$

subject to $\widetilde{c_{11}}\, t_1\, t_2 + \widetilde{c_{12}}\, t_1\, t_3 \leq 4,$

$t_1, t_2, t_3 > 0.$

5.1 Using Uncertainty Linear Distributions

$\widetilde{c_{101}} : L(40, 60),\ \widetilde{c_{102}} : L(40, 60),\ \widetilde{c_{201}} : L(600, 800),\ \widetilde{c_{11}} : L(0.7, 1.1),\ \widetilde{c_{12}} : L(1.7, 2.5).$

Hence the UCCMOGP problem reduces to

$$\text{Min} f_{10}(t) = \frac{50}{t_1 t_2 t_3} \quad 50t_2 t_3 \tag{5.2}$$

$$\text{Min } f_{10}(t) = \frac{700}{t_1 t_2 t_3}$$

subject to

$$(0.7(1-\alpha) + 1.1\alpha)t_1 t_2 + (1.7(1-\alpha) + 2.5\alpha)\, t_1 t_3 \le 4, \quad t_1, t_2, t_3 > 0.$$

Using the weighted-sum technique, the above UCCMOGP problem (5.2) becomes the single-objective geometric programming problem as follows:

$$\text{Min } f(t) = w_1\left(\frac{50}{t_1 t_2 t_3} + 50t_2 t_3\right) + w_2 \frac{700}{t_1 t_2 t_3} + \frac{50w_1 + 700w_2}{t_1 t_2 t_3} + 50w_1 t_2 t_3 \tag{5.3}$$

subject to

$$(0.7(1-\alpha) + 1.1\alpha)t_1 t_2 + (1.7(1-\alpha) + 2.5\alpha)\, t_1 t_3 \le 4, \quad t_1, t_2, t_3 > 0.$$

Here, Degree of Difficulty (DD) = 4 – (3 + 1) = 0.

Dual Problem of the above problem (5.3) is

$$\text{Maximize } d(\lambda) = \left(\frac{50w_1 + 700w_2}{\lambda_{01}}\right)^{\lambda_{01}} \left(\frac{50w_1}{\lambda_{02}}\right)^{\lambda_{02}} \left(\frac{(0.7(1-\alpha)+1.1\alpha}{4\lambda_{11}}\right)^{\lambda_{11}} \left(\frac{(1.7(1-\alpha)+2.5\alpha}{4\lambda_{12}}\right)^{\lambda_{12}} (\lambda_{11} + \lambda_{12})^{(\lambda_{11}+\lambda_{12})}$$

subject to

$$\lambda_{01} + \lambda_{02} = 1,$$

$$-\lambda_{01} + \lambda_{11} + \lambda_{12} = 0,$$

$$-\lambda_{01} + \lambda_{02} + \lambda_{11} = 0,$$

$$-\lambda_{01} + \lambda_{02} + \lambda_{12} = 0,$$

$$w_1 + w_2 = 1,$$

$$\lambda_{01}, \lambda_{02}, \lambda_{11}, \lambda_{12} > 0.$$

Solving the above system of linear equations, we have

$$\lambda_{01} = \frac{2}{3}, \lambda_{02} = \frac{1}{3}, \lambda_{11} = \frac{1}{3}, \lambda_{12} = \frac{1}{3}.$$

From the primal-dual relationship, we obtain

$$\frac{50w_1 + 700w_2}{t_1 t_2 t_3} = \lambda_{01} d(\lambda),$$

$$50w_1 t_2 t_3 = \lambda_{02} d(\lambda),$$

$$\frac{(0.7(1-\alpha)+1.1\alpha)t_1 t_2}{4} = \frac{\lambda_{12}}{\lambda_{11} + \lambda_{12}},$$

$$\frac{(1.7(1-\alpha)+2.5\alpha)t_1 t_3}{4} = \frac{\lambda_{12}}{\lambda_{11} + \lambda_{12}},$$

And the corresponding optimal solution is $t_3 = \left(\frac{w_1 + 25w_2}{4(1.7(1-\alpha)+2.5\alpha))w_1}\right)^{\frac{1}{3}}, t_2 = 2t_3, t_1 = \frac{2}{(1.7(1-\alpha)+2.5\alpha)t_3}.$

The optimal solutions of the UCCMOGP problem (5.2) under Linear Uncertainty Distributions are given in the following Table-1.

Table 1: Optimal solutions when using Uncertainty Linear Distributions.

α	Weights	Optimal solutions of primal variables			Optimal objective values	
		t_1^*	t_2^*	t_3^*	$f_{01}^*(t)$	$f_{02}^*(t)$
0.2	$w_1 = 0.1, w_2 = 0.9$	0.38	5.90	2.95	702.25	120.96
	$w_1 = 0.5, w_2 = 0.5$	0.79	2.88	1.44	178.10	244.18
	$w_1 = 0.9, w_2 = 0.1$	1.48	1.54	0.77	70.22	455.84
0.4	$w_1 = 0.1, w_2 = 0.9$	0.36	5.74	2.87	665.70	134.90
	$w_1 = 0.5, w_2 = 0.5$	0.74	2.80	1.40	170.59	275.79
	$w_1 = 0.9, w_2 = 0.1$	1.39	1.50	0.75	70.58	511.59
0.6	$w_1 = 0.1, w_2 = 0.9$	0.34	5.58	2.79	630.28	151.14
	$w_1 = 0.5, w_2 = 0.5$	0.71	2.72	1.36	163.20	304.60
	$w_1 = 0.9, w_2 = 0.1$	1.32	1.46	0.73	71.06	568.64
0.8	$w_1 = 0.1, w_2 = 0.9$	0.33	5.44	2.72	600.06	163.84
	$w_1 = 0.5, w_2 = 0.5$	0.67	2.66	1.33	158.39	337.51
	$w_1 = 0.9, w_2 = 0.1$	1.26	1.42	0.71	71.82	629.76

5.2 Using Uncertainty Normal Distributions

$\widetilde{c_{101}}: N(50, 5),\ \widetilde{c_{102}}: N(50, 5),\ \widetilde{c_{201}}: N(700, 70),\ \widetilde{c_{11}}: N(2, 0.2),\ \widetilde{c_{12}}: N(3,0.3).$

Hence the UCCMOGP problem reduces to

$$\text{Min}\, f_{10}(t) = \frac{50}{t_1t_2t_3} + 50t_2t_3 \tag{5.4}$$

$$\text{Min}\, f_{10}(t) = \frac{700}{t_1t_2t_3}$$

subject to $\left(2 + \frac{0.2\sqrt{3}}{\pi} log\left(\frac{\alpha}{1-\alpha}\right)\right)x_1x_2 + \left(3 + \frac{0.3\sqrt{3}}{\pi} log\left(\frac{\alpha}{1-\alpha}\right)\right)t_1t_3 \le 4,\ t_1, t_2, t_3 > 0.$

Using the weighted-sum technique, the above UCCMOGP problem (5.4) becomes the single-objective geometric programming problem as follows:

$$\text{Min}\ f(t) = w_1\left(\frac{50}{t_1t_2t_3} + 50t_2t_3\right) + w_2\frac{700}{t_1t_2t_3} + \frac{50w_1 + 700w_2}{t_1t_2t_3} + 50w_1t_2t_3 \tag{5.5}$$

subject to

$$\left(2 + \frac{0.2\sqrt{3}}{\pi} log\left(\frac{\alpha}{1-\alpha}\right)\right)t_1t_2 + \left(3 + \frac{0.3\sqrt{3}}{\pi} log\left(\frac{\alpha}{1-\alpha}\right)\right)t_1t_3 \le 4,\ t_1, t_2, t_3 > 0.$$

Here, Degree of Difficulty (DD) = 4 – (3 + 1) = 0.

Dual problem of the above problem (5.5) is

Maximize

$$d(\lambda) = \left(\frac{50w_1 + 700w_2}{\lambda_{01}}\right)^{\lambda_{01}} \left(\frac{50w_1}{\lambda_{02}}\right)^{\lambda_{02}} \left(\frac{\left(2 + \frac{0.2\sqrt{3}}{\pi} log\left(\frac{\alpha}{1-\alpha}\right)\right)}{4\lambda_{11}}\right)^{\lambda_{11}} \left(\frac{\left(3 + \frac{0.3\sqrt{3}}{\pi} log\left(\frac{\alpha}{1-\alpha}\right)\right)}{4\lambda_{12}}\right)^{\lambda_{12}} (\lambda_{11} + \lambda_{12})^{(\lambda_{11}+\lambda_{12})}$$

subject to

$$\lambda_{01} + \lambda_{02} = 1,$$

$$-\lambda_{01} + \lambda_{11} + \lambda_{12} = 0,$$

$$-\lambda_{01} + \lambda_{02} + \lambda_{11} = 0,$$

$$-\lambda_{01} + \lambda_{02} + \lambda_{12} = 0,$$

$$w_1 + w_2 = 1,$$

$$\lambda_{01}, \lambda_{02}, \lambda_{11}, \lambda_{12} > 0.$$

Solving the above system of linear equations, we have

$$\lambda_{01} = \frac{2}{3}, \lambda_{02} = \frac{1}{3}, \lambda_{11} = \frac{1}{3}, \lambda_{12} = \frac{1}{3}.$$

From the primal-dual relationship, we obtain

$$\frac{50w_1 + 700w_2}{t_1 t_2 t_3} = \lambda_{01} d(\lambda),$$

$$50w_1 t_2 t_3 = \lambda_{02} d(\lambda),$$

$$\frac{\left(2 + \frac{0.2\sqrt{3}}{\pi} log\left(\frac{\alpha}{1-\alpha}\right)\right) t_1 t_2}{4} = \frac{\lambda_{11}}{\lambda_{11} + \lambda_{12}},$$

$$\frac{\left(3 + \frac{0.3\sqrt{3}}{\pi} log\left(\frac{\alpha}{1-\alpha}\right)\right) t_1 t_3}{4} = \frac{\lambda_{12}}{\lambda_{11} + \lambda_{12}},$$

And the corresponding optimal solutions of primal variables are as follows

$$t_3 = \left(\frac{w_1 + 25w_2}{4\left(3 + \frac{0.3\sqrt{3}}{\pi} log\left(\frac{\alpha}{1-\alpha}\right)\right) w_1}\right)^{\frac{1}{3}}, t_2 = 2t_3, t_1 = \frac{2}{\left(3 + \frac{0.3\sqrt{3}}{\pi} log\left(\frac{\alpha}{1-\alpha}\right)\right) t_3}.$$

The optimal solutions of the UCCMOGP problem (5.4) under Normal Uncertainty Distributions are given in the following Table 2.

Table 2: Optimal solutions when using Uncertainty Normal Distributions.

α	**Weights**	**Optimal solutions of primal variables**			**Optimal objective values**	
		t_1^*	t_2^*	t_3^*	$f_{01}^*(t)$	$f_{02}^*(t)$
0.2	$w_1 = 0.1, w_2 = 0.9$	0.37	5.80	2.90	679.23	128.55
	$w_1 = 0.5, w_2 = 0.5$	0.76	2.84	1.42	174.36	261.02
	$w_1 = 0.9, w_2 = 0.1$	1.42	1.52	0.76	70.59	487.69
0.4	$w_1 = 0.1, w_2 = 0.9$	0.38	5.70	2.85	656.28	129.59
	$w_1 = 0.5, w_2 = 0.5$	0.78	2.78	1.39	167.84	265.42
	$w_1 = 0.9, w_2 = 0.1$	1.46	1.48	0.74	68.49	500.32
0.6	$w_1 = 0.1, w_2 = 0.9$	0.39	5.62	2.81	638.18	129.89
	$w_1 = 0.5, w_2 = 0.5$	0.79	2.74	1.37	163.64	269.77
	$w_1 = 0.9, w_2 = 0.1$	1.48	1.46	0.73	67.99	507.17
0.8	$w_1 = 0.1, w_2 = 0.9$	0.39	5.52	2.76	616.14	134.64
	$w_1 = 0.5, w_2 = 0.5$	0.80	2.70	1.35	159.52	274.35
	$w_1 = 0.9, w_2 = 0.1$	1.50	1.44	0.72	67.19	514.40

5.3 Using Uncertainty Zigzag Distributions

$\widetilde{c_{101}}$: $Z(20, 50, 60)$, $\widetilde{c_{102}}$: $Z(20, 40, 60)$, $\widetilde{c_{201}}$: $Z(600, 900, 1000)$, $\widetilde{c_{11}}$: $Z(0.9, 1.1, 1.2)$, $\widetilde{c_{12}}$: $Z(1.7, 2.1, 2.4)$.

When $\alpha < 0.5$

Hence the UCCMOGP problem reduces to

$$\text{Min } f_{10}(t) = \frac{42.5}{t_1 t_2 t_3} + 40 t_2 t_3 \tag{5.6}$$

$$\text{Min } f_{20}(t) = \frac{825}{t_1 t_2 t_3}$$

subject to

$$(0.9(1-2\alpha) + 1.1(2\alpha))t_1 t_2 + (1.7(1-2\alpha) + 2.1(2\alpha))\, t_1 t_3 \le 4, \quad t_1, t_2, t_3 > 0.$$

Using the weighted-sum technique, the above UCCMOGP problem (5.6) becomes the single-objective geometric programming problem as follows:

$$\text{Min f(t)} = w_1\left(\frac{42.5}{t_1 t_2 t_3} + 40 t_2 t_3\right) + w_2 \frac{825}{t_1 t_2 t_3} = \frac{42.5 w_1 + 825 w_2}{t_1 t_2 t_3} + 40 w_1 t_2 t_3 \tag{5.7}$$

subject to $(0.9(1-2\alpha) + 1.1(2\alpha))t_1 t_2 + (1.7(1-2\alpha) + 2.1(2\alpha))\, t_1 t_3 \le 4, \quad t_1, t_2, t_3 > 0.$

Here, Degree of Difficulty (DD) $4 - (3 + 1) = 0$.

Dual problem of the above problem (5.7) is

$$\text{Maximize } d(\lambda) = \left(\frac{42.5 w_1 + 825 w_2}{\lambda_{01}}\right)^{\lambda_{01}} \left(\frac{40 w_1}{\lambda_{02}}\right)^{\lambda_{02}} \left(\frac{(0.9(1-2\alpha)+1.1(2\alpha))}{4\lambda_{11}}\right)^{\lambda_{11}} \left(\frac{(1.7(1-\alpha)+2.1(2\alpha))}{4\lambda_{12}}\right)^{\lambda_{12}} (\lambda_{11}+\lambda_{12})^{(\lambda_{11}+\lambda_{12})}$$

subject to

$$\lambda_{01} + \lambda_{02} = 1,$$

$$-\lambda_{01} + \lambda_{11} + \lambda_{12} = 0,$$

$$-\lambda_{01} + \lambda_{02} + \lambda_{11} = 0,$$

$$-\lambda_{01} + \lambda_{02} + \lambda_{12} = 0,$$

$$w_1 + w_2 = 1,$$

$$\lambda_{01}, \lambda_{02}, \lambda_{11}, \lambda_{12} > 0.$$

Solving the above system of linear equations, we have

$$\lambda_{01} = \frac{2}{3}, \lambda_{02} = \frac{1}{3}, \lambda_{11} = \frac{1}{3}, \lambda_{12} = \frac{1}{3}.$$

From the primal-dual relationship, we obtain

$$\frac{42.5 w_1 + 825 w_2}{t_1 t_2 t_3} = \lambda_{01} d(\lambda),$$

$$40 w_1 t_2 t_3 = \lambda_{02} d(\lambda),$$

$$\frac{(0.9(1-2\alpha)+1.1(2\alpha))}{4} = \frac{\lambda_{12}}{\lambda_{11}+\lambda_{12}},$$

$$\frac{(1.7(1-2\alpha)+2.1(2.5\alpha)}{4} = \frac{\lambda_{12}}{\lambda_{11}+\lambda_{12}}.$$

And the corresponding optimal solutions of primal variables are as follows:

$$t_3 = \left(\frac{\frac{42.5w_1 + 825w_2}{40}}{4(1.7(1-2\alpha)+2.1(2\alpha))w_1} \right)^{\frac{1}{3}}, t_2 = 2t_3, t_1 = \frac{2}{(1.7(1-2\alpha)+2.1(2\alpha))t_3}.$$

When $\alpha > 0.5$

Hence the UCCMOGP problem reduces to

$$\text{Min } f_{10}(t) = \frac{42.5}{t_1 t_2 t_3} + 40 t_2 t_3 \tag{5.8}$$

$$\text{Min } f_{20}(t) = \frac{825}{t_1 t_2 t_3}$$

subject to

$$(1.2(2\alpha - 1) + 2(1 - \alpha)1.1)t_1t_2 + (2.4(2\alpha - 1) + 2(1 - \alpha)2.1)\, t_1t_3 \le 4, \quad t_1, t_2, t_3 > 0.$$

Using the weighted-sum technique, the above UCCMOGP problem (5.8) becomes the single-objective geometric programming problem as follows:

$$\text{Min f(t)} = w_1\left(\frac{42.5}{t_1t_2t_3} + 40t_2t_3\right) + w_2\frac{825}{t_1t_2t_3} = \frac{42.5w_1 + 825w_2}{t_1t_2t_3} + 40w_1t_2t_3 \tag{5.9}$$

subject to

$$(1.2(2\alpha - 1) + 2(1 - \alpha)1.1)t_1t_2 + (2.4(2\alpha - 1) + 2(1 - \alpha)2.1)\, t_1t_3 \le 4, \quad t_1, t_2, t_3 > 0.$$

Here, Degree of Difficulty (DD) $= 4 - (3 + 1) = 0$.

Dual problem of the above problem (5.9) is

Maximize

$$d(\lambda) = \left(\frac{42.5w_1 + 825w_2}{\lambda_{01}}\right)^{\lambda_{01}} \left(\frac{40w_1}{\lambda_{02}}\right)^{\lambda_{02}} \left(\frac{(1.2(2\alpha-1)+2(1-\alpha)1.1)}{4\lambda_{11}}\right)^{\lambda_{11}} \left(\frac{(2.4(2\alpha-1))+2(1-\alpha)2.1)}{4\lambda_{12}}\right)^{\lambda_{12}} (\lambda_{11}+\lambda_{12})^{(\lambda_{11}+\lambda_{12})}$$

subject to

$$\lambda_{01} + \lambda_{02} = 1,$$

$$-\lambda_{01} + \lambda_{11} + \lambda_{12} = 0,$$

$$-\lambda_{01} + \lambda_{02} + \lambda_{11} = 0,$$

$$-\lambda_{01} + \lambda_{02} + \lambda_{12} = 0,$$

$$w_1 + w_2 = 1,$$

$$\lambda_{01}, \lambda_{02}, \lambda_{11}, \lambda_{12} > 0.$$

Solution of the above system of linear equations, gives,

$$\lambda_{01} = \frac{2}{3}, \lambda_{02} = \frac{1}{3}, \lambda_{11} = \frac{1}{3}, \lambda_{12} = \frac{1}{3}.$$

From the primal-dual relationship, we obtain

$$\frac{42.5w_1 + 825w_2}{t_1 t_2 t_3} = \lambda_{01} d(\lambda),$$

$$40 w_1 t_2 t_3 = \lambda_{02} d(\lambda),$$

$$\frac{1.2(2\alpha - 1) + 2(1-\alpha)1.1}{4} = \frac{\lambda_{12}}{\lambda_{11} + \lambda_{12}},$$

$$\frac{2.4(2\alpha - 1) + 2(1-\alpha)2.1}{4} = \frac{\lambda_{12}}{\lambda_{11} + \lambda_{12}}.$$

And the corresponding optimal solutions of primal variables are as follows

$$t_3 = \left(\frac{\dfrac{42.5w_1 + 825w_2}{40}}{4(2.4(2\alpha-1) + 2(1-\alpha)2.1)w_1} \right)^{\frac{1}{3}}, \; t_2 = 2t_3, t_1 = \frac{2}{(2.4(2\alpha-1) + 2(1-\alpha)2.1)t_3}.$$

The optimal solutions of the above UCCMOGP problem under Zigzag Uncertainty Distributions are given in the following Table-3.

Table 3: Optimal solutions when using Uncertainty Zigzag Distributions.

α	**Weights**	**Optimal solutions of primal variables**			**Optimal objective values**	
		t_1^*	t_2^*	t_3^*	$f_{01}^*(t)$	$f_{02}^*(t)$
0.2	$w_1 = 0.1, w_2 = 0.9$	0.38	5.96	2.98	716.73	122.24
	$w_1 = 0.5, w_2 = 0.5$	0.78	2.90	1.45	181.16	251.53
	$w_1 = 0.9, w_2 = 0.1$	1.46	1.56	0.78	72.60	464.39
0.4	$w_1 = 0.1, w_2 = 0.9$	0.36	5.80	2.90	679.82	136.25
	$w_1 = 0.5, w_2 = 0.5$	0.74	2.82	1.41	173.49	280.38
	$w_1 = 0.9, w_2 = 0.1$	1.37	1.52	0.76	73.06	521.29
0.6	$w_1 = 0.1, w_2 = 0.9$	0.34	5.64	2.82	644.05	152.56
	$w_1 = 0.5, w_2 = 0.5$	0.70	2.76	1.38	168.29	309.43
	$w_1 = 0.9, w_2 = 0.1$	1.30	1.48	0.74	73.66	579.45
0.8	$w_1 = 0.1, w_2 = 0.9$	0.35	5.50	2.75	613.03	155.84
	$w_1 = 0.5, w_2 = 0.5$	0.72	2.68	1.34	160.08	319.07
	$w_1 = 0.9, w_2 = 0.1$	1.34	1.44	0.72	69.94	552.58

6. Conclusions

Multi-Objective Geometric Programming (MOGP) is a ground-breaking advancement method generally utilized for solving special types of nonlinear optimization problems, especially in engineering. Orthodox MOGP models expect that the parameters are deterministic and crisp. Still, the parameters or coefficients in real-life MOGP problems are frequently uncertain and subject to fluctuations. Therefore, we have moved toward the problem of formalizing and implementing imprecise and non-deterministic parameters using uncertainty theory. There exists plentiful writing on MOGP under uncertainty and its applications to problems (either chance-constrained or not) whose coefficients are fuzzy numbers, fuzzy variables or random variables. Notwithstanding, as far as we could possibly know, no past investigation has considered the formulation and/or solution of MOGP problems where the coefficients are given by uncertain variables (UVs). In this chapter, we have presented an Uncertain Chance-Constrained Multi-Objective GP (UCCMOGP) problem and proposed a solution procedure that applies to three of the most commonly used uncertainty distributions: we assumed the coefficients to be uncertain variables with linear, normal or zigzag uncertainty distributions. We showed that the corresponding uncertain chance-constrained multi-objective geometric programming (UCCMOGP) problem can be changed into conventional geometric programming problems with crisp coefficients and, thus, an optimal solution can be discovered using the duality theory. We have indicated the efficacy of the proposed UCCMOGP problem through three numerical examples. We believe that the framework proposed in this chapter contributes to revealing insight into the applications of MOGP to concrete problems, opening the way to further research in engineering and production management.

7. Acknowledgement

The author is thankful to the Department of Mathematics, University of Kalyani and DST-PURSE (Phase-II) Programme for supporting with financial assistance.

8. Future Research

From this chapter a lot of scope may arise for future research work.

- In this chapter we consider the coefficients of Multi-Objective Geometric Programming problem to be most commonly used uncertainty distributions i.e. linear, normal or zigzag uncertainty distributions but in future research other type of uncertainty distributions may be used.
- Also the proposed UCCMOGP problem may be applied to various operations research models and can be solved by the proposed solution techniques.

References

Antunes, C. H., Liu, G. P., Yang, J. B. and Whidborne, J. F. 2006. Multiobjective optimisation and control, Series: Engineering Systems Modelling and Control, Research Studies Press (2003), ISBN 0 86380 264 8.

Biswal, M. 1992. Fuzzy programming technique to solve multi-objective geometric programming problems. Fuzzy Sets. Syst., 51: 67–71.

Chankong, V. and Haimes, Y.Y. 2008. Multiobjective decision making: theory and methodology. Courier Dover Publications.

Das, P. 2014. multiobjective geometric programming and its application in gravel box problem. J. Glob. Res. Comput. Sci., 5: 6–11.

Dey, S. and Roy, T. 2015. Optimum shape design of structural model with imprecise coefficient by parametric geometric programming. Decis. Sci. Lett., 4: 407–418.

Ding, S. 2015. The α-maximum flow model with uncertain capacities. Appl. Math. Model., 39: 2056–2063.

Duffin, R. J. and Peterson, E. L. 1973. Geometric programming with signomials. J. Optim. Theory Appl., 11: 3–35.

Han, S., Peng, Z. and Wang, S. 2014. The maximum flow problem of uncertain network. Inf. Sci., 265: 167–175.

Islam, S. 2008. Multi-objective marketing planning inventory model: A geometric programming approach. Appl. Math. Comput., 205: 238–246.

Islam, S. and Mandal, W. A. 2019. Fuzzy Geometric Programming Techniques and Applications. Springer Nature, Singapore, ISBN-10: 9811358222, ISBN-13: 978-9811358227.

Li, S., Peng, J. and Zhang, B. 2013. The uncertain premium principle based on the distortion function. Insur. Math. Econ., 53: 317–324.

Lin, M.-H. and Tsai, J.-F. 2012a. Range reduction techniques for improving computational efficiency in global optimization of signomial geometric programming problems. Eur. J. Oper. Res., 216: 17–25.

Lin, M.-H. and Tsai, J.-F. 2012b. Range reduction techniques for improving computational efficiency in global optimization of signomial geometric programming problems. Eur. J. Oper. Res., 216: 17–25.

Liu, B. 2010a. Uncertain risk analysis and uncertain reliability analysis. J. Uncertain Syst., 4: 163–170.

Liu, B. 2010b. Uncertain set theory and uncertain inference rule with application to uncertain control. J. Uncertain Syst., 4: 83–98.

Liu, B. and Chen, X. 2015. Uncertain multiobjective programming and uncertain goal programming. J. Uncertain. Anal. Appl., 3: 10.

Liu, S.-T. 2006. Posynomial geometric programming with parametric uncertainty. Eur. J. Oper. Res., 168: 345–353.

Liu, S.-T. 2007. Profit maximization with quantity discount: An application of geometric programming. Appl. Math. Comput., 190: 1723–1729.

Liu, S.-T. 2008. Posynomial geometric programming with interval exponents and coefficients. Eur. J. Oper. Res., 186: 17–27.

Maranas, C. D. and Floudas, C. A. 1997. Global optimization in generalized geometric programming. Comput. Chem. Eng., 21: 351–369.

Miettinen, K. 1999. Non-linear Multi-objective. Optim. Kluwer's Int. Ser.

Ojha, A. and Biswal, K. 2014. Multi-objective geometric programming problem with∈-constraint method. Appl. Math. Model., 38: 747–758.

Ojha, A. K. and Das, A. 2010. Multi-Objective Geometric Programming Problem Being Cost Coefficients as Continuous Function with Weighted Mean Method. ArXiv Prepr. ArXiv10024002.

Ojha, A. K. and Ota, R. R. 2014. Multi-objective geometric programming problem with Karush- Kuhn- Tucker condition using ϵ-constraint method. RAIRO-Oper. Res., 48: 429–453.

Peng, J. and Yao, K. 2011. A new option pricing model for stocks in uncertainty markets. Int. J. Oper. Res., 8: 18–26.

Peterson, E. L. 2001. The fundamental relations between geometric programming duality, parametric programming duality, and ordinary Lagrangian duality. Ann. Oper. Res., 105: 109–153.

Sadjadi, S. J., Hesarsorkh, A. H., Mohammadi, M. and Naeini, A. B. 2015. Joint pricing and production management: a geometric programming approach with consideration of cubic production cost function. J. Ind. Eng. Int., 11: 209–223.

Samadi, F., Mirzazadeh, A. and Pedram, M. M. 2013. Fuzzy pricing, marketing and service planning in a fuzzy inventory model: A geometric programming approach. Appl. Math. Model., 37: 6683–6694.

Scott, C. H. and Jefferson, T. R. 1995. Allocation of resources in project management. Int. J. Syst. Sci., 26: 413–420.

Shiraz, R. K., Tavana, M., Di Caprio, D. and Fukuyama, H. 2016. Solving geometric programming problems with normal, linear and zigzag uncertainty distributions. J. Optim. Theory Appl., 170: 243–265.

Shiraz, R. K., Tavana, M., Fukuyama, H. and Di Caprio, D. 2017. Fuzzy chance-constrained geometric programming: the possibility, necessity and credibility approaches. Oper. Res., 17: 67–97.

Tsai, J.-F. 2009. Treating free variables in generalized geometric programming problems. Comput. Chem. Eng., 33: 239–243.

Tsai, J.-F., Li, H.-L. and Hu, N.-Z. 2002. Global optimization for signomial discrete programming problems in engineering design. Eng. Optim., 34: 613–622.

Tsai, J.-F. and Lin, M.-H. 2006. An optimization approach for solving signomial discrete programming problems with free variables. Comput. Chem. Eng., 30: 1256–1263.

Tsai, J.-F., Lin, M.-H. and Hu, Y.-C. 2007. On generalized geometric programming problems with non-positive variables. Eur. J. Oper. Res., 178: 10–19.

Wang, X., Gao, Z. and Guo, H. 2012a. Delphi method for estimating uncertainty distributions. Inf. Int. Interdiscip. J., 15: 449–460.

Wang, X., Gao, Z. and Guo, H. 2012b. Delphi method for estimating uncertainty distributions. Inf. Int. Interdiscip. J., 15: 449–460.
Wang, X., Gao, Z. and Guo, H. 2012c. Uncertain hypothesis testing for two experts' empirical data. Math. Comput. Model., 55: 1478–1482.
Wang, X. and Ning, Y. 2019. A new existence and uniqueness theorem for uncertain delay differential equations. J. Intell. Fuzzy Syst., 37: 4103–4111.
Wang, X. and Ning, Y. 2018. Uncertain chance-constrained programming model for project scheduling problem, Journal of the Operational Research Society, 69: 3, 384–39.
Worral, B. M. and Hall, M. A. 1982. The analysis of an inventory control model using posynomial geometric programming. Int. J. Prod. Res., 20: 657–667.
Yang, H.-H. and Bricker, D. L. 1997. Investigation of path-following algorithms for signomial geometric programming problems. Eur. J. Oper. Res., 103: 230–241.
Zhu, J., Kortanek, K. and Huang, S. 1994. Controlled dual perturbations for central path trajectories in geometric programming. Eur. J. Oper. Res., 73: 524–531.
Zhu, Y. 2010. Uncertain optimal control with application to a portfolio selection model. Cybern. Syst. Int. J., 41: 535–547.

CHAPTER 25

Optimal Decision Making for the Prediction of Diabetic Retinopathy in Type 2 Diabetes Mellitus Patients

Faiz Noor Khan Yusufi,[1,*] *Nausheen Hashmi,*[1] *Aquil Ahmed*[1] and *Jamal Ahmad*[2]

1. Introduction

Diabetes mellitus nowadays is one of the most prevailing non-communicable diseases throughout the world. The total number of patients suffering from this disease is increasing at an alarming rate. According to International Diabetes Federation (IDF Diabetes Atlas 2019), diabetes mellitus is a chronic metabolic condition in which the body is not able to control blood sugar levels and it is detected upon the incidence of high blood glucose levels (World Health Organization (WHO) 1999). There are more than 464 million people living with diabetes worldwide (IDF Diabetes Atlas 2019). Considering China, India and United States of America (USA), these countries form the majority of the diabetic population in the world (IDF Diabetes Atlas 2019).

Diabetes mellitus is of three types (WHO 1999): Type 1, Type 2 and gestational. Throughout the world Type 1 diabetes mellitus (T1DM) adds up to about 10% of the total diabetic population. In this type, there is no production or very little production of insulin.

Type 2 diabetes mellitus (T2DM) is much more spread out worldwide and is the most common type with 90% of the diabetic population suffering from this kind (Chen et al., 2012). With type 2 diabetes, the body either doesn't produce enough insulin or it resists insulin. The third type is known as gestational diabetes. It happens during pregnancy and affects both the mother and child (Hirst et al., 2012).

With so many complications developing because of diabetes, proper medications and lifestyle changes have to be implemented to prevent the incidence of the complications. Diabetic complications are: cardiovascular disease, kidney disease (diabetic nephropathy), nerve disease (diabetic neuropathy), eye disease (diabetic retinopathy (DR)), foot damage, leg amputation, skin infections, depression, dementia, hearing loss, pregnancy and oral complications (periodontitis) (WHO, 2016). The largest diabetic population is found in China (above 116 million), followed by India (above 77 million). The recorded quantities mentioned represent the number of T2DM patients.

2. Disease Risk Scores

Beyond its clinical and social burden, diabetes also has an impact on the consumption of economic resources. A complete and proper collection of patients' data is the basis of an effective set up of clinical improvement strategies. Health databases comprise of anthropometric, demographic and clinical variables including prescribed medications which are being used immensely in pharmacoepidemiologic studies (Ray, 2003). These variables are known as risk factors. Application of statistical techniques to form prediction risk scores or models is exceedingly beneficial in managing a disease.

Risk scores calculate the probability of disease incidence while knowing there is an absence of an exposure. In other words, risk scores determine the effect of an exposure on the occurrence of a disease (Miettinen, 1976). The outcome generally consists of death or incidence or prevalence of a disease, whereas an exposure can be explained by a risk factor, diagnostic test or a treatment.

Framingham heart study is the first and most famous predictive algorithm that has been developed, widely used for predicting cardiovascular, cerebrovascular, peripheral artery disease, heart failure and coronary heart disease events. The development of this study took almost ten years; it started in 1948 and published its primary findings in 1957 (Mahmood et al., 2014). It took an extended time to develop this score but finally the best recognized 10-year predictive algorithm known as the 'Framingham risk score for

[1] Department of Statistics & Operations Research, Aligarh Muslim University, Aligarh, U.P., India.
Emails: nausheenhashmi21@gmail.com, aquilstat@gmail.com
[2] Diabetes & Endocrinology Super Speciality Centre, Aligarh, U.P., India.
Email: jamalahmad11@rediffmail.com
* Corresponding author: fnkyusufi@gmail.com

coronary heart disease' was developed in 1998 (Wilson et al., 1998). This score has become the basis and foundation for all scores created in disease management. There are other scores that have been developed after the Framingham risk score (Conroy et al., 2003; Ridker et al., 2008; Hippisley-Cox et al., 2007).

3. Risk Scores for T2DM

It was understood that statistical models will provide a much cheaper option in addition to lifestyle management and preventive medical treatment in achieving a control over diabetes. Undiagnosed diabetes also known as prevalence is a scenario that is highly targeted (Lindstrom and Tuomilehto, 2003; Glumer et al., 2004; Mohan et al., 2005; Ramachandran et al., 2005; Sulaiman et al., 2018). The application of the created score on the same type of population is known as internal validation and if the population is from another country then it is termed as external validation. The set of patients on whom the statistical tests are applied, as well as the score is formulated is termed as a training set. The second set of patients that are used to validate the results obtained from the training set is named as a test set.

Predictive models for scenarios such as development of diabetes also known as incidence are one type of scores that have been created (Hippisley-Cox and Coupland, 2017; Hu et al., 2018). Scores for prediabetes, particular biomarkers and a combination of these cases have been constructed as well (Hertroijs et al., 2018). There are previous validation studies to find out the strength of the developed models (Stiglic et al., 2016).

Risk scores have not been created just for incidence and prevalence of diabetes but for the complications as well (Yang et al., 2008; Yu et al., 2018). The number of clinical parameters increase for such predictive models, emphasizing more on the blood biomarkers and not just on anthropometric and lifestyle habits. Validation of risk scores constructed for complications have also been performed (Van der Leeuw et al., 2015). There are different types of studies associated with the construction of risk scores and these are explained elsewhere (Riffenburgh, 2006; Rosner, 2010). This chapter aims at describing the procedures for developing and validating risk scores related to T2DM. The steps have been explained with the help of a case study for predicting the prevalence and incidence of DR in Indian T2DM patients (Yusufi et al., 2019).

4. Methodology

4.1 Development of Disease Risk Scores

A suitable sample size is calculated representing the population correctly. Next, the relevant demographic, anthropometric, clinical and medication variables that are required for the particular disease are to be decided, these will be termed as risk factors. The data can be collected through primary or secondary sources. The collected data is then divided into training and test sets. First of all, descriptive and graphical methods are used to find out the shape and trend of the data.

Various regression techniques are used on the training set to construct risk scores, providing a relationship between the study outcome, exposure and specific covariates of a group of patients. These regression methods are executed after their assumptions have been fulfilled. Commonly used regression methods are logistic, binomial, Poisson, Cox proportional hazards model, ridge, elastic net and lasso regression. The last three types of regression methods mentioned above are used in the case of overfitting and are known as regularization techniques or penalized regression methods. An overview of regression techniques is given below (Vittinghoff et al., 2011; Woodward, 2013; Rosner, 2010).

4.2 Linear Regression

A standard linear regression is used to model the average value of a continuous dependent variable based on a single or multiple independent predictors (total predictors being 'n'), given by Equation 1.

$$E[y \mid x] = \beta_0 + \sum_{i=1}^{n} \beta_i x_i \tag{1}$$

The right side of the above equation is a linear combination of β-coefficients and predictor variables. The outcome variable is of continuous form and follows normal distribution. 'y' is the vector of the corresponding outcome variable and 'x_i' ($i = 1,\ldots,n$) represents the respective predictor values for each individual observation.

4.3 Odds Ratio

If 'p' is the probability of occurrence of a disease, then the odds of the disease is given by Equation 2.

$$Odds = \frac{p}{1-p} \tag{2}$$

If 'p_1' and 'p_2' are the probabilities of occurrence of two separate diseases then odds ratio (OR) is the ratio of the odds of two diseases respectively, given by Equation 3.

$$OR = \frac{p_1 / 1 - p_1}{p_2 / 1 - p_2} = \frac{p_1 (1 - p_2)}{p_2 (1 - p_1)} \tag{3}$$

4.4 Logistic Regression

Logistic regression is used when the dependent variable has only two possible outcomes. The options in a dichotomous variable are usually represented by '*1*' and '*0*', and therefore, it is also known as a binary outcome. Presence of a disease or an event is coded as '*1*' and absence is denoted by '*0*'. Equation 1 shows limitations when used as a case of a binary outcome as it follows a binomial distribution. To rectify this, a logistic model estimates the probabilities of the dichotomous outcome. The second reason of inapplicability of the linear model is that the outcome is a probability. Therefore, the regression coefficients should result in such a manner that the calculated probability is restrained to a range between zero and one. A logistic model is shown in Equation 4.

$$P(y=1 \mid x) = \frac{\exp(\beta_0 + \sum_{i=1}^{n} \beta_i x_i)}{1 + \exp(\beta_0 + \sum_{i=1}^{n} \beta_i x_i)} \tag{4}$$

A logit transformation based on the calculation of odds is provided in Equation 5 and this forms a linear relationship between the log odds of the outcome and the predictor variables (Vittinghoff et al., 2011). A logistic model does not require the assumption of constant variance of the outcome variable to be followed.

$$\log\left[\frac{P(y=1 \mid x)}{1 - P(y=1 \mid x)}\right] = \beta_0 + \sum_{i=1}^{n} \beta_i x_i \tag{5}$$

Other alternatives for binary outcomes are an exponential model and a step function model. Logistic regression is used to predict the prevalence of a disease. On the other hand, incidence is predicted by using Cox proportional hazards regression model. This specific type of regression comes under survival analysis in which a time to event variable is included as a dependent variable.

5. Regularization Techniques

Obtaining suitable and optimal regression coefficients through the application of logistic or Cox regression is not that straightforward. Common issues arise due to bias and variance. Two scenarios, known as underfitting and overfitting, can occur, due to which the results will not be estimated correctly because the model fails to collect the exact facts being provided by the data. Underfitting arises when the predictive model is too simple and does not capture the true information. This situation is signified by high bias and low variance. However, low bias and high variance are the signs of overfitting and emerge when the model becomes more complex and fits the training set but not the test set (Lever et al., 2016). In the presence of underfitting, the model does not fit the training set properly. A model should fit both the training as well as test set satisfactorily in order to prove its accuracy for future prediction.

Increasing the sample size will reduce the bias and remove the issue of underfitting. Studies that involve large number of predictor variables will have a high possibility that the variables are correlated among themselves thus, making the model more complex. Presence of highly correlated variables is known as multicollinearity and it is the most vital reason for overfitting. The main drawback caused by overfitting (low bias and high variance) is that the error or noise of a model is included in the estimated regression coefficients. Therefore, the coefficients will result in values higher than their actual values. These higher values do not produce the correct results, and thus, must be reduced. Compared to underfitting, managing and rectifying overfitting is more complicated and is a stepwise procedure. It is essential to first of all know the techniques and methods that assist in the detection of overfitting. Variance inflation factor (VIF) and shrinkage estimation are the two methods used to test the presence of multicollinearity and overfitting.

VIF measures the amount of inflation in the variances of estimated regression coefficients and a factor greater than 5 signifies multicollinearity (Sheather, 2009). This is shown in Equation 6, in which R_d^2 is the coefficient of determination of a regression equation. The VIF for each predictor variable is measured and the specific variables that are the source of multicollinearity can be identified.

$$VIF = \frac{1}{1 - R_d^2} \tag{6}$$

Shrinkage is estimated using the likelihood ratio of a chi-square test denoted by Equation 7. If this equation results in a value less than 0.85, then it points to overfitting (Harrell et al., 1996).

$$Estimated\ Shrinkage = \frac{(Likelihood\ Ratio - number\ of\ predictor\ variables)}{Likelihood\ Ratio} \tag{7}$$

The next step is to find the best method that will reduce the predictor coefficients. This is accomplished by applying regularization techniques. There are three different regularization techniques available to rectify overfitting, which are also known as penalized regression techniques. They are ridge regression, lasso regression and elastic net regression. The main aim is to shrink the variable coefficients and reduce the variance so that the non-significant variables can be excluded from the study. Reducing the variance will result in actual coefficient values and will create a satisfactory predictive model however, the bias might increase by a small amount.

A linear regression model is used to explain the penalized regression methods. Let us assume a linear function '$\hat{f}(X) = X\hat{\beta}$', where '$X$' denotes the predictor matrix and '$\hat{\beta}$' is the regression coefficient matrix. While building predictive models, a prediction error will be present and it is shown in Equation 8 (James et al., 2013). An optimal model will have a low bias as well as a low variance. These two fundamental concepts can be controlled, however, 'σ^2' is the irreducible error and cannot be reduced or controlled. There will always be an error in a study even if the values of bias and variance are low. The sum of variance and square of bias results in an error termed as mean squared error (MSE) and its importance is utilized in the selection of appropriate regularization techniques.

$$Prediction\ Error = \sigma^2 + Bias^2(\hat{f}(X)) + Var(\hat{f}(X)) \tag{8}$$

The three penalized regression methods add a penalty term in the regression model so that the objective function minimizes the error (James et al., 2013). Introducing a small amount of bias in the model will reduce the variance significantly. Ridge regression adds a '*L2*' quadratic penalty term and is shown in Equation 9. The second term '$\lambda\sum_{i=1}^{n}\beta_i^2$' is the penalty term that will shrink the predictor coefficients. 'λ' is the tuning parameter and it will decide the amount of shrinkage.

$$min\left[\sum_{i=1}^{n}\left(y_i - \beta_0 - \sum_{i=1}^{n}\beta_i x_i\right)^2 + \lambda\sum_{i=1}^{n}\beta_i^2\right] \tag{9}$$

In the case of lasso regression (Equation 10), the penalty term changes to '$\lambda\sum_{i=1}^{n}|\beta_i|$' and is known as the '*L1*' absolute value penalty.

$$min\left[\sum_{i=1}^{n}\left(y_i - \beta_0 - \sum_{i=1}^{n}\beta_i x_i\right)^2 + \lambda\sum_{i=1}^{n}|\beta_i|\right] \tag{10}$$

Combination of '*L1*' and '*L2*' penalty produces elastic net regression. It is seen from Equation 11 that this specific category of penalized regression contains both penalty terms which are found in ridge regression and lasso regression. Both the penalties shrink the coefficient estimates by using maximum likelihood estimates (Goeman et al., 2018).

$$min\left[\sum_{i=1}^{n}\left(y_i - \beta_0 - \sum_{i=1}^{n}\beta_i x_i\right)^2 + \lambda_1\sum_{i=1}^{n}|\beta_i| + \lambda_2\sum_{i=1}^{n}\beta_i^2\right] \tag{11}$$

Equation 9, Equation 10 and Equation 11 depict a linear regression model. Penalized regression models can be obtained and similarly used for other regression techniques as well, such as logistic regression and Cox regression.

The shrinkage penalty terms in all three equations are the constraints for the minimization objectives and are included as additives with the help of Lagrange's multiplier 'λ'. The objective is to find the optimal value of 'λ'. This can be attained using crossvalidation. Various lambda values are tested and the one which produces the least amount of error is selected. This is usually the minimum lambda value. In ridge regression, the coefficients are reduced but are not set to a value of exact zero. Lasso regression on the other hand shrinks the parameters and reduces the non-significant predictor coefficients to zero. The difference between ridge regression and lasso regression is that ridge regression includes all the predictor variables in the final model and shrinks their coefficients. In lasso regression, less number of variables in the final predictive model are retained. Elastic net regression is categorized in between ridge regression and lasso regression. Lesser number of coefficients are reduced to zero as compared to lasso regression and more shrinkage penalty is administered than ridge regression. Out of ridge, lasso and elastic net regression techniques, the method which produces the minimum MSE is considered to be the best and most suitable method for a particular predictive model. Lasso regression and elastic net regression can be mentioned as variable selection techniques because they both remove non-significant predictor variables. In R software, '*glmnet*' package is available to implement penalized regression techniques (Friedman et al., 2010). This package is used to determine the optimal lambda value which is followed by the estimation of the penalized regression coefficients.

Once the proper and suitable regression techniques have been administered on the training set and the regression coefficients have been attained after implementing regularization methods if essential, the next procedure is to utilize these obtained coefficients. When formulating probabilities of an event, the 'β' values are substituted using formulas such as Equation 4. If the objective is to create scores, then 'β' values are modified and adjusted to construct simple and easily applicable risk scores. A risk score model for DR has explained the different weighing procedures that are available relating to the coefficients (Kulothungan et al., 2014). The most common and widely used method was developed by Glumer et al. in 2004. It was proposed that the achieved 'β' values be multiplied by 10 and then rounded off to the nearest integer. This is performed for each significant predictor variable and followed up with the addition of each variable score for every patient. A cut-off point for the score needs to be decided, above which the risk of the event increases.

6. Diagnostic Tests and Model Performance

Before the application of the obtained results on a separate test set, diagnostic tests are applied on the training group so that a cut-off can be established for the scores or probabilities. After this, the results are administered and validated on a separate test set. The reason for the validation is to remove the bias in the results achieved and to improve the accuracy and reliability of the model built. Another purpose for the application of diagnostic tests in a study is to avoid the occurrence of false positive and false negative errors (Woodward, 2013). Screening tests are conditional probabilities and can be estimated using Bayes' rule. The various diagnostic tests are sensitivity, specificity, positive predictive value (PPV), negative predictive value (NPV) and receiver operating characteristic (ROC) curve. ROC plot is interpreted by area under the curve, which ranges between zero and one. C-statistic or concordance statistic measures the goodness of fit of a model and it is equal to the value of area under an ROC curve. The more the area coverage of the curve, the better the reliability, predictability and accuracy of the model created. The curve is represented by plotting sensitivity versus (1 – specificity). The detailed description of diagnostic tests is available elsewhere (Woodward, 2013; Rosner, 2010). Equation 12 summarizes the diagnostic tests.

$$\left.\begin{aligned} &Sensitivity = P(positive\ test\ result \mid positive\ disease\ outcome) \\ &Specificity = P(negative\ test\ result \mid negative\ disease\ outcome) \\ &PPV = P(positive\ disease\ status \mid positive\ screening\ test) \\ &NPV = P(negative\ disease\ status \mid negative\ screening\ test) \end{aligned}\right\} \tag{12}$$

7. Youden's J Statistic (Youden's index)

Using sensitivity and specificity, Youden's index ranging between zero and one is computed by the formula '(sensitivity + specificity – 1)'. This index is estimated for each sample individual and the cut-off or screening value for the scores or probabilities is selected for those values of sensitivity and specificity which produce the maximum Youden's index (Equation 13) (Hilden and Glasziou, 1996; Kallner, 2014). Lastly, the test set is utilized to determine the number of patients with the disease that have their scores or probabilities above the cut-off value. This step is another validation of the created model.

$$Cut\text{-}off\ value = max(sensitivity + specificity - 1) \tag{13}$$

8. Markov Model

Logistic regression provides the probabilities for the time at which the data was collected. In other words, it predicts the present situation and not a future event (Equation 4). Suppose in a study, clinical readings of patients are collected and are substituted in a model which predicts the probability of prevalence of an event. The patients that have experienced the event are identified and the rest are separated. It is possible for a researcher to be interested in finding the incidence probabilities of the separated patients that remained event free. If the outcome is yearly based, then the incidence can happen anytime in the upcoming years. This is the disadvantage of using logistic regression. For this scenario a Markov model is useful.

The simplest type of model is a two state model in which there are two stages. A person being event free is the first state and developing an event is the second state. The objective is to determine the probability of developing an event by means of reaching the second state from the first state. The second state is an 'absorbing' state which means that further steps or transitions are not possible once this stage is attained. In terms of an event, it signifies that once a patient develops an outcome then it is impossible for him to return to his original healthy condition. Therefore, the first state is known as a 'transient' state. If the outcome is considered to be death then the first state is known as 'alive' state and 'death' is the second state (Andersen and Keiding, 2002). This model is expressed in Equation 14.

$$x^{(n)} = x^{(0)} P^n \tag{14}$$

In the above equation, the term '$x^{(n)}$' is the matrix form of Markov model predicting the probabilities for 'n^{th}' year. '$x^{(0)}$' signifies the available dichotomous options for an individual at the start of the study. The options presented are to either remain in the first category or reach an outcome by moving to the second state. The matrix '$x^{(0)}$' is defined as '$[1\ 0]$', where '0' is the coded form of the initial state and '1' is the coded form of the second state. 'P^n' denotes the transition matrix and for a two state model it will be a two by two matrix comprising of elements that estimate the probabilities of all possible steps between the two stages (Kampen, 1992). Memoryless property of a Markov chain is extremely beneficial in situations where only a single value such as the prevalence probability is available (Gagniuc, 2017). Due to this property only the current condition needs to be known and no prior information is required. The implementation of a Markov model to estimate yearly probabilities is a very helpful addition to the application of logistic regression.

A case study is presented next, explaining the application of above defined methods that are used to create probabilities which will predict the prevalence and incidence of DR complication in Indian T2DM patients.

9. Case Study

DR is a microvascular complication caused due to increased blood sugar levels. DR is the topmost reason for visual impairment across countries (Kempen et al., 2004). Its risk factors according to the Mayo Clinic (U.S.A.) consist of duration of diabetes, blood sugar level, high blood pressure, high cholesterol, tobacco use, pregnancy and race. Assistance in early prediction of a complication will allow for better control of the disease, as the presence of DR indicates dysfunction of various organs (Liew and Wong, 2009), resulting in the requirement of a more balanced lifestyle.

Apart from the score developed by Kulothungan et al., 2014, only two DR models have been developed worldwide. These scores were formulated in Iran and China (Hosseini et al., 2009; Wang et al., 2014) and they created models that will predict the prevalence of DR. Logistic regression was applied in both their studies to determine the regression coefficients and scores were created for the significant variables. All of the continuous variables were changed into categorical variables in both the Iranian and Chinese studies. However, the regression coefficients provided can be used to predict the incidence of DR in T2DM patients by estimating the probabilities.

The outcome of interest was DR and its presence was defined by numerous events. Patients that had mild and background type of DR were included. Additionally, patients that suffered from grade I, grade II, grade III and grade IV type of DR were also included as the target outcome.

The main objective of this case study was to validate the Iranian and Chinese DR models on Indian T2DM patients to find the better fit. The second objective was to evaluate probabilities that will assist in the prediction of current, annual, biannual or triannual risk of developing DR. To summarize, current prevalence and yearly incidence probabilities of DR were estimated.

10. Sample Size Determination and Collected Variables

At 95% confidence interval and 5% margin of error, the sample size was estimated as 388 T2DM patients and the required data was randomly collected through one to one interaction with the patients and also from their recorded patient files from Rajiv Gandhi Centre for Diabetes and Endocrinology (RGCDE), Jawaharlal Nehru Medical College (JNMC), Aligarh Muslim University (AMU), Aligarh, India. The Centre brings more than 400 patients weekly and this increases the diversity in the sample, resulting in a much more reliable model. The sample data was divided into two sets: 284 patients were part of the Indian training set and 104 patients were selected to be in the Indian test set. Training set was used to develop the probabilities and test set was used to validate the obtained results. The various risk factors collected were sex, age (years), height (cm), weight (cm), body mass index (BMI) (kg/m^2), blood sugar fasting (BSF) (mg/dL), glycosylated haemoglobin (HbA1c) (%), systolic blood pressure (SBP) (mmHg), diastolic blood pressure (DBP) (mmHg), total cholesterol (TC) (mg/dL), triglycerides (TG) (mg/dL), high density lipoprotein cholesterol (HDL-C) (mg/dL), low density lipoprotein cholesterol (LDL-C) (mg/dL), physical activity (yes/no), duration of T2DM (years), diet control (yes/no), hypertension (yes/no), history of antihypertensive drug treatment (yes/no), family history of diabetes (yes/no), waist circumference (cm), central obesity (yes/no) and history of DR.

Missing data was handled by using multiple imputation method. The variables height, weight and waist circumference were noted while the patients were in light clothes and without shoes. Diet control was decided on the basis of whether an individual was following the directed diet plan and not consuming more calories than advised. According to Ramachandran et al., 2005, having a waist circumference greater than or equal to 85 cm for males and 80 cm for females characterized these patients to be centrally obese. BSF was measured after at least 8 hours of fasting. BSF and lipid profiles were estimated by Biochemistry Analyzer, Lab Life Chem Master, India. HbA1c was measured using D10 Instrumentation, Bio-Rad, USA.

11. Iranian Risk Score and Chinese Risk Score

Equation 15 and Equation 16 represent the formulated DR risk score models by Hosseini et al., 2009 and Wang et al., 2014.

$$\begin{aligned} Hosseini's\ risk\ score = {} & (Age\ (15(\text{if } 40\text{-}49), 25(\text{if } \geq 50)) + BMI\ (10(\text{if } > 25) \\ & + Duration\ (20(\text{if } 2\text{-}4), 35(\text{if } 5\text{-}9), 50(\text{if } \geq 10)) \\ & + HbA1c(5(\text{if } 7\text{-}11), 10(\text{if } > 11)) + Sex(5\ (\text{if female})) \end{aligned} \quad (15)$$

$$\begin{aligned} Wang\text{'s risk score} &= (Age\ (5\ (if\ 45\text{-}64),\ 4\ (if\ \geq 65)) + Central\ Obesity\ (2\ if\ yes) \\ &+ Duration\ (9\ (if\ 1\text{-}5), 10\ (if\ 5\text{-}10),\ 14\ (if\ 10\text{-}15),\ 20\ (if\ \geq 15)) \\ &+ Hypertensive\ Treatment\ (7\ if\ yes)) \end{aligned} \tag{16}$$

Since, logistic regression was applied in the two studies, the scores represented the prevalence risk. Risk scores based on regression coefficients for the two models are shown in Equation 17 and Equation 18. These can be used to calculate the prevalence probabilities of DR.

$$\begin{aligned} Hosseini\text{'s model} = &(Age\ (0.471\ (if\ 40\text{-}49),\ 0.739\ (if\ 50\text{-}59),\ 0.763\ (if\ \geq 60)) \\ &+ Duration\ (0.638\ (if\ 2\text{-}4),\ 1.017\ (if\ 5\text{-}9),\ 1.781\ (if\ \geq 10)) \\ &+ HbA1c\ (0.187\ (if\ 7\text{-}9),\ 0.237\ (if\ 9\text{-}11),\ 0.348\ (if\ > 11)) \\ &+ BMI\ (0.347\ if > 25) + Sex\ (0.216\ if\ female)) \end{aligned} \tag{17}$$

$$\begin{aligned} Wang\text{'s model} = &(Age\ (0.49\ (if\ 45\text{-}64),\ 0.449\ (if\ \geq 65)) \\ &+ Duration\ (0.891\ (if\ 1\text{-}5),\ 1.039\ (if\ 5\text{-}10),\ 1.383\ (if\ 10\text{-}15), \\ &1.982\ (if\ \geq 15)) + Central\ Obesity\ (0.213\ if\ yes) \\ &Hypertensive\ Treatment\ (0.698\ if\ yes)) \end{aligned} \tag{18}$$

12. Statistical Analysis

12.1 Descriptive Characteristics

The number of patients that had DR were determined. The patients that did not have DR at the time of data collection were investigated further to assess whether they had developed DR over the upcoming years or not. These patients were used to calculate the incidence probabilities. For continuous variables, mean and standard deviation values were calculated and percentages were evaluated for categorical variables. These descriptive values were tested for differences based on the presence and absence of DR. Independent samples t test was applied for normally distributed continuous variables and Mann-Whitney U test was used for non-normally distributed continuous risk factors. Differences between the proportions of categorical variables were investigated using chi-square test. The descriptive characteristics and their respective p-values compared at 5% level of significance are shown in Table 1.

Table 1: Descriptive characteristics of T2DM patients.

Variables	With DR (n = 34)	Without DR (n=354)	p value
Age (years)	55.56 ± 10.49	51.57 ± 9.45	0.021
Sex: males	50.0% (17)	49.7% (176)	0.975
Duration of T2DM (years)	10.56 ± 5.38	5.51 ± 4.69	< 0.001
HbA1c %	9.38 ± 1.81	7.38 ± 1.16	< 0.001
BMI (kg/m^2)	26.09 ± 4.91	25.37 ± 4.35	0.481
TC (mg/dL)	172.47 ± 29.98	170.11 ± 29.89	0.401
TG(mg/dL)	150.29 ± 30.77	151.71 ± 37.11	0.965
BSF (mg/dL)	149.85 ± 46.72	114.24 ± 34.49	< 0.001
HDL-C (mg/dL)	40.50 ± 7.07	40.67 ± 7.13	0.921
LDL-C (mg/dL)	101.91 ± 18.13	99.12 ± 18.37	0.231
SBP (mmHg)	145.38 ± 12.48	132.71 ± 15.11	< 0.001
DBP (mmHg)	81.41 ± 5.01	81.55 ± 7.55	0.825
Waist circumference (cm)	96.03 ± 11.98	94.86 ± 12.51	0.602
Central obesity: yes	79.4% (27)	74.9% (265)	0.557
Hypertension: yes	91.2% (31)	80.2% (284)	0.119
History of antihypertensive drug treatment: yes	91.2% (31)	77.7% (275)	0.066
Diet control: yes	73.5% (25)	72.0% (255)	0.853
Family history of diabetes: yes	44.1% (15)	39.3% (139)	0.581
Physical activity: yes	67.6% (23)	64.1% (227)	0.682

12.2 Estimating Risk Scores

Risk scores were calculated by using Equation 15 and Equation 16 for each patient in the Indian training set. Scatter plots and ROC curves were plotted to observe the spread of scores and fit of the models. Cut-off point using Youden's index was evaluated for future application. The better fit between the two scores was determined.

12.3 Estimating Prevalence Probabilities of DR

Apart from scores, knowing the probabilities of having DR was essential. The prevalence probabilities were measured by using Equation 17 and Equation 18, followed up by utilizing Equation 4. The obtained probabilities lacked randomness and mostly resulted around the upper extreme. Scatter plots were and ROC curves were constructed. Since, prevalence probabilities were not satisfactory by the direct application of regression coefficients, a different procedure was applied.

It was not possible to obtain the original data, therefore, the complete data from the two countries was simulated. From the percentages, means, standard deviations and other characteristics presented for each variable in the Iranian and Chinese studies, data was simulated to be as close to the real data as possible. Two samples of size 3734 and 1869 respectively, with variables and statistical parameters provided in the two countries' studies were simulated. Simulation for continuous variables was implemented using the formula '$x = \mu + z\sigma$'. 'μ' and 'σ' were the respective mean and standard deviation values. This option removed the limitation of a specific distribution for each risk factor. However, duration of diabetes and TG factors followed lognormal distribution in the best way (Abrahamyan et al., 2010; Shanmugasundaram et al., 2014). If only median and interquartile ranges were available then mean and standard deviation were estimated according to the methods provided by Wan et al., 2014.

Logistic regression by enter method was performed on the two simulated samples. The new regression coefficients slightly differed from the original results because they were obtained from a simulated data. Continuous variables were not categorized and therefore, a single regression coefficient was sufficient for one variable in order to maintain the true information and statistical power (Streiner, 2002). Probabilities with the new achieved regression coefficients were calculated using Equation 4. Scatter plots and ROC curves were plotted once again. The C-statistic values for the probabilities obtained after simulation were acceptable. However, the spread of the scatter plots was not uniform and focussed mostly on extreme values. The reason was overfitting thus, regularization techniques were applied.

Optimizing the tuning parameter 'λ' shown in Equations 9–11 was needed to solve the trade-off between the prediction accuracy of training sample and prediction accuracy of the test sample. Using crossvalidation (Lever et al. 2016) by '*glmnet*' package in R software, optimal 'λ' and regularization technique was singled out from ridge regression, lasso regression and elastic net regression based on the smallest MSE. These techniques were applied for logistic regression by stating the family as binomial in the R command.

The regularized coefficients were applied on the Indian training set and prevalence probabilities of DR were determined once again using Equation 4. Risk scores to be substituted in Equation 4 were evaluated by the linear combination of coefficients and their corresponding variables. The obtained results were much more improved for the predictions of DR. Prevalence probabilities were plotted in scatter plots. The final conclusion about the better fitting model was verified by ROC curves. Youden's index was used to select an optimal cut-off point from the probabilities.

12.4 Estimating Incidence Probabilities of DR

Logistic regression produced prevalence probabilities and these were used to calculate the yearly incidence probabilities by the method of absorbing Markov chain (Andersen and Keiding, 2002). Considering a two state model, with 'No DR' and 'Developed DR' as the two states and supposing 'Developed DR' as an absorbing state, probabilities of DR determined from the regularized coefficients portrayed the probabilities of going from No DR state to Developed DR state. In other words, it signified taking a step from a transient state to an absorbing state (denoted as P(DR)). Figure 1 represents an absorbing two state Markov chain for DR. Once a patient develops DR, he will always suffer from it. Therefore, the probability of a patient remaining in that state was '1'. Since, the patient cannot go back to the initial state of 'No DR', the probability was '0'. The probability of staying event free and not developing DR was '1-P(DR)'. These probabilities were established on the characteristics of an absorbing state by using Equation 14. Presence of DR was indicated by a '*1*' and absence of DR was denoted by a '*0*'.

Substituting the coded values in '$x^{(0)}$' and all possible probabilities from Figure 1 in Equation 14, the following Equation 19 was obtained. Substituting integer values in 'n' will produce different yearly models. For example, '$n = 1$' estimated the probability

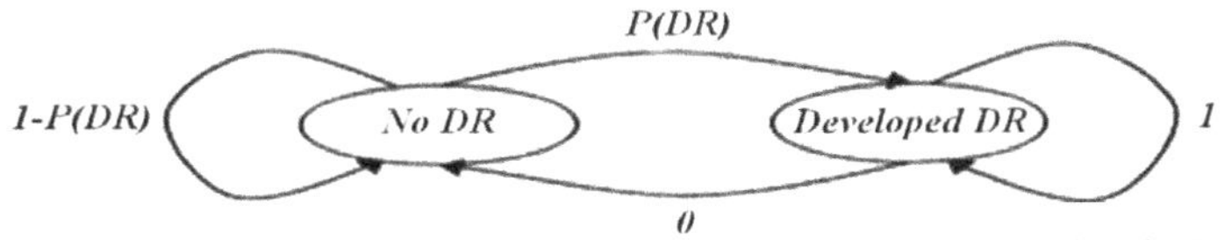

Figure 1: Two state Markov chain model with states No DR and Developed DR.

of having DR at the current time the data was collected, that is, prevalence probability. The value '$n = 2$' evaluated the incidence probability of DR for a duration of one year from the data collection. Similarly, substituting '$n = 3$' predicted the incidence probability for the second year, that is, after the completion of first year.

$$x^{(n)} = \begin{bmatrix} 1 & 0 \end{bmatrix} \begin{bmatrix} 1 - P(DR) & P(DR) \\ 0 & 1 \end{bmatrix}^n \tag{19}$$

Solving the above equation for each value of 'n' provided a more applicable form and it is shown in Equation 20. The prevalence probability of DR is substituted as '*Current P(DR)*'. It is to be noted that each year's DR probability uses the previous year's estimated DR probability.

$$\left.\begin{array}{l} I^{st}\ Year\ DR\ Probability = Current\ P(No\ DR) * Current\ P(DR) + Current\ P(DR) \\ II^{nd}\ Year\ DR\ Probability = Current\ P(No\ DR) * I^{st}\ Year\ P(DR) + Current\ P(DR) \\ III^{rd}\ Year\ DR\ Probability = Current\ P(No\ DR) * II^{nd}\ Year\ P(DR) + Current\ P(DR) \\ \vdots \qquad\qquad \vdots \qquad\qquad \vdots \qquad\qquad \vdots \\ \vdots \qquad\qquad \vdots \qquad\qquad \vdots \qquad\qquad \vdots \\ n^{th}\ Year\ DR\ Probability = Current\ P(No\ DR) * (n-1)^{th}\ Year\ P(DR) + Current\ P(DR) \end{array}\right\} \tag{20}$$

The achieved results from the better fitting country's simulated model were validated on the Indian test set. Incidence of DR for the first and second years was predicted. Diagnostic tests and ROC curve methods were applied to justify the results from this validation. Patients having any form of DR at the time of collection of data were part of the sample used to estimate the current prevalence probabilities. These specific patients were then excluded from further calculations of first year and second year incidence risk probabilities of developing DR.

13. Results

13.1 Descriptive Characteristics of the Study Sample

The sample of 388 patients comprised of 193 males (49.7%) and 195 (50.3%) females. The average age in years was 51.92 ± 9.59 (range: 28.00 – 80.00 years). On average, duration of diabetes was 5.95 ± 4.96 years (range: 0.05 – 25.83 years). The mean value of HbA1c (%) was 7.5 ± 1.3 (range: 5.0 – 13.4). The prevalence of DR was found in 34 (8.7%) patients at the time of data collection. The descriptive characteristics on the basis of presence and absence of DR in the form of mean, standard deviation, percentage and p-values of the variables are presented in Table 1.

13.2 Outcomes from the Application of Iranian and Chinese DR Scores

The application of the scores shown in Equation 15 and Equation 16 on the training sample produced scatter plots depicted in Figure 2 (a) and Figure 2 (b). The points in the two graphs were randomly distributed and did not form any clusters.

The areas under ROC curves of computed diagnostic accuracies attained were 0.815 (95% confidence interval: 0.765 – 0.859) for the Iranian score and 0.776 (95% confidence interval: 0.734 – 0.832) for the Chinese score. Considering risk scores, Iranian score was the better fit on the Indian data than the Chinese score.

Optimal cut-off point for the Iranian score was assessed as 65, sensitivity and specificity were 78.26% and 74.33% respectively. From the 23 patients which were diagnosed with DR, 19 (82.6%) recorded a DR risk score above 65. From the 261 patients that were not detected with DR, 163 (62.5%) patients noted a risk score below 65. This indicated that the fit of the Iranian risk score for predicting prevalence of DR is satisfactory. Table 2 shows some of the diagnostic test values of the tested screening points.

Table 2: Sensitivity (%), specificity (%), PPV (%), NPV (%) and Youden's index for the tested cut-off cases for the applied Iranian risk score.

Cut-off	Sensitivity	Specificity	PPV	NPV	Youden's Index
20	100.00	11.88	9.1	100.0	0.1188
50	91.30	45.21	12.8	98.3	0.3651
65	78.26	74.33	21.2	97.5	0.5259
75	60.87	85.82	27.2	96.1	0.4669
90	13.04	98.85	50.0	92.8	0.1189

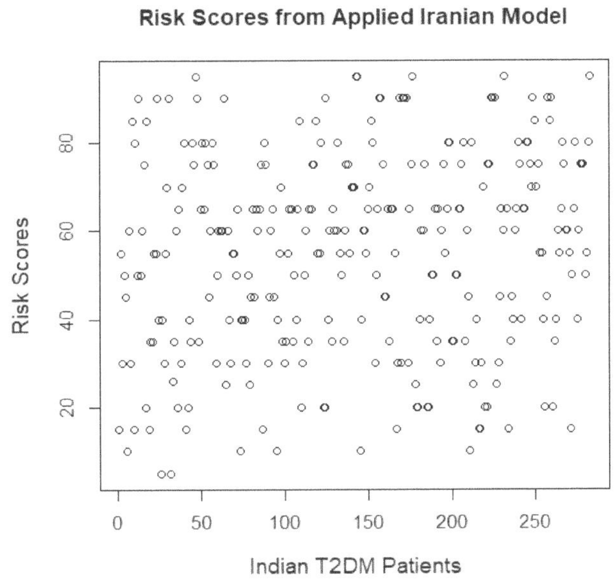

Figure 2(a): Scatter plot for applied Iranian risk score

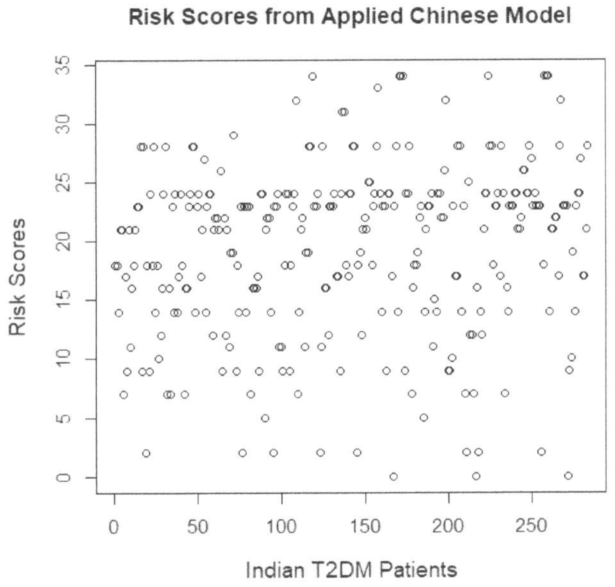

Figure 2(b): Scatter plot for applied Chinese risk score

13.3 Outcomes from the Estimation of Prevalence Probabilities of DR

Predicting the prevalence probabilities of DR by executing Equation 17 and Equation 18 in Equation 4 was the next objective. The Iranian model resulted in an area under ROC curve of 0.62 (95% confidence interval: 0.561 – 0.677). On the other hand, the Chinese model determined an area of 0.567 (95% confidence interval: 0.507 – 0.626). According to the C-statistic values, both of these models did not fit the Indian training sample well and the spread of the probabilities was not random and were mostly at the upper extreme.

To rectify the low accuracy criteria so that the predicted probabilities could be more spread out, two samples of size 3734 and 1869 were simulated and logistic regression was applied. TG, HbA1c and DBP variables were determined to be non-significant for the Iranian simulated model, while it was determined that diet control, central obesity, hypertension and regular physical activity variables were non-significant for the Chinese simulated model. Once again the probabilities were estimated and accuracy was determined by ROC curve. Application of simulated Iranian model produced an area under curve of 0.62 (95% confidence interval: 0.561 – 0.677). The simulated Chinese model had an area of 0.748 (95% confidence interval: 0.693 – 0.797). The C-statistic value of the simulated Iranian model clearly indicated that there was no improvement considering the fit, while the fit of the simulated Chinese model showed an improvement in the accuracy. However, the overall spread of predicted probabilities was not suitable and the values were still clustered at the upper extreme end.

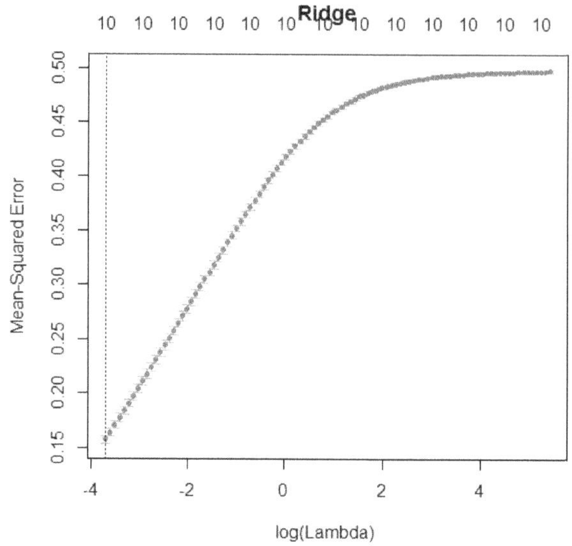

Figure 3(a): Various 'λ' values depicting different MSEs for ridge regression on Iranian simulated data.

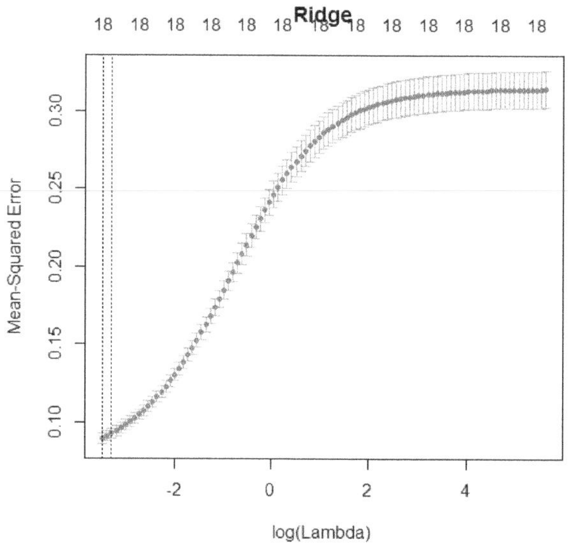

Figure 3(b): Various 'λ' values depicting different MSEs for ridge regression on Chinese simulated data

As mentioned previously, the reason for the clustered fit was overfitting and regularization techniques were used to remove multicollinearity. In this case, both the simulated Iranian and Chinese data achieved the smallest MSE for ridge regression, thus concluding it as the best method for removing overfitting. Selection of 'λ' is depicted in Figure 3 (a) and Figure 3 (b). In the graphs, the number of variables in the two models are written at the top in addition to the type of regularization technique applied. Optimal 'λ' was estimated as 0.0235 for the simulated Iranian ridge regression model and 0.0301 for the simulated Chinese ridge regression model.

For the estimated 'λ' values, ridge regression coefficients for the non-significant variables were reduced approximately to zero and this is portrayed in Figure 4 (a) and Figure 4 (b). Each individual line in the two plots represents the logistic regression coefficient of a risk factor being reduced.

Since ridge regression retains all variables, the newly estimated coefficient values and the ORs for full model are provided in Table 3 and Table 4. The obtained ridge coefficients have been minimized as compared to the previous coefficients. It is to be noted that some of the coefficients are approximately zero. These variables can be indicated as non-significant.

Using the evaluated ridge regression coefficients, Equation 4 was once again executed on the Indian training set to determine the prevalence probabilities of DR. Regularized Iranian model formed an ROC curve with an area of 0.648 (95% confidence interval:

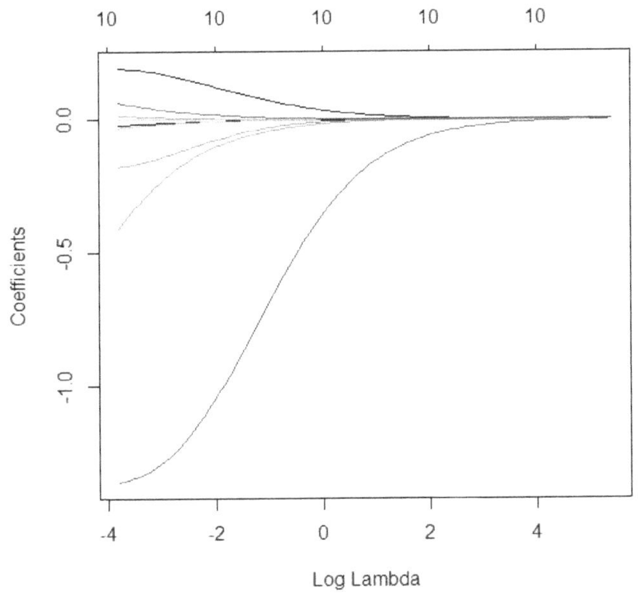

Figure 4(a): Ridge regression reduces the logistic regression coefficients of Iranian simulated data as 'λ' increases.

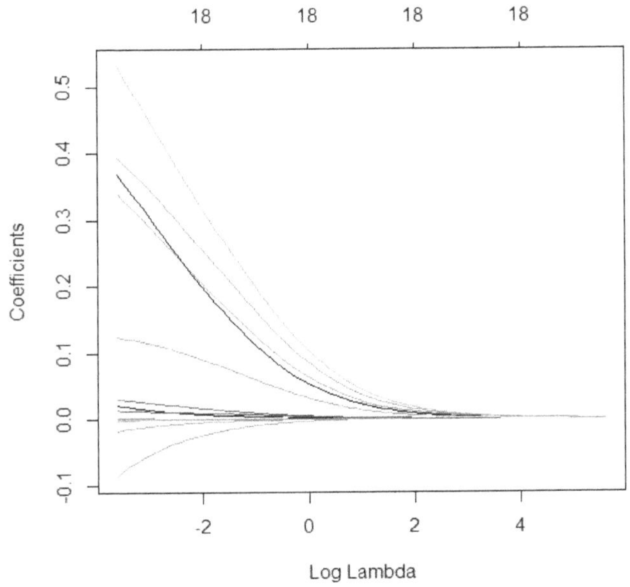

Figure 4(b): Ridge regression reduces the logistic regression coefficients of Chinese simulated data as 'λ' increases.

Table 3: Ridge regression coefficients for Iranian simulated data.

Variables	Ridge regression coefficient	OR
Sex: females	− 1.358	0.257
Age (years)	0.097	1.102
Duration of T2DM (years)	0.193	1.213
HbA1c (%)	− 0.175	0.839
BMI (kg/m^2)	− 0.383	0.682
TC (mg/dL)	− 0.019	0.981
BSF (mg/dL)	0.016	1.016
SBP (mmHg)	0.061	1.063
DBP (mmHg)	− 0.029	0.971
Constant	2.547	12.769

Table 4: Ridge regression coefficients for Chinese simulated data.

Variables	**Ridge regression coefficient**	**OR**
Sex: females	0.329	1.390
Age (years)	– 0.001	1.001
Duration of T2DM (years)	0.309	1.362
HbA1c (%)	– 0.004	0.996
BMI (kg/m^2)	– 0.064	0.938
Waist circumference (cm)	– 0.001	0.999
History of antihypertensive drugs: yes	0.481	1.616
TG (mg/dL)	0.027	1.027
Diet control: yes	0.135	1.145
Family history of diabetes: yes	0.118	1.126
Physical activity: yes	0.012	1.012
TC (mg/dL)	– 0.001	0.999
HDL-C (mg/dL)	0.003	1.004
LDL-C (mg/dL)	0.001	1.001
BSF (mg/dL)	– 0.001	1.001
SBP (mmHg)	0.017	1.017
DBP (mmHg)	– 0.013	0.987
Hypertension: yes	0.364	1.440
Constant	– 9.453	0.000078

0.588 – 0.702). On the other hand, regularized Chinese model produced an area under ROC curve of 0.784 (95% confidence interval: 0.731 – 0.830). Both these models improved the values of area under the ROC curves.

The spread of the probabilities for both models improved as well and these are represented in Figure 5 (a) and Figure 5 (b). A cluster near the higher probabilities is seen in Figure 5 (a), while Figure 5 (b) depicts a larger cluster of points near the lower range of probabilities. Concerning probabilities, the coefficients obtained from ridge regression applied on simulated Chinese data was a much better fit on Indian patients than that of the Iranian simulated data. Iranian ridge model was quite over predictive. Therefore, the area under the ROC curve and spread of probabilities (Figure 5 (b)) achieved for the Chinese ridge regression model confirmed it to be more applicable.

The optimal cut-off point for the regularized Chinese model from the Indian training set probabilities was assessed as 0.093 with sensitivity and specificity as 78.26% and 71.26% respectively. Diagnostic test values for the prevalence probabilities of DR are represented in Table 5. Out of 23 patients diagnosed with DR, 18 (78.3%) noted a DR risk probability greater than 0.093. From the remaining 261 T2DM patients that did not have DR, 185 (70.9%) patients observed a risk probability less than 0.093. Ridge regression retained all variables in the model, however, age, waist circumference, TC, LDL-C and BSF can be excluded if required due to the small impact of these variables.

14. Validation on Indian Test Sample

Out of a sample of 104 patients, 11 (10.5%) were diagnosed with DR. The ridge regression Chinese model was applied on the Indian test set. Prevalence probabilities of DR were determined for each patient using Equation 4 and the coefficients provided in Table 4. Out of the 11 patients who were diagnosed with DR, 10 (90.9%) patients scored a risk probability of over 9.3%. From the remaining 93 patients that did not have DR, 58 (62.4%) patients scored less than this cut-off value. ROC curve resulted in an area of 0.819 (95% confidence interval: 0.732 – 0.888). The curve is shown in Figure 6 (a). The spread of the probabilities obtained from the test set is represented in Figure 6 (b). An area of 81.9% and randomly distributed points across the scatter plot prove that the regularized Chinese model was a good fit on Indian T2DM patients for predicting the prevalence of DR.

Since 11 patients had DR presently, these were then excluded from the total test sample to allow a better estimation of one year risk probabilities. Excluding the 11 patients with DR, 9 (9.6%) patients developed DR within a year from the time of data collection. Further excluding the 9 patients to focus on the second year, 12 (14.2%) patients developed DR in the second year. The incidence

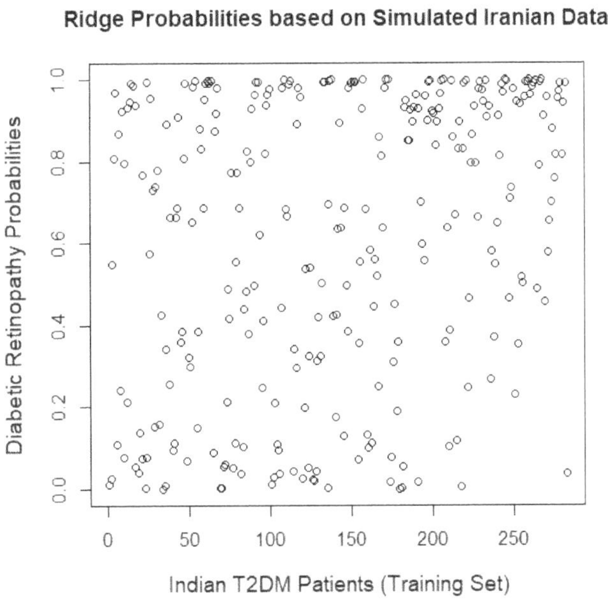

Figure 5(a): Scatter plot showing spread of Iranian ridge probabilities (Training set).

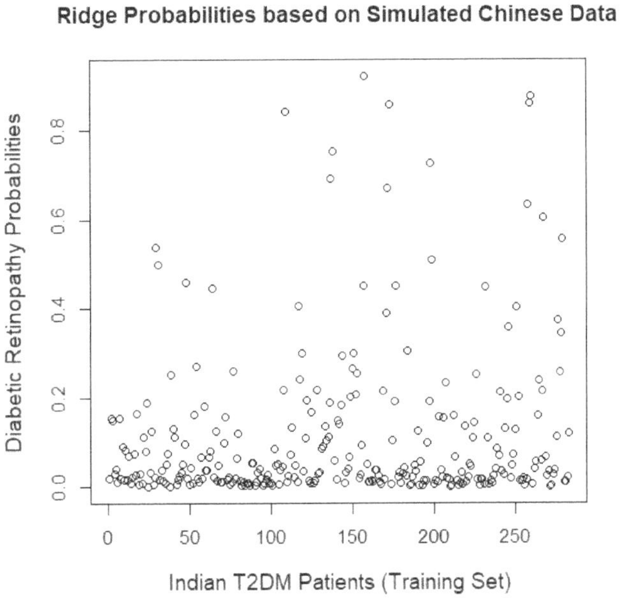

Figure 5(b): Scatter plot showing spread of Chinese ridge probabilities (Training Set).

Table 5: Sensitivity (%), specificity (%), PPV (%), NPV (%) and Youden's index to identify the optimal cut-off point for applied Chinese ridge regression model on the Indian training set.

Cut-off	Sensitivity	Specificity	PPV	NPV	Youden's index
0.0108	100.00	21.07	10.0	100.0	0.2107
0.0504	78.26	60.54	14.9	96.9	0.388
0.0934	78.26	71.26	19.4	97.4	0.4952
0.1443	65.22	79.69	22.1	96.3	0.4491
0.4458	21.74	94.64	26.3	93.2	0.1638

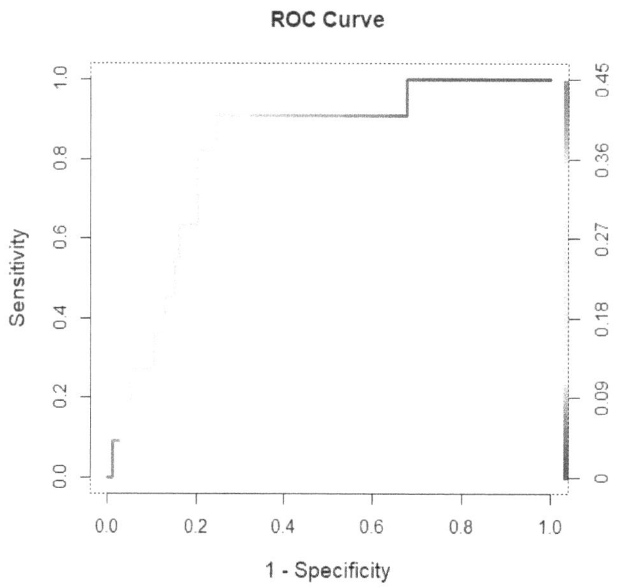

Figure 6(a): ROC curve for applied Chinese ridge regression model on Indian test set.

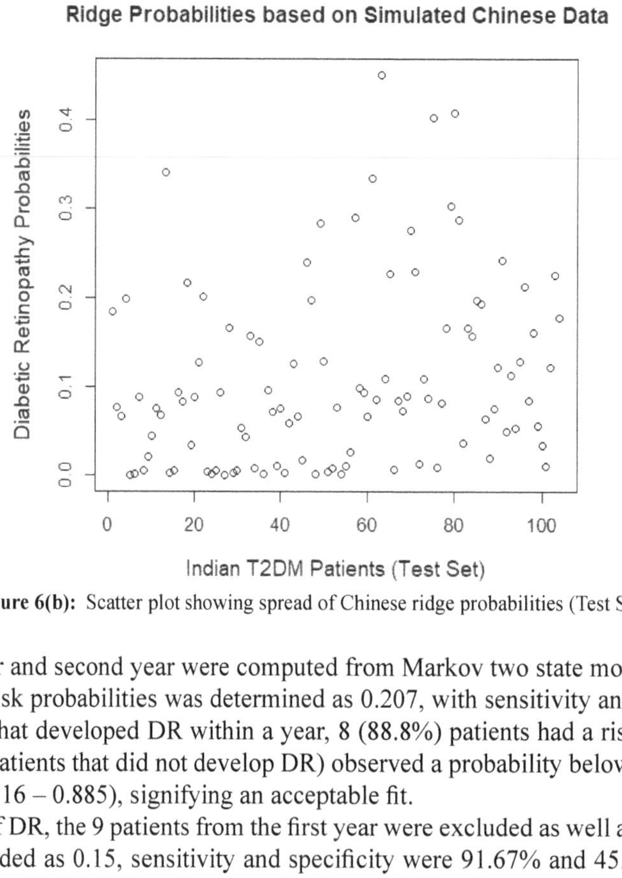

Figure 6(b): Scatter plot showing spread of Chinese ridge probabilities (Test Set).

probabilities of DR for the first year and second year were computed from Markov two state model using Equation 19 and Equation 20. The cut-off point for one year risk probabilities was determined as 0.207, with sensitivity and specificity as 88.89% and 76.19% respectively. Out of the 9 patients that developed DR within a year, 8 (88.8%) patients had a risk probability above 0.207, whereas 64 (76.2%) patients (out of the 84 patients that did not develop DR) observed a probability below 0.207. Area under ROC curve was 0.811 (95% confidence interval: 0.716 – 0.885), signifying an acceptable fit.

To study the second year risk of DR, the 9 patients from the first year were excluded as well and the probabilities were estimated. The optimal cut-off value was decided as 0.15, sensitivity and specificity were 91.67% and 45.83% respectively. The results were satisfying for the second year also as 11 (91.7%) patients noted risk probability above 0.15. ROC curve produced an area of 0.686 (95% confidence interval: 0.576 – 0.783). From the 72 patients that did not develop eye complication, 32 (44.4%) patients had their probability less than 0.15. Some of the diagnostic test values attained for the first and second year risk probabilities are provided in Table 6.

Table 6: Sensitivity (%), specificity (%), PPV (%), NPV (%) and Youden's index to identify the optimal cut-off point for first and second year DR risk probabilities on the Indian test set.

		First year risk probabilities			
Cut-off	**Sensitivity**	**Specificity**	**PPV**	**NPV**	**Youden's index**
0.0203	100.00	26.19	10.7	100.0	0.2619
0.1457	88.89	54.76	14.8	98.2	0.4365
0.2071	88.89	76.19	24.8	98.7	0.6508
0.2298	66.67	77.38	20.6	96.3	0.4405
0.3540	33.33	86.90	18.3	93.7	0.2023
		Second year risk probabilities			
Cut-off	**Sensitivity**	**Specificity**	**PPV**	**NPV**	**Youden's index**
0.0183	100.00	23.61	10.3	100.0	0.2361
0.0603	91.67	36.11	11.2	98.0	0.2778
0.1500	91.67	45.83	13.0	98.4	0.375
0.2939	41.67	79.17	15.0	93.9	0.2084
0.5178	16.67	91.67	15.0	92.6	0.0834

15. Discussion and Conclusion

An overview of diabetes mellitus and disease risk scores has been discussed in this chapter. Additionally, the methods required to create such models have also been discussed. The description has been made with the help of a case study (Yusufi et al., 2019). Iranian and Chinese DR risk scores, the only two models constructed on retinopathy, were applied on Indian patients and the Iranian score was concluded as the better fit. Several attempts were made to rectify the accurate prevalence probabilities from the two models. The ridge regression coefficients acquired from the Chinese simulated data satisfied the Indian data much better than the Iranian simulated data and the results were validated on a test set. Annual incidence probabilities of DR were estimated by Markov two state model and estimates were portrayed for the first year and second year from the time of data collection.

Consider an example of the Iranian score on the Indian data. A female patient of age 49 years has had diabetes for 9.42 years, HbA1c was 6.9%, BSF was 134 mg/dL and BMI was 24.65 kg/m^2. Such a patient gets a score of 55. Next, an example of the regularized Chinese model is explained. A male patient of age 60 years has had T2DM for 4.5 years, HbA1c was 7.6%, BMI was 19.03 kg/m^2 and waist circumference was 81 cm. Looking at the lipid profile, TG was 175 mg/dL, TC was 198 mg/dL, HDL-C was 47 mg/dL and LDL-C was 116 mg/dL. The patient was not following any diet control, was not physically active and had no family history of diabetes. The patient was hypertensive and was taking drugs for hypertension. SBP was 130 mmHg and DBP was 80 mmHg. The prevalence probability of DR for this patient was 0.072. First year incidence probability of DR was 0.139 and second year incidence probability scored was 0.202.

Even though the average values of BSF and HbA1c were higher in patients with DR, these variables did not have a strong impact in the ridge regression model. The reason was that these two factors were strongly penalized. This was a limitation of this case study. Another limitation of this case was that the original data for the two countries to apply regularization techniques on was not available. Nonetheless, the outcomes obtained were extremely satisfactory.

Standard errors and confidence intervals were not given in Table 4 because R software deliberately does not provide those parameters. The reason is that standard errors are not explanatory for highly biased estimates and ridge regression reduces the variance but increases the bias of the penalized estimates. It is difficult to calculate a precise estimate of the bias and bootstrapping can only produce an assessment of variance of the coefficients. It can be misleading to consider MSEs from this process as it ignores the inaccuracies caused by the bias. Thus, the confidence intervals from bootstrapping are affected as well, due to the fact that the calculation of these intervals are based on estimates of variances (Goeman et al., 2018).

Having DR prediction models available as a tool for prior assessment holds an advantage over expensive fundus photography. If risk scores are followed by doctors and patients then only the patients in the high risk criteria from the predictive model should be followed up for fundus examination. Such practices will.

This chapter described risk scores and the methods which are necessary to develop such scores. A case study is presented explaining the application of these methods needed to develop and validate prediction models for DR in T2DM patients. The study covered the difficulties and obstacles that arise in such prediction model building objectives. The discussed techniques can be used for the prediction of other diabetic complications and other diseases in order to provide self-awareness and management of both time and resources.

Abbreviations

AMU	Aligarh Muslim University
BMI	Body mass index
BSF	Blood sugar fasting
DBP	Diastolic blood pressure
DR	Diabetic retinopathy
HbA1c	Glycosylated haemoglobin
HDL-C	High density lipoprotein cholesterol
IDF	International Diabetes Federation
JNMC	Jawaharlal Nehru Medical College
LDL-C	Low density lipoprotein cholesterol
MSE	Mean squared error
NPV	Negative predictive value
OR	Odds ratio
PPV	Positive predictive value
RGCDE	Rajiv Gandhi Centre for Diabetes and Endocrinology
ROC	Receiver operating characteristic
SBP	Systolic blood pressure
T1DM	Type 1 diabetes mellitus
T2DM	Type 2 diabetes mellitus
TC	Total cholesterol
TG	Triglycerides
VIF	Variance inflation factor
WHO	World Health Organization

References

Abrahamyan, L., Beyene, J., Feng, J., Chon, Y., Willan, A. R. et al. 2010. Response times follow lognormal or gamma distribution in arthritis patients. Journal of Clinical Epidemiology, 63(12): 1363–1369.

Andersen, P. K. and Keiding, N. 2002. Multi-state models for event history analysis. Statistical Methods in Medical Research, 11(2): 91–115.

Chen, L., Magliano, D. J. and Zimmet, P. Z. 2012. The worldwide epidemiology of type 2 diabetes mellitus—present and future perspectives. Nature reviews Endocrinology, 8(4): 228–236.

Conroy, R. M., Pyörälä, K., Fitzgerald, A. E., Sans, S., Menotti, A. et al. 2003. Estimation of ten-year risk of fatal cardiovascular disease in Europe: the SCORE project. European Heart Journal, 24(11): 987–1003.

Friedman, J., Hastie, T. and Tibshirani, R. 2010. Regularization paths for generalized linear models via coordinate descent. Journal of Statistical Software, 33(1): 1.

Gagniuc, P. A. 2017. Markov chains: from theory to implementation and experimentation. John Wiley and Sons.

Glumer, C., Carstensen, B., Sandbæk, A., Lauritzen, T., Jørgensen, T et al. 2004. A Danish diabetes risk score for targeted screening: the Inter99 study. Diabetes care, 27(3): 727–733.

Harrell, Jr., F. E., Lee, K. L. and Mark, D. B. 1996. Multivariable prognostic models: issues in developing models, evaluating assumptions and adequacy, and measuring and reducing errors. Statistics in Medicine, 15(4): 361–387.

Hertroijs, D. F., Elissen, A. M., Brouwers, M. C., Schaper, N. C., Kohler, S. et al. 2018. A risk score including body mass index, glycated haemoglobin and triglycerides predicts future glycaemic control in people with type 2 diabetes. Diabetes, Obesity and Metabolism, 20(3): 681–688.

Hilden, J. and Glasziou, P. 1996. Regret graphs, diagnostic uncertainty and Youden's Index. Statistics in Medicine. 15(10): 969–986.

Hippisley-Cox, J. and Coupland, C. 2017. Development and validation of QDiabetes-2018 risk prediction algorithm to estimate future risk of type 2 diabetes: cohort study. bmj, 359, j5019.

Hippisley-Cox, J., Coupland, C., Vinogradova, Y., Robson, J., May, M. et al. 2007. Derivation and validation of QRISK, a new cardiovascular disease risk score for the United Kingdom: prospective open cohort study. Bmj, 335(7611), 136.

Hirst, J. E., Raynes-Greenow, C. H. and Jeffery, H. E. 2012. A systematic review of trends of gestational diabetes mellitus in Asia. Journal of Diabetology, 3(3): 5.

Hosseini, S. M., Maracy, M. R., Amini, M. and Baradaran, H. R. 2009. A risk score development for diabetic retinopathy screening in Isfahan-Iran. Journal of research in medical sciences: the official journal of Isfahan University of Medical Sciences, 14(2): 105.

Hu, H., Nakagawa, T., Yamamoto, S., Honda, T., Okazaki, H., Uehara, A. et al. 2018. Development and validation of risk models to predict the 7-year risk of type 2 diabetes: The Japan Epidemiology Collaboration on Occupational Health Study. Journal of Diabetes Investigation, 9(5): 1052–1059.

International Diabetes Federation. Available at https://www.idf.org/ aboutdiabetes/ complications.html. Accessed on 25 September, 2020.

International Diabetes Federation. IDF Diabetes Atlas, 9th Edn. Brussels, Belgium: Available at http://www.diabetesatlas.org. Accessed on 25 September, 2020.

James, G., Witten, D., Hastie, T. and Tibshirani, R. 2013. An introduction to statistical learning (Vol. 112, pp. 3-7). New York: springer.

Jelle, J. Goeman. 2018. Penalized R package, version 0.9–51.

Kallner, A. 2014. Laboratory statistics, handbook of formulas and terms. Chemistry International, 36(4): 23–23.

Kempen, J. H., O'Colmain, B. J., Leske, M. C., Haffner, S. M., Klein et al. 2004. The prevalence of diabetic retinopathy among adults in the United States. Archives of ophthalmology (Chicago, Ill.: 1960), 122(4): 552–563.

Kulothungan, V., Ramakrishnan, R., Subbiah, M. and Raman, R. 2014. Risk Score Estimation of Diabetic Retinopathy: Statistical Alternatives using Multiple Logistic Regression. Journal of Biometrics and Biostatistics, 5(5): 1.

Lever, J., Krzywinski, M. and Altman, N. 2016. Points of significance: model selection and overfitting. Nature Methods, 13(9): 703–704.

Liew, G., Wong, T. Y., Mitchell, P., Cheung, N. and Wang, J. J. 2009. Retinopathy predicts coronary heart disease mortality. Heart, 95(5): 391–394.

Lindstrom, J. and Tuomilehto, J. 2003. The diabetes risk score: a practical tool to predict type 2 diabetes risk. Diabetes Care, 26(3): 725–731.

Mahmood, S. S., Levy, D., Vasan, R. S. and Wang, T. J. 2014. The Framingham Heart Study and the epidemiology of cardiovascular disease: a historical perspective. The Lancet, 383(9921): 999–1008.

Mayo Clinic. Available at https://www.mayoclinic.org/diseases-conditions/diabetes/ symptoms-causes/syc-20371444. Accessed on 25 September, 2020.

Miettinen, O. S. 1976. Stratification by a multivariate confounder score. American Journal of Epidemiology, 104(6): 609–620.

Mohan, V., Deepa, R., Deepa, M., Somannavar, S. and Datta, M. 2005. A simplified Indian Diabetes Risk Score for screening for undiagnosed diabetic subjects. The Journal of the Association of Physicians of India, 53: 759–63.

Ramachandran, A., Snehalatha, C., Vijay, V., Wareham, N. J. and Colagiuri, S. 2005. Derivation and validation of diabetes risk score for urban Asian Indians. Diabetes Research and Clinical Practice, 70(1): 63–70.

Ray, W. A. 2003. Population-based studies of adverse drug effects. New England Journal of Medicine, 349(17): 1592–1594.

Ridker, P. M. Paynter, N. R., Rifai, N., Gaziano, J. M. and Cook, N. R. 2008. C-reactive protein and parental history improve global cardiovascular risk prediction: the Reynolds Risk Score for men. Circulation, 118(22): 2243–51.

Riffenburgh, R. H. 2006. Statistics in medicine 2nd ed.

Rosner, B. 2010. Fundamentals of biostatistics, Brooks/Cole, Cengage Learning. Inc., Boston, MA.

Shanmugasundaram, D., Jeyaseelan, L., George, S. and Zachariah, G. 2014. Analysis strategy for comparison of skewed outcomes from biological data: a recent development. Ann. Biol. Res., 5(12): 16–20.

Sheather, S. 2009. A modern approach to regression with R. Springer Science and Business Media.

Stiglic, G., Fijačko, N., Stožer, A., Sheikh, A. and Pajnkihar, M. 2016. Validation of the Finnish Diabetes Risk Score (FINDRISC) questionnaire for undiagnosed type 2 diabetes screening in the Slovenian working population. Diabetes Research and Clinical Practice, 120: 194–197.

Streiner, D. L. 2002. Breaking up is hard to do: the heartbreak of dichotomizing continuous data. The Canadian Journal of Psychiatry, 47(3): 262–266.

Sulaiman, N., Mahmoud, I., Hussein, A., Elbadawi, S., Abusnana, S et al. 2018. Diabetes risk score in the United Arab Emirates: a screening tool for the early detection of type 2 diabetes mellitus. BMJ Open Diabetes Research and Care, 6(1): e000489.

Van der Leeuw, J., Van Dieren, S., Beulens, J. W. J., Boeing, H., Spijkerman, A. M. W. et al. 2015. The validation of cardiovascular risk scores for patients with type 2 diabetes mellitus. Heart, 101(3): 222–229.

Van Kampen, N. G. 1992. Stochastic processes in physics and chemistry (Vol. 1). Elsevier.

Vittinghoff, E., Glidden, D. V., Shiboski, S. C. and McCulloch, C. E. 2011. Regression methods in biostatistics: linear, logistic, survival, and repeated measures models. Springer Science and Business Media.

Wan, X., Wang, W., Liu, J. and Tong, T. 2014. Estimating the sample mean and standard deviation from the sample size, median, range and/or interquartile range. BMC Medical Research Methodology, 14(1): 135.

Wang, J., Chen, H., Zhang, H., Yang, F., Chen, R. P. et al. 2014. The performance of a diabetic retinopathy risk score for screening for diabetic retinopathy in Chinese overweight/obese patients with type 2 diabetes mellitus. Annals of Medicine, 46(6): 417–423.

Wilson, P. W., D'Agostino, R. B., Levy, D., Belanger, A. M., Silbershatz, H. et al. 1998. Prediction of coronary heart disease using risk factor categories. Circulation, 97(18): 1837–1847.

Woodward, M. 2013. Epidemiology: study design and data analysis. CRC press.

World Health Organization. 1999. Definition, diagnosis and classification of diabetes mellitus and its complications: report of a WHO consultation. Part 1, Diagnosis and classification of diabetes mellitus (No. WHO/NCD/NCS/99.2). Geneva: World health organization.

World Health Organization. 2016. Global report on diabetes: executive summary (No. WHO/NMH/NVI/16.3). World Health Organization.

World Health Organization. Available at https://www.who.int/diabetes/ action_ online/basics/en/index3.html#:~:text=Microvascular%20complications%20include%20damage%20to,severe%20infections%20leading%20to%20amputation). Accessed on 25 September, 2020.

Yang, X., So, W. Y., Tong, P. C., Ma, R. C., Kong, A. P. et al. 2008. Development and validation of an all-cause mortality risk score in type 2 diabetes: The Hong Kong Diabetes Registry. Archives of Internal Medicine, 168(5): 451–457.

Yu, D., Cai, Y., Graffy, J., Holman, D., Zhao, Z. et al. 2018. Development and external validation of risk scores for cardiovascular hospitalization and rehospitalization in patients with diabetes. The Journal of Clinical Endocrinology and Metabolism, 103(3): 1122–1129.

Yusufi, F. N. K., Ahmed, A. and Ahmad, J. 2019. Modelling and developing diabetic retinopathy risk scores on Indian type 2 diabetes patients. International Journal of Diabetes in Developing Countries, 39(1): 29–38.

Index

Milton Keynes UK
Ingram Content Group UK Ltd.
UKHW020003121024
449327UK00055B/441

9 780367 618810